Pg 91 WKB B44

Geography

People and Places in a Changing World

Revised Second Edition

Paul Ward English is Professor of Geography and Middle Eastern Studies at the University of Texas at Austin. He has received a President's Associates Teaching Excellence Award and an Outstanding Teacher Award from *Change* magazine for his course in World Geography. He has also written a number of books and articles based on field research in Europe, Africa, and Asia under the auspices of the National Academy of Science and the Social Science Research Council. His special fields of interest are the Geography of Development, Human Geography, the Middle East, and Geography in Education.

Geography

People and Places in a Changing World

Revised Second Edition

Paul Ward English

Professor, Department of Geography and Middle Eastern Studies
University of Texas at Austin

with the assistance of
Brian Robinson
Elgin Independent School District

West Publishing Company
St. Paul   New York   Los Angeles   San Francisco

WEST'S COMMITMENT TO THE ENVIRONMENT

In 1906, West Publishing Company began recycling materials left over from the production of books. This began a tradition of efficient and responsible use of resources. Today, 100% of our legal bound volumes are printed on acid-free, recycled paper consisting of 50% new paper pulp and 50% paper that has undergone a de-inking process. We also use vegetable-based inks to print all of our books. West recycles nearly 27,700,000 pounds of scrap paper annually—the equivalent of 229,300 trees. Since the 1960s, West has devised ways to capture and recycle waste inks, solvents, oils, and vapors created in the printing process. We also recycle plastics of all kinds, wood, glass, corrugated cardboard, and batteries, and have eliminated the use of polystyrene book packaging. We at West are proud of the longevity and the scope of our commitment to the environment.

West pocket parts and advance sheets are printed on recyclable paper and can be collected and recycled with newspapers. Staples do not have to be removed. Bound volumes can be recycled after removing the cover.

Production, Prepress, Printing and Binding by West Publishing Company.

Design and production: Hespenheide Design
Principal cartography and map design: The DLF Group
Additional cartography: Mapping Specialists
Composition: American Composition & Graphics, Inc., Hespenheide Design
Artwork: Randy Miyake, Hespenheide Design

Printed with **Printwise**
Environmentally Advanced Water Washable Ink ∞

Acknowledgments

In my professional life, I have known several people who have dedicated their lives to education and to the production of good books. This book has benefitted from the contributions of such people.

The work of my colleagues and co-authors on the first edition of this text, Ed Lindop and Alice Schule, continues to focus and strengthen this new edition. Alice is remembered fondly, and this work is a credit to her commitment to excellence.

The new and expanded map program was refined through the efforts of Richard W. Wilkie, Ph.D. and Ute Dymon, Ph.D. (with the assistance of Mbobi E. Kiloson).

The comments and suggestions of Sarah Bednarz on the final draft of this manuscript deserve special thanks. Her expertise extends beyond geography to the unique demands of a textbook. Other individuals who made special contributions are Lois Helvey, Gerri Benigsen, and Sherry Henderson.

In addition to creating the new summaries, Quinton Priest developed the material in the teacher's wraparound edition, an extraordinary task. Carolyn McGovern did a meticulous editing job. Gary Hespenheide of Hespenheide Design is responsible for the beautiful design of this book, and for the extraordinary production work.

Many people at West Publishing Company have made invaluable contributions, but I would especially like to thank Carole Grumney and Lynn Bruton for their dedication, unfailing support, and endless hours of work.

—*Paul Ward English*

Reviewers/Consultants

The author and the publisher would like to thank the following reviewers/consultants for their valuable comments and suggestions throughout the development of this textbook:

Sarah W. Bednarz, Ph.D.
Texas A & M University

Debra Bickford
Hillside Junior High School
California

John Bickford
Sequoia Junior High School
California

Ute Dymon, Ph.D.
Kent State University
Ohio

Jeff Hackman
Western Mennonite High School
Oregon

Redginald R. Heth
Healdsburg High School
California

Lois Helvey
St. Bonaventure High School
California

Sherry Henderson
Northbrook High School
Texas

Evelyn R. Hill
Shadowlawn Middle School
Tennessee

Carol Laramore
Marianna High School
Florida

Barbara MacDonald
Glen Oaks High School
Louisiana

Joe Margraff
Incarnate Word Academy
Missouri

Dale W. Martens
Fairmont High School
Minnesota

Michael Martin
Maumee High School
Ohio

LTC Frank S. Meredith, Jr.
Missouri Military Academy
Missouri

Mary Jo Morton
Marianna High School
Florida

Quinton Priest, Ph.D.
Green Fields Country Day School
Arizona

Thomas L. Pumphrey
Marianna High School
Florida

John P. Reager
Samuel E. Shull School
New Jersey

Richard W. Wilkie, Ph.D.
University of Massachusetts (Amherst)

CONTENTS IN BRIEF

CONTENTS

Chapter 20
Tradition and Change in Africa

Chapter 26
Development, Disunity, and Change in South Asia

Maps and Figures

End-of-Chapter Skills

Vocabulary Skills are located at the end of each chapter following the Exercises and Skills.

The second edition of *Geography: People and Places in a Changing World* is a clearly written, carefully organized geography of the world. It tells the story of where we are now, and of where we seem to be heading. It outlines the two great challenges we face—providing for people and preserving our planet.

Geography: People and Places in a Changing World, Second Edition, is designed to interest you, to be easy to use, and to enlarge and enrich your appreciation of the world, a world you share with more than five billion other people. This opening statement explains how the book has been organized to make it up-to-date, relevant, helpful, and exciting.

How this Book is Organized

Geography: People and Places in a Changing World, Second Edition, is divided into twelve units. Unit 1 introduces geography and describes the environments of our planet.

Units 2 through 6 discuss the wealthiest and most powerful of the world culture regions: Western Europe, the Commonwealth of Independent States and Eastern Europe, the United States and Canada, Japan and the two Koreas, and the Pacific World.

These are five world regions in which science and technology have enabled a number of peoples with different backgrounds and environments to achieve and enjoy a level of well-being rarely known before on this planet. Most (the CIS and Eastern Europe excepted) have diversified economies, abundant access to resources, high levels of urbanization, slow-growing populations, and rising incomes.

Units 7 through 12 discuss the remaining six world culture regions: Latin America, Africa south of the Sahara, the Middle East and North Africa, China, South Asia, and Southeast Asia.

These six world regions have not fully achieved the high living standards that science and technology can provide, although great differences among countries and regions exist. Most have economies based on farming, poorly developed resources, slow rates of industrialization, rapidly growing populations, and stable or slow-growing incomes. These people are struggling to gain the levels of well-being enjoyed by their wealthier neighbors.

The characteristics of many regions are influenced by two contradictory forces: (1) *fusion,* the merging together into multinational units and (2) *fission,* the breaking apart of national units divided by race, religion, language, or economic disparities. Recent examples of *fusion* are the growth of the European Community (EC), the union of Canada, Mexico, and the United States in passing NAFTA (the North American Free Trade Agreement), and the spread of multinational corporations around the globe. *Fission* is seen in dramatic form in the collapse of the Soviet Union, civil war and disorder in Africa south of the Sahara, and religious and ethnic riots in India. *Fusion* directly aims at facing environmental and human challenges; *fission* makes progress more difficult to achieve.

How Units Are Organized

After Unit 1, each unit in *Geography: People and Places in a Changing World, Second Edition,* is devoted to one world culture region that is discussed in two or three chapters. The first chapter in each unit discusses the physical and human geographies of the region being studied. Detailed, accurate regional descriptions make up the bulk of these chapters. These chapters explain, for example, how the Industrial Revolution transformed Western Europe; how population pressure on the land plagues China; how India's unique environment creates problems of food production; and how Japan succeeds despite few natural resources.

The second chapter in each unit discusses the contemporary geography of the region being studied, including many geographical problems of current concern. These chapters explain, for example, why the U.S.S.R. collapsed; the retreat of Communism; Latin America's effort to cope with growing city populations; how the United States is changing from an industrial to a postindustrial society; and how environmental damage throughout the tropics threatens economic development and the global environment.

Unit and Chapter Features

All units are organized simply, directly and in a consistent manner for ease of use. At the beginning of each unit you will find six special elements. A **Unit Introduction** and a list of **Unit Objectives** provide a preview of what the unit covers and what goals you can achieve. **Skills Highlighted** lists the skills that will be emphasized throughout the unit. A wide variety of skills are treated, with special emphasis on map skills. The heart of the extensive map program in *Geography: People and Places in a Changing World, Second Edition,* is the **Miniatlas** at the beginning of each unit. The Miniatlas contains full-color physical, political and population maps of the region treated in the unit. This makes study and reference especially easy. Within each unit, detailed regional maps provide up-close information at a more detailed level, and additional special-purpose maps help enrich and clarify the study of geography. **At a Glance** contains basic statistics for all countries in the region under study in an easy to read table. **Keys to Knowing** is a brief sketch of the most essential information presented in the unit.

Each chapter contains three additional special features—**Close-Up**, **Discovering**, and **It's a Fact**. Close-Up articles are an in-depth look at an aspect of the region being studied. Discovering sections contain interesting information about geography from a human perspective. Each of these is linked to one of the themes of geography—location, place, people-environment relations, movement, and region. It's a Fact, found in boxes throughout the chapters, contain very short, memorable bits of information that will enliven your study of the region.

A multitude of sources have been examined to provide a current, accurate picture of the world. Most place names are consistent with those used in the *National Geographic Atlas of the World, Revised Sixth Edition*. Population and per capita income figures are based on the Population Reference Bureau's World Population Data Sheet. City populations are taken from many census and other sources; here estimates vary widely. Every effort has been made to use these sources to provide a clear presentation of the many changes in progress around the globe.

Testing and review methods that enable you to understand and remember geography have a high priority in this book. At the end of each unit there is a **Unit Review** containing exercises on human geography and physical geography, plus a writing activity to assist you in your studying and mastering of the unit material. The chapters are broken into easy-to-manage sections. Each section concludes with **Review Questions** and **Thought Questions**. At the end of each chapter, **Exercises** and **Skills** will help you apply chapter information and build your geography skills. Photo captions are informative and include thought-provoking questions that enrich the chapter content.

A very special feature of this text is the exclusive availability of the *CNN Geography Update*. This video update is produced by Turner Educational Services, using the resources of CNN, the world's first 24-hour, all-news network. West Educational Publishing is proud to be the exclusive educational partner of CNN for textbook/video integration in high school geography. With the *CNN Geography Update* you can enjoy the power of CNN, the network known for providing in-depth, live coverage and analysis of breaking news events, in your geography class. The *CNN Geography Update* is available exclusively with this text.

All the features included in this text will help you to understand the world. They will also prepare you to evaluate and analyze information as you master essential geography skills and concepts. They will prepare you to understand the challenges of today—providing for people and preserving the planet.

Geography

People and Places
in a Changing World

UNIT 1

WORLD GEOGRAPHY

UNIT INTRODUCTION Geographers study the Earth to gain a deeper understanding of where things are and why they are there. Five important themes in geography—location, place, people-environment relations, movement, and region—provide geographers with an organized understanding of the world in which we live.

This world has many different environments. *Environments* are created by forces inside and on the planet's surface that create and reshape land-forms, and by variations in climate, vegetation, and soils. The distribution of people on Earth is closely linked to the distribution of Earth's varied environments.

People change the world in many ways. Ten thousand years ago, during the Environmental Transformation, people developed agriculture and herding and built cities. The Scientific Transforma-tion culminated in the 1700s and 1800s with the development of modern industrial societies. The changes begun during these two periods of rapid transformation changed environments and ways of living. As technology and human thought contin-ue to evolve, new social, economic, and environ-mental conditions change our way of life.

The discoveries made during the Environmental and Scientific Transformations did not spread evenly across the Earth. The globe can be divided into eleven culture regions. In these culture re-gions, people face very different challenges and opportunities in their search for more secure and productive lives.

▲ *Many environments on our planet are thinly inhabited—deserts, mountains, tropical rain forests, polar regions. This has not changed in spite of population increases. What does this imply for the future?*

▼ *Increasingly, geographers are viewing the planet Earth as a single, environmental system. Name one issue related to this idea.*

◄ *Humans have radically altered the surface of Earth in places. What effect does the concentration of millions of people into huge cities have on the ecosystem?*

► *Tropical vegetation is crucial to the maintenance of our global atmosphere. Have you seen any programs or read any articles related to the importance of tropical vegetation?*

UNIT OBJECTIVES

When you have completed this unit, you will be able to:

1. State Earth's important characteristics, including its planetary movements that create day and night and the seasons of the year.

2. Name and define the five major themes in geography: location, place, people-environment relations, movement, and region.

3. Describe Earth's varied landforms and types of vegetation, and learn how they are developing and changing.

4. Understand that climate, characterized by temperature and rainfall, varies because of topography, latitude, and circulation systems in the atmosphere and oceans.

5. Explain the importance of the Environmental Transformation and the Scientific Transformation in changing the ways people live.

6. Relate how Earth today can be divided into a technological world of mainly industrialized, relatively prosperous countries and a developing world of mainly agricultural, relatively poor countries. Knowing about these variations within countries—including our own—is vital to understanding world geography.

SKILLS HIGHLIGHTED

- Reading Diagrams
- Comparing Diagrams
- Understanding Latitude and Longitude
- Using a Table
- Exploring the Unit Miniatlas
- Using a Physical Features Map
- Reading a Diagram of Vegetations and Soils
- Using a Climate Map
- Reading Graphs That Show World Population
- Using a Population Map of the World's Young People
- Vocabulary Skills
- Writing Skills

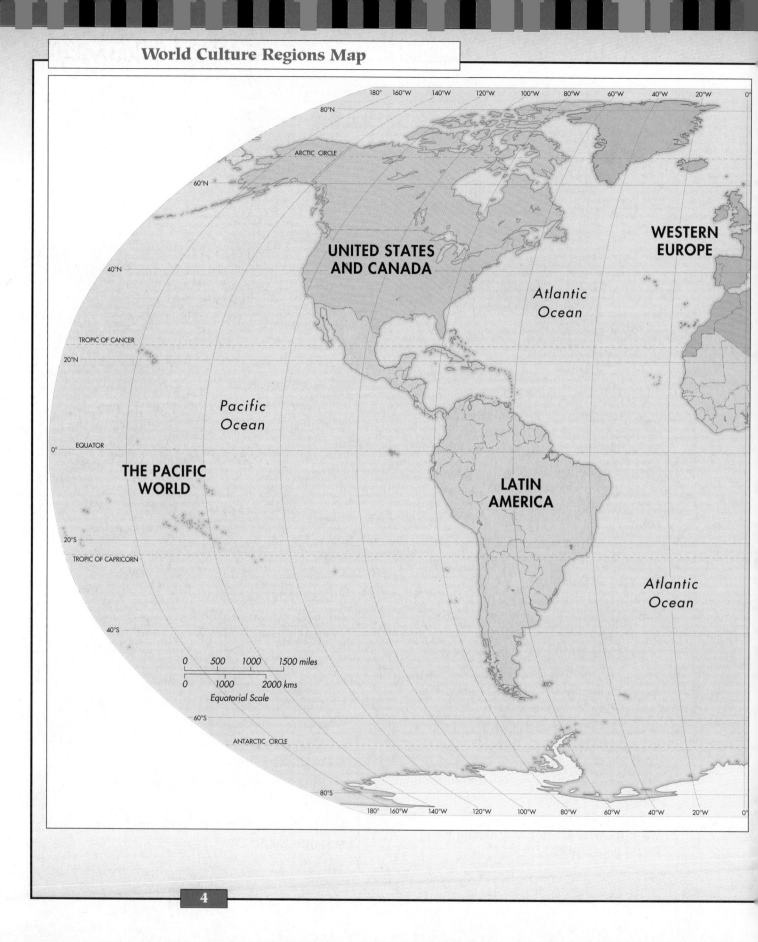

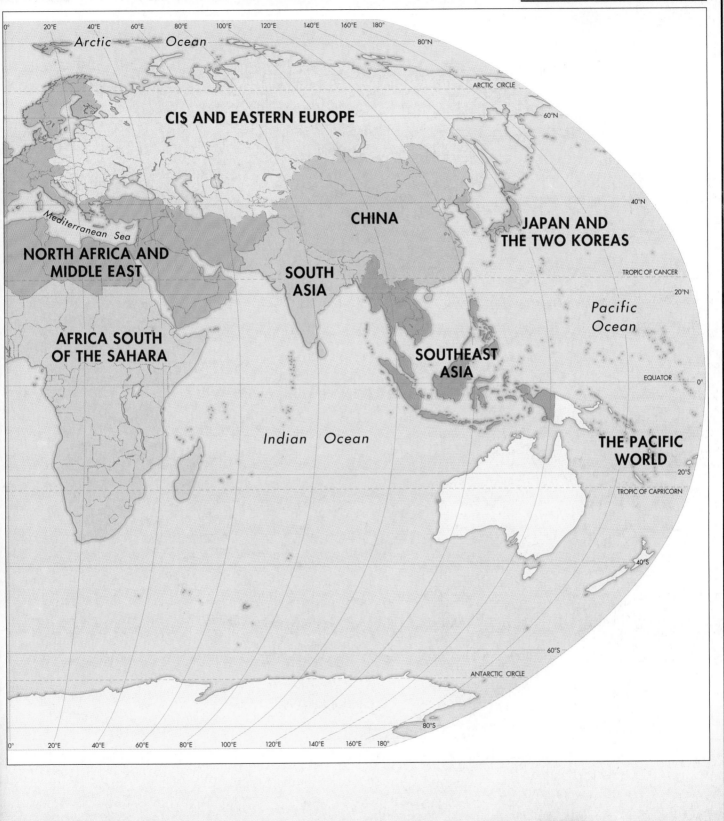

Arctic   Ocean

80°N

ARCTIC CIRCLE

60°N

CIS AND EASTERN EUROPE

40°N

CHINA

JAPAN AND
THE TWO KOREAS

Mediterranean Sea

TROPIC OF CANCER

NORTH AFRICA AND
MIDDLE EAST

SOUTH
ASIA

20°N

Pacific
Ocean

AFRICA SOUTH
OF THE SAHARA

SOUTHEAST
ASIA

EQUATOR   0°

Indian   Ocean

THE PACIFIC
WORLD

20°S

TROPIC OF CAPRICORN

40°S

60°S

ANTARCTIC CIRCLE

80°S

0°   20°E   40°E   60°E   80°E   100°E   120°E   140°E   160°E   180°

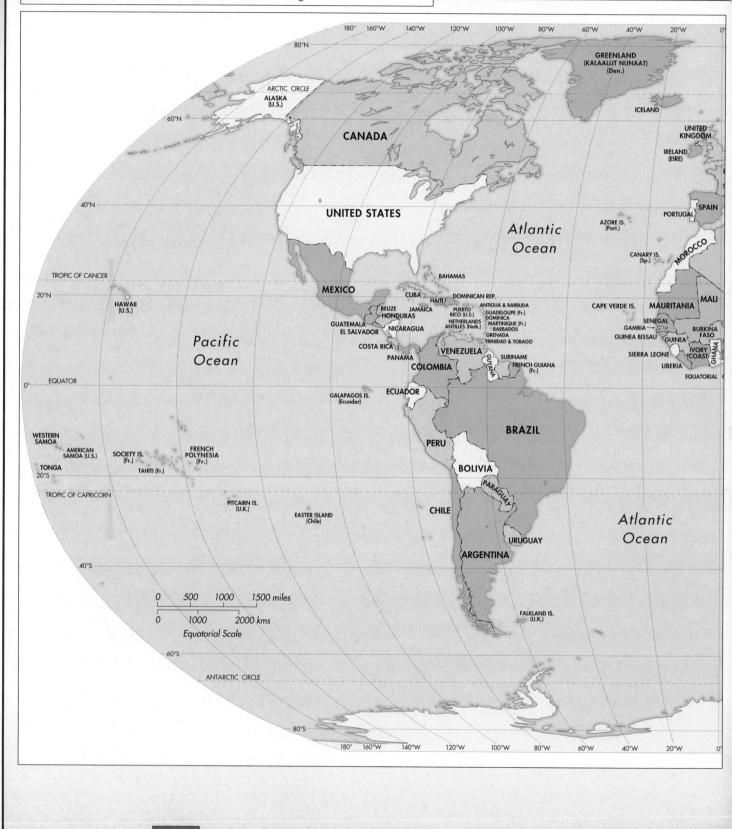

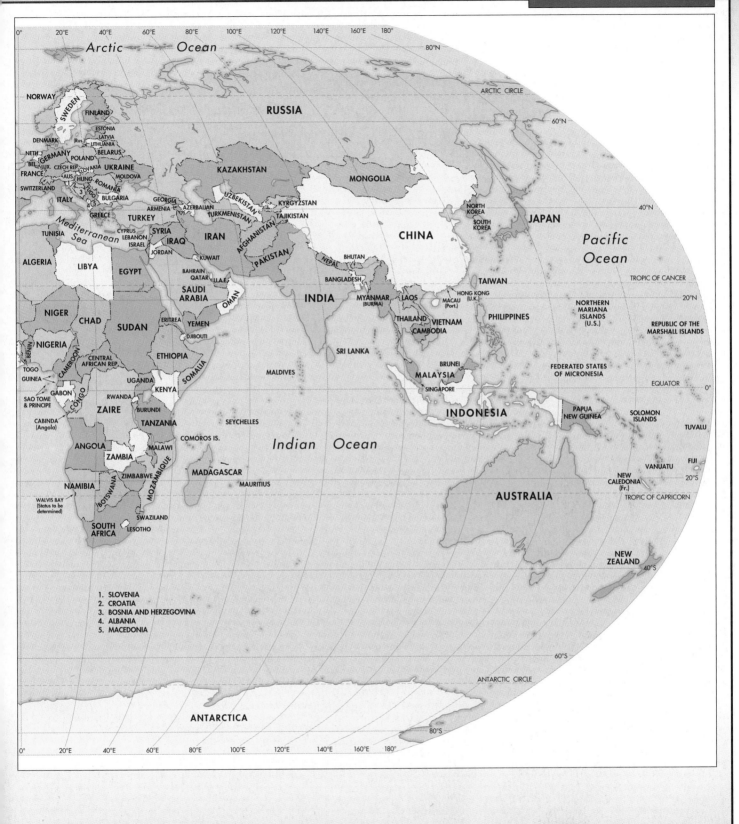

Arctic Ocean

ARCTIC CIRCLE

80°N

NORWAY

SWEDEN

FINLAND

RUSSIA

60°N

ESTONIA

DENMARK

(Rus.)

LATVIA

LITHUANIA

NETH.

GERMANY

BELARUS

BEL.

LUX.

POLAND

FRANCE

CZECH REP.

SLOVAKIA

UKRAINE

KAZAKHSTAN

MONGOLIA

40°N

SWITZERLAND

AUS.

HUNG.

ROMANIA

MOLDOVA

UZBEKISTAN

KYRGYZSTAN

JAPAN

ITALY

YUGO.

BULGARIA

GEORGIA

TURKMENISTAN

TAJIKISTAN

NORTH KOREA

GREECE

ARMENIA

AZERBAIJAN

TURKEY

SOUTH KOREA

CYPRUS

SYRIA

AFGHANISTAN

CHINA

Pacific Ocean

LEBANON

IRAQ

IRAN

ISRAEL

Mediterranean Sea

JORDAN

KUWAIT

PAKISTAN

NEPAL

BHUTAN

TAIWAN

TROPIC OF CANCER

TUNISIA

BAHRAIN

QATAR

BANGLADESH

HONG KONG (U.K.)

ALGERIA

LIBYA

EGYPT

U.A.E.

INDIA

MYANMAR (BURMA)

LAOS

MACAU (Port.)

20°N

SAUDI ARABIA

OMAN

THAILAND

VIETNAM

PHILIPPINES

NORTHERN MARIANA ISLANDS (U.S.)

NIGER

CHAD

SUDAN

ERITREA

YEMEN

CAMBODIA

REPUBLIC OF THE MARSHALL ISLANDS

NIGERIA

DJIBOUTI

SRI LANKA

BENIN

CENTRAL AFRICAN REP.

ETHIOPIA

BRUNEI

TOGO

MALDIVES

GUINEA

CAMEROON

UGANDA

SOMALIA

MALAYSIA

FEDERATED STATES OF MICRONESIA

EQUATOR

0°

SAO TOME & PRINCIPE

GABON

CONGO

KENYA

SINGAPORE

RWANDA

ZAIRE

BURUNDI

PAPUA NEW GUINEA

SOLOMON ISLANDS

CABINDA (Angola)

TANZANIA

SEYCHELLES

INDONESIA

TUVALU

ANGOLA

MALAWI

Indian Ocean

ZAMBIA

ZIMBABWE

MOZAMBIQUE

VANUATU

FIJI

NAMIBIA

MADAGASCAR

MAURITIUS

NEW CALEDONIA (Fr.)

20°S

WALVIS BAY (Status to be determined)

BOTSWANA

SWAZILAND

AUSTRALIA

TROPIC OF CAPRICORN

SOUTH AFRICA

LESOTHO

1. SLOVENIA
2. CROATIA
3. BOSNIA AND HERZEGOVINA
4. ALBANIA
5. MACEDONIA

NEW ZEALAND

40°S

60°S

ANTARCTIC CIRCLE

ANTARCTICA

80°S

0°   20°E   40°E   60°E   80°E   100°E   120°E   140°E   160°E   180°

World Physical Map

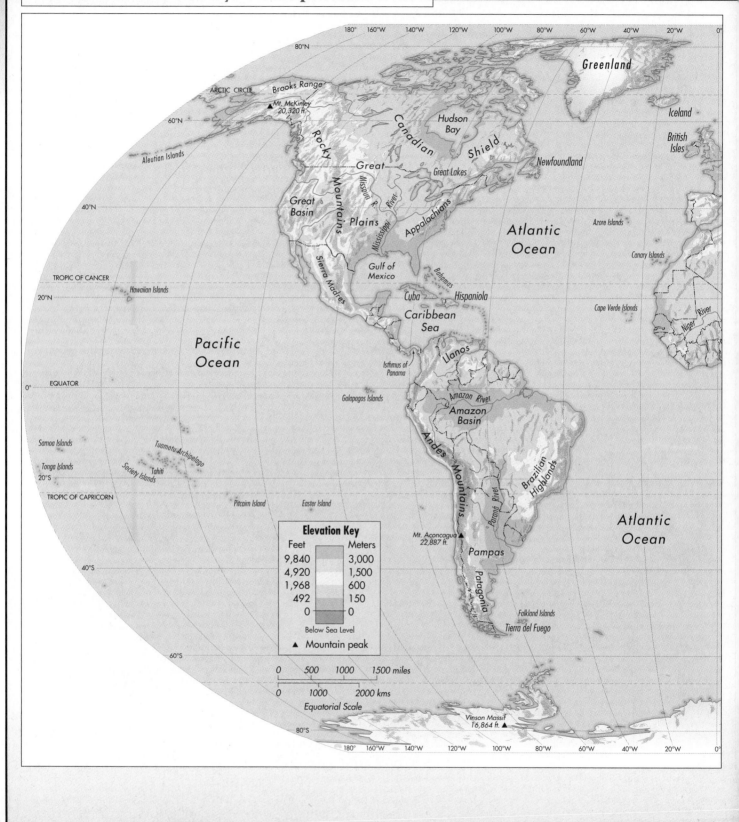

180° 160°W 140°W 120°W 100°W 80°W 60°W 40°W 20°W 0°

80°N

Greenland

ARCTIC CIRCLE

Brooks Range

Iceland

▲ Mt. McKinley
20,320 ft.

60°N

Canadian

Hudson
Bay

Shield

*British
Isles*

Aleutian Islands

Rocky

Great

Great Lakes

Newfoundland

Missouri R.

40°N

Mountains

*Great
Basin*

Plains

Ohio River

Appalachians

**Atlantic
Ocean**

Azore Islands

Mississippi

TROPIC OF CANCER

Sierra Madres

20°N

Hawaiian Islands

Gulf of
Mexico

Bahamas

Cuba *Hispaniola*

Canary Islands

Cape Verde Islands

**Caribbean
Sea**

Niger River

**Pacific
Ocean**

Isthmus of
Panama

Llanos

0° EQUATOR

Galapagos Islands

Amazon River

**Amazon
Basin**

Samoa Islands

Tuamotu Archipelago

Andes

*Brazilian
Highlands*

Tonga Islands

Society Islands Tahiti

20°S

TROPIC OF CAPRICORN

Pitcairn Island

Easter Island

Mountains

Paraná River

**Atlantic
Ocean**

Elevation Key

Feet		Meters
9,840		3,000
4,920		1,500
1,968		600
492		150
0		0

Below Sea Level

▲ Mountain peak

Mt. Aconcagua ▲
22,887 ft.

Pampas

40°S

Patagonia

Falkland Islands

Tierra del Fuego

60°S

0   500   1000   1500 miles

0   1000   2000 kms

Equatorial Scale

Vinson Massif
16,864 ft. ▲

80°S

180° 160°W 140°W 120°W 100°W 80°W 60°W 40°W 20°W 0°

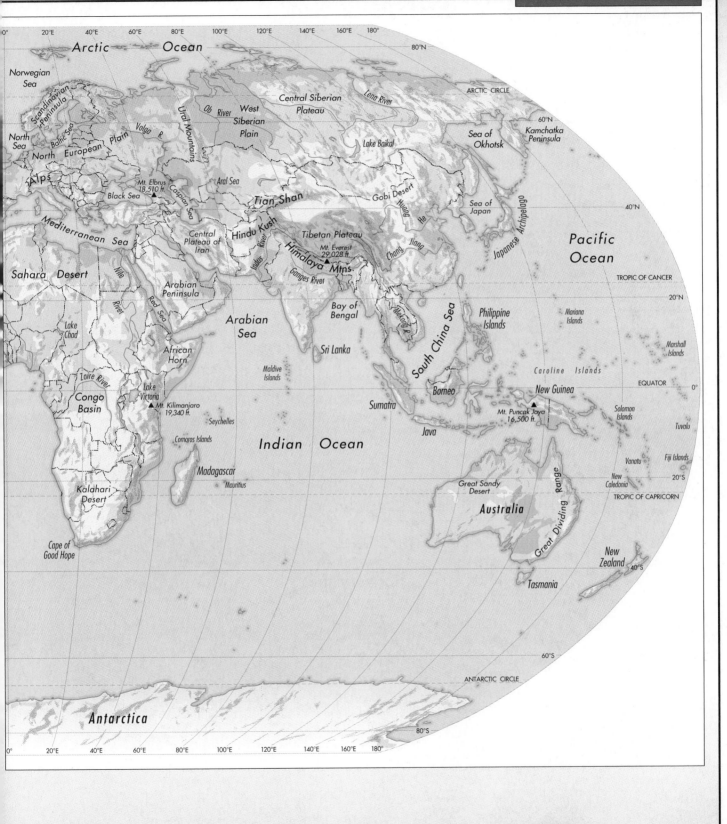

Arctic   Ocean   80°N
Norwegian   ARCTIC CIRCLE
Sea
                    Central Siberian   Lena River
Scandinavian   Plateau   60°N
Peninsula   Ob River West   Kamchatka
North   Siberian   Sea of   Peninsula
Sea   Baltic Sea Volga R. Plain   Okhotsk
North European Plain   Ural Mountains   40°N
Alps   Sea of
        Mt. Elbrus Aral Sea   Tian Shan   Gobi Desert Japan
        18,510 ft.   Japanese Archipelago   Pacific
Mediterranean Sea Black Sea Caspian Sea Hindu Kush Huang He   Ocean
                Central   Tibetan Plateau Chong Jiang
                Plateau of Himalaya Mt. Everest
Sahara Desert Iran   Indus River Mtns. 29,028 ft.   TROPIC OF CANCER
                Arabian Ganges River   20°N
        Nile Peninsula   Bay of   Mariana   Marshall
        River Red Sea Arabian Bengal Mekong R. Philippine Islands Islands
Lake   Sea   South China Islands
Chad   African   Sri Lanka   Sea   Caroline Islands   EQUATOR 0°
        Zaire River Horn   Maldive   Borneo New Guinea
Congo   Lake   Islands   Solomon
Basin   Victoria Mt. Kilimanjaro   Sumatra Mt. Puncak Jaya Islands
                19,340 ft. Seychelles   Java   16,500 ft.   Tuvalu
                Comoros Islands   Indian Ocean   Fiji Islands
Kalahari   Madagascar   Vanatu 20°S
Desert   Mauritius   Great Sandy   New
                                    Desert   Caledonia
Cape of   Australia   TROPIC OF CAPRICORN
Good Hope   Great Dividing Range   New
                                                          Zealand 40°S
                                              Tasmania

                                                          60°S

                                    ANTARCTIC CIRCLE
Antarctica   80°S

0°   20°E   40°E   60°E   80°E   100°E   120°E   140°E   160°E   180°

9

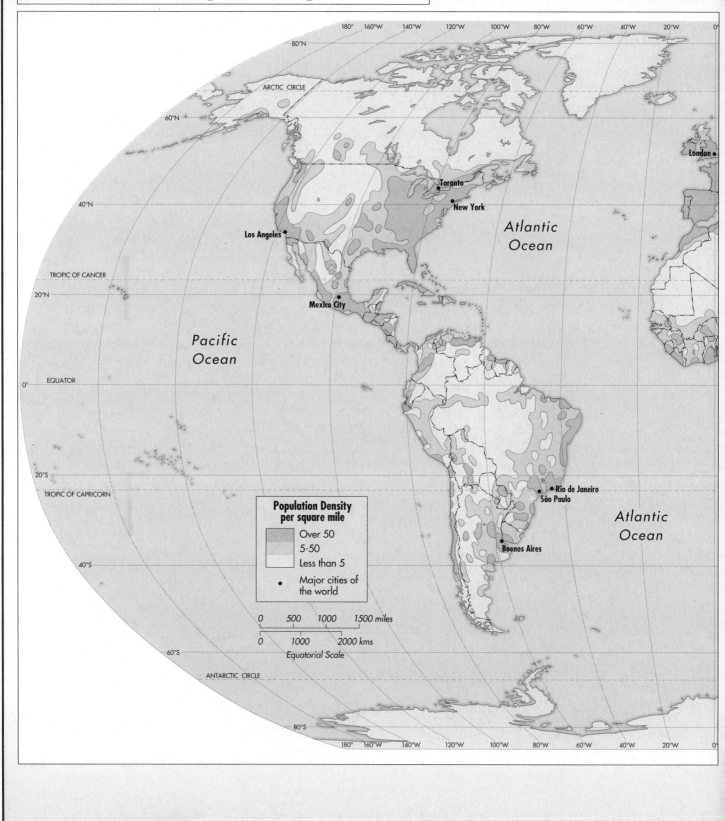

Population Density
per square mile

Over 50
5-50
Less than 5
• Major cities of
the world

0   500   1000   1500 miles
0   1000   2000 kms
Equatorial Scale

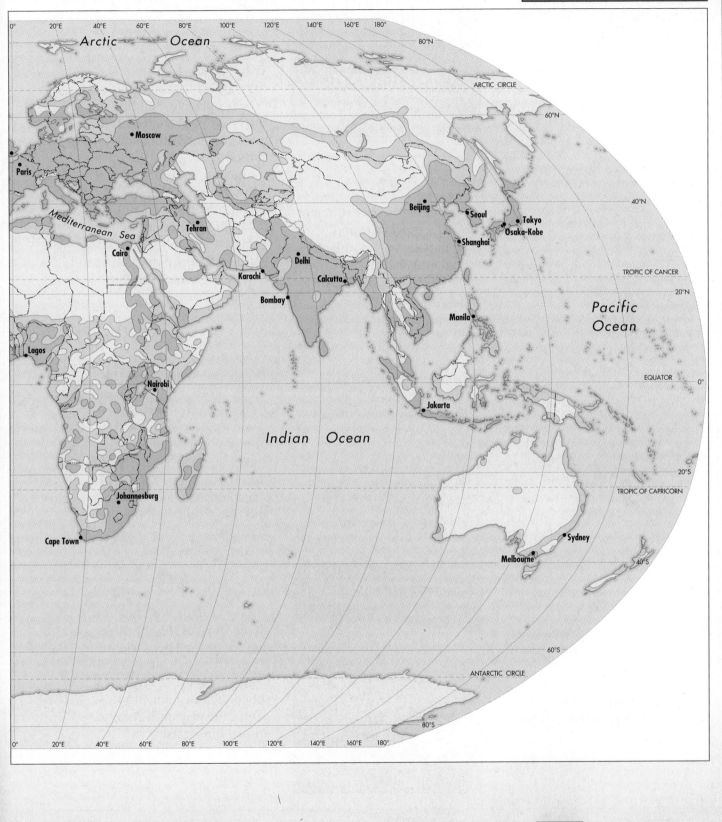

Arctic Ocean

80°N

ARCTIC CIRCLE

60°N

Moscow

Paris

40°N

Mediterranean Sea

Beijing •Seoul

Tehran •Tokyo

Osaka-Kobe

Cairo •Shanghai

Delhi TROPIC OF CANCER

Karachi

Calcutta 20°N

Bombay

Manila

Pacific Ocean

Lagos

EQUATOR 0°

Nairobi

Jakarta

Indian Ocean

20°S

Johannesburg TROPIC OF CAPRICORN

Sydney

Cape Town Melbourne 40°S

60°S

ANTARCTIC CIRCLE

80°S

0° 20°E 40°E 60°E 80°E 100°E 120°E 140°E 160°E 180°

KEYS TO KNOWING THE EARTH

1

Geography asks: Where are things located? Why are they located there?

2

Five major themes in geography are location, place, people-environment relations, movement, and region.

3

Earth is the only planet with an atmosphere that can support life as we know it. Its motions on its own axis and around the sun create day and night and the seasons of the year.

4

Maps using a grid of lines of longitude and latitude enable geographers to locate places on the planet.

5

Landforms on Earth's surface are created by tectonic plate movement, volcanic activity, folding, and faulting. Landforms are reshaped by weathering, erosion, and deposition.

6

Earth's climates have different annual temperature and rainfall patterns because of topography, latitude, and circulation systems in the atmosphere and oceans.

7

Earth's environments have distinct patterns of climate, vegetation, and soils that influence where people live.

▼ *The atmosphere which swirls around our planet is vital to life on Earth. What changes to the atmosphere have been in the news?*

8

Twice in human history, discoveries led to vital changes in how people live. Building on the knowledge and technology of earlier hunting and gathering societies, the Environmental Transformation involved the development of agriculture, herding, and urban life. The Scientific Transformation involved scientific and technological discoveries on which modern industrial life is based.

9

The technological world includes most of the countries and regions in five culture regions. These are Europe, the Commonwealth of Independent States (the former Soviet Union), the United States and Canada, Japan, and Australia and New Zealand in the Pacific world.

10

The developing world includes most of the countries and regions in six culture regions. These are Latin America, Africa south of the Sahara, the Middle East and North Africa, China, South Asia, and Southeast Asia.

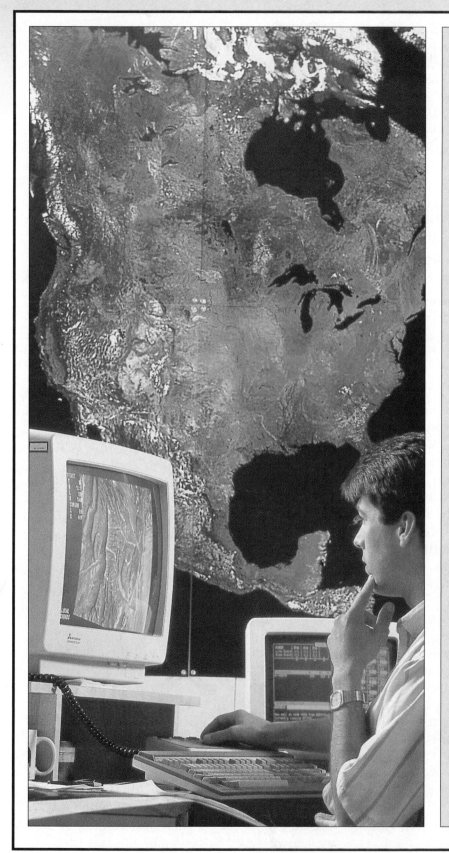

LEARNING THE GEOGRAPHY OF OUR PLANET

The photographs relayed to Earth by the *Apollo 12* spacecraft confirmed much of what we already knew about our planet. The Earth was nearly round. Its surface was primarily covered by water. In the satellite images, the oceans were blanketed by a canopy of shifting clouds that partially masked the boundaries between land and water. The atmosphere, a thin layer of air between the spacecraft and the planet, swirled above Earth's surface. Land, water, and air—the fundamental elements that make up all Earth **environments**—were actually seen as one unified reality.

◄ *Geographers use computers in analyzing locations and places. What satellite image is shown in this photograph?*

This image of the Earth as a single system reinforced our growing awareness of the need for global interaction to preserve and enhance planet Earth as our home. Earth is special among the planets that revolve around the sun. Its atmosphere and oceans provide water and air to support many forms of life. Its motions as it spins on its axis and moves around the sun create day and night and the seasons.

The word *geography* means "to describe the Earth." Geography—learning about the Earth—has fascinated people since time began. Today, geographers use five themes to organize the teaching and study of Earth: location, place, people-environment relations, movement, and region.

1. Planet Earth

The Earth in the Solar System

The Earth is a planet. A planet is a large body that revolves around and is illuminated by a star, in our case, the sun. This solar system is made up of 1 star, 9 planets, 42 known moons, some 50,000 asteroids, and billions of comets, meteors, and meteorites. More powerful telescopes are detecting new objects in our solar system constantly. All of these objects revolve around a star we call the sun. The sun is the source of all energy on Earth; it creates this energy by burning up 4 million tons of its mass every second. Earth receives this energy as heat and light.

The sun is nearly 93 million miles away from us. Light traveling from the sun at a speed of 186,000 miles per second reaches us in just over eight minutes. Light from the next closest star to Earth would reach us in 4.2 light years. An intergalactic journey in a modern spacecraft to this next closest star would take 75,000 years. Any message sent from the other side of the galaxy to Earth would take 60,000 years to reach us. For all practical purposes, then, we are confined to our solar system. Although huge, our solar system is really a small neighborhood in a galaxy made up of 100 billion stars. Beyond our galaxy lie billions of other galaxies that make up the universe.

We have identified only nine planets in our solar system. Their relative size and distance from the sun are shown on page 15. The four inner planets closest to the sun—Mercury, Venus, Earth, and Mars—are smaller, rounder, and more dense than the five outer planets of Jupiter, Saturn, Uranus, Neptune and Pluto. In fact, most planets are barren lumps of rock like Mercury or giant balls of gas like Jupiter. Venus and Mars are the two planets closest to Earth. Mars is perpetually frozen; Venus is a boiling inferno. Only Earth supports life as we know it.

The Earth Viewed from Space

A British economist, Barbara Ward, compared Earth to a spaceship. When astronauts travel through space, she noted, their spaceship must hold everything that they need to live. Earth has air to breathe, water to drink, and food to eat. It provides shelter from the freezing cold of outer space. Like a spaceship, Earth is a closed system that provides us with all the resources we need to stay alive.

You live at a time when people can actually see Earth as a spaceship because astronauts have been exploring outer space for more than twenty-five years. Satellite photographs of our planet give your generation a new vision by which to measure yourselves and your use of the planet on which you live. The Earth no longer seems infinite in size or boundless in resources. It is now the support system for 5.5 billion people.

The Structure and Size of Our Special Planet

Planet Earth is 7,926 miles in diameter when cut in half through the middle and 24,902 miles in circumference at the **equator**, an imaginary line that marks the midway point between the North and South poles. Earth is nearly a perfect sphere. The diameter at the poles is only 27 miles less than the diameter at the equator. It is also remarkably smooth. The highest point on Earth's surface is Mount Everest on the Nepal-China border in Asia at 29,028 feet. Farther west, K-2 (Godwin-Austen) in northernmost Pakistan is a close second. The greatest known depth below sea level is the 35,827-foot-

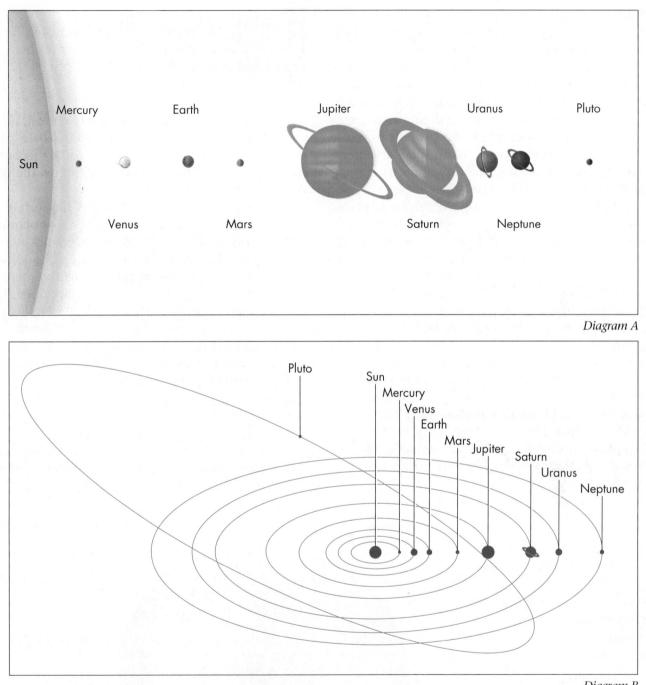

Diagram A

Diagram B

Our Solar System
The relative sizes of the nine planets in the solar system are illustrated in Diagram A.
The orbits around the sun are shown below, in Diagram B.

deep Mariana Trench east of the Philippine Islands in the Pacific Ocean. The difference between these extremes is less than 13 miles. Given its size, the surface of the Earth is smoother than that of a marble.

The fact that oceans and smaller bodies of water cover more than 70 percent of the Earth makes its surface even smoother. The largest bodies of water are Earth's four oceans—the Pacific, Atlantic, Indian, and Arctic. The Pacific Ocean alone is larger than the entire land area of the planet. Along with smaller bodies of water like the Mediterranean and Black Seas, these oceans are the source of all the Earth's available water. Evaporation from seas and oceans provides the moisture in the atmosphere that falls as rain and snow. The warm and cold ocean currents and air masses distribute heat energy from the equator to the poles. Without this air movement, Earth's climates would be boiling hot near the equator and perpetually frozen near the poles.

Although we have described the surface of planet Earth as smooth, its surface is constantly in flux—stretching, contracting, and moving. Stable as they may seem, the seven continents of Africa, Antarctica, Asia, Australia, Europe, North America, and South America and the tectonic plates under oceans are always moving. These movements are rarely noted by most people until earthquakes shake the foundations of human settlements or volcanoes spew fire and ash into the skies. Physical geographers, however, measure these events precisely. Hawaii is moving two inches closer to Japan each year, Australia is headed toward New Zealand at a rate of four inches a year. Parts of southern California are now sliding northward toward Canada. Stable as it may seem, the Earth's surface is elastic.

> **IT'S A FACT** The diameter of Earth is nearly 8,000 miles; the diameter of the sun is about 865,000 miles.

Internally, Earth is made up mostly of iron, nickel, and rock. These materials are arranged in four layers: (1) the crust, (2) the mantle, (3) the outer core, and (4) the inner core. Find them on the diagram below.

The **crust** is a band of solid rock at the surface of the planet. It is about 25 to 50 miles thick beneath the continents and from 5 to 10 miles thick under the oceans. This crust is divided into plates which slide and float on the **mantle**, a partly melted, white-hot inner layer of rock (called **magma**). This partly melted mantle extends for 1,800 miles toward the center of the Earth. Inside the mantle lies Earth's **outer core**, a 1,380-mile layer of still hotter molten iron and nickel. The **inner core** of the planet is made up of a solid 1,560–mile-wide ball of iron and nickel.

Earth is the only planet in our solar system capable of supporting many forms of life. Earth's atmosphere and oceans make this possible. The atmosphere moderates extremes of temperature. It filters sunlight and prevents dangerous radiation

Earth's Interior

We live on the Earth's crust, the thinnest of the four layers that make up our planet. The crust is in constant movement. Can you give an example of this movement?

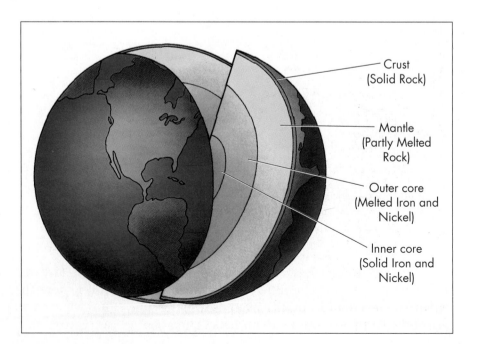

Crust
(Solid Rock)

Mantle
(Partly Melted
Rock)

Outer core
(Melted Iron and
Nickel)

Inner core
(Solid Iron and
Nickel)

from reaching the surface and provides the air that we breathe. Earth also has water, which is necessary for life. Most of this water is stored in the oceans, which accumulate excess heat during the day and release it slowly at night. Together with the atmosphere, oceans regulate temperature. These two elements make Earth livable.

The Earth Constantly Moves

The Earth moves quickly, rotating on its axis at a speed of more than 1,000 miles an hour at the equator. **Rotation** is the turning of the Earth on its axis. Earth's **axis** is an imaginary line that passes through the center of the planet from the North Pole to the South Pole. Earth's rotation is why we have night and day.

Earth also revolves around the sun, plowing through space at an average speed of 66,000 miles per hour. **Revolution** causes variations in the length of night and day at different places on Earth's surface; it also contributes to the seasons of the year. You have no sensation of these movements. You experience Earth as a solid platform because you are stuck to its surface by the force of gravity. The atmosphere is also held to Earth by gravity, just as you are. Although you cannot feel these motions, rotation and revolution vitally affect life on planet Earth.

Rotation Causes Night and Day

The earth rotates from west to east, as the diagram below shows. Each complete rotation takes twenty-four hours. This causes day and night. Day happens on the half of the Earth that is facing the sun and receiving sunlight. Night occurs on the dark half of Earth facing away from the sun. The imaginary line that separates the lighted from the darkened half of the planet is called the **circle of illumination**. This line moves as the Earth rotates. When the circle of illumination passes over us, we experience either dawn or dusk.

What would happen if the planet rotated more slowly and turned on its axis once every forty-eight hours instead of once every twenty-four hours? Our days and nights would be twice as long. If Earth spun faster, days and nights would be shorter. If Earth turned on its axis only once a year, the side facing the sun would always be hot and the other side would be perpetually frozen. As you can see, the rhythm of our life on the planet is closely tied to Earth's rotation.

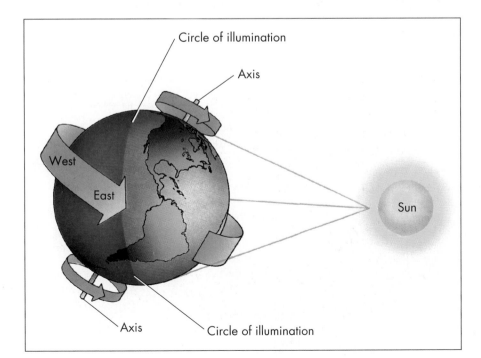

Earth's Rotation Causes Day and Night

The sun does not move around the Earth. The Earth rotates from west to east. One complete rotation takes twenty-four hours. As Earth rotates on its axis, half of our planet receives sunlight (daytime) and the other half does not (night). On this diagram, it is daytime on most of which two continents? Where would it be night on our planet? Is it morning or afternoon in California?

Revolution and Earth's Tilted Axis Cause the Seasons

Revolution is the annual movement of Earth around the sun. The fixed path that Earth follows as it moves around the sun is called its orbit. Each orbit around the sun takes 365¼ days, and this determines the length of the calendar year—365 days. What about the one-fourth day? Every four years, an extra day is added to the calendar in February. This added day every "leap year" takes care of the extra quarter day that Earth needs to complete its annual journey around the sun.

As Earth revolves around the sun, seasons come and go and days and nights vary in length. We have seasons because the Earth's axis is tilted at an angle of 23½ degrees as it moves around the sun. Because of this tilt, every place on Earth receives more sunlight when it is tilted toward the sun and so has a warm season, or summer. Each place receives less sunlight and has a cool season, or winter, when tilted away from the sun. Between these extremes, spring and fall are seasons with average amounts of sunlight and warmth for that location.

Look at the diagram below of Earth's annual revolution around the sun. This shows Earth at eight different times of the year. Notice that Earth's axis is always tilted at the same angle in all eight of the positions that represent the different times of the year, but the areas that receive sunlight differ. The North Pole axis is in total darkness in December and in

> **IT'S A FACT** Century years, such as 1800 and 1900, are only leap years if they are divisible by 400—as in the year 2000. This practice eliminates 3 leap years every 400 years—keeping the seasons aligned with the calendar.

total sunlight in summer. What do you think it would be like living in the Arctic or Antarctica?

Earth's tilted axis, then, creates our seasons. In June, the Northern Hemisphere is tilted toward the sun, so it receives more sunlight than the Southern Hemisphere. The North Pole has sunlight twenty-four hours a day; the South Pole is dark all the time. In December, the situation is reversed. The Southern Hemisphere is tilted toward the sun and receives more sunlight. At that time the North Pole is in complete darkness, and the South Pole is lit twenty-four hours a day. In March and September, the circle of illumination passes through the poles. Around March 21st and September 23rd (the dates may vary from year to year) are the only two days in the year when the entire planet has twelve hours of daylight and twelve hours of darkness.

If the Earth were not tilted on its axis, there would be no seasons. The equator would receive the most sunlight every day. The poles would receive the least sunlight every day. Days and nights would be equally long everywhere on Earth, all the time.

Earth's Revolution Around the Sun

Earth revolves around the sun once a year. Notice that the Earth's axis is always tilted in the same direction all year round. Some parts of Earth receive more sunlight and others less sunlight at different times of the year. Can you see where it might be warmest in June? In December?

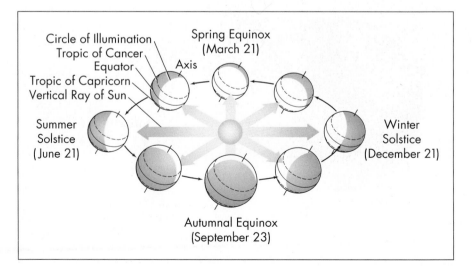

Equinoxes and Solstices Mark the Vertical Rays of the Sun

Days change in length from season to season, and this influences how much sunlight reaches Earth's surface at any given place. The longer the day, the more light arrives. But there is a second factor that influences the amount of total sunlight that reaches the Earth; this is the angle at which the sun's rays hit the surface.

Look carefully at the diagram on this page. This diagram shows the angle of the sun's rays as they reach Earth in June, when the Northern Hemisphere is tilted toward the sun. The vertical, or direct, rays of the sun strike Earth at an imaginary line known as the Tropic of Cancer (23½° N) on about June 21st. At the Tropic of Cancer, sunlight is very concentrated over a small area at the surface, because it pierces the atmosphere directly. The more direct the angle of the sun's rays are, the more heat arrives at Earth's surface. That is why direct sunlight in summer heats places on Earth's surface

to a greater degree than sunlight in winter. That is also why tropical regions between the Tropic of Cancer and the Tropic of Capricorn (23½° S) are warm. In contrast, the sun's rays in June strike the polar region in the Northern Hemisphere at a very sharp, or indirect, angle. Sunlight is spread over a larger surface area and must pass through a thicker layer of atmosphere. As a result, less heat reaches Earth's surface, which is why polar regions are cold.

The angle at which the sun's rays strike the Earth changes continuously during the year. Look again at the diagram that shows Earth's revolution around the sun. Twice a year on about March 21st and September 23rd, the vertical rays of the sun are directly over the equator. The circle of illumination passes directly through the North and South Poles. On these dates, day and night are of equal length everywhere on Earth. For this reason, these two dates are called **equinoxes**, from the Latin word meaning "equal night." In the Northern Hemisphere, March 21st is called the spring equinox. September 23rd is called the autumnal equinox. Do

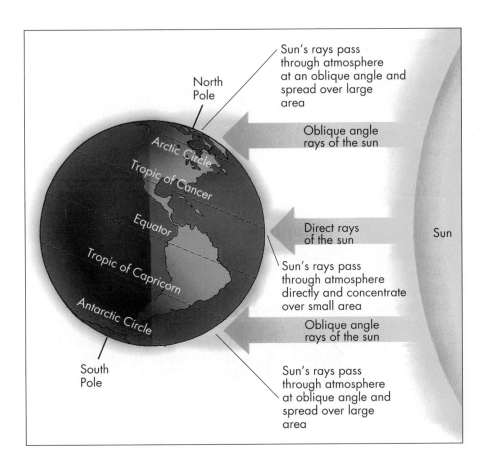

The Sun's Rays in June

In June, the Northern Hemisphere is tilted toward the sun. The direct rays of the sun hit Earth near the Tropic of Cancer. Remember that the more direct the sun's rays, the more heat reaches Earth's surface. Although the North Pole receives sunlight twenty-four hours a day in June, the oblique angle of the sun's rays reduces the amount of heat reaching Earth's surface at the pole. Note that in June, the South Pole receives no sunlight at all. Do you understand why it is warmer between the Tropics of Cancer and Capricorn than poleward of the Arctic and Antarctic Circles?

The sun appears to travel northward at a rate of 30 miles a day in spring.

you understand why the terms *spring* and *autumn* would be reversed in the Southern Hemisphere?

Between March and June, the vertical rays of the sun strike Earth farther and farther north of the equator. In the Northern Hemisphere, days lengthen and spring turns into summer. On about June 21st, the vertical rays of the sun reach their northernmost limit. The sun is directly above the Tropic of Cancer. Early observers of the sun's motions used the word **solstice** ("sun standing still") to describe that day when the sun appears to stop in its journey, pause, and turn back toward the equator. This day is called the summer solstice in the Northern Hemisphere because it is the longest day of the year. In the Southern Hemisphere, it is called the winter solstice because it is the shortest day of the year.

After June 21st, the vertical rays of the sun migrate back toward the equator. They pass over the equator on September 23rd (the autumnal equinox), and reach their southernmost limit over the Tropic of Capricorn (23½° S) on about December 21st. This day is called the winter solstice in the Northern Hemisphere, where it is the shortest day of the year. In the Southern Hemisphere, it is the longest day of the year. What do you think people in Australia or Argentina would call December 21st?

The Tropics, the Middle Latitudes, and the High Latitudes

Latitudes are imaginary lines drawn on globes to show distances north and south of the equator. The location of a place determines the amount of sunlight that that place receives. Five latitudes are especially important because they act as benchmarks that tell us the amount of sunlight reaching places along each line.

Find the equator in the diagrams on pages 18 and 19. The equator, latitude 0°, receives more direct sunlight than any place on Earth, which is why regions near the equator are usually hot all year. The Tropic of Cancer, north of the equator at 23½°

N, and Tropic of Capricorn, south of the equator at 23½° S, mark the farthest points away from the equator that receive vertical sunlight. This happens on the solstices. Find them on the two diagrams. The belt between these two lines of latitude, or **parallels**, is called the **low latitudes** or **tropics.**

Now locate on the two diagrams the Arctic Circle far north of the equator, 66½° N, and the Antarctic Circle far south of the equator, 66½° S. These are the parallels closest to the equator and farthest from the poles that have at least one day of complete sunlight and one day of complete darkness each year. This also happens on the solstices. North of the Arctic Circle and south of the Antarctic Circle are the **high latitudes** or **polar regions.** The high latitudes receive little heat from the sun, which is why they are cold. The North and South Poles have six months of daylight and six months of darkness each year. Two other belts are located between the low latitudes of the tropics and high latitudes of the polar regions. These are the **middle latitudes** or **temperate regions;** they are found between the Tropic of Cancer and the Arctic Circle and between the Tropic of Capricorn and the Antarctic Circle. Locate them on the diagram on page 21.

The terms for these belts of latitude are used to describe variations in the amount of sunlight that warms the surface of the Earth. Remember that these variations are caused by Earth's tilted axis. If the axis were not tilted, the sun's vertical rays would always strike our planet at the equator, days and nights would be equally long, and we would live in a world without seasons.

Mapping Our Round Planet

Modern mapmaking began with the Greeks. They were probably the first people to recognize that the Earth was round. They also developed the first grid system of lines of longitude and latitude that makes it possible to locate places accurately on the Earth's surface. Lines of **longitude,** or **meridians,** run north-south. They connect the North Pole to the South Pole and meet at the poles. The **Prime Meridian,** or 0° longitude, runs from the North Pole to the South Pole through the Royal Observatory at Greenwich, England, a suburb of London.

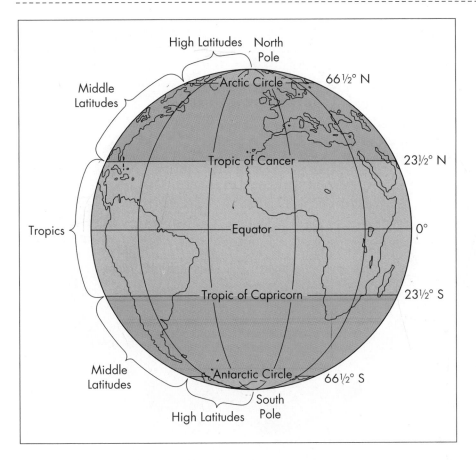

The Tropics, the Middle Latitudes, and the High Latitudes

The tropics are located between what two lines of latitude? What two lines of latitude form the boundaries of the high latitudes? The middle latitudes are located between what four lines of latitude?

Near the observatory, a brass line runs down a north-south street marking the "place where time begins," the Prime Meridian. From the imaginary extension of this line, meridians are labeled according to their location in degrees east or west of the Prime Meridian. At the equator, each degree of longitude is equal to slightly more than 69 miles or $\frac{1}{360}$ of the equator, which, being a circle, is divided into 360 degrees.

The equator is halfway between the poles. Lines of latitude, or parallels, are parallel to the equator. They connect points of equal distance north or south of the equator. Together, meridians and parallels form a grid that locates places exactly on Earth.

Reading Maps

Mapmakers, or **cartographers**, provide a great deal of information on each map—information that is explained in a **map key**, or **legend**. A **map scale**, which is sometimes in the key, is a line or bar that marks out how many inches on the map equal how many miles on Earth's surface. One inch may equal 100 miles or more on a map of the United States, but 1 inch may equal only 1 or 2 miles on a map of your town. Map scale is also expressed by using a representative fraction such as 1:62,500, which is frequently used because 1 inch on the map equals approximately 1 mile on the ground.

Various symbols on maps give you additional information. For example, you may be familiar with this symbol:

This directional indicator is often used to show where north is on maps. Another symbol sometimes used is a **compass rose**,

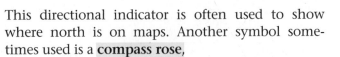

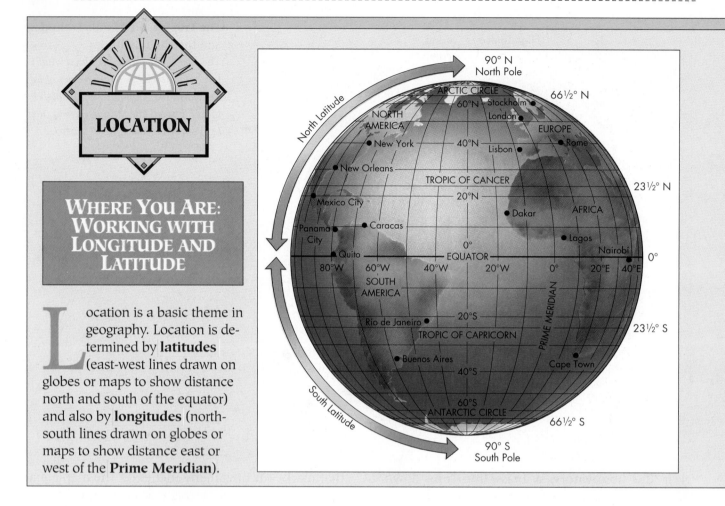

LOCATION

WHERE YOU ARE: WORKING WITH LONGITUDE AND LATITUDE

L ocation is a basic theme in geography. Location is determined by **latitudes** (east-west lines drawn on globes or maps to show distance north and south of the equator) and also by **longitudes** (north-south lines drawn on globes or maps to show distance east or west of the **Prime Meridian**).

which shows all four major compass points. In either case, if you know which direction north is, you also know where south, east, and west are on the map.

Other map symbols are usually explained in the map key. Line symbols mark boundaries of countries and states and show rivers, railroads, and streets. They also are used as contour lines to connect points on a map that have the same elevation. Point symbols mark the location of cities, oil wells, schools, and other places of interest. Area symbols use shading or color to provide information about areas that grow the same crops or have the same population density, climate, or elevation. Study the symbols that are used on the maps in the miniatlas at the beginning of this unit and explain why they are used.

When read correctly, a map will communicate much useful information. The language of maps is the language of symbols that the map key helps you to read. Developing map-reading skills is important to learning about the geography of our planet.

Map Projections

The only true map of the Earth is a globe, which is a model of Earth, but is hard to store and use. Its size also limits the amount of detail it can tell us. This creates a problem. How can mapmakers draw

IT'S A FACT   In Portugal, at the time of Columbus, selling a map to a foreigner was a crime punishable by death.

Using the picture of the globe and the information you have already learned, answer as many questions as you can about latitude and longitude:

🌐 QUESTIONS

1. What imaginary line is at 0° latitude?
2. What is the latitude at the North Pole?
3. What is the latitude at the South Pole?
4. What is the latitude at the Tropic of Cancer?
5. What is the latitude at the Tropic of Capricorn?
6. Are any parts of Europe or the continental United States in the tropics?
7. Are large or small parts of South America and Africa in the tropics?
8. What is the latitude of the Arctic Circle?
9. What circle lies at 66½° S?
10. Are New Orleans and Lisbon both in middle-latitude, or temperate, regions?
11. Is Mexico City in a middle-latitude, or temperate, region?
12. Is Buenos Aires in a middle-latitude, or temperate, region?
13. Which city in South America and which city in Africa are very near 0° latitude?
14. Is New Orleans or Rio de Janeiro farther from the equator?
15. Mexico City is about as far north of the equator as what South American city is south of the equator?
16. Which city in Europe is very near 0° longitude, or the Prime Meridian?
17. Which city in Africa is located at 20° E (longitude)?
18. Both Lagos and Dakar are in Africa. Which is east and which is west of the Prime Meridian (0° longitude)?
19. The location of what city is about 10° N (latitude) and 80° W (longitude)?
20. The location of what city is about 60° N (latitude) and 20° E (longitude)?
21. What is the approximate longitude and latitude of your home?
22. How did you do in this miniquiz?

the round surface of a globe on a plane such as a flat piece of paper? The truth is that they cannot, exactly.

A flat map, which is a representation of Earth, cannot show the exact shapes and sizes of lands and bodies of water, because there is no way to flatten the surface of a round globe without distorting shapes and sizes. Try to flatten the rind of half an orange and you will see that it tears and becomes distorted. This is a problem that cartographers must work with.

What they do is control the amount of distortion by creating different kinds of map projections that show either area, shape, distance, or direction correctly. Only degrees of longitude and latitude are correct on all maps. If you decide to become a cartographer, you will soon find out that it is impossible to show accurately area, shape, distance, and direction on the same map. This means that you must make a choice. You do this by your selection of map projection.

A **map projection** is a grid of lines projected onto one of several geometrical surfaces. Some map projections permit mapmakers to represent area in exact proportion to its reality on Earth. The shape may be distorted and so may direction, but any square or rectangle on the map has a correct area. These projections are called **equal-area projections**. Maps that show true shape for limited areas (it is impossible to do it for the entire map) are called **conformal projections**. Some projections are able to show distance accurately from one or two points, and these are called **equidistant projections**. There are also projections that show true direction.

Map projections are classified into three families, depending on the geometrical surface onto which

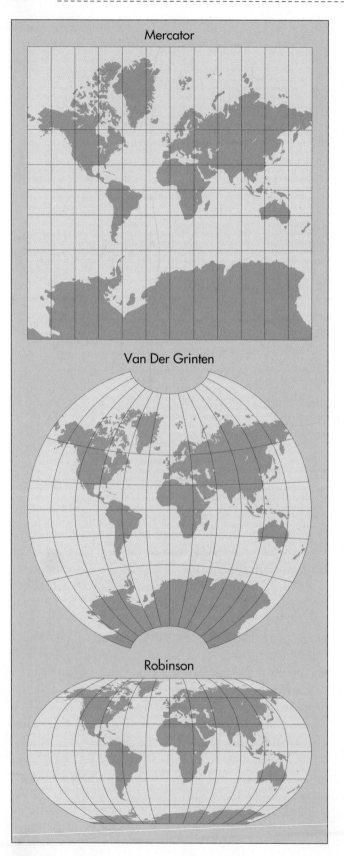

Mercator

Van Der Grinten

Robinson

the lines of longitude and latitude are projected. Planes, cones, and cylinders are the most frequently used geometrical surfaces, each of which distorts the representation of Earth in a different way.

Mapmakers are experts at knowing the exact amount of distortion of areas, directions, distances, and shapes on every map projection. As a result, they are able to choose the most suitable map projection for a given area or purpose.

The most famous map projection in history is the Mercator projection, a **cylindrical projection** named for its inventor, a Flemish geographer and cartographer of the 1500s named Gerhardus Mercator (see the diagram on this page). This man produced some of the best maps and globes of his time, but his most famous projection was special. Every straight line on a Mercator projection is a line of true direction, a **loxodrome**. This made it possible for navigators to plot straight-line, true-direction courses from one place to another on the surface of the Earth using a Mercator map. Imagine yourself a captain of a ship without modern navigational and communications equipment, and you will immediately realize how useful such a map would be.

Note, however, that when used as a world map, the Mercator projection greatly enlarges areas in the high latitudes or polar regions. That is why many of us think of Greenland as about the size of South America, even though it is only slightly larger than Mexico. And few map readers using a Mercator world map would realize that Australia is five times the size of Alaska, or that Scandinavia would fit into India four times.

Mercator's projection was suited to his time, but many others have been developed since then. In addition to cylindrical projections, azimuthal (point) and conic (cone) projections have been developed for special purposes (diagram page 25). **Azimuthal projections** measure equal distance from its central point to any other point on the map. **Conic projections** are useful to depict a **hemisphere,** or half the globe, or even smaller parts of the globe. Look at the difference that the choice of map projection makes in depicting the Commonwealth of Independent States (CIS) in the maps. Notice the differences in apparent distance between Moscow and Yakutsk on these three

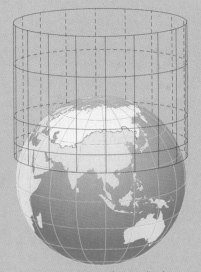

Earth projected on cylinder

Earth projected on azimuth (point)

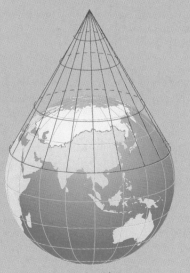

Earth projected on cone

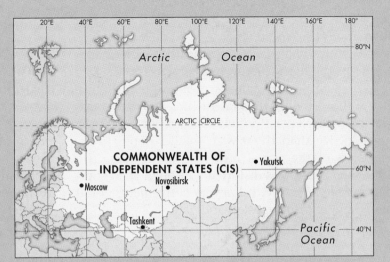

Cylindrical projection of the Commonwealth of Independent States

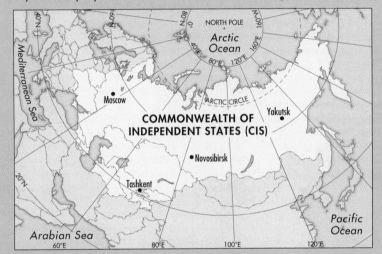

Azimuthal projection of the Commonwealth of Independent States

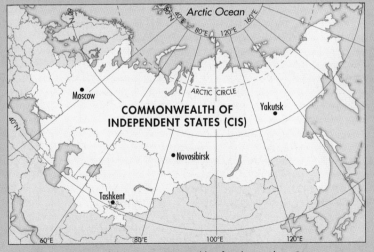

Conic projection of the Commonwealth of Independent States

projections. A cartographer would instantly understand how and why each projection created such different views. Would you?

Sometimes you may see a map that looks something like the figure below. In a **cartogram**, the size of each country, or state, is proportionate to the theme that is being illustrated. For example, countries with the largest populations would be shown larger in size than those with smaller populations, regardless of the actual area in square miles. The size of the country is distorted, although the general shape remains true. It is a very convenient, visual way to present information and data.

A New Map Projection for the National Geographic Society

The extreme distortion of high-latitude land-masses on Mercator world maps led the National Geographic Society to adopt the Van der Grinten projection for most of its world political maps.

In 1988, the National Geographic Society adopted a new map projection—a new view of the world—developed by the cartographer Arthur Robinson. The Robinson projection more accurately represents shapes and areas than the Van der Grinten projection.

Cartogram
The cartogram assigns to a particular region an area based on some value other than land-surface area. What value determines the sizes of states in this cartogram?

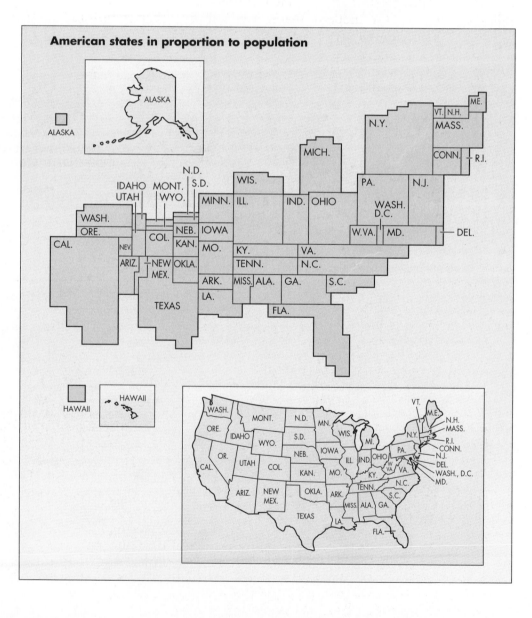

American states in proportion to population

Compare the Van der Grinten and Robinson projections in the diagram on page 24. On the Van der Grinten projection, Canada and Alaska are more than twice as large as they are in reality, and Greenland is six times its real size. On the Robinson projection, countries and continents more closely match their true size. As Gilbert M. Grosvenor, president of the National Geographic Society, notes, "The Robinson projection more accurately portrays round Earth on a flat surface."

Computer Cartography and Geographical Information Systems

Today, computer systems store much of the data that is presented on modern maps. **Computer** **cartography** creates base maps upon which data can be plotted, analyzed, and retrieved. Geographic Information Systems (GIS) are spatial databases of statistical information which are computer-based programs that enable you to display, store, and use data on various places in your county, state, or nation.

The types of information stored are called themes. In the figure below, data on four themes—cold, ruggedness, dryness, and poor soil—are presented for the Commonwealth of Independent States (CIS), the new name for the former Soviet Union. These data include the two essential features needed to create a GIS map: (1) what characteristics are connected to what places and (2) where these characteristics are located. Once this database exists,

Environmental Characteristics of the Commonwealth of Independent States

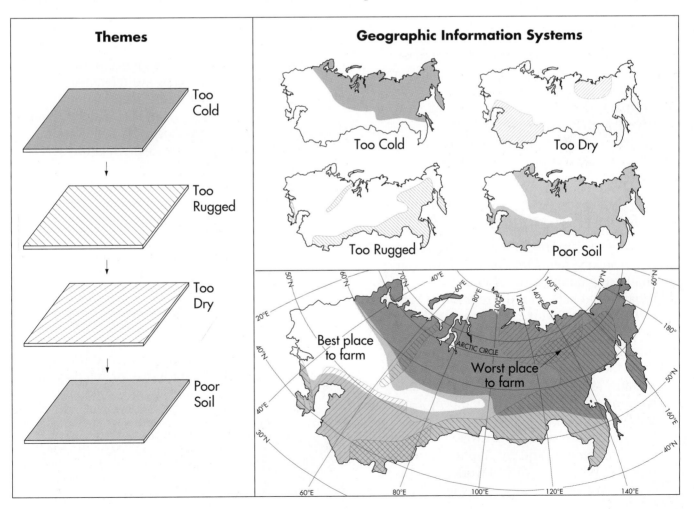

we can use it to answer interesting and practical questions.

In the diagram on the CIS, for example, the environmental characteristics would tell us where it would be wise or unwise to be a farmer in the CIS. The best place to farm would clearly be in the west, where none of the four negative conditions stored in our database occurs. The worst farming conditions in the CIS occur in a small area of northern Siberia that is cold, dry, rugged, and has poor soils. You can understand why utility companies, oil companies, and other firms that operate over large areas find GIS an important tool.

⊕ **REVIEW QUESTIONS**

1. What is a planet, and what planets are included in our solar system?
2. What are the four layers that make up the Earth? Each layer consists of what materials?
3. What does Earth's rotation cause? What does Earth's revolution around the sun cause?
4. Where are the (a) low latitudes or tropics, (b) high latitudes or polar regions, and (c) middle latitudes or temperate regions?
5. In what direction do lines of longitude run, and where do they meet?
6. On which projection is every straight line a line of true direction? Why is this useful to navigators?

⊕ **THOUGHT QUESTIONS**

1. Why are we confined, for all practical purposes, to our solar system? Do you think this will change in your lifetime? If so, how?
2. What would happen if the rotation of Earth were slower? What would happen if the rotation of Earth were faster? How might these conditions affect your body?
3. How does the tilt of the Earth cause our different seasons? Why do most North Americans think in terms of four seasons? Would people in central Brazil?
4. What questions could be answered using computer cartography and geographical information systems? Whom might you work for if this becomes your special field of knowledge?

2. Learning Geography: Five Themes

Five Themes in Geography

Geographers work to understand the differences and similarities among places on Earth. They study the "where" just as historians study the "when" of things. To gain this understanding, geographers focus on five main themes: (1), location, (2) place, (3) people-environment relations, (4) movement, and (5) region. By focusing on these five themes, geographers are able to gain a clear and organized vision of Earth. These themes help us to see our world in ways that satisfy our curiosity and build our understanding.

Location

Geographers study *where* things are located. For geographers, locations are reference points in space in the same way that dates are reference points in time for historians. **Absolute location** is the exact position of a mountain, river, lake, city, or town on Earth's surface. **Relative location** is the position of a location on Earth's surface in relation to other locations.

Absolute locations do not change in their degrees of longitude and latitude, just as dates do not change. This matters a great deal to astronauts, airplane pilots, and navigators who need to know precise distances and directions. Laser beams that bounce off satellites can now pinpoint absolute locations on Earth within an inch or two through what are called Geographic Positioning Systems (GPS).

The absolute locations of Philadelphia, Pennsylvania; Madrid, Spain; Ankara, Turkey; and Beijing, China, for example, are all the same distance north of the equator. Their locations in relation to the

IT'S A FACT Atlanta, Georgia, is closer to Detroit, Michigan, than it is to Miami, Florida.

Prime Meridian can be found on a world map. Absolute location does not tell us, however, that Philadelphia is an important industrial and port city with access to the Atlantic Ocean. It does not tell us that Madrid and Ankara are capital cities near the geographical centers of their countries. It does not tell us that China's capital city of Beijing is located on the North China Plain sixty miles from the Yellow Sea and just inside the Great Wall of China. Absolute location tells us only *where* these important cities are located, nothing more.

In the same sense, the absolute locations of your house, your street, your neighborhood, and your school are registered in local government offices. This is vital information for planners, property owners, and tax collectors. Absolute location does not, however, tell us much about where you live. We do not know, for example, whether you live in a central city, in a suburb, or on a farm. Nor do we know whether you live on a highway, on a side street, or on a rural road.

Knowing relative location gives us much more information. It tells us the ways in which a location is positioned relative to other locations on Earth's surface. Relative location, for example, tells us that the city of Philadelphia is located on the Delaware River, which flows into the Atlantic Ocean. The Schuylkill River, which joins the Delaware River in Philadelphia, gives the city access to the interior. Moreover, Philadelphia is one of a number of cities that line the East Coast of the United States. Like most of these cities, Philadelphia is located on a narrow coastal plain between the Appalachian Mountains and the Atlantic Ocean.

Similarly, the relative location of your home tells us where you live with respect to other locations in your area. It tells us whether you live in the center of a city or in the countryside, how far you travel to your school, and whether there are any movie theaters or parks near your home.

Relative locations change through time. Two hundred years ago, Philadelphia's site at the junction of the Schuylkill and Delaware Rivers was crucial to the city. Sailing ships from all over the world traded at docks along the Delaware River, and barges moving up and down the Schuylkill River gave Philadelphia access to farmland in the interior. Waterfalls on the river provided industry an inexpensive source of energy. Overland transport was more difficult and expensive than water transport— it was two days by stagecoach to New York City. Mail was so slow that news of the signing of the Declaration of Independence in Philadelphia took twenty-nine days to reach Charleston, South Carolina, about 500 miles away.

Today, Philadelphia's relative location has changed. Highways, railroads, and airlines connect the city with distant points in the United States and abroad. Forty miles of riverfront are lined with industries. Low-cost water transportation and direct access to international trade are still locational advantages. Railroads and roads across the Appalachian Mountains carry raw materials for manufacturing to and from Philadelphia. Overland trips that took days now are completed in hours. Philadelphia is still well connected to other locations, but its increased access to world resources and trade is an important change in relative location.

Geographers emphasize the theme of location because it is basic to understanding the world. Location helps to define the distribution of climates, vegetation, natural resources, and patterns of human settlement. Where are natural resources vital to industry located? Where is good farmland to be found? Where are new industries being built? What transportation and communication systems link specific locations with others? Which countries are located in areas where there is warfare? All of these questions require knowledge of absolute and relative location. They help us understand *where* and *why* events take place.

Place

A place is a particular city, village, or area with distinctive physical and human characteristics that distinguish it from other places. Geographers study places to understand human activity in a physical setting. Each place has distinctive landforms, bodies of water, climate, soils, and vegetation. Each is occupied by people with a particular religion, language, and set of political beliefs. These people build homes, stores, factories, places of worship, and roads. Each place's set of physical and human characteristics gives it meaning and character. In this sense, all places are unique.

Return for a minute to the example of Philadelphia. The city was founded in 1681 by William Penn and settled by English Quakers, Swedes, and Germans. It was a planned community two miles long and one mile wide set on a peninsula between the Schuylkill and Delaware rivers. A grid of north-south, east-west streets divided the city into blocks. Rows of brick townhouses lined the streets. A large square at the center was reserved for public buildings, and smaller squares were laid out as parks.

A hundred years later, Philadelphia was the largest city in the United States and our federal capital. Philadelphia boasted 40,000 residents, the best water supply of any American city, and the country's oldest continuously inhabited street. Its population of artisans and merchants made Philadelphia one of the most important cities in North America.

Its moderate climate, fertile farmland, river location, access to the sea, and talented people were significant factors in Philadelphia's growth and development.

Today, Philadelphia sprawls over many square miles and has a metropolitan population of nearly 6 million. Textile factories, shipbuilding facilities, oil refineries, and other industries occupy the riverfront and are situated along the main railroad lines that enter the city. Air, rail, and road networks connect Philadelphia with cities on the East Coast and far beyond the Appalachian Mountains. Residential areas spread outward from the city to smaller suburban communities in three states. Philadelphia is a different place now than it was when our country was founded. It has been shaped and reshaped by human ideas, aspirations, and activities.

▼ *Philadelphia, a grid-pattern town, was an early, important port city on the east coast of America.*
What factors in its relative location were important in its early development in the 1700s?

▲ *Modern Philadelphia differs in size, architecture, and relative location as compared with historic Philadelphia. What factors of relative location have changed over the last 300 years?*

People-Environment Relations

Geographers study interactions between people and environment to understand how human beings use the Earth. The word *environment* comes from the French word meaning "circle" or "surroundings." An environment may be as small as your neighborhood or as large as the world. **Environment** is the space you occupy, your place, your natural and social surroundings.

People modify environments to make them more or less productive or more or less comfortable. Each environment presents opportunities and limitations to human activity. How people change environments and adapt to them depends on their beliefs, ideas, economy, social organization, and technology.

The Great Plains of central North America show how interactions between people and environments explain why things are located where they are. In the early 1800s, this level, treeless grassland was incorrectly believed to be unfit for farming. One government map labeled the area "The Great American Desert." Bison-hunting tribes of Plains Indians occupied these grasslands. These Native

Americans were skilled horsemen and horsewomen who migrated northward in summer and southward in winter, living off of enormous herds of bison, some as large as 12 million animals.

Uncertain rainfall, scarcity of firewood, winter blizzards, and summer grass fires made the Great Plains unattractive to pioneer farmers crossing the Mississippi River and heading west. The farmers also incorrectly believed that land without trees was infertile. These settlers, therefore, bypassed the Great Plains and moved across the grasslands in wagon trains, headed for the tree-covered lands of Oregon and California. To farmers from the eastern United States, the Great Plains was an unfamiliar environment whose climate was very different from that back home. Thus the Great Plains remained thinly populated until after the Civil War in the 1860s.

When railroads penetrated the Great Plains, bison hunters, farmers, and cattle herders began to move onto the Plains Indians' hunting grounds. They soon destroyed the bison herds on which Plains Indian life was based, and cattle replaced bison on these grasslands. Soon after the slaughter of the bison, Native Americans living on the plains were driven off the land, rounded up and shunted onto reservations, or killed.

Once railroads could inexpensively transport beef to markets in Eastern cities, cattle were rounded up and driven to railhead cowtowns on the southern plains. Earlier, in 1862, Congress had passed the Homestead Act, which entitled every adult American to 160 acres of land in the Great Plains if they paid $10 and promised to produce a crop in five years. Encouraged by the construction of railroads, and lured by cheap land, thousands of settlers cut into the thick grassland sod and established farms and permanent settlements on the Great Plains. Slowly the environment of this region was transformed.

Final mastery of the Great Plains environment was accomplished by technology. The steel plow

▲ *The Great Plains of central North America were occupied by bison-hunting Native Americans.*
Now this environment is one of the great wheat-growing and cattle-raising areas of North America.
How do these two facts relate to the theme of people-environment relations?

cut through the thick turf, windmills pumped water to the soil's surface, and barbed wire fenced off cropland from animals. Mechanical reapers, binders, threshers, and other farm machines produced a torrent of grain from the Great Plains.

Today, this environment is one of the great wheat-growing and cattle-raising areas of the world. Generations of people transformed the Great Plains into a carefully managed human landscape divided into farms and ranches. Droughts in the 1880s and 1930s spelled disaster for many settlers and gave rise to the name "dust bowl" to describe the region. However, new techniques of animal raising and methods of dryland farming have again made the Great Plains a productive human environment. Changing people-environment relations on the Great Plains created new landscapes and new ways of living.

Movement

Movement is the study of interactions among people and other life forms located in different places and different environments. Three types of movement are of special interest to human geographers. These are migration, transport, and the spread of ideas.

Migrations of people on Earth's surface have changed patterns of living and altered environments around the globe. The large-scale migration of Europeans, Africans, Latin Americans, and Asians to the United States created a nation of immigrants who contributed their knowledge and strength to a growing country. Europeans brought labor and technology to help the United States become a leading industrial nation. Africans, imported against their will, provided labor to build the agri-

IT'S A FACT   **Every day seventy species of life on Earth become extinct.**

cultural economy of the South. Latin Americans, themselves migrants to North and South America, helped to develop the unique landscapes of the Southwest and California. Today, Asians form an increasing portion of the new immigrants to the United States. These movements of people brought together a rich blend of cultures and talents that form the fabric of America's human geography today.

Movement of goods by transport networks provides a second vital connection among places on Earth's surface leading to today's complex patterns of world trade. In the United States, the gradual building of turnpikes, canals, railroads, interstate highways, and airlines united the nation by providing more effective means of connecting people and resources.

The opening of the Erie Canal that linked the Hudson River Valley with the Great Lakes in 1825, for example, resulted in the growth of New York City, which is at the mouth of the Hudson River. New York City became the nation's largest port, trading the resources of the interior for other products. Transport networks to the fertile farmland and abundant mineral deposits of America's interior supported the growth of other large cities along the Atlantic coast, as well as newer cities in the central and western United States. Railroads made Chicago the hub of the Middle West. Natural transport routes like the Mississippi River channeled flows of goods between the north and the south. Spreading transportation networks led to the growth and expansion of places along those networks. Efficient, modern movement of goods is a reliable measure of a country's well-being. Few places can thrive without goods and services from other places and without allowing for economic specialization and the interdependence of regions.

The spread of ideas, attitudes, and discoveries through communication and trade is a third type of movement that geographers study to understand places. The spread of Christianity from the Middle East into Europe led to its establishment there, as well as in regions later settled by migrating Europeans. Technological and scientific discoveries that led to the Industrial Revolution in England in the 1750s spread to other lands, where they now form the underpinnings of industry in the countries of the technological world. New methods of farming changed landscapes. New modes of manufacturing led to the growth, decline, or relocation of industries.

No place can rely only on ideas developed at that location. The distinctive character of a place is often based on ideas that the people of that place have accepted and made part of their life and environment. A great number of these ideas developed in different places and different times and as they spread created cultural changes that are now seen as patterns of unity and variety worldwide.

Region

Regions are parts of Earth's surface that share one or more characteristics that distinguish them from surrounding areas. Geographers divide the world into regions to show similarities and differences among areas, just as historians divide time into historical periods.

Regions can be based on a single characteristic or on many. A region's unifying characteristics may be political, cultural, economic, or environmental. They can be as small as your neighborhood or as large as you choose. California, for example, is a political region within whose borders the laws of the state of California apply. The Great Plains, in contrast, is an environmental region whose landforms, climate, soil, and vegetation distinguish it from the Rocky Mountains to the west and the more humid parts of the Central Plains to the east.

Regions often describe very complicated combinations of culture, economy, and environment. This is the case when we use terms like *New England*, the *South*, *Scandinavia*, or the *Far East*. Such terms usually carry meanings that extend far beyond the location of these regions. They call to mind the customs of the people, the environments in which they live, their heritage, and certain economic activities.

Culture regions are defined by the language, history, and economy of large parts of Earth's surface. When we study Latin America, China, the Middle East, or Japan, we identify these regions as sharing common characteristics and ways of living that distinguish them from other culture regions. Geographers use regions to divide the world into more understandable segments that show patterns of unity and diversity.

The Geographer's Eye

Geographers view the world from a perspective that is different from that of historians, anthropologists, and other social scientists. Geographers' curiosity about location, place, people-environment relations, movement, and region reflects their conviction that it is important to understand patterns of human activity on Earth. The emphasis is always on *where* things are located and *why* they are located there.

Never has the geographer's eye been more important to the well-being of the people of the United States. The United States is now deeply entwined in a global economy and environment. Global interaction has become a reality, as events throughout the world affect our lives daily. Markets for our industrial and farm products extend around the globe. Our own dependence on oil and mineral resources from many countries involves Americans with faraway lands and peoples. Political changes on other continents transform our relations with foreign countries. The skills and talents of individual Americans are increasingly employed by multinational corporations that cross political boundaries. Interactions with peoples from other parts of the world occur frequently. As you learn about the different culture regions of this human world, you will gain perspective on forces that will affect your own life.

REVIEW QUESTIONS

1. What is the difference between absolute location and relative location?
2. How does a geographer define the word *place*?
3. What types of movement are of special interest to geographers?
4. On a global scale, how are culture regions defined?

THOUGHT QUESTIONS

1. What characteristics give each place a meaning that distinguishes it from other places? Describe those that distinguish your neighborhood, town, or city from others nearby.
2. Why do people modify environments, and what human factors determine how people change their environments? Look out the window and describe the changes people have made in the landscape you see. Have environmental changes in your town or city become issues of political debate?
3. Why do geographers divide the world into regions, and what characteristics can unify regions? What is the name of the region in which you live? What are its boundaries?

CHAPTER SUMMARY

The planet Earth is made up of land, water, and air. It revolves around the sun in the solar system. The Earth has special characteristics that define life as we know it: the Earth's daily rotation gives us day and night, the annual revolution around the sun gives us our year, and the tilt of the Earth's axis gives us the seasons. Geographers use many special terms, such as *solstice* and *equinox*, to describe the effects on the Earth of rotation, revolution, and tilt. Learning these terms will be useful, because you will encounter them often in your lifetime.

All map projections contain distortion because of the problem of projecting a sphere onto a flat surface. Many methods of projecting maps have been devised to make up for this distortion. Each method gives us a map with a special feature. Examples of these features are: true compass direction, equal areas, local accuracy, or accurate distance from one or two points. There is no one good map. To use a map correctly you need to know which features will serve your purposes and which will not.

The five fundamental themes of geography are location, place, people-environment relations, movement, and region. Know the five themes well, as we will use them throughout our study of geography.

EXERCISES

Find the Correct Term

Directions: One of the two terms in parentheses in each of the following sentences is correct. Choose the correct term and then write the complete sentence on your paper.

1. The sun is (a star, the largest planet).
2. The sun creates (meridians, energy).
3. Saturn is farther from Earth than (Neptune, Mars).
4. The circumference of Earth at the equator is about (15,000 miles, 25,000 miles).
5. Earth's outer core and inner core are made up mainly of (iron and nickel, gases and water).
6. The Earth's rotation produces (the seasons, day and night).
7. In June, the hemisphere that is tilted directly toward the sun is the (Northern Hemisphere, Southern Hemisphere).
8. In December, the (South Pole, North Pole) has sunlight twenty-four hours a day.
9. The movement of Earth around the sun is called (rotation, revolution).
10. Latitude is distance away from the (Prime Meridian, equator).
11. Lines of longitude connect the North Pole to the (South Pole, Prime Meridian).
12. The line on a map that shows how many inches on the map equal how many miles on Earth's surface is called the map (scale, key).

Complete the Sentences

Directions: Each sentence that follows has three possible endings. Select the correct ending and then write the complete sentence on your paper.

1. Map projections are projected onto (a) atlases and glossaries (b) planes, cones, and cylinders (c) diagrams and graphs.
2. Absolute location is more important than relative location in determining (a) relationships between cities and farms (b) regional traffic patterns (c) routes for mail carriers.
3. Relative location is more important than absolute location in (a) responding to fire alarms (b) deciding whether an airline will make stops at a particular city (c) determining the path of navigators.
4. The characteristic of a place that is most likely to change is (a) absolute location (b) relative location (c) climate.
5. Philadelphia, Madrid, Ankara, and Beijing all are (a) the same distance north of the equator (b) in the Western Hemisphere (c) capital cities.
6. Throughout history, people have modified (a) the inner core (b) ocean currents (c) environments.
7. An example of a political region is (a) the Rocky Mountains (b) California (c) the Great Plains.
8. The Industrial Revolution has had the strongest effects in (a) Southeast Asia (b) the United States (c) the Middle East.

Past or Present?

Directions: On your paper, write the word *past* if the following statement applied to Philadelphia two hundred years ago. Write the word *present* if it did not apply to Philadelphia two hundred years ago.

1. Philadelphia lies at the junction of the Schuylkill and Delaware rivers.
2. A grid of north-south and east-west streets divides the entire city into blocks.
3. Philadelphia has a metropolitan population of nearly 6 million.
4. Rivers with waterfalls provide an inexpensive source of energy for industry.
5. Charleston, South Carolina, is 500 miles from Philadelphia.
6. Railroads carry raw materials across the Appalachian Mountains to the city.
7. Philadelphia has a moderate climate and fertile farmland nearby.
8. Oil refineries occupy part of Philadelphia's riverfront.

CHAPTER 1

CHAPTER 1

Inquiry

Directions: Combine the information in this chapter with your own ideas to answer these questions.

1. What was the relative location two hundred years ago of the place where you live? One hundred years ago?
2. Today, what is the relative location of the place where you live?
3. Before the 1860s, why did the Great Plains of central North America remain thinly settled?

The Five Themes of Geography

Directions: You learned of the five themes of geography in this chapter. One or more of the five themes listed below applies to each of the following statements. Write each statement on your paper followed by the letters of the themes that apply. More than one theme may be used.

a. Location, relative and absolute
b. Place
c. People-environment relations
d. Movement
e. Region

1. Boulder is a city high in the Rocky Mountains of Colorado.
2. New York City replaced Philadelphia as the nation's main commercial port city on the Atlantic when the Erie Canal, running from Lake Erie to New York City, opened in 1825.
3. London, England, is 51.30 N and 0.10 W.
4. Japan is one of the United States' largest trading partners.
5. The Korean peninsula lies between Japan and China on the Sea of Japan.
6. South Korea is a peninsular country in East Asia with picturesque mountains to the east and fertile plains to the west.
7. The first transcontinental railroad was built in the 1860s to link the American West with the Central Plains and the East Coast.

8. City dwellers in Japan like to visit the island of Hokkaido in August because unlike the southern islands, it has cool summer weather, a flat-to-hilly landscape, hot springs resorts, and spectacular ocean views.
9. The so-called Muslim world includes the Middle East, North Africa, and Southeast Asia.
10. St. Louis, Missouri, became an important city because it served as a transportation link.

SKILLS

Reading Diagrams

Directions: A diagram is a drawing or design that explains something. In this skill lesson, you will read and compare several diagrams in Chapter 1. Look first at the two diagrams about our solar system on page 15. Then write the answers to these questions on your paper.

1. Which diagram shows the relative size of the planets?
2. Which diagram shows the relative distance of the planets from the sun?
3. Which is the largest planet?
4. Are Mercury and Mars smaller than Earth?
5. Which is the hottest planet?
6. Which is the coldest planet?
7. Is Jupiter colder or hotter than Earth?
8. Does Diagram B show that the Earth's orbit around the sun is longer or shorter than Saturn's orbit around the sun?
9. Which two planets have the longest orbits?
10. Which two planets have the shortest orbits?

Directions: Now look at the diagram of Earth's interior on page 16 to answer these questions:

11. How many layers of material are shown?
12. On which layer do people live?
13. Which layer consists of partly melted rock?
14. Is Earth's inner core solid metal or melted metal?
15. Which is the thinnest layer of material?

Directions: Next, study the diagram of Earth's rotation on page 17 and answer these questions:

16. What is the name of the imaginary line that runs from the North Pole to the South Pole?
17. What is the name of the imaginary line that separates the lighted half of Earth from the darkened half?
18. Does day occur on the half of Earth that is facing toward or away from the sun?
19. In this diagram, is Earth tilted?
20. In this diagram, is it day or night in most of North America and South America?

Directions: Study the diagram on the Earth's revolution around the sun on page 18 and answer these questions:

21. How many positions of Earth are shown in this diagram?
22. Does Earth's revolution around the sun take six months or a year?
23. At all times of the year, is Earth's axis always tilted at the same angle?
24. In June, does the Southern Hemisphere or the Northern Hemisphere receive more sunlight?
25. In December, does the Southern Hemisphere or the Northern Hemisphere receive more sunlight?

Comparing Diagrams

Directions: Write on your paper the title of the diagram that shows the following:

1. On June 21, the North Pole has sunlight twenty-four hours a day
2. Whether Uranus or Jupiter is larger than Earth
3. The Northern Hemisphere at the spring equinox
4. The relative width of Earth's inner core
5. The position of the Northern Hemisphere on March 21
6. The half of Earth that has night while the other half has day
7. The orbit of each planet around the sun
8. The type of material in Earth's outer core

Vocabulary Skills

I. Directions: Match the numbered definitions with the vocabulary terms. Write the term on your paper next to the number of its matching definition.

absolute location
azimuthal projection
cartogram
circle of illumination
compass rose
computer cartography
conic projection
crust
environment
inner core
legend
loxodrome
magma
mantle
map key
map scale
meridian
outer core
parallel
Prime Meridian
relative location

1. Layer of molten iron and nickel located between Earth's mantle and inner core
2. Any line of longitude
3. The imaginary line that separates the lighted half of Earth from the darkened half and that moves as the earth rotates on its axis
4. Information on a map provided by cartographers, sometimes called a map key
5. A line of true compass direction on certain map projections
6. The surrounding conditions within which an individual or organism lives
7. An ornamental design used on maps to show the direction north
8. Information on a map provided by cartographers, sometimes called a legend
9. The thick layer of hot, heavy liquid rock located between the Earth's crust and its core
10. Any line of latitude
11. The location of a place on Earth's surface in relation to the location of other places
12. A useful map to accurately depict hemispheres or other parts of the globe
13. Useful in creating base maps on which data can be plotted, analyzed and retrieved
14. Partly melted, white-hot rock

CHAPTER 1

15. Runs from the North Pole through Greenwich, England, to the South Pole and is represented on a map as zero degrees longitude

16. The exact position of a mountain, river, lake, city, or town on Earth's surface

17. A line, bar, graph, or representative fraction on a map that marks out how many inches or centimeters on a map equals how many miles or kilometers on Earth's surface

18. The innermost portion of the Earth

19. A map that measures equal distance from its central point to any other point on a map

20. A band of solid rock at the surface of the Earth that is 25 to 50 miles thick beneath continents and from 5 to 10 miles thick under oceans

21. Maps where relative size is based on some other measure than area, such as population

II. Directions: Select the correct word from the list that completes each sentence below. On your paper, write the word you selected next to the number of the sentence it completes.

axis	latitude
cartographer	longitude
conformal projection	low latitudes (tropics)
cylindrical projection	map projection
equal-area projection	middle latitudes
equator	(temperate regions)
equidistant projection	orbit
equinox	revolution
hemisphere	rotation
high latitudes	solstice
(polar regions)	

1. A globe, such as the Earth, represented on a flat surface is called a(n) ||||||||||.

2. The |||||||||| is a line of latitude (0°) that circles the Earth at an equal distance from the North and South Poles.

3. |||||||||| is the movement of the Earth around the sun.

4. A(n) |||||||||| occurs when the sun reaches its northernmost or southernmost limit. The ancient Greeks called it the "sun standing still."

5. |||||||||| is a term for a map that is able to show distance accurately from one or two points.

6. A(n) |||||||||| is an imaginary line from North Pole to South Pole that passes through the center of our planet.

7. |||||||||| is a line drawn on a map or globe that runs north and south and converges at the poles.

8. The area between the Tropic of Cancer and the Tropic of Capricorn is known as the ||||||||||.

9. The path the Earth follows as it moves around the sun is called an ||||||||||.

10. The Mercator projection is basically a ||||||||||.

11. The |||||||||| lie between the Tropic of Cancer and the Arctic Circle and the Tropic of Capricorn and the Antarctic Circle.

12. A mapmaker is sometimes called a ||||||||||.

13. A(n) |||||||||| shows true shape for limited areas on a map.

14. |||||||||| is a line drawn on a map or globe to show distances north and south of the equator.

15. |||||||||| is the turning of the Earth on its axis.

16. On a(n) ||||||||||, the shape or direction may be distorted but any square or rectangle on the map has a correct area.

17. The |||||||||| occurs two times a year when the vertical rays of the sun are directly over the equator, and day and night are of equal length anywhere on Earth.

18. The regions north of the Arctic Circle and south of the Antarctic Circle are called the ||||||||||.

19. The northern or southern half of Earth divided by the equator, or the eastern or western half divided by a meridian, is called a ||||||||||.

EARTH ENVIRONMENTS

Earth's varied landforms, climates, vegetation, and soils influence where people choose to live and how they earn their living. Many different environments exist on planet Earth. Some environments support large numbers of people, while others are thinly settled. But everywhere, environments form the stage on which human events unfold.

◄ *The intimate connection between climate and landforms is illustrated in this photograph of Mt. McKinley, the highest peak in North America. Like many of our scenic areas, it is located in a national park. Name two national parks in your region.*

1. The Surface of the Earth

Continents and Oceans

Viewed from space, the surface of Earth is made up of continents, islands, and oceans. Continents and islands make up 29 percent of the surface area of Earth, and oceans and smaller bodies of water cover the other 71 percent of Earth's surface.

Continents are the largest landmasses on Earth's surface. Seven continents are generally recognized: Asia, Africa, North America, South America, Antarctica, Europe, and Australia. Find them on the map on page 42 or 43. Continents vary considerably in size. Asia, the world's largest continent, is bigger than Antarctica, Europe, and Australia combined. It is six times larger than the smallest continent, Australia.

Five of the seven continents are connected to other continents. Only Antarctica and Australia are surrounded by water on all sides. North America and South America are connected at the Isthmus of Panama, and Africa is joined to Asia at the Isthmus of Suez. (An **isthmus** is a narrow strip of land that connects two larger landmasses.) Europe is a peninsula of Asia. (A **peninsula** is a land area that projects into and is nearly surrounded by a body of water.)

Oceans are the largest bodies of water on Earth. Find the Atlantic, Pacific, Indian, Arctic and Antarctic Oceans on the map on pages 8 and 9. The Pacific Ocean is larger than the entire land area of the planet, and its size equals the combined area of the four other oceans—the Atlantic, the Indian, the Arctic, and the Antarctic. These five oceans and other bodies of water cover nearly three-quarters of the surface of our planet. Some of the Earth's seas, such as the Mediterranean and Black Seas, can be looked upon as inland extensions of oceans.

The average depth of the oceans is much greater than the average elevation of the land. If Earth had a smooth surface, about a mile and a half of water would cover the entire planet. Great mountain ranges exist beneath the oceans. The Mid-Atlantic Ridge, for example, occupies about a third of the Atlantic Ocean and rises more than two miles from

The Continents Compared

	Total Area (square miles)	Percent of Total Land Surface	Highest Mountain (feet)	Largest Lake (square miles)	Longest River (miles)
Asia	17,297,000	29.9	Mt. Everest 29,028	Caspian Sea 143,240	Chang Jiang (Yangtze) 3,964
Africa	11,708,000	20.2	Mt. Kilimanjaro 19,340	Victoria 26,827	Nile 4,145
North America	9,406,000	16.3	Mt. McKinley 20,320	Superior 31,820	Mississippi 3,710
South America	6,883,000	11.9	Mt. Cerro Aconcagua 22,887	Titicaca 3,205	Amazon 3,999
Antarctica	5,405,000	9.4	Vinson Massif 16,864	—	—
Europe	3,835,000	6.6	Mt. Elbrus 18,510	Ladoga 7,092	Volga 2,193
Australia	3,287,000	5.7	Mt. Kosciusko 7,310	Taupo 234	Darling 1,700
World	57,821,000	100.0	Mt. Everest 29,028	Caspian Sea 143,240	Nile 4,145

the sea floor. In places, mountain peaks of the Mid-Atlantic Ridge rise above sea level to form islands or chains of islands like the Azores, Iceland, and the island of Surtsey, which was formed some 30 years ago. Locate the Mid-Atlantic Ridge and these islands on the diagram on page 47. The island of Surtsey is shown just south of Iceland. The depth profile beneath the map shows the ocean floor stretching across the North Atlantic Ocean from Boston, Massachusetts, to Gibraltar. Note the ruggedness of the Atlantic Ocean basin.

Earth's oceans and seas are the source of all of Earth's water. Evaporation of moisture from the surface of the oceans and seas form the clouds from which rain and snow fall. Without the movement of warm and cold water as currents in the oceans, Earth's climates would be very different.

Volcanism, Folding, and Faulting
Shaping the Surface of the Land

The shape of the land is created by two sets of processes. Tectonic forces like *volcanism*, *folding*, and *faulting* operate from inside the earth in the **lithosphere**, a zone which includes the crust and the uppermost part of the mantle. In this zone, these processes act to create landforms. On Earth's surface, gradational processes such as *weathering*, *erosion*, and *deposition* subdue these landforms by wearing them away. The gradual effects of wind, water, ice, air, and gravity work to reduce the relative elevation of the more prominent landform features created by tectonic processes.

Plate Tectonics

As you already learned, the Earth's surface is elastic. Physical geographers now know that Earth's crust is broken up into a series of huge, rigid slabs or **plates**, some of them as large as continents (page 43). The movement of these plates is called **plate tectonics**.

Two processes are occurring at the edges of these crustal plates. First, volcanic eruptions boil up along rifts in continents and ocean floors where the plates are spreading apart. Here, new surfaces are being created as, for example, along the Mid-Atlantic Ridge. Second, where plates are colliding together, one area of Earth's surface may slide beneath another into a vast trench. Over thousands of years, these

very slow movements of molten material between the mantle and the crust have reshaped landform features through the tectonic forces of volcanism, folding, and faulting.

Volcanism

Heat from the upper levels of the mantle forces its way to the surface at weak places in the Earth's crust leading to **volcanism**, the outpouring of molten rock onto the surface of the land. This liquid rock is called **lava** after it reaches the surface and magma when underground.

Sometimes lava piles up on the surface and forms a volcano like Mount Fuji in Japan or Mount Vesuvius in Italy. When a volcano erupts suddenly, devastation can occur. In May 1980, for example, the eruption of Mount St. Helens in the state of Washington killed more than sixty people in a matter of minutes. It spewed a layer of ash up to four feet deep across much of the state. It also blasted off the top of the mountain, leaving it 1,100 feet lower in elevation than before the eruption. More recently, in 1993, two newsmen had to be evacuated from the crater of the earth's most active volcano, Mount Kilauea in Hawaii, when it began to erupt. That same year, six scientists were fatally trapped in the sudden eruption of the 13,680-foot-high Galeras volcano in the Colombian Andes.

In other cases, magma boils up through cracks in Earth's crust and creates **lava flows** that spread out across large areas creating landforms such as the Deccan Plateau of central India. When the crust is sufficiently sturdy to keep the magma below ground, it may still warp the surface and create large **domes** such as those found in the Black Hills of South Dakota. After thousands of years of erosion, the magma from these domes may be exposed on the surface, as in the Rocky Mountains of Wyoming and Colorado. The Palisades, a sheer volcanic cliff that lines the west bank of the Hudson River across from New York City, is another striking example of a volcanic landform that has surfaced.

> **IT'S A FACT**   The Norse believed that earthquakes were quarrels between the gods.

REGION

VOLCANOES, THE "RING OF FIRE," AND PLATE TECTONICS

The largest belt of above-ground and underwater ridges and trenches in the world encircles the Pacific Ocean, an active volcano and earthquake zone sometimes called the *Ring of Fire* because scientists believe it may include 75 percent of the world's active volcanoes. This band runs southward along the west coast of North and Central America, then follows the Andes Mountains southward to the tip of South America, runs deep into the South Pacific, cuts northward through New Zealand, New Guinea, the Philippines, and Japan, and then swings eastward just south of the Bering Sea along the southern shores of Alaska. Mount Katmai in Alaska and Mount St. Helens, Mount Shasta and Mount Lassen in the continental United States are part of the *Ring of Fire*. The most spectacular features along this belt are the sizable Pacific-Antarctic Ridge between New Zealand and Antarctica and the depths of the Mariana Trench.

In the *Ring of Fire*, the huge Pacific plate and its small satellites (the Philippine, Juan de Fuca, Cocos, and Nazca plates) collide with the Australian, Eurasian, and American plates, as the plate map clearly shows. Compare these two figures and you will understand the relationship between the plates, earthquakes, and volcanic activity.

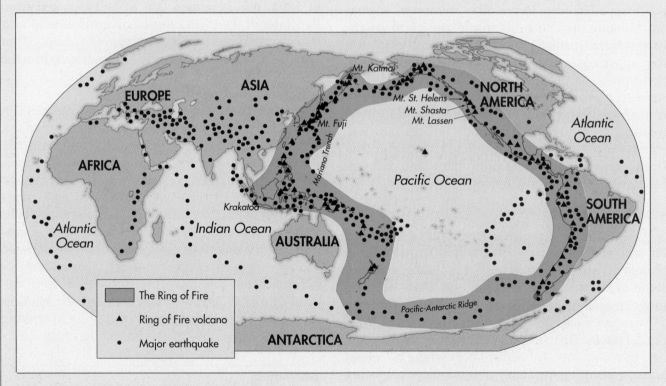

Volcanoes, Earthquakes, and the "Ring of Fire"

A second seismic belt (or area of earthquake activity) bisects the Atlantic Ocean from Iceland in the north to the ocean off Tierra del Fuego at the southern tip of South America. Called the Mid-Atlantic Ridge, this long series of towering peaks and deep gorges is one of the largest features on the planet, rising nearly two miles above the adjacent ocean floor. The volcanic island of Surtsey, one of the newer places on Earth, is at the northern tip of the Mid-Atlantic Ridge.

A third area of seismic activity stretches from the Alps of Europe across the Anatolian Plateau of Turkey to the Himalaya Mountains of Asia.

These plates float like pieces of a cracked eggshell above the hot, liquid rock in the mantle. As the theory of **plate tectonics** explains, the plates are constantly drifting apart and sliding together. Along their edges, which fit together like a jigsaw puzzle, volcanic eruptions and earthquakes occur most frequently. The restless movement of Earth's plates, grinding together and slipping apart, is where mountain building through the tectonic forces of **volcanism**, **folding**, and **faulting** usually occurs. But no one knows how many active volcanoes there are on Earth. In the spring of 1993, scientists discovered 1,133 underwater volcanoes in an area about the size of New York State 600 miles north of Easter Island in the South Pacific Ocean. Additional volcanoes have recently been found in Antarctica.

QUESTIONS

1. What is the relationship between tectonic plates, earthquakes, and volcanoes?
2. What landform would you expect to find in the region where tectonic plates are coming together? Why?

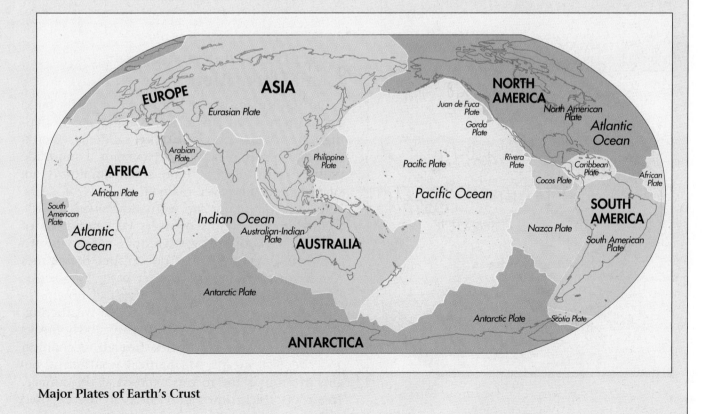

Major Plates of Earth's Crust

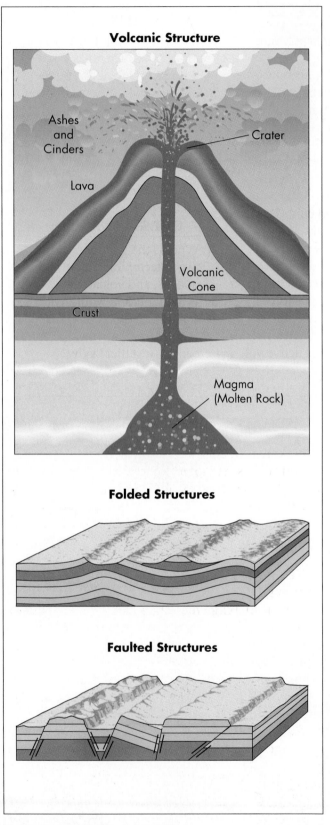

Volcanic Structure

Ashes
and
Cinders

Crater

Lava

Volcanic
Cone

Crust

Magma
(Molten Rock)

Folded Structures

Faulted Structures

Volcanism, Folding, and Faulting

Folding and Faulting

Folding and faulting create other landforms when forces inside the Earth cause rock in the crust to bend or break. When crustal plates move toward one another, they may compress the rock into a series of folds, like waves in the ocean (see the illustration on this page). In some areas where layers of rock with different levels of resistance to erosion are found, a series of parallel ridges and valleys may be seen—for instance, in the Appalachian Mountains.

Faulting takes place when tectonic pressures cause rock masses to push together or pull apart, and one mass rides up over or slides below the other. In some cases, rock masses slide sideways past one another. This happens along the San Andreas and other faults in southern California.

Volcanism, folding, and faulting occur as mountains are built. Remember that mountain building generally occurs at the edges of moving plates, where the crust is pulled apart or pushed together. Look carefully at the three examples of landform-building processes in the diagram.

Weathering, Erosion, and Deposition

On the surface of Earth, forces constantly build up and tear down landforms, shaping and reshaping the surface. This can happen suddenly when a tidal wave crashes against a shore or a mudslide plunges downhill, but more often landforms are shaped gradually over many thousands of years. Hills and mountains are worn down very slowly. **Deltas** at the mouths of rivers take hundreds of years to form. Rivers, over centuries, cut downward and form streambeds and floodplains. Sometimes they form new channels—water pathways—across Earth's surface. All of these processes take time.

Three forces are constantly at work: weathering, erosion, and deposition. **Weathering** slowly breaks rock down into finer particles through mechanical and chemical means. **Mechanical weathering** begins as soon as the rock is exposed at the surface. Tree roots reach down into cracks in the exposed rock and widen them, and other tensions chip away and create cracks in the rock. In cold climates, water trapped in these cracks expands when it

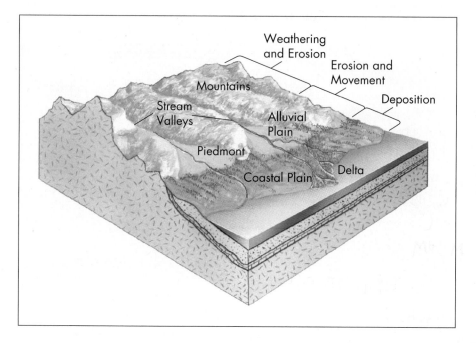

Erosion and Deposition Reshape Earth's Surface

The surface of the Earth is always changing. Erosion is most intense in uplands. Trace the deeply cut stream valleys on the diagram. Sand and small rocks are deposited on plains and form features like the delta at the mouth of the large stream. Between mountains and plains, materials are constantly being moved by wind and running water. Where would you expect to find boulders? Rocks? Sand? Fine silt? Why?

freezes into ice and breaks the rock into smaller pieces. These mechanical forces gradually cause rock masses to disintegrate.

Chemical weathering involves many processes that alter rock surfaces. These processes are most intense where heat and moisture are available; they usually operate more slowly in cold and dry climates. One of the most common of these is **oxidation**, the binding of oxygen to a mineral, which creates the green patina on copper and turns iron to rust. Weak acids in rainfall may dissolve rocks such as limestone through a process called **carbonation**. The phrase "hard as a rock" is misleading, because all rock breaks down through weathering. Some rocks just take longer than others.

Erosion is the movement of broken rock from one place to another by running water, wind, ice, or ocean currents. Running water cuts valley channels into the land surface. Over time, the down-cutting action of water creates a network of streams and rivers that drain water from the land.

Windblown sand and silt, or **loess**, is another form of erosion. Loess materials are believed to be carried long distances, usually by wind, and laid down as new soil or sand dunes in far-off places. Physical geographers, for example, recently discovered that air currents carried dust from the Sahara Desert in North Africa across the Atlantic Ocean

and renewed the fertility of the soils of the Amazon Basin in South America. Every year, windblown soil from the Gobi Desert in Mongolia drifts eastward across the plains of North China, giving that region a productive basis for agriculture.

In North America and Eurasia thousands of years ago, moving sheets of ice called **glaciers** plowed and scraped the landscapes of Canada, Western Europe, and Russia like huge earth-moving machines. They left behind landscapes with thin soils, disrupted drainage systems, and scattered rocks and rubble across the land. Smaller mountain glaciers are found in New Zealand and parts of South America. Along seacoasts, ocean currents, waves, and tides are constantly reshaping the shoreline. Each of these forces—water, wind, ice, and ocean currents—creates the distinctive landscape features that color human lives and livelihoods.

Deposition is the final result of erosion. Rocks, sand, and silt picked up by water, wind, ice, gravity, and ocean currents are deposited in a number of ways. Gravity can cause **mass movement**—the spontaneous downhill sliding of large amounts of

Gobi **is the Mongolian word for "desert."**

SURTSEY, ONE OF THE NEWER PLACES ON EARTH

▲ The volcanic isle of Surtsey came into existence during a volcanic eruption off the southern coast of Iceland in the Atlantic Ocean. Why is a volcano likely to form in this area?

On November 14, 1963, at 6:30 A.M., the small fishing vessel *Isleifur II* was about four nautical miles south of the Vestmannaeyjar, the southernmost islands off the coast of Iceland. An hour later the ship began to rock. Olafur Vestmann, the ship's cook, saw steam and fire erupt from beneath the sea. A 600-foot fissure, or crack, had split the ocean floor, tossing lava bombs as large as automobiles into the air. The crew rushed topside thinking the ship was on fire. But it was the sea that was on fire.

Lava flowed upwards from the seafloor for the next four years. This was the first volcanic island to push its way to the surface in the North Atlantic in the last 10,000 years. In a letter, the cook noted, "What we saw is now a matter of universal knowledge."

This fiery eruption in 1963 heralded the birth of a new island that was named Surtsey after the Norse fire god. A week after this volcanic eruption began on the northern tip of the Mid-Atlantic Ridge, the island of Surtsey was 700 square feet in area.

During the next four years, Surtsey grew, by way of more eruptions, to its present size of nearly

◄ These cinder cones and hills of volcanic soil are on Surtsey Island. Do you think there is any vegetation or animal life on this island?

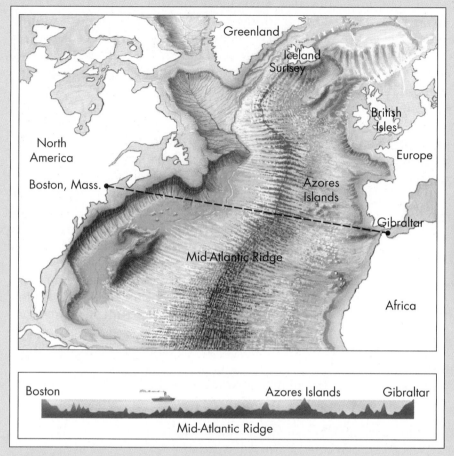

Greenland
Iceland
Surtsey
British Isles
Europe
North America
Boston, Mass.
Azores Islands
Gibraltar
Mid-Atlantic Ridge
Africa

Boston
Azores Islands
Gibraltar
Mid-Atlantic Ridge

Mid-Atlantic Ridge

Great mountain ranges exist beneath the oceans. The Mid-Atlantic Ridge occupies about one-third of the Atlantic Ocean and rises more than two miles from the sea floor. The depth profile beneath the map shows the ocean floor stretching across the North Atlantic Ocean from Boston, Massachusetts, to Gibraltar.

▼ *Although Surtsey was formed in 1963, some of these boulders look extremely old. What has happened to give them this aged appearance?*

one square mile with an elevation about 500 feet above sea level. Scientists keenly interested in how life would colonize this new landscape made the island into a laboratory. By the time the island had cooled enough for scientists to visit it, seaweed had already washed up on the black sand beach littered with boulders.

At the center of the island, two hills of sand and lava rose up, barren of living organisms. The first visitors to Surtsey were arctic seabirds that landed on the island while the eruption was still going on. They came to eat fish killed in the boiling water rising from the ocean. Now twenty-seven species of birds have come to the island and introduced a variety of insects and plants into this new environment. Six bird species now nest there. Every April, mother seals bear their young on the new black sand beaches.

Wind erosion has already reduced the size of Surtsey by a quarter, and reduced its elevation from 575 feet to 490 feet. Eventually, perhaps in centuries, this island will become smaller, and will finally disappear beneath the sea. By then, however, scientists will understand how life is initially established and ultimately destroyed on one of the newer places on Earth.

QUESTIONS

1. How does the island of Surtsey give us insight into the colonization of a "new" area by various life forms?
2. Surtsey is located at the northern tip of what seismic belt?
3. Describe how Surtsey provides an example of both tectonic and erosional forces.

material, such as a rock slide or a mudflow. Streams and rivers carry sand and silt downstream from uplands to lowlands, as the illustration on page 45 shows. They drop these materials wherever the volume or speed of the flow of water lessens, and the eroded material can no longer be carried.

These newly deposited soils form **alluvial plains** along stream beds, in the foothills, or **piedmont**, at the base of mountains, and on **coastal plains**. Where powerful rivers, such as the Mississippi, Nile, or Amazon Rivers, enter oceans, large deltas, or flat lowlands made up of these deposits, may extend far out to sea.

Landforms on Earth's Surface

Together, tectonic forces inside the Earth and gradational forces on its surface have produced four major types of landforms: (1) **mountains**, (2) **hills**, (3) **plateaus**, and (4) **plains**. The general distribution of these landforms is shown on the map on page 49, which helps us to visualize the relative elevation (or height above sea level) of the land surface in different places on Earth. These categories, however, are general and their boundaries are often difficult to identify. No one has ever, for example, satisfactorily defined the exact difference between a mountain and a hill in terms of relief. **Relief**, in this case, refers to variations in the elevation, shape, and forms of Earth's surface.

Mountains and Hills, Plains and Plateaus

Both mountains and hills are generally formed by volcanism, folding, and faulting. Mountains are generally higher in elevation and more rugged than hills, but the difference in height between one and the other is a matter of local opinion, although the differences can be expressed in terms of difference in relief.

Mountains tend to be more sparsely populated than hill lands and usually lie at 2,000 to 3,000 feet above sea level. The great mountain ranges are found where plate boundaries and earthquake zones are located on the land surface (see the maps on pages 42 and 43). Examples are the Rocky Mountains in North America, the Andes Mountains

▼ *The Andes Mountains of South America, like most great mountain ranges, are located where tectonic plates meet and earthquakes occur. What areas in the United States are prone to having earthquakes?*

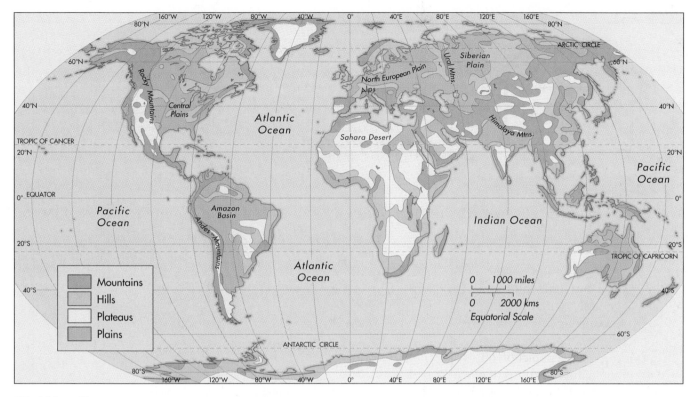

World Landforms

Plains are the most densely settled parts of Earth. Hills and mountains generally are sparsely populated.
What other factors do you believe might influence where people choose to live?

of South America, the Alps in Europe, the Ural Mountains of Russia, and the several ranges of the Himalayas that occupy much of central, east, and southeast Asia.

Plains and plateaus are also difficult to define precisely. Plains are relatively low-lying, level areas that in many cases are gently rolling but are sometimes deeply cut by stream valleys. The largest plains in the world are the Central Plains of North America, the Amazon Basin in South America, the North European Plain that extends deep into Russia, and the nearly flat plains of central Australia (see the map above). Plains located in river basins or along seacoasts in temperate climates are among the most densely settled areas in the world.

Plateaus are relatively level like plains, but they are located at higher elevations. Most plateaus have at least one steep side where a line of cliffs, an **escarpment**, separates the plateau from neighboring low-lying areas. Large plateau areas are located in the western United States, eastern Brazil, much

of Africa, central Spain, the Arabian Peninsula, central Asia, and western Australia.

Mapping Landforms

Geographers map landforms in two ways. Color may be used to show different types of landforms. On the map of world landforms above, green is used to show plains, yellow shows plateaus, and two shades of brown show hills and mountains. The key identifies what each color represents.

Topographic maps provide a more accurate method of showing elevation. Elevation is the height of the land above sea level. On topographic

IT'S A FACT Mount Everest, the world's tallest peak, was named by British mapmakers for a man who may never have seen the mountain.

LOCATION

TOPOGRAPHIC MAPS

A Topographic map shows landforms on Earth's surface by using contour lines that connect points of equal elevation. Topography is the depiction on maps of the natural features of a place or region. The view of the hilly island in the upper illustration shows how a topographic map is read. Imaginary horizontal lines spaced 100 feet apart cut through the island. Each of these lines forms a contour line of the same elevation wherever it touches the island's surface. The lowest contour line connects points along the shore. If you walked along the shore, you would stay at a constant elevation of zero. That is, you would be walking along the zero-foot contour line.

In the lower illustration, the hundred-foot contour lines and

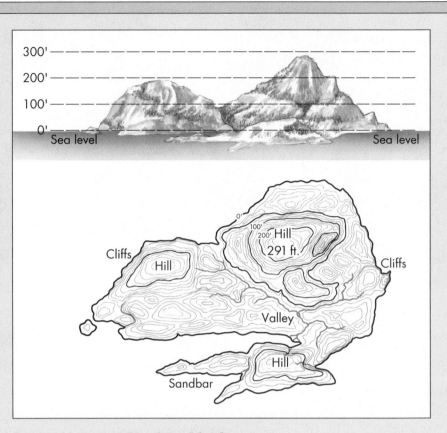

Contour Lines on an Imaginary Island

intermediate contour lines that are spaced 20 feet apart are transferred onto a flat piece of paper. This creates a topographic map. By looking at the pattern and spacing of contour lines, geographers can picture in their minds the shape of the island, with its two major peaks separated by a valley and with an outlying sandbar and smaller hill. The elevation of any point on the island can be

determined, because the elevation of each contour line is known.

The contour map uses contour lines representing distances 20 feet apart to show elevation. Every fifth contour (100 feet, 200 feet) is darkened to make the contour lines easier to read. The wide spacing of contours in the river valley indicates that slopes are gentle. One must travel some distance across the valley before

maps, points of equal elevation are connected by lines. These are called contour lines. If you walk along a contour line, you remain at the same elevation above sea level. Trace some of these contour lines on the map above.

The physical maps in this text, like the one on pages 8 and 9, are relief maps. These maps use color

to show differences in elevation. Areas of similar elevation are shown in the same color. Study the map key on page 8. Dark green is used to show land between zero (sea level) and 492 feet. Light green shows land elevations between 492 and 1,968 feet. Brown is used to show elevations above 9,840 feet. What do the other colors on this key show?

going up or down 20 feet in elevation. In contrast, the closely spaced contour lines along the coastal cliffs and at the edge of the hills indicate steep slopes and a difficult climb. Here, rapid changes in elevation occur within a very short distance.

The maps in the National Topographic Map Series of the U.S. Geological Survey (USGS) provide an invaluable source of information for geographers, regional planners, landuse specialists, and hikers, skiers, climbers, and other people engaged in outdoor recreational activities. Several map series are available for all or most of the United States at scales of 1:1,000,000 (1 inch equals 16 miles), 1:250,000 (1 inch equals 4 miles), 1:62,500 (1 inch equals approximately 1 mile), and 1:24,000 (1 inch equals 2000 feet). On these maps, contour lines and relief features are drawn in brown, water features in blue, vegetation of various types in green, and cultural features, such as railroads and buildings, in black. A complete listing of the extensive set of natural and man-made symbols used on these topographic map series is available from the USGS. You can probably buy one for your area. Compara-

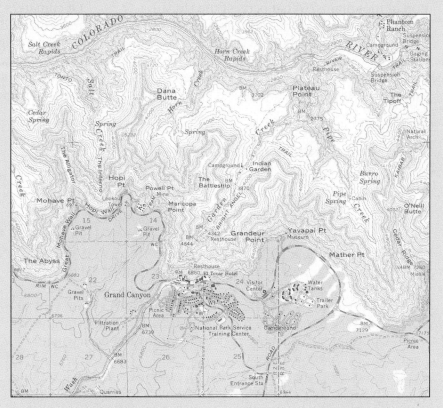

Topographic Map
The dramatic differences in elevation that appear on this topographic map of the Grand Canyon are striking. Do you understand why contour lines in the canyon are so close together?

ble topographic maps at various scales are available for Canada, Mexico, and many other nations.

QUESTIONS

1. What is a topographic map and how is the surface of Earth represented on these maps?

2. Where could you acquire a topographic map of your area?

3. Topographic maps are available at several scales. What scale would you choose to view a mountain range? A small valley? Your town?

⊕ REVIEW QUESTIONS

1. What is a continent, and which are the largest and smallest continents?
2. What three forces at the surface of the Earth shape and reshape landforms?
3. What is elevation, and how are points of equal elevation shown on topographic maps?

⊕ THOUGHT QUESTIONS

1. In what two ways are five of the seven continents connected to another landmass? Do you think these links between landmasses have any special importance? Have they affected the distribution of plants, animals and peoples? If so, give an example.

2. What natural processes account for the formation of mountains? What parts of the United States are seriously affected by these processes? How would you feel about living in one of these areas?
3. Why are mountains usually poorly suited to human settlement? Do mountains affect the distribution of farming? Of transportation routes? Of climate? Of vegetation? If you have visited mountains, describe how these areas were different from neighboring lowlands.
4. How does the topography of an area affect the lives of people who live there? Has topography had an effect on your life? If so, how?

⊕ SECTION 1 VOCABULARY SKILLS

A. Directions: Match the numbered definitions with the vocabulary terms. Write the correct term on your paper next to the number of its matching definition.

1. Huge, moving slabs that form the Earth's crust	carbonation
2. A form of chemical weathering in which oxygen is bound to a mineral to produce an oxide	chemical weathering
3. The action of a weak acid on rocks such as limestone to dissolve them	coastal plain
4. The study of large-scale movements of the Earth's crust	contour lines
5. Molten rock forced onto the earth's surface through volcanism	dome
6. The pushing up of surface rock by the intrusion of magma	faulting
7. Fertile, fine, windblown soil	folding
8. Lines on a topographic map that link points of equal elevation	glacier
9. A long, relatively flat topographic feature, created when deposition occurs along a coastline	hill
10. The physical effect of wind and water on the Earth's surface creating changes in the landscape	lava
11. River or sheet of moving ice	lava flow
12. The compression of rock into a series of folds by the movement of crustal plates	lithosphere
13. The flowing out of magma through crevices in the crust that spread out across large areas	loess
14. Caused when tectonic pressures cause one rock mass to ride up over or below or alongside another rock mass	mass movement
15. An elevated landform with more gentle slopes and less relief than mountains	mechanical weathering
16. A zone on the Earth that includes the crust and the uppermost layers of the mantle	mountain
17. The processes connected with volcanoes and volcanic activity	oxidation
18. A landform that is high in elevation, has steep slopes, and towers above surrounding areas	plates
19. Processes that alter rock surfaces especially in conditions of moisture and intense heat	plate tectonics
20. Variations in the elevation or shapes of features on the Earth's surface	relief
21. The spontaneous downhill sliding of large amounts of material	topography
22. The showing of the Earth's surface on maps	volcanism

continued

B. Directions: Fill in the blanks in the following sentences using the words in the list below. Write the completed sentences on your paper. Words may be used only once.

alluvial plain   peninsula
delta   piedmont
deposition   plain
erosion   plateau
escarpment   topographic map
isthmus   weathering

1. Foothills that form a transition between mountains and plains are called ‖‖‖‖‖‖.
2. The chemical, biological, and mechanical processes by which rocks are broken down are called ‖‖‖‖‖‖.
3. A(n) ‖‖‖‖‖‖ is described as a flat lowland made up of sediments dropped by a river at its mouth.
4. A(n) ‖‖‖‖‖‖ is a line of steep cliffs rimming a plateau.
5. A(n) ‖‖‖‖‖‖ is a narrow strip of land connecting two larger landmasses.
6. Geologists define ‖‖‖‖‖‖ as the laying down of rock fragments, sand, and silt by wind, running water, or ice.
7. A(n) ‖‖‖‖‖‖ is a level or gently sloping surface formed of sediments laid down by streams, generally during flooding.
8. ‖‖‖‖‖‖ is the breaking down and movement of rock particles by running water, ocean currents, wind, and ice.
9. A(n) ‖‖‖‖‖‖ is a relatively level highland that rises above surrounding areas.
10. A(n) ‖‖‖‖‖‖ is a relatively low, level area.
11. A(n) ‖‖‖‖‖‖ is an extension of land usually bounded by water on three sides.
12. A(n) ‖‖‖‖‖‖ is an accurate method of showing elevation (height above sea level).

2. Elements and Global Distribution of Climates

The distribution of climates on Earth is produced by four factors in addition to the Earth's rotation and revolution. These are (1) latitude, (2) circulation systems in the atmosphere, (3) variations in topography, and (4) location with respect to available sources of moisture. These factors affect weather and climate. **Weather** is the condition of the atmosphere at a given place and time. **Climate** is the average of these weather conditions over a period of years.

The Atmosphere

The **atmosphere** is the envelope of gases enclosing Earth. Dry, pure air contains about 78 percent nitrogen, 21 percent oxygen, and 0.03 percent carbon dioxide. Of all planets in our solar system, only Earth has an atmosphere that can sustain life as we know it. The atmosphere provides oxygen for breathing. It filters out harmful ultraviolet rays from the sun and also traps the sun's energy. It regulates temperatures, keeping them within the narrow limits that human beings can tolerate. The atmosphere is our envelope of life.

Energy from the sun warms our planet. During the day, land and water surfaces absorb solar energy and warm up, just as we warm up when lying in the sun. At night, the Earth's surface releases much of the energy back into space. Some of this radiant energy, however, is trapped for a while in the land and oceans and also in the atmosphere by carbon dioxide, a transparent gas in the air.

This trapping of energy is called the **greenhouse effect**. Like a greenhouse or a closed car left out in the sun, the atmosphere allows energy in but delays its rate of loss into space. The heat trapped in the atmosphere is distributed around the planet by global wind systems and ocean currents. Otherwise it would be extremely hot in the tropics, which receive much more solar energy than polar regions.

Circulation Systems in the Atmosphere and Oceans

The circulation systems in the atmosphere and the oceans redistribute energy. Differences in heating cause differences in the weight of air, which determines the **atmospheric pressure**. Warm air is lighter; cold air is heavier. When air is warmed, the

IT'S A FACT — About 99 percent of the weight of the 6,000-mile-deep atmosphere is compressed into its lowest 50 miles.

air expands, becomes lighter, and rises and areas of **low pressure** are created. As this air rises, clouds, rain, or storms often occur. When warm air rises because of surface heating of the land, the resulting rain is called *convection precipitation*. It is called *frontal precipitation* when a warm air mass rises above a colder air mass leading to rain, sleet, or snow. **Precipitation** refers to the condensed droplets of water vapor that appear as dew, rain, snow, sleet, or hail.

When air is cooled, it loses its ability to hold moisture and becomes dense and heavy. This cooler air creates **high-pressure** areas where the atmosphere tends to be very stable. Because air, like water,

▼ *The Earth's atmosphere traps energy on Earth through the greenhouse effect. What is the greenhouse effect?*

flows from areas of high pressure to areas of low pressure, heat is redistributed in the atmosphere.

The flow of air from high to low pressure occurs daily along coastlines, as the illustration on this page shows. These air flows, or winds, are caused by differences in the heating of the land and water surfaces. Winds are rivers of air that are named after the directions from which they blow.

Land heats up and cools down faster than water, and this causes differences in air pressure over the

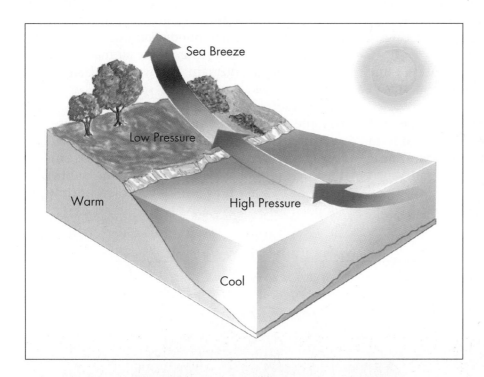

Land and Sea Breezes

Land and sea breezes are caused by differences in surface heating, which cause differences in air pressure. Air flows from areas of high pressure to areas of low pressure. This mixes the air and transfers energy. Sea breezes (top) flow in the daytime when land surfaces are warmer than neighboring water bodies. Land breezes (bottom) flow at night when land surfaces are cooler than neighboring water bodies. Try and think of the same principle moving air from high pressure areas to low pressure areas on the globe.

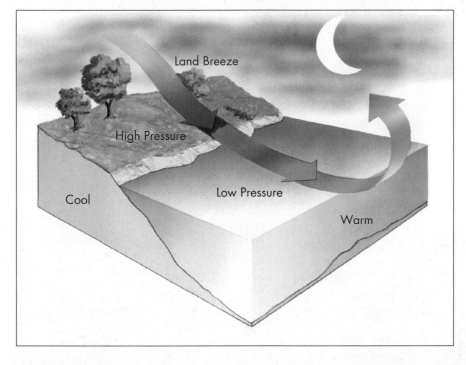

land and water. In the daytime, sunlight heats the land surface more rapidly than it heats the water. When the air above the land warms and rises, air from the slower-heating ocean flows inland from high to low pressure and creates a **sea breeze**. At night an opposite flow is set up, because the land loses its heat faster than the ocean. The air above the water warms, becomes lighter, and rises. The cooler air from the land flows seaward and creates a **land breeze.**

Pressure and wind systems carry air in the same way from high- to low-pressure areas and spread heat from the tropics to the poles. In the *tropics*, solar energy heats the Earth and creates a wide belt of low pressure called the **equatorial low** around the equator. Air masses from high-pressure areas near latitudes 30° N and 30° S (zones that are called the **subtropical highs**) flow steadily into this low-pressure area. These **trade winds** meet and create

rising columns of air near the equator and produce clouds, storms, and heavy rainfall. Other low-pressure belts, the **subpolar lows**, exist at latitudes 60° N and 60° S and are sandwiched between the subtropical highs and the **polar highs**. The **westerlies**, air that flows into these low-pressure regions from subtropical areas, and the **easterlies**, winds flowing toward the equator from the polar regions, collide in these middle latitudes. Here, heavy storms and rain occur when these air masses collide.

Note in the diagram below how these global wind systems function to spread solar energy throughout Earth's atmosphere. The trade winds that meet at the equator are warmed, they rise, and at higher elevations they spread out and flow poleward. This warm air from the tropics then sinks in the high-pressure belts located in the subtropics. Some of this air, however, flows poleward into the subpolar lows.

Global Exchange of Energy

Air is exchanged by winds flowing from areas of high pressure to areas of low pressure. Heat is spread from the tropics to the poles in this way. What would happen otherwise?

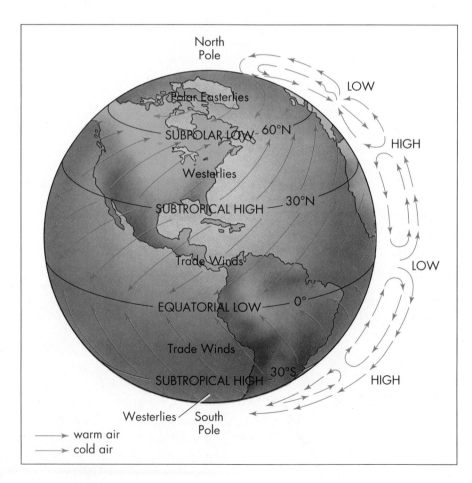

This exchange of energy is reinforced by ocean currents that are driven by these wind systems. The ocean currents, gigantic streams of seawater, carry warm water poleward from the tropics and cold water toward the equator from higher latitudes. The famous North Atlantic Drift (an extension of the Gulf Stream) carries water northward from the Gulf of Mexico and warms the southern shores of the British Isles and Northern Europe. Cool ocean currents carry this water back to the tropics where it is rewarmed to complete the energy cycle. These circular flows are called gyres.

Location and topography also influence the distribution of temperature and rainfall. Areas located near oceans usually have maritime climates with more moderate temperatures and more rainfall than places located in the interior of land masses, which have continental climates. These climatic patterns are also affected by topography. As you climb a mountain, temperature decreases at a rate of 3.5° F every thousand feet. This is why high mountains, even in the tropics, have snow-capped peaks.

Warm air that is forced to rise above a mountain range also cools. It loses its capacity to hold moisture, which then falls as rainfall or snow on the windward, or facing, slopes of mountains. This is called *orographic precipitation*. After mounting the crest, the air is dry. It sinks and creates a rain shadow on the leeward, or sheltered, slopes.

In Asia, for example, the Himalaya Mountains form a high wall that blocks warm, moist air from the Indian Ocean from entering Central Asia. Torrential rains fall on the southern or windward slopes of the Himalayas. On the leeward side, Central Asia's climates are steppes and deserts. In North America, the Rocky Mountains cast a rain shadow over large areas of the Great Plains. In Europe, the Alps steer the middle-latitude storms of the westerlies across the North European Plain into Russia.

Global Distribution of Climates

The world's climates are generally classified on the basis of whether they are (1) warm or cold, (2) wet or dry, (3) maritime or continental, or (4) highland. The most common global classification system of climates was first devised by the Russian-born geographer Vladimir Köppen and later revised with the assistance of Rudolf Geiger.

Köppen's classification system is based on the rainfall and temperature needs of certain types of vegetation. He identified five major climatic types (identified by capital letters): tropical (A), dry (B), temperate (C), continental (D), and polar (E). These five major climatic zones are further subdivided according to seasonal variations in temperature and rainfall during the year (identified sometimes by other capital letters but most often lowercase letters). Highland climates (H) are placed in a separate category because of the vast differences in temperature and rainfall that occur in areas of high elevation. The global distribution of these climates is shown on the map on pages 58 and 59.

Tropical Climates (A)

The tropics are located near the equator between the Tropic of Cancer (23½° N) and the Tropic of Capricorn (23½° S). Two types of climate are found here. Tropical rain forest climates (Af) are located at and near the equator. The letters *Af* in the Köppen system describe climates associated with *tropical rain forests*. Temperatures never cool in these regions: they are constantly hot and constantly wet. It rains almost every day. The great rain forests of the world include the Amazon Basin of South America, the Congo Basin of central Africa, and the vast forests of the islands and mainland of Southeast Asia.

Tropical monsoon and tropical savanna climates (Am, Aw) are also hot year-round, but unlike tropical rain forest climates, these areas have a pronounced dry season during the winter. In areas with high seasonal rainfall, like the coasts of India, monsoon forests and tropical monsoon climates (Am) are found. Where the dry season limits tree growth, forests gradually thin out and diminish in size and variety. Most savannas are occupied by tall grasses interspersed with trees. Tropical savanna climates (Aw) are located north and south of the tropical rain forests in South America, Africa, and Asia. In the United States, tropical climates are found in

World Climate Map

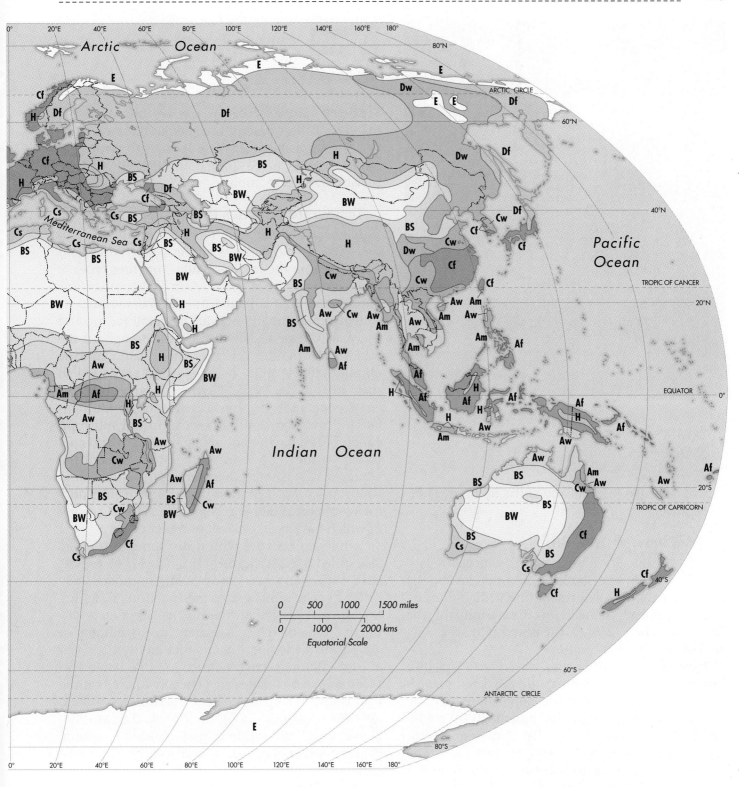

Arctic   Ocean

80°N

E

E   E

Cf   Dw   ARCTIC CIRCLE
H   Df   E   E   Df
Df   Df   60°N
Cf   Dw   Df
H   H
BS   BS   H
Cf   BW   BW   Df
H   Cs   Df   Cw   40°N
Cs BS H   BS   BS   Cf
Cs   BS H   Dw   Cw   Cf
Mediterranean Sea Cs   BS BW   H   Cf
Cs   BS   BS   H   Cf
BS   BW   BW   Cw   Cf
BW   H   BS   Cw Aw   Cf   TROPIC OF CANCER
H   Aw Cw Aw   Am   20°N
H   BS   Am   Aw Am Aw
BS   Am   Aw   Am   Pacific
H   Aw   Am   Ocean
Aw   H   BS   Af   Aw   Am
Am   Af   H   H   Af   Af   Af   Af
Aw   H   H   Af   H   Af   EQUATOR   0°
BS   H   Af H   Af   H   Af
Aw   Am   Aw   Af
Cw   Aw   Aw
BS   Aw Af   Indian Ocean   Am Aw
Cw   BS Cw   BS   BS   Cw   Aw   Af   20°S
BW   BW   BS   TROPIC OF CAPRICORN
Cs   BW   Cf
Cf   Cs   BS
Cs   BS
Cs   Cf   Cf   40°S
H

0   500   1000   1500 miles
0   1000   2000 kms
Equatorial Scale

60°S

ANTARCTIC CIRCLE

E   80°S

0°   20°E   40°E   60°E   80°E   100°E   120°E   140°E   160°E   180°

▲ *Tropical rain forests are located near the equator where the climate is hot and wet year round. Where are the largest rain forests located?*

the island state of Hawaii and at the tip of the Florida peninsula.

Dry Climates (BW, BS)

Dry climates are found in the subtropical latitudes just north and south of the tropics. Here the stable, sinking air of the subtropical high-pressure belts discourages rainfall. In the vicinity of the Tropic of Cancer (23½° S), the **desert climates** (BW) are hot and dry year-round.

Plant life is limited to species that are adapted to survive heat and drought, although where irrigation water is available dense populations may be found. The world's largest hot deserts (BWh) are located in the subtropics—in the American Southwest, North Africa, the Middle East, and the Arabian Peninsula, as well as the smaller Southern Hemisphere deserts of central Australia, southwest Africa, and southern South America. In Central Asia, deserts extend into the cooler, higher latitudes and are cold and dry year-round.

Steppe climates (BS), which are found on the margins of these deserts, have somewhat more rainfall. Short grasses cover these lands with steppe climates. Rainfall is too limited and unpredictable to sustain dense vegetation.

Temperate Climates (C)

Temperate climates are found in the middle latitudes where cold air masses from polar regions and warm air from the tropics collide. The three types of temperate climate all have warm and cool seasons but differ in their annual distribution of rainfall.

Temperate climates that have rainfall for most of the year (Cf) are widely distributed throughout the middle latitudes. Among these are **marine climates**. These mild, moist climates are found in such areas as the Pacific Northwest, much of Western Europe, most of New Zealand, and in southern and southwestern South America. **Humid subtropical climates**, having hot, muggy summers and cool to cold winters, occur in southeastern areas of the United States, China, South America, Australia, and South Africa. Most of these moist-climate areas originally supported dense middle-latitude forests, many of which have been cleared for human use.

Temperate climates with a dry summer (Cs) are also called **Mediterranean climates**. These regions have mild, wet winters and hot, dry summers. Only in this type of climate does precipitation fall during the winter rather than during the summer. A distinctive type of scrub vegetation called **maquis** or **chaparral** made up of plants that are well adapted to dry summers covers much of the landscape. Mediterranean climates are generally located on the west coasts of continents between latitudes 30° and 40°—for example, along much of the Mediterranean coast, in south central California, central Chile, southwestern Africa, and southwestern Australia.

The third type of temperate climate is characterized by dry winters and wet summers (Cw). These humid areas support luxurious mixed forests and grasslands. This type of climate is found in parts of southern Brazil, parts of the interior of southern Africa, northern India, and parts of both northern and southern China.

Continental Climates (D)

Continental climates have long, cold winters and short, hot summers because of their locations in the central and eastern parts of continents in the middle latitudes of the Northern Hemisphere. Temperature is the crucial variable in the productivity of these regions. The **humid continental climates** (Df), which are located on the southern margins of this belt, have hot summers that support some of the Earth's richest agricultural environments—those of the American Midwest, and Ukraine. The North China Plain has a continental climate with a dry winter. Farther north, humid continental climates merge into **subarctic climates** (Df) that have much colder winters and less rainfall. At their northern limits, subarctic climates have virtually no summer.

In Canada and Russia, these climates support a vast forest of coniferous trees called **boreal forests** (northern forests) or **taiga**, the Russian word for "swamp forest." These forests extend for thousands of miles across Canada and Eurasia in the Northern Hemisphere. Continental climates are not found in the Southern Hemisphere because neither South America, Africa, nor Australia extends far enough into the high latitudes.

Polar Climates (E)

Polar climates are located at the northern and southern extremities of the planet, north of the Arctic Circle (66½° N) and south of the Antarctic Circle (66½° S). Both climates in this category have extremely cold winters and very short summers. **Tundra climates** (E) have at least one month with an average temperature above freezing. During this brief warm period some plants can grow but generally tundra vegetation is limited to scattered patches of mosses, lichens, and ferns. Trees cannot survive in the many months of cold weather. Still further poleward, **ice cap climates** (E), which are located around the poles, never have temperatures above freezing. Temperatures are so low that the ground is permanently frozen. Virtually no plant life exists in these frozen areas. Both tundra climates and ice cap climates have low rainfall, be-

cause cold air is not able to hold much moisture. Some geographers view these icy wastes as great deserts because they in fact receive about the same amount of precipitation as the Sahara Desert.

Highland Climates (H)

The final category of the Köppen system of classification is called **highland climates** (H), which vary a great deal depending on slope and elevation. In general, temperatures are cooler at higher elevations. Rainfall is high, particularly on windward slopes. Vegetation changes with elevation. In tropical areas, rain forests give way to a variety of forests, then meadows, and finally permanent snow cover in high ranges like the Andes and the Himalaya Mountains. In middle latitudes, trees rarely grow above elevations of 10,000 feet.

REVIEW QUESTIONS

1. What are the characteristics of low-pressure areas? Of high-pressure areas?
2. Describe the Köppen system of climate classification. How is it useful?
3. What is climate, and how is it measured?
4. How has climate affected patterns of human settlement?

THOUGHT QUESTIONS

1. Why is the atmosphere called "our envelope of life"? Would life exist on Earth without the atmosphere? How could human beings survive on a planet without an atmosphere? Can you imagine a planet with a different kind of atmosphere?
2. Why is it colder at the top of a mountain than at the bottom? What does this tell you about how the Earth is heated?
3. Describe the climate in which you live. In what climate would you like to live? Why?

IT'S A FACT Antarctica and the Sahara Desert get about the same amount of precipitation annually.

⊕ **SECTION 2 VOCABULARY SKILLS**

A. Directions: Match the numbered definitions with the vocabulary terms. Write the correct term on your paper next to the number of its matching definition.

1. A belt of low-pressure area located along the equator
2. The temperature and rainfall conditions of any place on a given day
3. These winds flow steadily from high-pressure areas near 30° north and south latitudes toward the equator
4. The trapping of solar energy inside the atmosphere by carbon dioxide
5. The envelope of gases—mainly nitrogen, oxygen, and carbon dioxide—that encloses the Earth
6. The average of weather conditions over a period of years
7. A type of scrub vegetation adapted to dry summers and found in Mediterranean climates
8. Created when air above the ocean flows inland from high to low pressure
9. Facing away from rain-bearing air masses
10. An ocean current that carries water north from the Gulf of Mexico to the southern shores of the British Isles and Northern Europe
11. Facing the wind
12. Formed where air cools, loses its ability to hold moisture, and becomes dense and heavy
13. Forests in subarctic Canada and Eurasia
14. Created when air above a landmass heats up and flows out to sea from high to low
15. Falling products of condensation in the atmosphere, such as rain, snow, or hail
16. Circular flows of oceanic currents from warm water regions to cold water regions and back again
17. A semipermanent belt of winds that flow from subtropical high-pressure regions to the middle latitudes
18. A region in the lee of mountains that receives less rainfall than the region windward of the mountains
19. An area of light, warm air that tends to rise
20. The pressure exerted on Earth's surface by the weight of Earth's atmosphere
21. A semipermanent belt of high-pressure air cells that lie near latitudes 30° N and 30° S
22. A semipermanent belt of high-pressure air cells that lie along the Arctic and Antarctic Circles
23. A semipermanent belt of winds that flow from subpolar low-pressure regions to the middle latitudes
24. A semipermanent belt of low-pressure air cells that exist at latitudes 60° N and 60° S

atmosphere
atmospheric pressure
boreal forest or taiga
chaparral or maquis
climate
easterlies
equatorial low
greenhouse effect
gyres
high pressure
land breeze
leeward
low pressure
North Atlantic Drift
polar highs
precipitation
rain shadow
sea breeze
subpolar lows
subtropical high
trade winds
weather
westerlies
windward

continued

B. Directions: Write the first-level letter (A, B, C, D, E, or H) of the Köppen climate classification system after each climate zone. Then write a short description for each climate zone as described in your text.

1. continental climate
2. desert climate
3. highland climate
4. humid continental climate
5. humid subtropical climate
6. ice cap climate
7. marine climate
8. Mediterranean climate
9. steppe climate
10. subarctic climate
11. tropical rain forest climate
12. tropical savanna climate
13. tropical monsoon climate
14. tundra climate

3. Climates, Vegetation, and Soils

Climate, Vegetation, and Soils Make Up Distinctive Environments

Climate has a vital influence on the kinds of soils and vegetation found on Earth. Soil is the very top layer of Earth's rock crust that has been broken into tiny pieces by weathering, erosion, and deposition. Made up of rock particles and decayed plant and animal materials (organic matter or **humus**), soil is the loose material in which plants grow. *Vegetation* is a general term for all the plants that grow in an area.

Temperature and rainfall, the basic elements of climate, are the major factors in determining what kinds of plants will grow in an area. Some plants thrive in cold weather, while other plants cannot survive cold climates. Some plants need lots of water, and others can grow in dry climates. Vegetation loosens soils and adds organic matter to them.

Most plants, in turn, draw on the soil for their nutrients. Climate, vegetation, and soil make up distinctive environments that influence where people live and how they earn their livings.

The distribution of soils and vegetation around Earth is similar to the distribution of climates. Each climate has its own special kinds of vegetation and soils. One can identify the many different environments found on our planet by traveling northward or southward from the equator to the North or South Pole. Trace the journey northward on the diagram on pages 64 and 65.

Tropical Environments

Three types of environment are found in tropical latitudes: (1) *tropical rain forests*, (2) *tropical forests*, and (3) *tropical savannas*. Near the equator, **tropical rain forests** form a tall, thick mass of vegetation so dense that little sunlight reaches the ground. These forests are sustained by constant warmth and humidity. Temperatures are high and rainfall is abundant in these areas. Although an extraordinary variety of plant life thrives in this environment, rain forest soils are poor in nutrients. Low levels of sunlight limit plant cover at ground level. Heavy

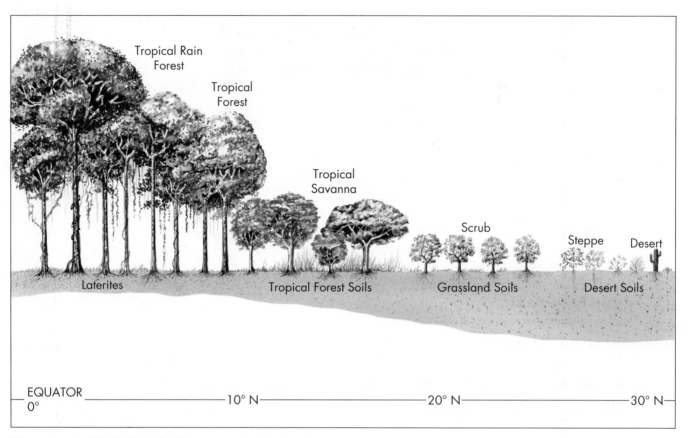

EQUATOR
0°

10° N

20° N

30° N

Vegetation and Soils from the Equator to the North Pole

Climate, vegetation, and soils change as one moves from the equator to the North Pole. Each of these environments poses different challenges and opportunities for human settlement. Can you identify some of the challenges? Opportunities?

rains wash through the soil and dissolve many elements that plants need to grow. Also, insects clear the forest floor of fallen leaves, which reduces the amount of organic matter or humus in the soil. The major components of these tropical rain forest soils are iron and aluminum; they form hard, compact, yellow-to-red soils called **laterite**, which in Latin means "brick." These soils are poor for farming because of their low fertility. Most of the evergreen trees in the rain forests gain their nourishment directly from the air and the water. Vast areas of rain forest in South America and Africa are now being cut down for timber, exposing these vulnerable soils

to erosion. An area larger than the state of Georgia is now being cleared in rain forests each year.

Moving northward from the equator, we see that tropical rain forests thin out to **tropical forests**, which have evergreen and deciduous trees, and then to savanna grasses as rainfall decreases in winter. Evergreen, or **coniferous**, trees hold their leaves year-round; **deciduous** trees drop their leaves in the dry season. **Tropical savannas** cover the surface where a long winter dry season occurs. Large areas of tall grass (some as high as ten feet) are interspersed with groves of trees, as the diagram on this page shows. Savannas, which are among the richest grazing lands on Earth, are found in Venezuela and Brazil, Central and East Africa, and northern Australia. Ranching and herding are the main human activities in the savannas.

With populations increasing worldwide, savanna grasslands are also being used for farming. The soils

IT'S A FACT Each decade the world is losing 7 percent of its topsoil.

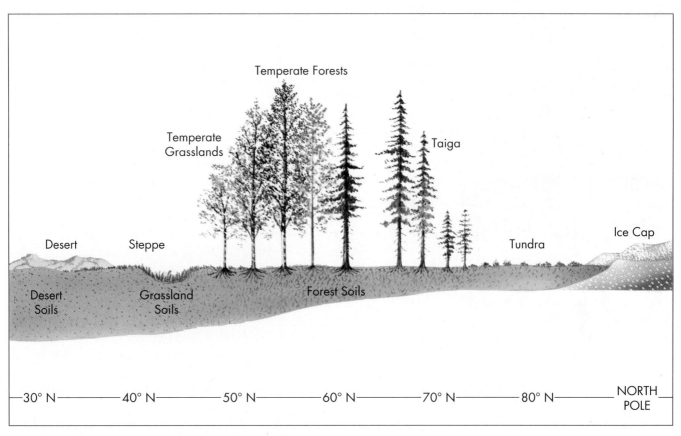

Vegetation and Soils From the Equator to the North Pole (Continued)

of the tropical savannas, like all grassland soils, are very fertile. Grass roots reach deep into the earth and break up soil particles, forming rich, dark soils. A single grass plant puts out as much as 350 miles of roots in a year. These roots decay and add plant nutrients and organic matter to savanna soils.

Dryland Environments

On the northern margin of the savannas, the tall grasses give way to short, thorny trees and shrubs as the drying effect of the subtropical high-pressure zone takes over. This **scrub** vegetation changes into short grass **steppes** and then deserts as rainfall decreases.

When rainfall decreases to below ten inches a year, plant cover almost disappears. Only cactus, thornbush, and other plants adapted to drought survive in such dry lands. These specially adapted plants are called **xerophytes**. The vast deserts of North Africa, the Middle East, Central Asia, south-

ern Africa, and South America are among the driest places on Earth. Some weather stations in the Sahara have never recorded any rainfall. In these areas, human settlement is confined to isolated sources of ground water and to river valleys like the Nile Valley in Egypt.

The soils of steppes, and to a lesser extent of deserts, are often rich in plant nutrients. These soils are rich in salts because there is little rainfall to wash the salts away. They have little organic matter, however, because vegetation is sparse. Productive farming areas do exist in steppes and deserts wherever water can be made available, such as in the Imperial Valley in southern California. In general, however, dry lands are thinly populated.

Middle-Latitude Environments

In the temperate and continental climates of the middle latitudes, vegetation and soils vary considerably. Areas with Mediterranean climates, such as

PEOPLE-ENVIRONMENT

BIOSPHERE 2, SPACESHIP EARTH

On September 26, 1991, eight volunteers (four men and four women) sealed themselves off within an airtight three-acre greenhouse with approximately 3,800 other plant and animal species from five of Earth's biological realms. The eight biospherians remained inside for two years. What they and the research teams who follow them learn from their experience may provide scientists with information to improve the quality of life on Earth and contribute to building a colony on Mars.

They tried to build a planet in a bottle. Located near Oracle, Arizona, at the foot of the Santa Catalina Mountains, Biosphere 2 was intended to be an island ecosystem that would test the ability of life forms to thrive in a sealed environment.

The private world inside Biosphere 2 contained five simulated environments—rain forest, savanna, marsh, desert, and ocean. The biospherians grew their own vegetables and ground their own grain. Fellow inhabitants included turtles, bees, domesticated animals, saltwater coral, an octopus (that stowed away), and numerous plants and insects.

Technology was vital to the project. Machinery cooled and circulated the air, created mist to maintain the humidity of the rain forest, supplied fog to quench the thirst of desert plants, and made waves that gently moved in the 900,000-gallon saltwater ocean.

This private venture fulfilled the dreams of a group of young ecologists. They worked to raise their food and sustain their lives, while noting how the various life forms were faring. They each had their own private apartment and access to the community library, gym, and observatory.

The risks were great. A proper balance of oxygen and carbon dioxide had to be maintained by the plants to keep the air breathable. Two great domes, called "lungs," acted as variable volume chambers. Connected to Biosphere 2 by sealed, underground passageways, they enabled hot air to flow out during the daytime so that it would not break the huge panes of glass in Biosphere 2's structure. Pest control was maintained by ladybugs and other predatory insects, and by crop rotation and plant selection. Pesticides or fertilizers would have quickly contaminated the water supply. An agricultural area provided fruit, grains, and vegetables, while a barnyard provided meat, milk, and eggs.

Initially hailed by the Smithsonian Institution, London's Kew Gardens, and other scientific organizations, Biosphere 2 captured the imagination of many. David Letterman even devoted a "top ten" segment to it. Soon, however, the project came under blistering criticism.

Within a year, scientists ridiculed it for stashing an eighty-nine-day supply of food "just in case." The project organizers were also forced to pump in fresh air

---

southern California, have a scrub vegetation composed of plant species adapted to drought. Middle-latitude grasslands, or **prairies**, are found in the interior of North America and Asia, where rainfall decreases because of the distance from the ocean and the rain shadow effect of mountains. The humid subtropical areas of the southeastern United States and China support temperate forests, al-though those of China have been almost completely cut down. Thick **temperate forests** made up of deciduous and coniferous trees also grow in the marine west coast climates of the Pacific Northwest and northern Europe. Northward, the cold continental climates support the vast coniferous or boreal forests that stretch across much of southern Canada and Russia where they are called taiga.

▲ *Biosphere 2 is a unique experiment designed to create a self-contained planet under glass. Would you consider spending a year or two of your life as a biospherian?*

and to remove carbon dioxide from its atmosphere. When the biospherians emerged after two years on September 26, 1993, it was discovered that too much organic matter in the 30,000 tons of soil had absorbed oxygen.

Its founder continued to maintain that "Biosphere 2 is a child of our Earth's biosphere, grown from the same flesh and genetic material, and born of the perspective gained with Apollo's distant images of Earth." One of Biosphere 2's administrators put it more simply: "Someone has to have the courage to start; we are the first willing to be wrong." A second experiment in human habitation in Biosphere 2 is planned.

QUESTIONS

1. What is the major goal or purpose of the Biosphere 2 experiment?
2. What were some of the risks involved for the biospherians?
3. What are some of the criticisms aimed at the project?
4. Do you think the scientists are justified in spending research funds on the Biosphere projects?

There are variations in soil and in vegetation cover depending on location, but in general, forested areas have less fertile soils than grasslands. Extremely thin acidic soils underly the coniferous forests of Russia. The grasslands of the central United States and Ukraine have deep fertile soils. Most of these middle-latitude environments are used for farming and ranching.

High-Latitude Environments

In the high latitudes, vegetation cover decreases because temperatures become quite cold and summers grow short. Trees gradually disappear, and only low shrubs, mosses, and some ground plants survive in tundra regions. Find 80° N on the diagram on page 65, and you will see how vegetation decreases.

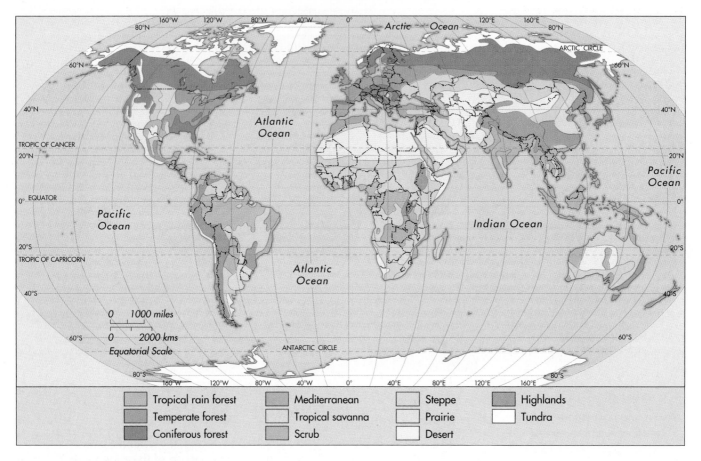

World Natural Vegetation

Compare the distribution of world natural vegetation regions above with the diagram showing vegetation and soils from the equator to the North Pole on pages 64–65.

Closer to the poles, much of the ground is frozen all year. This condition is called **permafrost.** Polar climates kill most life. The **tundra** and polar regions have little vegetation, very thin soils, and few people. Cold is a powerful environmental deterrent.

Environments and People

Most Americans now live in cities and experience the physical environment in which they live indirectly. Although blizzards, storms, droughts, and floods do occur, most people are insulated from direct contact with the environment by modern technology. However, environments still limit as well as present opportunities for human activities, as can be seen in the world map of population on pages 10 and 11. In the end, people live where they can make a living.

Large areas of our planet are almost empty of people. Small communities can be separated by many miles of unoccupied land. The tropical rain forests in Africa and South America and the deserts in North Africa, the Middle East, Central Asia, Australia, and the American Southwest (except for oasis cities like Phoenix and Las Vegas) support very few people. Also, few people live in the cold areas of Canada and Russia and in mountainous areas like the Andes of South America, the Rocky Mountains of North America, and the Himalayan ranges of Asia. Rapid population growth in the world has not led many people to move into these empty lands.

⊕ REVIEW QUESTIONS

1. What two basic elements of climate are major factors in determining what will grow in any region?
2. Which large deserts are among the driest places on Earth?
3. What vegetation once covered large areas of temperate and continental climates?
4. What causes vegetation to decrease in the high latitudes?

⊕ THOUGHT QUESTIONS

1. Tropical rain forests form a tall, dense mass of vegetation, yet rain forest soils, or laterites, are poor in nutrients. Why is this so? Does this make rain forests more vulnerable or less vulnerable to damage by human activities?
2. Why are ranching and herding the chief human activities in the savannas? Why aren't these regions intensively farmed? Do you think they might be in the future?
3. Why is living space not expanding to absorb the growing numbers of people on Earth? Why do people tend to cluster in existing areas of settlement rather than on frontiers? Would you like to live on a frontier like northern Canada or Alaska?

⊕ SECTION 3 VOCABULARY SKILLS

Directions: Match the numbered definitions with the vocabulary terms. Write the correct term on your paper next to the number of its matching definition.

1. Middle-latitude, treeless grasslands, characterized by low rainfall, found in the interior of North America and Asia	coniferous
2. Vast, nearly level, treeless plains of the Arctic regions of Europe and Asia	deciduous
	humus
3. A vegetative pattern characterized by short, thorny trees and shrubs, found on the margins of savannas due to low rainfall	laterite
	permafrost
4. Hard, compact, yellow-to-red soils that are poor for farming because of their low fertility	prairie
	scrub
5. A luxuriant evergreen forest found in the tropics where rainfall is abundant and there is no dry season	steppe *
	temperate forest *
6. Drought-tolerant plants, such as cactus and thornbush, that survive in dry lands and deserts	tropical forest
	tropical rain forest
7. Organic matter found in soil	tropical savanna *
8. A mixed evergreen and deciduous forest found in the parts of the tropics with a dry season	tundra *
	xerophyte

7. Organic matter found in soil
8. A mixed evergreen and deciduous forest found in the parts of the tropics with a dry season
9. Permanently frozen subsoil found in polar regions
10. Semiarid grassland found near deserts
11. A tropical grassland with a dry season in the winter and a wet season in the summer
12. Describes trees that drop their leaves in the cold or dry season
13. Describes trees that hold their leaves all year round
14. A forest of deciduous and coniferous trees found in a middle latitude climate

4. Changing Environments: An Altered Planet

While environments have placed limitations on human activities, people have also changed their environments to suit their needs. They have cut down woodlands for farming and encouraged the spread of some species of plants and animals at the expense of others. They have tried to make their environments more productive for themselves. In the past, this was always done in a piecemeal fashion.

But now people have an unprecedented level of scientific and technological power at their fingertips. Furthermore, the human population has increased dramatically, and a more intense use of environments has become common. As a result, people are rapidly changing Earth's land, air, and water to the extent that these basic elements of life have become issues of global concern.

This situation has posed new questions. How do we solve environmental problems that cross national boundaries? For example, neighboring countries may disagree about how to maintain and share the waters of rivers like the Rio Grande. Other problems may create conflict among countries of a whole region. Nations worry about water quality in large bodies of water like the Mediterranean Sea or the Great Lakes. *Acid rain* disturbs areas downwind from industrial centers. *Deforestation* or *desertification*—two processes of land degradation—in one country cause erosion and siltation in nearby countries. (Siltation is the depositing of fine-grained soil in the bed of a river, lake, or sea.) The condition of Earth's atmosphere and oceans now presents new challenges—and new concerns as well.

People Change Earth's Atmosphere

Human beings have attempted to influence the weather for thousands of years. Now several unplanned changes caused by human activity are changing Earth's atmosphere at an alarming rate. These are: (1) the increase of pollution and carbon dioxide in the atmosphere of the planet, especially in the Northern Hemisphere, and (2) the depletion of the *ozone* shield in Earth's upper atmosphere. The effects of these changes in the atmosphere may be severe. If the predicted damage occurs, it will be impossible to reverse.

As you know, carbon dioxide in the atmosphere traps energy and delays its return to space. Plants, particularly in tropical rain forests, remove carbon dioxide from the atmosphere and convert it into oxygen through the process of photosynthesis. The temperature of the planet is controlled by the amount of carbon dioxide in the atmosphere. More carbon dioxide means warmer temperatures, and less carbon dioxide means cooler temperatures. Carbon dioxide functions as Earth's thermostat.

Carbon dioxide is released into Earth's atmosphere by burning hydrocarbons like wood, coal, or oil. Factories and automobiles are major producers of carbon dioxide, and the atmosphere above the industrial areas of North America, Europe, and western Asia have high concentrations of this gas. Burning forests and grasslands create similar concentrations in the Amazon Basin.

Measurements show that the total amount of carbon dioxide in Earth's atmosphere has increased 30 percent since the Industrial Revolution began in 1750 and of that, 13 percent since 1959. This increase in carbon dioxide has raised Earth's temperature about 1° F. The correlation between these two events is shown in the figure on page 71.

How fast the atmosphere is undergoing this process of global warming is uncertain. All predictions are based on computer models. Most scientists expect Earth to be warmer in the next ten years than at any time in the last 100,000 years. Indeed, the four warmest years in the last century occurred in the 1980s, and 1990 was the warmest year ever recorded. But the eruption of Mount Pinatubo in the Philippines in June 1991 created a global volcanic haze that cooled world temperatures during the next several years.

Glaciers give solid evidence of global warming during the last century. Ice cores from Tibet show that temperature has risen between 2.5 and 4° F in the last hundred years. Moreover, alpine glaciers have shrunk nearly 50 percent since 1850. The evidence is not conclusive, but it cannot be ignored.

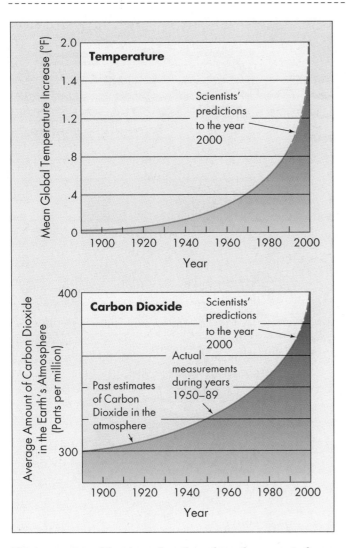

The temperature of the atmosphere depends on the amount of carbon dioxide in it. Both have been increasing steadily.

This warming would also change the distribution of rainfall on Earth as the map on page 74 shows. North Africa, Western Australia, India, and Mexico would be likely to receive more rainfall. The central portion of North America probably would lose rainfall as would the Commonwealth of Independent States (the former Soviet Union).

Many experts expect that warming will be greatest at the poles, melting the ice caps and raising sea levels. Already, enormous icebergs—one larger than the state of Rhode Island—have broken off the coast of Antarctica. A ten-foot rise in mean sea level (the mean is midway between the highest and low-est) would flood most of the world's coastal cities, as well as low-lying, densely populated areas like the Netherlands in Europe. A three-foot rise in mean sea level would also affect many large nations and leave 72 million people homeless in China, 11 million in Bangladesh, and 8 million in Egypt. In 1992, a group of thirty-seven small island nations expressed their concern at the United Nations over this potential threat to their countries.

The depletion of **ozone**, a form of oxygen found mostly in the upper atmosphere, is a second unexpected, human-made change in the world's envelope of life. Although ozone is a toxic chemical at lower levels of the atmosphere when highly concentrated, in the upper atmosphere ozone forms a protective shield that acts as an umbrella and blocks most ultraviolet light from reaching Earth's surface. Ultraviolet light causes skin cancer, damages the body's immune system, inhibits plant growth, and kills plankton, the microscopic organisms that are the basis of all marine food chains.

In the late 1980s, scientists discovered that the ozone shield above Earth had declined by 5 percent in only seven years. They also discovered an enormous hole, as large as the continental United States, in the ozone layer over Antarctica. By 1993, the ozone layer was thinner than it had been in the last fourteen years in both the Northern and Southern Hemispheres. That same year, scientists studying southern Argentina and Toronto, Canada documented the first large increase in ultraviolet radiation over populated areas of the world.

Ozone is destroyed by **chlorofluorocarbons**, or CFCs, a family of chemicals used in refrigerators, air conditioners, and aerosol cans. Scientists estimate that CFCs are being released into the atmosphere six times faster than they are being absorbed. Moreover, each CFC molecule that reaches the stratosphere—the second closest of five atmospheric layers to Earth—destroys an estimated 100,000 ozone molecules.

In 1987, twenty-four major industrial nations signed the Montreal Protocol, an agreement to reduce by half their production and use of CFCs by the year 2000. This was the first major success in global environmental diplomacy.

The ozone hole over Antarctica has continued to expand, and evidence now suggests that a similar

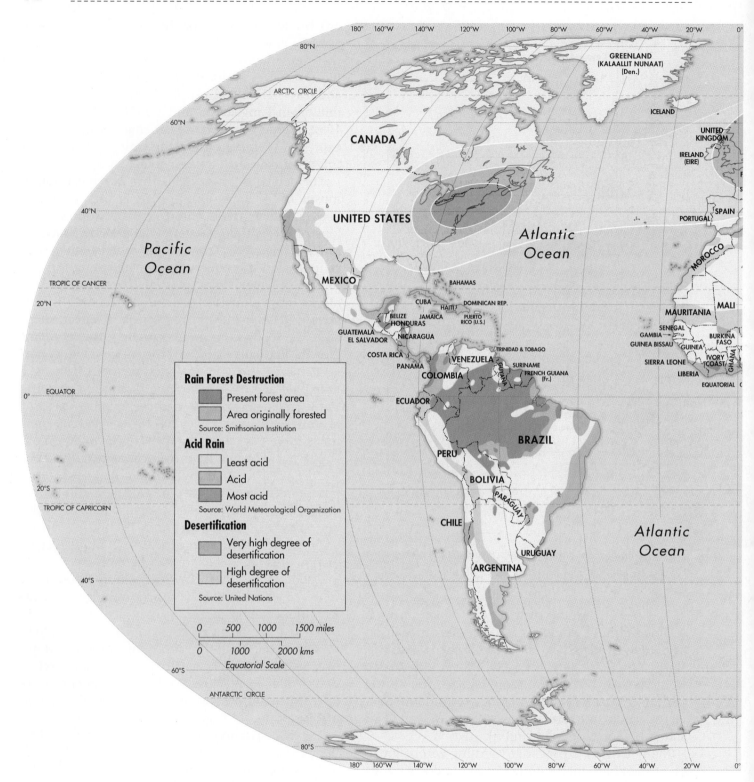

World Desertification and Acid Rain Map

1. SLOVENIA
2. CROATIA
3. BOSNIA AND HERZEGOVINA
4. ALBANIA
5. MACEDONIA

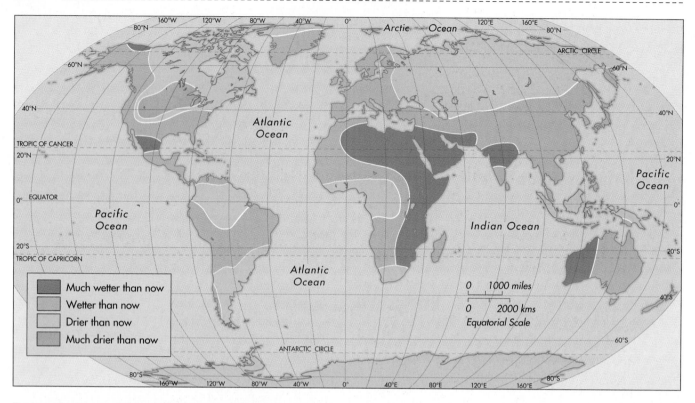

Projected Changes in the Distribution of Rainfall

Changes in the distribution of rainfall will occur as the carbon dioxide content of the atmosphere continues to increase.

depletion of the ozone layer is spreading over North America, Europe, and Russia. In 1992, eighty-seven countries met in Copenhagen and moved up the phaseout date to 1996. But the damage to the ozone layer has been done, and it will last one hundred years even if the CFC emissions are cut by 90 percent immediately. As one expert noted, so many tons of CFCs are already on their way to the stratosphere that the ozone layer will not recover until the middle of the next century.

Humans Change the Face of the Earth

On the land, growing populations are exerting pressure on land resources and creating other environmental issues of global concern. Among the most immediate of these problems are (1) the rapid destruction of tropical and high-latitude forests (**deforestation**); (2) a reduction in **biodiversity** (the number of species that exists in a given habitat and on the planet); (3) *acid rain*, which pollutes rivers and lakes and damages trees; and (4) soil erosion and *desertification*, which will reduce food production. Grain harvests and fish catches have already begun to decline on a per capita basis.

Tropical rain forests cover less than 7 percent of Earth's surface, yet they contain nearly half of all living species in our world. About 40 percent of these environments have been cut down. Rain forests are being plowed, burned, and logged over at an estimated rate of 40 million acres a year, an area roughly the size of the state of Florida (see map on pages 72 and 73).

Along with this deforestation is the elimination of species at a rate of seventy species per day, a reduction in the biodiversity of the planet's living species comparable to the age when dinosaurs became extinct 65 million years ago. Some scientists believe that if the forests of Central and South

America, Africa, and Southeast Asia continue to be cut down at current rates, a quarter or more of all species on Earth could become extinct.

Acid rain produced by sulphur dioxide and nitrogen oxides has also reduced the rate of forest growth and eliminated some species of trees. This type of pollution from the factories of the industrial Middle West in North America and the industrial heartland of Europe has caused large areas of the eastern United States, Canada, and Scandinavia to suffer major losses of timber and fish (see illustration on pages 72 and 73).

Soil erosion is also happening at an alarming rate in the world's agricultural zones. In the United States, losses of topsoil to erosion are occurring at six to nine times the rate that it takes new soil to form. Eroded soil flows out of the Mississippi River at a rate of 900 tons per minute. In South America, the reddish-brown eroded soil carried out of the mouth of the Amazon River stains the waters of the South Atlantic Ocean for hundreds of miles.

In addition, desertification—the degradation of productive rangeland, semiarid cropland, and irrigated farmland into desert—is occurring. Desertification is common on the margins of the world's largest deserts, as the map on pages 72 and 73 shows. It is also found where high population densities and poor land use practices have stripped the land bare. Worldwide, desertification has claimed 2 billion acres of productive land in the last fifty years.

⊕ REVIEW QUESTIONS

1. What causes global warming?
2. What is the ozone layer?
3. Where is desertification usually found?
4. What regions are affected by acid rain? Why?

⊕ THOUGHT QUESTIONS

1. What factors are causing damage to Earth's environments? What do you believe should be done about it? Are any clubs or groups at your school working to reduce environmental damage? What can you do as an individual to help?
2. Do you believe that recent conferences on the environment will have positive effects on changing attitudes and practices that damage the environment? Why or why not?

⊕ SECTION 4 VOCABULARY SKILLS

Directions: Use the following vocabulary terms in sentences that help to explain their meaning.

acid rain
biodiversity
chlorofluorocarbons (CFCs)
deforestation

desertification
global warming
ozone
siltation

⊕ CHAPTER SUMMARY

Three major components of Earth's environments are: landforms, climates, and vegetation. Earth's surface features are created by the interaction of forces of weathering, erosion, and deposition, and by movements of Earth's crust. These forces work on Earth's surface to create the variety of landforms that we take for granted: mountains, hills, plateaus, and plains.

Weather and climate affect every region on Earth. Weather varies daily in most regions of the world; climate is the cumulative pattern of daily weather events measured over a long period of time. Climate, when measured in terms of temperature and precipitation in a given region, is a major factor determining the types of soils and vegetation found on Earth.

Weather patterns, and therefore climates, are also influenced by a global system of semipermanent high- and low-pressure cells in the atmosphere. These cells serve to redistribute heat Earth receives from the sun by means of warm air currents from the tropics flowing to the poles and cool air from the poles flowing to the tropics.

Geographers have classified and described these global climate patterns. The Köppen classification system identifies five major climatic types based on the rainfall and temperature needs of types of vegetation.

Finally, you learned that humans have altered their environments in order to suit their needs. While this has often enriched our lives, or simply made them more comfortable, human activities have also caused air and water pollution and environmental degradation in many world regions. The environmental problems that have resulted from human activity are among the primary problems we face today and in the future.

EXERCISES

Complete the Sentences

Directions: Each of the following sentences has three possible endings. Find the correct ending and then write the completed sentence on your paper.

1. More carbon dioxide in the atmosphere causes (a) cooler temperatures (b) warmer temperatures (c) more rainfall.
2. Warm, light air that rises is associated with (a) low-pressure areas (b) high-pressure areas (c) strong volcanic action.
3. The North Atlantic Drift begins in the Gulf of Mexico and warms the shores of (a) Brazil (b) Japan (c) the British Isles.
4. Since the Industrial Revolution began in 1750, the amount of carbon dioxide in the atmosphere has (a) decreased 5 percent (b) stayed about the same (c) increased 30 percent.
5. In a recent period of seven years, the ozone layer in Earth's atmosphere has (a) decreased 5 percent (b) stayed about the same (c) increased 30 percent.
6. Climates are generally dry in areas (a) of low pressure (b) of high pressure (c) that are mountainous.
7. Tropical rain forest climates (Af) (a) have a dry season (b) are hot and dry all year (c) are hot and wet all year.
8. Desert climates are dry (a) all year (b) only in the summer (c) only in the winter.
9. Temperate climates are found mainly in the (a) low latitudes (b) middle latitudes (c) high latitudes.
10. Short grasses prevail in areas with (a) tropical rain forest climates (b) desert climates (c) steppe climates.
11. One area with a Mediterranean climate is (a) northern Europe (b) southern Canada (c) southern California.
12. Long, cold winters and short, hot summers are characteristics of a (a) Mediterranean climate (b) continental climate (c) desert climate.
13. Tundra climates have (a) no month with an average temperature above freezing (b) cold winters and long, hot summers (c) low rainfall.
14. Tropical rain forests have (a) rich soils (b) poor soils (c) dry soils.

Matching

Directions: The terms in Part A are described by the statements in Part B. On your paper, write the letter in front of each term next to the number of the statement it matches.

Part A

a. carbon dioxide
b. continents
c. erosion
d. humus
e. land and sea breezes
f. landforms
g. ocean currents
h. ozone
i. vegetation
j. volcanoes

Part B

1. The largest landforms on Earth's surface
2. The general term for the plant cover of an area
3. The organic matter in the soil
4. Landscape features on the surface of Earth
5. Usually located along the edges of plates
6. Released into Earth's atmosphere by burning wood, coal, or oil
7. Forms a protective layer in Earth's upper atmosphere
8. Caused by the differences in the heating of land and water surfaces
9. Giant streams of seawater driven by winds
10. Movement of rock fragments by wind, running water, ice, and ocean currents

Inquiry

Directions: Combine the information in this chapter with your own ideas to answer these questions.

1. Geologists predict that severe earthquakes may occur someday along the San Andreas fault in southern California, yet this area is heavily populated. Why do you think its population keeps growing?

2. On which type of landform—plain, plateau, hill, or mountain—do you live? Name the locations nearest to you where the other three types of landforms exist.
3. How would you describe the climate, including rainfall, of the place where you live? What types of vegetation are well suited to your area?
4. Describe two of the ways that people have altered Earth's environment. Why are these human alterations to planet Earth problems of a region, rather than problems of a single state or country?

SKILLS

Using a Table

Directions: A table shows important information at a glance. Near the beginning of each unit in this book, you will find a table that highlights the vital statistics of the regions discussed in that unit. Additional tables appear in many units.

Look at the table on page 40 and then answer these questions:

1. Which is the largest continent?
2. Which is the smallest continent?
3. Is South America or North America larger?
4. Is Africa or Europe larger?
5. Asia occupies what percentage of the total land surface?
6. North America occupies what percentage of the total land surface?
7. What is the highest mountain in the world? On what continent is it found?
8. Mount Everest is how much higher than Mount McKinley?
9. What is the largest lake in the world? On what continent is it found?
10. What is the largest lake in North America?
11. Does Europe have any lakes larger than 8,000 square miles?
12. What is the longest river in the world? On what continent is it found?
13. What is the longest river in South America?
14. What is the longest river in Asia?

15. Which continent has no recorded lakes or rivers?

Exploring the Unit Miniatlas

Directions: There is a miniatlas with important maps at the beginning of each unit in this book. You may refer to the miniatlas maps many times while you are studying a unit. Turn to pages 4–11 and answer these questions:

1. How many maps are in the miniatlas for Unit 1?
2. What are the titles of these maps?
3. Do all of these maps show the names of continents?
4. Which map shows the names of countries?
5. Which map shows the elevation of landforms?

Using a Physical Map

Directions: Look closely at the world physical map on pages 8 and 9 and answer these questions:

1. What color shows mountains over 8,200 feet high?
2. Does the eastern or western part of North America have the largest ranges of mountains?
3. Does the eastern or western part of South America have the largest ranges of mountains?
4. What is the name of South America's largest range of mountains?
5. Are there more ranges of mountains in Europe or Asia?
6. Is most of Australia mountainous?
7. Mount Everest in Asia is part of what mountain range?
8. Are there mainly lowlands east of the Mississippi River in North America?
9. Is Europe a continent composed chiefly of lowlands or of high mountains?
10. Are there wider stretches of lowlands in the northern or southern part of Africa?
11. Are there more large lakes in North America or South America?
12. Does Asia or South America have the most islands?

13. What oceans are shown on the map?
14. Is the Indian Ocean or the Pacific Ocean larger?
15. What ocean is located between Africa and Australia?

Using a Climate Map

Use the climate map on pages 58 and 59 to answer these questions:

1. What color on the map shows tropical rain forest climates?
2. What colors on the map show dry climates?
3. What colors on the map show continental climates?
4. What color on the map shows polar climates?
5. What color on the map shows highland climates?
6. Does the eastern or western part of North America have the largest areas with a continental climate?
7. Are areas with a dry climate located in the northeastern or the southwestern part of North America?
8. Are the large areas in North America with a highland climate located near the Atlantic Ocean or the Pacific Ocean?
9. In what part of South America do highland climates prevail?
10. Are the areas in South America with a tropical climate located north or south of the Tropic of Capricorn?
11. Are there areas with dry climates along part of South America's Pacific Coast?
12. Is the dominant climate of Western Europe temperate, tropical, or dry?
13. Notice the wide variety of climates in Asia. Are the largest areas of Asia with dry climates generally north or south of the areas with continental climates?
14. Are the areas in Asia with highland climates larger or smaller than the areas in Africa with highland climates?

15. Are most of the areas in Asia with tropical climates north or south of the Tropic of Cancer?
16. What is the prevailing climate of the islands near the southeastern coast of continental Asia?
17. Does Australia have any large areas with highland climates?
18. The central part of Australia has what kind of climate?
19. Is northern or southern Africa a large area with a dry climate?
20. Are most of the areas with a polar climate north or south of the Arctic Circle?

Reading a Diagram of Vegetation and Soils

Directions: Turn to pages 64 and 65 to answer these questions about the ways vegetation and soils change as one moves from the equator to the North Pole.

1. What is the name of the soils in areas that have tropical rain forests?
2. Are forest or grassland soils found in areas that have scrub vegetation?
3. Is there denser vegetation in tropical rain forests or temperate forests?
4. Do steppes generally have more or less vegetation than savannas?
5. Are taiga forests usually as dense as temperate forests?
6. Do taiga areas generally have more or less vegetation than tundra areas?
7. Are temperate forests or tropical forests nearer the North Pole?
8. Are tropical forests or tropical savannas nearer the equator?
9. Do trees generally grow taller in tropical rain forests or in taigas?
10. Do trees generally grow taller in temperate forests or in tropical savannas?

3

PEOPLE CHANGE THE WORLD

Geographers study the ways environments influence people's lives and the ways people influence their environments. What geographers have learned is that relationships between people and environment generally change slowly. Twice in human history, however, important discoveries led to major changes in how people live. These two periods of change, or transformation, are called the Environmental Transformation and the Scientific Transformation.

◀ *This beautiful image of Earth poised above the moon's horizon illustrates the oneness of the planet. The astronauts have commented on the lack of borders on Earth when viewed from space. How might this concept of global interdependence influence our uses of Earth's life-sustaining resources?*

1. Earth Environments and People—This Human World

This Human World

Today 5.5 billion people live on Earth. They all must have air and water to live, and they all need food, clothing, and shelter. They all must live in an earth environment that provides for their basic needs, and so people think, plan, build, and create. They also destroy.

This human world is what people have made of Earth. It includes all the people who live on the planet and all the different ways they use, and have used, their environments. As you know, people have profoundly changed landscapes all over the Earth; people are important shapers of their environments.

People Shape Environments Through Culture

How people use and modify Earth environments depends on their culture. Culture is the learned behavior of a society or nation. It includes people's knowledge, religion, law, language, technology, and ways of living, and it embodies all the attitudes, skills, and ideas that people learn from one another. You have been acquiring your culture from relatives and friends; from television and radio programs; from books, movies, classes, and computers; and most of all from your own life experiences. Moreover, each day you use your knowledge to change your immediate environment. In cold weather, for example, you turn up the heat indoors or wear heavier clothing outdoors.

Our culture in the United States enables people to modify environments on a large scale. Dams harness rivers to prevent floods and produce electric power. Earth-moving equipment is used to reclaim swamps, to build roads and canals, and to mine ores. In food-producing regions, people have replaced the natural vegetation with thousands of acres of crops like corn and wheat.

Technology Develops Resources

Technology includes all of the methods, tools, and knowledge that people use to obtain products they need or want. A group's technology tells us a great deal about the way of life of its members.

Members of societies with a simple technology often spend most of their time producing and collecting food. Their ability to change environments is limited because they do not have time for specialization. People with more advanced technologies have greater control over their environments. They are freer to pursue a variety of activities beyond just getting food. Fewer people are employed in agriculture, and they can instead specialize in manufacturing; in building towns, roads, or bridges; in creating works of art; in helping others; or in learning more about this human world.

Technology develops and creates resources, materials or human skills that can be used to meet a need. The word *resource* comes from the Latin word meaning "to spring up." Resources become resources when people apply their knowledge and tools to making use of them.

Petroleum, for example, did not become an important source of energy in the United States until the mid-1800s. At that time, new high-speed machines needed oil as a lubricant. This created a market for better drilling and pumping devices that could tap oil deposits deeper below the surface. In 1859, a man named Edwin Drake drilled a seventy-foot well at Titusville, Pennsylvania, and struck oil. Soon thereafter, the oil fields of Pennsylvania were producing 30 million barrels of oil a year.

Steel manufacturing developed on a large scale when Andrew Carnegie, a famous American industrialist, adapted the Bessemer process, a method of injecting oxygen into iron that purified and strengthened the iron into steel. After Carnegie applied new production techniques to create low-cost steel, a new mining technology was developed to dig up large quantities of iron and coal. Soon, canals and railroads were built to transport these heavy materials from mines to factories.

In the American South, the invention of the cotton gin, a simple machine that separated cotton fiber from cotton seed, led to a 1,000-mile-wide belt of cotton fields that stretched across the South from the Atlantic coast to central Texas in the 1800s. A relatively new resource today is uranium. Algae may be an important food resource in the future.

Human and Natural Resources

Geographers distinguish between human and natural resources. People are **human resources**. Their skills and knowledge are vital to our society and are used to create new technology. That is also why many countries invest heavily in educational systems and large companies support major research programs. You are a human resource that is being developed. Can you think of reasons taxpayers in your community should invest in increasing your knowledge and skills?

Natural resources—materials supplied in nature and available for human use—are renewable or nonrenewable. Forests are **renewable resources** that can be harvested, planted, and reharvested many times over. Soils that have been planted in crops are not renewable resources. They can, however, be maintained as stable resources if care is taken to prevent erosion and depletion of soil nutrients. Coal, oil, and minerals are **nonrenewable resources**. Only a limited amount of each of these elements exists on our planet, and they cannot be replaced once they are used up.

Looking Back in Time

For most of human history, people lived in a very different world than ours of today. Twenty thou-

IT'S A FACT Destruction of rain forests now threatens with extinction a quarter of all species of life on Earth.

sand years ago, small groups of men and women lived in scattered locations across the planet. They made a living scavenging the landscape as hunters, gatherers, and collectors. They acquired their food by locating and collecting plants, and by fishing, killing animals, and securing any food source within their reach. Their choices were few and their knowledge was limited. Although they made sophisticated tools, their use of Earth's resources was limited by their small numbers.

People lived briefly then; only a small number survived to an old age. The entire Earth probably supported a total population of only 5 million people. This is less than one-third the number of people living in the New York City area today.

That these early hunters and gatherers survived at all is remarkable. They faced hostile surroundings, unreliable food supplies, and dangerous animals. Life was uncertain. In order to survive, these people probably had to be intelligent, resourceful, and able to work cooperatively in groups.

Then two periods of inventive change occurred that created more secure and productive ways of living and transformed the geography of the Earth. During the first period, the Environmental Transformation, people changed their relationships with

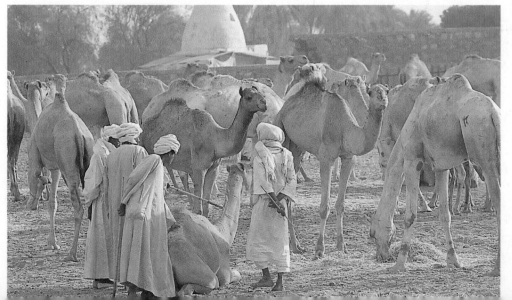

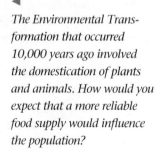

◄

The Environmental Transformation that occurred 10,000 years ago involved the domestication of plants and animals. How would you expect that a more reliable food supply would influence the population?

nature. About 10,000 years ago, they began to practice agriculture and tamed animals for their own use. They built cities, created civilizations, and developed new patterns of culture. By the 1500s, world population had grown to 500 million.

A second period of great change, the Scientific Transformation, began in Western Europe between 1500 and 1750. This revolution involved changes in technology that enabled Europeans to explore nearly all of the Earth, to create new inventions, and to harness natural resources. Scientific understanding of Earth increased beyond the imagination of people living just a few hundred years earlier. It continues to increase at an accelerated pace today. As this revolution in science and technology spread across the globe, the world came to be divided into the two regions that you will be studying: the technological world and the developing world.

⊕ REVIEW QUESTIONS

1. What necessities must all people have in order to live?
2. What is the meaning of the word *culture*, and what does it include?
3. Why was life so difficult for people who lived 20,000 years ago?
4. What is a natural resource? A human resource? A renewable resource? A nonrenewable resource?

⊕ THOUGHT QUESTIONS

1. For many centuries huge amounts of coal, iron, and oil lay untouched beneath Earth's surface. Why do you think they became resources only in modern times? What new resources are becoming available now? Do you think that computer technology, biotechnology, space exploration, or oceanography will create new resources? Explain.
2. When did the Environmental Transformation begin, and what major changes did it involve? What do you think it would be like to live by hunting, scavenging, and collecting?
3. When did the Scientific Transformation begin, and what major changes did it involve? These changes have clearly affected your life. Can you identify three or four examples of such changes?

2. The Environmental Transformation

Farming and Herding in the Middle East

During the Environmental Transformation, people developed agriculture, tamed animals, improved their technology, and settled down in villages. Plants and animals that were **domesticated**, or adapted for use by and made dependent upon humans, began to provide more reliable sources of food and clothing for growing populations of farmers and herders. These people were learning new ways to control their environments and improve their lives.

Early archaeological data suggests that agriculture probably was developed first in the Middle East about 10,000 years ago, although it appeared later throughout Asia, Africa, and North and South America. In the foothills surrounding the river valleys of Mesopotamia in what is now modern Iraq, people began to cultivate wheat and barley and to herd sheep and goats. This may have occurred as a result of a change in climate or an increase in population may have forced people to shift from collecting to planting. Whatever the reason, farmers soon moved from the foothills down into the valleys of the Tigris and Euphrates Rivers in Mesopotamia. Locate this area and these rivers on the map on page 83.

Agriculture and herding appeared in many other areas such as in Egypt, Southeast Asia, India, China, and along the shores of the Mediterranean Sea soon after its appearance in the foothills of the Middle East. Much later, these activities developed in Mexico and Peru. Exactly why human beings be-

IT'S A FACT The average height of hunters and gatherers at the end of the Ice Age was 5 feet, 9 inches for men and 5 feet, 5 inches for women, much as it is today.

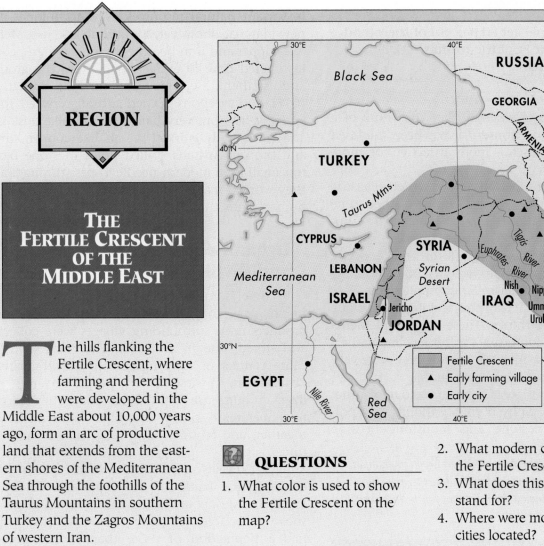

DISCOVERING

REGION

THE FERTILE CRESCENT OF THE MIDDLE EAST

The hills flanking the Fertile Crescent, where farming and herding were developed in the Middle East about 10,000 years ago, form an arc of productive land that extends from the eastern shores of the Mediterranean Sea through the foothills of the Taurus Mountains in southern Turkey and the Zagros Mountains of western Iran.

QUESTIONS

1. What color is used to show the Fertile Crescent on the map?

2. What modern countries share the Fertile Crescent today?

3. What does this symbol ▲ stand for?

4. Where were most of the early cities located?

came farmers and herders 10,000 years ago is still not known, but when they did, they changed the geography of Earth and they changed themselves.

This New Way of Life Involved Many Changes

Farming and herding gave people a more reliable food supply than had hunting and gathering. People's lives were no longer governed by the natural distribution of edible plants and animals, although these continued to be important food sources. But people could now produce food where they settled and store it in large quantities for future use. As a

result, more children survived, people lived longer, and population grew. Human beings moved into environments best suited to planting their crops and tending their animals. Life flowed with the seasons, with the year marked by festivals celebrating important annual events, such as planting and harvesting.

Village populations grew to 200 to 500 people, and these fixed places became the centers of daily life. Land near the villages was cleared to plant crops, and hillsides were terraced to add level land for farming. Forests were cut for firewood and for building materials. Dry lands were irrigated. These activities gradually modified natural landscapes to make them more productive. Can you think of a landscape near your home that has been modified to make it more useful, more attractive, or more productive?

Day-to-day living became more settled. Because people no longer had to spend most of their time moving about in search of food, they began to improve their technology to meet the demands of their new lifestyle. Some people became full-time artisans, and many improvements in technology took place in their villages. Stone implements were created to grind grain; clay pots and jugs were made to carry and store things. These early implements would have been too heavy for hunters and gatherers to carry from camp to camp, but they were very valuable to farmers living in one place.

Early agricultural tools were made out of limestone and sandstone by boring, grinding, and polishing. Thus, limestone and sandstone became

▼ *This ancient relief from the tomb of an Egyptian pharoah depicts an offering of the dead pharoah to the gods. How does this artifact support the belief that this is the product of an advanced civilization?*

important natural resources. New techniques and new materials, however, soon came into use. Metals like copper and iron became resources when people discovered how to extract them from the ground and use them. Metal tips were made to strengthen plows and weapons. Expert artisans created tools and jewelry from very hard materials like obsidian, and from easily worked metals like copper, silver, and bronze. Artisans drilled holes in stone beads too small for a modern needle to pass through, and they worked marble and alabaster into animal and human figures.

Urban Beginnings and Civilizations

In time, some villages with favorable environments grew into cities—large settlements where products from surrounding villages were collected and stored, bought and sold. The cities soon became centers of learning for the world's earliest civilizations. In these centers, people became "civilized" in our sense of the word today. They learned to cooperate, to solve the problems of living closely together, and to share knowledge in a formal setting.

By 5,000 years ago, a number of large cities existed along the banks of the Tigris and Euphrates Rivers in Mesopotamia, as the map on page 83 shows. The largest of these cities was Uruk, which had a population of more than 50,000 people. These centers of Sumerian civilization governed a rich farming region 10,000 square miles in area.

In the cities, priests and kings directed people's activities. Farmers, fisherfolk, and shepherds came to the cities to exchange their products for urban goods and services, and traders brought goods to the cities from distant places. Men and women were organized into labor groups to build temples and palaces. Writing was developed to keep records that scribes were taught how to keep.

Many negative aspects of civilization also appeared among the Sumerians. Slaves formed a large portion of the population; they built and repaired city walls, irrigation networks, and public buildings. Professional armies were organized to protect each city from enemies in neighboring cities, and warfare became so common that nearly a quarter of the

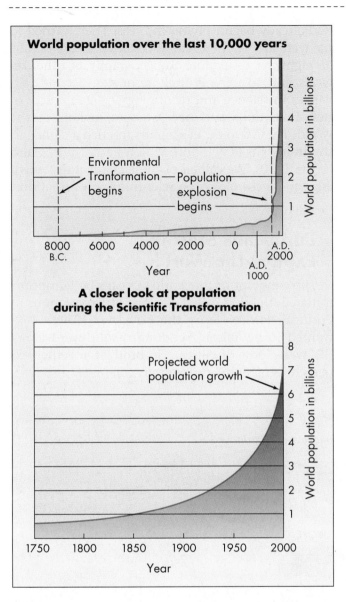

World population over the last 10,000 years

Environmental Tranformation begins

Population explosion begins

World population in billions

Year

8000 B.C.   6000   4000   2000   0   A.D. 1000   A.D. 2000

A closer look at population during the Scientific Transformation

Projected world population growth

World population in billions

1750   1800   1850   1900   1950   2000

Year

World Population Then and Now

For most of our history, human population grew slowly, as the top graph shows. The Scientific Transformation began a "population explosion" in the 1750s that has continued to accelerate. As the lower graph shows, the 5 billion mark for world population was passed in 1987.

and gathering. By A.D. 1500, the population of the planet had increased from 5 million to around 500 million people. Many of these people lived in sophisticated urban civilizations in the Middle East; in Western Europe; on the shores of the Mediterranean Sea; in the river valleys of Pakistan, India, and China; in West and East Africa; and in the Aztec and Inca empires of Mexico and Peru.

The human world came to be divided into distinct civilizations that occupied different parts of the Earth in A.D. 1500. Each civilization had its own religion, language, art, and economy.

⊕ REVIEW QUESTIONS

1. Where were people first able to get food without having to depend on the natural distribution of plants and animals? Agriculture and herding soon appeared in what other regions?
2. Why did the development of food production encourage the development of new technologies?
3. What were some of the major characteristics of Sumerian cities?
4. By A.D. 1500, where did sophisticated urban civilizations exist?

⊕ THOUGHT QUESTIONS

1. What new activities developed when people no longer had to spend most of their time finding food? How would your life have been different if you lived during the Environmental Transformation?
2. What do we mean when we say people became "civilized"? Think about the meaning of the word "civil." Do you believe that people who live in cities are more "civilized" than others? Why or why not?
3. Why did the population of the Earth increase from 5 million to around 500 million by A.D. 1500? Why were a person's chances of survival better? Would you prefer to have lived in an earlier time period? If so, when and where?

wealth of this civilization was spent on the military. Resources were used in great quantities to support palaces, rulers, and armies.

The Environmental Transformation Spread

The Environmental Transformation slowly spread from early centers of civilization like Mesopotamia into new environments. By A.D. 1500, agriculture, domesticated animals, and town life had changed the geography of a large part of the world.

The ability to produce food supported one hundred times more people on Earth than had hunting

3. The Scientific Transformation

The Scientific Transformation Began in Western Europe

The second major period of change, the Scientific Transformation, began with intellectual and geographical explorations in the 1500s and culminated in the Industrial Revolution in England in the 1750s. The scientific and industrial advances of this period made it possible for a few European countries to dominate many other civilizations on Earth.

Two developments in Western Europe between 1500 and 1750 led to one of the most exciting periods in human history. First, Europeans began exploring the coastlines of other continents, although earlier Arab, Indian, and Chinese traders had already begun plying the seas. They learned to navigate across open seas and started to use oceans as highways to explore the geography of other regions and to exploit their resources. Second, advances in science and technology created a host of new inventions and new ways of looking at the world. These two developments reached a climax in the Industrial Revolution, which began in England in the 1750s. We call all of these changes in knowledge and technology the Scientific Transformation.

Europeans Start to Explore the World

The conquest of the world's oceans by Europeans initiated a period called the Age of Exploration. In Portugal, Prince Henry the Navigator (1394–1460) wanted to be able to send ships southward along the west coast of Africa. He brought together experts in geography, cartography (mapmaking), navi-

The Age of Discovery: Early European Voyages

What does this map show? Read the two map keys at the bottom of the map to learn the answer.

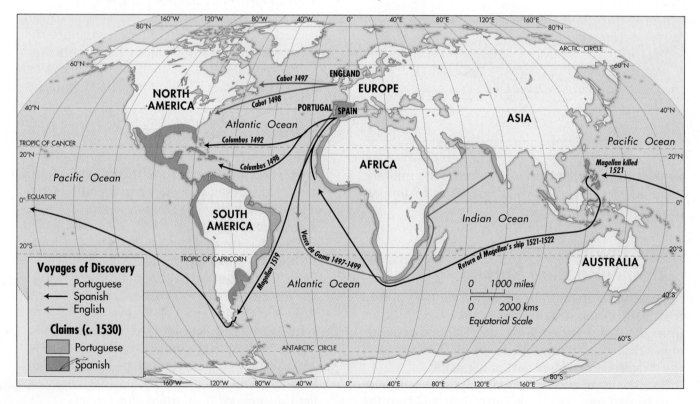

gation, and shipbuilding. These experts invented a new kind of ship, the **caravel**, which could sail in any direction, even into the wind. The caravel also was large enough to store provisions for long journeys. With these new ships, European sailors could cross the world's oceans and return safely to their home ports.

In the Italian port city of Genoa, one such sailor—Christopher Columbus (1451–1506)—read Marco Polo's fascinating account of his overland travels to the rich and wonderful empire of China in East Asia. Columbus, like many other people of his time, knew that the Earth was round well before his historic voyage to the New World. However, his miscalculation of the circumference of the planet caused him to think the wealth of Asia could be reached by sailing only 5,000 miles to the west, less than half the true distance to Asia. He would later explain this error in judgment by likening the shape of the Earth to that of a pear.

Sponsored by Queen Isabella of Spain, Columbus set sail across the Atlantic Ocean. On October 12, 1492, he landed on the island of San Salvador in the Bahamas, just south of Florida. Columbus arrived in the New World, and began an exchange of plants, animals, people, and wealth that would transform both the New World and the Old—a process now called the "Columbian Exchange." At the time, no one in Europe knew that two huge continents and many islands lay sprawled between Europe and Asia.

Columbus returned to Spain a national hero. He brought back to Spain the electrifying news that land lay only thirty-three days' sailing time to the west of Europe. Spain quickly sent Columbus with another, better equipped expedition to colonize these lands across the Atlantic. Because many believed that Columbus had reached Asia, the spices, silks, jewels, and gold of the Far East that Europeans desired so much now seemed to be within their reach. They soon discovered that the New World blocked their path.

Despite this, the race for the riches of Asia was on among those European countries that were able to sail the oceans of the world. In 1497, John Cabot (a navigator from Venice, Italy) sailed from England across the Atlantic to find a northwest passage around the landmass discovered by Columbus and in that way to get to China. Instead he ran into the rocky shores of modern Canada (follow his route on the map on page 86). That same year, the Portuguese explorer Vasco da Gama sailed eastward around the southern tip of Africa and reached the west coast of India.

In 1513, Vasco Núñez de Balboa of Spain crossed the Isthmus of Panama and became the first European to wade in the Pacific Ocean. Then, in 1519, Ferdinand Magellan, with his Portuguese crew, embarked on the first journey around the world. They sailed south and west around South America into the 10,000-mile-wide Pacific Ocean. Magellan was killed in the Philippine Islands, but eighteen of the sailors in his ship *Vittoria* succeeded in returning to Spain.

Other voyages of exploration filled in details on increasingly accurate maps. The Age of Exploration opened a new and larger world to the Europeans. For the first time in history, Earth was actually proven to be round, and almost every part of it could be reached by Europeans.

Explorations in Science and Technology

At this same time, a different world was being explored by European mathematicians and physicists like Nicolaus Copernicus, Johannes Kepler, Galileo Galilei, Francis Bacon, René Descartes, and Isaac Newton. These men were scientists, a word that was not to be invented until 1840, who observed the world with new inventions like the telescope and microscope. They measured things in the natural world, performed experiments, and recorded their results. This new process of observation and experimentation was called the **scientific method**. They changed people's understanding of the world.

Scientists discovered that Earth moves regularly around the sun. They learned that Earth is not, as previously believed, located at the center of the universe. New concepts of space, time, matter, and motion were discovered and debated, and a thirst for knowledge swept across Europe. When this knowledge was applied to improvements in

▲ *Galileo Galilei is often referred to as the first modern scientist. He conducted many experiments in an effort to discover how the laws of nature operated. He was one of the first scholars to use the scientific method. What is the scientific method?*

manufacturing, many new inventions, like those listed on page 89, began to change day-to-day life.

The Industrial Revolution

The two major forces underlying the Scientific Transformation—geographical and scientific exploration—led to the beginning of the Industrial Revolution in England in the 1750s. Geographical exploration opened up new sources of wealth, resource regions that Europeans could exploit, and new markets to sell goods from Europe. Scientific exploration uncovered the principles upon which a new technology could be based.

In the 1750s these forces began to reshape England into the world's first industrial society. Three factors were of great importance. First, population began to grow, forcing some people to leave the farms and migrate to cities. Improvements in agricultural production fed growing city populations. Second, science created a new technology that used the energy of coal and the strength of iron to increase the production of goods initially in the textile industry. England had ample supplies of both coal and iron. Third, cities, transportation networks, and industries grew quickly. People began to process raw materials into products on a large scale; this was called **mass production.** These three factors helped to create a new human environment, the industrial city.

Science Led to New Technology in England

England's growing population supplied laborers to work in the new factories that sprang up during the Industrial Revolution. These same people also bought the products manufactured cheaply in these new factories.

The factories were built with wealth brought to England by trade. Merchants and bankers who had made money in world trade became interested in making more money by applying scientific principles to productive technology. They invested **capital**—money and materials—in improvements in technology. They built factories and started businesses. In this way, England's capitalists encouraged and financed the ideas of talented scientists and inventors.

The new technology created by these inventors substituted coal and other sources of **inanimate** (nonliving) **energy** for muscle power and wood (animate sources). Technology also created machines that replaced human skills and increased the productive capacities of each worker. Workers now labored in large factories and mills rather than in their homes or small shops.

Two inventions that were crucial to the Industrial Revolution were developed at this time. James Watt's steam engine, modified for production in 1776, provided energy to run many different kinds of machines. Once coal (instead of charcoal from

**PEOPLE-
ENVIRONMENT**

MAJOR INVENTIONS OF THE 1700s AND 1800s

Inventions result in technological change and growth in an economy. Think, for example, how the invention of the computer is changing the way people make and do things today. Computers also are leading to the growth of new jobs and new businesses.

❓ QUESTIONS

1. Choose three of the inventions listed in this table and suggest three changes that each invention may have caused.
2. Name five of the more important inventions from the last 100 years that you would include on an updated list and give reasons.

Major British, European, and American Inventions of the 1700s and 1800s

Year	Invention	Country
1764	Spinning jenny	England
1769	Self-propelled steam vehicle	France
1776	Steam engine	England
1783	Puddling iron furnace	England
1785	Power loom	England
1786	Threshing machine	Scotland
1793	Cotton gin	United States
1802	Steamboat	United States
1811	Cylinder printing press	Germany
1824	Portland cement	England
1825	Steam locomotive	England
1831	Electric generator	England
1834	Reaper	United States
1839	Photography (daguerreotype)	France
1839	Vulcanization of rubber	United States
1844	Telegraph	United States
1845	Rotary printing press	United States
1850	Corn picker	United States
1851	Refrigerating machine	United States
1855	Bessemer process of steelmaking	England
1859	Gas engine	France
1859	Oil-well drilling	United States
1861	Passenger elevator	United States
1866	Open-hearth steel furnace	United States
1867	Reinforced concrete	France
1869	Railway air brake	United States
1876	Four-cycle gas engine	Germany
1876	Telephone	United States
1879	Incandescent light	United States
1882	Steam turbine	France
1884	Artificial silk (rayon)	France
1884	Linotype	United States
1884	Photographic roll film	United States
1888	Pneumatic tire	Ireland
1892	Diesel engine	Germany
1892	Electric motor (alternating current)	United States
1892	Gasoline automobile	United States
1893	Motion pictures	United States
1895	Wireless telegraphy	Italy
1900	Caterpillar tractor	United States

wood) was used to purify iron, this once scarce metal became more widely available and cheaper. Iron was soon used in making a variety of machines and tools. Factories no longer had to be situated on streams for water power, and iron forges no longer had to be located in forests near a supply of wood. Steam power and coal-purified iron meant that industries could move to areas where coal was found.

Cities, Transportation Systems, and Industrialization Changed England

England became the world's first urban society. People left the countryside and moved into industrial towns in search of jobs in factories. By 1800, two-thirds of England's population lived in urban areas (cities or towns). London attracted many of these people. Others moved to cities—like Manchester, Liverpool, and Birmingham—near the coal fields of the north, where the growing use of coal resulted in the construction of many factories and mines.

Cities were soon linked together by improved transportation systems. By 1775, canals had been built to connect every major city in England by water. Barges on these canals cut the costs of moving heavy, bulky materials like coal, iron, stone, and lime. By the middle of the 1800s, this network of canals was reinforced by a web of railroad lines that soon crisscrossed the English countryside. Steam locomotives greatly reduced the time and cost of transporting goods overland.

These transportation systems further encouraged the development of huge cities like London, Liverpool, Manchester, and Birmingham. Railroads not only carried great quantities of food and raw materials needed for manufacturing to these cities,

▼ *Steamships transformed oceans from barriers into highways and contributed to the creation of a global economy. Explain how this new technology increased contact among the peoples of the world.*

but also distributed urban-made manufactured goods back to the countryside.

The Spread of the Scientific Transformation

The spread of the Scientific Transformation from England to the continent of Europe and to other parts of the world took place in the 1800s and 1900s. Before the 1800s, colonies of Europeans had already settled in North and South America and in South Africa, and Europeans controlled some trading stations on the coasts of Asia and Africa. However, Europeans were not found in great numbers outside of Europe.

Then, beginning in the late 1700s, a wave of 53 million emigrants swept out of Europe and settled in North America, South America, South Africa, Australia, and New Zealand. Most of these regions were located in middle-latitude environments similar to those of Europe. Here, immigrants planted crops, started towns, moved into the interior, and dislodged or destroyed local peoples like Native Americans. In each of these regions the ties to European culture and language remained strong among immigrants long after new countries like the United States, Canada, Brazil, Argentina, Australia, and New Zealand became independent.

The immigrants brought with them skills in manufacturing and farming. In time, regions settled by European immigrants brought together the capital, resources, technology, and organization needed to become industrialized countries. These regions became a part of the technological world.

In the tropical regions of Latin America, Africa, South Asia, Southeast Asia, and East Asia, the effects of the Scientific Transformation were quite different. These regions produced tropical agricultural products and minerals wanted in the marketplaces of Europe; however, their environments were unappealing to most European settlers. In the tropics, Europeans formed a small but well-armed ruling minority. They started plantations and mines, but, as colonial rulers, they lived apart from native peoples.

Europeans provided the capital and technological know-how in tropical colonies, and local peoples provided the labor. As a result, colonies in these regions became dependent on the export of their raw materials to industrial countries in Europe. Japan was one of the few countries to escape European colonization.

Most of these colonies won political independence in the years following World War II in a world designed by and for Europeans. The former colonies never were fully able to become part of the technological world. Today, they remain the less developed, poorer nations.

Vast inequalities separate the "haves" of the technological world and "have-nots" of the developing world today. About one-quarter of the 5.5 billion people on Earth are relatively well-fed, healthy, and long-lived. Three-quarters are not. Today, competition for food, energy, and mineral resources is generating stress in world affairs. These problems are an important part of the study of this human world by modern geographers.

⊕ REVIEW QUESTIONS

1. What two developments in Western Europe led to the Scientific Transformation?
2. What three factors shaped England into the world's first industrial society?
3. Why were the steam engine and coal-purified iron process crucial to the Industrial Revolution?
4. What fraction of Earth's people today have neither enough food nor long life spans?

⊕ THOUGHT QUESTIONS

1. The caravel, a new kind of ship developed during the Age of Exploration, made it easier for sailors to travel across Earth's oceans. What types of even more efficient transportation have been built since the days of Prince Henry the Navigator? What effect do you think faster transport around the globe has had on your town or region? Have these changes affected you personally?
2. What were the environmental characteristics of regions permanently settled by European emigrants?

4. Our Human World Today

One World Divided

The Scientific Transformation did not spread evenly across Earth. Countries settled by emigrating Europeans had direct access to the knowledge and technology of the Industrial Revolution. In time, these countries (and later Japan) developed their human and natural resources and became technological societies. In these technological societies, population is growing slowly. People live longer, eat better, are healthier, have access to more energy resources, and are wealthier than people in developing countries.

In developing countries, many of which were European colonies, populations are growing rapidly. People do not live as long, their use of energy is low, living standards are rising slowly, and most people are poor.

To develop economically, these countries need to invest money in human resources and in expanding their production of natural resources. Many developing countries are already deeply in debt, yet their growing numbers of young people need and demand more schools, better medical care, and jobs. Most developing countries simply do not have enough money to pay for these long-term investments.

As a result, the world today stands divided between the rich and the poor, the haves and have-nots, or as they are called in this book, the people of the technological world and the developing world.

The Technological World

The technological world includes most of the countries located in five culture regions: (1) Western Europe; (2) the Commonwealth of Independent States (CIS), formerly the Soviet Union and parts of Eastern Europe; (3) the United States and Canada; (4) Japan; and (5) Australia and New Zealand in the Pacific. A **culture region** is a large area of the world unified by a common culture.

Each of the five culture regions of the technological world has a middle-latitude environment. With the exception of Australia and New Zealand in the Pacific World, all are located in the Northern Hemisphere. All but Japan, Korea, and small countries like Taiwan and Singapore are products of Western culture. Taken together, these regions are inhabited by less than a quarter of the world's people. They are frequently called *modern* or post industrial because of their ability to support large populations at a high **standard of living**—a measure of material goods and services consumed per person.

What defines these countries as members of the technological world is their wealth. Defining wealth is very difficult, but usually a country's wealth is measured by its **gross national product (GNP)**, the total value of all goods and services produced by a country in a given year. The GNP includes the value of all salaries paid, all goods bought and sold, and all resources produced. This figure is divided by the number of people living in that country to find **per-capita GNP** (per-capita means per-person). Every technological country has a high per-capita GNP, usually $7,500 per year or more. In contrast, the poorest countries in the world have per-capita GNPs of $1,000 per year or less.

Ten Characteristics of the Technological World

The countries of the technological world share a number of characteristics. Each country has developed its human resources to a high degree. Each country either uses local natural resources or imports raw materials to support its prosperous urban, industrial population. Most of the ten characteristics described below are found in each country of the technological world. As you think about these ten characteristics, see how well they agree with what you know about the United States as a technological society.

1. *People are relatively well off.* Per-capita GNP is high in technological countries. This wealth frees most people in the technological world from the basic struggle for food, clothing, and shelter. These countries also have wealth avail-

able for investment in new enterprises and activities.

2. *People live long lives.* Within the technological world, advances in medical science and health care enable people to have a life expectancy of seventy-five years or more. This average life span is considerably longer than ever before in history. Our own country has increasing numbers of senior citizens. What future policies will best enhance their ability to contribute to society?

3. *People are well educated and highly skilled.* **Literacy**, the ability to read and write, approaches 100 percent in technological societies. Educational systems are well developed. Corporations invest heavily in training systems, and young people spend many years in school preparing to take part in an increasingly complex society and economy. Many schools have already introduced computer technology into the classroom. Many new jobs are related to new technologies.

4. *Population is growing slowly.* Birth rates and death rates are low in technological countries. As a result, total population is growing slowly or not at all. In the United States, population is increasing at a rate of less than 1 percent a year. By the year 2000, the population of the United States will increase by only 20 million. In some Western European countries, total population is holding steady or declining slightly.

5. *Most people live in cities.* Technological societies are urban societies. A majority of their citizens live in cities and work in offices, factories, and businesses. Three-quarters of the people of Western Europe, the United States, Canada, Australia, New Zealand, and Japan live in cities, and two-thirds of all citizens of Russia and Eastern Europe are urbanites.

6. *Few people work in agriculture.* Technological countries require relatively few workers in agriculture. Countries that are self-sufficient in food, like the United States, Canada, Australia and New Zealand, have highly mechanized farms and ranches—factories in the fields—that literally manufacture food. These large farm businesses are able to feed their country's people and also produce food for export. Tech-

nological countries that import food, like Britain and Japan, have found it more profitable to exchange manufactured goods and services for food than to cultivate the land. In most technological countries, less than one-tenth of the people work in agriculture.

7. *Consumption of energy is high.* The countries of the technological world consume nearly 85 percent of the world's energy. The United States alone uses as much energy as the world's eighty least developed nations. Most work is done by machines powered by the major sources of energy—oil, natural gas, coal, and nuclear power. Many people in technological countries are freed from manual labor. They are instead trained to use machines to increase their productivity, which enables them to maintain a higher standard of living. Per-capita consumption of energy is generally considered one of the most reliable measures of modernization.

8. *Most people work in manufacturing or service industries.* An overwhelming majority of the people in technological countries work in factories, businesses, government offices, and organizations located in cities. Their activities produce the goods and services desired and needed in a modern mass-production, mass-consumption society. Except in the Commonwealth of Independent States (CIS) and Eastern Europe, urban populations produce industrial and consumer goods in large quantities for sale at home and abroad. They also provide the basic medical, educational, and social services required to maintain a high standard of living. The citizens of technological countries create the new ideas and products and develop the resources that will shape future economies.

9. *Transportation and communication networks are well developed.* To sustain their high-energy, industry-based economies, technological countries have well-developed transportation and communication networks. Roads, railroads, and sea and air transport connect urban centers, manufacturing areas, farmlands, port cities, and areas with natural resources. The costs of transport are low. Computer, satellite television, telephone, fax machines, and radio communications tie together most sectors of a technological society.

10. *Environmental control is high.* Knowledge of science and advanced technology gives people in the technological world a great deal of control over their environment. Resources are fully exploited. River systems are altered and forests are harvested and replanted. Urban centers of production and consumption operate on such a large scale that damage to the environment in the form of pollution is a matter of serious concern.

Most countries of the technological world share these ten characteristics in various degrees. Taken together, these characteristics describe wealthy, productive, politically powerful countries that control

◄

This aerial view of Vancouver, Canada, illustrates many characteristics of cities in technological societies. Describe and discuss three of these characteristics.

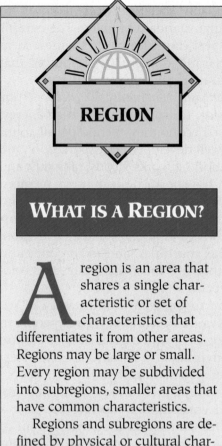

REGION

WHAT IS A REGION?

A region is an area that shares a single characteristic or set of characteristics that differentiates it from other areas. Regions may be large or small. Every region may be subdivided into subregions, smaller areas that have common characteristics.

Regions and subregions are defined by physical or cultural characteristics. Physical regions like the Rocky Mountains of North America, the Sahara Desert in Africa, and Amazonia in South America are defined by environmental conditions. Political regions are marked by boundaries. Every state has borders and every country has national boundaries.

Regions based on one characteristic often cut across regions based on other characteristics. The Rocky Mountains, for example, extend from Mexico, through the United States, and into Canada, crossing two international boundaries. The Sahara Desert extends across at least nine North African countries.

This book organizes our study of the Earth around eleven large **culture regions**, each of which is a complex area that shares common cultural and, in some cases, environmental features. The Middle East and North Africa (Unit 9), for example, is a culture region with a dry environment and a largely Islamic culture. The boundaries of China (Unit 10) enclose the world's oldest continuous culture.

All of these culture regions are complex. Within the culture region of the Middle East and North Africa, Israel does not share the predominant culture of the region. This is also true of the inner reaches of China, which are occupied by Tibetan, Mongol, and Turkish peoples. These are examples of subregions within a larger culture region. Indeed, most Tibetans view their region as independent of China, politically and culturally.

QUESTIONS

1. On the map on page 95, five of Earth's eleven culture regions are identified as belonging to the technological world. Six are members of the developing

---

much of the world economy and influence world affairs. Yet many also have serious environmental problems and pockets of poverty in important regions of their countries.

The Developing World

The developing world includes most of the countries located in six culture regions: (1) Latin America, (2) Africa south of the Sahara, (3) Middle East and North Africa, (4) China, (5) South Asia, and (6) Southeast Asia.

Each of the six culture regions of the developing world has a tropical or subtropical environment. All are crossed by either the Tropic of Cancer (23½° N) or the Tropic of Capricorn (23½° S). The equator cuts through Latin America, Africa, and Southeast Asia. Only China reaches into the middle latitudes to a significant extent. The southern tips of Latin America and Africa also have middle-latitude environments. Five of the six—Latin America being the exception—have non-Western cultures. More than three-quarters of the world's people live in these six culture regions.

Ten Characteristics of the Developing World

The countries of the developing world share many characteristics despite important differences in culture, environment, and history. Most of these countries are now struggling to develop their human resources by investing in education, health, and other social services. Many are attempting to

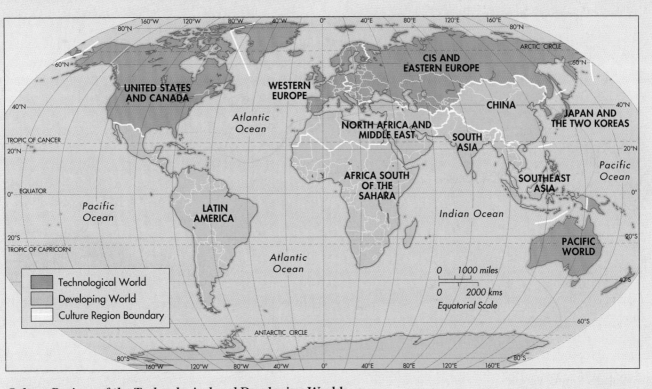

Culture Regions of the Technological and Developing Worlds

world. Locate each of these culture regions on the map. Do you see how subdividing the world into regions makes it easier to understand regions based on other characteristics?

2. Provide an example not given in the text of a region defined by one characteristic.

create new jobs for their rapidly expanding populations by exploiting natural resources and building industry. Others, afflicted by debt, natural disasters, or war, are becoming poorer each year. Some governments have used money from a high-priced resource like oil to modernize their countries. Few countries in the developing world, however, have enough money to pay for the staggering costs of modernization.

1. *People are relatively poor*. Most citizens in developing countries have low incomes. Even in the more prosperous of these countries, people's incomes equal only one-fifth of those of people in the technological world. Poverty breeds poverty. Because there is no money left over after paying for food and shelter, investment in

educating a son or daughter or starting a new business is often impossible. Only the oil-rich countries of the Organization of Petroleum Exporting Countries (OPEC) have money to invest. Modernization in the rest of the developing world involves great personal sacrifice or an infusion of foreign capital.

2. *People do not lead long lives*. People born in many countries of the developing world can expect to live (on average) into their sixties, about fifteen years less than people in the technologi-

IT'S A FACT   Four of every ten people on Earth live in India or China.

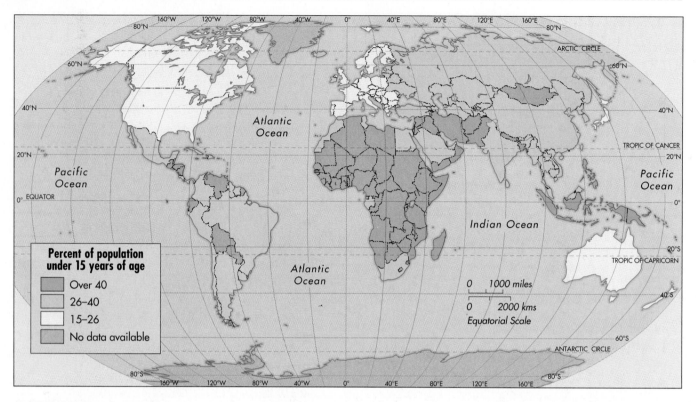

The World's Young People

Which culture regions have the highest percentage of their people under fifteen years of age? Which culture regions have the lowest percentage? How do you think this will affect life in the United States in 30 years? In Latin America?

cal world. In the poorest countries of Africa, Southeast Asia, and Latin America, average life spans barely reach fifty years. These relatively short average life spans result from high infant mortality, lives of toil, inadequate diets, poor health care, and too few doctors, nurses, and public health officials. In many countries, medical specialists are found only in capital cities. Almost every country is trying to improve health conditions; the success of these efforts depends on the country's environment, culture, and wealth.

3. *Most people are not well educated*. Only a limited number of people in the developing world have learned to read and write in their own language. Even fewer have learned European languages that would provide them with direct access to books on science and technology. Teachers, schools, and technical institutes are increasing in number, but educational systems in many developing countries are inadequate.

Because schools frequently exist only in cities and towns and because in many cases village children must begin work in the fields as soon as they can walk, many children have no opportunity to go to school.

4. *Population is growing rapidly*. Most countries in Latin America, Africa, and Asia have high birth rates and falling death rates. This results in rapid population growth—in many countries, at the rate of between 2 and 3 percent a year. This means that these countries will double their total population every twenty-five to thirty-five years. Between 1980 and 1990, more than 80 percent of the estimated 1 billion people added to the population of Earth were born in developing countries. Children under the age of fifteen now make up half of the population of many countries in the developing world. This places an enormous burden on educational, medical, and service facilities in the very countries least able to afford the cost.

5. *Most people live in rural areas or urban slums.* Developing societies usually are rural societies. Most people in India and China still live in villages, where they tend fields and herd animals. Almost all work is done by hand rather than by machine, and many workers are needed to produce a successful harvest. As population grows, however, some villagers are being forced to leave the land and move into cities. The cities have few jobs for these people because urbanization is happening faster than factories are being built. As a result, huge metropolises like Mexico City, Rio de Janeiro, Cairo, Bombay, and Calcutta are becoming centers of poverty. These cities are growing so rapidly that the number of people is overwhelming the available supplies of water, electricity, and housing. As a result, slum dwellers form a growing percentage of city populations.

6. *Many people work in agriculture.* A large portion of the population in many developing countries work to produce food. Some people work on plantations that grow commercial products like tea, coffee, and bananas for export. Still more people work to produce food for their families. Few countries produce enough to feed their own people. Today, the poorest countries regularly rely on international agencies for food. Malnutrition is now believed to be one of the most widespread health problems in the developing world.

7. *Consumption of energy is low.* Although the countries of the developing world include more than three-quarters of Earth's population, they use only 15 percent of the world's energy. The main source of energy in these countries is muscle power, because machines that run on coal, oil, and natural gas are too costly to buy, operate, and repair. Because people still work with their hands, their productivity and standards of living are low. The next time you see a road or a building under construction, notice that machines do the heavy work. Could the same project be completed without using these machines? How many people would be needed to do the job?

8. *Few people work in manufacturing or service industries.* Most developing countries have govern-ment programs to build more factories, mines, and plantations in an effort to become more prosperous. Few countries, however, have achieved this stage of economic development. Most developing nations are trapped in a situation in which populations are growing faster than jobs. As a result, many industrial products must be imported, consumer goods are in short supply, and social services cannot meet growing demand. The inability to provide people with ways to earn a decent living causes discontent in many countries.

9. *Transportation and communication networks are not well developed.* A major obstacle to using resources in the developing world is inadequate road, rail, and air networks to move people and goods from place to place. In Africa, Latin America, and much of Asia, railroads have just begun to open up landlocked regions, making it possible for local producers to sell their goods in urban markets. In most developing countries, plans are underway to create more effective networks of transportation and communication. Each new transportation link creates economic opportunities, just as railroads built in the late 1800s gave Texans the means to sell longhorn cattle to cities on the Atlantic seaboard in our country.

10. *Control of the environment is low.* Knowledge of science and technology is limited in the developing world. The costs of importing foreign specialists and technology are high. In many parts of the developing world, money is not available to build dams, clear land for farms, complete transportation projects, and protect people from natural hazards. People there live closer to nature. They are often not protected from droughts or floods.

The countries of the developing world share many of these ten characteristics, yet tremendous geographical and cultural diversity exists among these countries. Venezuela, Costa Rica, and Argentina, for example, stand as beacons of progress in Latin America when compared with the abject poverty of Haiti and Bolivia. In China, the economy is booming, but in parts of Africa, some countries have been engulfed by the southward

▲ *As in many less developed countries in the world, a majority of Indian people are village farmers. Why are so many people farmers in countries like India and so few people farmers in countries such as the United States?*

march of the Sahara Desert and other countries are engaged in civil war.

In Asia, excepting China, population growth threatens many plans for economic growth. In the oil-rich states of the Middle East, money is available, but the development of human resources is difficult in this arid, war-torn region.

Together, these ten characteristics of the developing world describe in varying degrees the poorer and less productive societies on Earth. Their inability to play significant roles in the world economy and world affairs is a constant source of frustration.

The Revolution of Rising Expectations

Nearly two-thirds of Earth's inhabitants still lead the difficult lives of their ancestors. The wealthy countries of the technological world have assisted the less developed countries through organizations like the World Bank and the International Mone-

tary Fund. This has helped only to a limited degree. The poor have remained poor, except in those countries where determined leadership or the discovery of new resources has made rapid development possible.

In a basic way, the people in the developing world have been changed by the "revolution of rising expectations." In Ethiopia, a plowman still uses the broken branch of a tree to cultivate the land, but he knows that with cash he could buy a metal plow and have larger harvests. In the Middle East, villagers live in homes similar to those built at the time of Jesus, but they know that their children must leave the village if they are to lead better lives. In Latin America, farmers leave their families to find work in the city because most good farmland is owned by wealthy landlords, but these farmers know that a better life is possible and that some people have such a life.

In a sense, all of these people—your fellow human beings—are enduring in the same way that their ancestors endured. However, they now have rising expectations for a better life. The knowledge of the wealth of other countries has made them feel poorer. The rising expectations of people in the developing world are central to the agenda of issues your generation will soon face.

An Agenda for Tomorrow

The problems that face the people of the developing world make up an important part of the study of this human world. These problems—including food shortages, persistent poverty, low standards of living, and growing populations—were created over the centuries by the ways the two great revolutions of the past, the Environmental Transformation and the Scientific Transformation, have affected people. In the technological world, land degradation, pollution, equality, equity, and quality of life are major problems.

Looking back, it is comforting to recognize that we human beings have accomplished much and endured more. Our attitudes toward modernization, environmental quality, pollution, and the preservation of human dignity will shape the future policies of our government. These policies can help forge a new world geography. The central message of the *Apollo 12* photographs from space showing planet Earth is that we live together on the same limited support system. Now more than ever before, each of us needs to view our lives on a global level. The human condition on Earth is the business of your generation of Americans—part of your agenda for tomorrow.

⊕ REVIEW QUESTIONS

1. What culture regions are included in the technological world?
2. What mathematical process do you use to find a country's per-capita gross national product?
3. What are the major sources of energy in the technological world? In the developing world?
4. What country uses as much energy as the eighty least developed countries on Earth?
5. What culture regions are included in the developing world?

⊕ THOUGHT QUESTIONS

1. Why do countries of the technological world need well-developed transportation and communication networks? Describe the transportation system of your state, city, or town. Has this ever been a factor affecting your family's travel plans, choice of job, or place to live?
2. How does the growth of population in the technological world compare with the growth of population in the developing world? Do you think that the number of people your age will have an effect on your job opportunities in future years? If so, how?
3. Why do many people in the developing world and few people in the technological world work in agriculture? Do you know any farmers? Would you consider farming as a future career? What factors would you consider? How has farming changed in your area in the last two generations?

⊕ CHAPTER SUMMARY

Geography includes the study of the ways that environments and people interact. Important discoveries during two periods in human history led to major changes in how people interacted with their environment. The first was the Environmental Transformation, in which the domestication of plants and animals paved the way for such changes as the growth of cities and towns, trade, and craft specialization. The second was the Scientific Transformation, which continues today. The Scientific Transformation included the development of the scientific method of learning how the natural world works through observation and experimentation, global voyages of discovery, and the Industrial Revolution. This revolution began in England and spread rapidly to other parts of the world.

The world became divided between the technological world and the developing world as a result of the Scientific Transformation. The Industrial Revolution meant that European countries sought additional natural resources as raw materials for their factories. When Europeans migrated to less developed regions, they developed natural resources there and sent them to industrial countries in exchange for manufactured products. As a result, developing countries had their economies transformed to meet the needs of the industrial countries of Europe. After these regions were taken over politically and made into colonies and subsequently became independent nations, this pattern of economic dependency continued.

The ten characteristics of the technological world are: (1) people are well off, (2) they live long lives, (3) they are well educated, (4) the population grows slowly, (5) most people live in cities, (6) few work in agriculture, (7) energy consumption is high, (8) most people work in manufacturing or service industries, (9) transportation and communication networks are well developed, and (10) people control their environments. The ten characteristics of the developing world are: (1) people are poor, (2) their life expectancy is shorter, (3) most people are not well educated, (4) population grows rapidly, (5) most people live in rural areas or urban slums, (6) most people work in agriculture, (7) energy consumption is low, (8) few people work in manufacturing or service industries, (9) transportation and communication networks are not well developed, and (10) people do not control their environment. Despite the difficult lives led by people in the developing world, their attitudes are being changed by the "revolution of rising expectations."

CHAPTER 3

EXERCISES

One Wrong Word

Directions: In each of the following sentences, the italicized word is wrong. Rewrite the sentences on your paper, substituting the correct word or words for each wrong word.

1. People in *India* created the Sumerian civilization.
2. The Scientific Transformation made it possible for *Asian* countries to dominate most of the world for many years.
3. The Industrial Revolution began in the country of *Mesopotamia* in the 1750s.
4. John Cabot searched for a northwest passage to *America*.
5. Vasco da Gama's expedition finally reached the west coast of *Africa*.
6. In England in the early 1800s, two-thirds of the population lived in *rural* areas.
7. In the 1700s, *oil* replaced wood as the major fuel for producing iron.
8. The countries of the technological world generally are located in the *tropics*.

Find the Correct Ending

Directions: On your paper, write the following sentences with the correct ending for each sentence.

1. The earliest people on Earth were (a) farmers (b) hunters and gatherers (c) temple builders.
2. The first major change in human culture was the (a) Scientific Transformation (b) Environmental Transformation (c) Industrial Revolution.
3. Agriculture probably developed first in (a) Mexico (b) Peru (c) Mesopotamia.
4. Cartographers are (a) mapmakers (b) navigators (c) priests.
5. Prince Henry the Navigator sponsored expeditions that sailed along the west coast of (a) North America (b) Northern Europe (c) Africa.

6. The first European explorer to sail from the Atlantic Ocean to the Pacific Ocean was (a) Cabot (b) Columbus (c) Magellan.
7. A country whose population grows at a yearly rate of 3 percent will double its population in (a) one hundred years (b) fifty years (c) twenty-five to thirty-five years.
8. A country whose population is growing rapidly is (a) France (b) India (c) the United States.
9. One of the most prosperous countries in Latin America is (a) Costa Rica (b) Haiti (c) Bolivia.
10. A large city that is a center of poverty is (a) Tokyo (b) Calcutta (c) London.
11. One of the regions in the developing world is (a) Western Europe (b) Japan (c) Latin America.

Inquiry

Directions: Combine the information in this chapter with your own ideas to answer these questions.

1. People who lived thousands of years ago faced many dangers that most of us don't face today. What were some of these dangers? What dangers do we now face that did not confront these earlier people?
2. When people settled down to farm, their lives became governed by the seasons. In what ways are your activities governed by the seasons?
3. Early people gradually changed landscapes in various parts of the Earth to suit their needs. In what ways are people still doing the same thing?
4. The death rate in Western Europe began falling sharply after 1700. What medical breakthroughs in recent years have caused the death rate to continue declining?

SKILLS

Reading Graphs That Show World Population

Directions: A graph is a type of chart that shows how two or more ideas are related. Graphs often can show relationships more effectively and more quickly than can words. Turn to page 85 and look first at the graph on the top of the page to answer these questions:

1. What does the horizontal line at the bottom of the graph show?
2. What does the vertical line at the right side of the graph show?
3. Was the world population more or less than 1 billion in A.D. 1000?
4. Did the world population grow faster between 8000 B.C. and A.D. 1000 or between A.D. 1000 and the present time?
5. Would you conclude from this graph that the world population has grown at an even rate or at an uneven rate since 8000 B.C.?

Directions: Now answer these questions about the bottom graph on page 85.

6. Does this graph show fewer or more years than the other graph?
7. Was the world population still less than 1 billion as late as 1800?
8. Did the world population grow at a faster rate between 1750 and 1800 or between 1900 and the present time?
9. Is the world population still growing at a rapid rate?
10. Would you conclude from this graph that the Scientific Transformation had a small effect or a large effect on population growth?

Using a Population Map of the World's Young People

Directions: Maps can be used for many special purposes. A special-purpose map in this chapter is the population map of the world's young people on page 96. First, study its key carefully, then answer the questions that follow. To do this exercise, you may also want to refer to the political map of the world on pages 6 and 7.

1. Does this map show the entire world or only a portion of it?
2. It tells about people in what age group?
3. The dark red on the map shows regions where young people under age fifteen make up what percentage of the population?
4. Is the percentage of young people in the population larger in South Asia or in Western Europe? You read in Chapter 3 that one of these two regions has had greater industrial growth. Which region is it?
5. About what percentage of the population of the United States consists of young people under fifteen years of age?
6. Find Mexico on the map. This is the nearest southern neighbor of the United States. Mexico has one of the fastest growing populations of any country in the world. What percentage of the population of Mexico consists of young people under fifteen years of age?
7. Notice that some parts of South America have at least 40 percent of the population under fifteen years of age. In most parts of South America, is the percentage of young people in the population larger or smaller than that of the United States?
8. Why do you suppose that regions with a relatively high proportion of young people have a more rapid population growth than regions with a relatively high proportion of older people?

Vocabulary Skills

Directions: Match the numbered definitions with the vocabulary terms. Write the correct term on your paper next to the number of the matching definition.

Bessemer process
capital
caravel
culture
culture region
domesticate
gross national
  product (GNP)
human resources
inanimate energy
literacy

mass production
natural resources
nonrenewable
  resources
per-capita GNP
renewable resources
resources
scientific method
standard of living
technology
urban

1. To tame plants or animals, especially by generations of breeding to live in close association with humans
2. Wealth in the form of money or property that is used to produce more wealth
3. Able to be restored or replenished with proper planning for its use
4. Energy derived from nonliving sources like coal or oil
5. Describing a city or a town as opposed to the country
6. A country's per-person gross national product

7. Not able to be restored or replenished once it is used
8. The learned behavior of a society or nation, including people's knowledge, faith, law, language, technology, and ways of living
9. A process of observation and experimentation that includes recording the procedures for others to test the results
10. The total value of all goods and services produced in a country in a year
11. The methods, tools, and knowledge used to obtain products needed or wanted
12. A small sailing ship of the 1400s and 1500s that could sail in any direction, even into the wind
13. People whose skills and knowledge are vital to a society and are used to create new technology, among other things
14. Of the quality of life of people, based on such measures as income and ownership of material goods
15. Materials or human skills that can be used to meet a need
16. The ability to read and write
17. Process of making steel that inexpensively cleans impurities from iron
18. A large area of the world unified by a common culture
19. Wealth supplied in nature and available for human use
20. The large-scale process of raw materials into products

Which Ending?

Directions: Each of the following sentences has three possible endings. Choose the correct ending, and then write each sentence on your paper.

1. Both Mercury and Venus are (a) colder than Earth (b) warmer than Earth (c) larger than Earth.
2. In December, the Northern Hemisphere is tilted (a) away from the sun (b) toward the sun (c) at an angle of about 14 degrees.
3. Today the people whose lives are most like those of their ancestors' are in (a) Ethiopia (b) England (c) Japan.
4. The Industrial Revolution began in (a) China (b) the United States (c) England.
5. Today, the distinctive character of most places is based mainly on ideas (a) developed recently at that location (b) handed down by the government (c) developed in different places and at different times.
6. Coal, oil, and minerals are (a) the only sources of energy (b) all removed from mines (c) nonrenewable resources.
7. Most of the people who left Europe in the late 1700s and 1800s migrated to regions in the (a) low latitudes (b) middle latitudes (c) high latitudes.
8. The invention of agriculture probably occurred first in (a) the Middle East (b) Mexico (c) Peru.
9. The chief effect of the Scientific Transformation on the tropical regions in Latin America, Africa, and Asia was to turn these regions into (a) exporters of raw materials (b) industrial societies (c) democratic societies.

Beginnings and Endings

Directions: Part A consists of sentence beginnings; Part B consists of sentence endings. Find the correct ending for each beginning, and write the complete sentence on your paper.

Part A
1. People invented agriculture and herding
2. The building of dams to harness rivers
3. During the Age of Exploration, Europeans started to use

4. Both the United States and Japan
5. The areas between the Tropic of Cancer and the Tropic of Capricorn are in
6. The Scientific Transformation in the 1700s and 1800s
7. Both China and Latin America
8. Regions with long, cold winters and short, hot summers
9. Polar regions are found in
10. Temperate forests are found in

Part B
a. the low latitudes.
b. the middle latitudes.
c. the high latitudes.
d. are part of the developing world.
e. are part of the technological world.
f. during the Environmental Transformation.
g. led to the establishment of modern industrial societies.
h. have continental climates.
i. oceans as highways to explore the Earth.
j. enables people to modify environments on a large scale.

Writing Skills Activities

Directions: Answer the following questions in essays of several paragraphs each. Remember to include a topic sentence and several sentences supporting your main idea in each paragraph.

1. What causes day and night and the different seasons?
2. Why is Earth's atmosphere very important, and what things are people doing that harm the atmosphere?
3. What are the major factors that determine the climate of a region?
4. What were the most important changes brought about by the Environmental Transformation and the Scientific Transformation?
5. What are the major differences between the countries of the technological world and the nations of the developing world?

2

WESTERN EUROPE

UNIT INTRODUCTION

Western Europe is a small world region, one-third the size of the United States, that is divided into eighteen countries and seven microstates. It is located at the western end of the huge landmass of Eurasia (Europe and Asia). Despite its small size, Western Europe was the most important center of political power and economic strength in the world in the 1800s and the early part of the 1900s.

After fighting two destructive world wars, however, Western Europe lost this dominant political and economic position. The continent of Europe emerged from World War II divided into the independent states of Western Europe and the Russian-dominated communist countries of Eastern Europe. Moreover, Western Europe's overseas colonies were rapidly becoming independent.

The countries of Western Europe—some of them former enemies—have faced this new situation by banding together as economic allies. Barriers to trade such as taxes and tariffs were removed among some Western European countries so that the region could build a larger and more powerful economy. They recognized that only through economic cooperation could Western Europe regain its high standards of living and its important place in the technological world.

Today, Western Europe is a major economic power whose eighteen countries produce three times the wealth of Japan and nearly as much as the United States. New circumstances, however, present new challenges. In 1989, the Berlin Wall was torn down, creating an avalanche of immigrants from the east into this region. In 1990, West and East Germany became one country. And shortly thereafter, the Soviet Union collapsed, freeing Eastern Europe from the yoke of communism after half a century. All of these events will affect the future course of Western Europe.

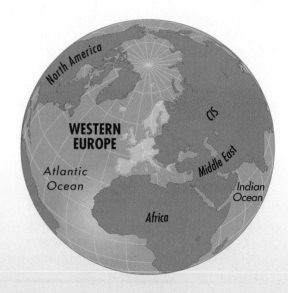

UNIT OBJECTIVES

When you have completed this unit, you will be able to:

1. Describe the major geographical features of Western Europe and how they have affected ways of living in this region.

2. Identify many of the different peoples and environments of Western Europe.

3. Relate that the Industrial Revolution began in Western Europe and that it had an enormous impact on this region and on many other parts of the world.

4. Describe the vast economic and political changes that occurred in Europe after two devastating world wars in this century.

5. Explain the serious new economic problems that confronted Western Europe after World War II and learn how this region dealt with these problems.

6. Point out the severe damage done to this highly industrialized region by water and air pollution, and describe the ways Western European governments are trying to solve these problems.

SKILLS HIGHLIGHTED

- Finding the Right Map or Globe
- Learning More about Hemispheres
- Understanding Chronology
- Vocabulary Skills
- Writing Skills

▲ The distinctive architecture of Stockholm, Sweden, suggests the many different peoples who live in Western Europe. What economic activities do you think might be important in Stockholm?

KEYS TO KNOWING WESTERN EUROPE

1

Western Europe is about one-third the size of the United States.

2

Despite its small size, Western Europe has a great variety of environments and people.

3

Europe, which is itself a peninsula at the western end of the Eurasian landmass, is composed of many islands and peninsulas.

4

Western Europe has four major subregions: the British Isles, Northern Europe, Continental Europe, and Mediterranean Europe.

5

The Industrial Revolution began in England in the mid-1700s and spread to the European continent in the 1800s.

6

Western Europe achieved economic and political domination over much of the world in the 1800s and early 1900s.

7

World War I (1914–1918) and World War II (1939–1945) shattered Europe's economic and political power.

8

The eighteen major countries of Western Europe have about 382 million citizens; most of these people have high standards of living.

9

The European Community (EC) is an economic union of twelve West European nations that has achieved prosperity by encouraging economic cooperation among member nations.

10

Western Europe's industrial areas are crowded and polluted; its agriculture is heavily supported by government subsidies.

◀ *Three quarters of all Europeans live in cities like Amsterdam in the Netherlands. Identify several examples of American influence on this busy street.*

Population Density per square mile

- Over 250
- 25-250
- Less than 25

Urban Population

- ● Over 1,000,000
- ● 500,000-1,000,000
- • Less than 500,000

Reykjavik

Norwegian Sea

ARCTIC CIRCLE

Atlantic Ocean

Helsinki

Oslo

Stockholm

Göteborg

Gulf of Bothnia

ESTONIA

RUSSIA

LATVIA

LITHUANIA

(RUS.)

Baltic Sea

North Sea

Edinburgh

Glasgow

Belfast

Newcastle

Middlesbrough

Manchester

Leeds-Bradford

Dublin

Liverpool

Sheffield

Nottingham

Birmingham

Coventry

Cardiff

Bristol

London

English Channel

Copenhagen

Malmö

Hamburg

Bremen

Berlin

Amsterdam

Hannover

POLAND

The Hague

Düsseldorf

Rotterdam

Essen

Antwerp

Wuppertal

Dresden

Cologne

Leipzig

Lille

Brussels

Bonn

Frankfurt

CZECH REPUBLIC

Weisbaden

Nürnberg

Mannheim

SLOVAKIA

Paris

Munich

Stuttgart

Vienna

HUNGARY

Zurich

Bern

ROMANIA

SLOVENIA

CROATIA

YUGOSLAVIA

Bay of Biscay

Lyon

Turin

Milan

Bologna

MOLDOVA

Black Sea

Bordeaux

Genoa

Florence

BOSNIA & HERZEGOVINA

ALBANIA

MACEDONIA

BULGARIA

Bilbao

Marseille

Adriatic Sea

Porto

Zaragoza

Madrid

Barcelona

Rome

Naples

Thessaloniki

TURKEY

Lisbon

Valencia

Seville

Palermo

Ionian Sea

Aegean Sea

Athens

Mediterranean Sea

Catania

MOROCCO

ALGERIA

TUNISIA

0   200   400 miles

0   400   800 kms

20°E

10°E

Country	Capital City	Area (Square miles)	Population (Millions)	Life Expectancy	Urban Population (Percent)	Per Capita GNP (Dollars)
Austria	Vienna	32,375	7.9	76	54	20,380
Belgium	Brussels	11,749	10.1	76	97	19,300
Denmark	Copenhagen	16,629	5.2	75	85	23,660
Finland	Helsinki	130,127	5.1	75	62	24,400
France	Paris	211,208	57.7	77	73	20,600
Germany	Berlin	137,857	81.1	75	85	23,650
Greece	Athens	50,942	10.5	76	58	6,230
Iceland	Reykjavik	39,768	0.3	78	91	22,580
Ireland (Eire)	Dublin	27,135	3.6	75	56	10,780
Italy	Rome	116,305	57.8	77	68	18,580
Luxembourg	Luxembourg	992	0.4	76	86	31,080
Netherlands	Amsterdam	14,405	15.2	77	89	18,560
Norway	Oslo	125,181	4.3	77	72	24,160
Portugal	Lisbon	35,552	9.8	74	30	5,620
Spain	Madrid	194,896	39.1	77	78	12,460
Sweden	Stockholm	173,730	8.7	78	83	25,490
Switzerland	Bern	15,942	7.0	77	60	33,510
United Kingdom	London	94,525	58.0	76	90	16,750

* Microstates not included

THE VARIED LANDSCAPES AND PEOPLES OF WESTERN EUROPE

Western Europe is made up of eighteen small countries. The largest country, France, would fit into the United States seventeen times. Indeed, the entire region is only twice the size of Alaska. After World War II, Western Europeans faced a bleak future; their lands had been ravaged by years of warfare. With the help of the United States, Western Europe recovered. Western Europe is now highly urbanized, much of the region being densely populated. Many of the cities are linked by efficient transportation systems. Its talented,

◀ *Vienna, once the capital of a very large empire, is still an extremely important city on the border between Western and Eastern Europe. It is now the capital of what country and located on what river?*

hard-working, and educated people now have one of the highest standards of living in the world.

This region still faces difficulties, however. Differences in wealth between the modern industrialized countries and the less developed ones in Western Europe are increasing. Even so, the independent states of Western Europe have been able to provide social services and opportunities to most of their people. Through cooperation, most of the countries of Western Europe enjoy stable and productive economies. The challenge for the future is to prevent high European manufacturing costs and economic competition with the United States and Asia from eroding Western Europe's industrial stability.

1. Regions and Countries of Western Europe

The Peninsula Continent

Most of Western Europe is made up of a series of peninsulas and islands that reach northward to the frigid coasts of the Arctic Ocean and extend southward to the sun-drenched shores of the Mediterranean Sea. The Atlantic Ocean and the North Sea bound this region to the west and north, and the Mediterranean Sea lines its southern margins. The people of Western Europe share a common heritage, although they speak different languages and occupy a variety of landscapes.

The British Isles lie off the west coast of the continent in the North Atlantic Ocean. East of these islands, the five countries of Northern Europe occupy three peninsulas. Norway and Sweden are located on the Scandinavian Peninsula, which extends southward from the Arctic Ocean. Denmark occupies a second peninsula, Jutland, that reaches northward from Germany and nearly joins Scandinavia. Together, these two peninsulas almost seal off the Baltic Sea from the Atlantic Ocean. Finland is located on the third peninsula on the eastern shore of the Baltic, opposite Sweden. Locate these countries, peninsulas, and islands on the maps on pages 107 and 108.

IT'S A FACT The British Isles include more than 5,500 islands.

Southern Europe also has three peninsulas. Spain and Portugal occupy the Iberian Peninsula, which faces the Atlantic to the west and the Mediterranean to the south. Italy is a second peninsula that projects deep into the Mediterranean. Farther to the east, Greece occupies the southern part of the Balkan Peninsula. Notice how these peninsulas give Western Europe a very long and uneven coastline.

The Role of the Sea

No place in Western Europe is more than 300 miles from the sea. France, the largest country in this region, has coastlines on both the Atlantic and the Mediterranean. Belgium, the Netherlands, and Germany line the shores of the North Sea. Of the Western European nations, excluding the microstates, only Switzerland, Austria, and Luxembourg are landlocked countries. A **landlocked** country lacks a seacoast and has no direct access to the sea.

Large rivers also connect Western Europe to the sea. The Rhine, for example, flows from landlocked Switzerland through several countries to empty into the North Sea. Few regions are as closely tied to the sea as Western Europe.

Western Europe is located much farther north than most people realize. Sweden is as far north as Alaska and reaches far beyond the Arctic Circle. London is near the same latitude as Newfoundland in Canada. If the United States were in Europe, its northern border would actually run through Paris. Yet temperatures in this region are much more moderate than one would expect at these latitudes. If this were not true, Western Europe would have less farmland and fewer people than it has now.

These moderate temperatures are caused by the influence of the Atlantic Ocean and Mediterranean

IT'S A FACT Five countries in Western Europe are smaller than Washington, D.C.: Liechtenstein, Malta, Monaco, San Marino, and Vatican City.

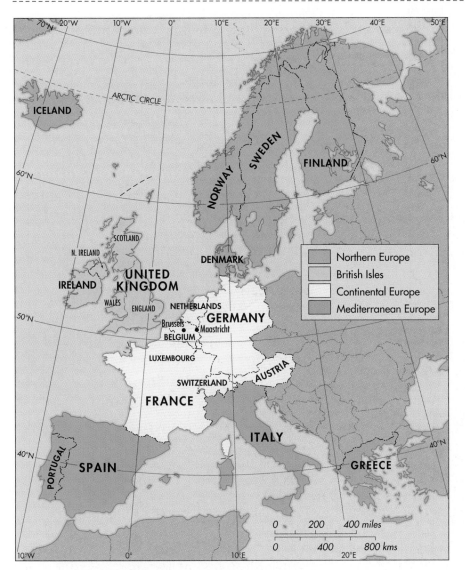

Subregions of Western Europe
Western Europe has four major subregions: (1) the British Isles, (2) Northern Europe, (3) Continental Europe, and (4) Mediterranean Europe.

This close relationship between land and sea has encouraged ocean trade, seafaring, and fishing in Western Europe. You can understand why navigation and shipbuilding skills developed in this region and why the Netherlands, Portugal, Spain, and Great Britain became sea powers. Europe's location in relation to other regions of the world was a great advantage once **maritime technology,** or systems for navigating the world's oceans, improved.

A Useful World Location

Western Europe lies at the center of that half of the globe called the land hemisphere. This **land hemisphere** contains 90 percent of the world's inhabited lands and 95 percent of the people on Earth. The other hemisphere is mostly water and is called the **water hemisphere.**

If you look at the two globes on page 114, you will see that most of the Earth's land area lies close to Europe. It is this central location and relatively easy access that helps to explain why Western Europe plays such an important role in world trade.

Sea, which reduce extremes of temperature, making winters warmer and summers cooler.

Westerly winds and a broad ocean current called the North Atlantic Drift (part of the Gulf Stream) push warm water across the Atlantic to the shores of the British Isles and Northern Europe. Palm trees can grow in southern England, which is slightly farther north than Québec, Canada. Southern Iceland and the coast of Norway also have farming.

The Mediterranean Sea moderates temperatures in Spain, southern France, Italy, and Greece. Winters in Rome, Italy, and Athens, Greece, are milder than those in U.S. cities like Philadelphia or Washington, D.C., which are at the same latitudes.

⊕ **REVIEW QUESTIONS**

1. What three large bodies of water form the boundaries of most of Western Europe on the north, west, and south?

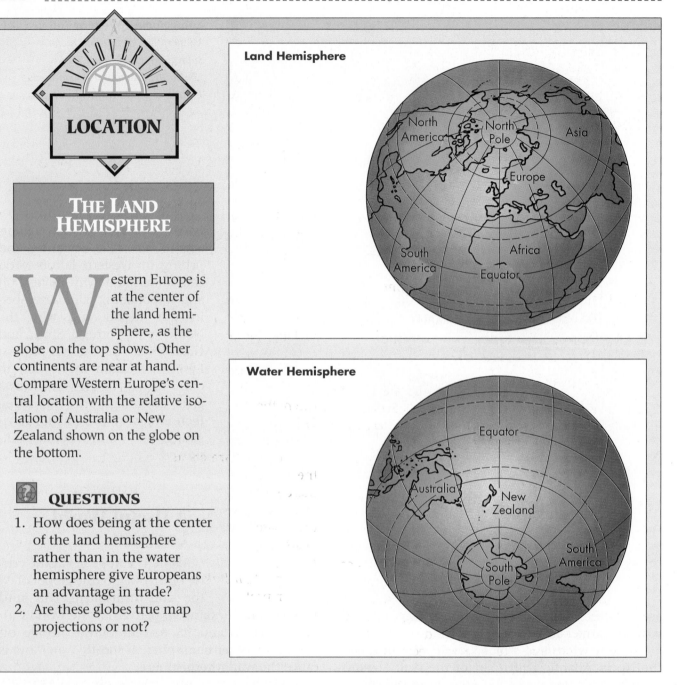

DISCOVERING LOCATION

THE LAND HEMISPHERE

Western Europe is at the center of the land hemisphere, as the globe on the top shows. Other continents are near at hand. Compare Western Europe's central location with the relative isolation of Australia or New Zealand shown on the globe on the bottom.

QUESTIONS

1. How does being at the center of the land hemisphere rather than in the water hemisphere give Europeans an advantage in trade?
2. Are these globes true map projections or not?

Land Hemisphere

Water Hemisphere

2. No place in Western Europe is more than how many miles from the sea?
3. Which European city is at the same latitude as the northern boundary of the United States?

THOUGHT QUESTIONS

1. Why are temperatures in Western Europe much more moderate than one might expect for a region at such high latitudes? How has this affect-ed land use and ways of living there? If your state had more moderate temperatures, what areas of your life and environment would be affected?
2. Why is it said that Western Europe has a useful world location? Is distance an important factor in economic development? Is this a case of absolute location or relative location? Give reasons for your choice.

2. The British Isles and Northern Europe

The British Isles

About one-sixth of all Western Europeans, some 61.6 million people, live on the British Isles. The British Isles include two large islands and many small ones close to the Western European mainland. The largest island is Great Britain; the second largest is Ireland.

The island of Great Britain is made up of the once separate lands of England, Scotland, and Wales. The country of the United Kingdom (often mistakenly called England) includes these three areas as well as Northern Ireland, which occupies the northern one-fifth of the island of Ireland.

The Republic of Ireland, or Eire, is a separate and independent nation that occupies the southern four-fifths of the island of Ireland. Locate Great Britain, the United Kingdom, England, Scotland, Wales, Northern Ireland, and Ireland (Eire) on the map on page 113.

The United Kingdom

The United Kingdom has 58 million people, most of whom live on the island of Great Britain. Highlands cover the northern and western parts of this island, and rolling lowlands are found to the south and east. The climate is moist. Winters are mild, and summers are cool. As you might expect, winters are harsher in the north and in the highlands. In fact, reindeer herding was recently introduced in northern Scotland.

Rugged uplands cover most of Scotland. These rock-strewn highlands are barren of trees. Villages dot the highlands and line the Scottish coasts. On the east coast, some towns are now booming because of the discovery of oil in the North Sea, which is described on page 123. Still, four-fifths of Scotland's 5 million people live in a narrow lowland that cuts across the island in southern Scotland. Glasgow, the country's most important industrial center and largest city (population of 1 million), an-

chors the western end of this coal-rich lowland corridor. Edinburgh, the capital of Scotland, is located at its eastern end. A strong movement for Scottish independence from the United Kingdom has grown in recent years. This is also true in Wales.

Most of Wales to the southwest is also covered by hills and mountains. The 3 million Welsh live along the southern coast and in upland valleys, where coal is strip-mined. They are particularly concentrated near the port city of Cardiff, whose heavy industry is based on nearby and inland coal fields. With the coal fields closing here (as they are in Appalachia in the United States), unemployment and poverty have forced many Welsh men and women to emigrate, most of them to neighboring England. The interior highlands of Wales, like those of Scotland, are sparsely settled.

The island of Ireland, to the west of Great Britain, includes Northern Ireland, or Ulster, which is part of the United Kingdom. The capital and port city of Belfast is Northern Ireland's major population center. Northern Ireland remained under British rule when the rest of Ireland gained independence in 1922. The Protestants, who form a majority in Northern Ireland, are the descendants of English and Scottish farmers and merchants who occupied the area in the 1600s. Centuries of conflict between these Protestants and native Catholics erupted into violence in the mid-1960s. The conflict is not entirely based on religious differences, however. At the heart of their conflict is the Protestants' desire to remain part of Great Britain with the economic advantages that link brings and the Catholics' desire to be reunited with the Republic of Ireland or to be granted equal access to jobs and housing in Northern Ireland. Bombings, riots, and killings have continued to plague Northern Ireland for nearly thirty years.

England, with its 48.5 million people, dominates the United Kingdom. In the north, mountains called the Pennines reach southward into England from Scotland. Large industrial cities grew up on the slopes of the Pennines during the Industrial Revolution. Liverpool and Manchester became important textile-weaving cities. Rich deposits of coal and iron are found here. Fertile farmlands nearby support a dense population. Birmingham, located just south of the Pennines, is the largest city in England's

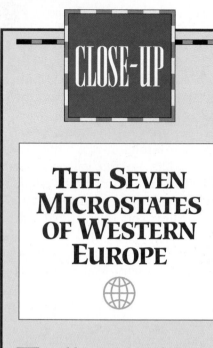

THE SEVEN MICROSTATES OF WESTERN EUROPE

I n addition to the eighteen major countries that make up Western Europe, seven microstates dot this region. Their locations are shown on the map on page 107. Five of these tiny places—Liechtenstein, Malta, Monaco, San Marino, and Vatican City—are independent countries. Andorra was in part regulated by the president of France and the Bishop of Urgel in Spain, until the country passed its first constitution in more than 700 years in the early 1990s. Gibraltar is a dependent British territory. As you can see, these microstates are survivors of past political arrangements. They prosper today in diverse ways.

Andorra (175 square miles in area) is located between France and Spain high in the Pyrenees Mountains. Monaco, which is 3 miles long, 1/2 mile wide, and half the size of Central Park in New York City, is located on the coast of the Mediterranean. Both countries have flourishing tourist economies. Exempted by France and Spain from paying taxes since 1867, Andorrans have become wealthy by selling tax-free perfumes, radios, cameras, and automobiles to visiting tourists. Each year roughly 10 million visitors climb steep mountain roads to ski and shop in Andorra, thereby providing a good living for its 54,000 citizens.

Monaco, with a population of 30,000, has based its economy on banks, beaches, stamps, and the world-famous Monte Carlo gambling casino. Small businesses, stamps, and tourism are its primary sources of wealth; the casino provides only 4 percent of the country's income. The country has been independent for some 800 years, but should the reigning prince of Monaco ever fail to have a male heir, the country would lose its independence and revert to France.

In contrast, Liechtenstein (62 square miles in area), situated in the Alps between Switzerland and Austria, and San Marino (24.1 square miles), located near the Adriatic coast of central Italy, have economies similar to neighboring regions. Both are small, clean, and prosperous countries. Much like Switzerland, Liechtenstein is shifting from alpine agriculture to precision manufacturing of items as varied as false teeth and engineering elements. This industry, in ad-

St. Peter's Cathedral is located at the heart of Vatican City in Rome. This small city is the center of what religion?

dition to the country's tax-haven regulations, has brought wealth to the 30,000 people of Liechtenstein.

Like surrounding areas in Italy, San Marino produces wine, leather, textile products, and building stone for export, although most of its revenue comes from tourism, finely minted coins, and stamp sales. The country was established in A.D. 301 and is the world's oldest and smallest republic. Unlike the other microstates of Europe, San Marino is not a tax haven. Yet, the tiny country perched on the summit of Mount Titano attracts as many as 3 million day tourists during the season. The country is now so wealthy that it pays three-quarters of the airfare of any Sammarinese (San Marino resident) living abroad to come home and vote in general elections. In 1992, it became the smallest full member of the United Nations.

Gibraltar (2.5 square miles), located on the southern coast of Spain, and the island republic of Malta (121.9 square miles), situated south of Sicily, were important British military bases in the Mediterranean. They guarded the passage of British ships to and from the Suez Canal in Egypt. Gibraltar, a tiny peninsula, guards the entrance to the Mediterranean. It has been a British crown colony since 1713 and remains so by choice. It still maintains a military garrison. Malta, with a population of approximately 400,000 gained its independence in 1964. It has a balanced economy based on farming, industry, trade, and tourism.

Vatican City (0.2 square miles) is the seat of the Roman Catholic Church and the surviving remnant of the papal states, which were powerful political entities ruled by the popes before the rise of the Italian state in 1870. Vatican City occupies a little over 100 acres, including St. Peter's Square in Rome, and has some 800 residents. You can walk around the entire place

▲ *This castle in Liechtenstein illustrates the beauty and age of this small country. The emphasis in Liechtenstein is switching from agriculture to what industry?*

in forty-five minutes. As the residence of the pope, the Vatican oversees the activities of one billion Catholics worldwide. Visited by millions of tourists each year, its grounds are protected by a contingent of Swiss guards, many of whom are not Catholic. Its primary local sources of revenue are tourism, coins, and stamps.

▼ *Monaco, on the coast of the Mediterranean, is an important banking and tourist center. For what else is it famous?*

? QUESTIONS

1. What is the major industry (source of income) in most of the microstates?
2. What country is the world's oldest and smallest republic?
3. Why do you think these microstates exist? Would you expect to see the creation of similar states today?

British Isles

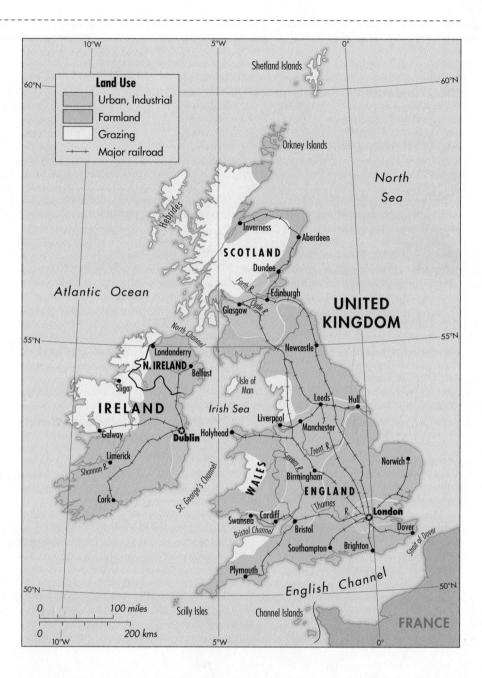

Land Use
- Urban, Industrial
- Farmland
- Grazing
- Major railroad

smog-bound "black country," the most important coal- and iron-mining district in the country.

The rest of southern England is a rolling plain dotted with neat farms, well-kept meadows, and hedged fields. This densely settled agricultural area is dominated by the city of London, which in the recent past was the largest city in the world. London's suburbs sprawl across the landscape. The city is located on the Thames River, forty miles from the river's mouth. London is the capital of the United Kingdom and its major port and richest market center. The 24-mile-long Channel Tunnel, an underwater tunnel recently built in the English Channel connecting Britain and France, will enhance this region's already intense communication links with the continent.

IT'S A FACT
Ireland is Western Europe's newest independent nation.

LOCATION

SITE AND SITUATION, A TALE OF TWO CITIES

London, England, and Paris, France, have been important European cities for many centuries, and today they are the largest urban centers in Western Europe. Why? Why do some cities flourish for many centuries while others prosper briefly or not at all?

Geographers believe that great cities are found where site and situation give them an advantage as compared to other places. **Site** is the actual location of a city. **Situation** is the relationship between this location and its surroundings. Let's see how the sites and situations of London and Paris have made them two of the world's greatest cities.

London is located forty miles inland on the Thames River. Its *site* is in a basin where the tidal flow from the Atlantic Ocean ends and where the river can easily be crossed. On this site, the Romans founded their city of Londinium and built a bridge near today's Tower Bridge. Londinium's strategic location gave the Romans control of movement up and down the Thames River. As one of the few crossing points on the Thames, the bridge was much used by land traffic.

London's *situation* is also superior. The city is located at the heart of the sunniest and most populous expanse of fertile farmland in Great Britain, and London became this area's most important market center. Because the mouth of the Thames River lies opposite the Rhine River delta, London was also superbly located to handle most of the trade between Great Britain and the continent.

Paris's site and situation are remarkably similar to those of London. The city was founded on a tiny island (the Ile de la Cité) in the Seine River, a *site* originally chosen for a fortress because it offered natural security. Because it was also located at a natural crossing place on the Seine, which flows into the English Channel, the city soon became

▲ *Two of London's most famous landmarks are shown in this photograph. Can you identify them?*

a crossroads of land and water traffic.

The *situation* of Paris at the heart of the magnificently fertile Paris Basin assured a steady flow of food, products, and people in the city. The Seine River served as a waterway for trade with a wider world, just as the Thames River expanded the reach of London's merchants.

QUESTIONS

1. Now that you understand the concepts of site and situation, try to apply them to the largest city in your state. Does it have a superior location? If so, why?
2. What factors have made this city grow?
3. Do you believe these factors will endure in the future?

The Republic of Ireland

The island of Ireland floats like a small bowl in the Atlantic Ocean to the west of Great Britain. The Republic of Ireland (Eire) covers most of this saucer-shaped land. Deep-cut rivers flow westward toward the North Atlantic through green meadows. Farm villages with thatched, whitewashed cottages line the river valleys, and fishing villages and towns along the coast are sheltered by low mountains. Dublin (with a population of 1 million) on the east coast is the main port and manufacturing center of the Republic of Ireland.

Today Ireland's economy is developing in spite of its lack of natural resources and an industrial base. The government is promoting high-technology industries and boasting of its highly literate workforce, but it is not clear whether these efforts will succeed in bolstering the economy, which is now the twenty-fifth wealthiest in the world. The country "which never had an industrial revolution," as the Irish claim, is attempting to achieve a postindustrial revolution. Yet in the 1990s some of its young, educated people are still migrating to other lands, while many of Ireland's 3.6 million people still keep alive their Celtic language, literature, and traditions.

Europe's Northern Fringe

Northern Europe is the most sparsely populated subregion in Western Europe. A quarter of this area lies north of the Arctic Circle, and the long winters, snow-covered mountains, wetlands, and poor soils limit settlement here. The five countries of Norway, Sweden, Finland, Denmark, and Iceland have a combined population of only 23.6 million people, in spite of the fact that Northern Europe embraces one-third of the total land area of Western Europe. Of these five countries, only Sweden has a population larger than metropolitan Chicago.

▼ *Clifden, in Western Ireland, is a fishing and tourist center. Its green landscape is typical of Ireland. Why do you think Ireland is called the "Emerald Isle"?*

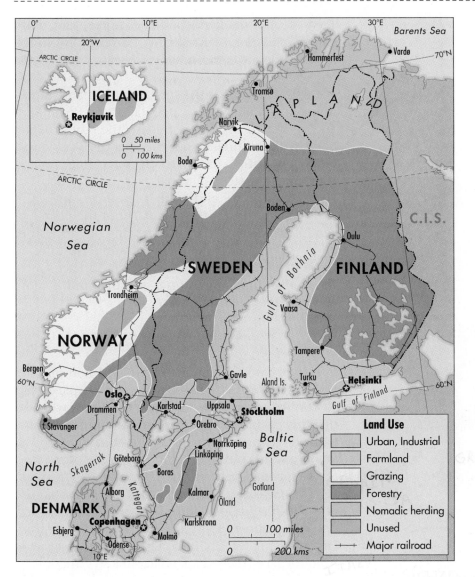

Northern Europe

Norway

Norway is a narrow, mountainous country with a coastline longer than that of the United States. It occupies the western half of the Scandinavian Peninsula, which points southward from the Arctic Ocean into the Atlantic Ocean. In the north, the cold is severe, winters are long, and vegetation is sparse. Most of the inhabitants of the far north are Lapps, who earn their living by herding reindeer. Farther south, Norway's mountain spine is densely forested. Lumbering and fishing are important in Norway.

Mountain glaciers, which are rivers of ice, creep down the mountain valleys to the sea, stripping the valleys clean of soil. Along the coast, these mountain glaciers have created Norway's best known geographical feature, its beautiful fjords. **Fjords** are long U-shaped valleys partly filled by the sea that have been carved out by glaciers. These deep inlets along Norway's west coast provide protected havens for fishing fleets when stormy gales rage across the North Atlantic.

Along the west coast of Norway, the climate is cool and wet. Farming and fishing settlements nestle at the heads of the fjords. The cities of Trondheim, Bergen, and Stavanger are located in these sheltered inlets. Many of these old fishing and trade centers are now booming as new industries connected to the development of offshore North

▲ *This is one of the numerous fjords, or U-shaped valleys, that line the coast of Norway. How are fjords created?*

Sea oil and natural gas are being built. South and east of these coastal cities, more than half of Norway's 4.3 million people live in or near Oslo, Norway's capital and largest city. Oslo is located in a region of fertile farmland and moderate climate.

Sweden

Sweden's 8.7 million citizens make it Northern Europe's largest and most populous country. Sweden occupies the eastern side of the Scandinavian Peninsula. The country's northern half is a very cold land covered by an enormous forest of pine, spruce, and birch trees. Small fishing and lumbering settlements are found here. Deposits of high-grade iron ore also are mined in this frontier area;

there are also smaller deposits of copper, lead, manganese, and zinc. Most Swedes, however, live in the more moderate environments of central and southern Sweden.

Central Sweden is the country's heartland. Stockholm, the capital, has a population of 1.5 million and is located on the east coast; the city of Göteborg, with its fine harbor, lies on the west coast. These two cities are linked by three huge inland lakes and a system of canals. Central Sweden has abundant mineral and lumber resources, and grain and dairy farms dot the area. Most of the people of central Sweden work in high technology industries and service jobs, and manufacturing is scattered throughout the towns of this region. Farther south, where the climate is warmer and the land is more level, are many wheat farms.

Finland

East of Sweden, across the Gulf of Bothnia, lies Finland. The Finns are a Central Asian people culturally unlike other northern fringe neighbors who settled this cold, swampy region a thousand years ago. Finland's climate is more severe than that of the Scandinavian countries. (Scandinavia is a name used to refer collectively to the countries of Denmark, Norway, Sweden, and sometimes also Iceland, the Faroe Islands, and Finland.) The winters are cold and dry, the summers short and hot, and soils are thin. Furs and timber are important products gathered from the lake-studded forests in Finland's interior.

Most of Finland's 5.1 million people live along the southern and western coasts. Even here, the winters are cold and long. The capital city, Helsinki, which has a population of 500,000, has frosts as late as June and as early as September. Its harbor is jammed with ice during the winter.

Denmark

Denmark is the smallest, warmest, and most evenly settled country in Northern Europe. This level land is carpeted by fields and farms that surround small villages and market towns. Three-quarters of all of Denmark's land is scientifically farmed. Farm management and organization are carefully planned to produce high-quality meat,

PEOPLE-ENVIRONMENT

STRIKING OIL IN THE NORTH SEA

It was front-page news in 1970 when an American company struck oil beneath the North Sea off the coasts of Great Britain, Norway, Denmark, and the Netherlands.

Most of the deposits belonged to the United Kingdom and Norway. Suddenly these two countries owned a giant oil and gas field, and they quickly made use of it. By the mid-1970s, oil and gas pipelines stretched from oil platforms in the North Sea to surrounding coasts. Norway now exports oil and natural gas. The United Kingdom has become self-sufficient in oil. No longer is Western Europe so heavily dependent on oil from the countries of the Middle East.

The technology needed to produce offshore oil from the floor of the North Sea is a marvel of advanced engineering. Floating platforms as large as several football fields are anchored to the sea bottom. Drilling towers rise 300 feet above these platforms, which are more than 100 miles from land. Beneath the towers, drill bits grind deep into the seabed to reach oil and gas at depths of 10,000 feet.

The hazards to human life and to the environment are enormous. Gales in the North Atlantic are fierce. Winds of 90 miles an hour and waves more than 60 feet high have toppled oil rigs, breaking them up. Drilling is expensive as well as dangerous. Despite these risks, North Sea oil is providing wealth and security to the countries that own the mineral rights.

▼ *This oil derrick is one of many that tap the large offshore fields located in the North Sea. Why would you expect North Sea oil to be relatively expensive to produce?*

QUESTIONS

1. Why was the discovery of oil and gas in the North Sea so important?
2. Why is drilling in the North Sea especially dangerous?

dairy, and poultry products, Denmark's leading exports.

The chief port and capital city of Denmark is Copenhagen, or "merchant's harbor." This city, which is the gateway between the North Sea and the Baltic Sea, is home to a quarter of Denmark's 5.2 million people.

Iceland

The island nation of Iceland is the westernmost country in Europe. Once a part of Denmark, Iceland became an independent republic in 1944. Many of its characteristics seem unfamiliar to people living in larger countries. The country's capital, Reykjavik,

IT'S A FACT
The names given to places can be misleading. For example, Iceland is green but Greenland is not. Greenland is largely covered by ice.

for example, has no jails. Its 260,000 people are listed in telephone directories by their first names. In Iceland, it's against the law to cut down a tree. Everyone is expected to learn how to read and write. Fishing in the North Atlantic is the basis of the country's economy.

This block of volcanic rock that is Iceland rises out of the frigid North Atlantic just south of the Arctic Circle. Iceland is often called the "land of fire and ice" because of its volcanic activity and large glaciers. Volcanic eruptions often occur, and new volcanic islands appear in the waters around Iceland, the most recent being the island of Surtsey (see pages 46 and 47). Hot springs and geysers provide **geothermal energy,** or heat from the earth's interior. In Iceland, geothermal energy warms the homes and powers the factories of Reykjavik, the northernmost capital city in the world, where 90 percent of all Icelanders live.

⊕ **REVIEW QUESTIONS**

1. What lands are part of Great Britain? Of the United Kingdom?
2. What are the five countries of Northern Europe?
3. What discovery in 1970 changed the economies of the United Kingdom and parts of Northern Europe?

⊕ **THOUGHT QUESTIONS**

1. Why have the people of Northern Ireland (Ulster) been plagued by troubles for many years? Look up this question in your library to discover how long antagonism between Irish Catholics and Protestants has existed. Do you know of conflicts between religious groups elsewhere in the world?
2. Why is the combined population of the five countries of Northern Europe so low? In what ways has the environment affected the distribution of people in these lands? Is this true in other northern regions as well?

3. Continental Europe

The Core of Western Europe

Continental Europe includes seven countries—France, Germany, Belgium, the Netherlands, Luxembourg, Switzerland, and Austria. These seven nations form the core of Western Europe. Locate them on the map on page 107.

The climates of this subregion originate in three different areas. From the west, the cool, wet, middle-latitude climate of the British Isles extends deep into northern France, Belgium, the Netherlands, Luxembourg, and Germany. From the north, the cold, continental climates of Scandinavia dip southward. And in the south, Mediterranean climatic influences reach northward into Continental Europe except where blocked by the Alps and the Pyrénées Mountains.

Nearly half of all West Europeans, 180 million people, live in Continental Europe—mostly in cities in an industrial zone that arches from northern France through Belgium, Luxembourg, and the Netherlands southward along the Rhine River. Farming is important on the cool, damp plains north of this industrial zone and in Mediterranean France to the south. Between these agricultural areas lies the economic heart of Western Europe.

France

On the north and west, French borders meet the North Sea, the English Channel, and the Atlantic Ocean. The Mediterranean Sea forms France's southern boundary. The Pyrénées Mountains separate France from Spain in the southwest. To the north and east, France borders on five countries—Belgium, Luxembourg, Germany, Switzerland, and Italy. The Rhine River forms part of the boundary between France and Germany. Extensions of the Alps separate most of France from Switzerland and Italy. If you look at the maps in the miniatlas, you will see that France is well located for trade with other countries.

France, Western Europe's largest country, is one-and-one-half times the size of Germany. It has im-

Continental Europe

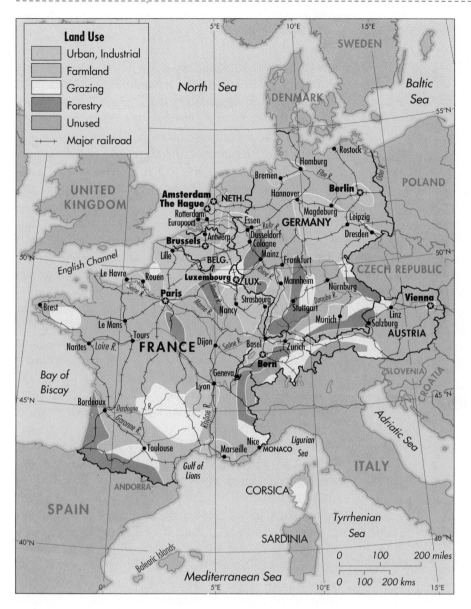

portant mineral deposits, fine forests, and a variety of farmlands. France is tied together by large rivers—the Seine, the Loire, and the Rhône—that transport goods to the sea as well as by a network of railroads. France also has a distinctive culture, a fertile land, and a durable identity.

The heart of France has long been the Paris Basin, a broad lowland area in the north that is partly rimmed by highlands. The Seine River and its tributaries flow through the rich farmlands of the Paris Basin into the English Channel. Paris, the capital, is located on the Seine, and is now the largest city on the mainland of Western Europe.

France has numerous and varied industries. Many of its heavy industries use iron mined from the played-out ore deposits of the northeastern province of Lorraine. Lille is the largest city in the industrial northeast. The city of Lyon, on the Rhône River, is a transportation center for the railroads, roads, and waterways that connect southern France with the Paris Basin.

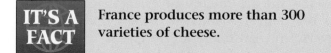

IT'S A FACT   France produces more than 300 varieties of cheese.

▲ *The Arc de Triomphe is a famous Paris landmark that commemorates Napoleon's victories. An eternal flame burns near the structure. What does this symbolize and where might you find another eternal flame?*

Because of France's well-developed industry and agriculture, most of the country's 57.7 million people enjoy a high standard of living. France, the largest food-producing country in Western Europe, is second only to the United States in food exports. France is Western Europe's largest producer and exporter of grain, milk, meat, and sugar. Northern France has a damp, cool climate. Hedges are used as fences to surround fields of wheat, barley, and sugar beets that grow well in this climate. Dairy cattle graze in meadows, and neat, well-tended fields and rolling green pastures ring the villages and towns of northern France. In the south, by contrast, the summers are long and sunny and the winters are mild. Vineyards and orchards dot the countryside of central and southern France much as they do in southern California. Resorts line the Mediterranean coast along the Riviera. Marseille (population of 1.1 million) is France's principal Mediterranean port.

Germany

In November 1989, the Berlin Wall, a hated symbol of Soviet domination over East Germany, was torn down. For twenty-eight years the wall had cut off the Russian sector from the rest of Berlin, located wholly inside East Germany. Germany's primary city had been divided into the American, British, French, and Russian sectors since the end of World War II. Built in 1961 to stop the flood of East Germans to West Berlin, the wall ended free travel between the sectors.

But in 1990, the map of Western Europe changed when West and East Germany were unified into one country. West Germany gained 42,000 square miles of territory and 17 million people. In 1991, the Germans established Berlin, located on its eastern borderlands, as the capital of a reborn Germany. Together once more, Germans began to face the challenges of change in Western Europe.

West and East Germany had been one country before 1945, but they were separated after World War II when part of Germany fell under Soviet control. In the process, Germany lost a third of its land area and a fifth of its people to the new East Germany. World War II also had destroyed many of Germany's cities and industries. With the help of the United States, the West Germans rebuilt their country; but the East Germans prospered little under Soviet rule. Today Germany is a strong, unified, and prosperous industrial nation of 81.1 million people. The problems and high costs of unification, however, may last well into the next century.

Germany, which is smaller than France, has fewer natural resources. Germany lies directly east of France, Luxembourg, Belgium, and the Netherlands. It reaches from the North and Baltic Seas southward to the Alps of Switzerland, Austria, and the Czech Republic. In the north, the gently rolling North European Plain stretches from France across Germany to Poland. Fishing villages and the great port cities of Bremen and Hamburg are located on the coast of the North Sea. Cool-weather crops like rye, oats, potatoes, and sugar beets are grown on the plain.

Hills and mountains cover southern Germany. As one moves southward toward the borders of Switzerland, Austria, and the Czech Republic, forested highlands give way to the towering peaks of the Alps and nearby mountain ranges. Valleys with level land are densely settled in this highland area. Small farms compete for land with busy cities, the largest of which is Munich. Because of its many mountains and hills and its cool climate, Germany produces only half of the food it needs.

Today, Germany is Western Europe's industrial leader, despite the high costs it must pay to rebuild the territories that were formerly part of East Germany. Most of its industrial districts are located in the west along the Rhine River, as the map on page 155 shows. Trace the course of the Rhine on the map on page 108 as it flows from Switzerland northward to the coast of the Netherlands on the North Sea.

The Rhine is one of the world's busiest rivers, with its steady stream of barges carrying goods to and from cities, mines, and industries along the river and its tributaries. In the south, the Saar industrial district's coal mines and heavy industry spread westward to the banks of the Rhine and the French border. Farther north lies the Rhine port of Bonn, the administrative capital of West Germany before German unification.

The largest industrial district in Western Europe, the Ruhr, is located along one of the many smaller rivers that flow into the Rhine. The Ruhr district

has coal and iron deposits, and its cluster of sixty industrial cities spreads out on both sides of the Rhine. Millions of workers in this area mine ores, produce iron and steel, and manufacture a wide variety of textile, chemical, and electronic products. Germany's industrial heartland has recently suffered many problems caused by competition from newer industrial plants in Japan and other developing countries. Moreover, the Germans have been noticeably reluctant to introduce high technology into their industries.

One bright spot, however, lies to the east in the province of Saxony, around the city of Leipzig. This area, which became heavily polluted when it was part of East Germany, is noted for highly specialized industries such as ceramics, textiles, and publishing. With its strong industrial tradition based on the raw materials of the Erzgebirge (Ore Mountains) on the border of the Czech Republic, Saxony is now emerging as the strongest economic region among the former East German states.

The Benelux Countries

Near the mouth of the Rhine River, the three small Benelux countries of *Be*lgium, the *Ne*therlands, and *Lux*embourg are located at the crossroads

▼ *Industrial districts line the banks of the Rhine River in Germany. Along what river that flows into the Rhine is the largest industrial district in Western Europe found?*

PEOPLE-
ENVIRONMENT

SUPER DIKE: THE DUTCH BATTLE THE SEA

The Polders

For more than a thousand years the Dutch have waged war against the sea. They have built dikes along the coast to hold back the waters of the North Sea. Low-lying lands reclaimed from the sea, or **polders**, make up half of the area of the Netherlands. For hundreds of years, the country's famous windmills have pumped salt water out of these polders to create usable land and freshwater lakes out of the sea bottom. Today many of these windmills have been replaced by electric pumps.

This storm surge barrier separates the North Sea from the mouth of the Eastern Schelde River, a branch of the Rhine. Super Dike is designed to protect some of the lowlands in the Netherlands from flooding during storms, while preserving the wildlife of coastal marshes. The dike is built to withstand battering from 20-foot waves. It completes a series of dikes, filling in the last gap in the 625 miles of coastal

barriers that protect the Netherlands from being flooded by the sea.

The 5.6-mile-long Super Dike took ten years to build at a cost of $2.5 billion. Concrete piers as tall as church towers, each weighing 20,000 tons, anchor the barrier in place. Two artificial islands were created so that Super Dike would not have to stretch an unbroken

5.6 miles. Huge 500-ton steel gates, 17 feet thick, can be opened or closed depending on weather conditions. In calm seas, these enormous gates hang suspended above the ocean, allowing the waters of the North Sea to flow into the mouth of the Eastern Schelde River. It is believed that this will preserve the abundant coastal wildlife in nearby

▲ *The 5.6-mile Super Dike in the Netherlands is designed to protect the country's reclaimed lowlands (polders) from flooding while preserving the wildlife of coastal marshes at the mouth of the Eastern Schelde River. Can you think of any other project of this scale in which preservation of the environment has been an important concern?*

marshes. When a severe storm threatens, however, the gates will be lowered to seal off the Eastern Schelde from the sea and to prevent flooding.

Reflecting a remarkable new Dutch attitude toward the environment, the government is buying up land and returning it to forests, wetlands, and lakes. Consistent with this new policy, the queen of the Netherlands noted at the dedication ceremony of Super Dike, "The Eastern Schelde has been made secure but not closed. Nature is under control but not disturbed."

QUESTIONS

1. Why is the Super Dike needed in the Netherlands?

2. In what ways is the government of the Netherlands demonstrating their commitment to preserving the natural environment?

3. Can you think of any other areas in the world that could benefit from reclaiming land?

of Western Europe. They are highly urbanized and intensively farmed and have a combined population of 25.7 million. Of these three countries, the Dutch have built their prosperity on the Netherlands' excellent trading location, situated where three rivers, including the Rhine, enter the North Sea.

The Netherlands (often called Holland) is the most densely populated country in Europe. Most of its 15.2 million people live near the capital city of Amsterdam and neighboring Rotterdam, both of which have populations of over 1 million. The land beneath these cities, in addition to half the land in the Netherlands, lies below sea level. This land, called **polder**, has been reclaimed from the sea. Some 4,000 miles of canals drain this low-lying country, where land is still so scarce that crops are grown between airport runways. Europoort, the new port at Rotterdam, was built after World War II and has become Europe's busiest trading center.

Neighboring Belgium has uplands and coal resources in the south. Its 10.1 million people work in heavy industry, manufacturing, and trade. Antwerp is Belgium's leading port city, but its capital is Brussels which is also the headquarters of the *European Community (EC)* that you will learn about in Chapter 5.

Tiny Luxembourg is smaller than Rhode Island. Luxembourg is a steel-producing country whose hill and plateau landscape supports fewer than half a million people.

Switzerland and Austria

These two small, landlocked countries that lie south and southeast of West Germany among the jagged, snow-capped Alps have had very different histories. Switzerland has long been an independent republic whose policy of neutrality has kept the Swiss out of Europe's many wars. It is not a member of the United Nations. In 1993, it voted to remain independent of the European Community (EC). By contrast, before World War I, Austria was the center of one of the largest monarchies in Europe, and its capital city of Vienna was a center of political power as well as culture.

The people of Switzerland speak four different languages: German, French, Italian, and Romansh

> **IT'S A FACT** Austria borders on more countries than any other nation in Western Europe.

(an offshoot of Latin). Although the country has few natural resources, it has beautiful scenery, abundant water power, and a determinedly independent people.

The Swiss, who number 7 million people, have made the most of what they have, achieving the highest standard of living in the world by skill, ingenuity, and hard work. Tourists come from all over the world to ski the Alps, vacation in their hotels and inns, and enjoy Switzerland's beautiful lakes and mountains. The Swiss make and export precision tools, watches, jewelry, and fine textiles. The dairy cattle pastured in the valleys and on the mountain slopes produce milk that is used to make the cheese and chocolate the Swiss export all over the world.

About three-quarters of the Swiss live on one-third of the land. The city of Geneva in the southwest is the center of French-speaking Switzerland. Romansh and Italian are spoken in the southeast. The main language spoken in the northern city of Zurich is German. Zurich is the largest city in the country and a center of international banking and trade. Switzerland depends on international trade for its high standard of living. Many Swiss products are shipped down the Rhine from the port of Basel to the Netherlands. The Swiss have built tunnels, roads, and railroads through the mountains to improve transportation and communication with the Mediterranean. Bern, the capital, is located on the country's central plateau.

Austria, a much younger and less stable country, has more resources and more agricultural land than Switzerland and is twice as large. Its standard of living, however, is lower mainly because of its history during this century. World War I reduced Austria from one of Europe's largest monarchies to a small country about the size of Maine. The country was occupied by the Germans during World War II and by the Allies until 1955. Today, Austria is an independent nation of 7.9 million people located at the geographic center of Europe.

▲ *The small, land-locked countries of Switzerland and Austria are dominated by what mountains?*

The country has two different environments. The climate and topography of the western two-thirds of Austria are similar to Switzerland's. Snow-capped peaks soar above the horizon; the scenery is spectacular. More than five hundred ski resorts cater to tourists drawn to this region. The eastern third of Austria is a lowland through which the Danube River flows on its way to the Black Sea. The capital city of Vienna, with a population of more than 1.5 million, is located on the Danube in Austria's most important farming region. Vienna is an international trading city and a bridge between west and east.

⊕ REVIEW QUESTIONS

1. What seven countries form the core of Western Europe? Which of these countries is largest?
2. What country in Western Europe produces the most food?
3. East Germany brought how much land area and population to the reunification?
4. What are the Benelux countries, and where are they located?

⊕ THOUGHT QUESTIONS

1. Why is Germany now the industrial leader of Western Europe? How do you think German unification will affect its economy in the short run and in the long run? Can you think of another nation where people are moving toward unification rather than separation?
2. Why have the Dutch had to wage war against the sea, and how have they done it? Do you think their efforts will be sustained over a very long period of time? What do you think of the Netherlands' new policy to restore parts of the natural environment?

4. Mediterranean Europe

IT'S A FACT The oldest unchanged national border in Europe is between Spain and Portugal.

Europe's Southern Fringe

Spain and Portugal, Italy, and Greece occupy three peninsulas that jut southward from Europe into the Mediterranean Sea. In the west, Spain and Portugal are located on the Iberian Peninsula, and Italy and Greece are farther east. Together, these four countries have a third of the land in Western Europe and about a third of its people.

The Mediterranean countries have delicate environments; their landscapes are rugged and their climates subhumid. As in southern California, most of the rain falls in winter rather than in summer. People have lived in these countries for thousands of years and have stripped the land of much of its vegetation. The remaining scrub is called *maquis* in this region. The deforestation, subhumid climate, sloping hills, and constant use by farmers and herders have eroded the soil and harmed many of the region's environments.

Spain and Portugal

Spain and Portugal share the Iberian Peninsula in the southwestern corner of Europe, facing both the Mediterranean and the Atlantic. This location encouraged these countries to look outward from their homelands to the sea. In the 1500s, they established large colonial empires overseas. In the process, Spain and Portugal destroyed native civilizations and reshaped the culture of Latin America. The Spanish also strongly influenced the heritage of parts of the United States.

Mediterranean Europe

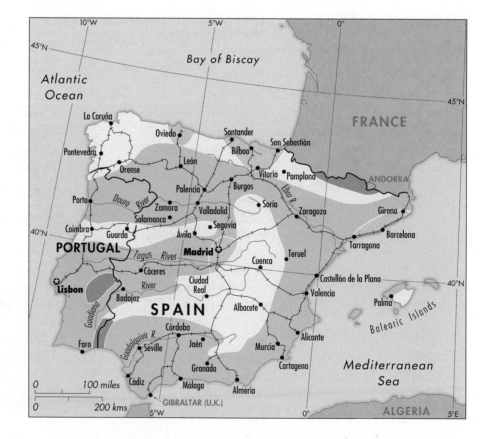

Spain is the second-largest country in Western Europe, but it is a thinly settled land. Some 39.1 million people live in this country, which is almost as large as France. More than half of Spain is a dry, rugged central plateau called the *meseta*. The meseta has few farming settlements because of thin soils, dry summers, uncertain rainfall, and sparse vegetation. Some wheat and cattle are raised where conditions permit. The major city on the plateau, with a population of 3.1 million people, is Madrid, Spain's capital and largest urban center.

Most Spaniards live along the margins of the *meseta* on coastal lowlands where grain crops, vineyards, olive groves, and orchards flourish. Spain's two major industrial zones are also located along coasts. Barcelona (population of 1.7 million), in the northeast, is Spain's leading Mediterranean port and most important industrial city. Iron and coal deposits along the Atlantic coast in the north form the basis of a steel industry in Spain.

Neighboring Portugal, a thin, rectangular country of 9.8 million people, imports four-fifths of its

▲ *Most of the rivers in Spain, like the Tagus River shown here curving past the city of Toledo, flow westward into the Atlantic Ocean. What large river in Spain flows eastward into the Mediterranean Sea?*

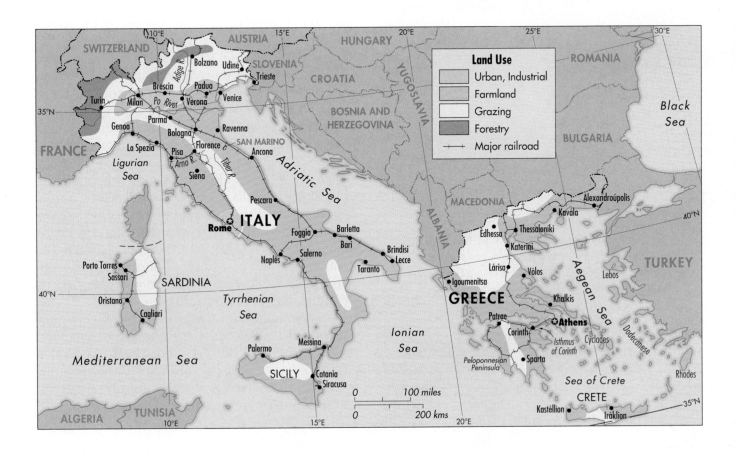

▲ *Venice is one of the most beautiful and unusual cities in the world, but its very existence is threatened by pollution. In what ways could the canals contribute to this problem?*

energy, nearly half of its food, and most of its technological goods. Joining the *Economic Community (EC)* in 1986 gave a boost to the Portuguese economy, particularly in supporting its marginal agricultural areas. But large-scale emigration has drained many of the country's most talented people.

The northern half of Portugal is hilly, green, and fertile. Wheat, vineyards, and orchards blanket broad river valleys where three-fourths of the Portuguese live. This region is famous for its fine wines. Fishing villages line the coast.

Southern Portugal, the Algarve, in contrast, has the dry summers that make it as attractive to European tourists as are the shores of other Mediterranean countries. Groves of olive trees and forests of drought-resistant cork oak trees furnish this area's major export products. The capital city of Lisbon (population of 2.1 million) is the most important port and manufacturing center in Portugal.

Italy

Italy occupies the 750-mile-long Italian Peninsula, which extends from the Alps in the north to deep into the Mediterranean Sea. Italy has an abundance of sun, art treasures, and history but few raw materials or energy resources. The peninsula, which is only 100 miles wide along most of its length, ends in the island of Sicily at the "toe" of the Italian "boot."

The fertile Po River Valley at the foot of the Alps in northern Italy is the economic heart of the country. This broad expanse of fertile, productive land is where 40 percent of Italy's 57.8 million people live and work. Irrigated farming in the Po Valley is highly mechanized, and this region produces most of the grain, olives, and wine that Italy exports.

The Po Valley is also Italy's most important industrial region. Milan, Turin, and the nearby port of Genoa (all cities with over a million people) are key centers of steel, machinery, automobile, and textile manufacturing. Venice, located in the Po delta, is a floating city built on hundreds of small islands clustered in a shallow lagoon. Although plagued by flooding and pollution, Venice's timeless beauty makes it an important tourist destination.

South of the Po Valley, the Apennine Mountains run the length of the peninsula. Here the land is rugged, broken, and dry; soils are thin and time-worn. Rainfall is low, and many streams are dry in summer. Towns are perched on the rocky crests of hills and mountains. Terraced fields planted in wheat, orchards, and vines spread out below these walled settlements.

The farmers of southern Italy—the *Mezzogiorno*—are poor. Many Italians have left this region to build new lives in the United States. Today they also migrate to the capital city of Rome (population of 2.8 million) or Naples which are located nearby, to industrial centers in the Italian north, and to manufacturing centers elsewhere in Western Europe to earn a living. Although the government is trying to strengthen southern Italy's economy, the nation remains divided: the north is prosperous; the south is poverty-stricken.

Greece

Greece, a country comparable in size to the state of New York, lies between the Ionian and Aegean Seas at the tip of the Balkan Peninsula. A mountainous country with water on three sides, Greece has been a seafaring nation for many centuries. Its deeply indented coastline is more than 100,000 miles long, and no place in Greece is more than 85 miles from the sea. Its merchant fleet is the fourth largest in the world. More than 2,000 islands in the surrounding Mediterranean make up one-fifth of the territory of Greece.

◄

This satellite photo showing part of Greece provides important clues about the country. What occupations and industries would you expect to be important? Would you expect agriculture to be a profitable activity here?

This nation of 10.5 million people is one of the poorest in Western Europe. Greece has little industry and few resources except the breathtaking beauty of its scenery. Northern Greece and the interior of the country are forested and mountainous, with several plains on which grain and tobacco are grown.

Southward from mainland Greece, the Peloponnesian Peninsula, or the Peloponnesus, extends into the Mediterranean like a hand with four fingers. Olives, citrus fruits, wheat, and other Mediterranean crops are grown on the Peloponnesus. Fishing villages line the coast and dot nearby islands.

The capital city of Athens (population of 3 million) lies east of the Isthmus of Corinth, which joins the Peloponnesus to the mainland. (As you have already learned, an **isthmus** is a narrow strip of land connecting two larger landmasses.) Now a thriving industrial city, Athens was the birthplace of democracy and is the most important center of Greek life.

⊕ REVIEW QUESTIONS

1. Mediterranean Europe includes what four countries?
2. What has eroded the soil in many parts of Mediterranean Europe?
3. Which country has a coastline over 100,000 miles long and also has more than 2,000 islands?

⊕ THOUGHT QUESTIONS

1. Why are Spain and Portugal much poorer countries today than they were in the 1500s? Do you think that the distribution of people is related to environmental conditions?
2. Why is the Po Valley very important to Italy's economy? What are the largest cities in the Po Valley?
3. Why do you think that Greece is a poor country now, even though it built a great empire in the past? Do you believe that Greece will rise to world prominence again? Why or why not?

⊕ CHAPTER SUMMARY

The countries of Western Europe occupy a peninsula continent on the western edge of the Eurasian landmass. Western Europe is closely tied to the sea, and many cities that are situated inland are connected to the sea by one of Europe's large rivers. While much of Western Europe lies farther north than most of the United States, its climate is very moderate due to the influence of the surrounding oceans and seas.

The continental peninsula, several smaller peninsulas, and islands are the basis of four major subregions of Western Europe: (1) the British Isles, which include Great Britain (England, Scotland, and Wales) and Ireland (the Republic of Ireland and Northern Ireland); (2) Northern Europe, a sparsely populated subregion that encompasses Norway, Sweden, Finland, Denmark, and Iceland; (3) Continental Europe, the core of Western Europe, which includes France, Germany, Belgium, the Netherlands, Luxembourg, Switzerland, and Austria; and (4) Mediterranean Europe, which consists of Spain and Portugal, Italy, and Greece on three peninsulas jutting out into the Mediterranean Sea.

While some Western European countries are comparatively large, many are small. Seven political units are so small that they are termed microstates: Andorra, Gibraltar (U.K.), Liechtenstein, Malta, Monaco, San Marino, and Vatican City.

Western Europe has many important natural resources. One that has been recently discovered is a large oil field beneath the North Sea. These oil deposits have made several countries self-sufficient producers of oil. In addition, Western Europe's highly developed industry and its productive agriculture have contributed to the high standard of living in most of its countries.

EXERCISES

Missing Terms

Directions: Select the correct word from the list below that completes each of the following sentences. On your paper, write the word you selected next to the number of the sentence it completes.

1. Most of Western Europe is composed of a series of |||||||||| and islands.
2. Western Europe lies at the center of the |||||||||| hemisphere.
3. The |||||||||| are found entirely in Italy.
4. Spain and Portugal share the |||||||||| in the southwestern corner of Europe.
5. The |||||||||| borders Western Europe to the south.
6. Norway and Sweden are located on a peninsula that extends southward from the ||||||||||.
7. Rivers of ice that creep downhill toward the sea in Northern Europe are called ||||||||||.
8. The deep inlets along Norway's west coast are called ||||||||||.
9. The |||||||||| separate France from Spain in the southwest.
10. Switzerland and much of Austria lie in the lofty ||||||||||.

Terms

Alps	land
Apennine Mountains	Mediterranean Sea
Arctic Ocean	mountain glaciers
fjords	peninsulas
Iberian Peninsula	Pyrénées Mountains

True or False?

Directions: On your paper write true or false next to the number of each of the following statements.

1. Today, only three cities of Western Europe have populations of 1 million or more.
2. The United Kingdom includes Ireland (Eire), Scotland, and Wales.
3. Mediterranean Europe has the highest standard of living in Western Europe.
4. The countries of Northern Europe are densely populated.
5. The mild climate in much of Western Europe is caused by the moderating influence of the Atlantic Ocean and the Mediterranean Sea.
6. France is the largest food-producing country in Western Europe.

Inquiry

Directions: Combine the information in this chapter with your own ideas to answer these questions.

1. For several years the United States has been importing more products from some countries of Western Europe than it has exported to these countries. What problems does this create for our own economy?
2. Nuclear power plants produce large amounts of the electricity consumed in several Western European countries. What do you think about the peaceful uses of nuclear power? List reasons for and against the use of nuclear power.

SKILLS

Finding the Right Map or Globe

Directions: Perhaps you have noticed that this unit has a variety of maps. It is helpful to be able to find the map that you want quickly. Skim through the unit and then list on your paper the page number of each of the following maps or globes:

1. The Land and Water Hemispheres
2. Western Europe Physical Map
3. The Polders
4. Western Europe Political Map
5. Western Europe Population Map

Directions: Number your paper from 6 to 20, and next to each number write the title of the map or globe you would use to find each of the following geographic features. You may use the same map or globe for more than one answer.

6. The areas of Western Europe that have more than 250 people per square mile
7. The location of Germany
8. The Alps
9. The least populated areas of Western Europe
10. Whether Norway or Sweden is more densely populated
11. The southernmost country in Western Europe
12. Whether Spain has any cities with more than 1 million people
13. The length of the Apennine Mountains
14. The parts of the Netherlands that have been reclaimed from the North Sea
15. The continents nearest to the center of the land hemisphere
16. The continents farthest from the center of the land hemisphere
17. The countries in Western Europe that border on the Mediterranean Sea
18. The capital cities of Western Europe
19. The distance from Paris to Berlin
20. The landlocked countries in Western Europe

Learning More About Hemispheres

You have already learned about some hemispheres—the Western and Eastern Hemispheres and the Northern and Southern Hemispheres. Remember that *hemisphere* is a term used to describe half of the Earth, or half of a globe. Geographers can make globes and maps dividing the Earth in various ways, and as long as each part includes half of the Earth, it is a hemisphere.

Directions: Look now at the globes of the land and water hemispheres on page 114 and answer these questions about them.

1. Is all of Europe within the land hemisphere?
2. Is all of North America within the land hemisphere?
3. Is all of Africa within the land hemisphere?
4. Is most or only a little of Asia within the land hemisphere?
5. What part of South America is within the land hemisphere?

6. Is Australia within the land hemisphere?
7. Does most of the Earth's land lie north or south of the equator?
8. What three continents do not appear in the water hemisphere?
9. Does an examination of the land and water hemispheres suggest that trade (and therefore travel) between Europe and North America have been heavy or light?
10. Do these hemisphere maps suggest that trade and travel between Europe and Australia have been heavy or light?

Vocabulary Skills

Directions: Match the numbered definitions with the vocabulary terms. Write the term you choose on your paper after the number of its matching definition.

fjord
geothermal energy
isthmus
land hemisphere
landlocked
maritime technology

meseta
mountain glacier
polder
site
situation
water hemisphere

1. The half of Earth that contains most of its land area
2. Heat from the Earth's interior that surfaces in the form of hot springs and geysers
3. A dry, rugged plateau in central Spain
4. The absolute location of a place
5. River of ice that flows down mountain valleys, removing their soils
6. Technology relating to oceanic navigation
7. The relative location of a place
8. U-shaped valleys partly filled by the sea that were carved out by glaciers
9. A narrow strip of land connecting two larger land masses
10. An area surrounded by land with no direct access to the sea
11. The half of Earth that contains most of its water area
12. Land lying below sea level, reclaimed from the sea and protected by dikes

THE RENEWAL OF WESTERN EUROPE

The Industrial Revolution began in Western Europe. The new and more complex ways of living that grew out of industrialization created a skilled and educated population. Modern methods of transportation, a wide variety of natural resources, and new technologies came into use. Western Europe rapidly became the world's richest region. Advanced industrial and military power enabled some countries in Western Europe to dominate other areas of the world.

Western Europe's dominance in the world did not last long, however. Twice during the 1900s, Europe was devastated by world wars. This region has

◄ *The new and old Kaiser Wilhelm Memorial Churches in Berlin radically illustrate tradition and change in Western Europe. How would you describe the architecture of these two buildings?*

had to overcome many difficult obstacles to re-emerge today as an important economic and industrial region.

1. The Industrialization of Western Europe

The Industrial Revolution Began in England

The Industrial Revolution began in England in the mid-1700s. There were several reasons that England was the starting point. First, England had many of the natural resources needed for modern industry, such as coal fields, deposits of iron ore, and rivers that provided water power. Second, because its population doubled between 1750 and 1820, England had enough people to work in the new factories and also to purchase the manufactured products. Third, money for investment was available in England among merchants who had become rich through trade.

Perhaps most important, the imagination and genius of England's inventors fostered the growth of industry in England. As they discovered new ways of making and transporting goods, one invention led to another. The steam engine, powered by coal, came into use as England ran out of wood for fuel. So coal production tripled in less than one hundred years. When coal miners dug deeper mines, they needed larger pumps to get rid of floodwater in the mine tunnels. The engines which ran these larger pumps were then used to run machines in new factories. Once started, England's inventors solved one problem after another, and their inventions created the industrial way of life.

IT'S A FACT When James Hargreaves developed the "spinning jenny" to spin yarn, he was driven from town by a mob of workers who thought this new technology would cost them their jobs.

Inexpensive methods of using coal to make iron and steel made these metals available for many uses—for instance, in railroad locomotives, railroad tracks, and factory machines. The cost of transport went down after George Stephenson invented his railroad locomotive, the Rocket, in 1829. Soon, more efficient and powerful steam engines were designed, allowing bulky raw materials like coal and iron to be moved cheaply from the countryside to city factories.

In the countryside, inventions dramatically increased farm production. New seeds, systems of crop rotation, animal breeding techniques, and farm machines changed traditional agriculture. Both agricultural and industrial production grew faster than the population.

New energy sources and technologies were creating a new human geography in England, in which

▼ *Although the Industrial Revolution gradually increased the wealth of the English people, many poor factory workers were forced to live in crowded city slums. One of the most infamous was Whitechapel in London's East End. What aspects of this illustration suggest poverty and hardship to you?*

factory towns and cities became the living environment of many people who once lived in villages. More and more farmers were forced to move to factory towns and cities in search of work, and this shift was often catastrophic for those who could not find jobs.

English cities were not prepared for the problems brought by this migration. Workers had to live in crowded city slums without clean water, sewage systems, lighting, or any method of waste removal. In the workplace, they had no rights; they worked long hours in dangerous and filthy conditions. Strikes and bitterness increased in English society. Although most people made slow but steady gains in income, education, health, and nutrition, it was many years before the tensions created during England's Industrial Revolution were eased.

The Industrial Revolution Spread to the Rest of Europe

Several geographical obstacles slowed the spread of the Industrial Revolution from England to the European mainland. First, Continental Europe is about twenty-five times larger than England. Railroads and roads had to span long distances to connect one place with another. Second, coal and iron deposits in Europe are widely separated from one

another. Third, physical barriers, like the Alps and Pyrénées Mountains, separate northern from southern Europe. In addition, many Europeans, particularly the French, disliked the industrial way of living. To them especially, the Industrial Revolution had destroyed a satisfying social order and disrupted traditional ways of living.

Industry, therefore, developed slowly on the continent during the 1800s. In 1850, most of the iron produced in France was still made in small charcoal-burning forges located in the forests. In Germany, iron smelters still used old techniques. Everywhere on the continent, textiles were woven in small, family-owned shops. In northern and southern Europe, **heavy industry**, which is the production of heavy goods like steel and machinery, was almost absent. Of all of the countries on the continent, only Belgium was keeping pace with England.

By 1850, a crisis was brewing in Europe. Populations were growing rapidly, but there were not enough factory jobs for these people. As poverty in rural areas increased, people began to leave their villages for cities like Paris, Rome, and Berlin. Strikes, riots, and uprisings alerted Europe's political leaders that something had to be done about the growing number of poor people in the cities. They began to associate the idea of progress with industrialization.

◄

The Industrial Revolution in England made it possible for many jobs previously done by hand to be done by machines, as this photograph of an English cotton mill illustrates. What contemporary new technology is now changing the way in which people work?

PLACE

FRANKENSTEIN'S MONSTER AND THE INDUSTRIAL AGE

Think of the Frankenstein monster. The image is unforgettable. A humanoid with a square head and lanky hair lurches forward dressed in rags. Stitched seams join together the parts of his body. Electric plugs jut out from his neck. The Frankenstein monster is a fiction of science, but also a symbolic creature of the industrial age.

The Frankenstein monster was invented by an eighteen-year-old woman named Mary Shelley. On a rainy night in Geneva, Switzerland, in June, 1816, she had accepted a dare to write a ghost story.

Her ghost story became the novel *Frankenstein*, in which a scientist named Victor Frankenstein discovers the greatest secret in the world—the secret of creating life.

In his laboratory, Dr. Frankenstein conducts scientific experiments in which he sews together limbs from dead bodies. When transplanting a brain into his creature, he accidently uses the brain of a criminal. Then, with bolts of electricity, he succeeds in injecting life into this creature, but he has unwittingly created a monster.

▲ *This is a photograph of Boris Karloff as Frankenstein's monster. Can you think of another book or film that deals with the theme of technology out of control?*

The scientist abandons the horror he has made. The monster wanders blindly in search of love, and in the end, the creature returns and destroys the man who created him.

The story of Frankenstein has lasted because of what Mary Shelley sensed about the Industrial Revolution. New powers were being unleashed, and science was revealing knowledge that could open up new worlds. Yet, as young Victor Frankenstein discovered, science could also produce an unexpected and uncontrollable "monster."

The young scientist created the monster out of pride to satisfy his own ambition. In today's world of nuclear power, biogenetics, and robotics, Frankenstein's powerful and enduring message has kept Mary Shelley's novel—and the fears behind it—alive.

QUESTIONS

1. According to this feature, what characteristics did Frankenstein's monster share with the Industrial Revolution?
2. What "modern Frankensteins" are being created by the use (or abuse) of science and technology?

The Construction of Railroads in Western Europe, 1850–1900

The most important factor in the industrialization of the continent was the building of railroads. Nearly 50,000 miles of railroad track were laid by 1880.

Railroads connected Europe's coal fields, iron ore mines, industrial centers, and port cities with one another. Distance and mountains were conquered by a web of steel tracks. European governments recognized that an industrial economy was needed to build their countries' prosperity, power, and prestige. Military strength now also depended on heavy industry.

Because the Germans were first to recognize the need to industrialize, railroads were built rapidly across Germany. The coal deposits of Germany's Ruhr Valley made it Continental Europe's greatest industrial center. The Ruhr Valley is located at the center of the *industrial triangle* on the map on page 155. By 1900, Germany surpassed England in railroad growth and iron production. The three graphs on page 144 show the close relationship between the building of railroads and the expansion of heavy industry in England, Germany, and France.

In France, railroads fanned out from Paris, the capital, like the spokes of a wheel. The railroad network was particularly dense in northeastern France, where heavy industry was based on iron ore from Lorraine and coal from the nearby Saar Basin on the German border. These important resource regions are also located inside the industrial triangle, close to Germany's Ruhr Valley. In spite of these resources, France's reluctance to accept industrialization caused it to fall further and further behind England and Germany.

Heavy raw materials were now moving across Western Europe on bands of steel. Everywhere on the continent, railroads broke through land barriers. Engineers built tunnels through the Alps and the Pyrénées to connect northern to southern Europe. Small industrial centers developed in the Po Valley of northern Italy and in northeastern Spain around Barcelona. Before 1900, Europe led the world in economic development.

Rapid Changes in Science and Technology, 1850–1900

In rapid order, more technologies were introduced that changed daily life in Western Europe. By the late 1800s, a second industrial age based on steel, oil, and electricity had spurred even higher levels of economic growth.

Steel, a material of great strength, hardness, and flexibility, became cheap and widely available. The **Bessemer process**, which cleans impurities from iron by injecting it with oxygen, made it possible to manufacture inexpensive steel. Because the Bessemer process made iron into steel in only twenty to

▼ *The industrial revolution gradually spread from England to the continent. By 1900, Germany was Europe's largest producer of iron. The Krupp Works in Essen, Germany, was one of the largest of these industrial operations. What are the connections between industrialization, city growth, and the existence of slums?*

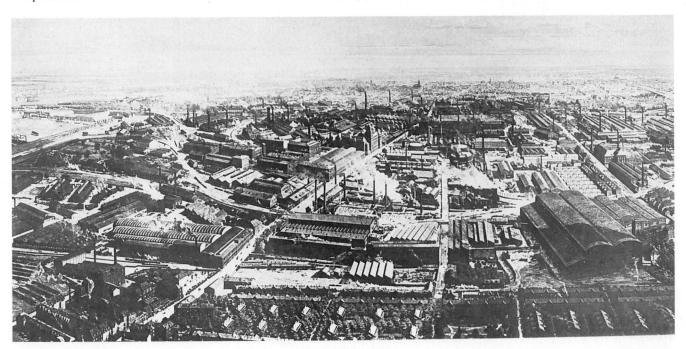

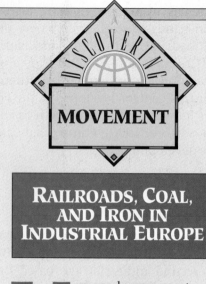

MOVEMENT

RAILROADS, COAL, AND IRON IN INDUSTRIAL EUROPE

You can learn a great deal about what happened in England, Germany, and France in the late 1800s by studying these graphs of railroad growth and of coal and iron production. They show that transportation lines were very important to the production of coal and iron in Europe.

England built railroads first. The graph on the left shows that England had more miles of track than either France or Germany until 1870. Don't forget, however, that Germany and France are much larger than England. Thus, the building of railroads slowed in England in the late 1800s largely because no more railroad lines were needed. The expansion of railroads in France and Germany, by contrast, continued well into the 1900s.

Railroads connected coal- and iron-producing areas with factories in industrial cities. These heavy raw materials could now be transported cheaply over long distances. The center graph shows that England was the largest producer of coal throughout the period. In Germany, coal and iron production expanded rapidly, as the center and right-

hand graphs show. By 1900, Germany was producing more iron than England. France, however, remained a modest producer of coal and iron; the growth of heavy industry was slowed down by French resistance to the Industrial Revolution.

QUESTIONS

1. Approximately how many miles of track were laid in Germany between 1850 and 1910?
2. What percentage of coal was produced by France in 1910 as compared to England in the same year?
3. In what decade did the greatest increase in iron production in Germany occur?

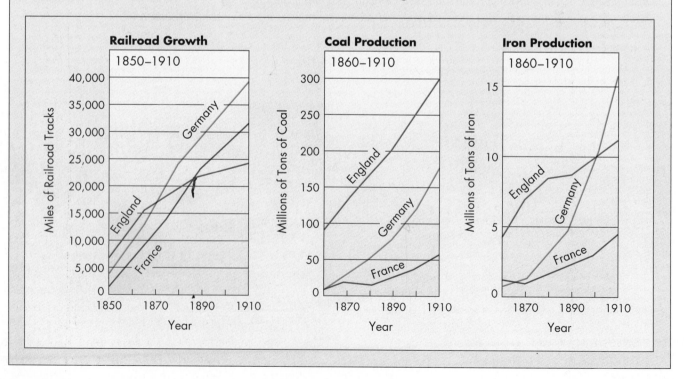

thirty minutes, the cost of steel dropped from $200 to $4 a ton.

Electric energy was also put to new uses. Coal was burned to create the electric power for lighting, telephones, telegraphs, and cable cars. Oil became an inexpensive alternative to coal. New products like typewriters, sewing machines, printing presses, watches, and mass publications changed people's day-to-day lives.

Scientific advances in chemistry created additional products. For example, chemical fertilizers increased crop yields on Europe's worn-out farmlands. Chemical dyes, drugs, inexpensive soap, and synthetic cloth like rayon were other products of this new science.

This explosion of science and technology led to competition rather than cooperation among the industrial leaders of Europe. Great Britain, Germany, France, and Belgium began to compete for foreign sources of raw materials and foreign markets in which to sell their products. In the end, they would almost destroy one another.

⊕ REVIEW QUESTIONS

1. Western Europe is divided into how many countries (excluding the microstates)?
2. What was the most important factor in the industrialization of Continental Europe? Which country first rivaled England as an industrial leader?
3. What is the Bessemer process, and why is it important?
4. What were some of the new uses for electric energy in the late 1800s?

⊕ THOUGHT QUESTIONS

1. Why did the Industrial Revolution first occur in England? What factors were most important? How did this change people's ways of living? If you had lived in this period, would you have changed occupations? Residence? Why or why not?
2. Why was the Industrial Revolution slow to spread to the European continent? Name several of the most important factors. Why do you think it spread swiftly to the United States? Can you identify ways this revolution of 250 years ago influences your life today?

2. A New International Economy

World Trade Increases Interdependence

Steamships made trade across the oceans quick and reliable, just as railroads had on land. These vessels carried manufactured products, raw materials, and passengers from one continent to another faster, cheaper, and in greater numbers than ever before. International trade, which in the early 1900s expanded tenfold, had important effects on economic activity in Europe.

Wheat shipped from the grasslands of the American Middle West began to be sold in Europe more cheaply than wheat grown in Europe itself. Inexpensive meat from Argentina and butter from Australia and New Zealand also were shipped to Europe. Gradually it became clear that European farmers, given their poorer environments, could not compete with the cheap price of these imported foods.

This fact changed the world economy. Western Europe began to sell manufactured goods abroad in return for food and other raw materials. Less developed suppliers outside Europe began to specialize in products that Europeans wanted in order to obtain European manufactured goods in return. Many of these suppliers were or had been European colonies. India grew tea; rubber came from Malaya; sugar from Cuba and the Philippines; and coffee from Brazil. From Egypt, the Ivory Coast, Nigeria, the Belgian Congo, South Africa, and other African countries came a variety of agricultural and mineral products.

Ocean trade routes became busy highways, and Europeans usually controlled both ends of these trade lines. European money paid for the establishment of mines, plantations, railroads, and ports in many developing countries in Latin America, Africa, and Asia. Some people in these countries "got rich quick" when Europeans bought their raw materials. In the long run, however, the less developed countries suffered because they depended on selling one or two products on world markets.

When the prices of these commodities went down, the people of these countries suffered.

Europeans Settle around the World

In the middle 1800s and the early 1900s, Europeans began to move about freely. They left their farms and moved to Europe's growing cities and from one European country to another. They also left Europe and migrated to other continents. Passports and visas were not needed then, and steamship companies offered cheap fares across the ocean.

In a **mass emigration** that began in the 1840s, millions of Europeans sought better lives and homes in the United States, Canada, southern Latin America, Australia, and New Zealand. When the potato blight destroyed Ireland's farm economy and life hit rock bottom there, tens of thousands of Irish people fled famine in their native land. From then on, each crisis in Europe triggered a new wave of emigrants. By the late 1800s, a half-million Europeans were leaving each year. After 1900, this figure swelled to 1 million emigrants each year.

All together, 34 million Europeans emigrated abroad between 1840 and 1910. Four out of five came to the United States. Irish, Scottish, and English people fled their land-poor countries throughout this period. Many Germans and Scandinavians also left Europe between 1850 and 1900; immigrants from Italy, Eastern Europe, and Russia, generally left Europe after 1880. Some of you may have great-grandparents who remember the hopes that these European immigrants brought to America. Many of your fellow Americans are descendants of these Europeans who took a chance on a new life in a new land.

⊕ REVIEW QUESTIONS

1. How did Europe's economy change when international trade expanded enormously in the early 1900s?
2. In the 1840s, what caused many thousands of Irish men and women to flee their native land?
3. Most of the 34 million European emigrants went to what country? What other countries and culture regions also attracted European emigrants?

⊕ THOUGHT QUESTIONS

1. How did economic dependence on Europe affect many developing countries in Latin America, Africa, and Asia? Are these trading patterns still factors in hindering economic development in lesser developed countries?
2. Why was the development of the steamship important? Imagine leaving the United States to migrate to a foreign land with only the goods you could carry. Do you know any people who did leave Europe for the United States? If so, ask them to describe their experiences.

▶

Ellis Island, at the mouth of the Hudson River, was a landing point for many immigrants from Europe. Recently, Ellis Island has been renovated as a historical monument. Information on all immigrants from Europe who passed through this gateway to America can be looked up on computers there. Where would you begin your search for your family history if your ancestors came from Europe in the 1800s or 1900s?

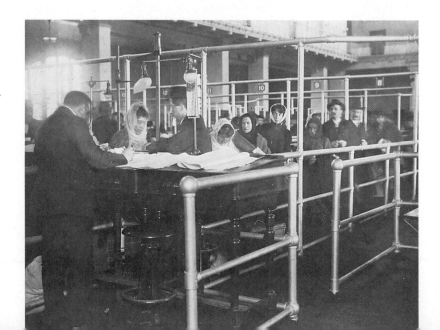

3. The European Scene Before World War I

Europe in 1900

Western Europeans were the first people to live in a modern industrial society—the first to experience industrial pollution, rapid technological change, and the tensions of modern life. Their countries competed for raw materials, markets, and national prestige all over the world. In Europe itself, two forces increased the pressures of daily life: population growth and the growth of cities.

A Crowded Continent Supports More People

Europe was quite a different place in 1900 than it had been in 1800. During these one hundred years, the number of Europeans doubled. Although overseas migration drained off some of this population surplus, competition for land and jobs increased throughout Western Europe.

The larger populations in industrialized countries were supported by higher levels of economic productivity. Britain with 42 million people and Germany with 50 million were the two most populous countries in Western Europe. Their methods of increasing economic productivity were quite different. Britain chose to sell manufactured goods abroad and to import cheap grain to feed its population since British farmers could not grow grain as cheaply. As a result, many British farmers were forced to leave the land and move to cities. German farmers, by contrast, increased grain yields by using chemical fertilizers and machine technology on their farms. The Germans soon were among Western Europe's most productive farmers, along with the Dutch, the Swiss, and the Danes.

France did not experience the population growth that occurred in the rest of Western Europe. In the early 1800s, France was the most populous country in Europe; thereafter, the French population grew more slowly than that of other large Western European countries. By 1914, there were three Germans for every two French. Although still an important power, France was rapidly falling behind Germany in both people and production.

The populations of Spain and Italy grew, but their economies did not keep pace. These countries had little coal or iron for industrial production so not enough new factory jobs were created in cities. Another factor was that most farmland was still in the hands of politicians, landlords, and the Catholic Church. Population growth on the remaining land, therefore, led to the breakup of small farms into even smaller farms, a decline in the standard of living, and rural poverty.

The Growth of Cities in Western Europe

In most of Europe, growing cities absorbed the increasing number of people. Country people moved to cities where factories and new businesses provided jobs. New cities sprang up near mines and factories, and port cities grew as trade with Europe's colonies increased. In 1800, 20 cities in Europe had more than 100,000 people; by 1900 there were 140 such cities.

By 1914, many people in Western European countries were **urbanites**, or people who lived in cities. About 80 percent of the British people, 60 percent of the Germans, and 45 percent of the French lived in cities. London was a sprawling urban center with 6.6 million people in 1900. Paris had a population of 3.7 million, Berlin had 2.7 million, and Vienna 1.7 million. Industrial centers like Manchester, Liverpool, and Birmingham in England; Hamburg in Germany; and Barcelona in Spain quickly grew into cities with more than a half-million people. So did port cities like Marseille in France, Copenhagen in Denmark, and Rotterdam in the Netherlands. Locate these important places on the map on page 109.

Europe's cities were ill equipped to handle this rapid increase in population. Slums grew up next to factories and near railroad yards. Badly constructed apartments were hastily built to shelter factory workers. Water supply systems were inadequate, and sewage and garbage removal systems nonexistent. Because crowding was intense, tension and anger grew. Throughout Europe, people were moving to cities, and few cities were ready for them.

The Quality of Life Improves

Despite these difficulties, the Industrial Revolution gradually improved the quality of day-to-day life for most people in Western Europe. Mass production made items like shoes, soap, cloth, and small appliances available to more people. In the cities, streets were paved and lit, and water supply and sewage systems were installed. Police and fire departments gave city dwellers more protection. Medicine provided cures for diseases that had been deadly just a few decades earlier.

Social leaders began movements for reforms to better the lives of the working poor. These leaders pushed for laws to abolish discrimination against religious and ethnic groups. They pressed for better working conditions in factories and mines, for public education, for equal rights for women, for an end to the abuses of child labor, and for the organization of labor unions. These efforts at reform eased tensions and probably spared much of Europe from bloody revolution.

⊕ REVIEW QUESTIONS

1. What two forces increased the pressures of daily life in Europe in the early 1900s?
2. By 1914, what percentage of the people in Britain, Germany, and France lived in cities? What were Western Europe's three largest cities in 1900?
3. What improvements began to better the lives of the urban working poor?

⊕ THOUGHT QUESTIONS

1. In what ways were Europe's cities not able to handle the increase in urban population? Why do urban slums still exist in most large cities after 250 years of industrialization? What can be done about it?
2. How did the Industrial Revolution gradually improve the quality of life in Western Europe? Who benefited most? Least? Do you think this pattern of gradual improvement is being repeated in many of the lesser developed countries of the world today?

4. The Decline of Europe

World War I, a Chain Reaction

World War I (1914–1918) left large parts of Europe in ruins and demolished its economy. Forces leading toward conflict had been steadily building when this war was triggered by an assassination. In June 1914, a young nationalist from Serbia (one of two republics left today in the Federal Republic of Yugoslavia after its breakup in 1991) assassinated the heir to the throne of the Austria-Hungary monarchy, Archduke Francis Ferdinand. At this time, the major countries of Europe formed two opposing camps allied against one another by military treaties. Thus, when Austria used this assassination to make demands on the government of Serbia, Serbia called on its ally, Russia, for help. Russia, in turn, invoked its alliance with France. On the opposing side, Austria-Hungary asked for assistance from its own ally, Germany. Because Germany's growing power worried Britain, Britain reluctantly joined on the side of Russia and France.

Soon most of the continent was involved in a chain reaction of treaty commitments that led to the most destructive war the world had ever seen. What we now call World War I was largely fought in Europe. The United States entered the war in 1917 on the side of Britain, France, and Russia.

This was the first war in which machine technology was used on a large scale. Armored tanks, airplanes, machine guns, submarines, and poison gas were all used. More than 9 million soldiers died, and civilian casualties also numbered in the millions. The cost of the war in blood and money was immense. Hundreds of billions of dollars were spent before Germany and Austria-Hungary were defeated. When the war was over in 1918, Europe lay shattered.

 IT'S A FACT   The distance between Rome and Paris is less than that between Miami and Atlanta.

◄

The use of modern military technology in World War I made this four-year struggle one of the most devastating in history. What were the major new elements of technology used in this war?

World War I Changes the Map of Europe

The Treaty of Versailles, which ended World War I, changed the boundaries of many European countries, as the maps on pages 150 and 151 show. Compare these two maps and note the boundary changes that followed this war. These changes are still influencing politics in Eastern Europe today.

Austria-Hungary, centered in Vienna, was carved up into Austria and Hungary and parts of Czechoslovakia, Poland, Yugoslavia, and Romania. Russia lost large pieces of land all along its western borders. Poland, Lithuania, Latvia, Estonia, and Finland became independent countries. Germany lost land on both its western and eastern flanks. France, Denmark, Italy, and Belgium gained land.

Britain, a victor in the war, had in fact lost a great deal. Markets for its manufactured goods in Europe and around the world had disappeared. Germany had lost most of its coal and iron deposits, as well as one-tenth of its population. As a loser, Germany was also forced to pay compensation to the winners for the costs of the war. These payments and lost resources soon resulted in Germany's economic collapse.

Meanwhile, economic leadership of the world shifted from Europe to the United States. The United States started exporting large amounts of manufactured goods, and Japan began taking over Britain's former markets in the Pacific. Revolution in Russia led to the establishment of a communist state there, the Soviet Union. Western European economic control of much of the world slipped badly.

The Western European countries struggled to rebuild their separate economies, but most were in debt and were tied into the powerful banking system of the United States. When the New York stock market crashed in 1929, precipitating the Great Depression, Europe was immediately affected. Prices and production fell, and unemployment soared. Banks in Austria and Germany closed. An economic depression settled over Europe.

World War II Creates a New Geography

Europeans' faith in democracy was badly damaged by the devastation of World War I. New tensions developed because of economic hardships. Bread lines and bankruptcies became common in most countries. As people lost faith in traditional

Europe Before World War I

Compare these two maps. What countries appear on the map on the opposite page that are not on this map? What two large states shown on this map shrink on the map opposite? Now compare the opposite map with the map of Europe today on page 107. What three countries that are on the opposite map have disappeared from the map on page 107? What happened to Germany as a result of World War I?

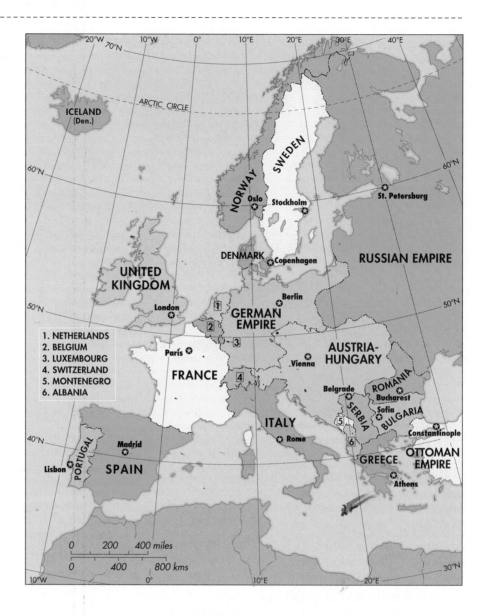

1. NETHERLANDS
2. BELGIUM
3. LUXEMBOURG
4. SWITZERLAND
5. MONTENEGRO
6. ALBANIA

institutions, political extremists gained power in Italy and Germany. In Italy, Benito Mussolini rose to absolute power promising Italians order and a powerful new nation led by his Fascist party (derived from *fasces*, a bundle of rods with an axe that was a symbol of authority in ancient Rome). Germany was taken over by Adolf Hitler's National Socialist, or Nazi, party. Hitler specifically tapped into the deep resentment felt by Germans over the harsh conditions of the Versailles Treaty.

When Nazi Germany invaded Poland, Britain and France declared war on Germany on September 3, 1939. Europe was plunged into World War II (1939–1945). Italy and Japan became allies of Germany. The Soviet Union signed a peace treaty with Germany and divided Poland between them in 1939. After Germany invaded the Soviet Union in June 1941, however, the Soviets joined forces with Britain and France. As you know, the United States entered the war after Japan bombed the U.S. naval base at Pearl Harbor, Hawaii, on December 7, 1941.

World War II saw the end of Western European countries as major world powers. After World War II ended in 1945, Europe faced a world dominated by the United States and the Soviet Union, the new su-

Europe After World War I

perpowers of the twentieth century. Western Europeans began to search for a new role in this rapidly
changing world.

⊕ REVIEW QUESTIONS

1. What event led to the outbreak of World War I?
2. In 1914, what countries were Serbia's chief allies? Who was Austria-Hungary's main ally?
3. Who were Hitler and Mussolini, and how did
   Hitler plunge Europe into World War II?
4. What two countries emerged as superpowers
   after World War II?

⊕ THOUGHT QUESTIONS

1. What happened to the economy of Western
   Europe as a result of World War I? How did the
   collapse of the American stock market affect
   Western Europe? Can you think of a recent example in which decisions or events elsewhere in
   the world have affected the economy of the
   United States?
2. War is expensive in both blood and money.
   What benefits, if any, are achieved by warfare?
   Please explain your answer in terms of gains and
   losses.

5. Building a New Economy in Western Europe

The Need to Rebuild Western Europe

For almost a hundred years, most countries in Western Europe gained wealth through international trade. They exported manufactured products throughout the world and imported the raw materials and food that they needed. After World War II, however, world conditions changed. Many countries in Latin America and Asia that had once imported European manufactured goods now had their own factories. The Soviet Union had constructed an "iron curtain" that closed off trade and other contacts between the industrial countries of Western Europe and the agricultural countries of Eastern Europe. The Western Europeans had to deal with these new developments if they were to rebuild their economies.

The most immediate problem was to recover from the war's devastation to the Western European landscape. In 1945, thousands of farms, factories, railroads, bridges, and entire sections of cities lay in rubble. Most countries had no financial reserves to pay for rebuilding. The cost of the war in lives and property had been staggering.

During this crisis, the United States provided grants and loans to Western Europe, including West Germany, to help rebuild cities, industries, and transportation systems under a program called the Marshall Plan. Americans believed that Germany's inability to recover from World War I had been a primary cause of World War II, and they wanted to help the country to become prosperous and stable.

As a result of this timely aid, most countries in Western Europe regained prewar levels of industrial production by the 1950s. But basic problems still remained. Western Europe was composed of eighteen small countries in a world dominated by the much larger superpowers—the United States to the west and the Soviet Union to the east. Twelve countries in Western Europe had populations smaller than the state of New York. Only four—West Germany,

the United Kingdom, Italy, and France—had more than 50 million citizens.

Because of their small sizes and limited populations, no single Western European country could build or staff the huge enterprises that were becoming common in U.S. industry and the government-sponsored industry of the Soviet Union. Clearly a new economic and political strategy was needed.

Cooperation Builds the European Community

Some forward-looking European leaders realized that cooperation rather than competition was the key to Western Europe's future. A bigger Western European market was needed to provide workers for factories and to buy the products these factories would produce so that Europe could compete in a rapidly changing postwar world.

In 1951, the six countries of West Germany, France, Italy, the Netherlands, Belgium, and Luxembourg created the European Coal and Steel Community, in which their most basic heavy industries were organized under a single commission. Six years later, in 1957, they founded the Common Market, or European Economic Community (EEC). In 1957, the European Atomic Energy Community (EURATOM) was established to control nuclear energy in these countries. These three groups were integrated into the **European Community (EC)** in 1967.

Barriers to trade, such as tariffs among the six countries, were gradually abolished so that goods and raw materials flowed freely from country to country within the European Community. As a result, trade among these countries tripled. At the same time, tariffs were imposed on goods coming into the European Community from other countries. Despite these tariffs, the European Community's trade with nonmember countries doubled because of its favored location, lingering ties with former colonies, and increasing prosperity.

In 1968, the European Community began to allow workers to move from job to job within member countries. This meant that people from depressed areas in Mediterranean Europe could seek jobs in the industrial areas in Continental Europe.

This free movement of labor encouraged more economic growth in Western Europe.

The success of the European Community attracted other countries. In 1973, the United Kingdom, Ireland, and Denmark were admitted, and in 1981, Greece entered the European Community. In 1986, Spain and Portugal increased the membership to twelve nations. In 1995, Austria, Finland, and Sweden became full members of the European Community. Norway and Switzerland are the only major Western European countries which are not members of the EC.

Today, the European Community has a larger population than the United States. Its educated people have rebuilt their industries. With far less farmland than our country, Western Europe is now self-sufficient in food, and also produces more steel than either the United States or Japan. Nearly half of all world trade passes through the European Community countries.

The Maastricht Treaty proposed in December, 1991, is named after a small Dutch town where the document was formulated. This treaty envisions a broader and larger "European union" as well as closer economic and monetary ties with a single currency, the European Monetary Unit (EMU). In addition, the European Community would regulate some working conditions, employer compensation,

European Community (EC)

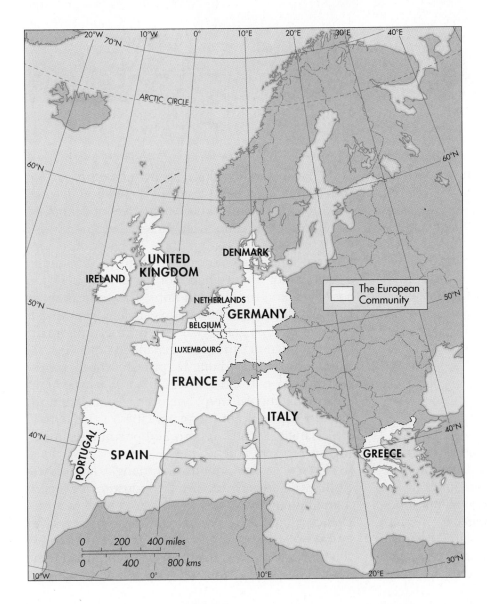

▶ *The headquarters of the European Community (EC) in Brussels, Belgium, is becoming an increasingly important center of power and influence. What 1991 treaty increased the authority of the EC?*

and taxes, as well as the distribution of money to less developed regions.

Needless to say, many Europeans see the Maastricht Treaty as an infringement on their national prerogatives; in fact, the Dutch, French, and British nearly vetoed it. In spite of these problems, the economic benefits of the EC are attracting the attention of most Eastern European countries, as well as Turkey and Morocco to the east and south. Attempting to merge the traditions and practices of this diverse group of countries, however, is a daunting task.

The Decline of Old Industrial Areas

Western Europe is one of the world's major industrial regions. The value of its industrial products is second only to that of the United States. It is a leading producer of steel, automobiles, machines, textiles, chemicals, and consumer goods. Western Europe's main manufacturing zone stretches from England through the Benelux countries, the French northeast, and the German Ruhr, southward to the Po Valley of northern Italy, as the map on page 155 shows. Half of Western Europe's population lives in this industrial zone, and three-quarters of its manufactured goods are produced there.

The greatest concentration of heavy industry in Western Europe is found in an **industrial triangle** composed of the French northeast, the Benelux countries, and the German Ruhr. This industrial region developed one hundred years ago when local deposits of high-grade coal and iron were brought into production. Industry expanded after canals and railroads were built to provide cheap transportation to port cities like Hamburg, Amsterdam, Rotterdam, and Antwerp. Engineering and chemical industries grew up in the major cities of this industrial triangle. Locate Western Europe's industrial triangle on the map on page 155.

Western Europe's industrial triangle and other coal-based centers of heavy industry are now changing rapidly. Century-old plants have become inefficient and have been closed. Coal mines have shut down because other sources of energy such as oil and nuclear power are being used. Imported iron is now cheaper and of better quality than local ores.

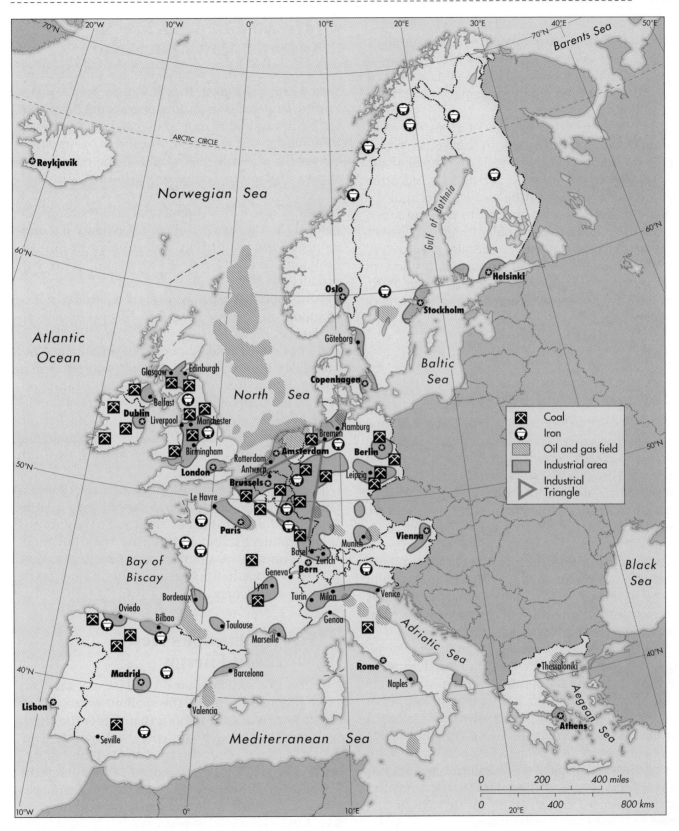

Resources and Industrial Areas in Western Europe

As a result, some older industrial areas, with their out-of-date plants, are unable to compete with modern steel centers. Like many steel-producing areas of the United States, these regions now have large numbers of unemployed people, as well as severe pollution problems.

In an effort to bring back older industrial areas, new industries are being built in these areas to provide jobs for the unemployed. Young men and women are being retrained for jobs in industries like precision manufacturing and high technology. Welfare programs are attempting to ease the suffering of the workers. But the older industrial districts in Western Europe still have serious problems. In the Welsh coal mines, the industrial districts of England, northeastern France, the German Ruhr, and Saxony, smokestacks are no longer a sign of prosperity.

New Industrial Patterns

In spite of this decline in Europe's old industrial areas, the European Community regularly produces 100 to 150 million tons of steel each year. New steel centers are being built on coasts so that inexpensive raw materials can be brought in from overseas. High-grade iron ore is imported from Sweden, Canada, and until recently, Russia; in fact, two-thirds of the iron ore used in Western Europe is now imported. New industrial ports have been built in cities on the coast of the North Sea, in Wales, in southern France, and in Italy. Coastal steel plants now account for one-third of the steel produced in the European Community.

Western Europe's automobile factories provide an important market for the steel produced in these coastal plants. The European Community produces nearly 9 million automobiles each year, and more than 60 million cars clog the highways of Western Europe. Industrial countries like Germany average one car for every five people. Less developed farming countries like Spain and Portugal have one car for every ten people. Western European auto manufacturers also sell cars to countries throughout the world, including the United States.

Increased Dependence on Imported Raw Materials

Western Europe's dependence on imported raw materials is increasing. In addition to iron, Western Europe imports large amounts of copper, chrome, lead, zinc, and other industrial metals and minerals. In addition, this region still must rely on imported energy. In the early 1990s, half of the energy consumed in Western Europe was imported. Although the eighteen countries of this region have only 8 percent of the world's population, they consume 20 percent of the world's energy. Demand has outstripped local production, even though the United Kingdom, Germany, and France are important coal producers.

Petroleum has replaced coal as the main source of industrial energy in Western Europe, and much of the petroleum it uses comes from the Middle East. Port cities like Rotterdam, Amsterdam, and Marseille are important refining centers where crude oil is transformed into gasoline, heating oil, and other petrochemical products. A 4,400-mile-long network of pipelines moves these fuels from port refineries to industrial centers throughout the region. Western Europe also buys natural gas from fields in the Russian Arctic.

Imported raw materials are expensive; Western Europeans must produce and export large quantities of manufactured goods and food to pay for the energy, minerals, and metals that they buy from other countries. In an effort to reduce these costs, Western Europe has invested heavily in developing new sources of energy.

Western Europe's New Sources of Energy

In the 1970s, major new oil and natural gas fields were developed beneath the shallow waters of the North Sea. Five of the nations that border the North Sea—the United Kingdom, the Netherlands, Germany, Denmark, and Norway—divided this body of water into zones to be developed by each country. Recent exploration has located more energy reserves as far north as the Shetland Islands, north of Scotland.

▲ *The steelworks in Taranto, Italy, is one of the new, efficient coastal factories built in Western Europe after World War II. Why is it advantageous to build steel plants in coastal cities?*

The North Sea fields already produce enough petroleum and gas to make the United Kingdom self-sufficient. Norway produces large amounts of hydroelectric power, so it sells its petroleum and natural gas to other nations in the European Community. Despite the North Sea discoveries, Western Europe needs to diversify its sources of energy. Many countries are developing their natural gas deposits—for instance, large gas fields in the Netherlands are just coming into production. Hydroelectric power is a major source of energy in Switzerland, Austria, Italy, and the Scandinavian countries.

In addition, Western Europe relies on nuclear power more than does any other world region. Today, over fifty nuclear power plants are operating

and sixty more are being built. These plants produce more than half of the electricity consumed in France, Belgium, and Sweden, and roughly a third of that used in Germany, Spain, Switzerland, and Finland.

The Shift to High Technology

Western European nations are working to create new jobs in the growing fields of high technology like automation, computers, information technology, aerospace engineering, and satellite communications. In the 1970s and 1980s, a gap in technology separated Western Europe and the United States. The United States was spending six times as much money on research and development in high technology as the European Community. Europeans were not as advanced in high-technology fields as either the United States or Japan. In the 1990s this gap is widening.

Part of the problem is that few European corporations are as large or as wealthy as industrial organizations in Japan and the United States. Also, 50,000 to 60,000 scientists and engineers leave Western Europe each year to work in universities and industrial laboratories in North America. With less research support and fewer trained people, Western Europe faces major difficulties.

To cope with these problems, the United Kingdom, France, and—to a lesser extent—Germany are dramatically increasing investment in high-technology fields. More scientists and engineers are being trained. Western Europe is developing its own space technology based on advances in high-energy physics. France is pouring funds into high-technology research designed to create centers of electronics and aerospace engineering. The United Kingdom is financing research in biotechnology and robotics. Private corporations in Germany are attempting to develop microchips, computers, and electronic processing equipment.

Today Western Europe has hundreds of small high-technology firms, yet few European companies have the resources to compete with the world's largest corporations. Despite European Community progress in coordinating advances in science and technology on the continent, American and Japanese products still dominate markets in Western Europe—a condition that must change if this region is to remain an important center of industry.

People on the Move

The growing wealth of the European Community has been a magnet to millions of immigrants from poorer regions of North Africa as well as the newly independent states of Eastern Europe. After the collapse of the Soviet Union, more than a million ethnic Germans left Russia, Poland, Romania, and elsewhere in Eastern Europe for Germany. Russians are also fleeing to Poland, and Romanians to

▶

A network of high-speed rail lines is being constructed to connect Europe's major cities. Europeans rely on public transportation to a great degree. Why is this less true in the United States?

Hungary and Austria. Albanians are being turned back from Greece. And Middle Easterners and North Africans are moving into France, Germany, Italy, and countries as far north as Sweden.

A swelling tide of immigrants is entering Western Europe. Between 1982 and 1990, an estimated 8 million migrants—mostly from North Africa, Turkey, and the former Federation of Yugoslavia—had already entered Western Europe. Now that the emerging countries of Eastern Europe have torn down the barriers that kept their people in, the countries of Western Europe are beginning to erect new barriers to keep them out.

The primary motivations for this mass migration, the largest since World War II, are clear. Dense populations, stagnant economies, and political repression have propelled Eastern Europeans, Middle Easterners, and North Africans to leave their countries. The normal flow of immigrants from these areas has doubled each year—and this does not include illegal immigrants. As many as 7 to 8 million citizens of the former Soviet Union are expected to head west into Western European countries that have not even begun to deal constructively with this unexpected problem.

Resentment of foreigners has led to violence, particularly in Germany. Most Western European countries are now severely restricting the number of people they accept as immigrants. But flight from famine, unemployment, and civil war are powerful motivating forces, and how the Western Europeans will deal with the staggering numbers of foreigners pleading for safe haven and a place to live is not clear. As one report notes, "migration is the visible face of social change." In other words, people vote with their feet.

Farming in Western Europe

After World War II, many Western European countries wanted to be as self-sufficient as possible in food production. Even countries with environments that were poorly suited for farming, such as Sweden and Switzerland, encouraged farmers to remain on the land and to produce as much food as possible.

IT'S A FACT   A farmer in the European Community sells wheat for $163 per ton. The market price is $99 per ton. The government subsidizes the remaining $64.

This was done by granting **subsidies,** or price guarantees, to farmers. Governments guaranteed the income of farmers by setting minimum prices on products like wheat, butter, eggs, and meat. Only the United Kingdom did not subsidize farm incomes at that time; the United Kingdom had decided in the 1800s to import inexpensive food from Canada, Australia, and New Zealand rather than to grow its own food at home.

This policy of subsidizing farming is encouraging farmers to stay on the land. Even farmers in dry, rugged, or cold regions that are difficult to farm continue to produce food. Many new farmers in the United Kingdom benefit from this European Community policy. One gentleman farmer in France puts it this way: "The more wheat I produce, the more I'm paid. It's as simple as that." But the cost of growing food on poor land is higher than the price this food can be sold for on world markets. Because governments make up the difference by subsidizing farmers, the landscape of Western Europe has remained broken up into small, inefficient farms.

No issue has vexed the European Community more than what to do about its farmers. Leaders of some countries want to eliminate food surpluses by lowering price supports paid for crops. However, member countries with large farm surpluses like Belgium, France, Italy, and the Netherlands insist that price supports be kept high. So the 12 million farmers in the twelve countries of the European Community have been guaranteed high food prices and protection against imports of lower priced food from countries outside the European Community.

As a result, tons of overpriced grain, butter, and beef are piling up in Western Europe. The cost of subsidizing farmers is a heavy financial burden on the European Community. But when price cuts are threatened, farmers revolt; in 1993, French farmers ringed Paris with tractors and brought traffic to a standstill in the capital. As a result of the subsidies, city dwellers in Europe pay two to three times as much for their food as people in other regions. More than two-thirds of the European Community's budget pays for the production of surpluses of butter, meat, wine, and other food products.

In spite of these government subsidies, some European farmers are leaving the land and moving to urban centers. Small, inefficient rural farms are gradually being abandoned, mostly in marginal farming areas. These would include regions in mountains where soils are thin, in Northern Europe where the growing season is short, and in Mediterranean areas where soils are exhausted from centuries of continuous use.

In the early 1990s, less than 10 percent of Western Europe's workforce is still engaging in agriculture. The rest of its workers live in the growing industrial cities of this prosperous region.

⊕ REVIEW QUESTIONS

1. In 1945, what was Western Europe's most immediate problem?
2. What is the European Community, and when was it started?
3. Where are the greatest concentrations of heavy industry in Western Europe?
4. How did the governments of Western Europe encourage farmers to remain in farming?

⊕ THOUGHT QUESTIONS

1. Why do the nations of Western Europe face difficulties in competing with Japan and the United States in the fields of high technology? Do you think the United States faces any of the problems Western Europe faces in these fields?
2. What problems have been caused by granting subsidies to farmers in Western Europe? Do you think that subsidies for growing crops are a good idea? Do any American farmers receive government subsidies in your state? How do they feel about it? How do you feel about subsidies?
3. What are some of the reasons people emigrate? How does large-scale immigration become a problem for wealthier countries?

6. Problems Caused by Urban Industrial Concentrations

As you have learned, Western Europe is an urban industrial region. Today, more than twenty cities in Western Europe have populations of more than 1 million people. All of them are capital cities, ports, or manufacturing centers.

With an industrial economy and 230 million people living in its cities, Western Europe has serious pollution problems. **Pollution** exists when the by-products of human activity accumulate faster than the natural movements of air and water can remove them.

Air Pollution

Air pollution is a problem in Western Europe. The increasing numbers of automobiles pour exhaust fumes into the atmosphere creating a smog dome over every major city in Western Europe. For this reason, most European governments have passed clean air laws.

The United Kingdom was the first to pass such laws. In 1952, smog settled over London and killed 3,500 to 4,000 people. The protests that followed led to a series of clean air laws. As a result, smoke levels and **emissions,** or discharges into the air, of pollutants from factories, vehicles, and home furnaces are now monitored and controlled throughout the United Kingdom. By the 1990s, levels of smog in the air over cities like London and Manchester had fallen by more than half. Today, the yellow pall of smog that blanketed British industrial centers has all but disappeared. In other countries like West Germany, France, Belgium, and

IT'S A FACT Pollution in England's "black country" caused a local species of moth to change its camouflage from green to dark gray to match the sooty tree trunks.

IT'S A FACT Air pollution was a nuisance in London, England, as early as four hundred years ago.

the Netherlands, comparable pollution control measures have been adopted. In what was formerly East Germany, no pollution controls were adopted under communist rule. A massive clean up is now planned by the German government.

Because Western Europe has eighteen countries crowded into an area about one-third the size of the United States, pollution from one country often affects other countries. Pollution is a global problem because of Earth's large-scale atmospheric and ocean circulatory systems. Air pollution from industrial nations like the United Kingdom has increased the amount of acid rain and snow that falls on Norway and Sweden, killing trees and stunting fish growth. Acid rain from the German Ruhr is now destroying the famous Black Forest, as well as the woodlands of Eastern Europe.

International cooperation is needed to solve these problems. The twelve-member European Community has already placed limits on the production and use of chlorofluorocarbons (CFCs), which destroy the ozone layer, and is also trying to create a unified environmental policy that will protect Western Europe's landscapes and wildlife.

Water Pollution

Most large cities in this region are located on rivers that provide water for household and industrial use. As the population of Western Europe has grown and become more concentrated, pollution from cities and industries has severely damaged the water quality of these rivers.

In France, rivers like the Seine, Rhône, and Loire are polluted, as are the country's major canals. In the Netherlands, water is not drunk or used on gardens without being filtered first. In England, clean water laws have made the Thames River once again safe for fish, but many inland waterways are still polluted.

Nowhere is the problem more acute than in the Rhine River, where pollution has reached dangerous levels. As the Rhine flows northward to the sea, it passes through a continuous band of cities and industries. By the time it reaches the Netherlands, the river is carrying almost 70,000 tons of solid waste a

▼ *This photograph shows the effect of acid rain and air pollution on some of the forest areas of Western Europe. What man-made structures are being damaged by acid rain in Europe?*

day, a staggering 25 million tons each year. This material is then deposited in the North Sea. The Meuse River arrives at Maastricht loaded with waste products from Luxembourg and Germany, although for twenty-five years the Dutch have pleaded with their neighbors to be more protective of the river.

Some of the loveliest mountain lakes in the Alps of Switzerland, Germany, Italy, and France have also been contaminated by acid rain and domestic sewage. The fish population in many of these lakes is rapidly diminishing. To the south, the beaches along the Mediterranean must be sprayed with disinfectant regularly to kill organisms from sewage that cause disease and to dissolve the tar released by the many oil tankers that pass through the Mediterranean Sea. In the 1990s a survey showed that hundreds of these beaches were in violation of the new environmental pollution laws enacted within the European Community.

Like the Mediterranean, the North Sea is also heavily contaminated. The rivers from nine West European countries dump waste materials into the North Sea. Salt from potash mines, phosphates from urban sewers, mineral wastes from steel mills, and pesticides from farms flow directly into the North Sea. Pollutants are also added by passing ships and by direct dumping of wastes from coastal cities.

More than forty international conferences have been held in Western Europe to discuss water pollution. Although the problem is far from being solved, the Western Europeans have demonstrated great imagination and skill in fields requiring international cooperation.

⊕ REVIEW QUESTIONS

1. What are the major causes of air pollution in Western Europe?
2. What are some of the effects of acid rain in Western Europe?
3. What has polluted the North Sea?

⊕ THOUGHT QUESTIONS

1. What cooperative efforts have been made by the various countries of Western Europe to decrease industrial pollution?

2. How are many Western European governments dealing with the problems caused by air and water pollution? Do you believe these measures will relieve the problem? Is air or water pollution a problem in your city, state, or region? What is being done about it? What can you do to help?

⊕ CHAPTER SUMMARY

The Industrial Revolution began in England and spread to Western Europe in the 1800s. The Industrial Revolution occurred first in England for several reasons: abundant natural resources, a large working population, money for investment, and a talent for invention. Industrialization caused many social and economic problems for the British government, some of which the country was able to solve over a long period of time. In mainland Europe, the Industrial Revolution spread more slowly because of resistance to change and because of Western Europe's larger geographic size and rugged terrain. Only when Western European countries faced a severe economic crisis did their governments recognize the value of industrialization.

Because of its new found industrial strength, Europe dominated most of the developing world politically and economically in the 1800s and early 1900s. But intense competition among Western European countries led to two devastating world wars which took from Europe much of its world power. The world emerged from World War II dominated by two superpowers, the United States and the former Soviet Union. European countries discovered that cooperation was their best alternative in such a world.

Today, many Europeans are hopeful that the European Community (EC), a union of fifteen Western European nations, will help the region maintain its prosperity and global economic position. As part of the technological world, most people in Western Europe today enjoy high standards of living. However, the highly industrialized countries of Europe are now struggling with the problems of pollution and crowding, slowing economic growth, and learning to live in a multicultural society.

CHAPTER 5

EXERCISES

Two Correct Sentences

Directions: Write the two correct sentences from each group of three sentences on your paper.

1. a. The Industrial Revolution began in England in the late 1800s.
   b. English inventors discovered new ways of making and transporting goods.
   c. England had many of the natural resources needed for modern industry.
2. a. The development of railroads reduced the costs of transporting goods and people.
   b. The terrible conditions in factories caused a great migration of Europeans from cities to villages.
   c. At the same time that new industrial inventions increased the production of factory goods, other inventions improved methods of farming.
3. a. By 1900, Germany surpassed England in railroad growth and iron production.
   b. In northeastern France, there is a zone of heavy industry based on iron ore from Lorraine and coal from the Saar Basin.
   c. By 1900, France overtook both England and Germany in industrial production.
4. a. The number of Europeans doubled between 1800 and 1900.
   b. Because of the workers' poor living conditions and the huge migration to the United States, the population of Europe declined in the 1800s.
   c. Germany had a larger population than England in 1900.
5. a. Europe's colonial empires were larger in 1945 than they had been in 1900.
   b. Chiefly as a result of two world wars, Europe's colonial empires dissolved.
   c. After World War II, Europe was faced with a world dominated by the United States and the former Soviet Union.

Name the Missing Country

Directions: Number your paper from 1 to 6 and complete the following sentences by writing on your paper the name of the appropriate country or region within a country. Country names can be used more than once.

Austria	Italy
Belgium	Luxembourg
Denmark	Netherlands
England	North Sea
Finland	Norway
France	Sweden
Germany	United Kingdom
(or West Germany)	Wales

1. The six countries that are the original members of the Common Market, which later became the European Community (EC), are |||||||||| , |||||||||| , |||||||||| , |||||||||| , |||||||||| , and |||||||||| .
2. Three Western European countries that in 1995 joined the European Community are |||||||||| , |||||||||| , and |||||||||| .
3. The three countries known as the Benelux countries are |||||||||| , |||||||||| , and |||||||||| .
4. Soon after the discovery of oil in the North Sea, |||||||||| , |||||||||| , |||||||||| , |||||||||| , and |||||||||| , the five nations that border the North Sea, divided this body of water into zones to be developed by each country.
5. Industrial waste is a problem in many European countries. In |||||||||| , for example, the Seine, Rhône, and Loire Rivers are polluted.
6. Clean water laws have enabled fish to live once again in |||||||||| 's Thames River.

Inquiry

Directions: Combine the information in this chapter with your own ideas to answer these questions.

1. Why do you think that the French were reluctant to accept the Industrial Revolution? Can you mention some of this revolution's negative features? Do you think they outweighed its benefits?

2. Do you believe that rail-roads are as important to a country's economy today as they were in the last century? Why or why not?

SKILLS

Understanding Chronology

Directions: As you study events that occurred in different parts of the world, you need to know which event came earliest in time and the *sequence*, or order, of other events. This is called understanding *chronology*. Once you have acquired this skill, many subjects that you study will become more interesting and meaningful. On your paper, write the numbers of the statements in each group in chronological order.

Group A
1. World War I was fought.
2. Western Europe was the most important center of political power and economic strength in the world.
3. The continent was divided into the independent states of Western Europe and the communist states of Eastern Europe.
4. World War II was fought.

Group B
1. Hitler and Mussolini came to power.
2. World War II began.
3. After the New York stock market crash, Western Europe suffered an economic depression.
4. The Treaty of Versailles was signed.

Group C
1. The Industrial Revolution started on the continent of Europe.
2. The Industrial Revolution started in England.
3. International trade expanded tenfold.
4. The United States provided grants and loans to Western Europe to rebuild.

Group D
1. Gasoline made from oil was used to run automobiles.

2. Coal was burned to create the electricity needed for telephones, telegraphs, and electric lights.
3. Coal was burned to run steam engines.
4. The steam-powered railroad locomotive was invented.

Vocabulary Skills

Directions: Select the correct word from the list that completes each sentence below. On your paper, write the word you selected next to the number of the sentence it completes.

Bessemer process	industrial triangle
emissions	mass emigration
European Community (EC)	pollution
	subsidy
heavy industry	urbanites

1. When crises such as famines or depressions occur, they may trigger waves of ||||||||||, changing population patterns in many regions.
2. Industry that manufactures goods like steel and machinery is called ||||||||||.
3. Pollutants that automobiles and manufacturing plants contribute to the air are called ||||||||||.
4. Farmers in many countries of the technological world receive ||||||||||, government grants of financial support.
5. Many highly industrialized countries suffer from ||||||||||, which is the contamination of air, water, or land by chemicals, gases, and other by-products of human activities.
6. People who live in cities are often called ||||||||||.
7. In Western Europe, the greatest concentration of heavy industry is found in the ||||||||||, an area composed of the French northeast, the Benelux countries, and the German Ruhr.
8. The |||||||||| made large quantities of inexpensive steel available for the first time in the late nineteenth century.
9. The fifteen member |||||||||| has a population larger than the United States, no trade barriers between member countries, and free movement of labor among these countries.

Human Geography

Directions: Select the correct ending for each sentence and then write the complete sentence on your paper.

1. The Industrial Revolution began in (a) France (b) England (c) Italy.
2. The Industrial Revolution started in the (a) 1600s (b) 1700s (c) 1800s.
3. The number of major countries in Western Europe is (a) fourteen (b) fifteen (c) eighteen.
4. The European Community is (a) an economic union (b) a political union (c) an armed alliance.
5. Western Europe has a population of about (a) 225 million (b) 382 million (c) 515 million.
6. Western Europe's agriculture is (a) not subsidized (b) slightly subsidized (c) heavily subsidized.
7. The most important factor in the industrialization of the continent was the building of (a) canals (b) roads (c) railroads.
8. Four out of five Europeans who emigrated in the late 1800s went to (a) Latin America (b) Australia (c) the United States.
9. After World War II, the rebuilding of Western Europe was accomplished with aid from (a) the United States (b) the United Kingdom (c) Japan and China.
10. In World War I, Britain and France were allied with (a) Germany (b) Austria-Hungary (c) Russia.
11. After World War II, European economic and political control over countries in the developing world (a) decreased sharply (b) increased significantly (c) stayed about the same.
12. After World War II, the superpowers were (a) Britain and France (b) Britain and the United States (c) the United States and the Soviet Union.

Physical Geography

Directions: There are two terms in each of the parentheses below. Write the correct term on your paper.

1. Western Europe is roughly one (third, half) the size of the United States.
2. Western Europe is composed of many islands and (isthmuses, peninsulas).
3. England's large deposits of (oil, coal) helped provide fuel for much of the manufacturing in the early years of the Industrial Revolution.
4. France, Italy, and Spain have coastlines on the (Baltic, Mediterranean) Sea.
5. London is near the same latitude as (Virginia, Newfoundland).
6. The influence of the Atlantic Ocean on much of Northern Europe makes winters (warmer, colder).
7. Western Europe is one of the most accessible areas in the world because it lies at the center of the (land, water) hemisphere.
8. Norway's fjords were carved out by (mountain glaciers, earthquakes).
9. In Iceland, homes are warmed and factories are powered by (nuclear, geothermal) energy.
10. West Germany's most vital trade link is the (Po, Rhine) River.

Writing Skills Activities

Directions: Answer the following questions in essays of several paragraphs each. Remember to include a topic sentence and several sentences supporting your main idea in each paragraph.

1. Why did the Industrial Revolution begin in England instead of on the European continent?
2. How has Western Europe been affected by air and water pollution, and what steps are being taken to deal with these problems?
3. How are Europeans coping with the following: (a) reducing trade rivalries, (b) obtaining sufficient energy, and (c) providing enough food for their people?

▲ *St.Basil's Cathedral in Moscow was built by Ivan the Terrible, Czar of Russia. How does it compare to Western European cathedrals like Notre Dame?*

UNIT 3

THE COMMONWEALTH OF INDEPENDENT STATES AND EASTERN EUROPE

UNIT INTRODUCTION

In 1992, the last empire on Earth, the Soviet Union, broke apart into fifteen independent republics. Twelve of these countries have now joined a new loose federation called the Commonwealth of Independent States (CIS). Three countries chose complete independence from the former Soviet Union.

For 75 years, the 300 million people in the Soviet republics lived under a Communist regime that reached from Eastern Europe across Asia to the Pacific Ocean. Its heartland was Russia; its capital city was Moscow.

Over the centuries, the Russians had conquered much of northern Asia to create the largest empire on Earth. But in 1900, Russia was still little more than a huge, backward agricultural land that was far less developed than Western Europe. Most Russians were poor farmers. Their lives were governed by a *czar,* an absolute ruler, and by rich landowners who controlled the country's farmland. In 1917, a Communist revolution ended centuries of total rule under the czars.

After this revolution, Communist leaders changed Russia's name to the Soviet Union or U.S.S.R., the Union of Soviet Socialist Republics. Many of these republics were occupied by non-Russian, captive peoples. The Soviet leaders believed that *communism,* with a state-controlled

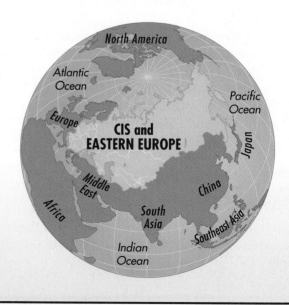

CIS and EASTERN EUROPE

North America

Atlantic Ocean

Europe

Pacific Ocean

Japan

Middle East

China

Africa

South Asia

Southeast Asia

Indian Ocean

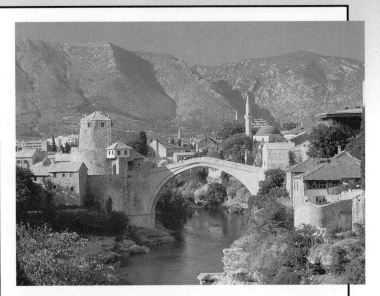

This bridge in Mostar, Yugoslavia, was built by Romans about 2,000 years ago. What modern events put this bridge in jeopardy?

UNIT OBJECTIVES

When you have completed this unit, you will be able to:

1. List the major geographic features of the Commonwealth of Independent States (CIS) and Eastern Europe and the ethnic, religious, and linguistic differences of the people who live there.

2. Describe the problems that existed in Russia under the czars.

3. Explain how the principles of communism differ from the principles of capitalism.

4. Identify changes that have occurred in Russia since the Communist revolution.

5. State how two world wars have affected national boundaries and ways of life in Eastern Europe.

6. Describe how the collapse of communism in the former Soviet Union resulted in the creation of the Commonwealth of Independent States in 1991.

SKILLS HIGHLIGHTED

- Using Historical Maps
- Locating the CIS's Neighbors
- Learning about Distances
- Using Physical Maps
- Understanding International Time Zones
- Using Maps to Compare Eastern Europe at Two Different Times
- Making an Outline
- Vocabulary Skills
- Writing Skills

economy, was a better social and economic system than capitalism. They rejected *capitalism,* or private ownership of the means of production, which had spurred industrialization in many Western European countries.

The Soviets' goal was to build a strong military country as quickly as possible. They accomplished this in a single generation. As a result, the Soviet Union emerged from World War II a global superpower. But in the process, the rights of many different ethnic groups in the Soviet empire were trampled by the new absolute rulers—the leaders of the Communist party—who had replaced the czars.

In 1990 the Berlin Wall was torn down, unifying Germany, and Eastern Europe secured its freedom from indirect Soviet rule. Within a year, the subject peoples of the former Soviet republics declared their independence from Moscow. Now most of these countries are attempting to build Western-style market economies; to reestablish their native religious, ethnic, and linguistic identities; and to repair the serious environmental damage that occurred under Soviet rule. Others, however, are striving to reassert communism, and the struggle between these two visions of the future is intense.

1

Russia, the largest republic in the Commonwealth of Independent States (CIS), is also the largest country in the world. It covers one-eighth of the land area of the planet.

2

The Commonwealth of Independent States contains more than 100 different ethnic groups; Russians make up only half of the commonwealth's population.

3

The Commonwealth of Independent States has three major subregions: (1) the North, (2) the Fertile Triangle (including the Baltic Republics, the Slavic Core, and the Asian Frontier), and (3) the Southern Rim, which includes the Transcaucasus and Central Asia.

KEYS TO KNOWING THE COMMONWEALTH OF INDEPENDENT STATES AND EASTERN EUROPE

4

The Commonwealth of Independent States occupies the world's largest plain, most of which lies at a latitude farther north than the border between the United States and Canada.

5

Russia's czars conquered this vast plain, created a huge Eurasian (European and Asian) empire, and imposed Russian law, language, and culture on many of its subject peoples.

6

The Soviet Union became the world's first Communist country after a revolution overthrew the last of the czars in 1917.

7

Three-quarters of the population, farmland, and industry of the former Soviet Union lie in the Slavic Core of the Fertile Triangle.

8

The Commonwealth of Independent States has the world's largest reserves of energy and a rich variety of industrial resources.

9

Eastern Europe, with its complex topography and varied climates, is located between Western Europe and the Commonwealth of Independent States.

10

The peoples and countries of Eastern Europe, now released from Soviet domination, are striving to build strong independent states.

▶

These Siberian women are wearing traditional dress and playing folk instruments. With what North American ethnic groups do they appear to share physical characteristics?

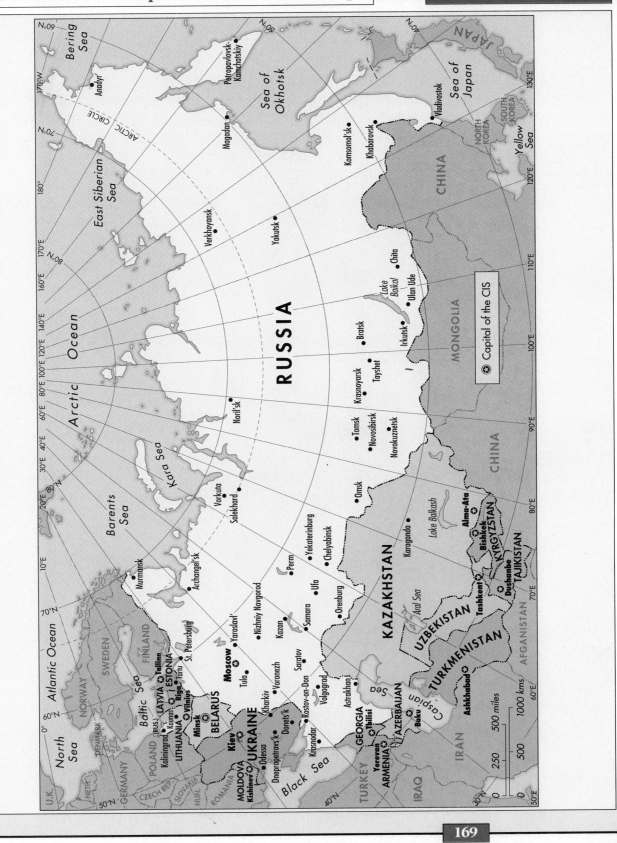

Commonwealth of Independent States Physical Map

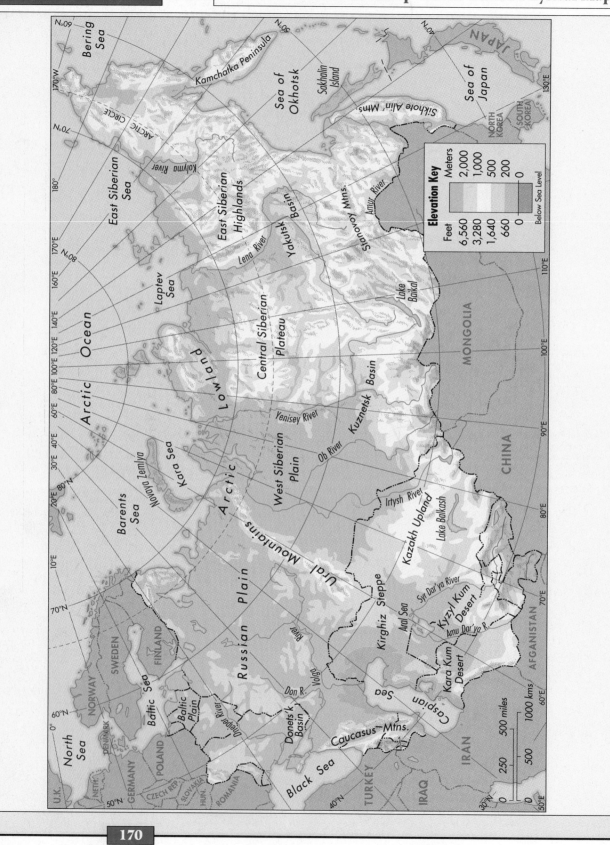

Bering Sea

Kamchatka Peninsula

ARCTIC CIRCLE

Sea of Okhotsk

Sakhalin Island

JAPAN

Sea of Japan

Sikhote Alin Mtns.

NORTH KOREA

SOUTH KOREA

East Siberian Sea

East Siberian Highlands

Kolyma River

Amur River

Stanovoy Mtns.

Elevation Key

Feet	Meters
6,560	2,000
3,280	1,000
1,640	500
660	200
0	0

Below Sea Level

Arctic Ocean

Laptev Sea

Lena River

Yakutsk Basin

Central Siberian Plateau

Lake Baikal

MONGOLIA

Lowland

Kuznetsk Basin

Yenisey River

Ob River

CHINA

Kara Sea

Novaya Zemlya

West Siberian Plain

Irtysh River

Kazakh Upland

Lake Balkash

Barents Sea

Arctic

Ural Mountains

Syr Dar'ya River

Kirghiz Steppe

Aral Sea

Kyzyl Kum Desert

AFGANISTAN

70°N

Russian Plain

River

Kara Kum Desert

Amu Dar'ya R.

SWEDEN

FINLAND

60°N

Baltic Sea

NORWAY

Dnieper River

Volga River

Don R.

Donets'k Basin

Caspian Sea

IRAN

Baltic Plain

North Sea

DENMARK

GERMANY

POLAND

NETH.

CZECH REP.

SLOVAKIA

HUN.

ROMANIA

Caucasus Mtns.

Black Sea

TURKEY

IRAQ

U.K.

1000 kms

500 miles

500

250

500

0

30°E

50°E

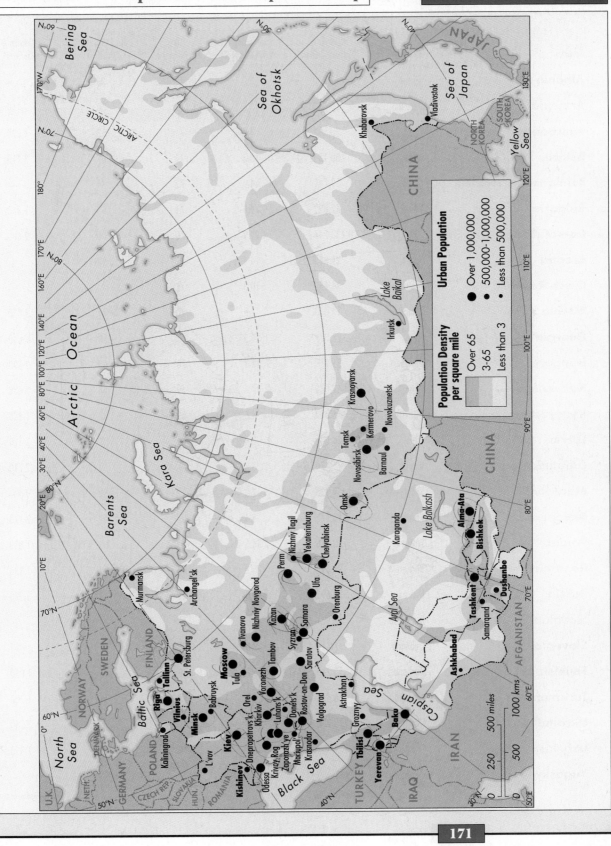

Urban Population
- ● Over 1,000,000
- ● 500,000–1,000,000
- • Less than 500,000

Population Density
per square mile
- Over 65
- 3–65
- Less than 3

Country	Capital City	Area (Square miles)	Population (Millions)	Life Expectancy	Urban Population (Percent)	Per Capita GNP (Dollars)
Albania	Tirana	11,100	3.3	72	36	—
Armenia	Yerevan	11,583	3.6	72	68	2,150
Azerbaijan	Baku	33,591	7.2	71	53	1,670
Belarus	Minsk	80,154	10.3	71	67	3,110
Bosnia and Herzogovina	Sarajevo	19,741	4.0	72	34	—
Bulgaria	Sofia	42,822	9.0	71	68	1,840
Commonwealth of Independent States	Minsk	8,433,247	294.9	70	66	2,680
Croatia	Zagreb	21,829	4.4	72	51	
Czech Republic	Prague	30,450	10.3	72	—	
Estonia	Tallinn	17,413	1.6	70	71	3,830
Georgia	Tbilisi	27,027	5.5	73	56	1,640
Hungary	Budapest	35,919	10.3	70	62	2,690
Kazakhstan	Alma-Ata	1,049,039	17.2	69	58	2,470
Kyrgyzstan	Bishkek	76,834	4.6	69	38	1,550
Latvia	Riga	24,942	2.6	70	71	3,410
Lithuania	Vilnius	25,174	3.8	71	69	2,710
Macedonia	Skopje	9,928	2.0	72	54	—
Moldova	Kishinev	13,127	4.4	69	47	2,170
Poland	Warsaw	120,726	38.5	71	62	1,830
Romania	Bucharest	91,699	23.2	70	54	1,340
Russia	Moscow	6,592,692	149.0	69	74	3,220
Slovakia	Bratislava	18,921	5.3	71	—	—
Slovenia	Ljubljana	7,719	2.0	73	49	—
Tajikistan	Dushanbe	55,213	5.7	70	31	1,050
Turkmenistan	Ashkhabad	188,418	4.0	66	45	1,700
Ukraine	Kiev	233,206	51.9	71	68	2,340
Uzbekistan	Tashkent	172,588	21.7	70	40	1,350
Yugoslavia	Belgrade	39,450	9.8	72	47	—

—Figures not available

172

THE LANDS AND PEOPLES OF THE COMMON-WEALTH OF INDEPENDENT STATES

The Commonwealth of Independent States (CIS) (formerly the Soviet Union) reaches halfway around the Earth, covering one-sixth of the land area of the planet. It is 2½ times larger than the United States, and nearly 100 times larger than Great Britain. The CIS is so large that it stretches across eleven time zones; an airplane flight across its vast expanse takes about twelve hours, and a train ride takes nearly a week.

◄ *This Tazjik man is demonstrating his strength by lifting weights during celebrations of the traditional Central Asian spring holiday. Why do you think his clothes appear to be a combination of ethnic costume and contemporary western-style dress?*

173

Most of the CIS lies north of the latitude of the United States–Canada border. Its two largest cities, Moscow (the capital of Russia) and St. Petersburg, are located as far north as central Canada. The cold temperatures in this northern region are one reason the CIS has trouble feeding its people. It also explains why, with more than 30,000 miles of coastline on three oceans, the CIS is icebound most of the year.

1. Environmental Regions

The Commonwealth of Independent States (CIS) has common borders with fifteen countries; nine of these are in Eastern Europe and six in Asia. Seven of the nine Eastern European neighbors have close economic ties with Russia. Three of its Asian neighbors have Communist governments. In the 1980s, the Soviet Union occupied one of these countries, Afghanistan, to try to support a Communist government. Its losses in men, equipment, and money

Area and Latitude Comparison of the Commonwealth of Independent States and the United States

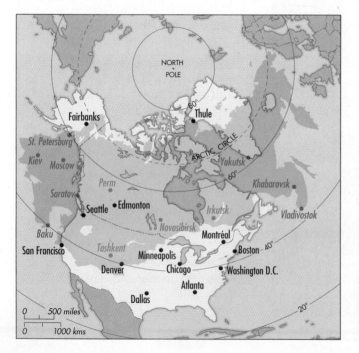

IT'S A FACT

Two-thirds of the farms in the CIS lie north of the United States–Canada border.

in the Afghan war, which lasted a decade, contributed to the 1991 breakup of the Union of Soviet Socialist Republics, or U.S.S.R., into fifteen independent republics.

Twelve of these republics are now members of the commonwealth, a loose federation that has replaced much of the U.S.S.R. The other three former Soviet republics—Estonia, Latvia, and Lithuania—are politically independent of the commonwealth, but are still heavily dependent on it for oil, natural gas, and other natural resources.

The commonwealth sprawls across the high latitudes of Europe and Asia. Minsk, the capital of Belarus and of the commonwealth, and Moscow, the capital of Russia, are located farther north than Edmonton, the northernmost city of any size in Canada. St. Petersburg is at the same latitude as Hudson Bay. Only the dry lands of Central Asia, located along the commonwealth's borders with Iran, Afghanistan, and China, reach as far south as Denver, Colorado, or San Francisco, California.

The three major subregions of the CIS are (1) the North, (2) the Fertile Triangle, and (3) the Southern Rim. These areas have difficult environments, which is why only 285 million people live in a country so vast in area.

The North

The North occupies much of an immense plain stretching 7,000 miles from Northern Europe, across Russia and the frozen wastelands of Siberia, to the Pacific Ocean. It is a cold region, open to northern winds from the Arctic Ocean but bounded on the south by a barrier of high mountains. Westerly rain-bearing winds from the Atlantic sweep across Western Europe and penetrate deep into this

IT'S A FACT   **Siberia means the "sleeping land."**

featureless plain. This Atlantic influence decreases as one moves farther into the interior of Siberia, the area east of the Ural Mountains. Here, far from the ocean, **continental climates** prevail, with their great differences in seasonal temperature. Summers are very hot, winters very cold; these ranges in temperature affect where and how people live.

Large vegetation belts stretch across the great plain that makes up much of the commonwealth. The North is a vast, frozen land—larger than either Canada or the United States—with penetrating cold temperatures that make it the least-developed area of the commonwealth. People are few and far between in this region. An estimated 5 million of them live in seclusion in this cold environment.

A **tundra** environment stretches from Europe to the Pacific along the rim of the Arctic Ocean and covers about one-tenth of the commonwealth. Winters are long and bitter. No month has an average temperature above 50° F in the tundra, even when the sun shines twenty-four hours a day in the summer.

The only trees that can grow in this cold, dry region are stunted birches. The rest of the vegetation consists of mosses, lichens, and ferns, plants that provide food for the reindeer that support small groups of herders who have long lived in these polar latitudes. Other people found in Siberia are oil drillers and miners who work at taking scarce and valuable resources from the frozen earth, and mili-

▲ *A Russian man uses his hatchet on a piece of lumber being used to build his house. Does Russia have many forests?*

tary personnel who live on bases. Prisoners sentenced to forced labor under the Soviet regime may still live in this barren land in labor camps, such as those vividly described by Alexander Solzhenitsyn in his book, *The Gulag Archipelago.*

South of the treeless tundra lies the world's largest forest, the taiga. The **taiga** is an unbroken

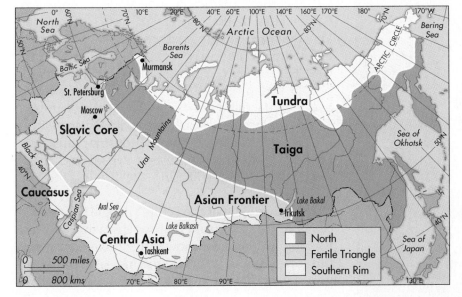

Environmental Subregions of the Commonwealth of Independent States

The Commonwealth of Independent States has three major subregions: (1) the North, (2) the Fertile Triangle, and (3) the Southern Rim.

swath of evergreen trees that stretches 4,000 miles across the country in a belt 1,000 to 2,000 miles deep. Few people live in the taiga. There is some lumbering of pine, spruce, and fir trees. Hunters trap fur-bearing animals like sable, ermine, and beaver that first lured people into the taiga. But for the most part, people stay out of the taiga. The area is so sparsely settled that forest fires sometimes rage for weeks in this wilderness before anyone notices.

The highly acid **podzol** soils of the taiga are among the poorest in the world. Much of the subsoil is permanently frozen, a condition called **permafrost**. In many places, standing bogs cover the frozen subsoils of this region. In other parts of the taiga, the land surface freezes during winter and thaws in summer, causing roads, railroad lines, and buildings to rise and fall until they break into pieces.

These conditions are moderated in the west in the Baltic states and the Slavic Core by winds from the Atlantic Ocean. Murmansk, located at a latitude equivalent to that of northern Alaska, is one of the commonwealth's only ice-free ports. Fish and forest products are shipped by rail from the Murmansk area to cities farther south.

▼ *The plains of northern Russia stretch from one horizon to another—a flat, featureless expanse of land. What part of the United States is similar to this region?*

It is estimated that two-thirds of the commonwealth's coal reserves, oil and natural gas fields, and potential hydroelectric power are located in this northern region. But the difficult environments and remote locations of the tundra and taiga make these resources expensive to develop. Tundra and taiga cover more than half of the country. In these latitudes, nature still commands.

The Fertile Triangle: The Slavic Core and The Baltics

The Fertile Triangle lies between the tundra and taiga to the north and the dry and rugged environments of the Southern Rim to the south. This is the **heartland**, the center of activity and core area of a culture, of Russia and of the commonwealth. The Fertile Triangle is broad in the west where moist Atlantic winds flow across Western Europe into Asia. It narrows eastward where the moderating effects of this Atlantic air decrease.

The most densely populated environment in the commonwealth, the Fertile Triangle is also the only large area that is suitable for farming, as the maps on pages 27 and 175 show. Other commonwealth regions are either too cold, too dry, or too mountainous to support dense farming populations.

Three-quarters of the people of the commonwealth live within the Fertile Triangle, which includes most of the four republics of the Slavic Core—Russia, Ukraine, Belarus, and eastern Moldova. The three politically independent, non-Slavic Baltic states of Estonia, Latvia, and Lithuania are also discussed in this section because of location, environment, and existing economic ties with the CIS. This combined region supports 215 million people and produces most of the country's food and manufactured goods. All five of the Commonwealth's major industrial centers are located within the Fertile Triangle. These industrial areas are Moscow (the Central Industrial Region), St. Petersburg, the Donets'k Basin, the Urals, and the Kuznetsk Basin.

The Slavic Core of the Fertile Triangle lies near the center of the great plain that sweeps across northern Europe and the commonwealth. It extends beyond the low, rounded Ural Mountains eastward to the northern borders of Kazakhstan. The Moscow subre-

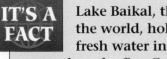

Lake Baikal, the deepest lake in the world, holds 80 percent of the fresh water in the CIS and more water than the five Great Lakes combined.

gion is a forested area that has poor soils, cold damp winters with six months of snow, and few natural resources. Moscow's location near important rivers, however, made it an early capital of Russia and later of the Soviet Union. Canals and rivers link Moscow to the Baltic, Black, and Caspian Seas, and railroads connect the Russian capital with all parts of the commonwealth. Food, raw materials, energy resources, and manufactured goods flow into the city along these transportation routes. Moscow has 9 million people, is a metropolis that serves 45 million people, and was the center of the communist world until 1991.

To the northwest, the three independent Baltic nations of Estonia, Latvia, and Lithuania occupy a northern land of bogs, lakes, and forests. This area is located so close to the Arctic Circle that the sun barely rises above the horizon in midwinter and barely sets in midsummer.

St. Petersburg, built by Peter the Great where the Neva River empties into the Baltic Sea, gives Russia a "window on the West." Rivers, canals, and railroads link this important port with Moscow to the southeast, Ukraine to the south, and the Ural Mountains to the east. St. Petersburg is a city of 5.2 million people in a subregion of 30 million. It is a manufacturing center with engineering, electronics, chemical, and metalworking industries.

Ukraine, which is also part of the Slavic Core of the Fertile Triangle, lies south of Moscow and the Baltic states. Ukraine has the most productive farmland in the commonwealth because of its moderate climate, its long growing season, and its fertile **chernozem** soils that are rich in humus. These environmental factors made Ukraine the breadbasket of the former Soviet Union. An important industrial center, the Donets'k Basin, is also located in southern Ukraine, where large iron and coal deposits are found. Nearly 52 million people live as farmers and industrial workers in Ukraine. Most are farmers who

grow potatoes and wheat and tend livestock in the north and west. The rest work in the industrial belt that encircles the Donets'k Basin. Most of these people are Ukrainians, who speak a Slavic language related to Russian and who form the second largest national group in the commonwealth.

The Fertile Triangle: The Asian Frontier

The Fertile Triangle narrows as it extends eastward from the Slavic Core. It crosses the plains of the Volga River, which form a transition zone between Europe and Asiatic Russia, then the triangle pierces the low ranges of the Ural Mountains, and extends deep into south central Siberia.

Energy and mineral resources attracted settlers to the Volga plains, the Ural Mountains, and the Kuznetsk Basin at the eastern tip of the Asian Frontier located entirely within Russia. These raw materials form the basis for industrial centers where more than 70 million people live and work. Vast hydroelectric dams built on the Volga River supply industrial energy in what was, until the collapse of the Soviet Union, the fastest growing industrial region in the country.

The vast array of minerals found in the Ural Mountains and the Kuznetsk Basin meant that industry grew more rapidly than farming. The most important minerals were high-quality iron ore in the Urals and abundant coal in the Kuznetsk Basin. A 1,200-mile-long railroad connecting the Ural iron fields and the coal fields of Kuznetsk was constructed, and steel factories were built at both ends of this railroad. Iron ore from the Urals is sent to steel mills in the Kuznetsk Basin. Coal from the Kuznetsk Basin is shuttled westward to fuel steel plants in the Urals.

On the southern margins of the Asian Frontier, in western Siberia, some settlers carved out farms on the wetter northern edges of the southern steppes. A **steppe** is a middle-latitude grassland.

The Trans-Siberian Railroad is the longest railroad in the world.

Newly built railroads made it possible for these farmers to transport and sell their grain and butter to city dwellers in the Slavic Core. But rainfall was too uncertain. As in parts of our American West, these grasslands suffered from periodic droughts and wind erosion, and frequent crop failures in this region forced many farmers to abandon their land.

Eastward, almost everyone lives within twenty miles of the only railroad that crosses eastern Siberia to the Pacific, the Trans-Siberian Railroad, which was completed in 1916. Cities based on coal and steel were built along this railroad in this frontier region, despite the great distances that separated this area from the country's main centers of population. In 1989, the 2,200-mile-long Baikal-Amur Mainline (BAM) railroad was completed to provide a politically secure, higher-speed rail link 200 miles north of the Chinese border across Russia's Far East and to open up this remote area to new development.

The Southern Rim: The Transcaucasus and Central Asia

Far to the south, the commonwealth is bordered by high mountains with peaks far higher than those of the Alps and by the deserts and steppes of the Southern Rim. In the west, between the Black Sea and the Caspian Sea, the turbulent Transcaucasus (the area south of the Caucasus Mountains) is a rugged region with fertile valleys occupied by a variety of non-Russian peoples. East of the Caspian lie the dry and rugged lands of the Southern Rim. The Southern Rim includes Georgia, Armenia, and Azerbaijan, as well as the five Muslim republics of Central Asia.

The snow-capped peaks of the Caucasus Mountains line the southern border of Russia between the Black and Caspian Seas. South of the Caucasus, thirty different ethnic groups live in this mountainous land called the Transcaucasus. Many, like the Armenians, who are now engaged in a bloody war with neighboring Azerbaijan over the Armenian enclave of Nagorno-Karabakh, have cultures far older than the Russians. An **enclave** is the territory occupied by a group of people belonging to one nation or nation-state that is located within another nation-state.

▲ *Three friends enjoy tea in a traditional tea-drinking establishment in Chaikhara, Central Asia. On the table are some round bread-like loaves. Do similar loaves appear in other cultures as well?*

The Transcaucasus, like most highlands, is a region of varied environments. In Armenia and Georgia, animal herds graze in mountain meadows below the snow line. Fields of corn dot narrow river valleys. Tea, citrus fruits, and vegetable crops are grown on south-facing mountain slopes. The mountains rise steeply from the shores of the Black Sea in Georgia and the Caspian Sea in Azerbaijan, where the climate is similar to that of central California, making this a delightful resort area. The oil fields around Baku on the Caspian Sea are of great importance to the new republic of Azerbaijan.

East of the Caspian Sea is Central Asia, a land of desert and grassland steppes occupied by Turkic- and Persian-speaking Muslim groups. These peoples were conquered by the Russians in the 1860s. Those who were nomads were forced to settle down on the land as farmers. Here, most people live at the foot of the towering mountains that separate the five independent republics of Kazakhstan, Kyrgyzstan, Tajikistan, Turkmenistan, and Uzbekistan from Iran, Afghanistan, and China to the south.

The Turkic-speaking Muslim peoples of Kazakhstan, Kyrgyzstan, Turkmenistan, and Uzbekistan, like the Persian-speaking people of Tajikistan, have kept their language and faith in spite of attempts to integrate them into Russian society. Fewer than 10 percent of the people of this subregion speak Russian. Those who do speak Russian are recent immigrants sent by the Soviet government to cities like Tashkent, the largest city in the subregion with a population of 2 million. Since the collapse of the Soviet Union, many of the Russians are returning to their original homes in the Slavic Core.

⊕ REVIEW QUESTIONS

1. The Commonwealth of Independent States (CIS) covers what fraction of the planet's land area? How much larger is it than the United States?
2. What are the tundra and the taiga, and what parts of the commonwealth do they cover?
3. Which part of the Fertile Triangle has the most productive farmland? Where is it located and what are its environmental conditions?
4. How do topography and vegetation differ in the two subregions of the Southern Rim?

⊕ THOUGHT QUESTIONS

1. How does the northern location of the Commonwealth of Independent States (CIS) help explain why it has trouble feeding its people? Do you think that environmental factors alone could explain this economic failure?
2. The Fertile Triangle is the heartland of the commonwealth. What factors would lead you to agree with this statement? Can you identify a heartland for the United States? If so, where is it located?
3. Why did the Asian Frontier experience rapid development in spite of its remote location?

2. The Creation of a Russian Empire

Russia's Czars Conquer a Vast Empire

For centuries Russia was ruled by a **czar**, an emperor or an empress whose personal power was total and absolute. In the 1300s, the Russian czars ruled only the city of Moscow and its surroundings, and governed a small state called Muscovy. Muscovy was the **culture hearth** of the Russians, the area where Russian civilization began. During the next 600 years, the czars and empresses of Muscovy spread Russian rule across Asia and created the largest empire in the world.

During these centuries, Russians "went east" just as Americans "went west." People were drawn to the east by the fur trade and by the lure of free land. In less than 100 years, Russian adventurers, explorers, soldiers, and settlers pushed 5,000 miles into the forbidding environments of northern Asia. They reached the shores of the Pacific Ocean in 1640. By 1700, Russian rule reached north to the Arctic Ocean, southeast to the mouth of the Volga River, and east across Siberia to the Pacific.

Under Peter the Great, a dynamic ruler who was determined to make Russia a great nation, Russia conquered land west of Moscow on the shores of the Baltic Sea. At enormous human cost, Peter built the city of St. Petersburg on this marshy coast. St. Petersburg enabled Russians to expand commercial and cultural contacts with West Europeans. Nearly all of present-day Russia, Ukraine, and Belarus were ruled by Peter's armies. The vast size of Russia when Peter died in 1725 is shown on the map on page 180.

By the time of the outbreak of the American Civil War in the 1860s, Russia's czar controlled an empire that stretched from the Baltic Sea in Europe

IT'S A FACT   Russia is the only large country that expanded from west to east instead of from east to west.

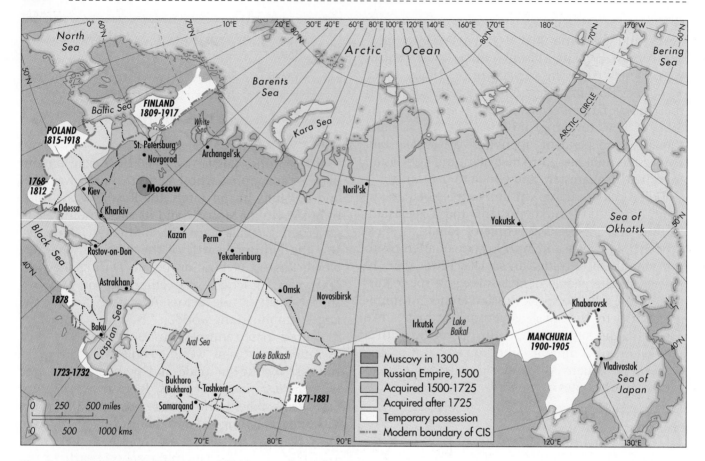

Territorial Expansion of Russia to 1914

across the entire continent of Asia to the shores of the Pacific Ocean. Many different ethnic groups with their own customs, religions, and languages lived within this empire. Less than half of the people spoke Russian, although many more knew a related Slavic language.

A Land of Poor Farmers

In 1900, Russia was a vast land which had not become as industrialized as the rest of Europe. Most Russians were poor farmers; their poverty was caused by frequent droughts, early frosts, wretched roads, high taxes, harsh landlords, and poor tools. Four of every five Russians lived in communal villages, or *mirs,* where village elders distributed land to farmers according to the size of their families.

Farming on these communally owned *mirs* was difficult, because climate was marginal for agriculture. Early frosts threatened harvests in the north.

In southern and eastern Russia, the chance of losing an entire crop to drought was three times greater than in Western Europe. Most of Russia's farmland was planted in grain crops, but food production was very low.

Russia's farming tools and techniques were inefficient. In most areas, farmers used wooden plows. Only one of every three Russian farmers owned a horse, and in many villages, men and women pulled the wooden plows because they had no animals. Harvesting and threshing grain were also done by hand. Crop yields were much lower in Russia than in Western Europe. Usually, Russian farmers ate only a little bread, cabbage, beets, and

IT'S A FACT Starting in 1500 A.D., Russia added territory to its empire at a rate of one Norway every seven years.

PLACE

ST. PETERSBURG, RUSSIA'S WINDOW ON THE WEST

Peter the Great (1682–1725) was a man of huge size, great strength, and incredible energy. More than any other czar, Peter was determined to westernize his nation's people, to change backward Russia into a modern European state.

Peter was chosen to be czar of Russia at birth, but he did not assume power until the age of twenty-two. He had an unusual education for a future ruler. During his teens, Peter devoted his energy to learning the martial arts and trained with groups of young men his own age in a series of war games. He was fascinated with ships and shipbuilding, and became a master carpenter and shipbuilder.

At age twenty-five, he disguised himself and traveled to Western Europe, where he learned of the great wealth and power that lay beyond Russia's borders.

Peter used his three major interests in warfare, ships, and Western modernization to create a stronger, more modern country. In 1709, he conquered territory along the Baltic coast, which gave Russia access to the sea. Peter founded a port city there called St. Petersburg; then he shifted the capital of Russia from the gloomy forests of Moscow in the interior to his new city on the coast.

Foreign experts were imported to build a great urban center in a swamp. Thousands of carpenters were drafted by royal decree to build St. Petersburg. Serfs from estates across Russia were brought in as laborers. Many

▲ *Peter I, known as Peter the Great, was one of many czars of Russia. Since the 1300s, the czars expanded their rule from a small area around Moscow to an empire larger than the former Soviet Union before its collapse. Does the term* **czar** *remind you of another ancient title for "leader"?*

thousands of these people died. A description of the new city captures the desperation of the builders of St. Petersburg: "On one side the sea, on the other sorrow, on the third moss, on the fourth a sigh."

St. Petersburg became a thriving port with a population of 40,000 by the time Peter died in 1725. Foreign merchants traded there, and the Russian Empire had its "window on the West." Great shipyards, personally supervised by Peter, were built along the docks to create a fleet of war and merchant ships. Named Leningrad during Communist rule, this great Russian city is once again called St. Petersburg.

In this city, Peter began his program of reforms. He modernized Russia's army and navy, established the Academy of Sciences, sponsored the first Russian newspaper, reformed the Russian alphabet, and forced landlords to break tradition and cut their beards and wear Western-style clothes. In making these changes, Peter presented Russia with a dilemma that still exists. How Western a country can, and should, Russia be?

QUESTIONS

1. What was the primary goal of Peter the Great?
2. Why did the location of St. Petersburg—its site and situation—ensure rapid growth?
3. What reforms did Peter the Great attempt to impose on his people?

In 1917 the Russian people rose up against the Czarist government. It was believed that communism would bring about a better life for all. In what major ways would life have been different after the revolution for this family?

cucumbers each day; such poor diets often led to illness and early death.

Despite these conditions, Russia's population grew to 130 million people in the early 1900s, because basic medical knowledge had trickled into Russia from the West. Village farmlands had difficulty supporting the increasing number of people, and Russia's farmers tried to escape hunger by leaving the villages. At least 4 million villagers fled to cities; millions more moved eastward along the Trans-Siberian Railroad in search of new land to cultivate. Most farmers, however, stayed in their home villages because official permission was needed in order to move, and permission was hard to get.

Hunger increased in the villages because of the czar's policy of exporting Russian wheat to Western Europe to pay for imported manufactured goods and technology. The czar's minister of agriculture declared, "We shall undereat but we will export." It was Russia's farmers who underate.

Russia Begins to Industrialize

Industry in Russia's empire began to expand in the 1880s. Railroads opened new areas for settlement and provided access to new resources. Wherever railroads were built, towns sprang up along the tracks, mines opened nearby, and land was put to the plow. Large areas that had once been isolated now became places where Russian manufactured goods could be bought, sold, or exchanged for crops.

Machinery to produce manufactured goods was imported from the West, and factories and mines were built with money borrowed from French, Belgian, and German bankers. A new center of heavy industry was built in the Donets'k Basin in southern Ukraine. The industrial center that Peter the Great had established in the Ural Mountains in earlier times was expanded. Moscow became an important trade center.

Russia was a budding industrial power by 1914 when World War I broke out. Cities like Moscow, St. Petersburg, and Kiev were growing rapidly, as Russians moved to cities in search of work. The cities, however, were poorly equipped to house and provide municipal services to these newcomers. Work-

IT'S A FACT   The average Russian farmer produces less than one-tenth as much food as the average American farmer.

ing conditions in Russia's factories and mines were dreadful, with work shifts lasting twelve hours or longer. Wages were very low. Discontent in the cities was growing fast.

After Russia was defeated by Japan in the Russo-Japanese War in the Far East in 1905, riots broke out in a number of cities. The first major defeat of a European power by non-Europeans was humiliating. Factory workers went on strike. Farmers burned manor houses, took over estates, and raided villages throughout Russia. This revolt was brutally put down by the government. Thousands of Russians were shot, and thousands more were exiled to Siberia.

Political oppression, a humiliating defeat, and economic hardship were the primary causes of these riots. In spite of the growth of factories and cities, Russia was still a country of poverty-stricken farmers; it lagged far behind Western Europe in per-capita wealth. As Russia's population continued to increase, farmers had less and less land to till. And there were not enough factory jobs to give them work.

▲ *The last of the czars, Nicholas Romanov (Nicholas II), and his family were executed in 1918 by the Bolshevik revolutionaries. Why do you think the entire family was killed?*

The Effects of World War I

These problems were made worse when Russia entered World War I in 1914 in what turned out to be an expensive disaster. Farm crops were seized by the government to pay for weapons and arms. When Russia lost battle after battle to the Germans, the czar's troops began to desert by the thousands. Hunger spread across the country.

Czar Nicholas II was trapped in an unwinnable war and an incomplete industrial revolution. As urban workers rioted in the streets and farmers seized the land in the countryside, Czar Nicholas, the last czar of Russia, was forced to abandon an empire that stretched from the Baltic Sea to the Pacific Ocean. In the midst of the chaos that followed, three exiled or imprisoned Communist leaders—V. I. Lenin, Leon Trotsky, and Joseph Stalin—returned to Russia in the spring of 1917.

⊕ REVIEW QUESTIONS

1. At the beginning of this century, the Russian Empire stretched from what sea on the west to what ocean on the east?

2. Who built St. Petersburg? Why was it called Russia's "window on the West"?

3. What were the *mirs,* and how was land distributed to farmers under the *mir* system?

4. What were the primary causes of discontent and rioting in Russia in 1905 and 1917?

⊕ THOUGHT QUESTIONS

1. Why was Peter the Great one of the most remarkable czars in Russian history? What were his principal personal interests? How did these interests affect his rule of Russia?

2. Why were most Russian farmers very poor under the rule of the czars? Do you think that a different government might have been able to overcome Russia's difficult environments? What would be the policies of such a government?

3. Why did the czar's government export wheat, even though many Russian people did not have enough food to eat? Was this a good strategy to modernize the country? Why or why not?

3. The Communists Take Control of Russia

Few suspected that Russia was on the brink of a revolution in 1917. The czar did not seem worried by Russian food shortages, protests, and demonstrations. On March 6, 1917, the British ambassador sent a cable to London from Russia stating: "Some disorders occurred today, but nothing serious." He was quite wrong. The 400-year rule of the czars over Russia was finished. A new revolution that would transform Russia into the world's first Communist country and a powerful modern state had begun.

The Communist Program for Change

Lenin returned from Switzerland, Trotsky from New York, and Stalin from a prison in Siberia. In only six months, these three Communist leaders, formed the Communist party and took over the government of Russia in the late autumn of 1917.

Russia's Communist party based its program for change on the teachings of a German political philosopher named Karl Marx. Marx believed that human misery was caused mainly by the exploitation of the poor by the rich. The capitalists, people who owned property and wealth, controlled the means of production, such as land and factories. Under **communism**, the government, acting for the people, owned the property and the means of production and decided what was produced. In theory, everyone would be given what they needed by the state. But in **capitalism**, open competition among government agencies and privately owned companies determines what is produced and who benefits.

According to Marx, capitalists used their economic power to enslave and impoverish workers.

IT'S A FACT Russia borders more countries than any other country in the world.

Marx thought that existing governments, like that of the czar, protected the power and property of the capitalists. If a better society was to emerge, a revolution was needed that would destroy the capitalistic system.

So Lenin, Trotsky, and Stalin were determined to use the absolute power of the Communist party to build a Communist state in Russia. By 1922, Russia had been renamed the Union of Soviet Socialist Republics (U.S.S.R.). A **soviet** is a committee chosen by the Communist party to run a local government. The Soviet Union was made up of a federation of fifteen independent republics run by such Communist committees.

Between 1917 and 1921, the Communist party under Lenin struggled to establish this program in Russia. First, the Communists signed a humiliating peace treaty with the Germans to get Russia out of World War I. They gave up claims to lands in Eastern Europe including Poland, Finland, and the Baltic provinces of Estonia, Latvia, and Lithuania

▼ *During the Russian Revolution, workers capture a military armored car. What colorful terms did the opposing sides use for their armies?*

that had taken the czars 200 years to conquer. Second, the Communists, or Reds, destroyed the followers of the now-dead czar, who were called Whites, in a civil war that raged for three years. By 1921, the Communists, although disliked by many Russians, had defeated the Whites, and Russia lay in ruins. Its industry, transport system, and agriculture had been shattered by seven years of foreign and civil war.

Russia Is Transformed into a Communist Country

Lenin, as leader of the Communist party, transformed Russia into a Communist country. Private property became the property of the state; factories, shops, and houses in the cities were taken over by the government. Farmland owned by the czar, the Orthodox church, and the landlords was confiscated by the government. Capitalism—private ownership of the means of production—was eliminated.

The initial impact of communism was disastrous. Factory and farm production in Russia plunged. Transportation was disrupted. Farmers began to eat more of their crops and refused to provide food to the cities. Bread was rationed in Moscow and Leningrad (the Communists' name for St. Petersburg), and factories had to be shut down because of a lack of fuel. In the face of these economic upheavals, Lenin temporarily retreated from communism in order to get Russia's crippled economy moving.

In 1921 he launched a New Economic Policy (NEP), which encouraged limited private investment in agriculture and industry. Farmers were permitted to own their own land, and they began to produce more food and to trade food in urban markets. Private investment in industry also increased. An economic revival took place in Russia because Lenin allowed some capitalistic practices that were contrary to communist principles.

Joseph Stalin: An Absolute Communist Dictator

Joseph Stalin took control of the Soviet Union after Lenin's death in 1924. Stalin believed that the Soviet Union was a country ringed by enemies who

▲ *A statue of Lenin gazes out from a crate at a world much changed from that which he helped to create. What are some of the things happening in Russia today that would not have occurred under Lenin?*

had consistently taken advantage of his country's backwardness. He was convinced that the Soviet Union had to become a strong world power immediately. To make the Soviet Union a strong military nation, Russian farmers had to be forced to feed workers in the cities and also to provide grain for export to pay for building heavy industry.

Stalin began to transform Russian agriculture and industry despite furious opposition by farmers and workers. By 1930, Stalin organized more than half of the farmland in the Soviet Union into **collective farms**, on which farmers were forced to pool their land and labor under the supervision of Communist government officials. Each farmer on the collective was paid a share of the harvest; on some state-owned farms, they were paid wages. Private or "capitalist" ownership of land, allowed under Lenin, was eliminated by Stalin. By 1937, 93 percent of all Russian farm families lived on collective farms.

The products from these farms were sold outside Russia to pay the costs of building heavy industry. About 25 million individual farms were consolidated into 250,000 collective farms closely supervised by members of the Communist party. People who resisted were relocated to work camps in Siberia. The determination of Joseph Stalin broke all opposition to collective farms.

Soviet Industry and Cities Grow

Stalin's goal of modernizing the Soviet Union under a communist system led to a similar collectivization of industry. All factories were placed under government control. Each factory was assigned a **production quota**—that is, the Soviet government ordered each factory to produce a certain amount, or quota, of goods each year. If the quota was not met, factory managers faced severe punishment.

The vast mineral resources of the Soviet Union were harnessed to provide the raw materials for industrial growth. By 1940, new coal and oil fields were brought into production in the area between the Volga River and the Ural Mountains. Hydroelectric power was developed. Minerals like aluminum, copper, zinc, tin, nickel, and magnesium were mined. New blast furnaces were built to produce more iron and steel. Some 20 million Russians moved into these industrial centers to work in expanding urban industries.

The Soviet Union Gains Land in World War II

When Lenin withdrew from World War I, Russia lost most of Poland, as well as Estonia, Latvia, and Lithuania. At the end of World War II, Soviet armies regained control over these countries from the defeated armies of Nazi Germany, and Stalin's government made Estonia, Latvia, and Lithuania republics of the U.S.S.R.

Other countries in Eastern Europe also fell under Soviet control. In addition to Poland, Soviet armies occupied Bulgaria, Romania, Hungary, Czechoslovakia, Austria, and the eastern part of Germany. In most of these countries, Stalin set up Communist governments taking orders from the Soviet Union. Austria regained total independence in 1955.

▶ *Ex-soviets mark the fortieth anniversary of the death of Joseph Stalin, the Communist leader of the Soviet Union from the late 1920s to early 1950s. What do historians say about Stalin's policies?*

Yugoslavia and Albania remained communist, but were not dependent on the Soviet Union. When Stalin died in 1953, the Soviet Union was an empire larger than that of the czars. It was also a military superpower, a dominant force in Europe and Asia, and one of the two most powerful nations on Earth.

⊕ REVIEW QUESTIONS

1. Who were the three main Communist leaders who gained control of the Russian government in the late autumn of 1917?
2. Russia's Communist party based its program for change on the teachings of what man? From what country did he come?
3. The Communist revolution led to a civil war between what two groups? Which side won?
4. Who took control of the Soviet Union after Lenin's death? Why did this man believe that farming and industry had to be reorganized? What did he do? Was he successful?
5. By the end of World War II, Soviet armies occupied what countries?

⊕ THOUGHT QUESTIONS

1. What did Karl Marx believe was the main cause of human misery, and how did he think this situation could be changed? How did these beliefs affect the actions of the Communists after they gained control of Russia?
2. Why were the Russian Communists willing to sign a humiliating peace treaty with the Germans in World War I? Do you think that these territorial changes might have affected developments in Eastern Europe after World War II? Why or why not?
3. In what way was Lenin's New Economic Policy contrary to communist principles? Why was it necessary?

 ## CHAPTER SUMMARY

The Commonwealth of Independent States, covering one-sixth of the Earth's land area, contains twelve of the fifteen republics of the former Soviet Union. The commonwealth republics can be grouped according to their environments, religious heritages, cultures, languages, and history into (1) the North, (2) the Fertile Triangle, and (3) the Southern Rim.

The three independent Baltic states of Estonia, Latvia, and Lithuania—formerly republics of the Soviet Union—are discussed as part of the Fertile Triangle. The difficult environments of the Commonwealth include the tundra and taiga belts of the North and the southern mountains and deserts of the Southern Rim. The Fertile Triangle lies between the tundra and taiga to the north and the rugged dry lands of the Southern Rim. The republics of Russia, Belarus, Ukraine, and Moldova are in the Slavic Core. The Southern Rim republics are Georgia, Armenia, and Azerbaijan in the Transcaucasus and the Muslim republics of Central Asia: Kazakhstan, Kyrgyzstan, Tajikistan, Turkmenistan, and Uzbekistan.

For centuries, Russia was ruled by czars who held absolute personal power and ruled this region from their capital in Moscow. Over a period of 600 years, the czars created the largest empire in the world by conquering neighboring lands and peoples. Under Czar Peter the Great, Russia started to industrialize. Peter conquered land to the west on the Baltic Sea and created the city of St. Petersburg to expand commercial contacts with Europe. Russia reached its greatest size when it stretched from the Baltic to the Pacific Ocean, embracing many different ethnic groups.

But Russia failed to develop its empire economically. The czars demanded more and more from rural people to build industrial centers, and participated in expensive, unwinnable wars. The people revolted in 1905 and again in 1917, when the cost of World War I drove them into desperate poverty. The Communists under Lenin, Stalin, and Trotsky took control after the 1917 revolution and a civil war and changed the name of the Russian Empire to the Union of Soviet Socialist Republics (U.S.S.R.) in 1922. Using the principles of communism developed by Marx and Lenin, the Communist party tried to industrialize the Soviet Union. Following World War II, much of Eastern Europe fell under Soviet control. The U.S.S.R. expanded to become a huge empire.

EXERCISES

Which Came First?

Directions: Two events in Russian history are described in each pair of sentences that follows. Number your paper from 1 to 5, and next to each number write down the letter of the event that occurred first.

1. (a) Czar Nicholas II ruled Russia. (b) Czar Peter the Great ruled Russia.
2. (a) St. Petersburg was built. (b) Moscow was built.
3. (a) *Mirs* were established. (b) Collective farms were established.
4. (a) The New Economic Policy was introduced. (b) Production quotas for factories were started.
5. (a) Soviet troops occupied countries in Eastern Europe. (b) The Reds defeated the Whites in the civil war.

True or False?

Directions: Write *true* or *false* on your paper for each of the following statements.

1. Russian explorers went east, just as American explorers went west.
2. Russian explorers were unable to cross the Siberian wasteland until the 1800s.
3. Since the early days of the Russian Empire, Moscow has been Russia's "window on the West."
4. Within the Russian Empire in 1860, there were many different ethnic groups, and less than half of the people spoke Russian.
5. In the days of the czars, the overwhelming majority of Russians were poor farmers.
6. Russian farmers had inefficient tools, which is one reason crop yields were much lower in Russia than in Western Europe.
7. Because of poor living conditions, Russia's population decreased in the early 1900s.
8. Before the Communist revolution, many Russian farmers tried to leave their villages and move to cities.
9. When hunger in the Russian Empire increased in the early 1900s, the government ended the export of wheat.
10. When Russia began to industrialize, most factory workers earned good wages and had satisfactory working conditions.
11. Beginning in the 1880s, railroads opened new areas for settlement and provided access to new resources.
12. The patriotism that flourished in Russia during World War I lessened the demands of the Russian people for better living conditions.
13. Russia lost much land to Germany in World War I.
14. Lenin's earliest efforts to impose communist principles in Russia met with widespread support from farmers and city workers.
15. By the end of World War II, the Soviet Union had imposed Communist occupation on many countries in Eastern Europe.

Inquiry

Directions: Combine the information in this chapter with your own ideas to answer these questions.

1. How have environmental factors shaped the distribution of people and economic activity in Russia?
2. Why can the Fertile Triangle be referred to as the heartland of the Commonwealth of Independent States?
3. In what ways was the *mir* system of landholding under the czars similar to the collective farms under the Communists?
4. When Karl Marx was living, most parts of the world had a small class of rich people and a large class of poor people. Today in the United States, Western Europe, Japan, and some other regions there is a large middle class that is neither very rich nor very poor. If this large middle class had existed in Marx's time, do you think it would have affected any of his ideas? Why or why not?
5. Russia, under its new Communist leaders, deserted the Western powers that were its allies

and surrendered to Germany before World War I ended. Do you think that this might have affected the attitudes of the Western powers toward Russia after the war? If so, how?

SKILLS

Using Historical Maps

Directions: Look at the map on page 180 showing how the Russian Empire has grown since 1300. This type of map, which shows changes that have occurred over a period of time, is called a *historical map*. Use it to answer the following questions.

1. What city on the map is located within the original part of Russia known as Muscovy?
2. Did the Russian Empire in 1500 stretch across to the Bering Sea?
3. After what year did the Aral Sea become a part of the Russian Empire?
4. What is the name of the temporary possession that covered the largest land area?
5. During what years was Finland a temporary possession of the Russian Empire?
6. Besides Moscow, what three other cities are located in the empire that existed in 1500?
7. Between 1300 and 1500, did the Russian Empire grow mainly to the northeast or to the northwest?
8. Is Kiev in the part of Russia that was acquired between 1500 and 1725?
9. Is Vladivostok in the part of Russia that was acquired between 1500 and 1725?
10. Was Manchuria acquired after 1725?

Locating the CIS's Neighbors

Directions: The following is a list of fifteen countries and seas. Without looking at a map, try to name the ten countries and seas on this list that border the Commonwealth of Independent States (CIS). After you have finished, use the political map on page 169 to check your work.

1. Great Britain
2. Sweden
3. Finland
4. Barents Sea
5. France
6. China
7. Afghanistan
8. Iran
9. Israel
10. Poland
11. Baltic Sea
12. Sea of Okhotsk
13. Caspian Sea
14. Mediterranean Sea
15. Black Sea

Learning About Distances

Directions: Use the political map on page 169 to answer the following questions about distances.

1. Is Moscow nearer to the countries of Eastern Europe or the countries of Asia?
2. Is Moscow or St. Petersburg nearer to the Baltic Sea?
3. Is Murmansk or Nizhniy Novgorod nearer to the Arctic Ocean?
4. Is Russia or Tajikistan nearer to Afghanistan?
5. Is Baku nearer to the Sea of Okhotsk or the Caspian Sea?
6. Is Vladivostok nearer to China or Finland?

Using Physical Maps

Directions: Use the physical map on page 170 to answer the following questions.

1. What is the largest lake in the Commonwealth of Independent States?
2. The Caucasus Mountains are located between what two seas?
3. Are the Eastern Siberian Highlands located in the European or the Asiatic part of the commonwealth?
4. The Volga River empties into what sea?
5. Is the Bering Sea located north or south of the Sea of Okhotsk?
6. Is the Kamchatka Peninsula in the European or the Asiatic part of the commonwealth?

7. Do the Eastern Siberian Highlands or the Caucasus Mountains spread over a larger section of land?
8. Is the Donets'k Basin nearer to the Yakutsk Basin or the Black Sea?
9. Does the Amur River flow through the European or the Asiatic part of the commonwealth?
10. Is the Kara Kum Desert found north or south of the Ural Mountains?

Vocabulary Skills

Directions: Match the numbered definitions with the vocabulary terms. Write the terms on your paper next to the numbers of the correct definitions.

5 capitalism
4 chernozem
2 collective farms
16 communism
13 continental climates
17 culture hearth
6 czar
11 enclave
8 heartland
3 mir
9 permafrost
14 podzol
15 production quota
12 soviet
9 steppe
1 taiga
10 tundra

1. The vast evergreen forests of the subarctic climate zones in North America and Eurasia
2. Farms on which workers pool their land and labor under government supervision
3. A communal village in pre-Communist Russia
4. Forest soil that is rich in humus
5. An economic system with open competition in a free market and private or corporate ownership of the means of production
6. The absolute ruler of Russia before the Communist revolution of 1917
7. Permanently frozen subsoil found in polar regions
8. The area where a culture or civilization originates
9. A middle-latitude grassland
10. A vegetation composed of mosses, lichens, and ferns that is located between the northern limit of trees and the polar regions in North America and Eurasia
11. The territory occupied by a group of people belonging to one nation-state that is located within another nation-state
12. A committee selected by the Communist party that was the local governing council in the former Soviet Union
13. Characterized by long, cold winters and short, hot summers caused by location in the interior of continents at middle to high latitudes
14. Highly acidic soils lacking nutrients for the cultivation of plants
15. The amount of goods assigned by the Soviet government that each factory must produce in a certain year
16. An economic system based on common ownership of property and the means of production
17. Center of activity and core area of a culture

ГОЛОСУЙТЕ
ЗА РЕФЕРЕНДУМ!
ЗА УЧРЕДИТЕЛЬНОЕ
СОБРАНИЕ!
ЗА НОВУЮ КОНСТИТУЦИЮ!
ЗА ПРЕЗИДЕНТСКУЮ
РЕСПУБЛИКУ!

THE COMMON-WEALTH OF INDEPENDENT STATES IN TRANSITION

I n the summer of 1991, the Union of *Soviet Socialist* Republics (U.S.S.R.) changed its name to the Union of *Sovereign* Republics. *Socialist* means government ownership of production, such as land, resources, and factories. **Sovereign** means independent, or having supreme power. This name change was an important signal of fundamental changes in the U.S.S.R.

Later that year, the U.S.S.R. broke apart. In its place, the Commonwealth of Independent

◀ *Moskovites march before the Ninth Extraordinary Congress bearing support for their leader, Boris Yeltsin, who was instrumental in bringing about the dissolution of the former Soviet Union. How many independent countries were formed from the old republics of the U.S.S.R.?*

States (CIS) eventually joined together twelve of the fifteen republics that had been part of the U.S.S.R. The Soviet empire was gone, the Communist party was discredited, and a new group of independent countries replaced the old republics of the Soviet Union. The last empire on Earth had collapsed.

1. The Collapse of the Soviet Empire

Until 1991, the Soviet Union was made up of the fifteen republics shown on the map below. Russia is the largest republic by far; most other republics are much smaller. Some non-Russian republics had become part of the Russian Empire after being conquered by the czars, others at the end of World War II. Many had been independent countries in their own rights for centuries. As a result of these conquests, more than 150 different ethnic groups lived in the Soviet Union. An **ethnic group** (1) shares common beliefs, language, and culture; (2) often lives in a particular territory; and (3) is tied together by a strong sense of national unity.

Ethnic Unrest

When the Communist revolution overthrew the last of the czars in 1917, these republics declared their independence and resisted Russian Communist rule. Wars between Russians and non-Russians raged along the shores of the Baltic Sea in European Russia and in the Asiatic realms of Central Asia and Siberia. In the early 1920s, the Russian Communists defeated these ethnic groups and reconquered the non-Russian regions of their former empire. These lands became "republics" within a federation called the Soviet Union.

In theory, each republic had the right to secede from the Soviet Union. But in fact, the Communist party and the Soviet army controlled life in the U.S.S.R. For seventy-five years the rights of non-Russian peoples were stifled. Non-Russian languages were discouraged. Russians were sent to non-Russian areas in the U.S.S.R. to spread Russian language, culture, and control. All religions were persecuted, and many churches were destroyed.

Early Communist leaders believed that ethnic differences would disappear under communism. But national, ethnic, and religious feelings remained very strong among the non-Russian peoples of the Baltic, the Ukraine, the Transcaucasus, and

Subregions of the Commonwealth of Independent States

Central Asia. These feelings were strengthened by living in a country dominated by Russian law, language, and customs.

In fact, Russians formed only half of the people of the Soviet Union. More than 40 percent of the people did not speak the Russian language as their mother tongue. Moreover, the birthrate among Russians was declining, while birthrates among the Muslim peoples of the Transcaucasus and Central Asia increased. By the end of this century, more than half of all commonwealth citizens will be non-Russians, and nearly a quarter will be Muslims. Baltic peoples are also growing in number.

Until 1991, the Soviet Union attempted to impose Russian language and culture on these people by forcing Russians to migrate to non-Russian areas of the country, such as the industrial regions on the margins of Central Asia that were viewed as politically unreliable. But state-controlled migration did not result in a blending of peoples. Russians and Muslims continued to live side by side but not together.

Language policy was another tool that the U.S.S.R. used to force non-Russians to give up their separate identities. Russian was introduced as a second language in all Soviet schools. A growing number of Ukrainians accepted Russian as their first language. But few non-Russians in the Baltic, the Transcaucasus, Central Asia, or Siberia abandoned their native languages despite the political and economic advantages to be gained through learning Russian. Indeed, when the Russians attempted to eliminate the teaching of some non-Russian languages in 1978 and 1979, people in the outer

Ethnic Makeup of the Commonwealth of Independent States

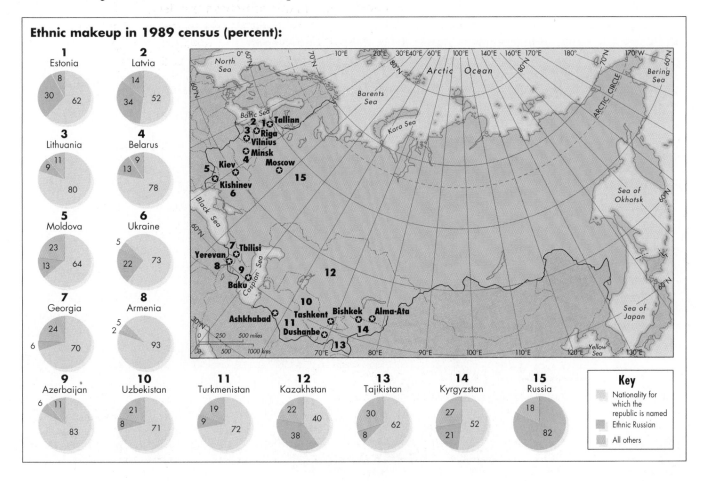

republics responded with protests and public demonstrations.

In a further effort to assimilate non-Russians, Uzbeks, Tatars, Azerbaijanis, Georgians, and other minority peoples were moved northward to work in the urban factories of European Russia. Tensions between Russians and non-Russians in these cities became intense. Similar difficulties were created by moving Russians into Estonia, Latvia, and Lithuania, which had previously been independent countries. Open conflict between Muslim recruits and Slavic-speaking officers in the Soviet armed forces that invaded Afghanistan emphasized the ethnic differences that separated the many peoples of the Soviet empire.

The Soviet Command Economy

The Communist government owned all natural resources, means of production, and financial institutions in the Soviet Union. The water, land, energy resources, minerals, factories, stores, means of transport, communication systems, and banks of the Soviet empire were state property. Private individuals, companies, and corporations played virtually no role in Soviet economic life.

The Soviet Union had a command economy. In the Soviet **command economy,** state planners and bureaucrats made all economic decisions about what was produced, what industries received resources, how much of any good was produced, and where it was produced. Planners decided on the prices of all products and how much workers would be paid to produce them. All economic decisions were made at the top in Moscow. Consumers played no role in these decisions, as in a **demand economy** like our own, in which the marketplace, or the demand for goods, determines what will be produced and how much workers will be paid.

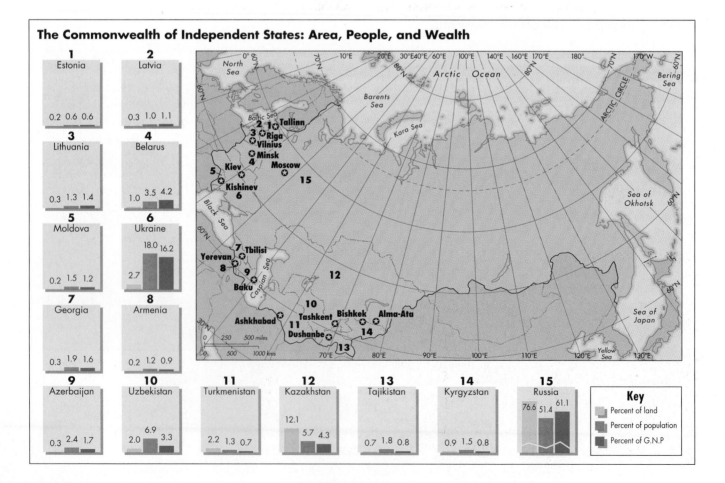

The Commonwealth of Independent States: Area, People, and Wealth

	1 Estonia	2 Latvia
	0.2 0.6 0.6	0.3 1.0 1.1
	3 Lithuania	4 Belarus
	0.3 1.3 1.4	1.0 3.5 4.2
	5 Moldova	6 Ukraine
	0.2 1.5 1.2	18.0 16.2 / 2.7
	7 Georgia	8 Armenia
	0.3 1.9 1.6	0.2 1.2 0.9

9 Azerbaijan	10 Uzbekistan	11 Turkmenistan	12 Kazakhstan	13 Tajikistan	14 Kyrgyzstan	15 Russia
0.3 2.4 1.7	2.0 6.9 3.3	2.2 1.3 0.7	12.1 5.7 4.3	0.7 1.8 0.8	0.9 1.5 0.8	76.6 51.4 61.1

Key
- Percent of land
- Percent of population
- Percent of G.N.P

> **IT'S A FACT**
>
> After you buy a car in the CIS, it takes eight years for delivery.

The resources of the Soviet Union were developed under the command system. The country became a military power because this was the principal goal of Soviet planners who perceived of themselves as being surrounded by dangerous enemies. Economic development was lopsided. Investment was made in heavy industry and energy resources for the military, and little attention was given to consumer industries that produced clothing, foodstuffs, appliances, television sets, and cars. Over time, the Soviet people became increasingly dissatisfied.

By the 1980s, economic growth in the Soviet Union had slowed down, because rigid control from the top reduced worker initiative and increased inefficiency. Lack of incentive and widespread alcoholism lowered the productivity of workers. Moreover, many Soviet products were of poor quality and could not be sold abroad. The country's woolen goods could not compete with British ones, its leather products were inferior to those from Italy, and its electronic goods were primitive by Japanese and American standards.

Soviet leaders, however, were unwilling to introduce the **profit motive,** or desire to make a profit, that spurs economic growth in Western Europe, Japan, and the United States. The profit motive implies that if people can gain more by working harder, they are motivated to be productive workers, but if the benefit to all workers is the same no matter how hard they work, they have no incentive to be creative and productive. The Soviets correctly believed that if they gave economic decision making to the people who run factories and farms, the government would also have to give up some political power. In addition, they were unwilling to admit that their command economy based on communist principles had failed to meet the needs of the people.

◄

One of the continuing effects of the breakup of the Soviet Union has been the continuation of the long lines to purchase food. Much of the food produced by farmers fails to reach stores. Why?

The Economy Declines; Ethnic Tensions Flare

When the economy of the Soviet Union began to fail in 1990 and 1991, anger against the Communist central government grew. Long-suppressed ethnic, religious, and economic resentments flared.

The Baltic republics of Estonia, Latvia, and Lithuania declared themselves independent of the Soviet Union. So did the Transcaucasus republic of Georgia and the republic of Moldova, along the Soviet Union's western border with Romania. The huge Russian Republic led by its president, Boris Yeltsin, soon followed. Resources, food, and tax money stopped flowing from the republics to the central government in Moscow.

Republics defiantly passed new laws that replaced those of the Communist central government. Some republics, including Ukraine, formed their own armies. Five republics chose to change their names to better reflect their ethnic identities. Byelorussia was renamed Belarus, Khirghizia became Kyrgyzstan, Tadzhikstan became Tajikistan, Turkmenia became Turkmenistan, and Moldavia became Moldova.

As the Soviet economy worsened and ethnic tensions increased, the central government used the Soviet army to try to keep the empire together. Soviet troops entered Lithuania and threatened to enter the two other Baltic republics of Estonia and Latvia. The Soviet army was sent to Armenia and Azerbaijan to halt ethnic fighting and to protect the lives and property of ethnic Russians. But discontent grew. Strikers closed coal mines and steel factories, and railroads were forced to shut down. Rural areas withheld food from cities, and food rationing was imposed in Leningrad for the first time since the Nazi invasion during World War II. Throughout the Soviet Union, thousands of workers stayed home. People toppled statues of Lenin in one city after another. Protestors demanded bread, soap, consumer goods, and a better standard of living.

Two Visions, Two Solutions: Mikhail Gorbachev and Boris Yeltsin

Two Soviet leaders proposed quite different solutions to the many problems facing the Soviet Union. For six years, Mikhail Gorbachev, the president of

▶

A Tadzhik Popular Front fighter in 1992 stands on an armored vehicle in Kolkhozabad, near the Tajikistan-Afghanistan border. What is he guarding?

the Soviet Union and head of the Communist party, had pursued a policy called *perestroika* or "reorganizing." Gorbachev, a dedicated Communist who joined the party as a young law school student, intended this policy to (1) make communism a more efficient economic system, (2) allow greater freedom of speech and action in the Soviet Union, and (3) keep all fifteen republics under the rule of the Communist central government.

After six years under *perestroika*, Gorbachev had failed. Membership in the Communist party declined from 20 million to 15 million in a single year, the demands of the republics for independence from Moscow increased, the standard of living of the Soviet people declined, and the empire was on the verge of ethnic and economic revolt.

An alternative vision was held by Boris Yeltsin, the newly-elected president of the Russian Republic, who saw communism as a failure. In his view, central planning did not work, and a free-market economy was needed if the Soviet Union was to solve its economic problems. In addition, he believed that the republics of the Soviet Union needed more independence from Moscow, and that those republics occupied by rebellious minorities should go their own ways. Yeltsin supported a smaller Russian Empire built around the huge Russian Republic, which, with a population of 150 million, was a superpower in its own right.

By the summer of 1991, living conditions had deteriorated and a cold winter and poor harvest were predicted. The Communist central government was losing control of the countryside, as non-Russian ethnic groups in the republics were on the verge of revolt. Gorbachev and Yeltsin attempted to solve these problems by compromise. The central government would grant greater economic and planning powers to the republics, who for their part would send less tax money to Moscow for national defense, transportation, and communications. The Russian Republic would no longer give 70 percent of its wealth to the central government, nor would

Kazakhstan send 93 percent of its commercial crops and minerals to the central government in Moscow.

Opposition to these policies was strong among the military and hard-line Communists. The military stood to lose the one-quarter of the budget of the Soviet Union that it previously controlled. The Communists, already defeated in local elections, faced a much larger defeat. If resources and money stayed in the republics, Moscow's government planners would lose their jobs.

In August of 1991, these groups attempted to overthrow Gorbachev and reimpose central Communist control on the Soviet Union. The coup failed, leaving communism at least temporarily in disgrace. Military leaders were replaced. The republics declared independence, and Yeltsin assumed the reins of power. In one of the most dramatic events of the twentieth century, the Soviet Union collapsed and a Commonwealth of Independent States (CIS) centered at the city of Minsk in Belarus was created.

⊕ REVIEW QUESTIONS

1. What is an ethnic group, and how many ethnic groups lived in the Soviet Union?
2. What is a command economy and how does this differ from a demand economy?
3. How did the Soviet Union attempt to control its ethnic groups?
4. What policies did Mikhail Gorbachev and Boris Yeltsin represent, and how did these policies differ?

⊕ THOUGHT QUESTIONS

1. What were the factors that contributed to the collapse of the Soviet Union? Had you been in charge of the Soviet Union, what actions would you have taken to prevent this from happening?
2. Given this background, what future do you envision for this region? Do you believe that communism will once again become the dominant political and economic system, or do you believe that a free-market economy will be established?
3. Are there other world regions where ethnic differences are becoming important? If so, what regions and in what ways?

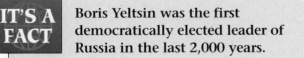

IT'S A FACT   Boris Yeltsin was the first democratically elected leader of Russia in the last 2,000 years.

2. The Commonwealth of Independent States

The Commonwealth of Independent States (CIS) is composed of twelve of the fifteen republics of the now-defunct Soviet Union, although the remaining three republics—Estonia, Latvia, and Lithuania—retain close economic ties with the CIS.

Basically, the fifteen former republics of the Soviet Union can be grouped into four cultural subregions: (1) the three independent Baltic republics of Estonia, Latvia, and Lithuania; (2) the four Slavic-speaking republics of Russia, Belarus, and Ukraine, and the eastern part of Moldova; (3) the three republics of the Transcaucasus region, Armenia, Azerbaijan, and Georgia; and (4) the five Central Asian republics of Kazakhstan, Kyrgyzstan, Tajikistan, Turkmenistan, and Uzbekistan, which form the Southern Rim of the commonwealth. Locate these fifteen republics that were once part of the Soviet Union on the map on page 169.

Differences Within the Former Republic

The essential basis for holding any union of peoples together is a common set of values. In many countries, such as France and Germany, values are based on a common history and language. The U.S.S.R., however, had no common language or set of values. As noted earlier, more than 150 different peoples lived in this empire. These peoples had different histories, languages, cultures, and religions. A large number were persecuted by the Russian army and the Communist party for seventy-five years. Most had been conquered by the Russians.

Early Communist leaders like Lenin and Stalin believed that ethnic, religious, and linguistic loyal-

 IT'S A FACT   The CIS is big enough to accommodate 2½ United States of Americas, 40 Frances, or 92 Great Britains.

ties would disappear under communism, as noted earlier. They believed that communism would create a "new Soviet citizen" to whom these cultural differences would no longer matter. This never happened. Until the end, passports in the U.S.S.R. identified every Soviet person as an "Armenian," a "Georgian," or a "Tajik" rather than as a Soviet citizen.

Today, the fifteen former republics of the Soviet Union, now independent countries, differ tremendously in area, population, and ethnic composition. Together, they form a sprawling multinational empire that embraces one-sixth of the land surface of the Earth. Russia alone extends across eleven time zones, and is more than twice the size of the United States. With a population approaching 150 million people, Russia is the dominant republic in the new Commonwealth of Independent States. By contrast, the three Baltic republics of Estonia, Latvia, and Lithuania are so small they could easily fit within New England. These three republics have a combined population of 8 million, smaller than that of the state of Florida.

Moreover, most republics in the new Commonwealth of Independent States have a variety of peoples within their borders. In nine of the fifteen former republics, the largest ethnic group makes up less than two-thirds of the total population. In Kazakhstan and Latvia, for example, Russians form a substantial portion of the population. Who should be accepted as a citizen in these new countries has already become an issue. The Russian Republic under Boris Yeltsin has declared that it is prepared to protect the 25 million ethnic Russians that live outside Russia in the "near abroad," as Russia refers to the fourteen other former Soviet republics. As border disputes and ethnic conflicts increase, ancient passions stifled by Communist military rule are rising to the surface in the subregions of the commonwealth.

The Baltic Republics of Estonia, Latvia, and Lithuania

The small Baltic republics of Estonia, Latvia, and Lithuania lie north and west of Moscow on the shores of the Baltic Sea. For several hundred years, these three republics were part of the Russian Empire under the czars. After World War I, they became independent countries. Their location, squeezed between Germany and Russia, however, meant their independence was short lived. In 1940, after less than twenty years of freedom, the Baltic states were taken over by the Soviet Union when Adolf Hitler signed a secret treaty with Joseph Stalin.

When Lithuania declared independence in 1990, street fighting broke out between the local militia and Soviet troops in the Lithuanian capital of Vilnius. Frequent demonstrations soon occurred in the neighboring capital cities of Tallinn in Estonia and Riga in Latvia. All three republics withdrew from the former Soviet Union in 1991, even though sizeable Russian minorities still live in these countries.

The Baltic republics are located on poor, swampy, glaciated land. Their well-developed economies are based mainly on farming, fishing, and shipbuilding. In recent years, increased trade and industry have raised standards of living in this subregion. With their borders now guaranteed by Poland to the west and Russia to the east, the Baltic states are among the best prepared former republics to develop independent economies and societies. They now, however, must import grain and oil from Russia at much higher prices than when they were part of the Soviet Union.

All three Baltic states have moved toward establishing market economies and closer ties with Western Europe. Part of this process includes redirecting their export-oriented economics to meet local needs. Tensions remain over the remnants of the Soviet army that still are stationed in this subregion, although by 1994 only 18,000 soldiers were

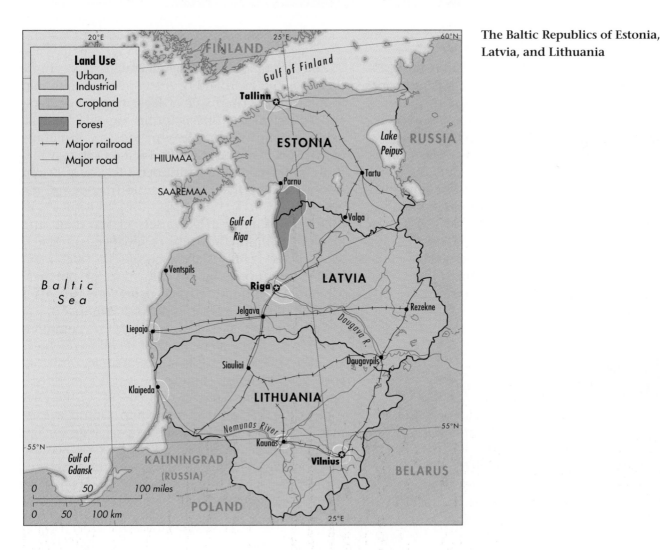

The Baltic Republics of Estonia, Latvia, and Lithuania

▶

The beautiful city of Riga is the capital of Latvia. What part did Latvia play in the collapse of the Soviet empire?

left in Estonia and Latvia. More difficult is the question of large minorities of Russian-speaking workers, many of whom settled this region. In some areas the Russians are now treated as second-class citizens and barred from citizenship until they pass difficult local language examinations. This issue is particularly acute since Russia is not only the Baltic states' greatest political concern, it is also potentially their most profitable trading partner. Antagonizing Russia is not to their advantage.

Although some attempts have been made at unifying the Baltic states, these small, vulnerable countries on the edge of Western Europe are very different places, each with its own language, culture, and history.

The Slavic Core: Russia, Belarus, Ukraine, and Moldova

The three Slavic-speaking republics of Russia, Belarus, and Ukraine formed the heartland of the Soviet Union. Together, these three republics have 211 million people. The commonwealth's most important industrial areas, all of its largest cities, and its major cultural centers are located in this Slavic-speaking region.

The Federation of Russian Republics (or simply Russia) is the heartland of the commonwealth and home to more than half of its people. The largest and richest of the fifteen former republics, Russia extends from its capital city of Moscow in Europe eastward across Siberia to the Pacific Ocean. A majority of the commonwealth's ethnic groups live in Russia, many of them in the Fertile Triangle and the North.

Russia includes three-quarters of the total land area of the Commonwealth of Independent States. Most of the Fertile Triangle occupies a huge plain briefly interrupted by the low rises of the Ural Mountains. Tundra and evergreen forests blanket northern Russia. Grasslands lie to the south.

This country produces half of the commonwealth's grain and steel. Most of its other resources are also located in Russia—three-quarters of its oil, coal, gas, diamonds, and gold. Russia's president, Boris Yeltsin, has asserted local control of these resources; he has also attempted to set up a free-market economy and maintains a separate foreign policy from the rest of the commonwealth. Because this country produces 70 percent of the wealth of the former Soviet Union, its independent control of these resources eliminated the power of the central government of the former Soviet Union. The economic and social changes now occurring in Russia echo throughout the region.

Belarus and Ukraine lie to the south and west of Russia. They have a combined population of 62 million people. The capital of Belarus is Minsk; this city is also the capital of the Commonwealth of Independent States. Minsk is an important industrial center and rail junction connecting the commonwealth with Poland. Much of Belarus has poor soils and few mineral resources, but major gas and oil pipelines passing through the country provide the raw materials for its oil refineries and petrochemical industries. Its location permits farming of flax and potatoes and some grazing. But it is still a cool, damp, marshy region. Because of its rural, traditional character, Belarus remains a communist state with a communist economy despite the changes initiated elsewhere in the commonwealth.

Farther south, Ukraine, which is a bit smaller than Texas, is the breadbasket of the Commonwealth of Independent States. It is the commonwealth's second most populous country and one of

The Slavic Core: Russia, Belarus, Ukraine, and Moldova

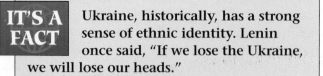

IT'S A FACT Ukraine, historically, has a strong sense of ethnic identity. Lenin once said, "If we lose the Ukraine, we will lose our heads."

its most important areas of heavy industry. Ukraine's fertile soils, long growing season, and mild winters make it a productive agricultural area. Ukraine grows a quarter of the commonwealth's food. Large iron and coal deposits located in the Donets'k Basin enable Ukraine to produce a quarter of the commonwealth's industrial output. These resources support one of the most important steel-producing centers in the commonwealth. Kiev, the capital of Ukraine, is the center of the Russian Orthodox Church. It is an important cultural, religious, and educational center. Large-scale demonstrations in Kiev forced Moscow to allow freedom of religion in the country in the early 1990s, but Ukraine has failed to deal with the economic changes brought by independence.

The failure of the central government to deal effectively with the nuclear disaster at the city of Chernobyl, north of Kiev, which spread radioactive material over a large area in 1986, adds to local unrest in Ukraine. Its government originally demanded destruction of all nuclear weapons located on Ukrainian territory but also asserts Ukrainian control of all of its resources. Ukraine has begun to issue its own currency and to establish its own armed forces. Because Ukraine claimed the former Soviet Union's navy in the Black Sea, the Russians had to negotiate to control half of it. Russia also now has right of access to the Crimea, which became Ukrainian territory in 1954. Although these issues have been partially resolved between Russia and Ukraine, the new republic has not been able to bring about major change. Its economy is declining, its external debt to Russia is substantial, and it is still dependent on Russia for energy and other raw materials.

Moldova, located north of the Black Sea on the border of Romania, has been the scene of ethnic violence. Much of this republic was part of Romania

▼ *This farm in the Ukraine was founded in 1989 and named "Iowa," after its owner returned from that state with American agricultural knowledge. Why is Ukraine so important to the people of Russia?*

until conquered by the Soviet Union in World War II. Many Moldovans are Romanians by birth and hope to rejoin Romania. East of the Dniester River, however, Russians and Ukrainians form a majority. Battles between ethnic Moldovans and small groups of Russians and Ukrainians have hampered development. Russian troops took over the army of this tiny republic in 1992 to stop the violence and to prevent Moldova from unifying with neighboring Romania.

The Transcaucasian Republics of Armenia, Azerbaijan, and Georgia

Armenia, Azerbaijan, and Georgia lie south of the Caucasus Mountains between the Black Sea and the Caspian Sea. This area, one of the world's most ethnically diverse because of complex geography, is home to twenty to thirty different ethnic groups. The three largest groups—Armenians, Azerbaijanis, and Georgians—have a combined population of 16 million. The Azerbaijanis are Turkic-speaking Muslims. The Armenians have their own ancient Christian church, as do the Georgians. Each group speaks its own language, and each group is fiercely independent. All three countries are members of the commonwealth.

Disputes over ancient religious and ethnic differences broke out in the Transcaucasus soon after the collapse of the Soviet Union. Armenians and Azerbaijanis are at war over the rights of Armenians living in Nagorno-Karabakh, an Armenian enclave in Azerbaijan. Russian troops have fought on both sides at different times. Violent incidents along the border between Azerbaijan and Iran are of great concern to the commonwealth. So is the high-grade oil that is produced at Baku, the capital city of Azerbaijan.

The republic of Georgia lies north of Armenia and Azerbaijan and east of the Black Sea. It often has been conquered by larger groups, such as the Romans nearly two thousand years ago and the Iranians in the 1600s. In the early 1800s, Georgia became part of the Russian Empire. It declared its independence at the time of the Communist revolution in 1917 but was reconquered in the 1920s. Throughout their history, Georgians have stubborn-

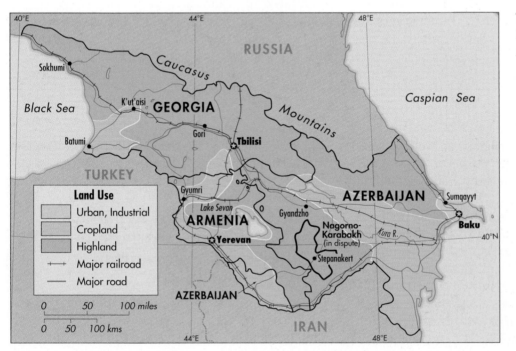

The Transcaucasian Republics of Armenia, Azerbaijan, and Georgia

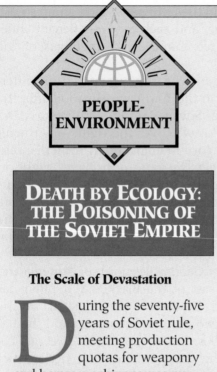

DEATH BY ECOLOGY: THE POISONING OF THE SOVIET EMPIRE

The Scale of Devastation

During the seventy-five years of Soviet rule, meeting production quotas for weaponry and heavy machinery was pursued without concern for protecting the environment. The extent of the ecological devastation that occurred is difficult to imagine. Three-fourths of the surface water in the Commonwealth of Independent States is contaminated, and half of the farmland has been damaged by salinization, chemical contamination, or erosion. In addition, approximately 12,000 containers of radioactive waste were dumped into the Arctic Ocean. Nearly half the children born in the former Soviet Union have some sort of birth defect. Conditions became so bad that rebellion against environmental degradation played an important role in the overthrow of Soviet rule.

Today, only 15 percent of the urban population lives in areas free of hazardous pollution; 175 million city dwellers do not. To save money, toxic chemicals are burned at one-half the temperature needed to destroy them, leaving cancer-causing agents in the air above 162 cities. This economy dominated by smokestacks has killed many and will kill more. As one former Soviet minister of health noted, "The way to live longer in Russia is to breathe less."

The Aral Sea

One extreme example of the Soviet planners' unreasonable focus on production quotas is the story of the Aral Sea, once the world's fourth largest lake. In the last thirty years, this sea has lost 60% of its water and almost all of its marine life. Only four species of fish remain capable of tolerating the salty water. The major coastal fishing town of Aral'sk, which used to be on this body of water, is now located 60 miles away from the lake's receding shoreline. Villages, fishing fleets, and large ships are stranded on giant salt flats that were located on the coast of the Aral Sea.

How did this happen? The planners in Moscow decided that the republics of Central Asia should produce cotton, so the

ly kept their culture and religion, as well as a language with its own alphabet; however, many Georgians speak Russian because they trade throughout the entire country. In 1991, Georgians voted overwhelmingly for total independence. Since then the country has collapsed into civil war and, in return for Russian military assistance, joined the commonwealth. Rebels in Abkhazia in western Georgia attempted to overthrow the central government at Tbilisi. In the north, Muslims along the Russian border started a rebellion, as did Kurds in the south. Georgia's larger neighbors—Turkey, Iran, and Russia—have interests in this conflict-ridden region, which has always been a place where rivals meet.

The Central Asian Republics of Kazakhstan, Kyrgyzstan, Tajikistan, Turkmenistan, and Uzbekistan

Central Asia stretches 2,000 miles along the southern border of the Commonwealth of Independent States from the Caspian Sea to the frontiers of

The Aral Sea has sustained extensive ecological damage as a result of Soviet policies. What has happened to it?

face a devastating shortage of water and arable land because of the environmental policies of the former Soviet regime.

QUESTIONS

1. What are some of the environmental effects of the Soviet policies over the last seventy-five years?
2. How has the relative location of Aral'sk changed in the last twenty years?
3. Knowing the probable consequences, why do you think the government of Uzbekistan continues to produce cotton?

two major rivers that fed the Aral Sea—the Amu Dar'ya and the Syr Dar'ya—were diverted by a maze of canals to irrigate cotton fields.

So cotton became the main cash crop of Uzbekistan, whose people were trapped in a cycle of wasteful production. Even after independence, Uzbekistan must continue to produce cotton al-though the cost is high. Half of Uzbekistan's land is deteriorating, as salt dust from the Aral Sea (75,000 tons of it each year) blows over it.

The damage has been done. By the end of the century, the Aral Sea will be a series of small lakes; the cotton fields will be desert. The peoples of Central Asia will

China. The people of this region are Turkic- and Persian-speaking Muslims. They were conquered by the Russian czars in the 1860s, and in the 1920s their territory was divided into five republics that are now independent countries—Kazakhstan, Kyrgyzstan, Tajikistan, Turkmenistan, and Uzbekistan. Kazakhstan, the largest of these republics, is nearly four times the size of Texas. Uzbekistan and Kazakhstan are the most populous countries in Central Asia. Over 50 million people live in this region.

Central Asia is a region that deeply concerns some members of the Commonwealth of Independent States because of its rapidly increasing population. The area is expected to have 100 million people twenty years from now. By that time, Russians will be a minority in the former Soviet empire. Muslims from Central Asia now make up a large portion of all soldiers in the Soviet army, but few Muslims are officers.

The Muslim religion is very important to the people of Central Asia. Mosques (Muslim places of worship) and religious schools have grown in number from 160 to over 5,000 in the last two years. Few Kazakhs, Kyrgyz, Tajiks, Turkmens, or Uzbeks

The Central Asian Republics of Kazakhstan, Kyrgyzstan, Tajikistan, Turkmenistan, and Uzbekistan

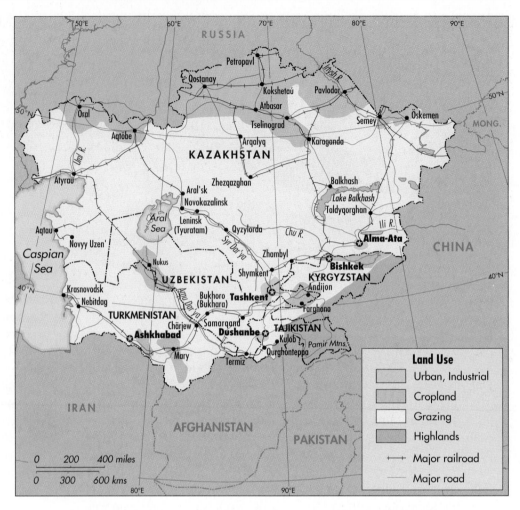

speak Russian. Although many Russians have migrated to industrial cities in Central Asia, particularly in Kyrgyzstan and Kazakhstan, this has not diminished the strong ethnic and religious feelings of the five groups that still make up a majority of the population of this region.

Deserts and steppes reach from the republics of Turkmenistan and Uzbekistan on the shores of the Caspian and Aral Seas north and eastward across Kazakhstan. Most people live in cities and villages in the fertile valleys of two major rivers. These rivers pour north and west down the slopes of the mountains of Tajikistan and Kyrgyzstan, which are located to the south and east, and empty into the Aral Sea. In the past, most of the people of the region were nomadic herders. But in the 1920s, they

were forced to settle down and work on collective farms where they grew cotton and other crops on irrigated land. Cotton irrigation severely decreased the amount of water reaching the Aral Sea, where the water level has dropped fifty feet in the last thirty years. Conflicts over water use are becoming more frequent. To make things worse, these republics are among the poorest in the Commonwealth of Independent States.

Ancient bazaars and mosques in trading centers like Bukhoro (Bukhara), Samarqand, and Tashkent contrast with modern factories and mines introduced into this region by the Russians. Kazakhstan has the commonwealth's most important copper mines, giant petroleum fields near the Caspian Sea, and the important coal deposits of Karaganda. This

▲ *Samarqand, Uzbekistan, was once the start of the famous Silk Road to China. Samarqand has been a part of many empires in history. Which empire gave it an Islamic heritage, as seen in the architecture in this square?*

republic has aggressively pursued contracts with Western corporations to develop these natural resources.

The Russians managed to keep an uneasy balance among the different ethnic groups in these five republics for many years. Now all five are independent, and ethnic fighting has engulfed the region. Riots between Uzbeks and Kyrgyz over land and water rights caused the Russians to send in troops to stop the violence. Tajikistan is in the middle of a civil war. Despite this region's poverty, cultural differences, and overpopulation, some of its leaders dream of a new Central Asian empire based on the Muslim religion. Whether this dream will become reality is highly uncertain.

⊕ REVIEW QUESTIONS

1. What countries are members of the Commonwealth of Independent States? Name three former Soviet republics that chose not to join this commonwealth.
2. What countries are located in the Transcaucasus? What is the major problem in this region?
3. Which republic is the largest and most important in the commonwealth? Why is this so?

⊕ THOUGHT QUESTIONS

1. What two factors caused the collapse of the Soviet Union? Are either of these important problems in the United States? Explain your answer.
2. The Soviet Union became home to many different ethnic groups, as did the United States. Why have the ethnic groups in the commonwealth reacted so strongly to reassert their cultural identities? Do you believe this could happen in the United States? Why or why not?
3. Geographers often debate whether Russia is European or Asian. What is your opinion and why?

WILL RUSSIA FALL APART?

At the core of the Commonwealth of Independent States (CIS) is the Russian Republic, which contains three-quarters of the area and half the population of the former Soviet Union. Despite the current independence of the fourteen other republics that formerly combined with Russia to make up the Soviet Union, a glance at the map shows that these former republics are located on the fringes of the old Soviet Empire in the Baltic, Transcaucasus, and Central Asian subregions. The center is still Russia, a country which is twice as large as the United States.

The Russian Republic was distinct from the beginning. Unlike the other fourteen Soviet Socialist Republics in the former Soviet Union, Russia was, and is, a huge multinational federation, with its own internal republics and smaller administrative units, 90 in number. The people in these republics and lesser administrative units are members of 100 minority ethnic groups, 31 of which have their own autonomous republics (states), *oblasts* (counties), *krays* (districts), or independent cities. This mix of peoples and patchwork of territories are the result of the growth of the Russian Empire by war and conquest. Russia today is a country among countries.

Moscow has lost economic dominance over the fourteen adjacent republics that made up the Soviet Union; communism has been at least partially discredited; ethnic minorities, like the Latvians, Estonians, and Lithuanians, have successfully declared their independence from Russian rule. Now the 21 interior republics are beginning to demand a greater economic share of their own resources and less Russian influence in cultural matters. They are making the same demand as the former republics of the Soviet Union.

Indeed in 1993, two new regions declared themselves to be republics—the Urals Republic located east of the mountains and centered in the city of Yekaterinburg, and the Maritime, or Far East, Republic centered on the Pacific port of Vladivostok. These political decisions bring the number of republics within Russia to 23. Hence the international concern that Russia, like the Soviet Union before it, may break apart into a group of small, ethnically-based independent countries.

The Republics Inside Russia

The republics located within Russia are widely distributed across its vast landscape. On the basis of geographical location, they can be described in three groupings: (1) the seven Republics of the Fertile Triangle, (2) the eight Republics of the Caucasus, and (3) the eight Republics of the North and East.

Republics of the Fertile Triangle

In the Fertile Triangle, across the Volga Plains and south of the Ural Mountains, lie the six republics of Mordvinia, Chuvashia, Mari-El, Tatarstan, Udmurtia, and Bashkortostan with a total population of 12.5 million. Also found in this region is the newly self-declared Urals Republic described above. Each of these areas is defined as the homeland of a non-Russian ethnic minority, but, given their location, all have a substan-

Dmitri Popov shows off his contest-winning national costume for a folk festival. Judging by the style of his costume, from what part of Russia does he originate?

Republics in Russia

tial minority of Russians who, until recently, controlled local cultures and economies.

Mordvinia, Udmurtia, and Mari-El are occupied by people who have ancient ties to Finnish-speaking peoples, although all of these peoples are outnumbered by Russians in their own republics. Mordvinia is principally an agricultural area, Mari-El is an important timber region, and Udmurtia is highly industrialized. None of these republics has made significant demands on Moscow.

Two of the remaining Fertile Triangle republics—Chuvash and Bashkortostan—are peopled by Turkic-speaking groups. The Chuvash are a majority within their re-

public; the Bashkort are not. The Chuvash are principally farmers and herders, whereas the economy of Bashkortostan is based on its huge oil-refining facilities and associated industries, as well as on a variety of natural resources ranging from timber to copper.

By far the most contentious republic in the Fertile Triangle, and indeed within Russia, has been Tatarstan, home of the descendants of the Mongols who swept across the plains of Russia nearly nine hundred years ago. The Tatars controlled much of Russia until they were defeated by Ivan the Terrible in 1552, losing their major city—Kazan. Now, these heirs of the Golden Horde are re-

claiming their culture and their land, though they make up only half of the population of Tatarstan.

As the Soviet Union disintegrated, the Tatars claimed sovereignty. They withheld tax revenues from Moscow and began controlling the export of nearly one-fifth of their annual 25-million-ton production of oil. Finally, in 1994, a landmark compromise treaty was signed between Russia and Tatarstan. It has been hailed as a model for normalizing Russia's unstable relationships with its 90 administrative subdivisions.

The treaty recognizes both the constitution of Russia and also the

continued on page 210

one adopted by vote by Tatarstan in 1992; local claims to sovereignty are dropped. In return, Russia recognizes Tatarstan's right to manage most of its natural resources, to conduct foreign trade, and to have its young men exempted from service in the Russian army. The treaty has apparently succeeded in appeasing local grievances, while avoiding broad demands for independence.

Republics of the Caucasus

Historically, the high mountains of the Caucasus have served as a refuge of peoples. The eight republics of this region have a total population of less than 8 million people, but their location on a narrow land bridge, between the Caspian and Black Seas, that connects Russia with Iran and Turkey (through the contentious CIS republics of Azerbaijan and Georgia) makes them important well beyond their size and population.

Six of the eight republics in the Caucasus are located in the high mountain valleys or on the slopes of the Caucasus Mountains. These are the states of Chenchenya, Ingushetia, North Ossetia, Kabardino-Balkaria, Karachayevo-Cherkessia, and Adyega. The two additional republics of Dagestan and Kalmykia border the Caspian Sea.

Chenchenya, Ingushetia, and North Ossetia have a complicated

history. The Chenchen and the Inguish, both Muslim groups, repelled Cossack attacks and Russian rule in the 1870s but were finally conquered. The two groups shared a republic until 1991, when Chenchen separatists attempted to take control of the republic.

At the time, the Inguish, who occupied high mountain valleys in the Caucasus, were being oppressed by neighboring Ossetians (themselves Muslims) who began a campaign of "ethnic cleansing" against their mountain neighbors in an effort to reunify historic land claimed by the Ossetian people.

In December 1994, tensions in this region deteriorated into open warfare. Russia sent tanks into Chenchenya, leveling the capital city of Grozny and killing an estimated 30,000 people. The Chenchenyan rebellion continues, however, taking the form of terrorist assaults on Russian territory in 1995 and 1996.

The mountain republics of Kabardino-Balkaria and Krachayevo-Cherkessia include a variety of minority groups including the Kabardinians, many of whom are of Cossack descent, the Turkic Karachay people, and the Christian Cherkess. Farther to the west, the Adygey Republic is occupied by people related to the Cherkess. All four of these republics are principally involved in primary activities like farming, herding, and logging.

The two remaining Caucasus republics of Kalmykia and Dagestan reflect the role of this region as a refuge area. Kalmykia is occupied

by nomads of Mongol descent who practice the Tibetan Buddhist religion, live in yurts, and are primarily livestock herders. Although the president of this republic demanded greater attention from Russia and other world powers, claiming to be "Europe's only Buddhist state," few have paid attention.

Dagestan, the last of the Caucasus republics, is occupied by many different national groups. It is important to Russia because of its location, abutting Azerbaijan to the south, and its substantial reserves of oil and natural gas.

The Republics of the North and East

The population of the eight republics of the north and east of Russia is somewhere between 6 and 10 million. Few people have chosen to live in these distant regions (voluntarily); the importance to Russia of these regions is either their strategic location or their resources. The three northern republics of Karelia, Komi, and Sakha (Yakutia) all have a Russian majority within their territories.

Karelia, located north of St. Petersburg and east of Finland, is a cold, forested and glaciated wilderness whose principal importance is its location on Russia's western border. Interestingly, the Karelians, themselves a Finnish-speaking people, make up only a tenth of the population. Similarly, the Komi Republic, which is located in the arctic region of Russia, west of the Ural Mountains, has a Finnish-speaking minority that amounts to roughly one-quarter of the population. Forest resources have been the principal industry of this region in the past, but coal, oil, and natural gas reserves have been discovered. The third northern republic of Sakha or Yakutia is an enormous region of harsh climate, poor soils, and permafrost. The Yakuts, a Mongol people, make up a third of the people of this area. Their control of large gold and diamond resources has brought them into conflict with Moscow, as was the case in Tatarstan.

Four of the five republics of the south and east (the self-declared Maritime, or Far East, Republic excepted) are historical remnants. Altaya, Khakassia, Buryatia, and Tuva (or Tannu Tuva) are home to small, non-Russian peoples of eastern origin and have a combined population of between 2 and 3 million people.

Altaya, a land of cattle breeders with a distinctive culture and society, is located where the borders of China, Kazakhstan, and Mongolia meet. To the north, Khakassia is occupied by Turkic-speaking nomads who tend livestock and farm the valleys in this stark environment. To the east, Buryatia, on the shores of Lake Baikal, was originally settled by Buddhist Mongols, but Russians now make up three-quarters of the population because minerals were found in this region. Tuva, famous in the past for its stamps, is occupied by Turkic-speaking pastoralists, who declared themselves independent of both China and Russia, but were annexed by Russia after World War II.

This class is studying dance at a cultural college in Salekard, Siberia. The school teaches music, art, and other vocations to Russian ethnic groups. Why would a school in Russia specialize in cultural courses for ethnic groups?

The declaration of self-rule by the Maritime, or Far East, Republic may be a significant development in Russia because of the region's strategic location on the shores of the Pacific. Russia's ongoing territorial dispute with Japan may be affected, as well as the export economy of this distant and previously inaccessible region.

Population, Resources, and Politics

The population of these diverse republics within Russia amounts to less than 30 million people, only a fifth of Russia's total. Moreover they are scattered across the huge spaces of the country, and are not unified by language, race, or religion. Despite these realities, which were also true of the republics that separated from the former Soviet Union, their tenacious resistance to Russian pressure, and their determination to retain their own culture and faiths, reflect weakness within this single largest remnant of the Russian Empire—Russia itself.

Accommodations are needed and are being made between Moscow and these small groups of previously conquered people. Although small in number, several occupy strategic locations, others control important resources. Beyond these practical concerns, however, most of these people simply do not want to be Russians. The ancient ties of blood, language, and faith are as strong in Russia as they have proved to be in the former Soviet Union and Eastern Europe.

QUESTIONS

1. How did all these diverse peoples come to be part of the Russian Federation?
2. How many interior republics are there? How many of them are self-declared?
3. What characteristics do these republics have in common? How do they differ?
4. How has the breakup of the Soviet Union affected the internal republics of Russia?

3. The Economy in Transition

Economic Collapse

The sudden collapse of the Soviet Union in 1991 threw its command economy into chaos. Stores had empty shelves and food was in short supply. Consumer goods were not available. Factories ground to a halt as transportation systems began to fail, and communications became uncertain. Paved roads had already deteriorated, railroads were in disrepair, and buildings were falling apart. The struggle between the central government and the fifteen republics for control of food and raw materials made matters worse. As economic activity stalled, aid poured in from Western countries to feed the Russian people.

For the first time, a survey of the economy was made for the entire commonwealth. The results were startling. Russia and the other republics were producing at a lower level than Brazil, Puerto Rico, or Uruguay. Per capita income in the commonwealth is estimated at $2,680 per year, about one-eighth that of the United States. Moreover, the evidence now suggests that the economy of the former Soviet Union had been stagnant for nearly twenty years, while those of many developing countries were expanding.

Since the early 1990s, major efforts have been made to replace the traditional centrally planned economy with a functioning market system, but these efforts have largely failed. The average incomes of the peoples of the commonwealth have fallen by one-third. Housing is in such short supply that one-fifth of the people have been on a waiting list for more than ten years. Telephones are a luxu-

 IT'S A FACT The CIS has only one-quarter the number of telephone lines of Western Europe and one-eighth the number of cars.

IT'S A FACT Nearly half of the Soviet Union has permanently frozen ground.

ry; they cost two years' average salary to install, and 15 million people are waiting to get telephones. Most products are of shoddy quality; television sets, for example, blow up with alarming regularity.

In sum, the commonwealth now lags far behind the technological world in almost every aspect of day-to-day living. The economy is so inefficient that it takes twice as much energy to produce one ton of steel in Russia as in Japan. This means that the commonwealth cannot sell manufactured goods to other countries because of their high costs.

As a result, the commonwealth economy is now similar to those of less developed countries. The commonwealth exports raw materials like oil, natural gas, diamonds, gold, furs, and caviar. It must import consumer goods, high-technology products, and food. Furthermore, inefficiency and high costs in agriculture and industry mean that little investment is made in protecting the environment. Health risks from pollution in large industrial areas are ten times greater than in the rest of the technological world. The wealth of the former Soviet Union was used to pay for weapons and to support the military, rather than to improve industry and agriculture and to raise the standard of living of the people.

Problems in Agriculture

The commonwealth was once the world's largest producer of wheat, and in the early 1990s record harvests as high as 240 million tons were expected. Experts believed that the commonwealth would become self-sufficient in food for the first time in many years. This did not happen. The legacy of seventy-five years of Soviet agricultural policies has badly damaged the farmers and their land.

As early as the 1950s, the Communist party announced a bold policy called the "Virgin Lands" plan to raise more food. Large wheat farms were established on the fertile unplowed lands east of the Volga River in Kazakhstan and western Siberia. This gigantic undertaking was a real gamble. The farms were located on an agricultural frontier in arid lands characterized by long winters, short summer growing seasons, and frequent droughts. At first this gamble appeared to succeed, as farm production in the Virgin Lands increased rapidly. Then the soils of the steppes began to lose fertility, droughts hit, and wind erosion scoured the new farms in Kazakhstan and western Siberia.

Failures like this forced Soviet leaders to face reality, as year after year they were forced to import thousands of tons of grain to feed their people. Gradually, restrictions against cultivating small, privately owned plots were eliminated. In 1990, private plots occupied only 3 percent of the farmland of the Soviet Union, but they produced 25 percent of Russia's farm crops and most of its eggs and green vegetables—a clear measure of peasant distrust and dislike of collective farming.

Other inherited problems persist. First, local diets are low in meat because there is not enough cattle feed to raise more animals. Second, yields per acre in the commonwealth are still much lower than those in other countries of the technological world. Third, one-fifth of the region's grain harvest is lost by late harvesting, improper storage, and problems in transporting crops to towns and cities. Finally, young people are leaving the farms and going to the cities—despite government restrictions on such movements—leaving older people to do the farm work. These reasons explain why the commonwealth still has food shortages, even though 25 percent of its people are farmers, compared to less than 5 percent in the United States.

The commonwealth has been forced to import grain and accept assistance from other countries since its inception in 1991. In a desperate attempt to recover crops, as many as 100,000 army troops have been assigned to harvest wheat. But each year about 40 million tons have rotted in the fields. Still more is eaten by rats in poorly constructed grain silos, or spoiled at railroad depots, waiting for trains to carry food to city markets. Nearly 60 percent of the fruit and vegetable harvest in the common-

◄

A farmer in central Bohemia works at his chores on his privately-owned farm. In what ways would his task be harder than that of people on a community-run farm? In what ways would it be easier?

wealth also spoils in the fields each year. Government representatives plead with people to harvest the fruit and vegetables, but few workers volunteer. In 1991, the central government raised the price of milk and lowered the price of grain, so that farmers fed grain to their cows instead of sending it to the cities, infuriating urbanites who ended up with milk but no bread.

In the mid-1990s, spare parts for agricultural machinery are no longer available. There are few batteries and no fuel to run farm machinery. One-fifth of the harvesting machines that do work have no drivers. A million people a year—most of them young men and women—are leaving poor farm villages to live in slightly less poor cities.

These circumstances have given more force to pressures for change. Turning agriculture over to private farms and doing away with collective farms

seemed the obvious solution. In November 1993, President Boris Yeltsin signed a decree allowing farmers to own and sell land freely for the first time since 1917. This is part of a larger land reform program that will grant the people who live on collective farms individual shares in their farmland. How this will work as long as fertilizer, pesticides, and tractors remain in government hands is unclear. As one farmer notes, "The yield today depends completely on chemical fertilizers. Without them, the land will give us nothing."

Failures in Soviet Industry

Decades of communist central planning created a difficult industrial situation when the fifteen republics became independent states. In many indus-

Oil Fields and Pipelines of the Commonwealth of Independent States

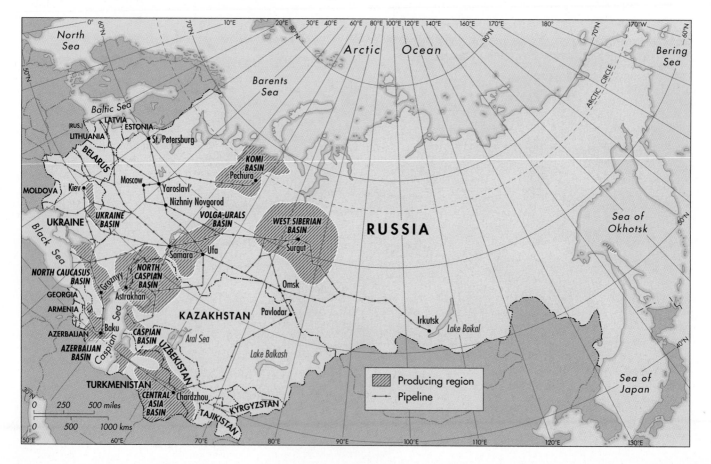

tries, one or two factories in the entire region produced individual items such as sewing machines, locomotive cranes, steel rails, machine parts, chemicals, and other basic industrial goods. When shortages occurred in one factory—the one that made steel rails, for example—the entire railroad system of the commonwealth was affected. Centralization created vulnerability.

These economic problems are not due to limited resources. The former Soviet Union has more than half the world's coal, a third of all natural gas, and greater petroleum reserves than any other country. Four major fields yield 60 percent of the country's coal: the Donets'k Basin in Ukraine, the Kuznetsk Basin in Russia, and two smaller fields located between the Urals and the Kuznetsk Basin. Today, coal provides 25 percent of the total energy consumed in the commonwealth, and coal is produced and used in greater quantities than in any other large industrial country.

Oil is equally plentiful. The early development of the Volga oil fields lessened the region's dependence on coal. Today, the neighboring Volga and Ural fields provide a third of the oil produced each year, much of which is piped westward to the industrial centers of European Russia and eastward to the Kuznetsk Basin. The recently discovered oil and natural gas deposits of western Siberia now produce more than half of the country's oil. These resources are piped 3,400 miles to countries in Eastern Europe and the borders of Western Europe.

Also rich in minerals and metals, the commonwealth is a top producer of iron (from Ukraine and the Urals), chromium, nickel, copper, and gold. Overall, it is one of the world's top three producers of nine different minerals.

Even these basic industries have been damaged by seventy-five years of central government control. Until 1991, oil was the Soviet Union's most important export, even though out-of-date technology and equipment caused much of the oil to be wasted. By 1993, oil production had declined by 40 percent. After the commonwealth was formed, economic specialists realized that the only factories that made oil-producing equipment were in Azerbaijan, an area engaged in a civil war with Armenia. When oil equipment breaks down, new equipment must be obtained from Azerbaijan or oil cannot be sent to refineries in Russia. If the oil cannot be refined, the fertilizer factories that use refined oil have to shut down. Without fertilizer, food production drops. The vicious circle continues.

◄

Mining is an important industry in mineral-rich Ukraine. Why do you think rail lines, like the one pictured here, are commonly found near mining operations?

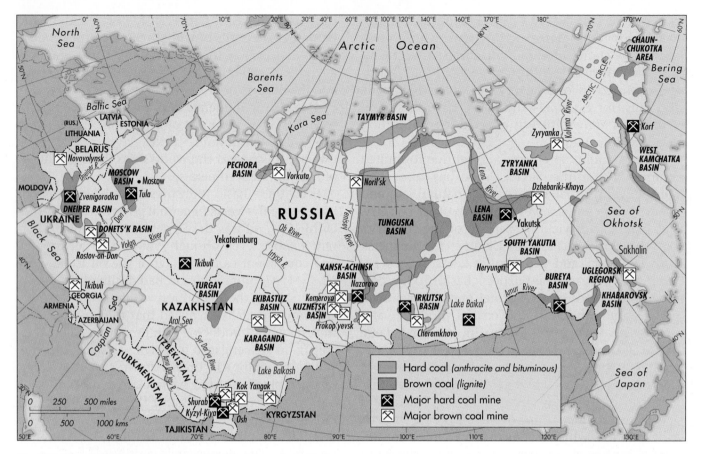

Coal Fields of the Commonwealth of Independent States

A cascade of shortages similar to this followed the coal strikes in Ukraine in 1989 and 1993. When the government initially refused to improve the miners' working conditions, coal shortages slowed down the commonwealth's entire economy. Goods began to flow more slowly between cities and villages. Many steel factories and machine plants were left without fuel. Now that the former Soviet Union has split into fifteen separate countries, the economic problems caused by worker protests, aging facilities, and poor management are severe. Although the commonwealth is now attempting to sell government factories to private citizens through a system of giving vouchers to workers, the creation of a market economy that will satisfy consumers' demands and create exports for trade will require considerable time and effort.

The Challenge of Change

Since the collapse of the Soviet Union, cooperation has increased between the two most powerful superpowers on the planet. The problem that now faces the United States is how to best help the former Soviet empire deal with its new countries, governments, and economies. Questions abound. Can the Commonwealth of Independent States form a meaningful economic union similar to Western Europe's European Community (EC)? Will the hardships of change encourage Russians to return to communism? Can or will the United States and other countries provide the resources needed to repair and reorganize the crumbling agricultural and industrial systems that exist in the commonwealth? Will the Soviet Union resurface as suggested by the

1996 agreement to form a common market among Russia, Belarus, Kazakhstan, and Kyrgyzstan? The sudden demise of the Soviet Union and at least for now communism has created new and unexpected questions and challenges.

⊕ REVIEW QUESTIONS

1. What are the major reasons economic growth in the Commonwealth of Independent States has slowed down?
2. Before the recent privatization of agriculture in the commonwealth, the 3 percent of the farmland occupied by private plots produced how much of the country's farm crops?
3. What percentage of the commonwealth's population must farm in order to feed the country? Are young people generally moving to or away from the farms?
4. Which areas in the commonwealth produce the most oil? The most coal?

⊕ THOUGHT QUESTIONS

1. In what ways was economic development in the former Soviet Union lopsided? How did this affect the people of the region? Do you believe this was a factor in the collapse of the country?
2. What problems are caused by a communist command economy? How are they avoided in a demand economy such as that of the United States?

⊕ CHAPTER SUMMARY

In 1991 the Soviet empire collapsed, and the Commonwealth of Independent States (CIS) was formed. The new republics of the commonwealth face many difficulties as they attempt to reform their old command economies into demand economies.

One of the causes of the Soviet Union's collapse was ethnic unrest. The government tried unsuccessfully to force a uniform Russian-dominated communist culture on the more than 150 ethnic groups in the Soviet empire. This was deeply resented by many non-Russians. A second cause was the failure of the Soviet command economy to satisfy the needs of the Soviet people. The government focused on heavy industry and energy for the military instead of on the manufacture of consumer goods. The government maintained rigid control from the top, which reduced worker incentive and initiative. These factors caused the slowdown of economic growth in the 1980s.

In the midst of the economic crisis, the republics began to declare their independence. At first, the Communist party under Mikhail Gorbachev tried to liberalize and reform the Soviet system. Gorbachev was replaced by Boris Yeltsin, and the republics became independent after the attempted coup by government hard-liners in August 1991.

What emerged was a confederation of twelve republics known as the CIS and three independent republics—Latvia, Estonia, and Lithuania—which still retain close economic ties with the CIS.

Perhaps the greatest challenge to the republics of the CIS is to replace their centrally planned economies with market economies. Numerous hurdles must be overcome, including waste and inefficiency, environmental degradation, equitable distribution of resources, and differing levels of natural and technological resources. New countries, new governments, and new economies are huge challenges to those members of the CIS who are reluctant to return to a communist system.

CHAPTER 7

Choose the Correct Ending

Directions: Part A lists the beginnings of nine sentences; Part B has the endings of these sentences. Find the correct ending for each beginning, and write the complete sentences on your paper.

Part A

1. The Fertile Triangle
2. St. Petersburg
3. Moscow
4. Ukraine
5. Vilnius
6. Kiev
7. The Kuznetsk and Donets'k Basins
8. Central Asia
9. Minsk

Part B

a. is the capital of Belarus and of the Commonwealth of Independent States.
b. is the capital of Ukraine.
c. is occupied by Turkic- and Persian-speaking peoples.
d. have large coal fields.
e. contains huge deposits of iron ore.
f. is a large plain that lies between the taiga in the north and the steppe grasslands in the south.
g. was called Leningrad by the former Soviet government.
h. is the capital city of Russia.
i. is the capital of Lithuania.

Missing Terms

Directions: Each of these sentences about the geography of the Commonwealth of Independent States is incomplete without a term from the list below. Find the correct missing term, and then write the complete statements on your paper (some words may be used more than once).

Central Asia	Lithuania
Estonia	Muslim
Kazakhstan	Transcaucasus
Kyrgyzstan	Ukraine
Latvia	Uzbekistan

1. A majority of the people who live in the republics of Central Asia are |||||||||| by birth.
2. The Baltic countries of ||||||||||, ||||||||||, and |||||||||| were among the first to declare their independence from the Soviet Union.
3. The republics of Armenia, Azerbaijan, and Georgia are all located in the |||||||||| region, south of Russia.
4. The Slavic-speaking Republic of |||||||||| grows a quarter of the commonwealth's food and accounts for a quarter of its industrial output.
5. The republics of Kazakhstan, Kyrgyzstan, Tajikistan, Turkmenistan, and Uzbekistan are located in ||||||||||.
6. The second largest republic in the Commonwealth of Independent States is ||||||||||.
7. Many Russians live in industrial cities in the Central Asian republics, particularly in |||||||||| and ||||||||||.
8. The Muslim religion is very important to the people who live in the republics of ||||||||||.

Inquiry

Directions: Combine the information in this chapter with your own ideas to answer these questions.

1. What factors do you think lower the productivity of workers in a communist system? Do you think that any of these factors is a serious problem in the United States? Why or why not?

2. Why do farmers in the Commonwealth of Independent States work much harder on private plots of land than on collective farms?

3. What steps is the government of Russia taking to transform its economy?

SKILLS

Understanding International Time Zones

Directions: Every traveler needs to know the local time of any place he or she is visiting. Because of the Earth's rotation on its axis, the beginning and end of a day and the hourly times are not the same everywhere. In one twenty-four-hour day, the Earth rotates 360°. Thus in one hour, the sun's rays pass over $1/24$ of 360°, or 15° of longitude.

The globe has been divided by international agreement into twenty-four time zones, one for each hour of the day. The boundaries between time zones are lines of longitude called *meridians*. One time zone was selected as the starting point for figuring time. That zone's middle line of longitude was used to establish noon for that zone. It is called the *Prime Meridian,* and it passes through Greenwich, England. Find the Prime Meridian on the map below. Time for every other zone is figured from the Prime Meridian. Time in a zone to the east of it is always later; time in a zone to the west of it is always earlier.

The line that is at 180° of longitude, or farthest from the Prime Meridian, is called the *International Date Line.* Find it on the map below. When it is noon at the Prime Meridian, it is midnight at the International Date Line. Travelers who cross the International Date Line either lose or gain a day. If they move westward across it, they lose a day; if they move eastward across it, they gain a day.

The figures at the top of the map show what time it is in the other twenty-three time zones when it is noon in Greenwich, England. Time zones have been adjusted somewhat to follow the boundaries of countries, states, and cities. If a city, for example, is on the boundary between two time zones,

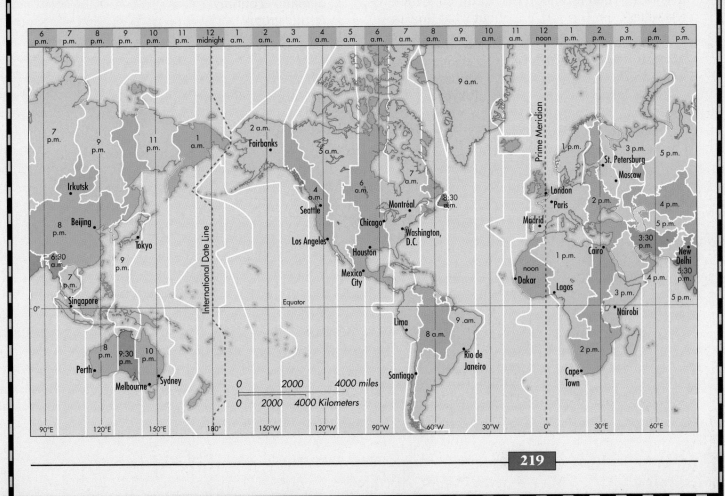

it would not be practical to have part of the city in one time zone and part in another. Some of these adjustments are shown on the map.

How Many Time Zones?

Directions: Use the map on page 219 to answer the following questions.

1. Not counting Alaska and Hawaii, how many time zones are there in the United States?
2. How many time zones are there in South America?
3. How many time zones are there in Africa?
4. How many time zones are there in Australia?
5. There are eleven time zones in Russia. What does this tell us about Russia?

Noon in Greenwich

Directions: When it is noon in Greenwich, which is a few miles from London, it is about what time in:

1. Los Angeles?
2. Washington, D.C.?
3. Madrid?
4. Rio de Janeiro?
5. Melbourne?
6. Paris?
7. Moscow?
8. Tokyo?

Crossing the International Date Line

Directions: Answer *yes* or *no* to the question of whether or not you would cross the International Date Line in each of the following trips. If you would cross it, tell whether you would lose or gain a day. Imagine that you are traveling:

1. Eastward from Houston to London
2. Westward from Moscow to Chicago
3. Westward from Los Angeles to Tokyo
4. Eastward from Phoenix to Tokyo
5. Westward from Washington, D.C., to Fairbanks
6. Westward form Rio de Janeiro to Melbourne
7. Eastward from Beijing to Los Angeles
8. Eastward from Montréal to Singapore

Vocabulary Skills

Directions: Write a brief paragraph defining each of the vocabulary terms below.

1. command economy
2. demand economy
3. ethnic group
4. *perestroika*
5. profit motive
6. sovereign

EASTERN EUROPE

E astern Europe is sand-
wiched between Western
Europe and the Slavic
Core of the Common-
wealth of Independent States
(CIS). Today this region, with a
combined population of nearly
125 million, is made up of
Poland, the Czech Republic,
Slovakia, Hungary, Romania,
Bulgaria, Albania, and the states
that declared independence
from Yugoslavia: Bosnia and
Herzegovina, Croatia, Macedonia, and Slovenia. The Federal
Republic of Yugoslavia is now
made up of the republics of
Serbia and Montenegro. Locate
each of these nations and republics on the four maps on
pages 226, 227, and 228.

Compare the political boundaries in Eastern Europe in 1914
with those that existed in 1990,
and those created after the fall
of the Berlin Wall and the
collapse of the Soviet Union.
You will see many countries

◀ *Prague was once a major cultural
center of Austria-Hungary. What
other Austrian city is famed for its
cultural life?*

with new boundaries. Until 1990, most of the changes occurred as a result of two world wars. Since then, new nations have been created by a revival of old identities among the peoples of Eastern Europe, a region of rich and varied cultures.

In the 1800s, most Eastern Europeans were ruled by the autocratic governments of czarist Russia, Prussian Germany, and Austria-Hungary. After World War II, Eastern Europe fell under the domination of the Soviet Union. In the 1990s, with strong and repressive governments no longer in control, the ethnic, religious, and linguistic differences erupted in political turmoil, as all sides fought to gain political power or disputed territory.

1. The Regions of Eastern Europe

Landscapes Are Complex

Eastern Europe reaches 1,000 miles from the Baltic Sea in the north to the Mediterranean Sea in the south. From west to east, it is 250 to 650 miles wide. Most important, it is located between Western Europe and the Commonwealth of Independent States (CIS).

Eastern Europe is usually divided into four subregions. The North European Plain, the first subregion, stretches through Poland along the coast of the Baltic Sea. Two major rivers flow northward across this plain. The Oder River forms Poland's western border with Germany and, along with the Vistula River, empties into the Baltic Sea. The area of rugged hills, mountains, plateaus, and upland basins that is located south of the North European Plain forms the second subregion of Eastern Europe. The highest of these ranges, the Carpathian Mountains, sweeps in an arc eastward through the Czech Republic, Slovakia, and southern Poland and circles back through Ukraine, then south and west into central Romania. Two fertile lowland areas along the Danube River, in the third subregion, are bounded by the Carpathian Mountains. These are the central plains of Hungary, Slovenia, and Croatia, and the plains of the lower Danube in Romania.

The Balkan Peninsula, the fourth subregion, includes the mountain ranges that fragment the lands of Yugoslavia (Serbia and Montenegro), Bosnia and Herzegovina, Macedonia, Albania, and Bulgaria. Study the map on page 223 and try to visualize the complex **topography**, the physical features of the landforms, of this region.

The rugged highlands south of the North European Plain and in the Balkans are less densely settled than the lowlands to the north. Over the centuries, minority groups driven from lowlands by larger groups of people often have taken refuge in these highlands. The mountains have not acted as barriers to migration in Eastern Europe, however. People, trade, and communications have flowed through the mountain passes of Eastern Europe and down its river valleys for centuries.

Climate Varies from North to South

Temperatures vary a great deal from north to south in Eastern Europe, because most of this region is located deep in the European interior far from the Atlantic Ocean. Temperatures in the north and central areas are not moderated by air masses from the ocean. Remember that climates that are far away from oceans and seas are called continental climates.

As a result, winters are long and cold in the northern parts of this region, much colder than those of Great Britain, which is surrounded by water. Snow lies on the ground for many months. Ice forms on rivers and frequently interferes with shipping in Baltic Sea ports along the coast of Poland. Summers are hot, with violent thunderstorms that bring sudden rains. Extreme temperatures are much more common in the interior than in coastal areas.

South of the Danube River, the climate of Eastern Europe warms up. Winters are milder. Summers are hot, but drier. Farther south, the coastlands of Croatia, Yugoslavia, and Albania have a warm Mediterranean climate. Winters are mild and sunny with moderate rainfall. During the long summers, almost no rain falls. The Adriatic coast is sunny while the plains of Hungary farther north are blanketed in snow.

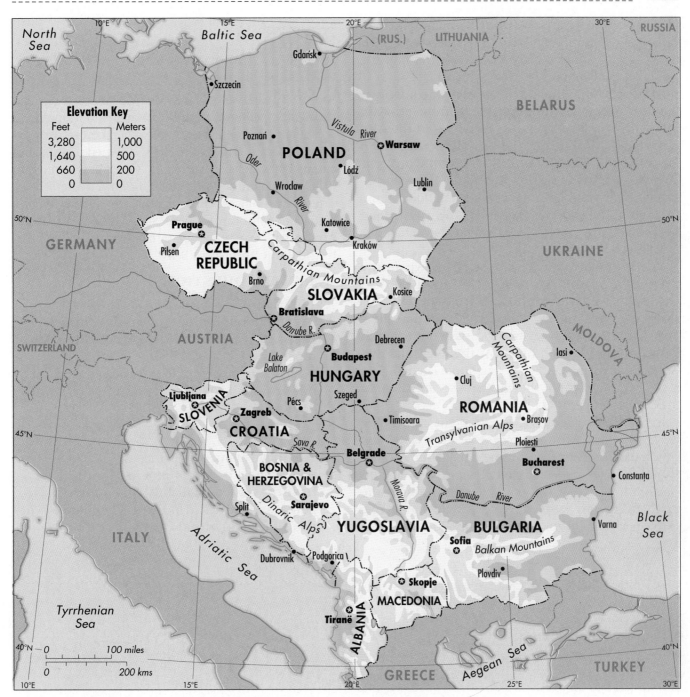

Eastern Europe

The Danube River

The Danube River is as important a highway of trade and communications in Eastern Europe as the Rhine River is in the industrial heartland of Western Europe. The Danube is 1,777 miles long, the longest river in Eastern Europe. It flows from Germany through Austria, along the border between Slovakia and Hungary, and then southward through the central plains of Hungary. It continues through Yugoslavia and along its border with Romania, and then forms the border between Romania and Bulgaria before emptying into the Black Sea. Great cities like Belgrade, Budapest, and Vienna line the banks of the Danube.

Although the Danube is a highway of commerce, the river has rapids and **narrows** (narrow passages) where it flows through steep-sided gaps in the Alps Mountains in Western Europe and the Carpathian Mountains in Eastern Europe. The most dangerous of these narrows lies along the border between Romania and Yugoslavia where the Danube flows through a tremendous gorge in the Carpathian Mountains. The gorge's sides are so steep that this place is called the Iron Gate.

The Iron Gate is a stretch of rapids where the river has worn through solid rock. The valley sides are nearly vertical, so building roads or railroads along the banks of the river has been impossible. This place has been a strategic pass dotted with fortresses since Roman times. At the river's narrowest point, Trajan's Tablet, a memorial etched into rock, commemorates the victorious campaign of the Roman emperor Trajan 1,800 years ago.

In the 1970s, Yugoslavia and Romania built a 200-foot-high dam with locks at the Iron Gate, the largest dam of its type in Europe. The dam has created an artificial lake that stretches behind the Iron Gate and evens out the flow of water through the gorge, making this stretch of the Danube **navigable**, or deep and calm enough for ships to pass, year-

IT'S A FACT Yugoslavia means "land of the southern Slavs."

round. The power station at the dam provides hydroelectric power to both Yugoslavia and Romania.

The Czech Republic and Slovakia, before they became separate nations in 1993, finished a system of canals and locks, more than 300 miles long, that links the Danube with the Elbe and Oder Rivers. These rivers flow through Poland and Germany into the Baltic and North Seas. Ships and barges now carry people and materials from the Black Sea to the Baltic and North Seas.

⊕ REVIEW QUESTIONS

1. Which countries make up Eastern Europe?
2. Eastern Europe extends north and south between what two seas? How wide is Eastern Europe?
3. What is the longest river in Eastern Europe? It flows through which countries and along which borders?

⊕ THOUGHT QUESTIONS

1. Why have many people in Eastern Europe experienced great political turmoil?
2. Have the rugged highlands south of the North European Plain acted as barriers to migration or refuges for minorities? Discuss.

▶

This picturesque photo was taken in Romania. Why do you think it unlikely that this castle is situated near the capital city of Bucharest?

2. A Mixture of Peoples

The Peopling of Eastern Europe

Eastern Europe is occupied by peoples who speak many different languages, practice different faiths, and belong to various ethnic groups. Eastern Europe has been invaded again and again—from the west by the Germanic and Latin peoples of Germany and Italy, from the east by the Slavic peoples of Russia, and from the south by the Greeks and from the southeast by Muslim Turks.

During the past 1,000 years, peoples from these neighboring regions have swept into Eastern Europe. Each group carved out a territory for itself and fought other groups to protect and expand its territory. This fighting is still going on.

Slavic peoples moved in from the east. These Slavs included a number of separate groups, such as the Poles, Czechs, and Slovaks, each of which spoke its own Slavic language. The Asiatic Magyars, or Hungarians, also arrived from the east and occupied the rich plains of the middle Danube River. Farther south, other Slavic peoples, such as the Serbs and the Bulgarians, entered the region.

Later, the Germans and Swedes moved in from the west. Although the origins of the Latin-speaking Romanians are still not certain, the Romanians believe that they are descendants of Roman soldiers who once occupied their country during the time of the Roman Empire.

Varied Religions in Eastern Europe

Christianity filtered into Eastern Europe from two different sources, from Rome to the west and from Constantinople (modern Istanbul) in Turkey to the south. People in the western part of Eastern Europe, therefore, are mostly Roman Catholics or Protestants. Roman Catholicism is particularly strong in Poland, the home of Pope John Paul II. Most people to the east are members of the Greek Orthodox faith. In the western part of this region

most groups use the Roman, or Latin, alphabet. People in the east use the Cyrillic alphabet, an alphabet of Slavic origin.

Muslims also brought their religion to Eastern Europe. The Ottoman Empire of Turkey ruled the southern part of the Balkan Peninsula for several hundred years. When the Ottoman Empire broke up in the late 1800s and early 1900s, it left behind Muslim minorities in Yugoslavia, Bulgaria, and Romania. Albania is the only country in Europe where Muslims form a majority of the population.

The Jews were another group that settled in Eastern Europe. Many Jews established homes in Poland, whose kings valued their skills and encouraged them to develop the Polish economy. By the 1500s, Poland was an important center of Jewish life and culture in Eastern Europe. These once-strong Jewish communities in several Eastern European countries were destroyed in World War II. Fewer than 100,000 Jewish people still live in the region today.

Eastern Europe in 1914

In 1914, the German and Russian Empires ruled the northern part of this region. Poland did not exist as an independent country, although it had been an important kingdom in earlier times. In the central region, the modern-day Czech Republic, Slovakia, Austria, Slovenia, and Hungary, as well as parts of Croatia, Bosnia and Herzegovina, Yugoslavia, and Romania, were ruled by the dual monarchy of Austria-Hungary. A number of small states that had once been governed by the Ottoman Empire existed on the Balkan Peninsula, among them Romania, Bulgaria, Serbia, Montenegro, and Albania. Trace Eastern Europe's political boundaries in 1914 on the map on page 226.

Now look at the map of Eastern Europe after World War I on page 226 to see how World War I completely changed the map of Eastern Europe. Self-determination was encouraged by Woodrow Wilson and other leaders of the time. Various small, weak states were established by breaking up Germany, Russia, and Austria-Hungary. Poland was carved out of Germany and Russia. Czechoslovakia

Political Boundaries in Eastern Europe Before World War I

Political Boundaries in Eastern Europe After World War I

and Yugoslavia were created largely out of the lands from Austria-Hungary. Hungary was separated from Austria, and both countries were reduced in size. Romania grew bigger, and Bulgaria became smaller. Montenegro and Serbia disappeared from the map. Only Albania, which became independent in 1912, remained unchanged. These small countries attempted to become self-sufficient states in spite of undeveloped economies, varied cultures, and limited experience in self-government.

Eastern Europe After World War II

Twenty years later, the map of Eastern Europe was again changed by war, as the map on page 227 shows. Germany, under Hitler, seized most of Eastern Europe at the beginning of World War II. Near the end of the war, in 1945, Soviet armies pushed the Germans out of the region. The Soviet Union then occupied many of the independent countries that had been created in Eastern Europe, as well as Austria and the eastern part of Germany.

Land nearly 60,000 square miles in extent was taken from eastern Poland, Czechoslovakia, and Romania and absorbed into the Soviet Union. Poland was given German lands along its western border to compensate for lands lost in the east. In effect, the entire country of Poland was picked up and moved westward. Germany was split into two countries, West Germany and East Germany. East Germany became a Communist country dependent on the Soviet Union.

The displacement of people was enormous. It is estimated that one Eastern European of every four was forced to move to a new home, often in another country. Eleven million Germans fled westward out of Yugoslavia, Czechoslovakia, and the parts of Germany that were taken over by Poland and the Soviet Union. Slavic groups moved in to take their places. The million or so Jews who survived the Holocaust left Eastern Europe, many of them

IT'S A FACT | **Albanians speak an Indo-European language much older than Latin.**

Political Boundaries in Eastern Europe After World War II

Eastern Europe After 1990

The collapse of the Soviet Union led to further changes in the borders of Eastern Europe. After the Berlin Wall fell in 1989, West and East Germany were integrated into a single nation. In 1993, the Czech Republic and Slovakia separated peacefully, but economic tensions between these two new nations remain strong.

In Yugoslavia, however, a full-scale ethnic war broke out between the Croats, the Muslims of Bosnia and Herzegovina, and the Serb armies of Bosnia and Herzegovina and Yugoslavia. Tens of thousands of people have been slaughtered in this fighting, and the economy of the region has been destroyed. Slovenia in the north escaped this turmoil after a ten-day war with Yugoslavia and declared its independence in 1992.

Macedonia to the south also declared itself independent of Yugoslavia. Macedonia's new status has been accepted by the world community in spite of

moving to Israel. The Soviet Union dominated Eastern Europe. Unsuccessful revolts against the Soviet Union broke out in East Germany in 1948, in Hungary in 1956, in Czechoslovakia in 1968, and in Poland in 1972, 1976, 1980, and 1984. Only Yugoslavia after 1948 and Albania after 1967 were able to escape direct Soviet rule.

Six Communist countries of Eastern Europe (East Germany, Poland, Czechoslovakia, Hungary, Romania, and Bulgaria) became linked with the Soviet Union in an economic alliance after the war. This economic association, known as the Council for Mutual Economic Assistance, or COMECON, organized trade among its members. When Eastern Europe gained independence from Soviet rule as the Soviet Union faced collapse, COMECON broke apart. Natural resources, industries, and collectivized agriculture, however, were developed under communist principles during the fifty-year period of Soviet rule. This meant that each country was told how to develop its agriculture and industry, what to produce, and where to export its products. Today this legacy is hampering economic development in much of Eastern Europe.

▼ *Warsaw, Poland, a city which suffered heavy damage at the beginning of World War II, was also the site of the first forced Jewish internment. Approximately how many survivors of the holocaust left Eastern Europe?*

Political Boundaries in Eastern Europe Today

the Greeks' complaint that Macedonia is the name of a neighboring Greek province. In the Balkans small differences have created large wars.

⊕ REVIEW QUESTIONS

1. The Slavic peoples who moved into different parts of Eastern Europe belonged to what separate groups?
2. What are the dominant religions in the western part and in the eastern parts of Eastern Europe?
3. What happened to the Jews in Eastern Europe after World War II?
4. Unsuccessful revolts against Soviet control of Eastern Europe broke out where and when? When did Eastern Europe become free of Soviet rule?

⊕ THOUGHT QUESTIONS

1. Why would it have been nearly impossible to form one single nation in Eastern Europe?
2. Describe the breakup of Yugoslavia, and discuss the factors that led to this major war. What countries were created? Is this type of violence occurring in other parts of Europe and Asia formerly under Soviet rule?

3. Peoples and Resources in Eastern Europe

Resources and Development Before World War II

Eastern Europe has some important energy and mineral resources. The map on page 229 shows important oil fields in Romania, but Eastern Europe still must import oil and natural gas from the commonwealth. Poland has coal fields along its southern borders. Iron is found in the western part of the Czech Republic, although iron is also imported from the commonwealth. Other Eastern European countries, particularly in the south, have fewer resources. Notice the location of resources and major industrialized regions in Eastern Europe on the map on page 229.

These resources were poorly developed before World War II. During the time when many of the countries of Western Europe were becoming industrial, the governments of Eastern Europe were unstable. Rivalries between empires and changing political boundaries discouraged investment in long-term industrial projects. Although Austria and Germany developed some industries in areas where natural resources were available, Eastern Europe generally remained economically backward.

The Recent Growth of Industries

After World War II, the Soviet Union encouraged the growth of industries in Eastern Europe. Equally important, modern industrial technologies from the West became available to some countries in the region. Levels of development vary from country to country.

The oldest centers of industry are in the more developed countries of the north—in Poland and the Czech Republic. The iron deposits of the western part of the Czech Republic were first developed by

IT'S A FACT   The waiting period to get an apartment in Poland is ten to fifteen years.

Germans. Industries were developed in Poland's important coal mining area near Kraków. The oil fields of Romania were the largest source of oil in Europe before the discovery and development of North Sea oil in the 1970s.

Elsewhere in Eastern Europe, industries are located in major cities. Each Eastern European country built new industries and expanded and modernized old ones under the direction of Communist government planners. Heavy industry, petrochemicals, and machinery were emphasized, and consumer goods received less attention.

In general, industrial development in Eastern Europe is greatest in the north and least in the south. Poland and the Czech Republic are still the leading industrial countries in Eastern Europe now that Soviet rule has come to an end. Their per capita industrial production is close to that of Western Europe. In contrast, the southern countries of Eastern Europe are less industrialized, partly because of a scarcity of resources and partly because of war. Albania, the only remaining Communist country in Europe, is one of the most isolated, least developed countries in the world. In most of Eastern Europe, basic minerals and energy resources are still imported from the Commonwealth of Independent States.

Today, the independent growth of national economies is hampered by reliance on the former Soviet Union for raw materials, especially because the prices of these raw materials have increased dramatically. The more industrialized countries of the north, particularly Poland, are importing advanced Western technology, which has increased the Polish debt to the United States.

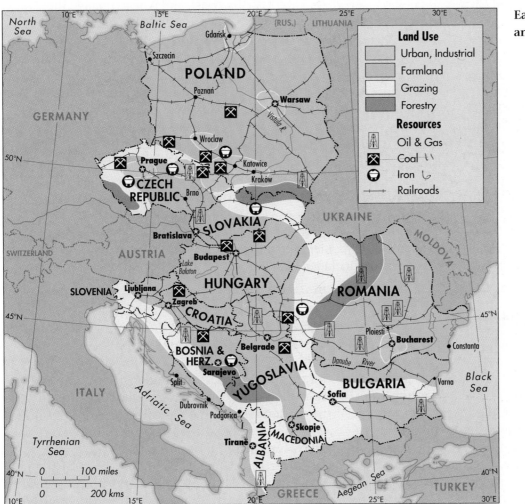

Eastern Europe Land Use and Resource Map

PLACE

WAR AND PEACE IN BOSNIA

Conflict in the Balkans

For centuries, the Balkan Peninsula has been an area of ethnic and religious warfare where peace is usually a pause between wars. Over the years, various religious and ethnic groups moved into this region from the Germanic regions to the west, the Slavic lands to the east, and Turkey in the southeast.

At this crossroads of cultures, ethnic and religious groups developed deep attachments to local territories. They fought ongoing battles with other groups to maintain or expand their territories.

The Disintegration of Yugoslavia

In 1991, the USSR disintegrated when, one after another, fifteen of its ethnically defined republics became independent countries. Communist Yugoslavia was also composed of ethnically defined republics—Serbia and Montenegro, Bosnia and Herzegovina, Croatia, Macedonia, and Slovenia. Like the USSR, Yugoslavia began to crumble.

In 1991, Slovenia and Croatia declared independence. In 1992, Bosnia voted for independence, and all three of these former republics were admitted to the United Nations. By 1993, Macedonia also was an independent nation. This left Serbia and Montenegro as a rump state that still referred to itself as Yugoslavia or land of the southern Slavs. (See map p. 228.) Except for Bosnia, however, each of the new countries had a dominant majority.

The Multiethnic State of Bosnia

In 1991, Bosnia (pop. 4.6 million) was a multiethnic state whose people included Bosnian

Muslims (40%), Orthodox Serbs (32%), and Catholic Croats (18%). The Bosnian Muslims are descendants of Serbs or Croats who, in order to own land, converted to Islam sometime during the 500 years that this region was part of Turkey's Ottoman Empire. The Serbs (defeated by the Muslims in the battle of Kosovo in 1389) are Orthodox Christians who use the cyrillic alphabet and identify with fellow Orthodox Slavs to the east. The Croats are Roman Catholics who use the Latin alphabet, relate to the people of the Germanic lands to the west, and who supported the Nazi occupation of this region in World War II.

These three groups speak the same language and are all Slavs. For several decades, the lives of Bosnian Muslims, Orthodox Serbs, and Catholic Croats were "woven together by marriage and culture."

War in Bosnia

War started in April 1992 as soon as Bosnia declared its independence. The Serb and Croat minorities rebelled against inde-

Changes in Agriculture

Eastern Europe has much fertile farmland, and agriculture is a major activity in most countries. Farming is particularly important in the southeast, where roughly half of the people live in the countryside. Before World War II, this region was the breadbasket of Europe, producing 25 percent of Europe's wheat, 50 percent of its rye, and 70 percent of its corn. Potatoes and rye are raised in the cool climates in the northern regions of East Germany and Poland. Wheat and corn are grown on the plains of Hungary and Romania, and until the 1990s war, both were grown in Yugoslavia.

This region is no longer an important exporter of food, however, for two reasons. First, the population of Eastern Europe has grown so rapidly that most food is consumed locally. Second, Communist governments organized farmland into collectives whose productivity has been low.

Collectivization, a system in which all the land is owned by the group that works it under government supervision, was forced on most countries in

The Dayton Agreement Map

pendence fearing the Muslim majority. These minorities were supported by the military forces of Croatia and Serbia, although Croatia had formed an alliance with Bosnia in 1994 to oust Serbs from occupied Croatian territory. As the war progressed, Bosnia became a killing field.

In the next four years, the country was devastated by savage fighting among Serbians, Bosnians, and Croats. The conflict deteriorated into "ethnic cleansing," the forcible removal of members of one ethnic group from a local area to make it "ethnically pure" for citizens of the aggressor group. All three groups engaged in ethnic cleansing by killing local populations, destroying their property, and occupying their land. Most of the accusations of massacres have been leveled at the Serbs; most of the victims were apparently Bosnians.

As a result, an estimated 200,000 people, most of them civilians, have been killed in the war. An additional 2.7 million people, more then half the country's population, are refugees as a result of this most recent episode of violence in the Balkans.

Peace in Bosnia

In early 1996, 20,000 American soldiers landed in Bosnia as part of a NATO (North Atlantic Treaty Organization) military expedition to enforce a peace treaty. The treaty, initialed in Dayton, Ohio, and signed in Paris on December 14, 1995, envisions the preservation of a single multiethnic state, Bosnia, subdivided into two entities—a Muslim-Croat Federation with 51% of the land and a Serb Republic occupying 49% of the land (see map). Sarajevo will remain the nation's capital. Nearby Pale will serve as the capital of the Serb Republic, although its military headquarters is at Banja Luka. The presidency of the country will rotate among groups; the central government will be weak. In addition, the country will have three different armies.

This settlement is an attempt to end the bloodiest fighting in Europe in the last 50 years. As long as the NATO forces remain in Bosnia, the peace will probably hold. But neither Croatia nor Serbia has given up the idea of incorporating Bosnian territory into their countries. But if this were to happen, Bosnia would be reduced to a small Muslim "statelet" bracketed between much larger, hostile neighbors. As one Bosnian Serb negotiator put it, "the Muslims are going to be like walnuts in a Serbo-Croat nutcracker."

QUESTIONS

1. How was Bosnia different from the other countries created from Yugoslavia?
2. Name three ways in which Bosnia will differ politically from the U.S.

Eastern Europe in the 1950s. By the 1960s, 95 percent of the land in Eastern Europe was organized into state or collective farms, except in Poland and Yugoslavia. In these two countries, resistance to collectivization was so intense that most of the land remained in private hands. Elsewhere in Eastern Europe, farmers were allowed to own only small private household plots and some domestic animals.

Collectivization of farming in Eastern Europe produced a situation similar to that in the Soviet Union. Production on collective and state farms was much lower than that on privately owned farms in Western Europe. In fact, in Poland and Romania, the food situation deteriorated so badly in the late 1980s that meat and bread were rationed.

The causes of low production were the same in Eastern Europe as in the Soviet Union. Farm management was inefficient, farmers were discontented, and the policy of providing industrial workers in cities with low-priced food at the expense of the farmers created deep resentment. Now that they are free of communist rule, most Eastern European

IT'S A FACT Hungary has the highest suicide rate in the world.

countries are trying to develop private industry, more private and more efficient farms, and closer ties with Western Europe.

A Hopeful Future

The countries of Eastern Europe have changed from agricultural countries into more industrialized societies. More people live and work in cities. The standard of living of most Eastern Europeans is slowly improving, but the costs of importing technology from the West and raw materials from the Commonwealth of Independent States have driven most Eastern European countries deep into debt.

The hopes of consumers for standards of living comparable to those of Western Europe, however, remain high because independence has given Eastern European countries a great degree of **economic autonomy**—the freedom to decide what resources they will develop and who will benefit. Except in the warring south, the new countries of Eastern Europe are now able to shape their own futures, which, given the history of this region, is an unusual opportunity.

⊕ REVIEW QUESTIONS

1. Where are the major mineral and energy resources located in Eastern Europe?
2. Most countries in Eastern Europe depend on the Commonwealth of Independent States for what raw materials?
3. What are the chief crops grown in Eastern Europe?

⊕ THOUGHT QUESTIONS

1. What natural resources helped the Czech Republic and Poland become the leading industrial countries in Eastern Europe?
2. Why is Eastern Europe no longer an important exporter of food?
3. What economic progress has been made by the countries of Eastern Europe in the last forty years? What economic problems still exist in this region?

⊕ CHAPTER SUMMARY

The region of Eastern Europe contains a mixture of peoples and a variety of climates and resources. Located between Western Europe and the Commonwealth of Independent States, Eastern Europe can be divided into four subregions: (1) The North European Plain is occupied by Poland; (2) Slovakia, the Czech Republic, southern Poland, and central Romania are located in the rugged mountains and hills to the south of this plain; (3) the plains formed by the Danube River are the location of Hungary, Slovenia, Croatia, and Romania; and (4) the Balkan Peninsula includes Yugoslavia, Bosnia and Herzegovina, Macedonia, Albania, and Bulgaria. Among these subregions, climates vary from continental to Mediterranean.

Topography, migration, and history have brought a mixture of peoples to Eastern Europe. Slavs from the east, Magyars from Asia, Germans and Swedes from the west—all made Eastern Europe their homeland and defended it from one another over subsequent centuries. Christianity and Islam also had their influences. From Rome, Catholicism took hold among people in the western part of Eastern Europe. From Constantinople, the Greek Orthodox faith attracted people in the eastern part of the region. Ottoman Turks brought Islam to the Balkan Peninsula, and today Muslim minorities live in Yugoslavia, Bulgaria, and Romania. Albania is a Muslim country. Jews also migrated into Eastern Europe, forming strong communities in many countries before World War II.

Ethnic and religious tensions are important themes in the history of Eastern Europe. The region has been dominated by Western Europe and by Russia and has been the location of many wars. As a result, the borders and names of Eastern European countries have changed many times, most notably in 1914, 1945, 1990, and 1991.

Industrialization came late to this region. When the Soviet Union gained control of the region after 1945, Moscow encouraged and also managed industrialization in Eastern Europe. In general, industrialization has been greatest in the north and least in the south. Poland and the Czech Republic are the region's leading industrial nations.

CHAPTER 8

EXERCISES
Cause and Effect

Directions: The statements in Part A are causes, and the statements in Part B are the effects that resulted from the causes. Number your paper from 1 to 8, and match each cause with its effect by writing the letter of each effect next to its matching number.

Part A: Causes

1. Temperatures in the northern and central areas of Eastern Europe are not moderated by oceanic air masses. *F*
2. The coastlands of Croatia, Yugoslavia, and Albania have a Mediterranean climate. *E*
3. Christianity filtered into the western part of Eastern Europe from Rome. *C*
4. Christianity filtered into the eastern part of Eastern Europe from Constantinople. *A*
5. Yugoslavia and Romania built a 200-foot-high dam with locks at the Iron Gate. *G*
6. Austria-Hungary was broken up after World War I. *D*
7. The Ottoman Empire ruled the southern part of the Balkan Peninsula for hundreds of years. *B*
8. There are major energy and mineral resources in Poland and the Czech Republic. *H*

Part B: Effects

a. People in this region are mostly members of the Orthodox church.
b. Today there are Muslims in Yugoslavia, Bosnia and Herzegovina, Bulgaria, Romania, and Albania.
c. People in this region are mostly Roman Catholics or Protestants.
d. Czechoslovakia and Yugoslavia became independent countries.
e. Winters in this region are mild and sunny with moderate rainfall.
f. Winters in this region are long and cold.
g. An artificial lake evens out the flow of water and makes a once-hazardous stretch of the Danube River navigable year-round.
h. Industrial development is greatest in the northern part of Eastern Europe.

Inquiry

Directions: Combine the information in this chapter with your own ideas to answer these questions.

1. Why is it said that World War I completely changed the map of Eastern Europe?
2. The Soviet Union exploited the political vacuum that was left in Eastern Europe after the Nazis were defeated in World War II. What does this statement mean?
3. Why were there so many revolts in Eastern Europe after 1948? Have these revolts succeeded?

SKILLS
Using Maps to Compare Eastern Europe at Two Different Times

Directions: The map on the left on page 226 shows Eastern Europe before World War I, and the map on the right on page 226 shows the same region after World War I. Study both maps and then answer the following questions:

1. What political state ruled Budapest and Prague in 1914?
2. Which one of these three countries did not exist in 1914—Bulgaria, Poland, or Romania?
3. Did Germany gain or lose territory after World War I?
4. The country of Czechoslovakia, which appears after World War I, was carved out of what pre-World War I political state?
5. Did Montenegro and Serbia in the southern part of Eastern Europe exist as countries after World War I?

Making an Outline

Directions: Making an outline helps you to organize your thoughts. The first step in making an outline is to identify the main ideas in the material that you are reading or the talk to which you are listening. When you find the first main idea, make a note of it in your own words. Next, find the supporting

233

details or facts that help explain the main idea. List these details under the main idea. Then, find other main ideas and the supporting details that go with them.

The main ideas become main headings in your outline; list them using Roman numerals (I). The supporting details become subheadings in your outline; list them using capital letters (A). You often will find additional facts that help explain the supporting details; list them using Arabic numerals (1).

Here is a sample outline based on the section "Climate Varies from North to South" on page 222. Read this part of the textbook again and then study the following outline.

I. Climate in the northern part of Eastern Europe
   A. Located in interior, away from ocean air masses
   B. Winters are long and cold
      1. Many months of snow
      2. Ice on rivers interferes with shipping
   C. Summers are hot, with violent thunderstorms
   D. Extreme temperatures are common

II. Climate in southern part of Eastern Europe
   A. Warmer south of Danube River
      1. Winters mild
      2. Summers hot but drier
   B. Coastlands of Croatia, Yugoslavia, and Albania have a Mediterranean climate
      1. Winters sunny with moderate rainfall
      2. Summers with little rain

Using an Outline

Directions: Complete the following sentences by writing the sentence on your paper with the correct ending.

1. Outlining helps you to (a) improve spelling and punctuation (b) organize ideas (c) distinguish between fact and opinion.

2. An outline can be used (a) only for reading (b) only for listening (c) both for reading and for listening.

3. Outlining helps you to (a) distinguish between main ideas and supporting details (b) find sources of information besides your textbook (c) remember every word that you read.

4. In an outline, the main ideas are listed using (a) capital letters (b) Roman numerals (c) Arabic numerals.

5. In an outline, the supporting details are listed using (a) capital letters (b) Roman numerals (c) Arabic numerals.

Now read page 225 again and outline the section called "Varied Religions in Eastern Europe."

Vocabulary Skills

Directions: Fill in the blanks with the correct vocabulary terms from the list below by writing the completed sentences on your paper.

> collectivization
> economic autonomy
> narrows
> navigable
> topography

1. |||||||||| refers to the relief features or surface configuration of an area or region.

2. When a nation can set its own production goals and standard of living, without interference from a more powerful nation, it is said to have ||||||||||.

3. Rapids that occur in the |||||||||| of rivers are especially dangerous because boats caught in them can be easily dashed.

4. |||||||||| is the process by which workers pool their land and labor under government supervision.

5. Before the development of modern means of transportation, the most important method of transporting people and products in many European countries was by |||||||||| rivers.

Human Geography

Directions: This exercise will give you the chance to be an editor. Find what is wrong in each of the following sentences. Then rewrite these sentences correctly on your paper.

1. Peter the Great started the city that is now called Moscow.
2. China was the first Communist country.
3. Stalin was the first head of the Soviet Communist government.
4. Three-quarters of the Russian people, farmland, and industry are located in the North.
5. The countries of Eastern Europe have been part of the Communist system since the end of World War I.
6. When Russia entered the twentieth century in 1900, its economy was chiefly industrial.
7. In 1900, four out of every five Russians lived in communal villages called steppes.
8. The Communist party based its program on the teachings of a Russian named Karl Marx.
9. The leading industrial countries are in the southern part of Eastern Europe.
10. Today Eastern European countries have large surpluses of food crops.

Physical Geography

Directions: Each sentence that follows has three endings, only one of which is wrong. On your paper write the numbers 1 to 9. Then after each number write the letters of the two correct endings for each sentence.

1. Russia (a) is the largest country in the world (b) has the largest population of any country in the world (c) covers one-eighth of the land area of the planet.
2. Eastern Europe has (a) varied climates (b) a Mediterranean climate along the coastlines of Croatia, Yugoslavia, and Albania (c) very cold winters in regions moderated by ocean air masses.
3. The Commonwealth of Independent States has (a) few natural resources (b) one of the best re-source bases of any industrial country (c) large deposits of iron, coal, and oil.
4. Most of the farmland in the commonwealth is (a) cultivated less efficiently than farmland in the United States (b) is located in Kazakhstan and western Siberia (c) was organized into collective farms until the collapse of the Soviet Union.
5. Moscow is a major (a) transportation hub (b) ocean port (c) industrial center.
6. The tundra (a) is a vast, frozen, treeless land (b) has large forest resources (c) has mosses, lichens, and ferns.
7. The taiga (a) is an evergreen forest (b) covers an area larger than the United States (c) is north of the tundra.
8. The Fertile Triangle has (a) the most farmland in the commonwealth (b) the country's only major deserts (c) the nation's major industrial centers.
9. Ukraine (a) lies north of Moscow (b) has fertile soils and a long growing season (c) produces much of the wheat in the commonwealth.

Writing Skills Activities

Directions: Answer the following questions in essays of several paragraphs each. Remember to include a topic sentence and several sentences supporting your main idea in each paragraph.

1. Why has climate played an important role in the economic development of the Commonwealth of Independent States?
2. What problems in Russia led to the Communist revolution?
3. How is communism different from capitalism?
4. Why did Communist governments fail to provide all they promised to the people of the former Soviet Union and Eastern Europe?
5. What changes came to Eastern Europe after World War I, after World War II, and after the collapse of the Soviet Union?

THE UNITED STATES AND CANADA

UNIT INTRODUCTION Four-fifths of the continent of North America is made up of two huge countries, the United States and Canada. Canada is the second largest country in the world, and the United States is the fourth largest. About 5 percent of the world's population lives in this world region. The United States has a population of 258 million; Canada's is about 28 million. They share the longest undefended border in the world.

The United States and Canada also share a common heritage. Both countries were originally settled by Native Americans, and both became British colonies. Later each country gained independence, and extended its borders westward across North America. People from all over the world came to join in building these young countries. Today, the United States and Canada are among the wealthiest nations in the world. Rich environments, abundant resources, talented people, and democratic forms of government have enabled both countries to develop strong economies.

As mature industrial societies, the United States and Canada face increased world economic competition. At home, urban and environmental problems pose challenges. Both countries now stand on the threshold of major change. *Postindustrial economies* based on services and high-technology manufacturing are developing in both countries to complement the rich farming and industrial traditions of the United States and Canada.

▲ *The Ambassador Bridge, which crosses the Detroit River, is one of oldest international bridges in North America. What two countries does it connect?*

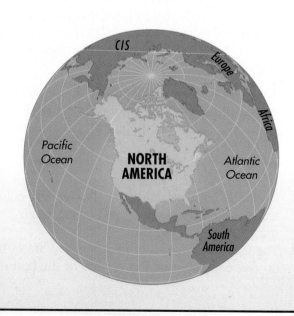

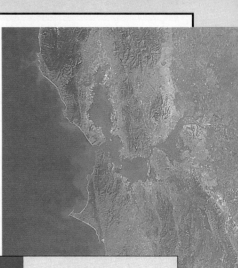

This satellite photograph of San Francisco Bay illustrates why this city became one of the most important ports on the west coast of North America. Why do you think this happened?

SKILLS HIGHLIGHTED

- Using a Transportation Map
- Reading Bar Graphs
- Making a Graph
- Using a Political-Physical Map
- Reviewing Latitude and Longitude
- Vocabulary Skills
- Writing Skills

▼ The Grand Canyon is one of America's most dramatic landform features. In what state is it located?

UNIT OBJECTIVES

When you have completed this unit, you will be able to:

1. Identify the roles that climate and landforms have played in the settling of the United States and Canada.

2. Relate that the United States and Canada began as colonies that later achieved independence, became two of the world's leading democracies, and have maintained close ties with one another.

3. Name some of the abundant natural resources found in the United States and Canada and describe how the United States faces the problem of running out of some of these resources.

4. Describe how the United States and Canada, nations of immigrants, have been enriched by cultural diversity.

5. Explain how the United States and Canada, very productive industrial and agricultural countries, are rapidly becoming postindustrial societies adjusting to major economic changes.

KEYS TO KNOWING THE UNITED STATES AND CANADA

1

The United States and Canada are two of the four largest countries in the world.

2

The United States and Canada share major landforms, such as the Atlantic Coastal Plain, the Appalachian Mountains, the Central Plains, and the Rocky Mountains.

3

The United States is a middle-latitude country with varied climates. Much of Canada is located in the colder high latitudes.

4

The United States and Canada began as colonies, gained independence, and became democratic countries.

▼ *On a clear day from the top of the Empire State Building in New York City, you can see several states. Using the political map in the miniatlas for this unit, can you guess which states?*

5

The United States and Canada have a combined population of about 285 million. Most Canadians live within 100 miles of the shared border.

6

The United States and Canada are nations of immigrants from many countries.

7

The United States and Canada have abundant and diverse natural resources.

8

The United States and Canada together form the most productive urban industrial area in the world. They also lead the world in agricultural exports.

9

The United States and Canada generally are wealthy societies with high standards of living for many of their people.

10

The United States and Canada are rapidly becoming postindustrial societies.

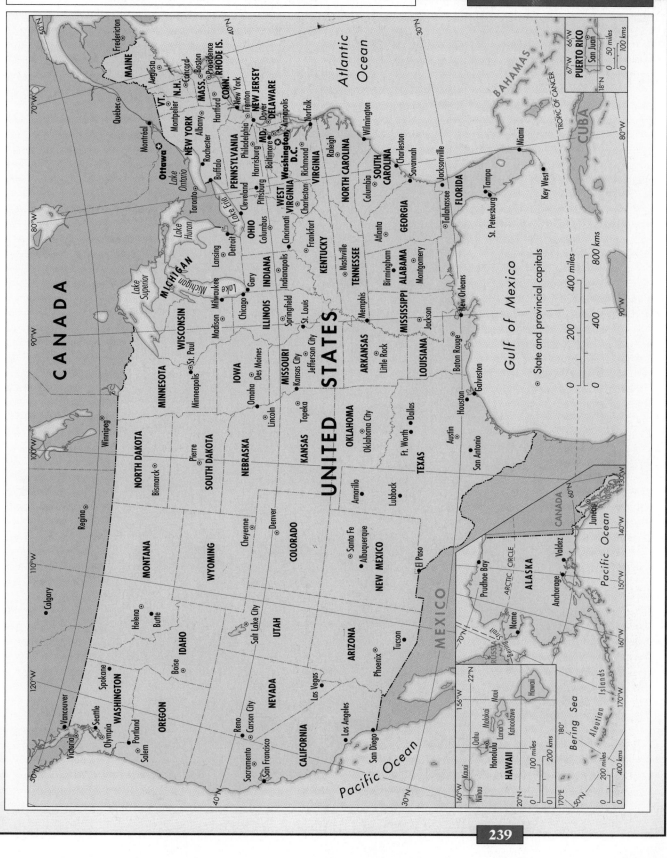

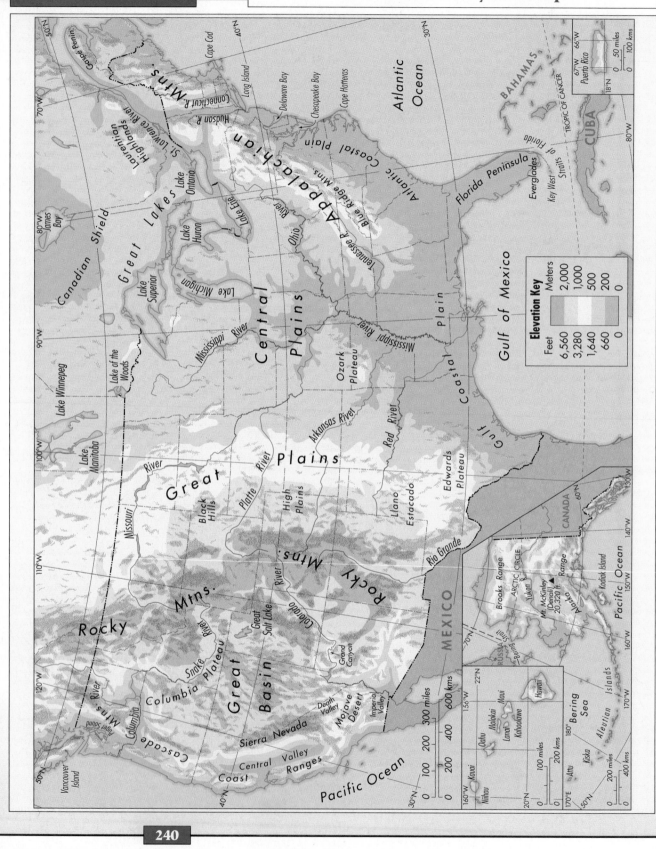

Gaspé Penin.
Cape Cod
Long Island
Delaware Bay
Chesapeake Bay
Cape Hatteras

Atlantic Ocean

BAHAMAS
TROPIC OF CANCER
CUBA

Connecticut R.
Hudson R.
St. Lawrence River
Laurentian Highlands
Appalachian Mtns.
Blue Ridge Mtns.
Atlantic Coastal Plain
Florida Peninsula
Everglades
Key West
Straits of Florida

Lake Ontario
Lake Erie
River
Ohio
Tennessee R.
Mississippi River

Great Lakes
Lake Huron
Lake Michigan
Lake Superior
Canadian Shield

James Bay

Central Plains

Coastal Plain
Gulf of Mexico

Gulf

Elevation Key

Feet	Meters
6,560	2,000
3,280	1,000
1,640	500
660	200
0	0

Lake Winnepeg
Lake of the Woods
Lake Manitoba

Mississippi River
Ozark Plateau

Arkansas River
Red River
Edwards Plateau

Great Plains
Plains
High Plains
Llano Estacado

Platte River
Black Hills
River
Missouri

Rio Grande

MEXICO

Rocky Mtns.
Mtns.
Rocky

Great Salt Lake
Colorado
Grand Canyon
River

Rocky

Columbia Plateau
Snake River

Great Basin

Death Valley
Mojave Desert
Imperial Valley

Columbia River
Cascade Mtns.
Puget Sound
Vancouver Island

Sierra Nevada
Central Valley
Coast Ranges

Pacific Ocean

CANADA
Brooks Range
ARCTIC CIRCLE
Yukon R.
Mt. McKinley (Denali) 20,320 ft.
Alaska Range
Kodiak Island
Pacific Ocean

RUSSIA
Bering Strait
Bering

Hawaii
Oahu
Molokai
Maui
Lanai
Kahoolawe
Kauai
Niihau

Bering Sea
Aleutian Islands
Attu
Kiska

300 miles	600 kms
200	400
100	200
0	0

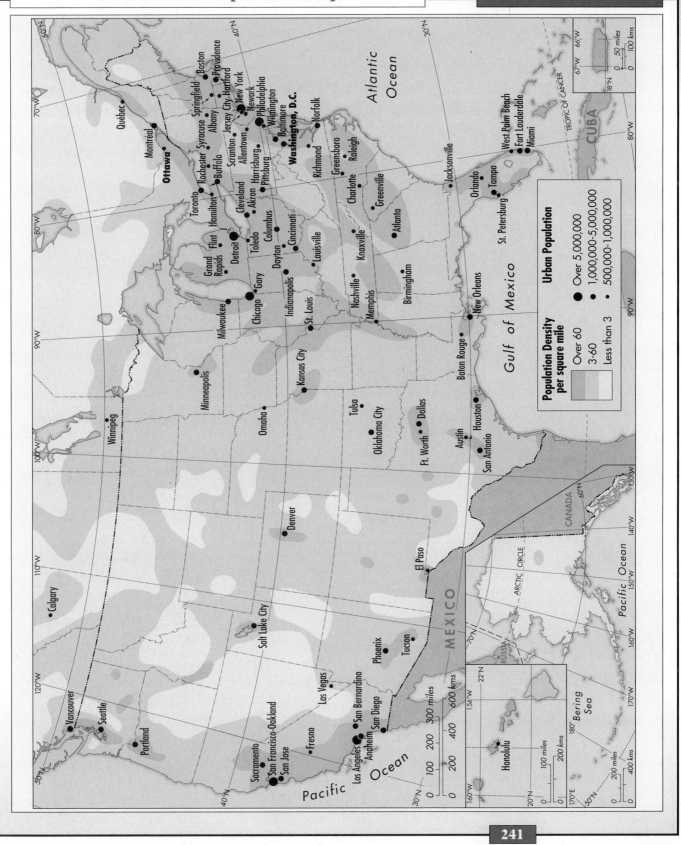

Country	Capital City	Area (Square miles)	Population (Millions)	Life Expectancy	Urban Population (Percent)	Per Capita GNP (Dollars)
Canada	Ottawa	3,851,792	28.1	77	77	21,260
United States	Washington, D.C.	3,615,104	258.3	75	75	22,560

ENVIRON-MENTS AND ECONOMIES OF THE UNITED STATES

The United States is bordered on the north by Canada and on the south by Mexico. Between these two neighboring countries, 258 million Americans live in the most populous nation founded on democratic principles in the world. President Theodore Roosevelt used to say that to govern this country properly, every president should have a "sense of the continent."

Today, the United States is facing the challenges of shifting from an industrial (or

◀ *This cattle round-up on the Big Bend Ranch in Texas illustrates a major economic activity of the Great Plains. Do you think the Great Plains are thinly populated or densely populated as compared with other regions in America?*

manufacturing) economy to a **postindustrial** economy (based on services and high technology). This change is causing economic hardship in some parts of the country and creating new economic opportunities in others. Americans are meeting these challenges in the same way they adapted to the forces of urbanization and industrialization in years past. The poet Archibald MacLeish once noted that no one can come to the Pacific coast of this country and feel that the end of anything has been reached. The American journey goes on; its society is constantly changing.

1. Many Regions in a Large Land

The Importance of Location

The United States is blessed with a location that is fortunate in several ways. The Tropic of Cancer (23½° N) runs just south of the Florida peninsula. The Arctic Circle (66½° N) passes through our northernmost state, Alaska. Most of the country lies in the middle latitudes, which means it has a variety of moderate climates and productive environments.

Our Atlantic coast faces the technologically advanced nations of Western Europe; our Pacific shores face Japan and the emerging countries of Asia. A large volume of trade flows across the Atlantic and the Pacific Oceans in this age of inexpensive ocean transport. Our northern and southern boundaries are shared with the friendly nations of Canada and Mexico.

The Role of Landforms and Climate

Landforms and climate are important factors in defining the subregions of the United States. Large-scale environmental features such as the Rocky Mountains, the Central Plains, the Gulf Coastal Plain, and the Atlantic Coastal Plain define our subregions. Understanding the varied environments of

IT'S A FACT   More than half of all Americans live within an hour's drive of an ocean beach.

the United States is essential to understanding the country. Basically, the United States is made up of three seacoasts, two major mountain ranges, and a central plain.

The Atlantic Coastal Plain of the eastern United States is where colonial America began in the 1600s and 1700s. In the north, New England's cold winters, short summers, and poor soils discouraged farming. But mild winters and fertile soils made the south an enormously productive agricultural area. The Pacific Coast has many different environments. California's variety of landscapes and climates is an important reason it is the most populous state. Southern California has a Mediterranean climate, with mild, wet winters and dry summers. The forests of northern California and the Pacific Northwest grow in cool, humid environments year-round. Five states border the warm Gulf of Mexico, our third coast.

The Appalachian Mountains in the east and Rocky Mountains of the west frame the Central Plains of the United States. The Appalachians, which stretch 1,500 miles north-south from Canada to Alabama, are an old, worn-down range of folded mountains whose ridges and valleys form a rippled surface. The steep and rugged Rocky Mountains in the west, by contrast, are a young mountain range. They are 3,000 miles long and have elevations twice those in the Appalachians. Many of their crests are capped by snow.

Between these two ranges lies a vast plain that covers a third of the nation. In this area, which is drained by the Mississippi-Ohio-Missouri River system, even small changes in rainfall or temperature can mean changes in types of farming and ranching. Crops and animals are bred to withstand bliz-

IT'S A FACT   North America has more lakes than any other continent.

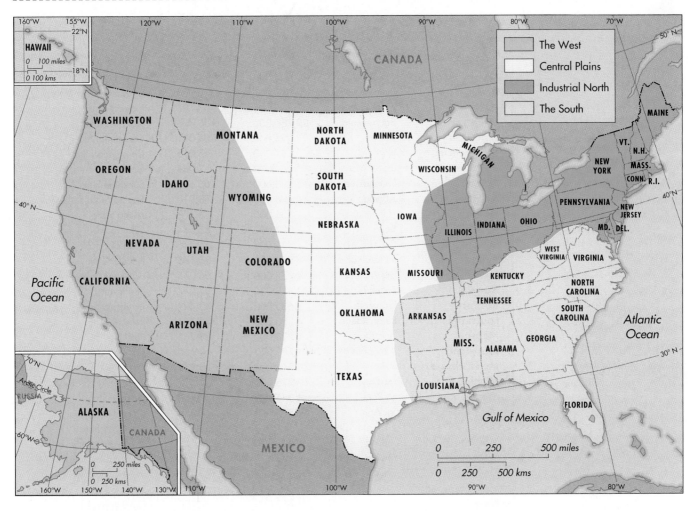

Subregions of the United States

The United States has four major subregions: (1) the Industrial North, (2) the South, (3) the Central Plains, and (4) the West.

zards in the north and to survive hot summers in the south. Ample rainfall occurs in the east, but it diminishes in the west.

In these environments, Americans established their settlements and created their economies. As you learn the characteristics of the Industrial North, the South, the Central Plains, and the West, pay close attention to the ways Americans have creatively used the varied landscapes of this huge country.

⊕ REVIEW QUESTIONS

1. Most of the United States lies in what belt of latitude? How does this affect climates?

2. What two mountain ranges frame the Central Plains of the United States?

3. How do climates differ in southern California and northern California and the Pacific Northwest?

⊕ THOUGHT QUESTIONS

1. In what ways is the United States fortunate in its location? Do you think location has been an important factor in the economic growth of the United States?

2. How would the United States be different today if areas like the South, Texas, and California had become independent countries?

The Fifty States

State	Capital City	Population	Area (sq. mi.)	Nickname
Alabama	Montgomery	4,041,000	51,705	Heart of Dixie
Alaska	Juneau	550,000	591,004	Great Land
Arizona	Phoenix	3,665,000	114,000	Grand Canyon State
Arkansas	Little Rock	2,351,000	53,187	Natural State
California	Sacramento	29,760,000	158,706	Golden State
Colorado	Denver	3,294,000	104,091	Centennial State
Connecticut	Hartford	3,287,000	5,018	Constitution State
Delaware	Dover	666,000	2,044	First State
Florida	Tallahassee	12,938,000	58,664	Sunshine State
Georgia	Atlanta	6,478,000	58,910	Empire State of the South
Hawaii	Honolulu	1,108,000	6,471	Aloha State
Idaho	Boise	1,007,000	83,564	Gem State
Illinois	Springfield	11,431,000	56,345	Land of Lincoln
Indiana	Indianapolis	5,544,000	36,185	Hoosier State
Iowa	Des Moines	2,777,000	56,275	Hawkeye State
Kansas	Topeka	2,478,000	82,277	Sunflower State
Kentucky	Frankfort	3,685,000	40,409	Bluegrass State
Louisiana	Baton Rouge	4,220,000	47,751	Pelican State
Maine	Augusta	1,228,000	33,265	Pine Tree State
Maryland	Annapolis	4,781,000	10,460	Old Line State
Massachusetts	Boston	6,016,000	8,284	Bay State
Michigan	Lansing	9,295,000	58,527	Great Lakes State
Minnesota	St. Paul	4,375,000	84,402	Gopher State
Mississippi	Jackson	2,573,000	47,689	Magnolia State
Missouri	Jefferson City	5,117,000	69,697	Show Me State
Montana	Helena	799,000	147,046	Treasure State
Nebraska	Lincoln	1,578,000	77,355	Cornhusker State
Nevada	Carson City	1,202,000	110,561	Silver State
New Hampshire	Concord	1,109,000	9,279	Granite State
New Jersey	Trenton	7,730,000	7,787	Garden State
New Mexico	Santa Fe	1,515,000	121,593	Land of Enchantment
New York	Albany	17,990,000	49,108	Empire State
North Carolina	Raleigh	6,629,000	52,669	Tar Heel State
North Dakota	Bismarck	639,000	70,703	Flickertail State
Ohio	Columbus	10,847,000	41,330	Buckeye State
Oklahoma	Oklahoma City	3,146,000	69,956	Sooner State
Oregon	Salem	2,842,000	97,073	Beaver State
Pennsylvania	Harrisburg	11,882,000	45,308	Keystone State
Rhode Island	Providence	1,003,000	1,212	Ocean State
South Carolina	Columbia	3,487,000	31,113	Palmetto State
South Dakota	Pierre	696,000	77,116	Mt. Rushmore State
Tennessee	Nashville	4,877,000	42,144	Volunteer State
Texas	Austin	16,987,000	266,807	Lone Star State
Utah	Salt Lake City	1,723,000	84,899	Beehive State
Vermont	Montpelier	563,000	9,614	Green Mountain State
Virginia	Richmond	6,187,000	40,767	Old Dominion
Washington	Olympia	4,867,000	68,138	Evergreen State
West Virginia	Charleston	1,793,000	24,231	Mountain State
Wisconsin	Madison	4,892,000	56,153	Badger State
Wyoming	Cheyenne	454,000	97,809	Equality State

2. The Industrial North

Economic Ties Bind the Industrial North

The Industrial North stretches from the Atlantic seaboard across the Appalachian Mountains to the southern shores of the Great Lakes. This subregion includes the Atlantic Coastal Plain of New England and the middle Atlantic states, the fertile farmland and industrial cities of the Great Lakes Interior, and that part of the Appalachian Mountains that lies between these two densely settled areas. Locate the boundaries of the Industrial North on the map below.

Most of our factories, mills, and businesses are located in the Industrial North. No part of the United States is as densely settled or has as many large industrial cities as the Industrial North. Railroads and highways crisscross the region, carrying goods to and from these cities. Suburbs sprawl outward from these centers of industry. Between many cities and suburbs are farms. Parts of this huge and varied region, however, are still forested.

New England, Birthplace of the Industrial North

New England, which is made up of six states—Maine, New Hampshire, Vermont, Massachusetts, Rhode Island, and Connecticut—has a population of 13.2 million people. Vermont is the only one of

The Industrial North

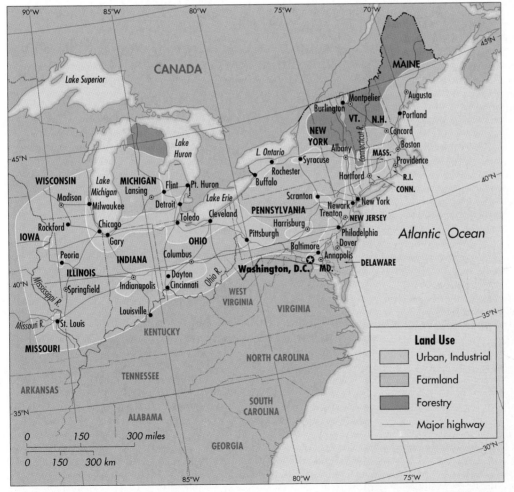

New England is bigger than old England.

The coast of Maine is 3,400 miles long.

these states that does not border the ocean. Locate New England's six states on the map on page 247.

New England's environment discourages farming. The short **growing season**, the period between the last frost in spring and the first frost in autumn, makes farming a risky venture. Winters are too long; summers are too short. Also, the sheets of ice, or glaciers, that swept over New England thousands of years ago scraped much of the soil off the land and left behind a jumbled, rock-strewn landscape

▼ *The landscape of New England is dotted with small villages whose economies are based on farming and tourism. This village in Vermont is shown during what season of the year?*

that supplies plenty of material for the stone walls that are so common in this area.

Away from the seacoast, the forested ridges and valleys of the northern Appalachians crease the landscapes of Vermont, New Hampshire, and western Massachusetts. Much of the land in these states is still forested. The beauty of its forests, lakes, and mountains attracts tourists year-round to New England. In the Connecticut River Valley, the only large area of level and fertile farmland in New England's interior, dairy farms, tobacco fields, vegetable gardens, and suburbs compete for space.

The core of New England is its narrow coastal plain. Most New Englanders live within thirty miles of the ocean. In the south, New England's coastal plain is part of a **megalopolis**, a belt of overlapping cities, that extends from Boston to Washington, D.C. (see the map on page 249). Boston, with its excellent harbor, is the major port and urban center of southern New England. The city is also known for high-technology research because of its many universities. Today, Boston and its suburbs house nearly 5 million people, over a third of all New Englanders, which is why the city is often called "the hub of New England."

Megalopolis, the Urban Core of the Industrial North

A megalopolis is a continuous urban strip made up of a number of cities that have expanded and merged with one another. About 48 million Americans live in cities in Massachusetts, Connecticut, Rhode Island, New York, Pennsylvania, New Jersey, Delaware, Maryland, and the District of Columbia. The largest of these cities is New York City, some of whose 18.1 million people live in suburbs that spill out over three states. Towering groves of skyscrapers on Manhattan Island in the Hudson River mark the city's center.

New York City handles half of the country's overseas air freight and passenger traffic. It is the nation's leading port and its most important manufacturing and service center. Incredible as it may seem, forty-seven of our states have fewer people than does the New York metropolitan area. Only the states of California, Texas, and New York have more.

South of New York, the metropolitan area of Philadelphia, Pennsylvania, has a population of nearly 6 million people. Philadelphia, like many eastern coastal cities, is located on the **fall line**, which is a topographic break connecting points where the Appalachian highlands meet the Atlantic Coastal Plain. Along the fall line, rivers from the highlands drop suddenly from upland slopes to the flatter lowlands and create waterfalls. These waterfalls provided power for the early economic growth of many East Coast cities including Philadelphia. Today, Philadelphia has a balanced economy based on international trade, oil refining, manufacturing, and service industries.

Baltimore, Maryland, a seaport city located at the head of Chesapeake Bay, serves both the Industrial North and the South. It has been called "the most southerly" northern city and "the most northerly" southern city. Today Baltimore produces steel using iron imported from abroad and coal from the Appalachian Mountains. The city's factories also process crops grown in the upper South. Its deep-water harbor serves ships from all over the world.

Nearby Washington, our country's capital, is located in the District of Columbia on sixty-nine square miles of once swampy land bordering the Potomac River—land that was originally part of the state of Maryland. The French engineer Pierre L'Enfant planned Washington to be a spacious city with broad vistas, impressive government buildings, and handsome parks. Wide streets spread like spokes in a wheel from the Capitol at the city's center. As the seat of government for the entire United

Megalopolis

The urban centers in the megalopolis line the Atlantic Coastal Plain from Massachusetts to Virginia. Many of the largest cities have good harbors and are located on streams, two factors that encouraged their early growth.

States, Washington, D.C., is the place where the most important political and economic decisions, many of which affect our personal and national welfare, are made.

The Appalachian Highlands of the Industrial North

The central Appalachian Highlands lie between the coastal plain of the megalopolis and the interior plains south of the Great Lakes. The Appalachians are rugged and sparsely populated, even today. Steep-sloped ridges and narrow valleys cut across the area, and in the valleys farming and mining communities line the banks of fast-flowing streams. Coal resources are the economic link between the highlands of the central Appalachians and the manufacturing areas of the Industrial North.

Pittsburgh, Pennsylvania, the only large city (population of 2.2 million) in the central Ap-palachians, is an iron and steel center. The city is located at the junction of three rivers. Coal is shipped in from fields in the Appalachian Highlands, and iron is brought in by barge through the Great Lakes from the Mesabi Range mines in Minnesota.

Pittsburgh's steel industry was developed by an immigrant from Scotland, Andrew Carnegie. Describing his booming industry, Carnegie bragged that he could produce a pound of steel at a penny a pound out of "two pounds of iron from Lake Superior, a pound and a half of coal from Appalachia, and a half pound of lime" shipped to Pittsburgh. This was the recipe for Pittsburgh's steel-based wealth.

Now Pittsburgh's iron and steel industries, as well as those elsewhere in the Industrial North, are struggling to survive. High labor costs, aging plants, and out-of-date equipment have made it difficult for these industries to compete with imported steel produced in more modern steel factories in foreign countries.

▼ *Air pollution from the mills and mines of Appalachia is intense because the foul air tends to get trapped in the narrow valleys in this region. What major tectonic force contributed to the formation of alternating ridges and valleys in the Appalachian Mountains?*

The Great Lakes Interior, the Western Rim of the Industrial North

The Industrial North reaches across the central Appalachians to include a band of manufacturing cities on the southern shores of the Great Lakes. These cities produce metal and machine goods, process farm products from surrounding areas, and trade with world markets through the St. Lawrence Seaway. The Seaway enables ships to journey 2,350 miles from the Atlantic Ocean into the center of the continent. It extends from the St. Lawrence River in eastern Canada through the Great Lakes to the western end of Lake Superior. Locate the St. Lawrence Seaway and the five Great Lakes on the map on page 302. There are a large number of cities along their southern shores.

In the Great Lakes Interior, many cities are well known for specializing in certain industries. Rochester, in upstate New York near Lake Ontario, is a center of high technology, and, along with Buffalo on Lake Erie, processes food products from neighboring farms. Farther west, Cleveland, Ohio, on Lake Erie, is a diversified steel-producing center like Pittsburgh. Detroit, Michigan, is the automobile capital of the United States. Its automobile plants assemble components made in many smaller cities and towns in Michigan, Ohio, and Indiana. Farther west, Chicago, Illinois, is a railroad hub whose lines of steel gather the farm products of the Central Plains for processing and transport. Milwaukee, Wisconsin, on Lake Michigan is noted for a variety of machine industries. The twin cities of Minneapolis and St. Paul on the upper Mississippi River developed as wheat milling centers.

These cities in the Great Lakes Interior are surrounded by farmland. Orchards of apples, pears, peaches, and cherries, as well as huge vineyards, line the southern shores of Lake Erie and Lake Ontario. Flowers and nursery plants are grown along the southern shore of Lake Michigan. Farther west, farmers raise wheat, corn, and hay and keep dairy cattle and poultry. Farming is an important complement to urban industries in the Great Lakes Interior.

▼ *One of the largest cities in the Great Lakes Interior is located on the southern tip of Lake Michigan. What city is this?*

⊕ REVIEW QUESTIONS

1. What smaller subregions are included in the Industrial North?
2. The megalopolis that stretches southward from Boston includes what other large coastal cities?
3. What waterway provides a trade route between port cities on the Great Lakes and world markets?

⊕ THOUGHT QUESTIONS

1. Why is New York such an important city?
2. Why are the iron and steel industries in the Industrial North struggling today?

3. The South

A New and Varied Economy

The South includes (1) the Coastal Lowlands, a broad plain that spreads along the shores of the Atlantic Ocean and the Gulf of Mexico; and (2) the Inland South, which includes the foothills, slopes, and forested ridges of the southern Appalachian Mountains and the Ozark Plateau.

Once the world's greatest cotton-growing region, the South today is changing rapidly. Southern cities are processing the region's natural resources, and farmers grow a variety of crops. Population is soaring because many Americans are moving to the South, attracted by its mild climate and growing economy.

The South includes a huge twelve-state area with a combined population of roughly 60 million people. Eight of these states border either the Atlantic Ocean or the Gulf of Mexico. Virginia, North Carolina, South Carolina, and Georgia form a north-south tier of states along the Atlantic, reaching inland across the Atlantic Coastal Plain to the Blue Ridge Mountains of the Appalachians.

Farther south, Alabama, Mississippi, and Louisiana border the Gulf of Mexico. Northern

The South

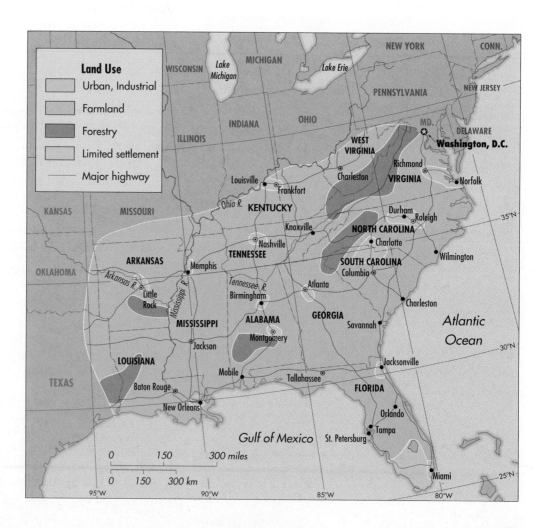

Alabama extends northward into the Appalachians, where there are rich deposits of coal and iron. Separating these Atlantic and Gulf states, the Florida peninsula juts like a 400-mile-long thumb into the Caribbean Sea. All of these coastal states have warm, humid climates with long growing seasons.

The remaining four inland states of West Virginia, Kentucky, Tennessee, and Arkansas are more rugged. Eastern Kentucky and Tennessee lie in the folds of the Appalachians. The western portions of these states then reach westward across rolling plains. The Ozark Highlands spread over northern Arkansas. West Virginia is so mountainous that a local saying claims that "workers can rest anywhere in the state by just reaching out and leaning on a slope." The climate in these highlands varies from place to place, but winters are colder and summers are shorter in the northern parts of the Inland South because of elevation, latitude, and distance from neighboring oceans.

The Coastal Lowlands

From the Great Dismal Swamp in Virginia to the **bayous**, or sluggish streams, of Louisiana, much of the southern coastal plain is poorly drained. Swamps and marshes, like the Everglades of Florida, cover large areas. Part of this coastal plain is farmed; part of it is an uninhabited wilderness.

Many of the South's port cities are located at the mouths of rivers. Along the Atlantic coast, port cities like Savannah, Georgia, are trading centers for farm products grown on the lowland plain. Cotton, tobacco, and peanuts are important crops because of the subtropical climate of this area. Timber is harvested from the pine forests found here. Sandbars, reefs, and islands line much of the coast. In Florida, these islands are called **keys**, from the Spanish word *cayo* meaning "shallow" or "reef." The most famous of these reefs is Key West at the southern tip of the Florida peninsula.

▼ *Miami, Florida, is one of the South's largest and most important port cities, but as recently as 1900 it was a small fishing harbor. Many places in the South are now growing rapidly. What factors do you believe are attracting more Americans to this region?*

HIGH TECHNOLOGY AND HIGHER EDUCATION IN NORTH CAROLINA AND NEW ENGLAND

High-technology research and manufacturing requires large numbers of well-educated, professional employees. The exchange of knowledge or information processing is the major product of these business-es. Both North Carolina and New England have created educational environments that are now attracting high-technology firms.

North Carolina's famous Research Triangle Park can be found in the area of the cities of Raleigh, Durham, and Chapel Hill. Businesses in the Research Triangle Park draw employees and consultants from three university communities in the area—Duke University, the University of North Carolina, and North Carolina State University. These universities are an important rea-son many companies decided to locate in Research Triangle Park. Government offices, military research, and development companies are also found here. Research Triangle Park is one of the most important centers of high-tech research and development in the United States.

New England, like central North Carolina, has a large pool of well-educated people; in fact, the state of Massachusetts alone has eighty-seven four-year colleges and universities. The availability of skilled graduates of these centers of higher education attracts firms specializing in electronics, robotics, and other high-technology fields to New England. As in North Carolina, the skills of New England's people are creating new opportunities on the cutting edge of change.

▼ *Duke University, one of three major universities in North Carolina's Research Triangle, is one of the premier institutions of higher learning in the South. What contribution has Duke made to the location and creation of the Research Triangle?*

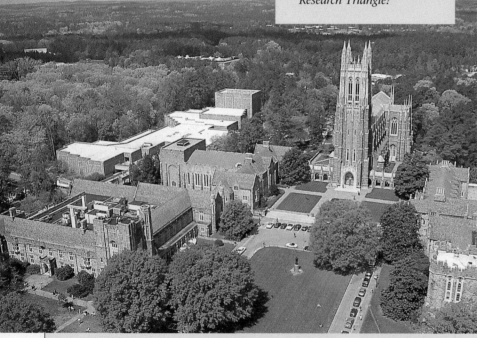

QUESTIONS

1. Heavy industry required raw materials like coal and iron to produce steel. What resource or resources are required in high-technology research and manufacturing?
2. What factors do New England and North Carolina have in common that attracted high-technology industries to these locations?
3. Do you intend to work in a high-technology industry? If so, what fields of knowledge will you study, and where in the United States might you live?

Inland, where the coastal plain meets the Appalachian **piedmont**, or foothills, many cities grew up along the fall line, the line of a sudden break in elevation where the hills meet the plains. Here, waterfalls provided power for textile mills in cities like Richmond, Virginia; Raleigh, North Carolina; and Columbia, South Carolina. Today, each is a manufacturing center, and Raleigh is part of North Carolina's high-technology Research Triangle (see page 254). Locate these fall-line cities as well as the South's major Atlantic ports on the map on page 252.

Florida is a gently rolling peninsula dotted with many lakes and swamps. Beneath parts of the peninsula lies **limestone**, a type of rock that dissolves in water. When limestone dissolves, it leaves behind caverns, underground streams, and **sinkhole lakes**, which are hollows melted out of the limestone that then fill with water. In central Florida, the influence of these sinkhole lakes and surrounding oceans, or the **maritime** influence, moderates extremes of temperature and protects citrus groves from frost damage in most years. Florida's warm climate attracts many retired people, millions of tourists, and enough permanent residents to make it the fastest growing large state in the United States in the 1990s. Two of its largest cities, Miami and Tampa, are centers of international finance and trade.

The Gulf Coastal Plain reaches westward along the shores of the Gulf of Mexico through the Florida panhandle, Alabama, Mississippi, and Louisiana. A **panhandle** is a narrow strip of land attached to another, larger piece—the "pan"—and here the "handle" refers to the northern band across Florida at the border with Georgia and Alabama. This area has hot and humid summers, mild winters, and ample rainfall throughout the year. Farmland and forests extend deep into its interior. Locate this region on the map on page 252.

The Mississippi River spills into the Gulf of Mexico in Louisiana and spreads a huge delta out into the Gulf of Mexico. A **delta** is land made up of sediments dropped by a river at its mouth. New Orleans, Louisiana, stands at the gateway to the Mississippi River Valley. Like many delta cities, New Orleans is surrounded by earthen embankments called **levees.** These levees protect New Orleans

> **IT'S A FACT**
>
> The drainage basin of the Mississippi River is twice the size of Alaska.

(much of which is below sea level) from flooding, as they did during the massive flood of 1993.

New Orleans is a busy international port; it also processes the grain, rice, sugar, and cotton grown on the Gulf Coastal Plain and shipped from all along the Mississippi River system. Its petrochemical industries process nearby deposits of oil and gas into airplane fuel, gasoline, chemicals, and other petroleum-based products. These oil and gas deposits are large enough to make Louisiana one of the leading energy-producing states in the country.

The Inland South

The Inland South is made up of the southern Appalachians, which cover large areas of West Virginia, Kentucky, Tennessee, and North Carolina, as well as parts of the Ozark Plateau farther west in Arkansas. These uplands have very different economies from the coastal lowlands that cover most of the South.

The foothills, or piedmont, of the Appalachian Mountains were settled by hard-working farmers on small farms early in our country's history. Dairy farms and orchards worked by families still cover much of the piedmont. West of the piedmont lie the steep-sloped ridges and narrow valleys of the southern Appalachians. Now called Appalachia, this region is a blighted land lacking any resource except a diminishing amount of coal that is expensive to mine. Its people are mainly farmers whose land is too poor to produce high yields and out-of-work miners, many of whom cannot find jobs to support their families. They are now leaving Appalachia in large numbers, abandoning their homes to search for steady work and better lives elsewhere in the United States.

West of the Appalachians, the lush bluegrass country of Kentucky and Tennessee supports rich farmlands. Tobacco is grown and horses are raised around the cities of Louisville and Lexington, Kentucky. Nashville, Tennessee, a recording center for country music, is this region's largest city.

In the 1930s on the southern margins of the Appalachians, the federal government created the Tennessee Valley Authority (TVA) in a seven-state area drained by the Tennessee River. The TVA was designed to improve the lives of people in this river valley. Thousands of small dams and twenty-one large ones were built along the Tennessee River to control floods, reduce soil erosion, and restore the fertility of this worn-out land. These dams generate electricity, which provides light and heat to homes in the Tennessee Valley, and powers factories in cities like Knoxville and Chattanooga.

The two largest cities of the southern Appalachians are Atlanta, Georgia, and Birmingham, Alabama. Atlanta's central location has made it a hub of railroad, highway, and airline transport. It has become the banking and trade center of the Southeast. Birmingham is a center of heavy industry because of local supplies of coal and iron. It is the only steel-producing center in the United States that is located on a coal field.

Across the Mississippi River, the Ozark Plateau in Arkansas has valuable mineral resources. Mining of coal, **bauxite** (aluminum ore), and iron is important in the Ozarks. Dairy farms and orchards are found in valleys. In the highlands, magnificent forests are carefully managed to support a growing lumber industry. Little Rock, the largest city serving the Ozark Plateau, processes the products of its mines, farms, and forests. Little Rock is located on the Arkansas River, which flows into the Mississippi River.

⊕ REVIEW QUESTIONS

1. Which southern states border the Atlantic Ocean, and which border the Gulf of Mexico?
2. How has climate helped the growth of farms and forests on the Gulf Coastal Plain?
3. What states are part of the Inland South?

⊕ THOUGHT QUESTIONS

1. Why is the South now a region that is changing rapidly?
2. Why is Florida attracting large numbers of people from other states?
3. Why are people leaving Appalachia in large numbers?

4. The Central Plains

The American Heartland

The heartland of the United States is a vast plain that reaches 3,600 miles northward from Texas to Canada. It extends 1,200 miles from the Appalachian Mountains in the east to the Rocky Mountains in the west. Eleven states lie completely within the Central Plains, and seven more are partly in this region. The Central Plains have the most productive grain fields and the most important cattle- and sheep-raising areas in the United States.

Climate divides the Central Plains into two subregions. Rainfall is greater in the east than in the west. Farming is important in states like Indiana, Illinois, and Iowa that are located in the wetter eastern half of the Central Plains. Cattle and sheep raising as well as wheat growing are more important economic activities in the drier western plains.

The transition zone between farming and ranching in the Central Plains occurs along the 100th meridian of longitude. Locate the 100th meridian (100°W) on the map on page 257. To the east of the 100th meridian, the plains average twenty inches or more of rainfall a year. This is enough to grow corn and wheat without irrigation. West of the 100th meridian, in the area often called the Great Plains, rainfall is frequently less than twenty inches, which makes crop raising more of a gamble. Time and again, farmers have bet on rain in this subregion and lost. Droughts, dust storms, and erosion have ruined their farms.

Temperatures also vary on the Central Plains. The length of the growing season is very important to both farmers and ranchers. In the long northern winters, bitterly cold polar air masses sweep down from Canada and bring subzero temperatures and raging blizzards to Montana, Wyoming, and the Dakotas. The growing season here is rarely more than 120 days long, which severely limits farming. Sometimes these Arctic air masses or "northers" reach as far south as central Texas and Florida, freezing crops in the fields.

In summer, the Central Plains are warmed by hot, humid air flowing northward from the Gulf of

Mexico. The growing season in Texas and Oklahoma lasts for 240 days or more. This makes it possible to raise almost anything wherever soils are fertile and water is available. The states of Texas and Oklahoma also have rich resources of oil and natural gas; pipelines transport these important fuels to cities on the eastern seaboard.

Farming and Ranching on the Central Plains

Variations in climate strongly influence the distribution of farming and ranching on the Central Plains. A series of farm belts exists in this region. A farm belt is an area where the climate encourages the cultivation of one or more commercial crops.

The northernmost of these farm belts is the wheat belt that stretches from the Canadian border of North Dakota as far south as Kansas. Two types of wheat are grown: in North Dakota, spring wheat is planted in April or May and harvested in September; farther south, in Kansas, winter wheat is planted in autumn and harvested in spring. Winter wheat cannot survive the freezing temperatures of the far north.

East of these wheatlands, a belt of dairy farms spreads across the glaciated areas of Wisconsin and Minnesota. South of the dairy belt, the corn belt stretches from Ohio, Indiana, Illinois, and Iowa

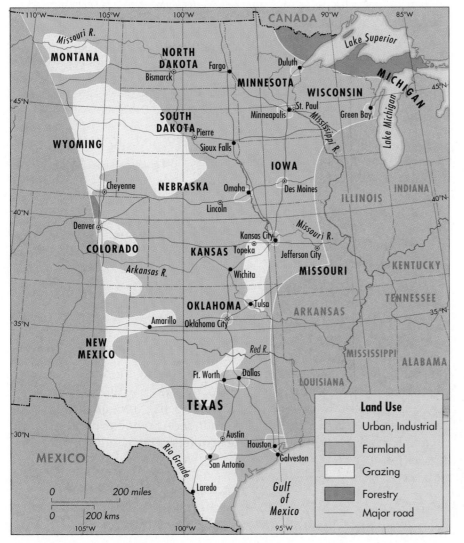

The Central Plains

PLACE

DO YOU KNOW YOUR STATES?

AMERICAN STATES:
A Miniquiz

1. What state was named after a foreign and a non-English king?
2. What two state names mention the same color?
3. What state was named after a mythical island?
4. What state was named after a real European island?
5. What state was named after a foreign country?
6. What state was named after a French province?
7. Which states were named after women?
8. How many states were named after American presidents?

9. Which state was named after a holy day?
10. Which state was once called "Deseret"?
11. Which state was named after another state?
12. Which states are not officially "states"?
13. What state still flies the Union Jack?
14. What state was for years an independent republic?
15. Which state is located in the tropics?
16. Which state touches the most other states?
17. Which states are completely rectangular?
18. Which state was once a kingdom?
19. Which state extends from ocean to ocean?
20. What states share a peninsula?
21. In which state does the geographic center of the United States lie?
22. Which state lies in two hemispheres?
23. Which island nation is closest to the United States?

Developed by Professor Emeritus Philip Wagner, University of British Columbia.

Answers

1. Louisiana
2. Rhode Island and Colorado—both "red"
3. California
4. New Jersey
5. New Mexico
6. Maine
7. Virginia, West Virginia, and Maryland
8. One—Washington
9. Florida, a Spanish term for Easter
10. Utah
11. West Virginia
12. The Commonwealths: Massachusetts, Pennsylvania, Virginia, and Kentucky
13. Hawaii (as part of its own flag)
14. Texas
15. Hawaii—Florida and Texas just barely miss
16. Tennessee and Missouri, meeting eight neighbors each
17. None—look closely at Wyoming and Colorado borders!
18. Hawaii
19. Alaska
20. Delaware, Maryland, and Virginia—"Delmarva"
21. South Dakota
22. Alaska—the Aleutians extend to about 175°E
23. Bahamas—nearer than Cuba

westward through South Dakota, Nebraska, and Kansas. Corn thrives in this area where summer temperatures are warm and summer rainfall usually totals about eleven inches a year. Given this environment, these states are covered by a carpet of corn fields. The corn is used to fatten cattle and hogs, which are then shipped to markets in the eastern United States.

West of the 100th meridian, rainfall decreases sharply. Huge herds of cattle and sheep roam on fenced ranches from Texas to Montana. Scientific animal breeding, controlled pastures, and special diets have increased the size and quality of these herds. Cattle are rounded up by cowhands in pickup trucks and helicopters. The "cattle kingdom" of

 IT'S A FACT The average elevation of Kansas is higher than that of Alaska.

◄

Farms blanket the eastern part of the Central Plains, the heartland of the United States. What is a farm belt?

the past, where riders on horseback tended cattle, is disappearing, even though its heritage remains alive in books, movies, and on television.

Nowhere is this memory of cattle kingdoms more vivid than in Texas. Yet Texas, because of its large size, is one of the most varied states in the Central Plains. In West Texas, cotton is irrigated in the panhandle, although the supply of underground water is limited. In southern Texas, some of our nation's largest oil and gas fields line the coast of the Gulf of Mexico. Citrus crops are grown in the Rio Grande Valley. And a group of large cities—Dallas, Fort Worth, Austin, San Antonio, and Houston—are major centers of urban living in the state.

Cities of the Plains

Most cities on the Central Plains are processing and service centers for the farms and ranches in this huge region. Chicago, Illinois, a city of the Industrial North, is also the largest transport center in the north plains. The city's transportation network fans out to cities in the wheat, dairy, and corn belts.

Smaller cities serve as collecting and processing points for beef packing, flour milling, and the preparation of pork products.

St. Louis, Missouri, is located just below the junction of the Missouri and Mississippi Rivers. Through this river port, food products and manufactured goods travel up and down the Mississippi River from New Orleans to Minneapolis-St. Paul. Denver, Colorado, an important center of commerce located at the foot of the Rocky Mountains, is the major city serving cattle and sheep ranches on the northern plains. Dallas-Fort Worth, one of the largest metropolitan areas in the United States, plays a similar role for the southern plains.

Houston, Texas, one of the largest cities in this region, has a diversified economy based on international trade, oil and natural gas production, and high-technology research. The space program of the National Aeronautics and Space Administration (NASA) is located in Houston. Although 25 miles or

IT'S A FACT Worldwide, forty-nine out of fifty tornadoes hit the United States.

REGION

WHERE OUR WATER SUPPLY IS RUNNING OUT

A quarter of the water used in the United States comes from underground deposits. This underground water is held in **aquifers,** which are porous layers of rock in which water is held and through which it moves. The total amount of groundwater in the United States is forty-five times the volume of Lake Michigan. It is equal to the amount of water that has flowed down the Mississippi River into the Gulf of Mexico during the last two hundred years.

Water in underground aquifers collects over thousands of years but can be quickly used up. Every day the United States takes 21 billion gallons more water from under the ground than trickles back into its aquifers. Our consumption of groundwater has doubled in the last thirty years. In parts of California and on the Central Plains, depletion of underground water has reached a critical stage.

The nation's most serious shortage of groundwater is occurring on the Central Plains in Texas, Oklahoma, New Mexico, Kansas, Nebraska, and Colorado. For years this fertile area, covering some 98,000 square miles, has drawn its water from the massive Ogallala aquifer, which has as much water as Lake Huron. Despite this aquifer's size, however, water is being pumped from it so rapidly that its level is falling between two and ten feet each year. At this rate, it will be effectively exhausted in forty years.

Farmers on the Central Plains are finding it difficult to stay in business. One of their problems is the soaring costs of pumping water. New technologies are temporarily dealing with the problem. For instance, lasers measure fields to make them absolutely flat to prevent any runoff of water. Efficient but expensive drip and trickle irrigation systems are replacing huge sprayers to conserve water. Cotton is grown on the plains in Texas and Oklahoma because this crop uses less water than others.

Despite this advanced technology, however, farming is being abandoned; an estimated 3.5 million acres will go out of production by the year 2000. Even successful plains farmers realize that in forty years it is likely that the only water available in this area will fall from the sky.

QUESTIONS

1. Where is water shortage a major problem in the United States? Can you think of another area in our country, not mentioned in this essay, that has a similar problem?
2. Describe some of the measures that farmers on the Central Plains have taken to try to conserve water.
3. Is water conservation an issue in your community? What organizations are involved? Does this issue interest you? Why or why not?

◄ *In the corn belt, drought can severely reduce the size of harvests, which results in lower farm incomes. Do you think the government should subsidize farmers? Give reasons for your answer. What is a subsidy?*

▲ *The modern skyline of Houston, Texas, rises dramatically from the state's coastal plain. What famous space agency is located in Houston?*

so inland, Houston is also an important port. The Houston Ship Channel, which enters the Gulf of Mexico at Galveston, brings much trade and business to the city.

⊕ REVIEW QUESTIONS

1. Where is the transition zone between farming and ranching in the Central Plains?
2. The corn belt stretches through which states?
3. How does the economy of Texas vary from area to area?

⊕ THOUGHT QUESTIONS

1. How do temperatures affect the length of the growing season in the northern and southern parts of the Central Plains? What effect does this have on crop patterns?
2. How do Chicago, St. Louis, and Denver serve the economic needs of the Central Plains?

5. The West

A Vast and Varied Region

The West is a vast region that extends from the snowy crests of the Rocky Mountains across plateaus and basins to lower mountains along the Pacific shores of California, Oregon, and Washington. It skips across western Canada to include Alaska, and over 2,000 miles of the Pacific Ocean to include Hawaii.

The West defies any simple description. It has towering mountains and vast deserts. Yet parts of this region are densely settled, and large cities are located along the Pacific coast. A fifth of all Americans, some 47 million people, live in the mountain and Pacific states in this region. Two of every three Westerners live in California, which has the largest population of any state in the United States. The population is clustered along the coast; the interior West is sparsely populated.

Contrast is the key to the West. This region includes our country's highest mountain, its deepest valley, its largest desert, its major glaciers, and its only temperate rain forest and subtropical rain forest. The economies of the West are diverse; they are based on, for example, sophisticated high-technology firms like those in central California's Silicon Valley near San Francisco, farms and orchards in the Pacific Northwest, fruit processing factories in Arizona and California, and frontier homesteads and oil in Alaska.

The Rocky Mountains

The jagged peaks of the Rocky Mountains rise steeply from the Central Plains, reaching elevations of 6,000 to 14,000 feet in the mountain states of New Mexico, Colorado, Wyoming, and Montana. Westward, they spread into neighboring Arizona, Utah, Nevada, and Idaho. To the south, they reach into west Texas. The series of ranges that make up the Rockies are the highest ranges in North America. They form a belt of rugged highlands several hundred miles wide.

Several great rivers begin in the melting snows of the Rockies and flow westward. Other rivers like the Missouri, Platte, Arkansas, Red, and Rio Grande flow eastward. The valley of the Columbia River in Washington and Oregon is the most densely settled region in the Pacific Northwest. The Snake River cuts through Wyoming and southern Idaho. The Colorado River flows 1,400 miles from the Rockies to empty into the Gulf of California in Mexico. As the river flows through Arizona, it cuts deeply into layers of rock to form the famous Grand Canyon. Locate these large rivers on the map on this page.

Rivers are extremely important to Western settlement because over half of this region is dry. Some of these drylands have very fertile soils when watered. Old-timers say, "Spit on the desert and a flower will bloom." So where dams like the Grand Coulee Dam on the Columbia River or the Hoover Dam on the Colorado River provide irrigation water, millions of acres of land are farmed, and electric power is produced for homes and factories. Like many other things in the West, these dams are huge. The Hoover Dam is so large that if gigantic refrigerators had not been used to cool and harden the concrete, it would have taken one hundred years to set.

It was riches, not rivers, that first attracted settlers to the Rocky Mountains. Gold and silver lured thousands of prospectors to these uplands, and later copper, lead, zinc, and uranium deposits were found. Towns grew up as trading centers for miners and for surrounding ranchers and farmers, but very few places grew into large cities.

Denver, Colorado, however, did grow because it serves both mining communities in the Rocky Mountains and ranches on the Central Plains. Its

The West

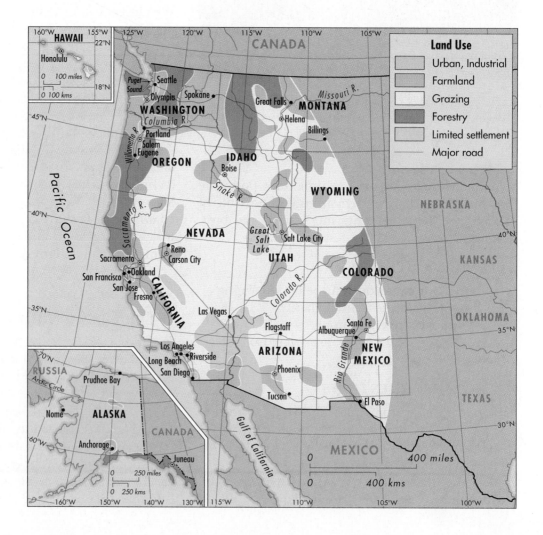

Lake Mead stretches out behind the Hoover Dam on the Colorado River. Why are dams so important to many settlements in the western part of the United States?

location at the foot of passes through the Rocky Mountains makes it an important transportation center, as well as the headquarters for many companies. Tourists further enrich the region's economy. But over large areas in the Rocky Mountains, people are few and far between and that is why there are few cities.

The Intermontane Basins and Plateaus

West of the Rocky Mountains lies a series of basins and plateaus that make up the largest area of sparse population in the United States. A **basin** is a relatively level lowland surrounded by higher land. A **plateau** is a relatively level highland that rises above surrounding areas. This dry, broken land that extends from the Rocky Mountains westward to the Sierra Nevada and the Cascade Mountains is called **intermontane**, which means "between the mountains," because of its location. It includes nearly all of Arizona and Nevada and parts of the states of Washington, Oregon, California, Idaho, and New Mexico.

The basins and plateaus of this intermontane region are underlain by soft **sedimentary rocks**, which are made up of ancient layers of sediment that were once deposited under water. Streams have carved spectacular, deep canyons into this soft rock. The deepest canyon in the United States, the Grand Canyon, was formed when the Colorado River cut completely through sedimentary rock to much older and harder rock below. Except where lakes form behind dams, like Lake Mead behind Hoover Dam on the Colorado River, surface water is hard to find in the intermontane West.

Farming areas are located where irrigation water is available or where dry farming techniques can be used. **Dry farming** techniques involve conserving water by plowing but not planting the land every year. The unplanted land is plowed to receive and store water more effectively. In the north, on the Columbia Plateau of eastern Washington and Idaho, dry farming has made this region the wheat belt of the West. Water from the Grand Coulee Dam on the Columbia River also irrigates more than a million acres of the Columbia Plateau. Spokane, Washington, the largest city in the area, uses this dam's abundant electric output for household and industrial purposes.

The Great Basin, south of the Columbia Plateau, has one of the driest climates in the United States.

This bowl-shaped area includes most of Nevada and parts of Utah, California, Idaho, Wyoming, and Oregon. Rain rarely falls in the Great Basin, because it is located in the rain shadow of the Rockies; in fact, in some years, no rain at all is recorded in the nearby Mojave and Sonora Deserts. For this reason, settlement in the Great Basin is limited to oases. An **oasis** is a fertile, watered area in the midst of a desert.

The largest oasis in the Great Basin is Salt Lake City, Utah, where the land is watered by streams flowing down from the Wasatch Range of the Rocky Mountains. This valley was originally settled by members of the Mormon faith who came to this isolated region to escape religious persecution in the 1840s. Other cities, like Las Vegas and Reno, Nevada, were early mining centers. Now they are tourist attractions for gamblers and for those who enjoy this area's hot, dry climate and stark beauty. Tucson and Phoenix in Arizona are other oasis cities that attract Americans in search of a location with a desert climate. Along with Albuquerque, New Mexico, they are among the fastest growing cities in the United States today.

The Landscapes of California

California is a land of spectacular vistas. The Coast Ranges along the Pacific rise sharply from the ocean, then the land falls abruptly into a series of interior valleys, and rises again to elevations of nearly 15,000 feet in the Sierra Nevada and the Cascade Ranges. The contrast is dramatic. The lowest elevation in the United States, Death Valley, is near Mount Whitney, the highest mountain in California. Deserts cover the southern interior; forests blanket the well-watered slopes of northern California. Nowhere in the United States do climate and topography change so rapidly over short distances.

California is where wet polar air from the north meets dry tropical air from the south. The seasonal shifts of these air masses give California its distinctive climates. Along the coasts of northern California, moist Pacific air dominates, rainfall is ample throughout the year, and temperatures are moderate to cool. In the northern interior away from the maritime influence of the ocean, however, winters can get quite cold. Southern California experiences dry air during the summer. Most of the rainfall in southern California falls in winter, a characteristic of Mediterranean climates, as you learned in Chapter 2.

As much as seventy inches a year may fall on the western or windward slopes of the rugged Sierra Nevada. The Sierra Nevada, in Spanish the "Snowy Mountains," provide water resources that are vital to California's people. This water is trapped, stored, and moved through a complicated system of dams, reservoirs, and canals. It is used to irrigate orchards and vegetable farms in the Central Valley of California and the Los Angeles Basin. Water from the Sierra Nevada also is used by the 16 million people who live in the Los Angeles Basin. Because of the subhumid and dry climates in the south, Californians have spent billions of dollars moving water over rugged terrain to places where it is needed.

Los Angeles, California's largest city, grew from a trade center for fruit and vegetable farmers in the early 1900s into a sixty-mile-wide sprawling metropolis today. One bonanza after another enriched this city: the first was oil, then motion pictures, next the aerospace industry, and now high technology. Word spread across the United States that in California jobs were plentiful and life was comfortable. So thousands headed west and more came from all over the world each year to live in cities like Los Angeles and San Diego, and to enjoy southern California's warm climate, lovely beaches, and attractive scenery.

Los Angeles became a city built for people with automobiles; in fact, cars outnumber people in this city. It has been called "twenty suburbs in search of a city," although Los Angeles is actually made up of sixty-four incorporated cities and towns linked together by freeways.

Los Angeles, incredible as it may seem, produces more goods and services than all but nine countries in the world. Today, unplanned growth as well as environmental hazards like fires, mudslides, and earthquakes are endangering the very features that attracted people to southern California in the first place. Air pollution is a critical problem, suburban sprawl has increased commuting distances to and from work, and smog has blurred the scenic vistas.

To the south, San Diego, with an important naval base, is another major urban center and port and is a city that links the United States and

Mexico. Inland from this city, the Imperial Valley is an irrigated farming area where cotton, alfalfa, date palms, vegetables, and fruits are grown. The All-American Canal brings water to this valley from the Colorado River.

San Francisco is the most important city of northern California, an area that has a cooler climate and more rain than southern California. Located on a magnificent bay with one of the few deep harbors on California's coast, San Francisco developed early as the country's most important Pacific port. Today the San Francisco Bay area forms an immense urban space where over 6 million people live. Cities in the Bay area are centers of international trade, heavy industry, shipbuilding, and high technology rivaling Los Angeles and Seattle in economic importance. This urban region is also the gateway to the Central Valley, which is one of California's largest and most productive farming areas.

The Pacific Northwest and Alaska

The states of the Pacific Northwest, Oregon and Washington, share two ranges of mountains and a coastline along the Pacific. The rocky coast of Oregon and Washington has a mild, moist climate with rainfall year-round. Storms from the Pacific carry moist air inland. When this moist air rises over the

◀

California is a large and populous state. The variety of climates and landscapes in California is striking. Can you identify some of the state's geographical features from this satellite photo?

LOCATION

THE ISLAND STATE OF HAWAII

Hawaii is a group of islands in the middle of the Pacific Ocean with a lush, subtropical environment and a unique blending of peoples. Mark Twain described Hawaii as "the loveliest fleet of islands that lies anchored in any area."

These islands are the tops of towering volcanoes that rise out of the Pacific Ocean about 2,400 miles west of San Francisco. They have to rise three miles from the ocean floor to reach the ocean's surface. Until recently, the Hawaiian Islands were isolated in the vastness of the world's largest ocean. Now, their growing population of more than a million Americans use enhanced transportation systems to make these islands a bridge between the United States and Asia. Many tourists visit Hawaii because of its tropical climate and astound-

▼ *Honolulu, Hawaii, is a world-famous destination for tourists. Where is it located?*

ing beauty. Waikiki Beach in Honolulu, shown in the photo, is a world-famous tourist destination.

Average temperatures in January are in the 70s, and summers are a little warmer. The vegetation is lush, but there are dry rain shadow areas as well. Hawaii boasts more than 900 species of flowering plants that bloom on sand beaches, sea cliffs, and forest-clad volcanic slopes. Its extremely fertile volcanic soils also support pineapple, coffee, and sugar plantations. Mauna Loa, the tallest active volcano in the world, spews out lava about once every four years.

The strategic location of Hawaii in the mid-Pacific has deeply influenced life on these islands. About 20 percent of the people of Hawaii are employed at military installations, like the huge naval base at Pearl Harbor. Moreover, its growing Asian-American population and economic links across the Pacific are providing a bridge between different world regions.

QUESTIONS

1. Describe the physical environment of Hawaii. What environmental factors would interest you in visiting or living in this state?
2. In what ways does Hawaii provide linkages across the Pacific between the United States and East Asia?

Cascade Range in the interior, large amounts of rain pour down its western slopes. Between the Cascades and the Coast Ranges, this water flows along the Willamette Valley of Oregon, an area famous for its fruit orchards. The Puget Sound area in Washington is also noted for orchards, as well as for dairy farms and vegetable production.

The first settlers in the Pacific Northwest were looking for farmland, but they discovered when they arrived that most of the land was rugged and densely forested. Settlements are framed against a background of evergreen forests in the Pacific Northwest. About 70 percent of Washington's land and half of Oregon's is covered by dense stands of Douglas fir, cedar, pine, and spruce trees—the most valuable forests in North America. These forests and the sea are important sources of wealth in Oregon and Washington. Today, the fishing industry of the northern Pacific has a catch forty times more valuable than that of the Atlantic.

Towns and cities in Oregon and Washington are trade centers for lumbering, fishing, and farm products from the interior. Portland, near the mouth of the Columbia River, is the largest city and busiest port in Oregon. Seattle, the dominant city on Puget Sound in Washington, is an important port for trade with Asia and Alaska, a center of aerospace en-

IT'S A FACT	The Alaska Highway crosses 129 rivers.

gineering and technology, and the economic hub of the Pacific Northwest.

Most of Alaska is a huge peninsula reaching westward from Canada into the Pacific and Arctic Oceans. The Brooks Range, an extension of the Rocky Mountains, sweeps across the northern part of the peninsula. South of the Brooks Range are the vast snowfields of the Yukon. The southern part of Alaska, known as the panhandle, stretches along the Pacific Ocean. The southern coast is bordered by the Alaska Range, another imposing chain of mountains where Mount McKinley (Denali) is located. A towering 20,320 feet in elevation, it is the highest peak in North America.

Alaska's largest city, Juneau, is located on the southern coast. As you would expect, this part of Alaska has the state's mildest climate. It is similar to that of the Pacific Northwest, with abundant rain and a fairly long growing season. Alaska's most important farming area is located in a valley near Anchorage.

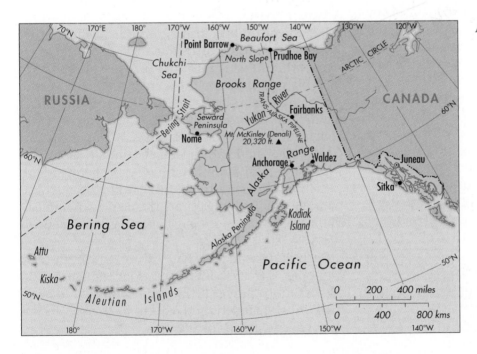

Alaska, Our Northernmost State

▲ *Anchorage is one of Alaska's largest cities. Is it located in the northern or southern part of the state?*

No railroad reaches Alaska from Canada or the rest of the United States. Only one highway travels to Alaska from the "lower forty-eight." Indeed, Juneau, the capital of Alaska, is cut off by mountains from all land contact. Ships reach the city by using the Inside Passage between Puget Sound in Washington and southern Alaska. The Inside Passage is a channel between mainland North America and the many islands that line this coast.

Gold brought miners to Alaska at the end of the 1800s, but timber and fish proved to be more important resources. Now oil and natural gas production is many times more valuable than all other products. Alaska contains nearly one-third of our country's total oil reserves. Vast deposits of oil and natural gas have been found on the frozen plain called the North Slope that sweeps northward from the Brooks Range to the Arctic Ocean. Developing and transporting these energy resources are expensive, however. The soil is permanently frozen deep beneath the surface, a condition called permafrost. The Alaska Pipeline that transports this oil and nat-

IT'S A FACT   Alaska, our largest state, has fewer people than Rhode Island, our smallest state.

ural gas from the Alaskan Arctic across the Brooks Range, the Yukon flats, and the Alaska Range, goes through some of the most forbidding country in the world to the ice-free port of Valdez.

⊕ REVIEW QUESTIONS

1. What resources first attracted settlers to the Rocky Mountains?
2. Where is the largest area of sparse population in the United States?
3. Why is the Sierra Nevada Range so important to California?
4. What are the two most important sources of wealth in Oregon and Washington?

⊕ THOUGHT QUESTIONS

1. Why are rivers especially important in the West? Do you think competition for water will create disputes in California in the future?
2. Why does California have varied climates? How have these climates benefited California's people? What problems do they pose?

⊕ CHAPTER SUMMARY

The United States is composed of many regions, each distinguished by its own landforms and climate. Moving east to west one first encounters the Atlantic Coastal Plain, then a series of foothills called the piedmont. Rising from the piedmont, the Appalachian Mountains stretch through Pennsylvania south and west into Alabama. Beyond the Appalachian Mountains, the Central Plains meet the Rocky Mountains far to the west. The Central Plains are divided according to climate along a transition zone that mostly runs along the 100th meridian. East of the meridian, adequate rainfall allows for wheat and corn farming, and west of the meridian, cattle ranchers raise beef, sheep, and wheat on the prairies. Still further west are the arctic, subtropical, mountain, and desert lands known as the American West.

Each of these regions has a history, a culture, and economic activities developed from the natural resources found there. Yet all subregions are increasingly interdependent and subject to changes in the national and global economy.

EXERCISES

Find the Right Word

Directions: In the following sentences, select the right word in the parentheses and write the complete sentence on your paper.

1. The Appalachian Mountains are (higher, lower) than the Rocky Mountains.
2. The most densely settled part of the United States is (the South, the Industrial North).
3. Appalachia is now one of the (richest, poorest) sections of the United States.
4. Evergreen forests cover much of the (Central Plains, Pacific Northwest).
5. Caverns, underground streams, and sinkhole lakes are found in Florida because of large (limestone, oil) deposits.
6. The Tennessee Valley Authority constructed many (dams, low-income houses).
7. The most productive grain area in the world is found in the (Great Basin, Central Plains).
8. The region with the highest mountains and largest deserts is the (Ozark Plateau, West).
9. One example of an oasis city is (Phoenix, St. Louis).
10. Many tourists visit (the Grand Canyon, Hawaii) because of its year-round moderate climate.
11. The most valuable industry in Alaska today is (lumbering, oil and natural gas production).

Cities in the United States

Directions: Match each numbered city with its correct description. On your paper write the letter of the correct description after each number.

1. Boston
2. Baltimore
3. Washington, D.C.
4. Pittsburgh
5. Detroit
6. St. Louis
7. New Orleans
8. Atlanta
9. Little Rock
10. Houston
11. Denver
12. Los Angeles

a. The automobile capital of the United States
b. A city planned by Pierre L'Enfant before it was built
c. A Gulf of Mexico port and the gateway to the Mississippi River Valley
d. A huge metropolis made up of sixty-four cities and towns
e. Serves the cattle and sheep ranches on the northern plains
f. The hub of New England and a center of high-technology industries
g. The most "southerly" northern city and the most "northerly" southern city
h. The largest city serving the Ozark Plateau
i. The banking and trade center of the southern Appalachians
j. The iron and steel capital of the United States
k. The home of NASA's space program
l. A busy port that serves the Mississippi and Missouri rivers

Inquiry

Directions: Combine the information in this chapter with your own ideas to answer these questions.

1. Are there any centers of high technology near your home? If so, where are they located and what do they produce?
2. In which region of the country do you live? What are its major landforms? Its major economic activities? Is the area in which you live similar to its description in the chapter? In what ways? In what ways is it different?
3. If you had your choice, in what state would you live? Why?

SKILLS

Using a Subregion Map

The subregion map on page 245 shows the land use in each of the four regions of the United States. Using this subregion map, answer the following questions.

1. Name the four regions of the United States.
2. Which of the four regions has the least amount of land devoted to urban industrial use?
3. Each region shows three, four, or five types of land use. Only the West has five. How would you explain the greater variety of land use in the West?

Vocabulary Skills

Directions: Write the term on your paper next to the number of the sentence it best completes.

aquifer	levee
basin	limestone
bauxite	maritime
bayou	megalopolis
delta	oasis
dry farming	panhandle
fall line	piedmont
farm belt	plateau
growing season	postindustrial
intermontane	sedimentary rock
keys	sinkhole lake

1. A(n) |||||||| is an earthen embankment that protects surrounding areas from flooding.
2. |||||||| is a water-soluble rock deposit that lies beneath the Florida peninsula.
3. The small islands that lie off the coast of Florida are called |||||||| after a Spanish word meaning "shallow" or "reef."
4. A(n) |||||||| landform region is located between two mountain ranges.
5. |||||||| regions are located on or close to the sea, and therefore have more moderate climates.
6. In any region, the |||||||| is determined by the average number of days between the last frost of spring and the first frost of winter.
7. The |||||||| is an imaginary line connecting points where highlands meet a coastal plain at which rivers drop suddenly as waterfalls.
8. On the east coast of the United States, a |||||||| was formed when a number of cities expanded and formed a continuous urban belt.
9. A(n) |||||||| is formed where an area is devoted primarily to the growing of a single crop.
10. |||||||| in areas of low rainfall conserves water by plowing but not planting the land every year.
11. A(n) |||||||| is a fertile, watered area in the midst of a desert.
12. New Orleans is located on a ||||||||, a flat lowland made up of sediments dropped by a river at its mouth.
13. Near the Atlantic coast lie the ||||||||, which are foothills that form a transition zone between mountains and plains.
14. Louisianans call a slow-moving, sluggish stream a ||||||||.
15. A(n) |||||||| is a relatively level highland that rises above surrounding areas.
16. |||||||| is a mineral ore out of which aluminum is made.
17. Many countries of the technological world are |||||||| societies, in which most people work in service occupations.
18. A |||||||| is a relatively level lowland surrounded by higher land.
19. |||||||| is composed of layers of sediments originally deposited under water.
20. A(n) |||||||| is a porous layer of rock in which water is held and through which it moves.
21. A(n) |||||||| is formed when water dissolves limestone deposits to form a cavern which collapses.
22. A(n) |||||||| is a narrow strip of land attached to a larger area.

THE UNITED STATES IN CHANGE

The principle of freedom took root and grew in the United States, nurtured by people whose heritage is as rich as the lands they settled. Many of the people came to our country as laborers, servants, or slaves from Europe, Latin America, Africa, and Asia. These new Americans took advantage of the opportunities that vast spaces, empty lands, and ample resources provided them. Today, people of different races and religions are members of a strong, independent society. In three hundred years, our country has become the world's most prosperous nation.

◄ *Computer education is now widely available in America's elementary and secondary schools. Is a working knowledge of computers important? Why or why not?*

1. The Geography of Settlement

The Settlement of a Growing Country

Twenty years after independence, the United States was growing fast. In 1803, before many settlers reached the Mississippi River, the United States doubled its land area by buying all the land between the Mississippi River and the Rocky Mountains from France—the Louisiana Purchase. Thirteen new states and some of the richest land in America were added to the Union through this land purchase. Other great chunks of land were added to the United States, and new states were created. Texas (1845) and California (1850) joined the Union after winning wars with Mexico in the 1840s. Oregon was added after negotiations with Great Britain. In 1853, the Gadsden Purchase of southern Arizona and southern New Mexico rounded out the frontiers of the continental United States, as the map on page 274 shows.

▼ *Steam railroads were vital in conquering the huge distances that separate America's regions. When was the first transcontinental railroad completed?*

Building Transportation Networks

In the 1800s, Americans built roads, canals, and railroads to tie the country together. Roads were built westward through the Appalachian Mountains to link the cities of the Atlantic coast to settlements in the interior.

Water transport also became very important. Canals were built to connect bodies of water and the towns and cities located on them. The Erie Canal, which was completed in 1825, connected the Hudson River and New York City with Lake Erie. In the Middle West, canals joined the Ohio and Mississippi river systems with the Great Lakes. Steamboats could transport goods thousands of miles along these inland waterways. By 1860, a thousand steamboats carried goods and people up and down the Mississippi, Missouri, and Ohio Rivers. Locate these important rivers on the map on page 240.

Steam railroads, however, were needed to master the huge distances that existed in the United States, and by 1860, railroad lines bridged the Appalachian Mountains and connected the Atlantic coast and the Middle West. Chicago became the railroad hub of the interior plains. The first transcontinental railroad was completed to California in 1869.

These improvements reduced the time and cost of transportation in the United States. Each region could now specialize in its own way of making a living because, with all of the regions tied together by efficient and inexpensive transportation, it was possible to exchange goods produced in one region for those produced in another. This specialization led to huge increases in agricultural and industrial production. A landscape of mills and factories was created in the North, and in the South, cotton plantations stretched from Georgia to central Texas. The rich soils of the Central Lowlands or Middle West were tilled by more and more farmers. Distinctive economies emerged in the Great Plains, Oregon, California, and Texas. Each of these areas was connected to the rest of the United States by a transportation network.

Regional Economies

The Industrial North

Industry grew rapidly in the Industrial North for several reasons. First, these Americans were experienced in technology and loved inventing things. Second, the rich farmland farther west produced grain more cheaply than farms in the cold climates of New England, so many New England farmers left the land and either moved farther west or became factory workers in places like Boston. Third, the cities of the northern seaboard had been centers for trade with many parts of the world for some time, and had access to foreign markets for their manufactured goods.

So industry flourished in the mill towns and port cities of the Northeast, which produced two-thirds of the country's manufactured goods in the 1800s. The Industrial North also led the world in precision goods like rifles, clocks, and sewing machines—the high-technology products of that day.

The South

In the South, cotton was king in the 1850s. After the cotton gin, a hand-driven machine, solved the problem of separating the cotton fiber from cotton seeds, thousands of acres of land were planted in cotton from Georgia to Texas. The South's environment was excellent for cotton farming, because cotton grows well wherever rainfall reaches twenty-five inches a year and the growing season is longer than 200 days. Cotton became the most valuable single crop grown in the United States. In fact, the American South produced three-fourths of the world's cotton. Southern plantations sent bales of raw cotton to be woven into cloth in textile mills in the Industrial North and in Britain. Because the fertile cotton lands of the South were too valuable to be used in any other way, many southern plantations bought their food from farms in the Middle West.

The expansion of cotton cultivation increased the demand for cheap labor in the South. One field hand was needed to work every five acres of cotton land. Because slaves were the cheapest laborers, the total number of slaves in the South reached 4 million in the 1860s. In southern cities, a strong, healthy field hand sold for $1,800, a very large sum of money for the time.

The Central Plains

In the northeastern part of the Central Plains, also called the Central Lowlands or the Middle West, pioneers built farms in forest clearings and began raising foodstuffs to sell to the Industrial North and to the South. A stream of wheat, flour, corn, pork, and beef soon flowed out of this area on the roads, canals, and railroads shown on the map on page 275. Cities like Cincinnati, St. Louis, and Chicago emerged to process these agricultural products.

◄

Cattle were collected in the stockyards of Chicago for transport to markets in the eastern United States. How was this accomplished?

MOVEMENT

TYING THE UNITED STATES TOGETHER

By 1860, Americans had a country that extended from the Atlantic to the Pacific. Later, two more territories were added: Alaska was purchased from Russia in 1867, and Hawaii was annexed in 1898. Alaska and Hawaii both became states in 1959.

These two maps show (1) when land was acquired and states were admitted to the Union and (2) the major roads, canals, and railroads that tied the country together in 1860. On the map below can you find when your state was admitted to the Union? When was its territory acquired? Was it one of the original thirteen states? Part of the Louisiana Purchase?

On the map opposite, you can see how transportation routes linked different parts of the United States with one another by 1860. In 1860, did a road, canal,

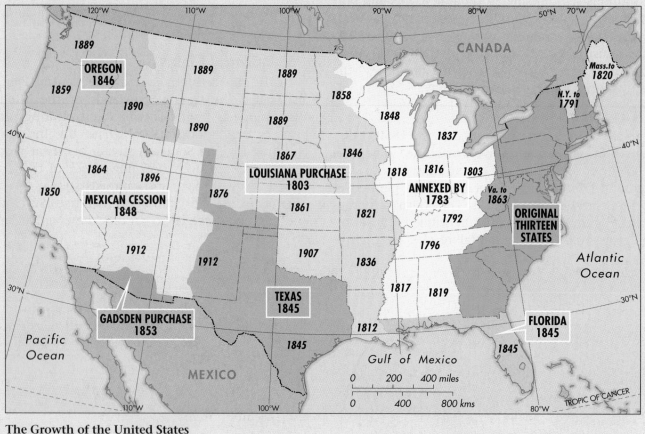

The Growth of the United States

These cities also served a growing population of immigrant farmers who had come to the United States looking for fertile land. They had seen advertisements claiming that cornstalks in the Central

IT'S A FACT Congress imported two boatloads of camels into Texas in the 1850s.

or railroad exist near where you live now?

QUESTIONS

1. Can you name a famous canal or a famous railroad that was particularly important? Where is it located?

2. How long did it take for the first forty-eight states (not including Alaska and Hawaii) to become members of the United States? Did states in the East become members first or states in the West?

3. Where was the transportation network most dense in 1860? Least dense?

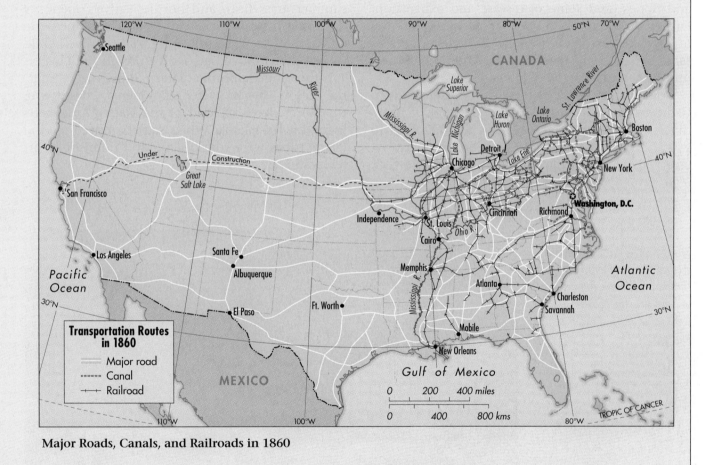

Major Roads, Canals, and Railroads in 1860

Plains grew as thick as tree trunks, cucumber vines grew fast enough to strangle unwary farmers, and the apples of Ohio's orchards weighed so much they could "sink the British navy." Although many headed west for land, others took industrial jobs in eastern cities or worked building railroads.

Mastery of the Central Lowlands was eventually achieved by machines. John Deere invented a steel plow that cut deep furrows through the grassland soils. Cyrus McCormick developed mechanical reapers in 1834 that could harvest large fields of grain quickly. Better ways of drilling wells provided water for the fields. With this machine technology, farmers produced a golden torrent of grain, and the Central Lowlands, or Middle West, became the breadbasket of America.

Oregon and California

Other Americans settled on the Pacific shores of the Oregon Territory and California. Wagon trains followed the Platte River westward across the Great Plains to the Rocky Mountains on trails that were lined with graves. Some travelers died in blizzards and floods, and some of disease and exhaustion. Others were killed by Native Americans attempting to defend their lands. Despite the dangers, thousands of pioneers made the nine-month, 2,000–mile trek to the Oregon Territory.

Gold nuggets were discovered at Sutter's Mill in California in 1848, and the next year 25,000 "forty-niners" booked passage on ships to California to search for gold. Another 55,000 people rode in covered wagons across the continent. In a single year, San Francisco mushroomed into a city of 25,000 residents. Soon California was producing half of the world's gold.

The Great Plains

The last fertile region in the American West to be settled was the Great Plains, the western half of our Central Plains. The Great Plains region, which covers almost a quarter of the continental United States, is an enormous semiarid grassland located between the 100th meridian of longitude and the Rocky Mountains. These rolling plains stretch from Texas into Canada.

Uncertain rainfall, scarce fuel, winter blizzards, summer grass fires, and hostile Native Americans discouraged settlers. So this region remained the domain of the bison-hunting Plains tribes while wagon trains moved across it to reach the Oregon Territory and California. After railroads were built, the bison herds were slaughtered, destroying the basic source of food of the Native Americans living on the plains. When settlers moved in, the Plains tribes were rounded up and imprisoned on reservations.

▶

Many immigrants to America crossed the continent in wagon trains to settle in California and the Oregon Territory. What were some of the dangers of this journey?

◄

Homesteaders who settled the Great Plains often lived in dugouts made from thick slabs of grass cut from the treeless prairies. They were called sodbusters because they frequently cut the sod using an axe and a sledgehammer. What 1862 government act encouraged people to settle the Great Plains?

On the southern Great Plains, longhorn cattle were driven from Texas northward to railroad towns like Abilene and Dodge City in Kansas for shipment to eastern cities. Soon, the wide-open spaces of the Great Plains were crisscrossed by railroads and staked out by the barbed-wire fences of farmers and ranchers.

The farmers settled the Great Plains after Congress in 1862 passed the Homestead Act, which gave Americans the opportunity to own 160 acres of land in the Great Plains if they paid $10 and promised to produce a crop within five years. When those pioneer farmers who previously had grown crops only in partially forested regions reached the grasslands, they faced an entirely new environment. Massive plows drawn by as many as fourteen people were needed to cut through and turn the thick grassland turf. Poorer people used sledgehammers and axes to cut the sod (they were called "sodbusters"). Wood for houses and for fuel was scarce on the treeless plains, and so were sources of water.

Inventions like the steel plow that cut through sod, windmills that pumped water from underground, and barbed wire that kept animals out of the grain fields helped farmers conquer the Great Plains. An area as large as Britain and France combined was transformed into farms and ranches.

⊕ REVIEW QUESTIONS

1. What purchase added thirteen new states and some of the country's richest farmland to the United States? From whom did we purchase this land?
2. What three types of transportation systems did Americans build to tie the country together?
3. By 1850, the South produced how much of the world's cotton?
4. What problems did pioneers traveling to the Oregon Territory face?

⊕ THOUGHT QUESTIONS

1. Why did improved transportation lead to great increases in agricultural and industrial production?
2. Why did New Englanders shift from farming to industry in the 1800s? Do you think this process is happening elsewhere in the United States today?

2. The Development of an Industrial, Urban Nation

The Growth of an Industrial Nation

Having mastered the land, Americans soon transformed their country from one based on farming to one based on industry. By 1900, the United States was already a manufacturing giant with an industrial heartland that stretched from New England to the Great Lakes Interior.

There were several reasons for this rapid industrial growth. First, abundant natural resources were discovered. Second, a flood of inventions—such as the telephone, the telegraph, and new machines—changed ways of living in the nation. Third, an ever-expanding road, waterway, and railroad network connected resources with people in the United States. This combination of resources, inventions, and cheap transportation triggered explosive industrial growth in the United States.

In Pittsburgh, Andrew Carnegie owned an industrial empire that included steel mills, iron and coal mines, limestone quarries, barges, and railroads. J. Pierpont Morgan, who bought out Carnegie, founded United States Steel, the first billion-dollar corporation and most important steel maker in the United States. John D. Rockefeller of Cleveland brought the same organizational genius to the oil industry. At one point, his corporation, Standard Oil, owned 95 percent of the oil-refining capacity of the United States.

The size and organization of these companies were astonishing. Each year, the amount and variety of manufactures increased steadily. Advances in technology, the driving energy and organizational ability of business leaders, and the hard work of immigrants and native-born laborers including the de-

Huge steel factories based on iron from the Mesabi Range in Minnesota and coal from the Appalachian Mountains emerged in the Industrial North. Do you think when this steel mill was photographed in 1937 many people were concerned about pollution?

scendants of African slaves, made the United States into a growing industrial nation.

The United States Becomes an Urban Nation

Cities grew along with industry in the United States as laborers piled into factory towns where there was work. By the early 1900s, 30 million Americans lived in cities. The largest of these cities had special sites and situations that encouraged growth. New York City, Philadelphia, and Boston were leading ports on the Atlantic Ocean. San Francisco, Los Angeles, and Seattle became the most important ports on the Pacific. Pittsburgh and St. Louis were located at the junctions of important rivers. New Orleans and Minneapolis-St. Paul anchored opposite ends of the Mississippi River. Chicago was the rail hub of the Central Plains.

Many of the residents of these cities were immigrants who had come to the United States in search of a better life. Hard times in their homelands had forced many of them to seek their fortunes in the United States. The Irish came because potato famines ruined their country. Crop failures and harsh rulers in central and eastern Europe encouraged Germans and Slavs to cross the Atlantic in the late 1880s. Jews escaped massacres in Russia in the 1800s. Millions more fled the worn-out farmlands of southern Europe. They came to the United States by the hundreds of thousands, and most of them ended up living in cities and working in factories.

By the 1900s, four out of five immigrants were factory workers who lived in cities; more than half of the industrial workers in the United States were foreign born. Russian Jews and Italians toiled in the garment trade in New York City; Poles, Greeks, and Syrians joined the Irish in New England's textile mills; and Irish, Slovak, Polish, and Italian miners worked the coal fields of Pennsylvania. At this same time, African Americans were migrating from the South northward to find jobs in the country's industrial heartland.

 IT'S A FACT In 1900, the United States had less than 500 miles of paved roads.

 IT'S A FACT California is the most urbanized state in the nation.

Conditions of Urban Life

Cities in the United States were overwhelmed by these immigrants. New York City had twice as many Irish as Dublin, more Italians than Naples, and probably the world's largest Jewish community. Chicago was the third-largest German city. One of every three persons in Milwaukee, Detroit, and Cleveland were immigrants. Desperate for work, these people took jobs in urban factories and moved into tenement districts nearby.

This enormous influx of people worsened conditions of housing, sanitation, crime, and corruption that already existed in many cities. The need for adequate health services and housing was severe. In 1900, only two out of every five infants survived their first years in one Chicago slum. In New York City, one out of every twenty people lived in a cellar. Sewage facilities were so primitive in cities like Philadelphia that epidemics of cholera were common. The urban crime rate skyrocketed. In the slums, living conditions were barely endurable.

In the early 1900s, some of these urban problems began to ease. Police and fire departments were established in many cities. Regular garbage removal was started, and sewage systems were installed. Streetcars became so popular that 850 trolley lines with 10,000 miles of tracks were built in the early 1900s. Cities began to spread, relieving crowding. New York City and Boston went underground and built subway systems to move people to and from work. Electric, gas, and telephone lines were installed in many cities in the United States. Aqueducts and reservoirs brought clean water to city dwellers.

Prosperity and Depression

Untouched by World War I, the United States continued to develop its industrial and agricultural resources. The country became as rich as all of Europe combined and the most powerful nation in the world.

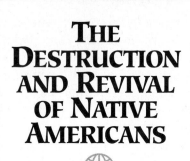

THE DESTRUCTION AND REVIVAL OF NATIVE AMERICANS

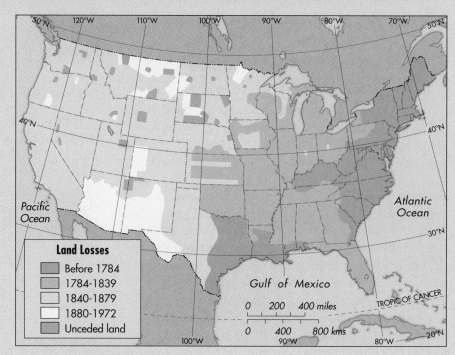

Native American Land Losses

N ative Americans, who number 8.7 million people, are the tenth-largest ancestry group in the country in the 1990s. For the first time, the validity of some of their claims to land and wealth are being treated seriously by the American government. They are now being allowed to decide on the uses of money derived from oil and gas royalties, grazing fees, and other income from their land.

Despite these efforts, Native Americans who live on reservations are still extremely poor, and one-third of their people live below the poverty line. Unemployment levels are between 50 and 80 percent on the reservations. The Native American story is one of the saddest in our country's history.

Between 1784 and 1880, the American government signed and broke 370 treaties with Native American tribes. As the frontier moved west in the 1800s, millions

of acres of land—two-thirds of a continent—were taken, as the map on this page shows. Today, Native American reservations in the United States amount to 50 million acres, or 2.2 percent of the nation's total area.

In 1860, some 250,000 Native Americans lived on the Great Plains. The strongest and most warlike groups were the Sioux, Blackfoot, Crow, Cheyenne, and Arapaho in the north, and the Comanche, Apache, Ute, and Kiowa in the south. The Plains tribes lived as nomadic hunters, migrating northward in summer and southward in winter, living off enormous herds of bison (often referred to as buffalo).

Some bison herds included as many as 12 million animals, on

which the tribes relied for food, shelter, and clothing. As the railroads penetrated the plains, bison hunters, farmers, and cattlemen began to destroy the herds, encroaching on hunting grounds officially granted by treaty to the Native Americans. By 1873, only one large bison herd was still intact; by 1876, no more large herds were left.

In desperation, the Native Americans fought back in fierce battles. One by one, the groups were subdued and confined on reservations (see the map on page 281). The Native American population was drastically reduced by raids, disease, and hunger. Long before the pitiless massacre of the last of the Sioux at Wounded Knee in 1891, the extinction of the bison

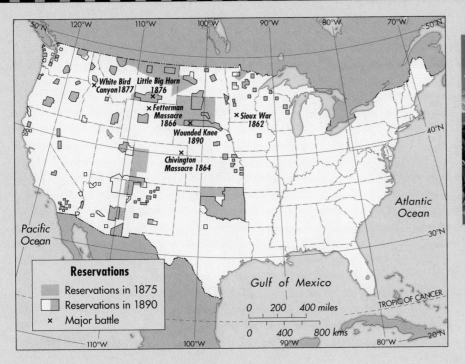

Reservations

█ Reservations in 1875
□ Reservations in 1890
× Major battle

White Bird Canyon 1877
Little Big Horn 1876
Fetterman Massacre 1866
Sioux War 1862
Wounded Knee 1890
Chivington Massacre 1864

Pacific Ocean

Atlantic Ocean

Gulf of Mexico

0   200   400 miles
0   400   800 kms

TROPIC OF CANCER

Native American Reservations and Major Battles

▲ *The Indian Gambling Regulatory Act of 1988 allows Native Americans to run bingo games such as this. What are some of the ways income from these gambling operations is being used?*

had sealed their destiny. The tragedy of their fate was poignantly summarized by Chief Joseph of the Nez Percé, who is reported to have said, after a fighting retreat over 1,500 miles of mountain and plain, and surrendering just short of asylum in Canada on October 5, 1877, "I am tired of fighting. . . . My heart is sick and sad. From where the sun now stands, I will fight no more forever."

Now prospects are beginning to change. In 1988, the Indian Gaming Regulatory Act was passed to reduce the poverty of Native Americans by allowing them to operate gambling games not prohibited by state law. Some seventy groups have initiated gambling. The once-poor Oneida tribe of central New York is opening a

huge casino east of Syracuse that will pay for free medical care and free education for every tribal member. In Connecticut, the Mashantucket Pequot tribe is running a gambling enterprise that has become a major employer in a depressed region. In other states, the status of such enterprises is pending.

In the West, Native American water rights are now generating substantial income. Native-American groups along the lower Colorado River are considering selling water to thirsty cities like Las Vegas, and possibly even Phoenix or Los Angeles. And in one of the largest land settlements in recent history, the United States is awarding 400,000 acres of land in Arizona to the Hopi tribe to ter-

minate its long dispute with neighboring Navajos.

The new relationship between the government and America's first citizens was symbolized in November 1992 by changing the name of the Custer Battlefield National Monument in Montana to the Little Bighorn National Monument.

❓ QUESTIONS

1. What law passed in 1988 opened up new economic opportunities for some Native Americans? Do you approve of this law?

2. Do you think these efforts to improve living conditions of Native Americans and grant them more respect is a good decision by the United States? Why or why not?

▶ *By the 1900s, many of the people who lived in America's largest cities were immigrants who tended to cluster together into ethnic neighborhoods. What is an ethnic neighborhood?*

Mass production was the economic key to this prosperity. The automobile industry provides the best example of this American system of industrial production. In 1908, Henry Ford produced his first Model T automobile, and by 1915 his corporation had produced 15 million of the sturdy black cars.

This astounding feat was made possible by **assembly-line** techniques of mass production, in which machines, equipment, and workers were arranged in line so that one operation followed another until the product was manufactured. This process eventually reduced the time required to build a Ford automobile from 12.5 hours in 1914 to 1.5 hours in 1925. As a result, the price of a Model T Ford fell from $600 to $290, and more people were able to buy a car. Soon mass-production techniques were applied to other industries, and the United States became the leading industrial nation of the world.

There were, however, economic problems in the United States after World War I. Old industrial areas like New England were beginning to decline. Times were difficult for coal miners in Appalachia as mines played out and oil became a popular alternative fuel. Farmers who bought tractors to feed millions of Europeans during World War I no longer had customers for their crops. And neglected minorities like Native Americans, African Americans, and Hispanic Americans were denied equal access to their country's freedom and prosperity.

In 1929 the stock market collapsed and the Great Depression that followed in the 1930s left the American people with devastating problems. In one year, 1,300 banks closed. Factories, business offices, and shops also closed. Millions of able-bodied people lost their jobs. So farmers burned crops they could not sell, and unemployed war veterans sold apples on street corners. Bread lines lengthened, and **shantytowns** made of makeshift shelters filled up with people who had lost their bank savings and their homes.

Huge federal construction projects were started to create jobs for the unemployed. Thousands of people—including the father of the author of this book—were put to work breaking rocks for government-sponsored roads and dams. In the South, the Tennessee Valley Authority was created to revive eroded farmlands and to supply electricity to rural areas in seven states. Federal jobs by the thousands provided unemployed Americans with work.

Building on Our Strengths

The Great Depression of the 1930s forced Americans to view their country from a different perspective, and they responded magnificently. The federal government began to take responsibility for the cost of public welfare. Americans recognized that poverty and unemployment could occur even in the most affluent societies. Government intervention, which had been seen as destroying individual freedoms, was now viewed by many as the only way of restoring them. The United States dedicated itself, in President Franklin D. Roosevelt's words, to "social values more noble than mere monetary profit."

This willingness to face up to injustice, misfortune, and inequality and to do something about it is our heritage. The commitment of the United States to personal freedom, political liberty, and justice has been reaffirmed in many different ways in the years that followed.

⊕ REVIEW QUESTIONS

1. What factors helped to promote industrial growth in the United States?
2. In the early 1900s, what proportion of the immigrants lived in cities?
3. What manufacturer developed assembly-line techniques? Describe how they made industrial production more efficient.

⊕ THOUGHT QUESTIONS

1. What difficult conditions existed in American cities in the early 1900s? How and to what extent were they remedied?
2. What severe effects were caused by the Great Depression that began in 1929? Do you agree that the thinking of President Franklin D. Roosevelt in the 1930s still applies today, or do you believe that the federal government should play a more limited role in the lives of Americans?

3. The Changing Economy of the United States

The American Way of Life

Today, Americans are among the most prosperous people in the world, although a large disparity in wealth exists between and among groups and regions. Many families own television sets and stereos, leisure-time equipment, one or more cars, and a range of appliances beyond the imagination of past generations. In addition, a variety of insurance, medical, and retirement programs provide security for many families. With a gross national product (GNP) of more than a trillion dollars, the United States surpasses every other country in wealth and power.

The Shift from an Industrial to a Postindustrial Society

Yet the United States faces problems growing out of basic economic changes. First, we are shifting from a manufacturing or industrial economy to a service or postindustrial economy. You are more likely to get a job as a teacher, computer operator, bank clerk, or salesperson than as a worker in automobile or textile manufacturing.

Second, our national economy is more integrated with the international economy than ever before. The names of many Japanese, French, German, and English products are familiar to you today, but most of these products were not part of the American scene when your grandparents were growing up. Third, we face competition from other industrial nations like Japan, Germany, and Canada, and from rapidly developing countries like Mexico, Chile, and Taiwan. No longer are the products of our country unchallenged in world markets.

By the 1950s the United States was the world's largest industrial machine. Its abundant resources had been harnessed by huge corporations to increase industrial production. In the 1990s, these **multinational corporations** control half of the industrial assets of the United States and employ hundreds of thousands of workers. They spend and earn

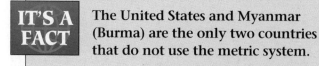

IT'S A FACT — The United States and Myanmar (Burma) are the only two countries that do not use the metric system.

1990s, more than four-fifths of all research and development efforts by corporations are concentrated in these high-technology fields. Most new jobs created in the 1990s are for well-educated, skilled workers in technical fields.

billions of dollars each year in dozens of locations in the United States and in other countries, which is why they are called *multinational*. They use advanced technologies to produce and sell goods and services to people and governments worldwide.

The shift to a postindustrial economy is based on new discoveries in computers, electronics, space exploration, and satellite communications. In the

Competition with other industrial nations is intensifying. The United States now produces only 15 percent of the world's steel, compared with 60 percent forty years ago. Likewise, American automobiles now face stiff competition from Asian and European car makers. Many basic products like automobiles, however, are joint creations made of parts from Japan, South Korea, Taiwan, and many other parts of the world, and then assembled in the

Resources and Industrial Areas of the United States

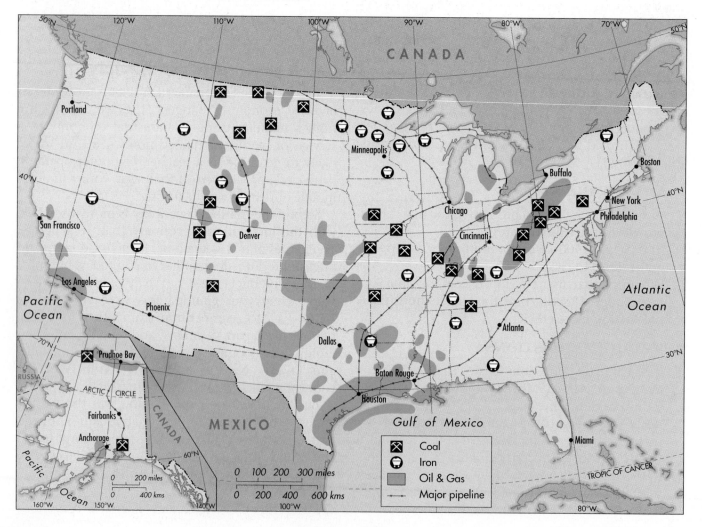

▲ *Strip mining results in great ecological damage if the area is not restored when mining operations are completed. Do you think mining companies should be allowed to use this method? Should mining companies have to pay the costs of restoring the land?*

United States or neighboring Mexico. This growing involvement in the international economy is changing the industrial landscape of the United States.

The federal government is attempting to control these economic forces. To accomplish this, our government has, at one time or another, instituted policies to ease inflation, and to help the troubled aerospace, automobile, and railroad industries, as well as agriculture.

These problems are part of a shift from an industrial economy based on manufacturing to a postindustrial economy based on service jobs, as you have already learned. Economic growth in industries like steel making is limited by rising costs of energy, by high labor costs, and by foreign competition. Americans are becoming aware that success in world markets is crucial to their own prosperity, and also that their own natural resources are limited.

America's Energy Resources

The United States has nearly half of the world's reserves of coal, and a quarter of the country's energy comes from coal. It is our cheapest, dirtiest (because it gives off many pollutants), and most abundant fuel. Our largest coal deposits lie near the surface and are strip mined. **Strip mining** removes surface layers of soil with gigantic earth-moving equipment to expose coal deposits. Because it is inexpensive, strip mining is used in both of our major coal fields: (1) the Appalachian coal field, which stretches from Pennsylvania to Alabama; and (2) the Eastern Interior coal field, which covers much of Illinois, as the map on page 284 shows. Already, 2 million strip-mined acres have become a scarred landscape of hills and gashed slopes. Laws have been passed recently to limit strip mining and to force mining corporations to restore the land to its natural state.

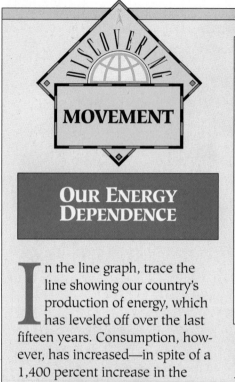

MOVEMENT

OUR ENERGY DEPENDENCE

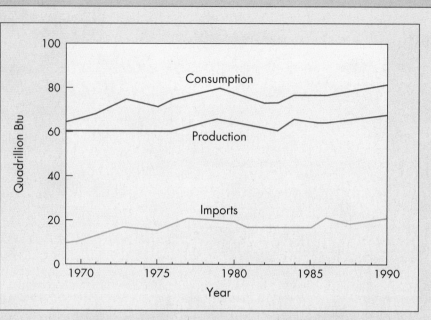

In the line graph, trace the line showing our country's production of energy, which has leveled off over the last fifteen years. Consumption, however, has increased—in spite of a 1,400 percent increase in the price of oil since 1973. As the bottom line shows, about 20 percent of our energy (and 40 percent of our oil) is now imported.

QUESTIONS

1. Why does the United States import so much foreign oil? What effect does this have on our trade deficit?

2. Why do you think Americans consume so much energy as compared with other industrial nations? For example, what is our attitude toward cars, air conditioning, and heating? Do you know how this differs from the attitude in other nations?

3. What effect does the cost of energy have on your life, if any? To conserve energy, what might you do?

Until 1975, the United States was the world's largest producer of oil and natural gas, which provide half of our total energy needs. Now we are the world's largest consumer and importer of these energy sources. This situation is an important factor in increasing our trade deficit because of the high cost of importing oil. About 90 percent of the oil and natural gas produced in the United States comes from three fields west of the Mississippi River: (1) the Gulf Coast Fields in Texas and Louisiana, which account for 60 percent of the oil and 70 percent of the natural gas produced in the United States; (2) the Midcontinental Field in Kansas, Oklahoma, and Texas; and (3) the California Field near Los Angeles. Locate these oil and natural gas fields on the map on page 284. Although these fields still have ample supplies of oil and natural gas, these resources are expensive to develop. That is why imports of less-expensive oil are increasing.

Our greatest undeveloped oil deposits are 200 miles north of the Arctic Circle in Alaska. Here, on the North Slope of the Brooks Range, an estimated 10 billion barrels of oil lie 10,000 feet below the surface. In a remarkable feat of engineering, this oil field was drilled, and a 799-mile-long pipeline across Alaska began transporting oil from this area in 1975. If demand for energy increases, expensive North Slope oil may be developed in spite of the massive oil spill in Prince William Sound in 1989.

Additional offshore deposits of oil and natural gas exist in the Arctic Ocean, the Gulf of Alaska, and the Gulf of Mexico, and off New England and

MOVEMENT

OUR DEPENDENCE ON IMPORTED MINERALS AND METALS

The United States is well endowed in minerals and metals, but our very complicated industries require large quantities of many different types of raw materials. We now import minerals and metals from neighboring countries like Canada and Mexico, and from faraway places like Zaire and Australia. Do you know why these metals and minerals are important?

QUESTIONS

1. What effect does the import of minerals and metals have on our economy?

2. Do you think that our dependence on many countries for minerals and metals will cause us to become more or less involved in international affairs? Support your view.

3. Why do you think the United States allowed itself to become dependent on foreign countries for oil, minerals, and metals?

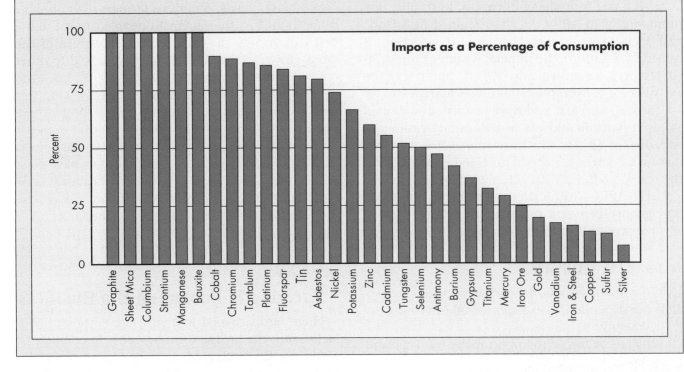

Imports as a Percentage of Consumption

California. Moreover, huge areas in Colorado, Utah, and Wyoming contain oil-bearing shale. Thus far, the low price of oil caused by the world glut in the late 1980s and early 1990s has discouraged efforts to develop these potential fuel reserves. Also, the development of these natural resources faces stiff opposition from Americans who believe that local environments will be damaged. What do you think should be done?

We Have Metal Ores, but We Need More

The United States is an important producer of metals like iron, copper, lead, and zinc, which are used in heavy industry. Recently, however, rising demand has outstripped our own production, so we have depended more on imports. As the graph above shows, the United States now imports more than half of its supply of a substantial number of important minerals and metals.

Iron ore production illustrates how production of metals in our country has changed. Iron ore deposits at the western end of Lake Superior in Minnesota produce 80 percent of the iron ore mined in the United States. This ore is shipped from Lake Superior through the Great Lakes to steel-making cities like Chicago, Gary, Cleveland, and Buffalo. The quality of these iron ores has declined, which means that higher-quality iron ore must be imported from other countries. So high-grade iron ore is shipped in from Canada, Venezuela, and Liberia to steel centers on the Great Lakes by way of the St. Lawrence Seaway.

Minnesota, Arizona, and Utah produce half of the country's minerals and metals. The United States is a major producer of copper, lead, zinc, uranium, and gold, all of which are mined in Arizona and Utah. Bauxite for aluminum production is mined in Arkansas. The mineral wealth of the United States is extensive and varied, but it can be exhausted if consumption continues to increase.

Think about what you have learned. The United States imports oil and gas, as well as many minerals and metals. This is why wise management of our natural resources is important. Good relations with the many countries with whom we trade is also critical. The 1994 NAFTA (North American Free Trade Agreement) between Canada, the United States, and Mexico is a step in this direction.

⊕ **REVIEW QUESTIONS**

1. What is a multinational corporation?
2. Is our dependence on imported oil increasing or decreasing?
3. Why are we forced to import many minerals and metals?

⊕ **THOUGHT QUESTIONS**

1. In what ways has America's economy become more involved in the international economy? Do you think this is good for the nation or not? Explain.
2. How do you think the shift from an industrial to a postindustrial economy might affect you personally?

4. Americans on the Move

The Flight from the Farm

The international economy is a major reason American farmers are facing problems. Some 23 million people lived on farms in 1950, 5.6 million Americans in 1980, and 4.6 million in 1990. Abandoned family farms are common sights in the farm belts of the United States. The graphs on page 289 dramatically show the changes taking place in farming in our country.

There are several reasons for these changes. First, improved farming methods, better seeds, and greater use of fertilizers and pesticides have increased crop yields. American farmers now produce more food than the American people can eat, more crops than they can use. Second, one-third of the crops grown in the United States were sold to other countries twenty years ago. In the 1990s, however, more countries are able to feed themselves, and many more countries are exporting food. These countries sell crops at low prices on world markets. Third, farm costs are going up faster than food prices in the United States. Many farmers have gone deeply into debt in order to stay in farming. Government farm-aid programs have not solved the problem. Small farmers are being pushed into other ways of earning a living.

Farming Is Becoming Big Business

Huge farms owned and run like corporations—factories in the fields—seem best able to compete in world markets today. Many of the largest of these farms are agribusinesses. An **agribusiness** is a huge farm under corporate management that raises, processes, and sells food products nationwide. As the right-hand graph on page 289 shows, farms more than doubled in size between 1950 and 1990 in the United States. Today, more than half of our farmland is cultivated in units larger than 1,000 acres. A quarter of all farms in the United States are owned by companies that process and sell food products.

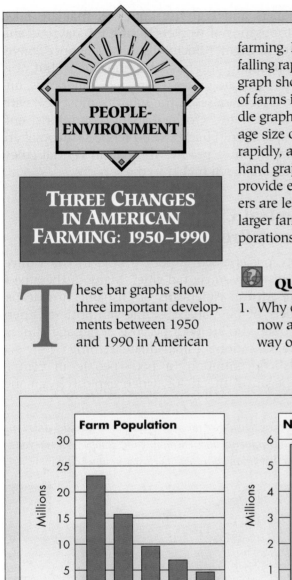

PEOPLE-ENVIRONMENT

THREE CHANGES IN AMERICAN FARMING: 1950–1990

These bar graphs show three important developments between 1950 and 1990 in American farming. First, farm population is falling rapidly, as the left-hand graph shows. Second, the number of farms is decreasing, as the middle graph shows. Third, the average size of farms is growing rapidly, as shown in the right-hand graph. These three graphs provide evidence that small farmers are leaving agriculture, while larger farms, many owned by corporations, are growing.

QUESTIONS

1. Why do you think people are now abandoning farming as a way of life and choosing to work and live in towns or cities? Do you know anyone who has done this?
2. Describe the type of schools (including universities), shops, hospitals, and cultural facilities (theatres, dance halls, and music clubs) you might expect to find in a rural farming area?
3. Why do you think the American government continues to subsidize farming when many farms are owned by large corporations? Do you remember the problem farm subsidies are creating in Western Europe?

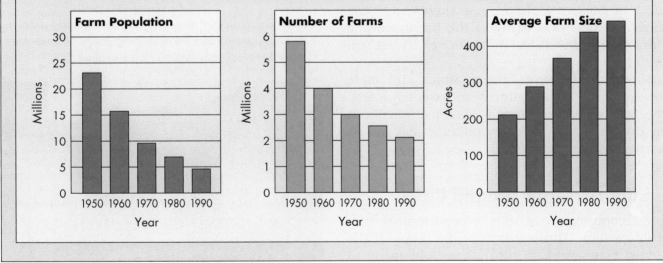

These agribusinesses produce huge amounts of crops and animals. They can invest large sums of money in modern farm machinery for planting, weeding, irrigating, and harvesting crops—technology that is too expensive for small farmers with limited acreage. As a result, the country's farms are fewer in number and are much more concentrated in specific areas.

In 1950, farming was practiced almost everywhere in the United States except in the forested mountains of the Pacific Northwest, the dry Southwest, and rugged areas like the Rocky Mountains, the Appalachians, and most of New England. In the 1990s, this distribution has changed dramatically. Farming is now practiced only in the most fertile areas in the Central Plains, the South, and the West.

Specialty farming of vegetables, fruits, herbs, and other products still is found near large cities, but in general poorer land that can no longer compete is being abandoned, and farmers with small acreage are leaving the land.

Americans Move to the "Sun Belt"

Changing economic realities are causing many Americans to move from one place to another in our country. Many families have moved from rural areas to towns and cities.

Others have left declining factories in the Industrial North and moved to the "sun belt," a cluster of southern and western states that extends from North Carolina, Georgia, and Florida through Texas and the Southwest to California. These states have mild climates, low energy costs, well-developed inland waterways, efficient transportation networks, and growing port cities that reach out to the world. Thousands of new and relocated businesses have created job opportunities in these states. Defense, aerospace, and high technology industries are clustered in California, Texas, and Florida—the "big three" of the 1990s. Many of the fastest growing cities in the United States are located in the sun belt.

Resorts and tourist centers attract still more Americans to these southern and western states. Tourists, as well as thousands of retired Americans, are drawn to the sun belt by mild climates and attractive living environments.

Suburbs Grow around Cities

A second important shift in population is from city to suburb. Three-quarters of the people of the United States work in urban areas, but many of these people live in city suburbs. Suburban areas reach out from city centers as far as commuters are willing to travel. Two-thirds of the population growth in the United States in the 1990s is happening in suburbs that ring the central business districts of large cities. Today three Americans live in suburbs for every two that live in central cities.

Automobiles allow most Americans to live far from their jobs. Suburbs spread out in widening circles around large cities. City centers are clusters of skyscrapers and high-rise buildings that are filled with thousands of workers during the daytime and deserted at night. Midcity districts were once convenient, pleasant, and busy neighborhoods, but now most of them are places where poor Americans, unable to afford houses in the suburbs, must deal with high crime rates, crowded living conditions, and aging apartment buildings and schools. Incomes are low among most people who live in central cities; incomes are much higher in the suburbs.

Central Cities Have Many Problems

Sixty percent of the poor people in the United States now live in central cities. Many of these people are African Americans or Hispanic Americans, members of our largest ethnic minorities. Others are people who are poor because of old age, illness, limited education, or inadequate training.

This concentration of poor people in central cities began in the 1940s, when explosive demand

▼ *The aging downtown areas of America's large cities are home to many minority and new immigrant groups. Why do you think new immigrants tend to cluster in inner cities?*

MOVEMENT

WE ARE ALL CHILDREN OF IMMIGRANTS

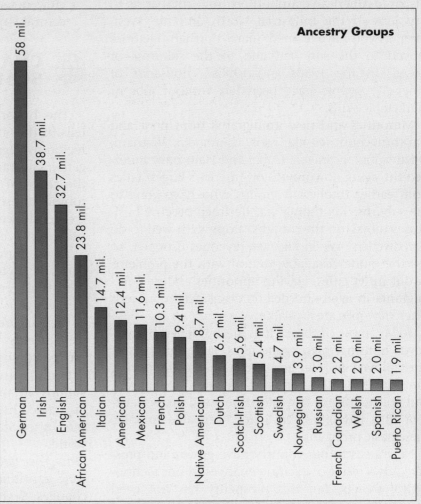

Ancestry Groups

German 58 mil.
Irish 38.7 mil.
English 32.7 mil.
African American 23.8 mil.
Italian 14.7 mil.
American 12.4 mil.
Mexican 11.6 mil.
French 10.3 mil.
Polish 9.4 mil.
Native American 8.7 mil.
Dutch 6.2 mil.
Scotch-Irish 5.6 mil.
Scottish 5.4 mil.
Swedish 4.7 mil.
Norwegian 3.9 mil.
Russian 3.0 mil.
French Canadian 2.2 mil.
Welsh 2.0 mil.
Spanish 2.0 mil.
Puerto Rican 1.9 mil.

The twenty largest ancestry groups as census respondents identified themselves in 1990.

In the 1990 census, the U.S. government asked people to identify their ancestry. There were 197 different responses. The largest groups, as the bar graph shows, claimed German, Irish, or English descent. African Americans were the fourth-largest group. Among the smallest groups were those from Uganda, Singapore, and Nepal. The government found out in this census what President Franklin D. Roosevelt noted years ago, that "we are all children of immigrants."

Today, new immigrants are attracted to America. According to the 1990 census, one-third of these new immigrants are Asian Americans from South Korea, Taiwan, China, Vietnam, Laos, and Cambodia, and more than one-third are Hispanic Americans from Mexico, Central America, and South America. If this trend continues, these ethnic groups will soon be named among the leading ancestry groups of people living in the United States.

Always a nation of immigrants, the United States is once again gathering in new ethnic groups from different parts of the world. As these immigrants, our "new pilgrims," recreate the American dream, the benefits to our country could be enormous.

QUESTIONS

1. What are the three largest ancestry groups named on the graph? When do you think most of them came to America?

2. Are you descended from one of the groups shown on the graph? How would you identify your ancestry?

3. Do you think new immigrants from Asia and Latin America should be allowed to enter the United States? Why or why not?

for factory workers lured African Americans, Hispanic Americans, and others from rural areas to city jobs in the Industrial North and the West. Later, as you have learned, many northern factories moved to the sun belt and to the suburbs—or closed—in the 1980s and 1990s. Thousands of inner-city people have been left without jobs or marketable skills.

Minorities and new immigrants from poor and war-torn countries like Laos, Cambodia, Vietnam, El Salvador, Nicaragua, Cuba, and Haiti have inherited the aging downtowns of America's largest cities from earlier immigrant groups who have gone to the suburbs. Tax money has declined because businesses moved to the suburbs, along with well-to-do city dwellers. When city tax revenues dropped, so did the money available to deal with the problems in our inner cities, leaving minorities and new immigrants in aged and deteriorating neighborhoods where few jobs are available.

Today, most of these cities are burdened by run-down housing, high crime rates, large numbers of unemployed people, and the high cost of paying the salaries of thousands of teachers, firefighters, and police officers. Downtown renewal projects, except in a few cities, have not been able to solve these basic problems.

New areas in our country have grown and prospered as a result of exciting changes in our technological society, but this prosperity has not been shared by many Americans who live in central cities. It is now the responsibility of all Americans to help them to participate in our country's new challenges and opportunities.

⊕ REVIEW QUESTIONS

1. What three basic economic changes does the United States face today?
2. What is an agribusiness?

⊕ THOUGHT QUESTIONS

1. What are the main problems of American family farmers today?

2. Why are the downtowns of America's largest cities deteriorating? Would you prefer to live in a central city or suburb? Why?

⊕ CHAPTER SUMMARY

The historical geography of the United States tells us that the American economy was developed region by region, and that these areas were tied together by transportation links. While each part of the country developed according to its own environment, resources, and unique features, the country developed an integrated economy by the late nineteenth century.

At the same time, the population of the United States began to become more concentrated in urban industrial centers in the East and Middle West. Farmers who were displaced by the mechanization of agriculture joined European immigrants who were fleeing wars and prejudice, as both groups moved to the cities in large numbers looking for better opportunity. American inventions and innovations in manufacturing, such as the assembly line, aided industrial growth and by 1900 propelled the United States into the forefront of the industrial world.

In the twentieth century, geographic patterns were continuously altered by domestic and global changes. Such factors as war, the automobile, radio and television, computers, and emerging industrial economies in places like Japan and Korea have dramatically changed the way we live.

Americans have moved out of city centers into the suburbs now that the automobile has given them greater mobility. Satellite cities, industrial parks, and shopping malls ring the older urban centers, and the population works and shops away from the inner city. The inner cities are now home to the urban poor and recently arrived immigrants. This means that the inner city no longer has a strong tax base, and urban decay and poorly funded schools add to urban problems. Today, the older inner cities are one of the biggest of the country's problems that are awaiting successful solutions.

EXERCISES

Select the Ending

Directions: Three possible endings are given for each of the following sentences. Select the correct ending and then write the complete sentence on your paper.

1. America was tied together by the construction of (a) steel mills (b) tenements (c) railroads.
2. The Erie Canal linked Lake Erie with (a) Lake Michigan (b) the Mississippi River (c) the Hudson River.
3. The railroad hub of the interior plains was (a) St. Louis (b) Detroit (c) Chicago.
4. By the 1850s, the most valuable single crop grown in the United States was (a) cotton (b) corn (c) rice.
5. The region that farmers had the most difficulty cultivating was the (a) South (b) Middle West (c) Great Plains.
6. Americans had the opportunity to buy land cheaply as a result of the (a) Gold Rush (b) Homestead Act (c) Louisiana Purchase.
7. By 1900, the industrial heartland of the United States stretched from New England to (a) the Great Lakes Interior (b) Texas (c) California.
8. The Great Depression (a) reduced government intervention in the economy (b) increased government intervention in the economy (c) was one of the causes of World War I.
9. The Tennessee Valley Authority was created to (a) revive eroded farmlands and provide electricity to seven states (b) turn the Tennessee Valley into an urban area (c) link the agricultural South to the Industrial North.
10. Between 1913 and 1925, the price of a Model T Ford (a) nearly doubled (b) stayed about the same (c) decreased by more than 50 percent.
11. In 1901, United States Steel (a) was sold by J. Pierpont Morgan to Andrew Carnegie (b) was the first billion-dollar American corporation (c) imported most of its iron ore from Western Europe.

Inquiry

Directions: Combine the information in this chapter with your own ideas to answer these questions.

1. In the past, the physical environment of a region strongly influenced the occupations of most people. Do you think this is still true today? Why or why not?
2. What are the chief reasons the United States became an industrial giant?
3. "In the 1930s government intervention, which had been seen as destroying individual freedoms, was now viewed by many as the only way of restoring them." How would you interpret the meaning of this statement?

SKILLS

Using a Transportation Map

Directions: The map on page 275 shows how roads, canals, and railroads linked parts of the United States by 1860. Use the map to answer these questions:

1. By 1860, were there more canals or railroads?
2. By 1860, was the West linked by roads to both St. Louis and Memphis?
3. Does the map show any canals west of the Mississippi River?
4. Was the first transcontinental railroad completed by 1860?
5. By 1860, was Mobile linked to Atlanta and Charleston by railroad?
6. Did more canals flow into Lake Erie or Lake Huron?
7. Were there more railroads in New England or Florida?
8. St. Louis gained fame as an important port for steamboats on the Mississippi River. By 1860, could people and products also reach St. Louis by railroad?

Reading Bar Graphs

Directions: Graphs can provide much valuable information at a glance. In this lesson you will be working with four graphs. First, turn to page 289 and look at the three bar graphs that show changes in farming in the United States. Then answer these questions:

1. What subject is covered by each graph?
2. On the farm population graph, are the population figures in hundreds, thousands, or millions?
3. Was the farm population in 1990 less than one-third of what it was in 1950?
4. Has the number of farms declined in every ten-year period since 1950?
5. Has the size of the average farm decreased in every ten-year period since 1950?
6. Did the size of the average farm double between 1950 and 1990?

Directions: Now look at the bar graph on page 287, which shows our country's dependence on imported minerals and metals, and answer these questions:

7. The United States imports 100 percent of how many of the raw materials shown on the graph?
8. The United States imports at least 75 percent of how many of the raw materials shown on the graph?
9. Does the United States import more or less than half of its iron ore?

Making a Graph

Directions: What are the most serious problems in the United States today? The class can discuss these problems. List them on the chalkboard and vote on which problem students think is the most serious. Add the votes and then calculate the percentage of the class that voted for each problem. Show the results in the form of a bar graph drawn on poster paper. Each 5 percentage points could be represented by one inch on the graph.

Vocabulary Skills

Directions: Match the numbered definitions in the second column with the vocabulary terms. Write each vocabulary term on your paper next to the number of the correct definition.

agribusiness
assembly line
multinational corporation
shantytown
strip mining
sun belt

1. The removal of surface layers of soil with earth-moving equipment to expose mineral deposits located near the surface
2. A huge farm under corporate management that raises, processes, and sells food products nationwide
3. A section of a city or town characterized by crudely built houses or shacks
4. A method of arranging machines, equipment, and workers in a line in which one operation follows another until a product is manufactured
5. The cluster of southern and western states that extend from Florida to California
6. A business that operates in more than one country

CANADA

Canada is a huge land with few people. Canada is equal in area to the entire continent of Europe, yet the country has a population of only 28 million. Climate, landforms, and resources dictate where settlements are found in Canada. Three-fourths of all Canadians live in large cities strung out like beads along their border with the United States. Few Canadians live in the snowbound environments of their northern frontier, but Canada's people have used the resources of the north to create a prosperous country. Today Canada is one of the ten wealthiest industrial countries in the world.

◄ *Vancouver is Canada's largest port on the West Coast. Remembering the satellite photo of San Francisco Bay in Chapter 9, what do you think these two port cities have in common?*

1. The Building of Canada

The peace treaty after the American Revolutionary War specified that the northern half of North America would remain separate from the southern half. Britain kept its northern colonies, which later joined together to form Canada. Canada's southern boundary ran east-west across the continent through rugged wilderness and across high mountains. The British colonies in Canada were separated from one another by formidable geographical obstacles.

Unifying Canada

In the 1840s, the English-speaking province of Upper Canada (Ontario) and French-speaking Lower Canada (Québec) were united. By 1850, a majority of Canadians were English-speaking people who lived on the Atlantic coast and in Upper Canada, or Ontario.

When pioneers surged westward in the United States in the 1850s, Canadians recognized that they had better develop their own western territory as quickly as possible. By 1867, a continentwide federation of provinces and territories began to take shape in Canada. In the east there were six colonies: Upper Canada (Ontario), Lower Canada (Québec),

Political and Physical Map of Canada

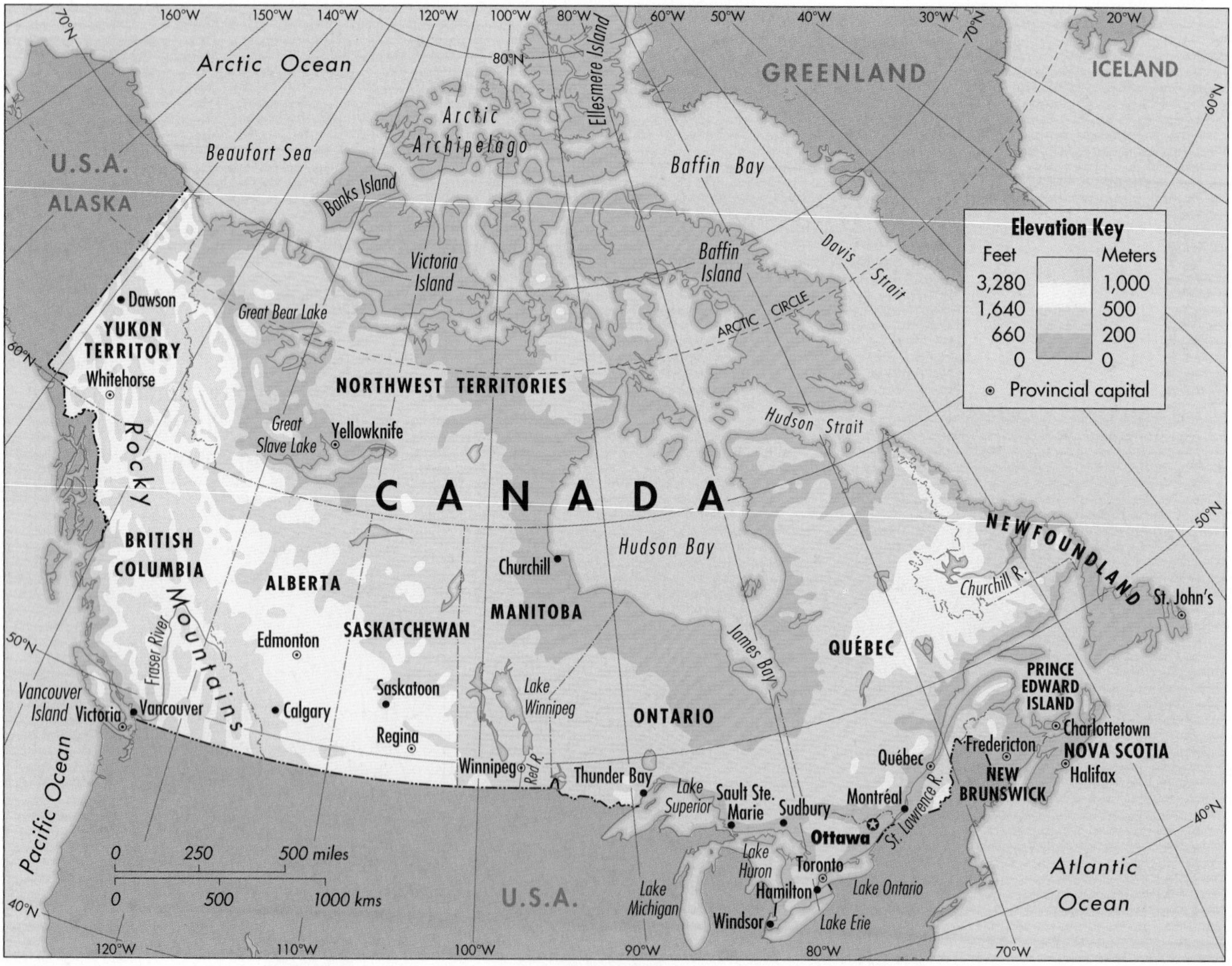

Elevation Key

Feet		Meters
3,280		1,000
1,640		500
660		200
0		0

⊙ Provincial capital

Newfoundland, Nova Scotia, New Brunswick, and Prince Edward Island. British Columbia was located in the far west. Between British Columbia and Upper Canada, British settlers occupied the Red River Valley, which later became the province of Manitoba. Still later, immigrants filled in the area between British Columbia and Manitoba, farming and settling the area that became Alberta and Saskatchewan. Throughout this settlement process, Native Canadians were treated with more respect and consideration than the Native Americans in the United States.

Gradually, Canada became more independent of Britain. The British gave their colonies in Canada self-government as **provinces** within a **dominion**. This dominion maintained close political and economic ties to the British government. Canada broke away from direct British rule peacefully.

Building a Prosperous Country

In the 1900s, Canadians moved westward across the fertile lands along their southern border. Millions of immigrants, many of them from Eastern Europe, started farms on the lush grasslands of south central Canada. Wheat became an important source of Canadian wealth. Gold was discovered in the Yukon Territory to the north, and additional mining strikes were made of nickel, iron, lead, and silver in ore-rich Canada. Logging and fishing also made valuable contributions to Canada's growing economy.

Meanwhile, in the St. Lawrence Valley and on the northern shores of the Great Lakes, manufacturing centers like those of our Industrial North began to transform Canada into an urban, industrial nation.

⊕ **REVIEW QUESTIONS**

1. What resources did the Canadian North contribute to the country's economy?
2. Describe how the separation of the United States from Britain differed from that of Canada.

⊕ **THOUGHT QUESTIONS**

1. Why do you think so many Canadians live near the border between Canada and the United States?
2. In what part of Canada do you think you would most like to live? Why?

2. The Regions of Canada

A Vast and Empty Land

Canada stretches over 3,200 miles from east to west, extending across five and one-half of the world's time zones. It reaches 2,900 miles northward to within 500 miles of the North Pole. Ringed by three oceans—the Atlantic, the Pacific, and the Arctic—Canada has one of the world's longest coastlines.

Most of Canada is empty of people. The northern half of the country is a wilderness of rocks, dense forests, lakes, and bogs. Native Canadian hunters and herders live in small settlements here. Mining centers, trading posts, lumber camps, and military outposts are among the larger settlements in the Canadian North. This region occupies the polar half of North America.

The cold climates and hostile lands of this polar wilderness have, as one Canadian notes, "squeezed Canadians like toothpaste" along their border with the United States. Three-quarters of Canada's people live within 100 miles of the border of the

Subregions of Canada
Canada has five major subregions: (1) the Canadian North, (2) the Maritime Provinces, (3) the St. Lawerence Lowlands, (4) the Prairie Provinces, and (5) British Columbia.

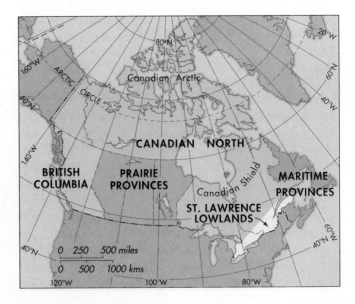

IT'S A FACT Canada is forty times larger than Great Britain.

United States; nine out of ten Canadians live within 200 miles. In contrast, only one of every ten Americans lives within 100 miles of Canada.

A Difficult Geography

Writers have often described Canada's geography as difficult. Natural barriers separate Canada's regions from one another, which makes national unity a difficult challenge. Most Canadians are confined to the southern margins of their country by the cold climates and the rugged lands of the Canadian North.

In the east, the Appalachian Mountains slice northward through Canada. These mountains separate the primarily English-speaking people of the Maritime Provinces of Nova Scotia, New Brunswick, Prince Edward Island, and Newfoundland from the primarily French-speaking people of Québec. Newfoundlanders must sail across the Gulf of St. Lawrence to reach the Canadian mainland. Indeed, Newfoundland did not become a province of Canada until 1949.

West of the St. Lawrence River, rocks, swamps, and forests make east-west travel between Ontario and the Prairie Provinces difficult. These provinces,

in turn, are blocked from easy communication with British Columbia on the Pacific coast by the jagged mountain ranges of the Canadian Rockies.

Canada is a united country because it has overcome the challenges of its difficult geography. In spite of many barriers that separate east and west, Canadians have created a prosperous country that reaches from the Atlantic to the Pacific. Within this country there are five major subregions: (1) the Canadian North, a frozen land that produces many mineral resources vital to the nation's economy; (2) the Maritime Provinces, whose relation to the sea is similar to that of Maine; (3) the St. Lawrence Lowlands of Ontario and Québec, which form the heartland of Canada's population and economy; (4) the Prairie Provinces, a vast grain- and oil- producing region that is an extension of our Central Plains; and (5) British Columbia, Canada's Pacific coast province. Each of these Canadian regions has a difficult landscape and climate. Over time, these regions have developed distinctive economies.

The Canadian North

The Canadian North is largely uninhabited. Under most of this region is a **shield**, a massive block of very old, very hard rock. The Canadian Shield covers about half of Canada's land, nearly 1.6 million square miles. Blanketed by coniferous forests, the shield's glaciated landscape is made up of bare rocks, swamps, shallow lakes, and thin soils.

▶

Banff National Park in the province of Alberta is located in a northern extension of the Rocky Mountains in Canada. How have these mountains affected the communication between Canada's eastern provinces and British Columbia?

The Canadian Shield begins in the arctic region of central Canada, swings southward to encircle Hudson Bay, and forms the rock base of virtually all of eastern Canada except for the St. Lawrence Lowlands north of Lake Ontario and Lake Erie. It reaches south of the Canadian border to northern Minnesota, Wisconsin, and Michigan.

North of the shield is the Canadian Arctic. This region is made up of hundreds of ice-covered islands that sweep nearly a quarter of the way around the Earth from Banks Island in the west to Ellesmere Island near Greenland. Locate these huge Arctic islands on the map on page 296. Winters here are so long and cold that only one month has an average temperature above freezing. Trees will not grow in this climate. The Arctic's tundra vegetation is made up of mosses, lichens, and ferns. Fewer than 20,000 people live in the Canadian Arctic; many are **Inuit** (Native Canadian) seal and whale hunters.

Only 5 percent of all Canadians live in the harsh environments of the Canadian North, but the resources of this region are vital to the Canadian economy. The shield is rich in mineral resources—iron, nickel, zinc, copper, uranium, gold, and silver ores—and is the source of half of Canada's mineral production. Canada leads the world in the production of nickel, zinc, and silver. More mineral finds

Resources of Canada

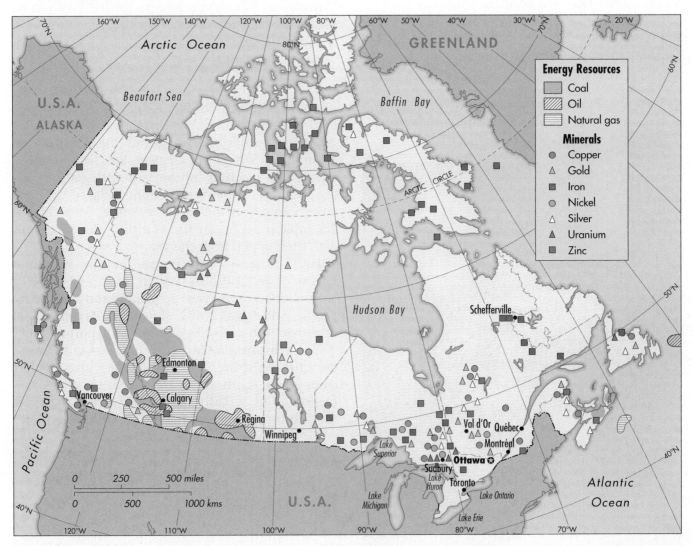

in the Canadian Arctic await development in the future.

Logging is also important in the Canadian North wherever timber, pulp, and paper mills are able to transport their products to the outside world. Hydroelectric power is generated from streams dropping off the southern edge of the shield. These streams provide inexpensive electricity to Ontario, Québec, and parts of northern New England. In the mid-1990s, however, a major dispute has developed between local residents, who see their ways of life being destroyed by a massive hydroelectric project, and the government in northern Québec.

In 1991, the Canadian government granted political domain of 770,000 square miles (one-fifth of Canada's area) of the eastern reaches of the North-

west Territories, an area called Nunavut to the 17,500 Native Canadians living there. Included in the agreement was a large cash settlement and limited mining rights in this frozen region. In 1997, Nunavut will probably become a full-fledged territory with its own government. Other Native Canadians such as the Cree and the Mohawks are equally interested in securing territorial rights.

Except for the Inuit, most people in the Canadian North live in mining camps, logging settlements, and military outposts. Even these small settlements must struggle with the environmental extremes of this hostile land. Many mines must be heated before minerals can be extracted. The permanently frozen soils that underlie much of this region make road building extremely difficult. Cold climates limit farming to a few favored locations. Virtually all food and supplies must be flown in to settlements in the Canadian North.

The Maritime Provinces

The Maritime Provinces of Nova Scotia, New Brunswick, Prince Edward Island, and Newfoundland line the Atlantic shores of Canada. They are located in the northern ranges of the Appalachian Mountains, which in Canada are low, forested mountains and hills interspersed with narrow valleys. Most settlements are fishing villages and port cities located on the rocky coastline. Some of the 2.5 million people in the Maritime Provinces still earn their livings from the sea.

Four of every five people in the Maritime Provinces are of British descent and speak English. Nova Scotia means "New Scotland," and some of its people still speak with a Scottish accent. New Brunswick has a large French-speaking minority.

The Maritime Provinces, as the gateway to Canada, carry on a lively Atlantic trade with seaboard cities in the northeastern United States. After completion of the St. Lawrence Seaway, which bypassed the Maritime Provinces, trade with Europe declined.

▼ *Logging is a major economic activity in British Columbia. Have you heard of any environmental issues related to forest clearing?*

◄

The coastal port of St. John's in New-foundland is one of the oldest cities in North America. Do most people in the Maritime Provinces speak English or French?

The region's major cities, Halifax in Nova Scotia and St. John's in Newfoundland, are small cities whose economies depend on logging, farming, fishing, tourist-related activities, and retail trade.

Logging is practiced in the hills and farming is important in protected valleys, but fishing once dominated the economy of the Maritime Provinces. Fishing fleets catch lobster and other shellfish along the coast; however, fishing for cod in offshore waters is now severely limited. The Grand Banks, once one of the world's richest fishing grounds, which lies off the coast of Newfoundland, has been badly overfished.

Increasingly, people in the Maritime Provinces supplement their incomes by serving tourists who visit the region to enjoy its wooded countryside and lovely coasts. Tourism is not sufficient, however, to raise levels of living very much. As a result, the Maritime Provinces are Canada's poorest region.

The St. Lawrence Lowlands

The St. Lawrence Lowlands, the heartland of Canada, are located entirely within the two provinces of Ontario and Québec. These lowlands extend from Québec City on the St. Lawrence River along the northern shores of Lake Ontario and Lake Erie to Lake Huron. The Canadian Shield is located north and west of these lowlands; to the south lies

the United States. In between, the St. Lawrence Lowlands form a narrow zone of fertile agricultural land that is the most densely settled, economically important area in Canada.

Nearly 16 million people, 60 percent of all Canadians, live in the St. Lawrence Lowlands. Canada's two largest cities are located in this area. Montréal, with a population of more than 3 million, is the largest French-speaking city in North America. Québec City to the east is the historic center of French culture in Canada. Toronto, Canada's largest city with a population of nearly 4 million, is the major center for the province of Ontario's English-speaking population. Canada's capital city of Ottawa is situated between Toronto and Montréal, in the border zone between English-speaking and French-speaking Canada.

The St. Lawrence Lowlands region is the economic core of Canada. Hydroelectric power is provided by waterfalls where rivers drop off the edge of the Canadian Shield. Raw materials from the mines and forests of the shield are used in industry. Three-quarters of Canada's manufactured goods are pro-

IT'S A FACT   Montréal, Canada, is the largest French-speaking city in the world except for Paris.

duced in the industrial cities that line the northern shores of the upper St. Lawrence River, Lake Erie, and Lake Ontario. Canada's most important publishing houses, financial institutions, and corporations are located in these cities.

The St. Lawrence Seaway gives these lowlands direct access to the Atlantic Ocean. An expensive and somewhat disappointing project built by the United States and Canada, this waterway reaches 2,300 miles into the North American interior, linking fifty-six port cities with the outside world. Its fifteen locks lift ships some 600 feet above sea level in Lake Superior, as the diagram below shows.

But the seaway was outdated almost as soon as it was completed in 1959. Its locks are too small to accommodate large, modern, oceangoing vessels, and the waterway is frozen three months of the year. Railroads can deliver most goods from middle America to the East Coast more quickly, so only bulk cargoes of wheat, coal, and iron ore are carried on the seaway. The 300-mile trip between Toronto and Montréal now takes about two days.

Fertile soils and moderate climates have enabled the St. Lawrence Lowlands to become one of Canada's major food-producing regions. Dairy farms, market gardens that specialize in vegetables, fruit orchards, and a variety of other farms provide food for nearby cities. This densely settled farming area hugs the shores of the Great Lakes; the rugged surface of the Canadian Shield prevents it from extending northward into the Canadian interior.

The Prairie Provinces

Another important farming area lies along the southern border of the Prairie Provinces of Manitoba, Saskatchewan, and Alberta. Here, 4.5 million Canadians live on the fertile grasslands that extend

The St. Lawrence Seaway

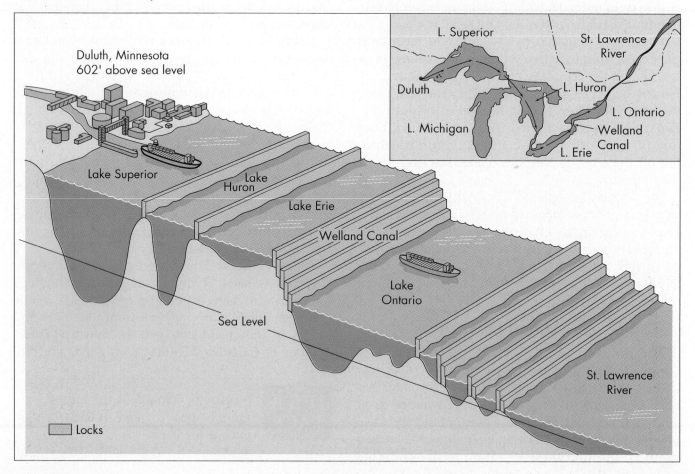

northward from our Central Plains into Canada. Large fields of spring wheat cover much of this region. The enormous productivity of these flat, fertile prairies has made Canada the world's second largest exporter of grain, after the United States. As a result, the Prairie Provinces are frequently called Canada's "breadbasket."

Discoveries of huge deposits of oil were made in Alberta in the 1940s. The oil and gas fields of Alberta shown on the map on page 299 now produce 85 percent of Canada's energy and petrochemicals. The cities of Calgary and Edmonton became oil boom towns and important petroleum service centers. These cities are relatively small, as are Winnipeg, the capital of Manitoba, and Regina, the capital of Saskatchewan. The oil fields, mines, ranches, and farms of the Prairie Provinces do not require the services that large city populations provide.

British Columbia

British Columbia faces the Pacific Ocean and is separated from central and eastern Canada by a wall of towering mountain ranges, one of which is part of the Rocky Mountain system. The snow-capped peaks, great rivers, and glaciers of this region are a sharp contrast to the flat landscape of Canada's Prairie Provinces. British Columbia is so distant from the rest of Canada that in the 1800s it considered joining the United States.

The economy of British Columbia is similar to that of the Pacific Northwest in our country. This subregion's forested slopes, rivers, and mountains produce a rich harvest of timber and hydroelectric power. Fishing and mining add to the wealth of this beautiful province. Vancouver, with a population of 1.5 million, is the largest city and most important port in British Columbia. This city is Canada's window on the Pacific.

⊕ REVIEW QUESTIONS

1. What is the Canadian Shield? What resources are found there?

2. Name the Maritime Provinces, and tell where they are located.

3. What region in Canada has the most people and the most economic activity?

4. Why are the Prairie Provinces called the "breadbasket" of Canada?

⊕ THOUGHT QUESTIONS

1. What difficulties of geography have made Canada's national unity challenging to build and maintain?

2. How are the St. Lawrence Lowlands the heartland of Canada?

◀

Wheat fields cover much of the landscape in this southern area of the Prairie Provinces in Canada. What three provinces are included in the Prairie Provinces subregion?

3. Canada's Future

A Bilingual Country

From its beginning, Canada's population has been divided in language, culture, and tradition between the French-speaking settlers of Québec and the English-speaking people of Ontario. Canada has had to overcome these cultural differences, as well as geographical obstacles, in order to create a unified nation.

Today, 41 percent of all Canadians trace their origins to the British Isles, 24 percent to the original settlers of New France, and 25 percent to German and Eastern European immigrants who moved to Canada in the 1900s. The French are concentrated in Québec, where nearly 80 percent of the people are fluent in this language. The government recognized these cultural differences by declaring both French and English to be official languages of Canada. Instruction in both languages is required in schools.

These language differences have not yet been resolved. In 1990, in response to French-speaking people's demands for an independent Québec, the Meech Lake Accord was developed by the government as a plan to preserve French language and culture. Many Canadians, however, believed the agreement gave French-speaking Canadians priority over other ethnic groups and overwhelmingly rejected this agreement in a national referendum. In October 1995, the province of Québec nearly seceded from Canada, and only by a narrow vote remained Canadian.

Like the United States, Canada is a nation of immigrants. Millions of Europeans became Canadians after World War II. Most of these people were refugees from Germany and Eastern Europe, and many settled in the Prairie Provinces and developed the resources there. Since 1945, Canada's population has doubled to over 27 million. In the process, new cultural elements have been added to Canada's mosaic of people. Recently, Jamaicans, Hong Kong Chinese, and Vietnamese refugees have established homes in Canada. Today, one of every six Canadians was born outside Canada.

Canada and the United States

Canada's immigrants have helped to build the country's economy, which is based on its wealth of natural resources. These resources make Canada an important trading nation. Today, Canada exports a quarter of its total resource production, much of it to the United States.

Canada's economic future is closely tied to that of the United States. Canadian hydroelectric power, minerals, oil, and wood products flow southward into the United States. Coal, iron ore, manufactured goods, and high-technology products move northward into Canada. Many companies in Canada are jointly owned by Canadian and U.S. corporations.

This close relationship with the United States is a matter of concern among Canadians. Economic problems in the United States frequently influence Canada; U.S. trade practices affect Canada's exports. Acid rain from factories in the United States and Canada kills forests and fish in eastern Canada. Declining industrial employment in the Industrial

▼ *The border between the United States and Canada is the longest open border in the world. Does a citizen of the United States need a passport to travel in Canada?*

PEOPLE-ENVIRONMENT

ACID RAIN

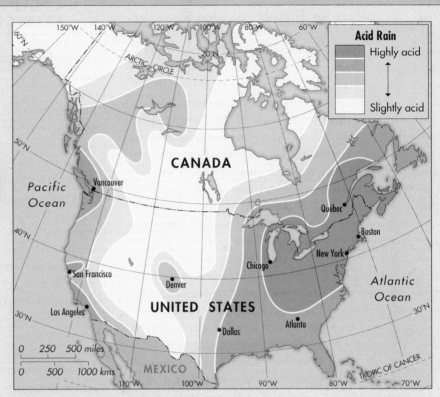

Acid Rain

Highly acid

Slightly acid

Pollution created when sulfur dioxide in smoke from coal- and oil-burning factories and nitrogen oxide from automobile emissions combine with water vapor in the atmosphere is called **acid rain**. This chemical reaction makes rain, sleet, and snow as acidic as vinegar in areas downwind from major industrial regions. Acid rain is widespread in Western and Eastern Europe and is now causing problems between Canada and the United States.

Acid rain contains a destructive chemical that kills fish, slows the rate of tree growth, damages buildings, eats the paint off of automobiles, and endangers human health. As the map above shows, the incidence of acid rain is particularly intense in the Industrial North, Appalachia, southeastern Canada, and New England—areas that are located near or downwind from densely populated manufacturing centers.

As many as 50,000 lakes in New England and eastern Canada are already so polluted that fish have been killed or reduced in number. Tree yields are expect-

ed to decline 15 percent in the next twenty years. Increasing evidence suggests that acid rain also causes bronchitis, asthma, and other lung diseases, particularly in children.

The people of eastern Canada are concerned that acid rain from the United States is damaging their environment. One Canadian conservationist commented, "They're the ones spewing it out. They're the ones who ought to pay for it." Some progress has already been made. Expensive pollution controls on factory smokestacks in the United States have reduced emissions by half. Canada has done less to reduce industrial pollution, but it produces far fewer pollutants. At

Canada's request, the United States has agreed to do more research on acid rain.

? QUESTIONS

1. Are Americans responsible for the acid rain that falls downwind? Should we pay for cleaning it up?
2. Looking at the map, do you live in an area with a high level of acid rain? If so, what effect could this have on your town, state, or country?
3. Do you believe all pollution, such as acid rain, should be eliminated, or should there be a balance between industrial costs and the level of pollution? Support your view.

North of the United States triggers similar changes across the border in Canada. As one Canadian prime minister pointed out, "When a mouse is in bed with an elephant, however well intentioned the elephant, the mouse must be conscious of its every twitch." Canadians often feel that the United States is not aware of its impact on Canada.

The United States and Canada are close trading partners and close allies. They have cooperated in many activities, including the construction and operation of the St. Lawrence Seaway, the nuclear defense of North America, and the environmental cleanup of the Great Lakes. Free trade between Canada and the United States has brought these nations even closer together. Canada and the United States form the largest trading partners in the world. As President Franklin D. Roosevelt remarked, we are good neighbors because "we discuss our common problems in the spirit of common good."

Canada has extended this neighborly spirit to many countries through active participation in world affairs. The Canadian military plays an important role in United Nations' peacekeeping forces, and Canada contributes social and economic assistance to developing countries. Recently, Canadians accepted refugees from the war-torn lands of Southeast Asia and Afghanistan into their country. For these reasons, Canada is held in high esteem among the world of nations.

⊕ REVIEW QUESTIONS

1. What is the major cultural division in Canada?

2. How much has Canada's population grown since 1945?

3. What are the chief products that Canada sells to the United States, and what are the main products that Canada buys from the United States?

⊕ THOUGHT QUESTIONS

1. How has Canada dealt with its cultural differences?

2. What is meant by the statement that Canada is "a nation of immigrants"?

⊕ CHAPTER SUMMARY

Canada, like the United States, was originally settled by Native Canadians like the Inuit and then by European immigrants. Today, the majority of Canadians are of English and French descent. The French were the original European settlers in Canada and they now comprise 80 percent of the population of Québec. English-speaking settlers followed and colonized Upper Canada. In the 1840s, the two Canadian colonies were united. While Canadian settlers moved westward and established themselves at about the same times as American colonists, they did not have to fight a war with Britain for independence. Canada was organized into provinces and received dominion status within the British Commonwealth.

Geographically, Canada can be divided into five major subregions: the Canadian North, the Maritime Provinces, the St. Lawrence Lowlands, the Prairie Provinces, and British Columbia. Natural barriers separate Canada's regions from one another. The cold northern climates confine most Canadians to within 100 miles of the U.S. border with the majority in the St. Lawrence Lowlands. Canada's industrial heartland is located along the St. Lawrence River, along with its largest cities. Logging and mining are the main economic activities of Northern Canada, and tourism, logging, and fishing support Canadians in the Maritime Provinces. Wheat farming in the Prairie Provinces makes that subregion Canada's "breadbasket." Forest products, hydroelectricity, fish, and minerals are British Columbia's major industries. Increasingly, British Columbia is becoming involved in the economy of the countries that ring the Pacific Ocean.

To create a unified nation, Canada needs to overcome the cultural differences that have traditionally divided the French-speaking settlers of Québec from the English-speaking settlers of the rest of Canada. Canada and the United States are working together to solve the problems caused by acid rain. Canada's economy, and hence its future, is closely linked to that of the United States.

EXERCISES

Two Are Correct

Directions: In each of the following groups of three statements, two are correct. Write the correct statements on your paper.

1. a. The United States is larger than Canada.
   b. Canada is the second largest country in the world.
   c. Canada is equal in size to the entire continent of Europe.
2. a. Climate, landforms, and resources dictate where settlements are found in Canada.
   b. Most Canadians live on their northern frontier.
   c. Three-fourths of all Canadians live near the border of the United States.
3. a. The largest cities in Canada are Montréal and Toronto.
   b. In the Maritime Provinces, mining is the most profitable economic activity.
   c. Québec is the main French-speaking province in Canada.
4. a. Canada leads the world in the production of nickel, zinc, and silver.
   b. The Canadian Arctic has hundreds of ice-covered islands that sweep nearly a quarter of the way around the Earth.
   c. Farming is more important than logging and mining in the Canadian North.
5. a. The Canadian Shield has greatly limited population growth and farming in the St. Lawrence Lowlands.
   b. Three-quarters of Canada's manufactured goods are produced in the cities along the northern shores of the St. Lawrence River, Lake Erie, and Lake Ontario.
   c. The St. Lawrence Seaway gives the St. Lawrence Lowlands direct access to the Atlantic Ocean.
6. a. The flat, fertile Prairie Provinces have helped Canada become the world's second largest exporter of grain.
   b. The chief grain raised in the Prairie Provinces is corn.

   c. Huge oil discoveries have been made in the province of Alberta.

Finish the Sentences

Directions: A word or term is needed to finish each of these sentences. Find the correct words or terms in the list that follows, and then write the complete sentences on your paper.

1. The largest French-speaking city in North America is
2. The largest city in Canada is
3. The large city that is Canada's window on the Pacific is
4. The capital city of Canada is
5. A city in the Prairie Provinces that experienced large growth because of oil and natural gas discoveries is

Vancouver	Calgary
Montréal	Toronto
Ottawa	

Inquiry

Directions: Combine the information in this chapter with your own ideas to answer these questions.
1. What factors divide Canada geographically, economically, and culturally?
2. In what ways are Canada's climate and landforms partly responsible for the closeness of economic and cultural ties between the peoples of Canada and the United States?

SKILLS

Using a Political-Physical Map

Directions: You have studied both political maps and physical maps in previous chapters. The map of Canada on page 296 combines both political and physical features. Use it to find the correct endings to each of the following sentences, then write the complete sentences on your paper.

CHAPTER 11

1. The elevation of eastern and central Canada is generally (a) about 3,280 feet (b) above 1,640 feet (c) below 1,640 feet.
2. A city on the St. Lawrence River is (a) Vancouver (b) Montréal (c) Calgary.
3. The capital of Canada is (a) Ottawa (b) Québec (c) Montréal.
4. Canada's highest mountains are in the (a) southeast (b) northeast (c) west.
5. Regina is a city in the province of (a) Alberta (b) Saskatchewan (c) Ontario.
6. The provincial capital of British Columbia is (a) Vancouver (b) Victoria (c) Ottawa.
7. In the northeast, Manitoba Province is bordered by (a) Hudson Bay (b) the Pacific Ocean (c) the Atlantic Ocean.
8. The Canadian Yukon is near (a) the Great Lakes (b) Greenland (c) Alaska.
9. Newfoundland is north of (a) Nova Scotia (b) Hudson Strait (c) Great Bear Lake.
10. Ellesmere Island lies in (a) the Pacific Ocean (b) Hudson Bay (c) the Arctic Ocean.
11. Most of Nova Scotia is (a) a low plain (b) mountainous (c) an archipelago.
12. The Rocky Mountains (a) are only in the United States (b) are near the Great Lakes (c) extend into British Columbia.
13. Toronto is a port city on (a) Lake Michigan (b) Lake Erie (c) Lake Ontario.
14. The land surrounding Hudson Bay is generally (a) a lowland (b) a highland plateau (c) mountainous.
15. A city near the Canada-United States border is (a) Edmonton (b) Winnipeg (c) Churchill.
16. The provincial capital of Nova Scotia is (a) Charlottetown (b) Halifax (c) St. John's.
17. Windsor lies on the Canadian side of (a) Lake Superior (b) Lake Toronto (c) Lake Erie.
18. The distance between Montréal and Toronto is (a) shorter than the distance between Vancouver and Calgary (b) about equal to the distance between Vancouver and Calgary (c) longer than the distance between Vancouver and Calgary.

Reviewing Latitude and Longitude

Directions: Use the political-physical map for this exercise. Write the true statements on your paper.

1. Most of Canada lies between 60° W longitude and 140° W longitude.
2. Most of Canada lies between 40° N latitude and 50° N latitude.
3. The Tropic of Cancer passes through Canada.
4. The Arctic Circle passes through Canada.
5. Winnipeg and Vancouver both are near 50° N latitude.
6. Winnipeg and Vancouver both are near 100° W longitude.
7. Montréal is at about 45° N latitude and 85° W longitude.

Vocabulary Skills

Directions: Match the numbered definitions with the vocabulary terms. Write each vocabulary term on your paper next to the number of its matching definition.

acid rain
dominion
Inuit
province
shield

1. Large platform of very old, very hard rock
2. A self-governing nation of the British Commonwealth—for example, Canada
3. Polluted rain, snow, or mist created when sulfur dioxide from coal and oil burning and nitrogen oxide from automobile emissions combine with water vapor in the air
4. The Canadian equivalent of a state in the United States
5. Native Canadians who are whale and seal hunters

Human Geography

Directions: Complete these sentences by writing on your paper the term or phrase that belongs in the blank.

1. No part of the United States is as densely settled or has as many large industrial cities as the ||||||||||.
2. Most of the newest immigrants to the United States come from the culture regions of |||||||||| and ||||||||||.
3. By the 1900s, four out of five immigrants lived in ||||||||||.
4. Americans have been leaving the occupation of |||||||||| for many decades since large corporations began providing for the needs of the country.
5. Because of the ||||||||||, many Americans are able to live in suburbs that surround industrial cities.
6. Most people of French descent in Canada live in the province of ||||||||||.
7. Generally, Canada is held in high esteem among nations because of its ||||||||||.

Physical Geography

Directions: Some of the following sentences are true statements with which you would agree. Write these true statements on your paper.

1. The United States, next to China, is the largest country in the world.
2. The economy of the United States is changing from an industrial to a postindustrial one, and competition in world markets is increasing.
3. California consists almost entirely of desert land that has been irrigated to produce citrus crops.
4. A megalopolis stretches along the Atlantic coast from Boston to beyond Washington, D.C.

5. Today, most urban workers in the United States live in inner cities because they want their homes to be near their work.
6. The average size of farms in the United States is increasing today, and the number of farms and farmers is decreasing.
7. Less than one-fourth of all Canadians live in cities.
8. Some regions in Canada are strongly affected by acid rain from factories in the United States.

Writing Skills Activities

Directions: Answer the following questions in essays of several paragraphs each. Remember to include a topic sentence and several sentences supporting your main idea in each paragraph.

1. Explain how, throughout its history, the United States has taken advantage of its excellent location, natural resources, and talented labor supply.
2. The United States has many diverse subregions. Select any two of these subregions and describe the ways in which they are different.
3. How has the economy of the United States been changing in recent years, and what problems has this caused? Discuss how farming has been affected.
4. Why does Canada have so few people in such a large country?

5

JAPAN AND THE TWO KOREAS

UNIT INTRODUCTION
Japan is an *archipelago*, or a group of islands, located off the east coast of mainland Asia. Together, Japan's four largest islands are about the size of California in area. On these islands, Japan crams a population half the size of that of the United States. With 124.8 million people, Japan is one of the most densely populated small countries and has twice as many people as any country in Western Europe except Germany.

During the last one hundred years, Japan changed from an agricultural country into an industrial nation and became a leader in world commerce. Improved technology, increased business, and foreign trade have made Japan an industrial giant in the technological world. Japanese cars, cameras, television sets, and CD players are found throughout the world.

Yet Japan's prosperity is fragile. The country must buy most of its energy and raw materials, as well as some of its food, from other countries. In a rapidly changing world, Japan's reliance on foreign trade means that the country's economic relationships with other nations are vital to its well-being.

The two Koreas represent experiments in communism and capitalism. Although they share a peninsula, they were divided by war in the years following World War II. North Korea is one of the few remaining countries that pursue communism both politically and economically. South Korea has become an industrial nation of increasing importance and has had the fastest growing economy in the world during the last two decades.

▲ *The torii, or gateway, off the coast of Hiroshima is an important Japanese cultural symbol. The contrast between this traditional symbol and the modern skyline of Hiroshima suggests the rapid rate of change that has occurred in Japan. Can you identify another harbor that is home to an important symbol?*

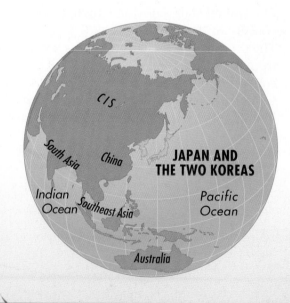

CIS

South Asia

China

JAPAN AND THE TWO KOREAS

Indian Ocean

Southeast Asia

Pacific Ocean

Australia

◀ *This crowded street in Seoul, South Korea, suggests the dense population that is characteristic of much of Asia. In what American cities might you expect to see such crowded conditions?*

SKILLS HIGHLIGHTED

- Using the Library's Catalog
- Using a Line Graph
- Using the Map of the Japanese Megalopolis
- Vocabulary Skills
- Writing Skills

UNIT OBJECTIVES

When you have completed this unit, you will be able to:

1. Describe the chief geographical features of Japan and how they have affected the Japanese people's ways of living.

2. Name some of the ways Japan has modernized.

3. Explain how Japan has dealt with the problems of an island nation without sufficient natural resources for industries.

4. Contrast the spectacular economic growth that Japan has enjoyed since the end of World War II with other Asian countries.

5. Identify some serious problems that Japan faces both now and in the future.

6. Explain how communism in North Korea and capitalism in South Korea have led to very different standards of living.

▲ *Mountains cover 80 percent of the Japanese landscape. Mount Fuji, a dormant volcano, is one of Japan's most famous tourist attractions. Why are lowlands in Japan more intensively occupied than highlands?*

311

1

Japan is an archipelago located off the east coast of Asia.

2

The Japanese homeland is composed of four major islands with a combined area slightly smaller than California.

3

Because forested mountains cover four-fifths of Japan, level land is scarce.

4

A government of *samurai* warriors opened Japan to the world in 1868 and began the process of modernization.

5

Japan accelerated its empire building in East Asia in the 1930s. This empire disintegrated after World War II, when Japan surrendered to the United States in 1945.

6

Japan's cities, industries, and farms are squeezed onto small coastal plains that make up only 11 percent of Japan's total land area.

7

Creative economic planning by government and industry is an important reason why Japan has been one of the fastest growing economies in the technological world.

8

Japan's industries are based on minerals, metals, and energy resources imported from foreign countries.

9

Because of rapid economic growth in a confined space, Japan faces difficult problems of air pollution, water pollution, and congestion.

10

The Korean Peninsula is divided politically and economically into communist North Korea and capitalist South Korea, but it is occupied by one people with the same history, language, and culture.

◀ *The festival in Sapporo, Hokkaido, is famous worldwide for its snow and ice sculptures. This scene of Japan's northernmost island is most suggestive of what part of the United States—Texas, California, or Maine?*

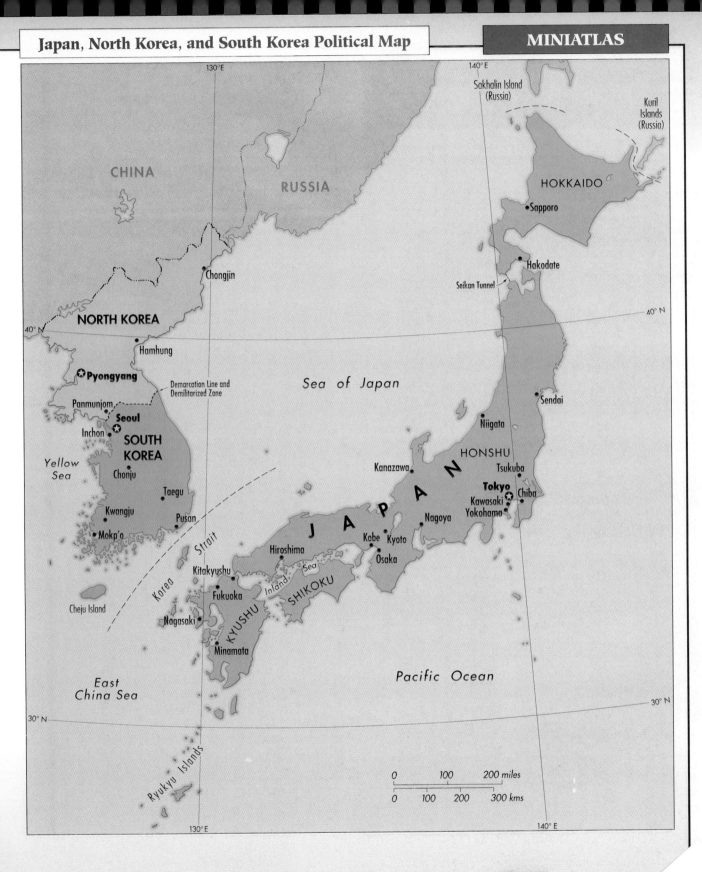

CHINA

RUSSIA

HOKKAIDO

•Sapporo

Sakhalin Island
(Russia)

Kuril
Islands
(Russia)

130°E

140°E

•Chongjin

Hakodate

Seikan Tunnel

40° N

40° N

NORTH KOREA

•Hamhung

Sea of Japan

Sendai

✪ **Pyongyang**

Panmunjom

Demarcation Line and
Demilitarized Zone

Niigata

•Seoul

Inchon

**SOUTH
KOREA**

HONSHU

Kanazawa

Tsukuba

*Yellow
Sea*

Chonju

J A P A N

Tokyo ✪ Chiba

Kawasaki

Taegu

Yokohama

Kwangju

Pusan

Nagoya

Mokp'o

Strait

Korea

Kobe Kyoto

Hiroshima

Osaka

Inland Sea

SHIKOKU

Kitakyushu

Fukuoka

Cheju Island

KYUSHU

Nagasaki

Pacific Ocean

Minamata

*East
China Sea*

30° N

30° N

Ryukyu Islands

130°E

140°E

0		100		200 miles

0	100	200	300 kms

Japan, North Korea, and South Korea Physical Map

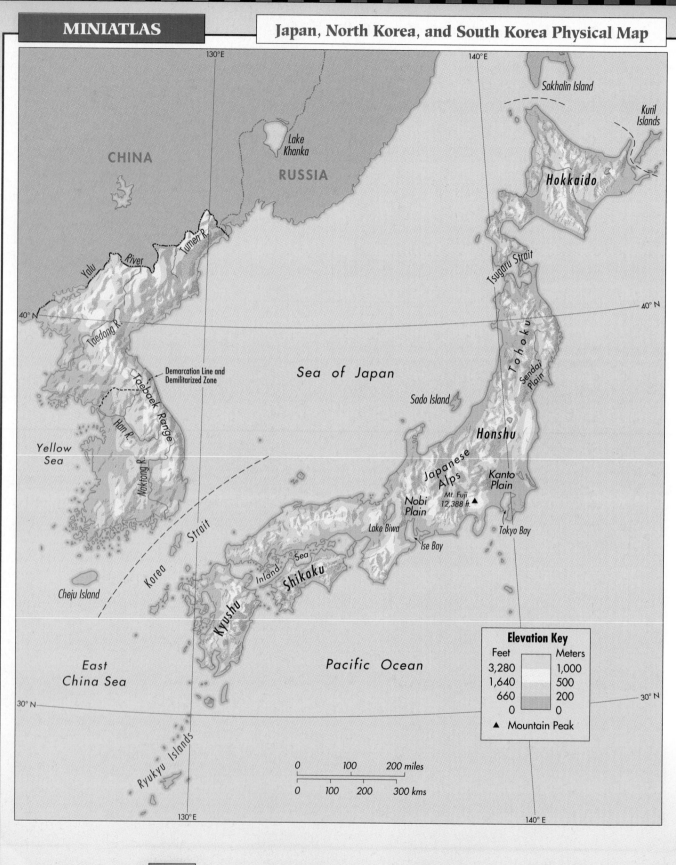

CHINA

RUSSIA

Lake Khanka

Sakhalin Island

Kuril Islands

Hokkaido

130°E

140°E

Tumen R.

Yalu River

40° N

Taedong R.

Tsugaru Strait

40° N

Sea of Japan

T o h o k u

Sendai Plain

Demarcation Line and Demilitarized Zone

Taebaek Range

Han R.

Sado Island

Honshu

Yellow Sea

Naktong R.

Japanese Alps

Kanto Plain

Nobi Plain

Mt. Fuji 12,388 ft.

Lake Biwa

Tokyo Bay

Ise Bay

Korea Strait

Inland Sea

Shikoku

Cheju Island

Kyushu

Pacific Ocean

East China Sea

30° N

30° N

Ryukyu Islands

130° E

140° E

Elevation Key

Feet		Meters
3,280		1,000
1,640		500
660		200
0		0

▲ Mountain Peak

0	100	200 miles
0	100	200	300 kms

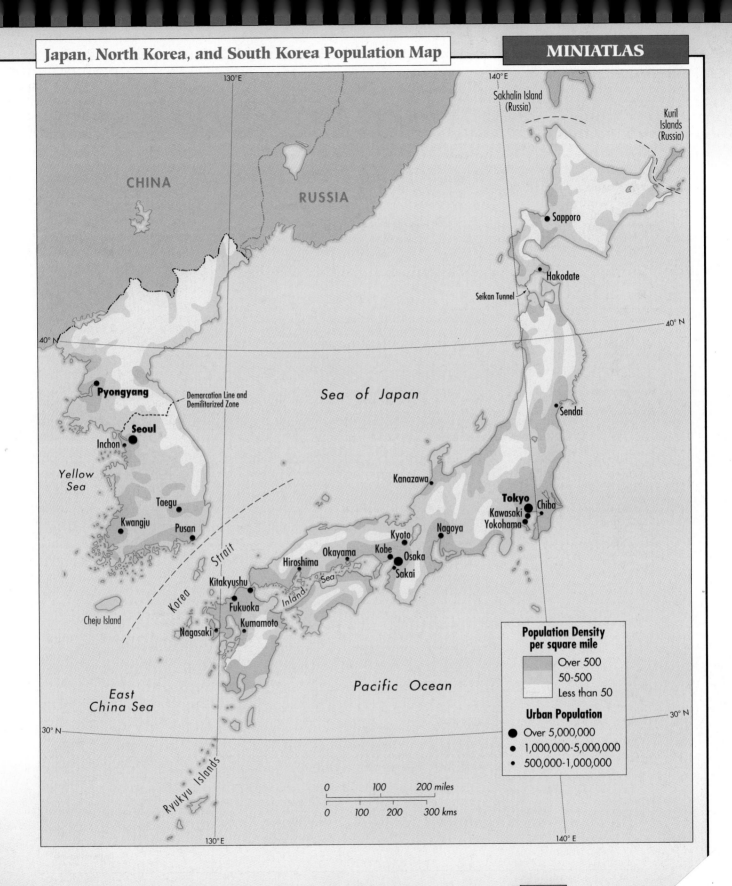

CHINA

RUSSIA

130°E

140°E

Sakhalin Island
(Russia)

Kuril
Islands
(Russia)

Sapporo

40° N

Hakodate

Seikan Tunnel

40° N

Pyongyang

Demarcation Line and
Demilitarized Zone

Sea of Japan

Sendai

Seoul

Inchon

Kanazawa

Yellow
Sea

Tokyo

Chiba

Kawasaki

Taegu

Yokohama

Kwangju

Pusan

Kyoto

Nagoya

Okayama

Kobe

Korea

Hiroshima

Osaka

Sakai

Strait

Inland

Sea

Kitakyushu

Cheju Island

Fukuoka

Kumamoto

Nagasaki

**Population Density
per square mile**

Over 500

50-500

Less than 50

Urban Population

● Over 5,000,000

● 1,000,000-5,000,000

• 500,000-1,000,000

East
China Sea

Pacific Ocean

30° N

30° N

Ryukyu Islands

0   100   200 miles

0   100   200   300 kms

130° E

140° E

Country	Capital City	Area (Square miles)	Population (Millions)	Life Expectancy	Urban Population (Percent)	Per Capita GNP (Dollars)
Japan	Tokyo	143,749	124.8	79	77	26,920
North Korea	Pyongyang	46,541	22.6	69	60	—
South Korea	Seoul	38,023	44.6	71	74	6,340

—Figures not available

CHAPTER 12

JAPAN ENTERS THE MODERN WORLD

In 1868, Japan abandoned the policy of isolation that it had imposed on itself for 250 years and established a new government led by young *samurai* warriors. These *samurai* wanted to make Japan a leading nation in the modern world, and by the 1930s, the Japanese had achieved world respect and military power.

Because Japan was a nation with limited resources, it wanted to control a resource-rich empire by conquering neighboring countries. The Japanese expanded their empire to include Manchuria, China, and Southeast Asia. Then Japan's attack on Pearl Harbor in Hawaii in 1941 led to war with the United States and defeat in World War II.

◀ *Japan became one of the world's most powerful economies in only a century. Why do you think Japan was successful in modernization, while other countries were not?*

317

1. The Regions of Japan

Regions and People

Japan consists of four large islands plus other smaller ones that curve along the eastern coast of Asia for a distance of 1,300 miles. Hokkaido, the northernmost of Japan's large islands, is located as far north as Maine. The southern tip of Kyushu lies at the same latitude as Atlanta, Georgia. The Sea of Japan separates the Japanese homeland from the Korean Peninsula, China, and Russia, a member of the Commonwealth of Independent States (CIS) (the former Soviet Union), on the Asian continent.

Look closely at this small island nation on the map on page 313 in your miniatlas. Japan is an **archipelago**, a group or chain of closely connected islands. Find the northernmost of the large islands.

Subregions of Japan

Japan has three major subregions: (1) Hokkaido, (2) Northern Honshu, and (3) the Heartland which includes the Japanese Megalopolis.

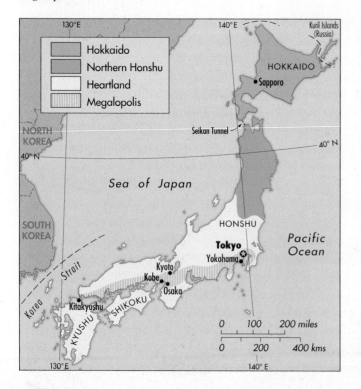

Which is the largest island? On what island is the capital city of Tokyo located? Find the Inland Sea, which is encircled by the islands of Honshu, Shikoku, and Kyushu. All four of Japan's large islands are the tops of submerged volcanic mountains that rise from the floor of the Pacific Ocean.

Three tectonic plates converge just south of Tokyo, so that this huge city—like Los Angeles—is anticipating a major earthquake. The people of Tokyo estimate this possibility on the basis of the "seventy year rule," because during the last 350 years earthquakes have struck Tokyo in 1633, 1703, 1782, 1853, and 1923—the last killing an estimated 140,000 people. For the estimated 34 million people who live on the Kanto Plain surrounding Tokyo, the 1990s are a tense period.

Landforms influence where Japan's 124.8 million people live. Forested mountains cover four-fifths of Japan, reaching elevations above 10,000 feet in the Japanese Alps in central Honshu. Most Japanese live on small coastal plains where mountain streams flow downslope to the sea. Here, along Japan's 18,487-mile coastline, farms and factories compete for level land. The cities and industries of Japan are located in these coastal areas.

Climate also influences human settlement in Japan. Temperatures vary just as they do along the Atlantic coast of the United States. The south is a warm, subtropical region of terraced rice fields and mulberry and tea plantations. The northern island of Hokkaido has a short growing season and long, snowy winters. Virtually all of Japan is well watered. Late summer brings great storms called **typhoons**, or hurricanes, with high winds and heavy rains.

Hokkaido, Japan's Northern Frontier

Hokkaido is remote from the centers of Japanese life. Settlement has been slow here because Hokkaido has a cold climate similar to that of northern New England. The island is blanketed by snow during its long winters.

 IT'S A FACT No place in Japan is more than 100 miles from the sea.

IT'S A FACT | Japan is roughly half the size of Texas.

Until over one hundred years ago, most people in Hokkaido lived by fishing and hunting. Their small villages were clustered on the coast of Hokkaido's southern peninsula. But when military bases were built in Hokkaido in 1869 to defend the island from Russian attack, large numbers of Japanese moved there. Settlers were sent in to cut down forests for lumber and to prospect for minerals. After World War II, the Japanese government encouraged more colonists to settle on Hokkaido. This island is still the most sparsely settled region of Japan, however. Its only important urban center is Sapporo, a city of 1.7 million people.

Farming is carefully adapted to the environment in Hokkaido. Pioneer Japanese farmers raised hay, potatoes, and oats. When cold-resistant and early-ripening strains of rice were developed in the early 1900s, this crop was introduced to Hokkaido. Rice is now grown on about one-fifth of the island's farmland. Other cold-tolerant crops, like potatoes, oats, beans, and sugar beets, are quite important. Beef and dairy cattle are now being raised on Hokkaido.

Natural resources are Hokkaido's other source of wealth. Coal is mined and oil fields are being tapped; further development is planned. A thirty-four-mile undersea railroad tunnel (the Seikan Tunnel) connecting Hokkaido with northern Honshu was completed in the 1980s. The longest undersea tunnel in the world, it has cut travel time between the two islands from four hours to thirteen minutes. Trains move through it at speeds of 270 miles per hour.

▼ *Rice has become a major crop on Japan's northern island of Hokkaido despite its cold climate. Compare and contrast this scene of rice cultivation with the one on page 336. What differences are apparent? Why?*

IT'S A FACT

Tokyo is Japan's seventh capital city.

Northern Honshu

The northern third of the island of Honshu is warmer than the Japanese north but cooler than the subtropical south. In this hilly and mountainous region, level land is scarce. The climate of northern Honshu is moderate when compared to that of Hokkaido, but winters are still long and only one crop a year can be grown.

Northern Honshu is a rural area. More than half of its people are involved in farming, fishing, or logging of the pine forests of the highlands for lumber from which charcoal is made. Only 9.5 million Japanese live in the region; most of them are clustered in settlements located in small, fertile upland basins.

On the western coastal plain, facing Asia, the climate is warm enough for rice growing. Often called "the rice bowl of Japan," the area produces one-fourth of Japan's grain.

▼ *Yokohama, near Tokyo, is one of Japan's largest and most important port cities. Why are virtually all of Japan's largest cities located along the coast?*

The Heartland of Japan

The heartland of Japanese civilization is located in central and southern Honshu and on the neighboring islands of Shikoku and Kyushu, which encircle the Inland Sea. The **heartland** of a culture is its place of origin and fullest expression. More than four-fifths of the Japanese live in central and southwestern Japan. Most of the country's industries and businesses are located in this region.

In central Honshu, mountains known as the Japanese Alps rise two miles above sea level. The famous and beautiful Mount Fuji (12,388 feet) is one of these volcanic peaks. From the Japanese Alps, mountain ranges extending along the shores of the Inland Sea form the mountain backbones of southern Honshu and the islands of Shikoku and Kyushu.

The Japanese heartland has a rich variety of environments. The region's warm climate and plentiful rainfall allow farmers to raise crops on plains and on terraced hillsides throughout the year. The landscape includes small villages, rice fields, orange groves, large cities, and industrial areas. Above the coastal plains rise forested mountains.

Varied environments and rapid economic growth result in striking variations in landscape similar to those in California. In the north, forested mountains dive dramatically into the sea as in northern California. Farther south, the huge smog-ridden metropolis of Tokyo in Japan, like Los Angeles in California, is among the most densely settled urban areas on Earth. The modern cities and suburbs of Tokyo, Nagoya, and Osaka contrast with nearby forested mountains and farms.

Level land is the key to understanding where people live in the heartland of Japan. Tokyo, for example, is located on Japan's largest plain. This plain is blanketed by industrial areas that provide jobs for the 12 million city dwellers of Tokyo and the neighboring city of Yokohama (population 3.2 million). The fields and farms that feed these city dwellers surround the urban centers and suburbs. Approximately 34 million people live on this small plain.

A narrow passageway (the Tokaido Corridor) runs through the mountains west of Tokyo, Japan's largest city, to a series of small delta plains located on the Pacific coast of Honshu. Along this narrow corridor, which links Tokyo with southwestern Honshu and the Inland Sea, the climate is mild and subtropical, and rainfall is plentiful. Tea and mandarin oranges are raised, as they have been for centuries.

The city of Nagoya lies roughly midway between Tokyo and the Inland Sea. Pine- and cedar-covered mountains form a backdrop to this thriving city. Located on Japan's second largest plain, Nagoya has more than 2 million residents. The city's industrial areas and suburbs are located in the center of the plain, while rice fields, vegetable plots, and orchards blanket the slopes of hills nearby.

Further west, the Tokaido Corridor reaches the large industrial concentration at the head of the Inland Sea. Here, the twin cities of Osaka and Kobe, both industrial ports, are encircled by lowland rice fields and terraced gardens. Inland lies Kyoto, the ancient center of Japanese civilization.

A number of smaller industrial urban areas are scattered along the northern shores of the Inland Sea. Population is dense along the coast, where factories, houses, and farms are packed tightly together. The oldest industrial center in Japan grew up in the coal-mining district of northern Kyushu.

⊕ REVIEW QUESTIONS

1. What are the names of Japan's four large islands?
2. How many people live in Japan?
3. The world's longest undersea railroad tunnel connects what two Japanese islands? How much travel time is saved by using this tunnel?
4. What is the heartland of Japan, and what proportion of the Japanese population lives there?

⊕ THOUGHT QUESTIONS

1. How do landforms in Japan influence the location of people, farms, and factories?
2. Hokkaido has a cold climate similar to that of northern New England. How has farming been adapted to this environment?

2. The Modernization of Japan

The Drive for Modernization

In 1868, the new *samurai* leaders of Japan intended to make the country a respected world power. They wanted their small group of islands to become a modern military and commercial nation. The main goal of **modernization** was to gain international respect, not, as is often the case today, to improve standards of living. The Japanese did not want to be humiliated again by Western military power as they had been in the 1850s.

They wanted to accomplish modernization as rapidly as possible. The *samurai* code of discipline, loyalty, sacrifice, and selfless labor was to provide the inspiration needed to achieve this goal. European and American experts in science and technology were brought in to teach Western methods. Delegations of Japanese visited the United States and Europe to find out which technical and social systems were best suited to the "new Japan."

Japan's new leaders also reorganized Japan's lands. Hereditary landlord control over feudal estates was abolished and these domains were divided into government-controlled counties. Japanese lands and villages were consolidated under the laws and leadership of the central government.

Communication Systems Unify Japan

By 1914, Japan had constructed 7,000 miles of railroad track, in spite of the challenges of building in the country's many mountains. These railroads made it possible to move goods and people across rough terrain. Communications were also improved, as telegraph and telephone lines linked the four main islands. Japanese merchant ships were built to handle increased trade in Japanese waters. More roads were built. All of these improvements in transport and communications helped to tie Japan together.

LOCATION

JAPAN'S LOCATION AND THE MISSION OF COMMODORE PERRY

Japan's location astride the trade routes between North America and China led to a confrontation with the United States in 1853. For centuries Japan had isolated itself from the world of nations, but when Commodore Matthew C. Perry steamed into Edo (Tokyo)

▼ *Commodore Matthew C. Perry meets with Japanese imperial commissioners in Yokohama in Japan in 1853. Why was Perry sent to Japan at the request of President Millard Fillmore?*

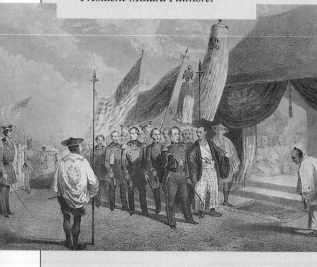

Bay with a squadron of four armed gunboats, Japan's self-imposed isolation was broken forever.

Enough Western knowledge had filtered into Japan to give Japanese military experts respect for the firepower of the American vessels. They were aware of the disastrous defeat that British troops had inflicted on China eleven years earlier in the First Opium War.

For a long time, other countries had attempted unsuccessfully to trade with Japan. Commodore Perry's well-armed ships were intended to force Japan to provide shelter for shipwrecked American sailors, to aid American vessels seeking shelter and supplies, and to open trade with the United States. After a brief stay, the American fleet departed, leaving the Japanese to consider the situation. This show of force left the Japanese with few options.

A year later, Commodore Perry returned to Japan with eight vessels. Expensive gifts were presented to the Japanese ruler. A miniature railroad was set up on the shore, and Japanese officials whirled around its tracks. A telegraph link was established between Perry's ship and the royal palace. Telescopes and armaments were displayed. The technological and military significance of these gifts was not

lost on Japan's leaders. Angry *samurai* demanded that the government "expel the barbarians," but Japanese leaders recognized that this was impossible.

Because of its location on major Pacific trade routes to China, Japan's two centuries of seclusion came to an end. Two ports were opened to American vessels. Trade agreements were soon signed with Britain, Russia, France, and the Netherlands. Japanese silk, tea, and copperware were sold to foreigners in exchange for cotton textiles, sugar, and ironware. Trade increased rapidly.

So did the opposition of the *samurai* to the government. They strongly opposed foreign intervention in Japan and believed that Japan's government was turning the country over to the West. Fifteen years after Matthew Perry's first mission to Japan, the Japanese government was overthrown by these *samurai*. The country's new leaders decided that the only way to cope with the West was to adopt its military technology, so they embarked on a policy of modernization designed to make Japan a powerful military state.

QUESTIONS

1. What were the three main objectives of Commodore Perry's visit to Japan in 1853?
2. Why do you think Commodore Perry demonstrated so many examples of American science and technology to the Japanese leaders?

The class barriers of feudal times were also eliminated. All Japanese men—not just the *samurai*—were required to serve in the military. New laws made education mandatory. By 1900, the Japanese were literate and united and were becoming a strong military power.

Farming Provides the Money for Development

The *samurai* leaders knew that developing Japan into a modern country would be expensive. Building factories, buying Western technology, and developing a strong and well-armed army and navy cost a great deal. These costs had to be paid by increasing farm production. The government gave farmers more land, but it also forced them to accept a strict system of tax collection. Improvements in Japanese agriculture helped to increase crop production. Rice harvests increased 80 percent between 1880 and 1920, and farmland expanded by one-third.

Japan's foreign trade led to changes in the crops grown in Japan—for instance, foreign imports of sugar and cotton eliminated local cultivation of these crops. Rice, tea, fruit, and vegetables occupied most of Japan's farmland. Tea and silk became the main exports.

In the end, Japanese farmers paid for the modernization of Japan. They grew the food for people living in Japan's industrial centers. Villages also supplied the workers for industry. Farmers produced the silk and tea for export and were responsible for about 90 percent of Japan's revenues during the early stages of modernization.

Japan Industrializes

The Japanese government encouraged *samurai* families to start up the heavy industries needed to supply a modern army and navy. Shipyards, machine works, and steel mills were built near the coal fields of northern Kyushu, and factories were built in the port cities of Osaka and Tokyo.

The government also established prototype factories as examples of how to produce cloth, silk, iron, paper, and glass efficiently. These government-sponsored industries served as models of Western methods of production. The government then sold these model factories to *samurai* business leaders but retained ownership of railroads as well as telegraph and telephone facilities. Today, many large Japanese firms trace their origins back to these early years of modernization.

Problems of Rapid Modernization

High land taxes and low food prices caused financial problems for Japan's farmers. Young men from the villages were being drafted into military service. Children were forced to go to school, so these youngsters could not help on the farms. Many farm families went bankrupt in the 1880s, and others migrated to islands in the Pacific and to North America.

Life for factory workers in Japan was just as bad as it was in Europe during the early stages of the Industrial Revolution. Twelve-hour work shifts in factories produced manufactured goods around the clock; production was high, but such a schedule tired the workers. Many teenage women were hired as low-paid workers in textile factories. Low wages for workers meant high profits for factory owners—profits that were invested in building more factories. Crowded slums where workers lived sprang up near factory districts, just as they did in cities in Europe and the United States.

In spite of these hardships, the Japanese approached modernization with a strong sense of national purpose. Most citizens accepted—and acted on—the belief that personal sacrifice was necessary to make Japan a powerful industrial nation.

REVIEW QUESTIONS

1. How were transportation and communications improved to modernize Japan?
2. The cost of modernization in Japan fell largely on what class? What products did this class provide?

THOUGHT QUESTIONS

1. Why do you think the *samurai* code helped in the modernization of Japan?
2. How did rapid modernization create hardships for farmers and factory workers?

3. Expansion, Empire, and Defeat

Japan Builds an Empire

Japanese leaders began to build an empire in the late 1800s. Their desire to take over Korea (then called Chosen), a peninsula long under the control of China, led in 1895 to war between China and Japan. In this war Japanese armies crushed Chinese forces. After their defeat, the Chinese were forced to give Japan the large island of Taiwan (then called Formosa) and to grant independence to Korea.

The peace treaty with the Chinese also gave the tip of the Liaodong Peninsula in southern Manchuria to Japan. The Russians were building a railroad there, however, and they seized this peninsula. The Japanese attacked the Russians in 1904, defeated them a year later, and then took over the Liaodong Peninsula.

This Russian defeat, in what became known as the Russo-Japanese War, sent shock waves through Europe. A European people had not been defeated by a non-European people in centuries. A humiliated Russia was forced to give the southern half of

The Japanese Colonial Empire

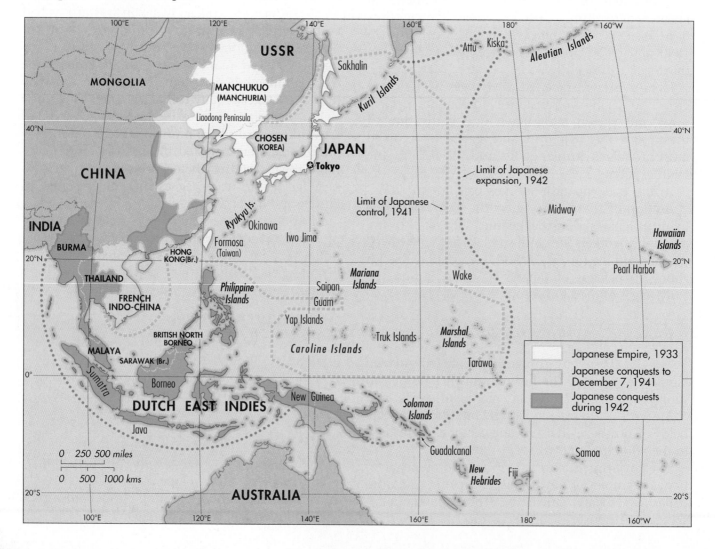

PEOPLE-ENVIRONMENT

JAPAN'S *SAMURAI* CODE

The *samurai code* played an important role in the modernization of Japan. Many early industrialists were *samurai*, and *samurai* started the largest corporations of modern Japan. These rules of the founder of one large multinational corporation show how old values have been applied in modern times.

Founder's Rules

Article 1. Do not be preoccupied with small matters but aim at the management of large enterprises.

Article 2. Once you start an enterprise, be sure to succeed.

▲ *Modernization began in Japan when* samurai *leaders took control of the country in the 1860s. What were the major goals that the* samurai *wanted to achieve in pursuing modernization?*

Article 3. Do not engage in speculative enterprises.

Article 4. Operate all enterprises with the national interest in mind.

Article 5. Never forget the pure spirit of public service and *makoto* (sincerity, fidelity).

Article 6. Be hardworking, frugal, and thoughtful.

Article 7. Utilize proper personnel.

Article 8. Treat your employees well.

Article 9. Be bold in starting an enterprise but meticulous in its prosecution.

❓ QUESTIONS

1. Which article might be interpreted as "Don't get lost in the details"?
2. Which article might be interpreted as "Pay attention to the details"?
3. How would you explain this apparently conflicting advice?

Sakhalin Island to Japan. Five years later, in 1910, Japan annexed Korea. In less than half a century, Japan had won the place it had so desperately wanted among the powerful nations of the world.

Why the Japanese Believed They Needed an Empire

Japan's industries grew so rapidly that they had to import large amounts of coal, iron, oil, rubber, and other raw materials to keep their factories running. Japan also relied on imported food from its colonies in Korea and Formosa (Taiwan). By the 1930s, Japan, like Britain, had to trade to prosper.

Expanded trade and economic growth increased Japan's wealth. Then, in the 1930s, a worldwide economic depression destroyed much of Japan's trade-based economy. Exports fell 50 percent in two years. Silk prices dropped by two-thirds and those for rice by one-half. Many young Japanese officers believed that Japan's dependence on world trade had ruined their country. They also felt uneasy surrounded by European colonies, such as French Indo-China, British Malaya, and the Dutch East Indies.

Their answer to Japan's problems was to create a self-sufficient Japanese empire in East Asia. Western influence would be eliminated from this empire. Japan would conquer areas rich in natural resources so that it would never have to rely on outsiders again.

The Japanese military set out to achieve this ambition. Without government orders, the Japanese army in 1931–1932 captured the iron and coal fields of Manchuria from China and renamed the area Manchukuo. Coal, iron ore, and steel soon flowed from Manchuria to Japan. In 1937, Japanese armies invaded central China to gain control over East Asia's largest nation. At home, the generals increased military spending and prepared the Japanese people for a major war to expand the empire.

Japan in World War II

In September 1939, Japanese armies moved southward into Vietnam, which was then part of a colony in Southeast Asia called French Indo-China. In Europe that same month, France and Britain declared war on Germany. Thus Japan's invasion of France's colony in Vietnam meant that Japan was also at war with France. Japan quickly signed treaties with Germany and Italy and joined their side in World War II.

The United States immediately banned exports of war materials—such as scrap iron, tire rubber, steel products, and oil—to Japan. This threat to their industries caused the Japanese to risk everything. They decided to gain economic independence by driving the Americans, as well as the Europeans, out of Southeast Asia and the Pacific Ocean.

On December 7, 1941, Japan launched a surprise attack on the Pacific fleet of the United States at Pearl Harbor in Hawaii. The attack was intended to destroy the U.S. Navy in the Pacific and to give Japan time to consolidate its hold on China and Southeast Asia. It was a gamble that failed. The American navy was damaged but not destroyed, and the Japanese gained a lethal opponent in the Pacific as the United States joined the French and British in the war.

The Japanese, however, were able to take over much of Southeast Asia by 1942, as the map on page 324 shows. It took more than three years for the United States and its allies to defeat Japan. In August 1945, the Japanese finally surrendered after atomic bombs were dropped on the cities of Hiroshima and Nagasaki. A month later the victorious American army began an occupation of Japan that lasted seven years.

⊕ REVIEW QUESTIONS

1. What large island did Japan receive after defeating China in 1895, and what peninsula did Japan capture from Russia in 1905?
2. What raw materials did Japan import in order to expand its industries?
3. When did the war between the United States and Japan begin, and when did the Japanese surrender?

⊕ THOUGHT QUESTIONS

1. In the 1930s, why did Japan's leaders want to create a self-sufficient empire in East Asia?
2. What was the Japanese strategy behind the attack on Pearl Harbor on December 7, 1941?

🌐 CHAPTER SUMMARY

Japan is a mountainous chain of islands with a series of densely settled coastal plains. The majority of people live on these plains in the major cities of Tokyo, Yokohama, Nagoya, Osaka, Kobe, and Kitakyushu. Around 1870, a group of *samurai* warriors began modernizing Japanese society. In fifty years, these warriors made Japan a powerful military and economic nation through empire building.

In two wars, Japan defeated first China and then Russia. As a result, Japan gained the island of Formosa (Taiwan) from the Chinese in 1895 and annexed Chosen (Korea) in 1910 after defeating the Russians. In the 1930s, Japan began to expand into China. Japan and the United States clashed over Japanese policies in East Asia and fought one another in World War II for more than three years. Japan lost the war that ended in the bombing of many of the country's cities, including the destruction of Hiroshima and Nagasaki by atomic bombs.

EXERCISES

Find the Correct Ending

Directions: Write on your paper the sentence beginnings that follow, completing them with the correct endings.

1. Japan is an island (a) estuary (b) archipelago (c) reef.
2. One reason for American interest in Japan in the 1850s was (a) the desire to obtain beaver pelts and other furs (b) Japan's nearness to Hawaii (c) the need to provide shelter for shipwrecked American sailors.
3. In the early days of Japanese trade with foreign nations, the chief products obtained from Japan were silk, copperware, and (a) coffee (b) tea (c) rubber.
4. In the early days of Japanese trade with foreign nations, the chief products obtained from other countries were cotton textiles, sugar, and (a) lumber (b) citrus fruits (c) ironware.
5. When the landlords lost control over their lands to the government, their estates were divided into (a) separate states for each island (b) government-controlled counties (c) three large states and many small colonies.
6. Japan captured the Liaodong Peninsula from (a) China (b) India (c) Russia.

Two Correct Sentences

Directions: There are two correct sentences in each numbered group of three that follows. After each number, write the correct sentences on your paper.

1. (a) Most Japanese farming is done on small coastal plains. (b) Shikoku is the island that provides most of Japan's potatoes, oats, and sugar beets. (c) Japan's chief cities and industries are located mainly on coastal lowlands.
2. (a) Japan's islands are the volcanic peaks of submerged mountains. (b) Mount Fuji is located in the Japanese Alps. (c) Tokyo is located in a beautiful mountain basin.

3. (a) Kyoto is the ancient center of Japanese civilization. (b) Nagoya is the largest city on Hokkaido. (c) Osaka and Kobe are both industrial ports.

Before or After?

Directions: Use the map of the Japanese colonial empire on page 324 to tell whether Japan made each of these conquests before or after December 7, 1941, when the Japanese attacked Pearl Harbor in Hawaii.

1. Borneo
2. Caroline Islands
3. Chosen (Korea)
4. French Indo-China
5. Malaya
6. Manchukuo (Manchuria)
7. Philippine Islands
8. Sumatra
9. Thailand

Inquiry

Directions: Combine the information in this chapter with your own ideas to answer these questions.

1. The primary goal of modernization in Japan was achieving international respect. In countries that are still developing today, do you think that the chief goal of modernization is gaining international respect or a higher standard of living? Give reasons for your answer.
2. Why would Japan's modernization have progressed much more slowly if the Industrial Revolution had not already occurred in Western Europe and the United States?

SKILLS

Using the Library's Catalog

The catalog in the library can help you do research. Before computers, all the books in a library were cataloged on index cards. The computerized library catalog allows you to call up information about books on a screen instead of looking through cards filed A to Z in a drawer.

Whether you search for books by computer or in card files, each nonfiction book is listed in three

CHAPTER 12

different ways—by subject, by title, and by the author's name. Computerized catalogs can be searched in other ways as well.

An important part of each entry is the book's *call number*. This number, which libraries assign to a book according to its subject, is in the upper left-hand corner of the card or at the top of a computerized listing. The call number tells you where this book is located in the library.

Directions: Look at the information on this card from the catalog and then answer the questions that follow.

JAPAN	Davidson, Judith. *Japan, Where East Meets*
952	*West*. Minneapolis, Minn.: Dillon Press,
D	©1983.
	142 p.; illus.
	Bibliography, p. 139. Index. An introduction
	to the history and culture of Japan, including
	a discussion of Japanese in the United States.

1. Has this entry been found under the book's title, author, or subject?
2. Who is the author of this book?
3. What is the title of the book?
4. What is the book's call number?
5. What is the copyright (publication) date of the book?
6. How long is the book?
7. What company published the book?
8. Does it have any pictures?
9. Does it have both a bibliography and an index?
10. Does it contain information about Japanese Americans?

Which Book?

Often many books are listed under the same subject in the catalog. You need to look at the titles of these books to tell which ones are likely to be helpful to you.

Directions: The following is a list of research topics. After the name of each topic are the titles of three

books. From its title decide which one of these books probably will have the most information on this topic. Write this book's title on your paper.

1. *Topic:* Urban customs and conditions in Japan
   Cities and Metropolitan Areas in Today's World
   Living in Tokyo
   The History of Settlement of Hokkaido

2. *Topic:* Subjects and symbols often used by Japanese painters
   Shapes and People: A Book about Pictures
   Enjoying the World of Art
   The Art of Japan

3. *Topic:* How Japan conquered large parts of East Asia
   Japan: The Years of Triumph; from Feudal Isolation to Pacific Empire
   Japan: Its History, Arts, and Literature
   Japan before Buddhism

Vocabulary Skills

Directions: Match the vocabulary terms with the correct definitions. Write the term on your paper after the number of its correct definition.

archipelago	*samurai*
heartland	*samurai* code
makoto	slum
modernization	typhoon

1. The process by which a country tries to improve its standard of living and gain respect among nations
2. A way of living for a certain class in Japan that stresses discipline, loyalty, sacrifice, and selfless labor
3. A place where a culture originates and achieves its fullest expression
4. A windstorm that comes to Japan in late summer, called a hurricane in the Americas
5. A group of islands, usually forming a chain
6. An area of a large city where the very poor live
7. The warrior class of Japan that led efforts to modernize Japan in the nineteenth century
8. The expression of sincerity and fidelity (loyalty) to a person or a company in Japan

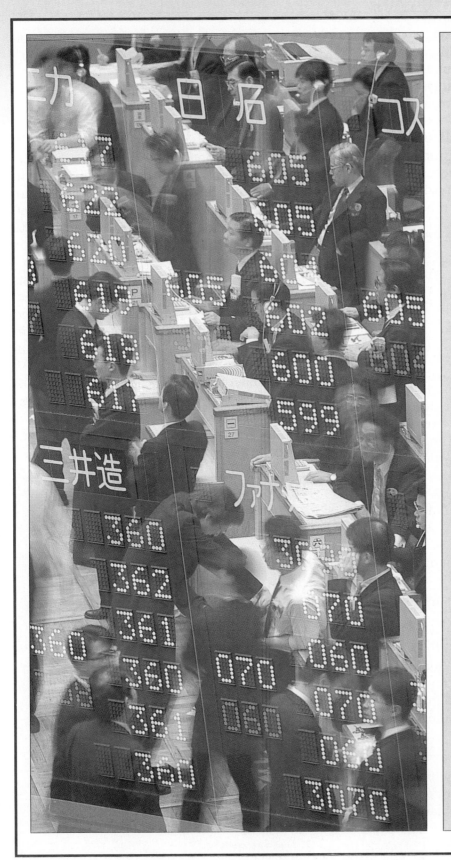

JAPAN AND THE TWO KOREAS TODAY

Few could have guessed that Japan would become a leader of the technological world so quickly after World War II. The country had lain in ruins. Two-thirds of Tokyo and Osaka had been burned to the ground. Hiroshima and Nagasaki were atomic wastelands. Japan's industry, power plants, and transportation networks had been destroyed. Yet, in less than forty years, the Japanese fully recovered from their wartime defeat. They built a new economy that made Japan one of the wealthiest nations in the industrialized world.

At the end of World War II, Korea was split into two

◀ *The Tokyo stockmarket is one of the most active centers of trade in the world. How do you think such a small, mountainous country could accumulate such great wealth? Can you think of another small mountainous country that has also accumulated great wealth?*

countries. North Korea chose to pursue a communist program of development. South Korea embarked on a capitalist drive for modernization. These countries share a common history, language, and culture, but their policies and programs during the last fifty years have led to striking differences in their standards of living. North and South Korea provide an excellent comparison of the effectiveness of capitalism and the ineffectiveness of communism in increasing a country's well-being.

1. Japan Becomes an Industrial, Urban Giant

The Benefits of Creative Economic Planning

After World War II, Japan became an industrial leader. The Japanese did this through creative economic planning, with government and industry working together closely. This planning program involved phasing out declining industries in favor of newer, more profitable ones. Workers in declining industries were retrained to perform new tasks. By shifting from textiles to heavy industry, and then from heavy industry to high technology, Japan has maintained a high rate of economic growth. It is now one of the richest countries in the world.

The special relationship between government and industry in Japan goes back to the late 1800s when *samurai* families were entrusted with building an industrial state. In recent years, this connection has made it possible for Japanese industrialists to plan for the future. Japan has scrapped old industries that cannot compete in world markets and has instead invested in high technology in the

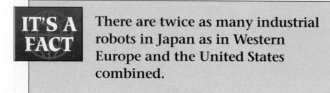

IT'S A FACT There are twice as many industrial robots in Japan as in Western Europe and the United States combined.

computer, communications, electronics, and robotics industries. Economic planning in Japan is coordinated by the Ministry of Internal Trade and Industry (MITI).

From Textiles to Heavy Industry

In the 1950s, Japan realized that its textile industry could not compete with the lower labor costs in neighboring countries like South Korea and Taiwan. Japanese companies making cotton, rayon, and synthetic fibers were encouraged to switch to the production of steel for ships, automobiles, and heavy machinery.

Steel facilities were constructed on the coasts where ships carrying imported iron ore, coal, and oil could unload. By the 1980s, Japan was second only to Russia in steel production and had surpassed the United States. Nearly half of Japan's in-

Creative Economic Planning
In the 1960s, textiles were important exports. In the late 1960s, the Japanese shifted to the export of steel products and color televisions, and then cars in the 1970s. Today the Japanese are specializing in high-technology exports.

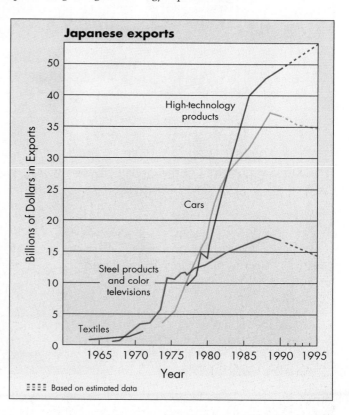

Japanese exports

High-technology products

Cars

Steel products and color televisions

Textiles

Billions of Dollars in Exports

1965 1970 1975 1980 1985 1990 1995
Year

Based on estimated data

dustrial workers were in the metal, chemical, and engineering industries. But by the 1990s, Japan had already largely shifted away from heavy industry into high technology.

Japan's Limited Resource Base

One reason for the shift from heavy industry is the high cost of imported raw materials. Because Japan does not have enough local energy or mineral resources, it must buy raw materials on world markets. As a result, Japan is very vulnerable to disruptions in world trade.

Take, for example, the problem of energy. Until 1950, half of Japan's energy was provided by its own coal fields in northern Kyushu. As the country's energy needs increased, Japan began to import more and more oil. The oil crises of 1973 and 1979, which increased oil prices by 1,400 percent, made the Japanese realize that their country was too dependent on oil imports.

Japan is now working with Russia to develop natural gas fields north of Hokkaido, but four islands without resources north of Hokkaido that were granted to Russia after World War II have created a dispute of pride between the two countries. Japan has signed a treaty with China to mine a rich coal deposit in the Chinese Northeast (Manchuria). Some 1,500 hydroelectric plants in Japan now supply electricity to Japan's largest cities, and nuclear power stations are another source of local energy. A strict energy conservation program has also been started. Yet Japan still depends on energy imports more than any other industrial nation.

Japan's supply of basic minerals and metals is also limited. Japan imports 90 percent of the iron ore and 80 percent of the coal used in the country. In addition, all of the nickel, copper, cobalt, and aluminum that Japan uses must be purchased abroad. To ensure future supplies, Japan is now investing in joint development of mineral deposits in other countries, including Australia.

From Heavy Industry to High Technology

The high cost of imports of raw materials and lower labor costs in heavy industry in neighboring countries are creating a second change in Japanese industry. Japan has shifted into high technology fields like advanced computers, robotics (the making and using of robots), and semiconductors (the building blocks of computers) and is already an exporter of memory chips and semiconductors. The MITI is masterminding a program to encourage Japanese industrialists to create new products. Japan is fast achieving a position of world leadership in high technology.

The shift to high technology has relocated industry into the Japanese countryside. The government is building high technology industries in rural areas, where there is clean air and clean water, because the slightest contamination makes a computer chip unusable. Mountain villages in northern Honshu now produce one-fifth of Japan's integrated electronic circuitry. Highland valleys in central Honshu have attracted manufacturers of components for videotape recorders, televisions, and stereo systems. The old steel center of Kyushu is now called "Silicon Island," because it produces 40 percent of Japan's semiconductors. **Silicon**, as you may know, is the raw material used to make computer chips. The government also plans to build nineteen **technopolises**, or "science cities," in Japan, similar to Tsukuba near Tokyo.

▼ *Japan increasingly is investing resources in basic research, as this photograph of a compact disc research facility near Tokyo illustrates. Do you believe it is essential that the United States continue to invest in basic research? What areas do you think will be important in the future?*

The Cities of Japan

The growth of Japanese industry also changed the distribution of the population. Farmers from rural areas have moved to coastal cities, where industrial jobs are available. Since the 1980s, three out of every four Japanese have been living in these sprawling urban centers.

The largest Japanese cities are located on plains along the Pacific or eastern coast. Four-fifths of all urbanites live in the small portion of Japan that is relatively flat.

A Japanese Megalopolis

By some definitions, Tokyo is now the largest city in the world. The sprawling suburbs of Tokyo and Yokohama hold an estimated 34 million people, more than a quarter of Japan's total population. Farther west, the city of Nagoya, with 2 million people, covers a smaller plain. The twin cities of Osaka

IT'S A FACT A piece of land the size of this page would cost $50,000 in central Tokyo.

and Kobe, with a combined population of 12 million, have spread inland to Japan's oldest center of civilization, Kyoto. With the exception of Kyoto (population 1.5 million), every large city in Japan is located on the coast and is equipped with port facilities.

Geographers today refer to the continuous urban strip that runs from Tokyo through Nagoya and the cities of the Inland Sea to Kitakyushu in northern Kyushu as the "Japanese megalopolis." A **megalopolis** is a large, sprawling urban area where the built-up centers and suburbs of various cities meet and overlap as, for example, along the coast of the American Northeast between Boston and Washington, D.C. (the Bos-Wash Corridor). Along

▼ *Tokyo is one of the most densely settled cities in the world. It also has the highest cost of living. What factors do you think might contribute to this situation?*

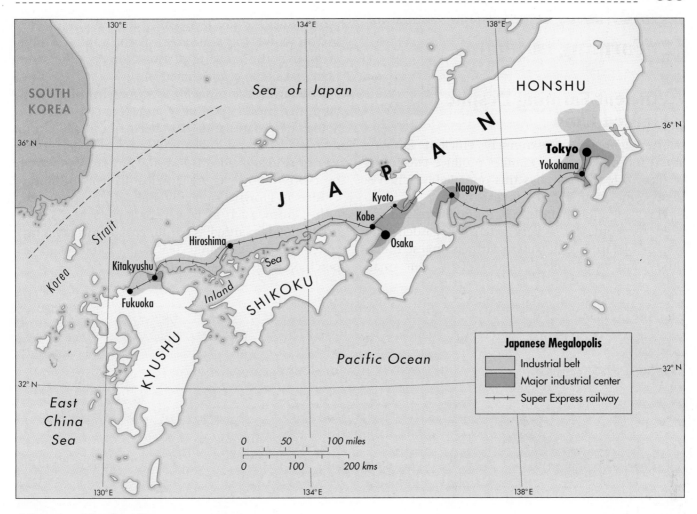

A Japanese Megalopolis
What is a megalopolis? What city anchors the western edge of the industrial belt? What city anchors the eastern end? Can you think of a comparable example in North America?

the Japanese megalopolis, highways and a super express railway connect the cities of Tokyo, Nagoya, Kyoto, and the Inland Sea ports.

This megalopolis contains most of Japan's people, industrial activity, and wealth. As in the American Northeast, the largest cities—Tokyo, Nagoya, Osaka, and Kobe—are important ports as well as industrial centers. People, industry, and agriculture compete fiercely for space. Problems of congestion are severe. Worse, only 65 percent of Japan's roads are paved, compared with virtually all those of Britain and Germany. The Japanese megalopolis continues to grow in spite of government efforts to spread a variety of industries, along with high technology firms, into Japan's outlying regions.

⊕ REVIEW QUESTIONS

1. By the 1980s, where did Japan rank among the world's largest steel producers? What percentage of its iron ore and coal does Japan import?
2. What kinds of products are being made by Japan's high technology industries?
3. The Japanese megalopolis stretches between which two cities?

⊕ THOUGHT QUESTIONS

1. How did creative economic planning by government and industry change Japan's economy after World War II?
2. How has the shift to high technology brought more industry into the Japanese countryside?

2. Farming in Japan

Efficient Farming Despite Limited Land

The Japanese are extremely efficient farmers. Today 3.8 million farm families produce two-thirds of Japan's food. How has this remarkable achievement come about?

The geographic handicap of limited farmland has been overcome by a very intensive, garden-style system of farming. Japanese farmers invest much time, hard work, energy, and skill in their small plots of land to produce some of the highest yields in the world. In warm areas in the south, **double cropping**, the growing of two crops on the same field in a single year, is practical.

The farmlands of Japan are a patchwork of lowland rice fields (paddies) and terraced hillsides. Some new farmland has been reclaimed from marshes along the coast, and farms also have been started on Hokkaido. These gains in farmland have been offset, however, by land lost to growing cities.

Rice is grown on nearly half of the farmland. Other crops, such as wheat, barley, soybeans, and potatoes, are planted as winter crops or in areas not suited to rice growing. Tea and mulberry groves perch on hilly slopes.

Land Use in Japan

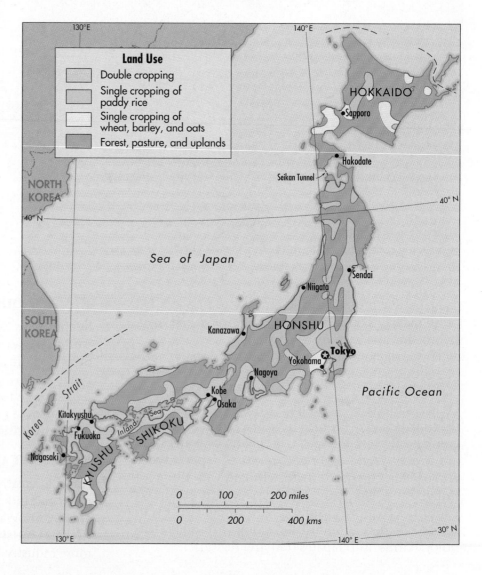

Important Changes in Farming

The United States carried out a land reform program in Japan after World War II. Land owned by landlords was bought by the government and then sold at low rates of repayment to the tenant farmers who had worked on it. One-third of Japan's farmlands changed hands in this way. As a result, most Japanese farmers now own the land they cultivate.

The land reform program also created larger, more rectangular fields that can be farmed with machines. Now small tractors, tillers, planters, and harvesters are used in nearly every Japanese rice field. Mechanization, better drainage, and improved irrigation (supplying crops with water by artificial means) have increased rice harvests dramatically. Moreover, the government has established agricultural experiment stations to provide farmers with the most recent information on seed varieties, fertilization practices, and methods of mechanizing the use of their land.

Japanese farmers now are well-off for two reasons. First, the price of rice in Japan is kept high by government subsidies, or grants, that guarantee farmers a high price for their rice—a price much higher than the rice would bring on world markets. Second, small factories are relocating into farming areas. In the 1990s, four of every five Japanese men living on farms work part-time in these factories to increase their family incomes.

⊕ REVIEW QUESTIONS

1. What crop is grown on nearly half of Japan's farmland? What other food crops are raised?
2. What is double cropping?
3. What has been done to increase rice harvests dramatically in recent years?

⊕ THOUGHT QUESTIONS

1. How did the growing of rice help to make the Japanese a group-oriented people?
2. In Japan, the price at which rice is sold is kept at a high level by government subsidies. Do you believe that the payment of government subsidies to farmers in the United States is a good policy? Why or why not?

3. Problems of a Technological Society

Social Changes and Economic Growth

After World War II, the Japanese faced the task of rebuilding their country. The traditional Japanese values of discipline, self-sacrifice, and obedience all helped to spur economic growth. But the same values that made Japan rich are now being challenged by the changes that success has brought.

Three social practices that have been important to Japan's success in manufacturing are (1) guaranteed lifetime employment, (2) salaries based on length of service with a company, or seniority, and (3) cooperative company unions. In the 1990s, these traditional industrial policies are breaking down, and the future structure of Japanese industry is becoming less certain.

Guaranteed lifetime employment is being affected by several factors. First, in industries that are declining, this policy means that large numbers of employees will have little work to do at a time when foreign economic competition is intense. Second, automation and robots are gradually reducing the number of workers needed in some industries. Third, more and more Japanese women are seeking jobs in industry and fewer are having children, increasing the number of available workers.

The seniority system, which rewards workers on the basis of number of years of service, is also being challenged. Some executives are impatient with the seniority system. While in the past they waited for salary increases and advancement, now they leave guaranteed jobs to start their own businesses. Other Japanese executives are working for industries in nearby developing countries such as South Korea and Taiwan.

| IT'S A FACT | Unless you have an off-street parking place, you cannot register a car in Tokyo. |

THE JAPANESE RICE CYCLE

A rhythm of life in rural Japan centers on the **rice cycle**—the planting, growing, and harvesting of rice plants. Seeds are sown in the spring in special seed beds. A few weeks later, the seedlings are set out by hand in flooded paddy fields. **Paddies** are embanked plots of irrigated land on which rice is grown. These fields are fertilized and tended by hand. When the rice is ripe, the paddy fields are drained. After the ripe plants are cut, they are separated into rice, straw, bran, and hulls. Each of these by-products is fully used and eventually returned to the soil as fertilizer.

Rice is eaten in various forms in Japan. Boiled rice is the main food at every meal. Rice pounded into dough is made into rice cakes, which are used as offerings at shrines and eaten on holidays. Rice is also made into crackers; the dough is compressed, cut into thin pieces, dried, grilled, and flavored with soy sauce. Powdered rice mixed with water and season-ings is a kind of candy, and fermented rice malt becomes sake, or rice wine.

Rice straw is used as livestock feed, fertilizer, and fuel. Floor mats, ropes, baskets, clothes, toys, and ceremonial religious objects are also made from rice straw. The ash from burned straw is used in firing pottery.

The bran, which contains vegetable oil, proteins, and B vitamins, is made into products like cooking oil, pickling ingredients, and chicken and livestock feed. Wrapped in a little cloth pouch, bran is used for skin care. The hulls are used as fertilizer and as

▼ *Rice farming is of great cultural and economic importance to the Japanese. The Japanese government guarantees its farmers high prices for rice grown in Japan. Why do you think this is an important trade issue between the United States and Japan?*

Rice paddies are found in deep valleys in the Japanese Alps. Why do you think farming is confined to valleys in this part of Japan?

QUESTIONS

1. Other than for food, in what ways is the rice plant used?
2. The Japanese dependence upon rice production helped to develop cultural values of co-operation and harmony. Recalling what you know about the American pioneers of the old West, what individual characteristics or qualities would have been most helpful in that time and place?

packaging for fragile objects. The ash is used to make a special pottery glaze.

The rice cycle regulates life in many Japanese villages. Raising wet rice requires that people work together in the paddy fields. Harmony is necessary to maintain complicated irrigation and drainage systems. Historically, these needs have helped to make the Japanese a group-oriented people, who see personal concerns as subordinate to family and village welfare. The entire system is based on order, continuity, and a thrifti-ness that permits no waste. Now that so few Japanese are farmers, and most of these farm only part-time, village Japan is beginning to disappear both physically and culturally.

In a rural area of Honshu, two merchants are making rice cakes. One pounds the rice with a wooden mallet while the other adds necessary ingredients. How does this scene differ from one you would expect in an American bakery or supermarket?

The Changing Role of Women in Japan

Since the 1980s, Japanese women have been playing new and different roles in society. Birthrates are falling so that fewer women need to stay home to care for children. A growing number are taking jobs outside the home and working in the country's shops, factories, and businesses.

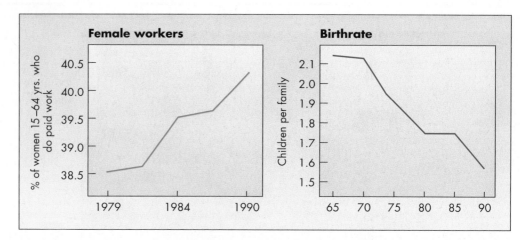

At the same time, company unions, which have cooperated with management in the past, are beginning to demand more benefits for their members. Strikes and work slowdowns, steadily increasing in the 1990s, threaten to raise the cost of Japanese export goods.

Creativity Is the Key to the Future

The shift to high technology is also bringing new challenges. Japan's success has been achieved by importing and modifying technology earlier developed in the United States and Western Europe. In the last thirty years, some thirty thousand agreements gave the Japanese the right to use research findings from the United States.

Today Japan's sophisticated export products are mainly improved versions of this borrowed technology. But in the 1990s, now that Japan has caught up with the West, it can no longer depend on other nations to fund basic research and provide scientific breakthroughs. Therefore, research and development costs are rising sharply in Japan. Indeed, Japan is now a leader in the kind of research that makes history, not just profits.

The Japanese government recognizes that basic research can no longer be borrowed and that creativity will be the key to future economic success. The number of engineers trained has tripled each of the last ten years. The government has built an entire science city, Tsukuba, forty miles northeast of Tokyo, with two major universities, more than fifty

scientific institutes, and a growing population of 250,000.

Pollution of the Environment

Japanese cities, industries, and people are crowded onto small coastal plains, as you have learned. This concentration of human activity on Japan's lowlands has caused a high level of environmental pollution. Complaints to the government about pollution began in the 1960s, and by 1990, conditions were so bad that the Japanese Diet (parliament) discussed moving Japan's government out of Tokyo.

Air pollution has created smog domes in the atmosphere above every large city in Japan. A **smog dome** is a canopy of polluted air that settles over an urban area. In Tokyo people buy tanks of oxygen in order to breathe on days with high smog levels. The use of high-sulfur petroleum from the Middle East increases the amount of pollutants in the atmosphere. Automobiles—now numbering 33 million vehicles—also contribute to Japan's air pollution.

Water pollution is a problem in Japan as well. Sewage from coastal cities and pollutants from

IT'S A FACT   The people of Japan eat more fish per capita than the people of any other nation.

Virtually every Japanese city suffers from severe air pollution and is covered by a smog dome. What American cities have serious air pollution problems?

industry are contaminating the waters along the coastal shores of Japan and of the Inland Sea. In the late 1960s, when Tokyo Bay was declared the most polluted body of water in Japan, fishing and

bathing were banned in the bay. People who ate fish caught in Minamata Bay on the west coast of Kyushu were stricken by diseases caused by heavy metal poisoning. On the northwest coast of Honshu, a similar case of mercury poisoning from industrial pollutants led to a large fine against a local industry.

These cases of human suffering from pollution have caused public outrage. More than 80,000 Japanese people have been officially diagnosed as suffering from pollution-related ailments; they are now receiving compensation from the government.

The Japanese government has passed strict antipollution laws based on a policy of making polluters pay for pollution. Legal standards of minimum air quality have been set up. Monitoring stations test the air in industrial-urban areas. Because of efforts to reduce pollution in coastal waters, Tokyo Bay again has a fishing industry.

Japan still has pollution because so much human activity is compressed into such a small area, but it has begun to export pollution-producing industries to less developed neighboring countries. The costs of environmental conservation inside Japan are very high, and moving industries away from the coast into the mountainous interior would be incredibly expensive and possibly make the pollution problem worse. The Japanese will have to reduce their rate of industrial growth and implement better pollution controls to really improve the quality of their environment. Perhaps Japan can apply the same creative vision used to rebuild its industries in tackling its severe environmental problems.

REVIEW QUESTIONS

1. What three social practices have been important to Japan's success in manufacturing?
2. What is the centerpiece of Japan's information economy?
3. What has air pollution created in the atmosphere above every large city in Japan?

THOUGHT QUESTIONS

1. What is meant by this statement: "Creativity will be the key to Japan's future economic success"?
2. How is Japan trying to deal with its pollution problems?

MOVEMENT

FOREIGN TRADE AND JAPAN'S HIGH-TECH VISION

Foreign trade is essential to Japan if the country is to continue as a world economic leader. To maintain their status, the Japanese are turning to knowledge-based industries. The production of steel and cars is being replaced by computer manufacturing, telecommunications, robotics, optics, and **biotechnology**—using living organisms to develop products for use in fields such as medicine and environmental science.

Many other countries share this high-tech vision of the future, but Japan has already taken enormous strides toward achieving success in these industries. Through focused, cooperative investment, government and industry are forging a major role for Japan as a trading nation.

The centerpiece of Japan's information economy is its computer industry, which is second in size only to that of the United States. Japanese computer companies are the only serious global competitors to U.S. companies. Because the Japanese domestic market is small, leading Japanese computer makers rely heavily on sales abroad.

The Japanese specialize in products with a high degree of uniformity, which allows maximum use of their well organized and disciplined production techniques and keeps prices low. Some of the most popular software programs ever written for computers come from Japan, including many of the better known programs used in video games.

The Japanese are investing in ideas and technologies that will shape the future. Scientists are working on a project to create computers with "artificial intelligence." Nineteen high-tech cities, called **technopolises**, are being planned for Japan; one such city, Tsukuba, has already been built. Given its dependence on trade, on the international movement of goods, Japan will need to keep transforming its high-tech vision into reality to meet the challenges that lie ahead.

QUESTIONS

1. What is meant by the term *knowledge-based industries*?
2. What would probably happen to Japan's computer industry if the movement of goods out of the country were stopped?

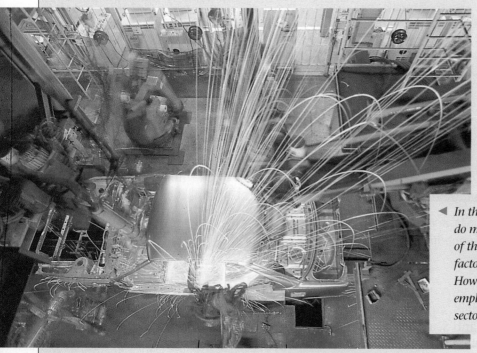

◀ *In this automobile factory in Japan, robots do most of the welding. Indeed, Japan is one of the few countries in the world that has factories where robots make more robots. How do you think this might affect future employment opportunities in the industrial sector as compared to the service sector?*

4. The Two Koreas: Communism and Capitalism Share a Peninsula

Korea: The Bridge Between

For centuries, the Korean Peninsula has been a bridge between China and the island world of the Japanese. The peninsula's location made it a natural corridor through which Chinese learning and trade flowed to Japan. Korea became a buffer state between these two strong nations. A **buffer state** is a small country located between two larger, more powerful countries. The Japanese periodically invaded Korea as a stepping-stone to China. More often, the Chinese influenced Korea's political and economic policies.

Through it all the Koreans retained their own culture, language, and sovereignty. From the 1600s onward, they adopted a policy of isolation and became known as the "hermit kingdom." Japanese diplomatic and commercial pressure broke Korea's seclusion in 1875, and twenty years later Japanese forces occupied and colonized the entire region. In 1910, they annexed Korea into a new and expanding Japanese empire.

Japanese Rule in Korea

At that time, Korea's population of 10 million was primarily engaged in agriculture. Villagers were concentrated in the lowlands of the southwest, whereas settlement in the uplands of the north and northeast was relatively sparse. Villagers were brutally driven to higher levels of production by the Japanese during the occupation, but this productivity did little to improve the life of the Korean people. Japanese and Koreans still harbor much bitterness and resentment toward one another.

The Japanese doubled the agricultural production of Korea during their fifty-year occupation by increasing irrigated rice land in the south. They extended water-control facilities, reclaimed coastal land for cultivation, and introduced mechaniza-

tion. In the hills, commercial crops, like cotton, tobacco, and commercial fruit orchards, were planted. The Korean fishing fleet was modernized, and catches increased substantially. Wood was harvested from the uplands, causing problems of soil erosion today. Production was geared to supply Japan's expanding economy and empire.

Roads and railroads were constructed to connect the agricultural areas of the south and the mineral centers of the north with port cities. Anthracite coal and iron deposits were mined for use by Japanese

The Korean Peninsula

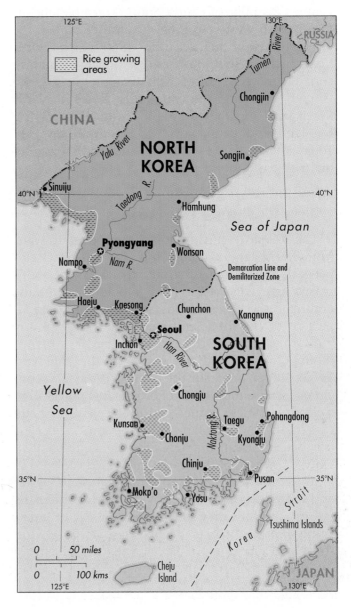

factories, and local hydroelectric plants were constructed to facilitate manufacturing.

As in Japan's other conquered colonies in Taiwan and Manchuria, a remarkable rate of economic development was achieved under Japanese rule. But in 1945, when Korea's population reached 28 million, most Koreans were living in poverty and need. The quality of life for the average Korean was worse in 1945 when Japanese occupation ended than it had been in 1900.

After the defeat of the Japanese in World War II, the Korean Peninsula was politically broken up into two countries. North Korea became a communist state with strong links to the Soviet Union and the People's Republic of China. South Korea became affiliated with Western capitalist countries.

This division hardened in 1950 when North Korea invaded South Korea, triggering a devastating conflict between United Nations forces led by the United States and the communist forces of North Korea and China. The war lasted until 1953, leaving the land and people of both countries ravaged. Today, the economies of North and South Korea remain on a war footing.

The Korean Peninsula

The Korean Peninsula juts 500 miles southward from the Asian mainland into the Pacific Ocean between northern China and the Japanese archipelago (see page 341). Hills and low mountains dominate the Korean landscape, covering some 70 percent of the total land area. The topography is most rugged along the northern border, where a mountain mass and the gorges of the Yalu and Tumen Rivers separate the Korean Peninsula from China. A mountain spine of broken low ranges runs from the Chinese border southward paralleling the east coast. Lowlands fan out to the south and west on the peninsula, but few are very large in area. Only 17 percent of the land is cultivated.

In the north, the hills and mountains are higher and more rugged, and are covered with coniferous forests. The hills dive steeply into the Sea of Japan along the east coast. In the west, swift-flowing streams course down narrow valleys and empty into tidal estuaries in the Yellow Sea. Climate is severe and continental in character.

Population and relief are closely related. In the north, hardy crops like wheat, millet, and barley are grown on hill slopes, and rice in sheltered valleys. The population of North Korea (currently 22.6 million) is smaller than that of South Korea (44.6 million) because its topography and climate are difficult for farming. A major level plain along the west coast straddles the thirty-eighth parallel that is now the border between the two countries. This divided agricultural region supplies food to the capital cities of both North and South Korea, Pyongyang and Seoul.

In the south, the mountains become hills, valleys broaden out, and the climate is more moderate. Unlike the north, the Korean south has a growing season (the period between frosts) long enough that farmers can practice double cropping. Irrigated rice is grown using the intensive garden-cropping Chinese system, so that level valley floors and adjacent terraces are covered with paddies. In the hills, wheat and barley are the primary crops.

The most densely populated areas are located on plains adjacent to the western and southern coasts in the vicinity of the capital city of Seoul, the cities of Kunsan and Mokp'o farther south, and the port of Pusan on the east coast.

The Two Koreas: Two Paths to Development

After their political division, each country attempted to pursue economic development, and both have undergone substantial change in the last fifty years. Today, the peninsula is occupied by 67.2 million Koreans united in language and culture but divided politically into two hostile states. North Korea has pursued modernization under communist rule and until the collapse of the Soviet Union in 1991 was aligned with that country. South Korea, by contrast, has a capitalistic society and is an ally of the United States; 35,000 American troops under the control of the United Nations are still stationed there, patrolling the DMZ, or demilitarized zone, along the thirty-eighth parallel. The results of these two pathways toward modernization have been strikingly different. South Korea is a wealthy nation and the people of North Korea are poor.

North Korea: Communism

From 1948 to 1994, North Korea, under the hard-line leader Kim Il Sung, pursued a distinctive form of communism reflecting the Korean concept of *juche*, or national self-reliance. North Korea pulled away from cooperation with foreign governments and foreigners, becoming one of the most isolated countries in the world. North Korea also became a police state and developed a highly centralized, planned economy.

Farmland was collectivized and farmers were regrouped into large communes in the 1950s. This meant that farmers worked communal fields and lived together in collaborative groups. The country has a food deficit in the 1990s. Indeed, North Korea faces a growing economic crisis as well as worsening food shortages, and discontent is mounting.

The major emphasis of the centralized economy, however, has been on the growth of heavy industry

IT'S A FACT Kim Il Sung, who died in 1994, headed his country longer than any other world political leader.

rather than on agriculture. The capital and leading industrial city of Pyongyang (with a population of 2 million) is supplied with coal and iron ore from important deposits in its **hinterland**, which is the region that serves an urban area. A second industrial zone with a similar concentration of minerals is located in the far northwest near the Chinese border.

Overall, North Korea yields four-fifths of the mineral and metal production of the peninsula. Despite this advantage, North Korea has not experienced the rate of economic growth of the South, a fact most observers attribute to the inflexibility and inefficiency of its communist economic system.

▼ *American troops are still stationed along the DMZ which separates North Korea and South Korea. What conflict led to this situation?*

North Korea has spent heavily on the development of a military that is enormous. Even more alarming, North Korea in 1993 threatened to become the first country to withdraw from the Nuclear Nonproliferation Treaty, which was designed to limit the spread of nuclear weapons. With so much of its resources going to the military, then, it is no wonder that North Korea has a GNP estimated at less than $1,000 per person, compared to the South Korean per capita GNP, which is six times larger. The North Korean economy is, in fact, on the verge of collapse. Indeed, by 1996 the country was forced to seek foreign aid in the form of oil and food from Japan, South Korea, and the United States.

South Korea: Capitalism

In sharp contrast to the north, South Korea has become a rapidly growing industrial power known as one of Asia's "four dragons." Like the other "dragons" (Hong Kong, Taiwan, and Singapore), South Korea has experienced extraordinary rates of economic growth since the 1960s. Personal income rose from $105 in 1965 to $6,340 in 1993. In many ways, South Korea has exhibited the most successful transition from an agricultural to an industrial economy in recent decades.

South Korea is now an urban country. Three-quarters of its people live in cities. Seoul, the capital, with a population of 11 million, is one of the largest cities in East Asia. It is linked by rail and superhighway with Inchon, its port on the Yellow Sea. In the south, Pusan is a **metropolis**—a large sprawling city—of nearly four million. The inland center of Taegu has nearly three million people. South Korea's people have retained their rural roots, however. Although Britain's farms are 60 times larger and America's 150 times larger, South Korea produces enough rice and other agricultural products for its own use and also for export to neighboring nations in East and Southeast Asia.

South Korea depends on local hydroelectric sources and imported oil for industrial energy but plans to expand coal mines on the east coast to in-

▼ *This electronics factory in South Korea reflects the rapid economic growth of this country during the last two decades. Why has the South Korean economy grown more rapidly than the North Korean economy during this period?*

crease domestic energy production. South Korea has five nuclear plants in operation and four under construction. Still, like Japan next door, South Korea must trade for raw materials to survive in the global arena. Its largest cities also share many of the same problems of pollution as those in Japan, Taiwan, and China.

South Korea's capitalist economic system includes strong government investment in large private enterprises. As in Japan, these businesses are run by an intimate circle of financial and industrial leaders. The economy has focused on the production of steel, petrochemical products such as rubber and fertilizers, textiles and clothing, consumer electronics, automobiles, and shipbuilding. Tourism, an important and growing local industry, was given a big boost by the 1988 Olympics, held in Seoul. South Korea exports workers as well as machinery and finished goods. Private enterprises in many Asian and Middle Eastern countries hire Koreans as workers in the merchant marine, the construction industry, hospitals, hotels, and restaurants. Moreover, South Korea is investing in industrial and commercial industries abroad.

Today's questions in South Korea center on the liberalization of the political and economic system. Students demonstrate and workers strike for a greater role in the growth of their country. They reflect the need for South Korea's human aspirations to catch up with its rapid material progress.

Movements toward the reunification of the "two Koreas" appeal to young people in both the north and south despite the divergent paths their countries have taken toward development. In 1991, North and South Korea signed an agreement renouncing force. But in 1993, North Korea was preventing international inspection of its nuclear facilities, creating deep concern among many nations, especially South Korea. Their separate military alliances and economic systems would seem to discourage any immediate unification of the Korean people, yet discussions among many Koreans continue at personal as well as political levels.

⊕ REVIEW QUESTIONS

1. The Korean Peninsula has been a bridge between what two countries for hundreds of years?

2. What happened to Korea after Japan was defeated in World War II?

3. What former country had North Korea mainly used as a model in developing its economy?

4. What are South Korea's chief exports?

⊕ THOUGHT QUESTIONS

1. The output of Korean farms and mineral products increased while Korea was occupied by Japan, but the quality of life of the average Korean did not improve. Why was this so?

2. How has each of these factors helped make South Korea more productive than North Korea: (a) geography; (b) climate; (c) the economic system?

⊕ CHAPTER SUMMARY

After World War II the Japanese rebuilt their country into a leader of the technological world. Cooperation between government and business was very important to the rebuilding of their economy. Today, the majority of the Japanese people live in the major cities along the eastern coast and the Inland Sea. These cities are so large that they are virtually connected, forming what is called a "megalopolis." However, despite Japan's strong industrial base, its economy is vulnerable because the country has few natural resources and must depend on global trade to gain the resources it needs, such as oil and raw materials. Japan is working with other countries to develop its own oil deposits and to promote nuclear energy. Japanese farmers have been forced to become very efficient cultivators because of limited availability of farmland.

After gaining its independence from Japan at the end of World War II, Korea was divided at the thirty-eighth parallel. Later this division gave rise to two countries, communist North Korea and capitalist South Korea. The creation of a global economy among capitalist nations and strong government efforts have favored South Korea's economic development, while the global collapse of communism and the closed society of the North Korean government have led to that country's near collapse.

EXERCISES

Cause and Effect

Directions: Each statement in Part A is a cause that made something happen. Each statement in Part B is an effect or change brought about by one of the causes. Number your paper from 1 to 9. Then write the letter of each effect after the number of its cause.

Part A: Causes

1. Hokkaido is the farthest north of Japan's large islands.
2. Japan could not compete with the lower labor costs in Korea and Taiwan.
3. For centuries, Korea was a buffer state between China and Japan.
4. Clean air and clean water are essential in making computer chips.
5. South Korea has fewer mountains and a milder climate than North Korea.
6. Japan uses high-sulfur petroleum from the Middle East for energy and is increasing its use of personal automobiles.
7. The crises of 1973 and 1979 increased the cost of a product by 1,400 percent.
8. Japan has very limited amounts of land on which to plant crops.
9. South Korea is a highly urbanized country; three-quarters of its people live in cities.

Part B: Effects

a. South Korea is more agricultural than North Korea.
b. The Japanese realized that they were too dependent on oil imports.
c. Air pollution is in the atmosphere above every large city in Japan.
d. It has a cold climate much like that of northern New England.
e. Japanese farmers invest much time, hard work, and skill in their small plots of land.
f. Korea was influenced, and periodically invaded, by its neighbors.
g. The government is locating many computer chip factories in the Japanese countryside.
h. In the 1950s, Japan began to switch from the production of textiles to heavy industry.
i. Urban dwellers in South Korea suffer from air pollution as do those in Japan.

True and False

Directions: If the statement below is true, write *true* on your paper; if it is false, write *false*. Then rewrite each false statement to make it a true statement.

1. Japanese government and industry historically have had a special relationship.
2. The special relationship between government and industry in Japan has slowed the country's economic development.
3. The Japanese economy recovered rapidly after World War II because Japan has abundant natural resources.
4. Japan's largest cities are located on plains along the Pacific coast and the Inland Sea.
5. Japanese farmers employ a very intensive, garden-style farming system because of limited farmland.
6. Rice farming in Japan has resulted in a population of individualists with little group identity.
7. The Japanese computer industry is one of the strongest in the world, second only to the American computer industry.
8. The South Korean economy is very weak due to the lack of economic development.
9. Seoul, Pusan, Taegu, and Pyongyang are four large cities on the Korean peninsula.

Inquiry

Directions: Combine the information in this chapter with your own ideas to answer these questions.

1. The Japanese government has worked with industry to phase out declining businesses in favor of more profitable ones. Do you think that a government should direct its nation's economy? Why or why not? What are some

ways the U.S. government influences the American economy?

2. No large cities in Japan are located in the mountains except Kyoto. Have mountains also affected the distribution of cities in the United States?

3. A megalopolis has certain advantages for the people who live in it. What are some of these advantages? What are some of the disadvantages of living in a megalopolis?

SKILLS
Using a Line Graph

The line graph on page 330 uses lines to show changes over time—specifically, how certain exports changed between 1965 and 1990 and estimated changes to 1995. Study this line graph and then write on paper the answers to the following questions.

1. What does the horizontal line at the bottom of the graph stand for?
2. What do the vertical lines on the graph stand for?
3. The diagonal lines show the value of exports from four major industries in different years. What are these four industries?
4. What was the approximate value of cars exported in 1990?
5. What was the approximate value of steel products and color televisions exported in 1990?
6. What was the approximate value of high technology products exported in 1990?

Directions: Certain important conclusions can be made based on this line graph. On your paper, write *yes* for each of the following conclusions that can be drawn from this graph. Write *no* for statements that are not conclusions that can be based on this graph. Then, answer questions 12 and 13.

7. By 1975, textiles no longer were Japan's leading export.
8. The export value of steel products and color televisions had its largest increase between the years 1980 and 1990.

9. The export value of cars nearly tripled in the five-year period between 1975 and 1980.
10. The export value of high technology products was more than twice the export value of steel products and color televisions in 1990.
11. The export value of cars and high technology products were both about $10 billion in 1976.
12. What industries have seen a decrease in exports since 1988?
13. Based on the trends indicated by the graph, what will be Japan's leading export in the year 2000?

Using the Map of the Japanese Megalopolis

Directions: Turn to the map on page 333. Then find the correct ending to each of the following sentences and write the complete sentences on your paper.

1. The Japanese megalopolis extends about (a) 300 miles (b) 550 miles (c) 1,000 miles.
2. The city at the northeastern end of the Japanese megalopolis is (a) Tokyo (b) Kyoto (c) Hiroshima.
3. The city at the southwestern end of the Japanese megalopolis is (a) Tokyo (b) Nagoya (c) Fukuoka.
4. The Super Express railway extends across (a) a small part of the megalopolis (b) about half of the megalopolis (c) the entire megalopolis.
5. An island that is not included in the megalopolis is (a) Shikoku (b) Kyushu (c) Honshu.
6. The nearest large city to Tokyo is (a) Osaka (b) Yokohama (c) Nagoya.
7. Osaka and Kobe are ports on the (a) Inland Sea (b) Sea of Japan (c) East China Sea.
8. Kyushu is southeast of (a) South Korea (b) Shikoku (c) Honshu.

9. The city nearest Kitakyushu is (a) Nagoya (b) Hiroshima (c) Fukuoka.
10. Most cities in the megalopolis are (a) on high plateaus (b) inland (c) coastal ports.
11. The Japanese megalopolis is located along the shores of the (a) East China Sea (b) Sea of Japan (c) Pacific Ocean.
12. The Japanese megalopolis is oriented (a) north-south (b) east-west (c) northwest-southeast.

Vocabulary Skills

Directions: Select the term from the following list that means the same as the italicized word or phrase in each sentence. Write each sentence on your paper substituting the correct term from the list.

biotechnology	metropolis
buffer state	paddies
coniferous	petrochemicals
DMZ (demilitarized zone)	rice cycle
double cropping	silicon
hinterlands	smog dome
irrigation	technopolis
juche	tenant farmers
megalopolis	

1. A *strip of cities and suburbs* stretches from Tokyo to Kitakyushu in northern Kyushu.
2. The Japanese *"science city"* is devoted to the manufacture of high technology components.
3. *This material* is used for manufacturing semi-conductors and computer chips.
4. In Japan, a farmer's life follows the *seasonal agricultural rhythm*.
5. *The artificial flooding* of the rice paddies is necessary in rice growing.
6. Before land reform in Japan, *they* worked the land for the landlord.
7. Japan is making use of *living organisms to develop products* for use in fields such as medicine and environmental science.
8. A *canopy of polluted air* has settled over many large cities.
9. *A small country between two large countries* often helps to keep the peace between countries.
10. North Korea is covered with forests of *evergreen* pine trees that grow high in the mountains.
11. North and South Korea have created *a "no man's zone" that divides the peninsula*. It is patrolled by troops from the two Koreas and the United Nations to prevent another war.
12. Unlike South Korea, which trades its products with many countries, North Korea has attempted *economic self-reliance* in its effort to produce all its own products.
13. Urban centers, such as Pyongyang, rely upon *these areas* for trade goods and natural resources.
14. Pusan is a *large city with several small towns and suburbs linked to it* on the southwest coast of South Korea.
15. The Japanese cultivate rice in *embanked plots of irrigated land*.
16. *Products made from oil*, such as rubber and fertilizer, are important to the South Korean economy.
17. *Growing two crops on the same field in a single year* is practical in warm areas in the south of Japan.

Human Geography

Directions: Are the following statements about Japan and Korea true or false? Write the true statements on your paper.

1. Hokkaido is the least settled part of Japan.
2. By the 1980s, the Japanese produced more steel than the United States.
3. Today, Japan's high technology exports account for more revenue than its textile products.
4. For its sources of energy, Japan no longer depends heavily on imports.
5. Today, three out of four Japanese people live in urban centers.
6. More than a quarter of the entire Japanese population lives in the cities of Tokyo and Yokohama and their suburbs.
7. Japan's largest cities are important ports as well as centers of industry.
8. Because so many farmers have moved to the cities, Japanese crop yields have decreased sharply.
9. Japanese farmers now raise more wheat than rice.
10. In 1895, Chinese forces occupied the entire Korean Peninsula and made it a part of China's expanding empire.
11. In 1950, United Nations forces came to the defense of South Korea when it was invaded by North Korea.
12. North Korea used the United States as a model in developing its economy.

Physical Geography

Directions: Complete these sentences by writing on your paper the terms or phrases that belong in the blanks.

1. Japan is an island |||||||||| located off the east coast of Asia.
2. The Japanese homeland is composed of |||||||||| major islands.
3. Tokyo is located on ||||||||||, the largest Japanese island.

4. The large, northernmost Japanese island is ||||||||||.
5. Kitakyushu is an important city on ||||||||||, Japan's large, westernmost island.
6. About four-fifths of Japan is covered by forested ||||||||||.
7. Japan's cities, industries, and farms are squeezed onto small coastal ||||||||||.
8. Japan's islands are the tops of underwater ||||||||||.
9. Along the northern border, high rugged mountains separate North Korea from the People's Republic of ||||||||||.
10. Rugged topography and cool climates limit the number of crops that can be raised in |||||||||| Korea.
11. The DMZ (demilitarized zone) that separates the two Koreas is along the thirty-eighth ||||||||||.
12. Seoul is the capital of ||||||||||.
13. Pyongyang is the capital of ||||||||||.

Writing Skills Activities

Directions: Answer the following questions by writing essays of several paragraphs each. Remember to include a topic sentence and several sentences supporting your main idea in each paragraph.

1. Japan has had certain advantages and disadvantages because it is an island nation. What are some of these advantages and disadvantages?
2. In the industrial age, how has Japan tried to compensate for its limited supply of natural resources?
3. How do you account for the fact that although Japan was devastated at the end of World War II, today it is one of the leaders of the technological world?
4. What serious problems does Japan face today and in the foreseeable future?

6

THE PACIFIC WORLD

UNIT INTRODUCTION The Pacific world includes Australia, New Zealand, and thousands of small islands dotted across the vast area of the Pacific Ocean. In the late 1700s, Australia and New Zealand were settled by British citizens, some of them convicts who chose "transportation" to the opposite side of the Earth, half a world away from Europe, over imprisonment at home. Australia's nearest neighbors are the island nations of Indonesia and Papua New Guinea, which lie to the north. The trading partner most important to Australia is Japan. The futures of Australia and New Zealand are linked to the rapidly developing Pacific Ocean region. These two countries are European in culture, but Pacific in location.

Like the Europeans who occupied North America, the British who settled these Pacific lands found thinly populated lands that were easily colonized. Only 300,000 native people were living in Australia and about 200,000 in New Zealand when the British arrived. Like the United States, Australia and New Zealand became British colonies and later independent nations.

Although many of the smaller islands of the Pacific world have retained their traditional economies and cultures, some have changed dramatically because of contact with Asian and Western societies.

▲ *The return of Ayers Rock in Australia's Northern Territory to the aborigines symbolizes the country's efforts to compensate native people for past losses. A similar process is occurring in the United States. Do you know of an example in your state?*

SKILLS HIGHLIGHTED

- Using a Rainfall Map
- Locating Australia's Mineral and Energy Wealth
- Working with the Map of New Zealand
- Using Encyclopedias
- Vocabulary Skills
- Writing Skills

UNIT OBJECTIVES

When you have completed this unit, you will be able to:

1. Locate the important geographical features of Australia and New Zealand and give examples of the ways the people there have adapted to their physical environments.

2. Relate the ways of life of the native peoples of Australia and New Zealand and the history of their relations with the British.

3. Explain how the economic development of Australia and New Zealand has made them nations of the technological world.

4. Analyze the growing relationship between Australia and New Zealand and their Asian neighbors.

5. Show why Australia and New Zealand cannot support much larger populations than they now have.

6. Describe how contact with other societies has changed ways of living in the islands of the Pacific world.

► *The harbor of Wellington, New Zealand, reflects the strong British influence on the history and development of this country. What elements in this photograph suggest European influences?*

▲ *This volcanic island in Tahiti is the top of a submerged volcano. How does the environment of high islands differ from that of low islands in the Pacific?*

351

KEYS TO KNOWING THE PACIFIC WORLD

1

Australia, with an area four-fifths that of the United States, is the world's smallest continent.

2

Australia is the world's flattest continent. Its interior is dry, but its coasts are humid.

3

Northern Australia lies in the tropics, and southern Australia in the middle latitudes.

4

Australia and New Zealand were made British colonies in the 1800s and eventually became independent countries.

5

Australia's aborigines and New Zealand's Maori suffered greatly from European contact. Now both countries are attempting to compensate these native peoples for past losses.

6

Most Australians live in cities and work in industries that are located on the coast. The interior is virtually uninhabited.

▼ *Sheep-raising stations such as this one in the shadow of the southern Alps are a common sight on South Island, New Zealand. What areas in the United States or Canada does this dramatic contrast in topography suggest?*

7

Australia is a vast storehouse of mineral and energy resources.

8

New Zealand is a humid, mountainous, forested land with few mineral or energy resources.

9

Australia and New Zealand are wealthy, literate societies with high standards of living and small populations.

10

The three main groups of islands in the Pacific world are (1) Melanesia in the southwest Pacific; (2) Micronesia to the north of Melanesia and the south of Japan; and (3) Polynesia, which reaches across the Pacific Ocean from Australia to South America.

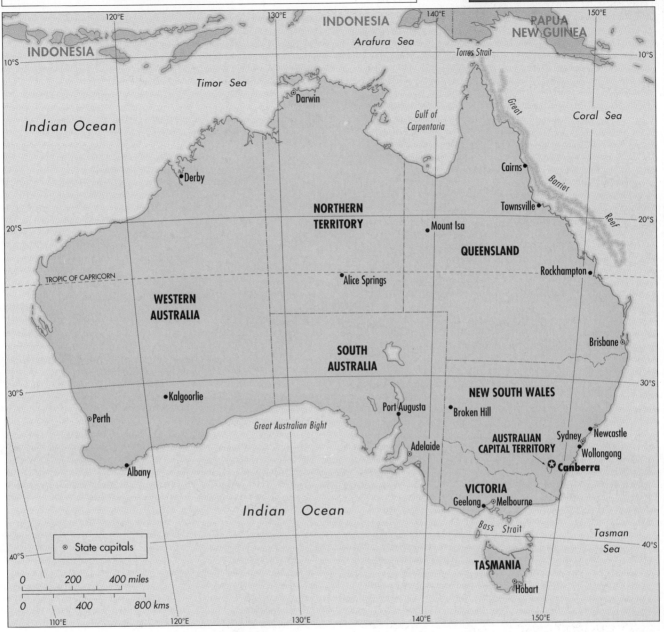

INDONESIA

Arafura Sea

Torres Strait

PAPUA NEW GUINEA

Timor Sea

10°S

120°E

130°E

140°E

150°E

INDONESIA

⊙ Darwin

Indian Ocean

Gulf of Carpentaria

Coral Sea

Great

• Derby

NORTHERN TERRITORY

Cairns •

Barrier

20°S

Townsville •

• Mount Isa

QUEENSLAND

Reef

TROPIC OF CAPRICORN

Rockhampton •

WESTERN AUSTRALIA

• Alice Springs

SOUTH AUSTRALIA

Brisbane ⊙

30°S

• Kalgoorlie

NEW SOUTH WALES

⊙ Perth

• Port Augusta

• Broken Hill

Great Australian Bight

AUSTRALIAN CAPITAL TERRITORY

Sydney • • Newcastle

Wollongong

• Adelaide

✪ **Canberra**

• Albany

Indian Ocean

VICTORIA

Geelong • ⊙ • Melbourne

Bass Strait

Tasman Sea

40°S

⊙ State capitals

110°E

120°E

130°E

140°E

150°E

TASMANIA

• Hobart

0   200   400 miles

0   400   800 kms

Australia Physical Map

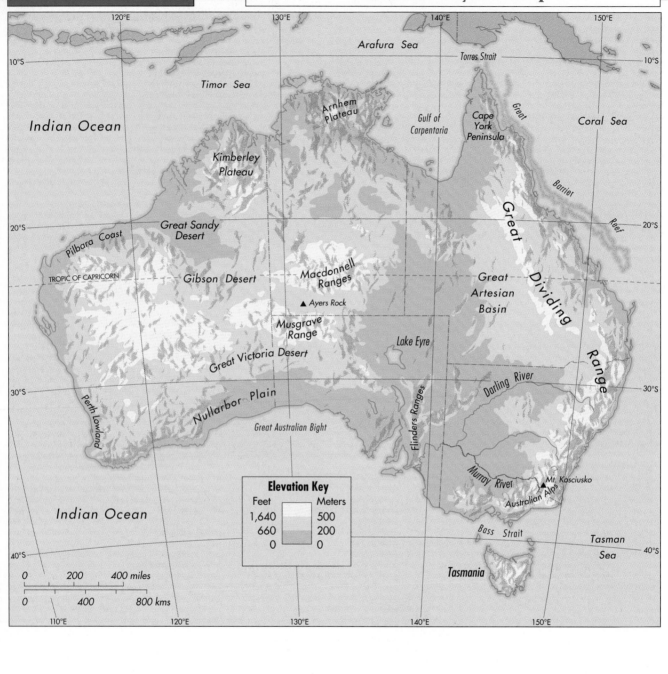

Indian Ocean

Timor Sea

Arafura Sea

Torres Strait

10°S

Arnhem Plateau

Gulf of Carpentaria

Cape York Peninsula

Coral Sea

Kimberley Plateau

Great

Barrier

20°S

Great Sandy Desert

Pilbara Coast

TROPIC OF CAPRICORN

Gibson Desert

Macdonnell Ranges

▲ Ayers Rock

Musgrave Range

Great Artesian Basin

Great Dividing Range

Reef

20°S

Lake Eyre

Great Victoria Desert

Nullarbor Plain

Darling River

30°S

Perth Lowland

Great Australian Bight

Flinders Ranges

30°S

Murray River

Mt. Kosciusko

Australian Alps

Indian Ocean

Bass Strait

Tasman Sea

40°S

Elevation Key

Feet		Meters
1,640		500
660		200
0		0

Tasmania

40°S

0   200   400 miles

0   400   800 kms

110°E   120°E   130°E   140°E   150°E

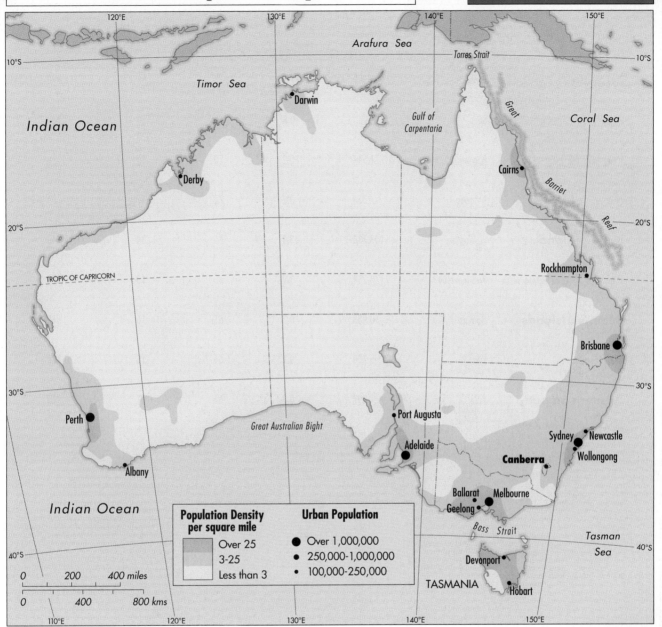

Arafura Sea

Timor Sea

Darwin

Indian Ocean

Torres Strait

Gulf of Carpentaria

Great

Coral Sea

10°S — 10°S

Derby

Cairns

Barrier

20°S — 20°S

TROPIC OF CAPRICORN

Reef

Rockhampton

Brisbane

30°S — 30°S

Perth

Port Augusta

Sydney • Newcastle

Albany

Adelaide

Canberra

Wollongong

Indian Ocean

Great Australian Bight

Ballarat • Melbourne

Geelong

Bass Strait

Tasman Sea

40°S — 40°S

Devonport

TASMANIA

Hobart

Population Density per square mile

Over 25
3-25
Less than 3

Urban Population

● Over 1,000,000
● 250,000-1,000,000
• 100,000-250,000

0   200   400 miles

0   400   800 kms

110°E   120°E   130°E   140°E   150°E

355

Country	Capital City	Area (Square miles)	Population (Millions)	Life Expectancy	Urban Population (Percent)	Per Capita GNP (Dollars)
Australia	Canberra	2,967,909	17.8	77	85	16,590
Fiji	Suva	7,065	0.8	64	39	1,830
French Polynesia	Papeete	1,544	0.2	70	65	—
New Caledonia	Nouméa	7,358	0.2	72	59	—
New Zealand	Wellington	103,736	3.4	75	84	12,140
Papua New Guinea	Port Moresby	178,259	3.9	55	13	820
Solomon Islands	Honiara	10,983	0.3	65	16	560
Vanuatu	Vila	5,700	0.2	65	18	1,120
Western Samoa	Apia	1,097	0.2	66	21	930

—Figures not available

CHAPTER 14

AUSTRALIA AND NEW ZEALAND IN THE PACIFIC WORLD

Australia and New Zealand lie deep in the South Pacific, as the map on page 374 shows. It is significant that these wealthy, Western societies are located next to large, overpopulated Asian countries. Many changes will affect these two countries and the islands of Melanesia, Micronesia, and Polynesia as the Pacific region grows in importance. More than most European countries of the technological world, Australia and New Zealand must adapt to changes not only in the Pacific world, but also in the developing world.

◀ *This ranch, or station, in the Australian outback suggests the vast expanses of open spaces. Each station is as independent as possible. In what ways would stations try to be self-sufficient?*

1. The Subregions of Australia

A Flat Continent with Few People

The continent of Australia is a somewhat rectangular island 1,980 miles long and 2,500 miles wide. It is four-fifths the size of the United States, yet only 17.8 million people live in Australia—fewer people than live in the metropolitan area of Mexico City, Tokyo, New York City, or London.

Australia is not only the smallest continent, but also the flattest continent. Its central plateau covers more than three-quarters of its total area. The Great Dividing Range borders this plateau on the east. Although these mountains are low, (they average less than 3,000 feet in elevation), they proved to be a serious barrier to inland movement in Australia much as the Appalachians did in the United States.

Climate Is Important

Climate, especially rainfall, has played the most important role in where Australians have chosen to live. People are clustered in the temperate **middle-latitude** regions of the south and east. You may recall from Chapter 2 that the middle latitudes lie between the Tropic of Cancer and the Arctic Circle in the Northern Hemisphere and the Tropic of Capricorn and the Antarctic Circle in the Southern Hemisphere. The middle-latitude climates of these parts of Australia are well suited to ranching and farming. The dry desert and steppe lands of the Australian interior, by contrast, are home to very few people because they are for the most part too dry to support ranching and farming.

The distribution of climates in Australia is determined by the continent's size and location. Australia stretches from the tropics to the middle latitudes. Hot tropical climates are found in northern Aus-

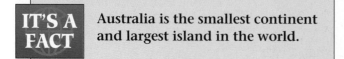

IT'S A FACT — **Australia is the smallest continent and largest island in the world.**

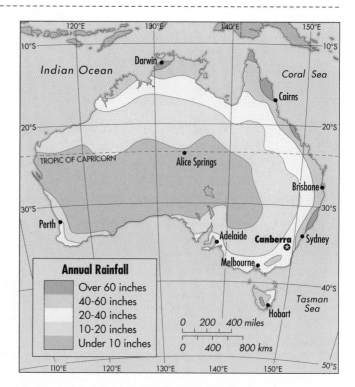

Annual Rainfall in Australia
Note the contrast between the wet coasts and the dry interior. Also note that the North has tropical climates. Do you know why?

tralia. Warm ocean winds near the equator bring heavy rains to these areas, particularly in the summer, so that tropical rain forests are found along Australia's northern coasts. As one moves inland to subtropical areas, rainfall decreases and rain forests give way to mixed woodlands and **savannas**, fields of tall grasses dotted with bushes and trees. Deeper in the interior, short grass steppes and deserts are found.

The coastal areas, where the majority of Australians live, have the most dependable rainfall in Australia. Middle-latitude westerly winds bring winter rains to parts of Australia's southern coast, just as they do in central California. The southwest near Perth and the southeast around Melbourne and Sydney receive as much as forty inches of rain each year. Australia's east coast gets rain year-round.

Between the tropical north and the middle-latitude south lies Australia's dry core, the third largest desert (if one counts Antarctica as well as the Sahara) in the world. This flat, stony area receives

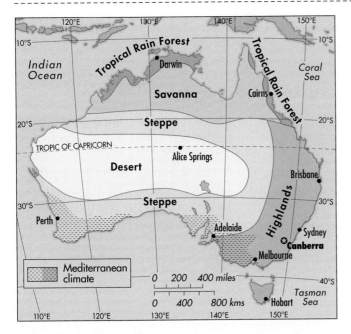

Subregions of Australia

*Australia's two major subregions are based on environment:
(1) the densely settled coasts have mediterranean, highland,
and tropical rain forest environments, and (2) the thinly
peopled core has savanna, steppe, and desert environments.*

The volume of water in all of
Australia's rivers is less than half
that of the Amazon.

desert, steppe, and tropical rain forest environments discouraged early European settlers. Most new immigrants from Britain were drawn to the middle-latitude environments of the southern and eastern coasts because they were familiar—environments that cover only a small area of Australia.

Second, Australia is remote from Europe. Most immigrants leaving the British Isles went to Canada and the United States. Relatively few wanted to make the dangerous 12,000-mile voyage to Australia. Third, ranching and mining were the bases of Australia's economy in the 1800s. Because neither of these industries requires many workers, few jobs were available. Finally, most Europeans came to Australia relatively recently. This means that natural population growth caused by births has not had enough time to build up.

almost no rainfall. On its margins the grass is so thin that it takes several square miles to provide graze for a single sheep. Australians refer to this vast, inland region as the **outback.**

In most areas, rainfall is low and unpredictable. Nearly two-thirds of Australia receives less than twenty inches of rainfall a year. Only 11 percent of the country gets more than forty inches of rain. One cattle ranch, for example, received ten inches of rain in a single day and then no rainfall for an entire year. Such uncertainty in the availability of water makes permanent settlement very risky. Severe droughts have caused great losses of sheep and cattle eight times since 1900. As you can understand, Australia's climates have strongly influenced where people live and work.

Why Australia's Population Is Small

Australia is one of the world's smaller countries in terms of population. There are several reasons Australia is so sparsely populated. First, Australia's

Australia's Densely Settled Coasts

Most Australians live on the eastern and southern coasts of the continent where the major port cities are located. In fact, eight of every ten Australians live within fifty miles of the sea. The only large city located away from the coast is Canberra, the national capital. Australia's two largest cities, Sydney and Melbourne, each wanted to become the capital city, and neither would give in, so the Australian capital was established at Canberra, halfway between Melbourne and Sydney, as a compromise.

Population is concentrated around the five mainland state capitals, all of which are seaports. Locate these on the map on page 355. Fewer than 1 million Australians live farther than 300 miles away from these important cities. The only exception is on the southern island of Tasmania, where half of the people live near the state capital of Hobart.

Australia is the only continent
that is a single country.

Australia's Thinly Populated Core

Small but important communities are found in the interior. Kalgoorlie, Broken Hill, and Mount Isa are centers of mining and ranching. (Settlements also dot the 1,000–mile coastline of Queensland.)

Australia's interior is a harsh, stony, and barren region that does not have enough water resources to support farming. Water must be piped hundreds of miles to the mining towns that are scattered in the desert. Railroads and roads were built to connect coastal cities (and eventually world-wide markets) with distant mining settlements and ranches in the interior. Raw materials are transported by truck or rail to the coast, and goods are shipped by the same routes from cities to people living in the interior.

Huge **road warriors**, double-decker trucks with three trailers, carry as many as 200 head of cattle to urban markets on dirt roads. These roads are important links between Australia's thinly populated interior and the coastal cities. Transport connections are especially important to Australia's econo-my, because a rich storehouse of minerals, metals, and energy resources is found in central Australia, as you can see on the map on page 367.

⊕ REVIEW QUESTIONS

1. How does the size of Australia compare with the size of the United States? What is Australia's population?
2. How much of Australia is covered by its central plateau?
3. What subregions of Australia are the most heavily populated? What subregion is sparsely populated?
4. What is Australia's capital city? It is halfway between what two seaports? Why is it located there?

⊕ THOUGHT QUESTIONS

1. How has climate played an important role in determining where Australians live?
2. Why is Australia one of the least densely populated countries in the world?

▼ *Sydney, Australia's largest city, is an important coastal port. Note its famous opera house in the foreground. Why do most Australians live along the coast rather than in the interior?*

LOCATION

IN AUSTRALIA, EVERYTHING IS UPSIDE DOWN

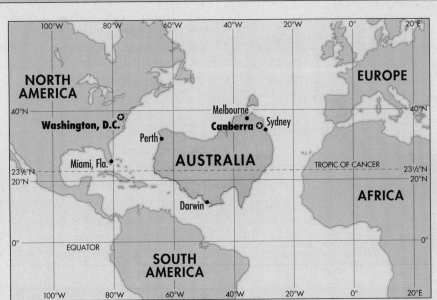

Australia on the Other Side of the Earth
This map shows where Australia isn't, but where its geographically opposite position would be. Note that the Australian "north" is closest to the equator.

The practice of drawing maps with north at the top is so common that it has influenced our way of thinking about directions on the globe. We often say "up north" or "down south." We use terms like "upper" Michigan and "lower" California. Australia is "down under."

We all know that a globe—like any sphere—has no up or down. And there is no reason maps could not be oriented in some other way. But we are accustomed to north as "up" and south as "down."

This mapping practice causes special problems when we think about Australia, mainly because we live in the Northern Hemisphere and we expect southern areas to be warm and northern areas to be cool. We think December will be cold and June will be warm. In Australia, however, everything is upside down, because Australia is in the Southern Hemisphere.

In the map on this page Australia has been turned upside down and placed in the North Atlantic in the Northern Hemisphere in a geographical position exactly opposite its real location on the globe. Looking at this map, we can see that northern Australia (including the city of Darwin) reaches into the hot tropics. In contrast, Australia's southern coast lies in the cooler middle latitudes. (Remember, however, that latitude is not the only factor that influences climate, as you have learned.) The city of Melbourne, for example, is located at the same latitude as Washington, D.C., except that Melbourne is located in the Southern Hemisphere.

Remember that this map shows where Australia is *not*. Australia is actually in the *Southern Hemisphere*.

QUESTIONS

1. Under what conditions would the term *upriver* or *upstream* mean moving in a southerly direction?
2. Find the city nearest your own home town on the political map in the Unit I Miniatlas. Locate a city in the Southern Hemisphere that is near the same latitude. Use the text index to determine if the climates are similar.

REGION

THE GREAT BARRIER REEF, AUSTRALIA'S EIGHTH WONDER OF THE WORLD

For sailors and explorers in the past, the Great Barrier Reef that lies off the northeast coast of Australia was a danger to their ships. For marine scientists and scuba divers today, the Great Barrier Reef is a living laboratory. It is the largest mass of coral in the world.

The Great Barrier Reef is made up of hundreds of islands ranging from tiny islets to the tops of submerged mountains. It extends 1,240 miles along the northeastern coast of Queensland and varies in width from 9 to 186 miles. Between the Great Barrier Reef and the mainland is a narrow water passage where Captain Cook, an early British explorer, wrecked his ship. Today, the reef's sparkling waters, teeming underwater life, and spectacular seascapes are the reasons that Australians call it "the eighth wonder of the world."

The architects of this natural wonder are **coral polyps**. These are tiny sea creatures that live in a protective shell of lime made from their own secretions. Generation after generation of coral polyps live and die, building on the skeletons of their ancestors. Over millions of years, these tiny marine creatures created the **coral reefs** that make up the Great Barrier Reef.

Can you imagine what it would be like to dive in the sparkling waters between the mainland and the Great Barrier Reef? Hundreds of species of coral present a dazzling array of colors. More than a thousand species of fish abound in the reefs. Here you come face to face with one of Earth's last frontiers, and like all frontiers, it holds dangers. Sharks, stingrays, and venomous jellyfish lurk beneath the sea. Shallow reefs test the skills of even experienced sailors. For less adventurous people, underwater observatories are available. Which option would you choose if you had a chance to visit this spectacular natural wonder? In the United States, similar underwater landscapes can only be viewed off the southern coast of Florida.

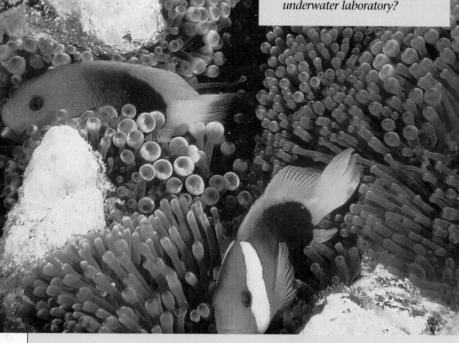

▼ *Tropical fish abound in the waters surrounding the Great Barrier Reef off the east coast of Australia. What creatures created this magnificent underwater laboratory?*

? QUESTIONS

1. How are the coral reefs created?
2. What are some of the dangers associated with coral reefs?

2. Australia's Changing Economy

IT'S A FACT   What we call the Far East, Australians call the Near North.

The Exploration and Settlement of Australia

In 1787, a fleet of eleven ships sailed from Britain carrying 1,030 people; 736 were convicts who chose "transportation" to Australia rather than prison terms in British jails. The fleet landed at Botany Bay and started the first British settlement nearby, calling it Port Jackson. Gradually, these prisoners established farms, and more settlers arrived from Britain. This initial settlement on the east coast grew into Sydney, Australia's largest city.

As in eastern North America, the low mountains of the Great Dividing Range kept early colonists confined to the coast for some time. By the 1800s, however, British colonists settled the fertile lands of the Murray and Darling River systems, which flow westward down the inner slopes of the Great Dividing Range. New settlements were soon established on the Australian coast, eventually forming the cities of Brisbane on Australia's east coast, Melbourne and Adelaide in the southeast, and Perth in the southwest. Locate these important cities on the map on page 353.

Australians Turn to Ranching and Farming

Australians began to import European plants and animals that were adapted to Australian environments. Their first success came with merino sheep. These animals from the dry plains of Spain flourished on the grasslands in Australia's interior. As the wool industry grew in importance, sheep-raising ranches, or **stations**, were established on the inner slopes of the Great Dividing Range. After wells were drilled to provide permanent sources of water, cattle were raised.

Wheat farming was established near Perth in Western Australia and at various points along Australia's eastern and southern coasts near Brisbane, Sydney, Melbourne, and Adelaide. Some wheat farms failed where rainfall was too low, too unpredictable, or both. Wheat did become an important crop in the Murray River Basin inland from Adelaide, on the inland slopes of the Great Dividing Range, and in the vicinity of Perth.

Australian ranches and farms were profitable businesses. They fed local people and produced wheat, meat, wool, and hides for export. New types of wheat that could resist drought were developed and grown on Australia's dry lands.

Railroads opened the interior to farming and ranching and also provided the means to ship products to the coast. Equally important, refrigerated ships made it possible for Australians to sell meat and milk products to Europe and North America. These early economic successes gave Australians the money to invest in gold finds in New South Wales and Victoria in the southeast. As rapid economic development continued throughout Australia, each Australian state demanded—and was given—the status of a British colony.

Finally, in 1901, the six Australian colonies of Queensland, New South Wales, South Australia, Victoria, Tasmania, and Western Australia joined together as states under a federal government, the Commonwealth of Australia. By 1914, this new "Western" country in the Pacific had a population of 4.9 million and a flourishing economy based on ranching, farming, and mining.

From Primary Production to a Postindustrial Society

Before World War I (1914–1918), Australia's economy was based on **primary production**—work in which people make a living directly from the land. In Australia, raw materials from farming, ranching, and mining were produced for export.

After the war, manufacturing and processing industries grew in Australia. The country's raw materials were being changed into products like textiles,

ANTARCTICA, THE ICE CONTINENT

The continent of Antarctica, which makes up 10 percent of the world's land area, is the most remote, least known, and least inhabited continent on Earth. In the past its glistening wastelands attracted only whale and seal hunters, explorers, and adventurers. Now scientists, military observers, tourists, and environmentalists from many nations endure harsh conditions year-round to study life and living conditions in this frozen land.

For nearly six months a year, the sun never rises over this enormous landmass, bigger than China and India combined, that is completely covered by a vast sheet of ice. This ice sheet is so thick that 70 percent of the world's fresh water is captured within it. In this land, temperatures regularly fall to 120° below zero (F). Winds of 200 miles per hour and sudden storms can be lethal. Moreover, almost no precipitation falls in Antarctica because of the extreme cold. For this reason, some observers refer to the

▲ *Penguins are one of the few creatures that thrive along the icy shores of Antarctica. Who owns Antarctica?*

continent as one of the world's great deserts.

Scientific work is mostly done during the relatively short summer which begins on September 21 when the sun rises. Living quarters in the scientific stations that dot the shores of Antarctica are cramped and primitive. Neverthe-

IT'S A FACT The Ross Ice Shelf, afloat at Antarctica's shore near McMurdo Station, is as large as Texas.

less, twenty-five nations conduct research in Antarctica on subjects like global warming and the ozone layer. In 1993, 4,000 scientists and technicians—the largest number yet—studied Antarctica.

This continent belongs to no one. In 1959, twelve nations signed the Antarctic Treaty that has governed this icy land with no permanent population. This treaty, now accepted by forty nations, bans all military activity or nuclear research in Antarctica and permits all nations to do scientific research there. Of the original signatories, seven nations—Argentina, Australia, Britain, Chile, France, Norway, and New Zealand—have claimed rights over Antarctic territory, as shown on the map. The United States and Russia have made no territorial claims on this continent and recognize no claims

▼ *Specially designed ships are required to provision the many research stations located in Antarctica. In what ways do you think the vessel in this photograph is adapted to the extreme cold of this continent?*

made by other nations. Overall, however, this treaty has proved to be a successful example of international cooperation.

In 1991, the treaty was extended considerably when twenty-four nations agreed to ban all oil and mineral exploration in Antarctica for the next fifty years. Environmentalists, who would like to turn Antarctica into a "world park," spearheaded this movement. As the British minister of the environment noted, "This is the protection of the last great wilderness."

❓ QUESTIONS

1. Why do some observers consider Antarctica one of the world's largest deserts?
2. If the environment of Antarctica is too harsh and inhospitable for settlement, why are so many countries interested in the continent?

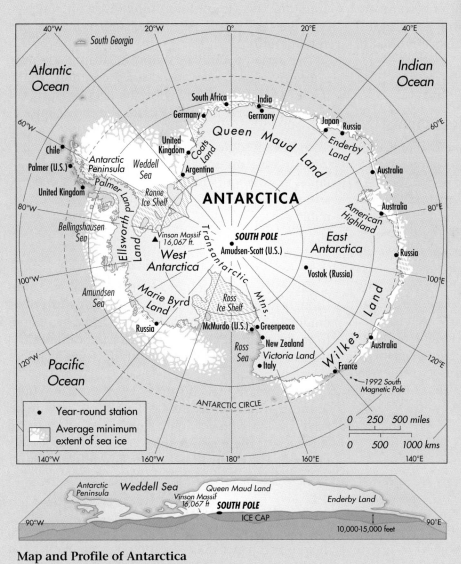

Map and Profile of Antarctica

automobiles, and machines. This is **secondary production,** in which people work in urban factories rather than on the land.

Today Australians are shifting to service occupations, or **tertiary production,** the hallmark of a **postindustrial society.** In a postindustrial society, most work is based on knowledge and information rather than raw materials. People earn a living as doctors, teachers, computer programmers, travel agents, and salespeople. In short, they provide services rather than make things.

Today, agriculture and mining produce only 10 percent of Australia's wealth and manufacturing provides 20 percent. Service industries account for the rest. Australia is rapidly becoming a postindustrial society, like other nations in the technological world.

Modern Farming and Ranching

About 40 percent of Australia's land is still used for farming and grazing; animals account for two-fifths of Australia's earnings from exports. About 15 percent of the world's sheep live in Australia, and they outnumber people eight to one. Sheep graze on land that is planted in wheat during wet years. Most of the wool produced on these ranches is exported to Japan, Western Europe, and Russia.

Cattle require more water and better grazing land than sheep. They are raised on the eastern slopes of the Great Dividing Range and in the well-watered Murray and Darling river basins. Recent developments in animal breeding, pasture management, and dry-weather varieties of wheat explain why Australia is the world's largest exporter of beef and an important exporter of wheat.

A Storehouse of Mineral Resources

Australia supplies minerals—such as iron ore, coal, bauxite (for aluminum), nickel, tin, manganese, lead, silver, zinc, and copper—to many countries in the technological world. Mining produces one-fifth of Australia's exports.

Australia is the world's leading exporter of iron ore. Its four largest mines in Western Australia operate at only two-thirds of capacity because the world steel market cannot absorb more iron ore at this time. Reserves of coal, a second important resource used in steel production, line the east coast of Australia, awaiting development.

In most Australian mining districts, various ores are found together. The earliest and most famous of these mining areas is Broken Hill, west of the Darling River in New South Wales. Now Broken Hill is one of the world's largest producers of lead and zinc, as well as an important source of copper and uranium. The Mount Isa mining district in Queensland produces silver, lead, zinc, copper, manganese, and uranium.

The array of minerals found in Australia is truly impressive, and new discoveries continue to be made. Recently, what may be the largest deposit of nickel was found near Kalgoorlie, the early center of Western Australia's gold-mining region. Income from these resources has helped to pay for Australia's growth into an urban, industrial society. Today, Japan buys about one-third of Australia's exports, and this Japanese connection has been very profitable indeed.

Rich Sources of Energy

Australia is also rich in oil and natural gas, even though Australia still imports about one-third of its crude oil. Most of its own oil and natural gas comes from fields in the Bass Strait between Tasmania and the mainland and from deposits off the coast of Western Australia.

Huge additional offshore reserves have been found in these same areas. In addition, new fields have been found in South Australia and Western Australia. If these new sources of oil and gas are developed, Australia will be self-sufficient in energy.

Manufacturing Develops

Today, Australia has a wide range of industries. Textile and paper mills, food-processing plants, and clothing factories are found in every large city. Firms that make precision instruments and electronics are

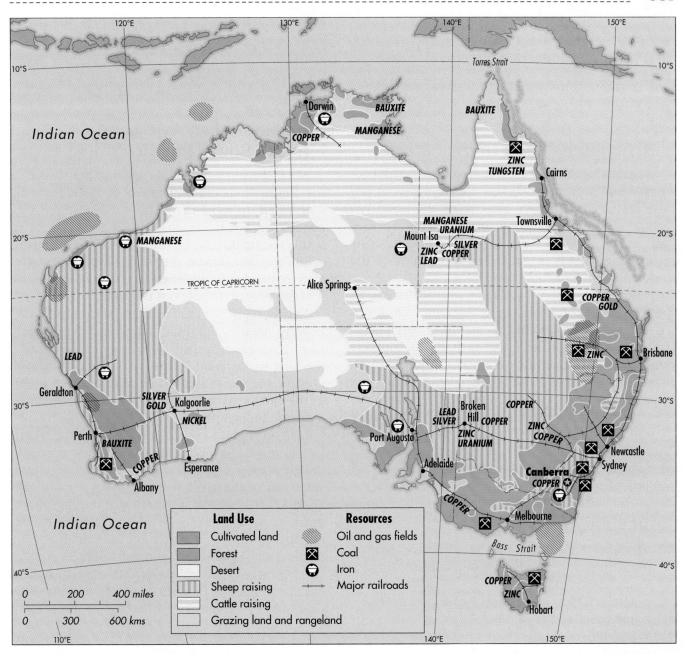

Australia's Mineral and Energy Resources

increasing in number. Other industries assemble automobiles, steel products, and machinery. Oil refineries and petrochemical plants convert Australian crude oil into gasoline, plastics, and fertilizer.

Now, one of every four Australians works in manufacturing. Most of these industries are located in Australia's largest cities, which, as you have learned, are also ports. The iron and steel industry is centered around Sydney. As the map above shows, Sydney is close to large deposits of coal and iron. Notice, also, the network of railroads in this densely populated area.

IT'S A FACT There is a 1,000-mile highway in Australia that has no curves.

Australians Look to the Future

Australia is rapidly becoming a postindustrial society. Most Australian workers are employed in service industries, and 85 percent of the country's people live in cities. Australians earn about $16,600 per person per year. These facts make Australia one of the most highly urbanized and wealthiest of the industrialized countries.

Australia faces some problems in maintaining its high standard of living, however. One example is the high cost of producing manufactured goods, which makes it difficult for Australia to compete with other manufacturing countries. Another problem is that Australia's small population limits the domestic market for its own products. A recent agreement joins Australia and New Zealand in a South Pacific "common market" that should increase trade between the two countries. However, this common market will still sell goods to only 21 million people.

⊕ REVIEW QUESTIONS

1. What portion of Australia's wealth comes from agriculture and mining? From manufacturing? From service industries?
2. Australia is the world's largest exporter of what animal product?
3. What mineral products does Australia export? Australia is the world's leading exporter of what mineral product?
4. Where are the fields that provide most of Australia's own oil and natural gas?

⊕ THOUGHT QUESTIONS

1. Why don't Australians mine all of their iron ore and coal deposits?
2. Why do you think Australia does not limit itself to producing and selling only raw materials to other countries?

3. New Zealand's Climate and Resources

A Middle-Latitude Land in the South Pacific

New Zealand's two main islands are shown on the maps on pages 369 and 374. Although they look small on most maps, these islands are in fact larger than Great Britain. New Zealand lies in the South Pacific directly in the path of winds blowing across the ocean from the west. This location gives New Zealand a cool climate and ample rainfall distributed evenly throughout the year.

New Zealand is a land of high mountains, swiftly flowing streams, mountain lakes, and dense forests. Early Polynesians called New Zealand the "land of the long white cloud." This name described the impressive fog banks created by land and sea breezes in the valleys, as well as the snow-capped peaks and dramatic topography.

Like Australia, New Zealand is part of the technological world. New Zealand is peopled by a wealthy, skilled, society that is using high technology to develop its industries. Some 80 percent of New Zealand's 3.4 million people live in cities and earn more than $12,000 per person per year. New Zealand's humid climate, mountainous landscape, forested slopes, and few mineral resources have led to a pattern of economic development that is different from that of its giant neighbor to the west, Australia.

New Zealand's Environment

Mountain ranges that slice across the country from northeast to southwest dominate New Zealand's North and South Islands. On both islands, topography strongly influences local climate. Patterns of settlement and ways of making a living are affected by the lay of the land.

North Island is somewhat smaller than South Island and has lower mountains and more level land. Its central highlands reach elevations of more than 8,000 feet. South Island is much more rugged.

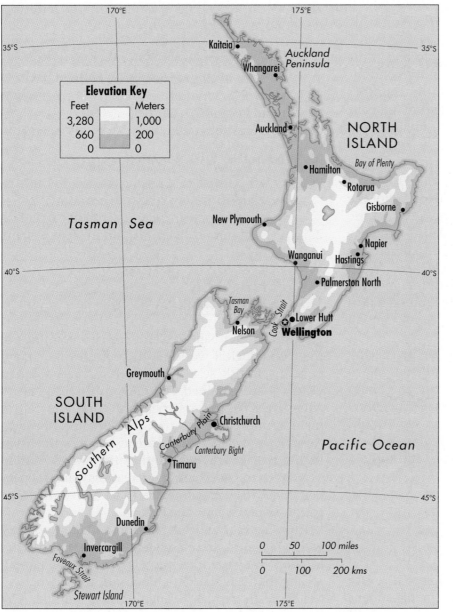

New Zealand

(the slopes facing *away from* the direction which the wind is blowing) of these mountains receive as little as thirteen inches of rain in some years. They lie on the dry side in the "rain shadow" of New Zealand's Southern Alps, just like areas east of the Rockies in North America.

Because no part of New Zealand is very far from the ocean, temperature extremes are rare. Forests and pasture land cover much of the country. The population is concentrated on level land along the coasts, just as in Japan (although population densities are much lower).

New Zealand Develops Differently

New Zealand's development was different from that of Australia. Its two islands were smaller, more mountainous, wetter in climate, and located 1,200 miles farther away from Britain. Economic development progressed rapidly on South Island after gold was discovered on the west coast in the 1860s.

This gold strike brought immigrants to New Zealand. Gold also attracted investment money to build roads and railroads, which in turn opened up to cultivation the best wheat-farming land in New Zealand, the Canterbury Plain south of Christchurch. Beyond the plains, in New Zealand's mountain interior, pioneer settlers grazed sheep.

The snowy Southern Alps are a spectacular mountain range whose highest peaks rise two miles above the sea.

Rainfall in New Zealand varies from 12 to 315 inches a year. The droughts that Australians suffer do not occur here. The western slopes of the Southern Alps receive an enormous amount of precipitation each year. When warm, moist air masses from the west reach the coast, they are forced to rise up the rugged **windward** slopes (the slopes facing the direction *from* which the wind is blowing) of the Alps and to drop their moisture. The **leeward** slopes

Sheep and Cattle Raising in New Zealand

For two hundred years, raising sheep and cattle has been the backbone of New Zealand's economy. Wool, meat, and dairy products are the country's most important exports. Only 3 percent of New Zealand's land is cultivated, and the principal crop is animal fodder.

New Zealand's high standard of living is derived from the country's 60 million sheep and 9 million cattle. Magnificent animal pastures cover half of the country. The most important lowland areas are the Auckland Peninsula of North Island and the Canterbury Plain inland from Christchurch on South Island. Here fertile soil, ample rain, and a long growing season support lush pastures and rich farms. In these areas, mixed farming, fruit orchards, and vegetable fields are interspersed among meadows.

Few workers are needed in the highly mechanized, scientifically run sheep- and cattle-raising industries. Futhermore, competition in the dairy industry is intense and profits are small. Therefore, most of New Zealand's people work in service jobs and processing industries in New Zealand's three largest cities—Auckland, Christchurch, and the national capital of Wellington.

How Can New Zealand Diversify Its Economy?

New Zealand, like Australia, is trying to diversify its economy, so the country will not be dependent on sheep and cattle raising. However, New Zealand has few of the mineral resources that industry needs. Hydroelectric power is the only important source of energy.

Economic diversification in New Zealand is quite difficult. The country's distance from world markets, the small scale of local industries, the limited number of people to buy products, and high labor costs discourage investment in new enterprises. New Zealand must import machinery and energy resources. It has been unable to develop advanced manufactured goods or electronic products that can compete on world markets. At this time, New Zealand is far behind Australia in its industrial development.

⊕ REVIEW QUESTIONS

1. What kind of climate does New Zealand have?
2. What type of landform dominates both of New Zealand's islands?
3. How much does rainfall vary in New Zealand?

⊕ THOUGHT QUESTIONS

1. How is it possible that raising sheep and cattle is the backbone of New Zealand's economy, yet most of the people in New Zealand live in cities?
2. Why is economic diversification in New Zealand difficult?

🌐 CHAPTER SUMMARY

Australia and New Zealand are two countries of the technological world. Both countries were colonized by British settlers in the last three hundred years. They remained colonies until this century when they became independent commonwealths, maintaining close ties to Great Britain. However, Australia and New Zealand are also part of the Asian Pacific world. This means that the two countries' economies are becoming increasingly interdependent with the technological and the developing countries of Asia and the Pacific.

Australia is physically a saucer-shaped country with hills and mountains near the coast and a flat interior. The interior's very unpredictable rainfall means that it is sparsely populated; those who live there mostly work in ranching and mining. The southern and eastern coasts, which receive more abundant rainfall, are where most Australians live and where most of Australia's large cities and industries are located. Australia is rapidly becoming a postindustrial society, with many jobs in information processing and service industries rather than in manufacturing.

New Zealand is an island country with a mountainous interior. It has a smaller population, fewer resources, and is less well developed technologically than Australia. The New Zealand economy is dependent primarily on sheep ranching and dairy farming.

EXERCISES

Missing Terms

Directions: Each of the statements below cannot be completed without including the missing term. Find the missing term in the list below, and write it on your paper next to the number of the sentence to which it belongs. There are more terms than blank spaces.

1. Hot tropical climates are found on Australia's |||||||||| coast.
2. The most dependable rainfall occurs in Australia's |||||||||| regions.
3. Australia has a great oceanic laboratory off its eastern coast, called ||||||||||.
4. |||||||||| is the capital of Australia.
5. Two economically important inland areas of settlement in Australia are the |||||||||| and |||||||||| river basins.
6. One of Australia's most important industries, accounting for one-fifth of its exports, is ||||||||||.
7. |||||||||| is a sparsely populated continent belonging to no one and to everyone.
8. New Zealand's primary economic activity is ||||||||||.
9. The capital of New Zealand is ||||||||||.

Terms

the Great Barrier Reef	Sydney
Canberra	mining
Ayers Rock	Darling
coastal	southern
northern	inland
Antarctica	Australia
Auckland	Murray
ranching	manufacturing
farming	Wellington

Inquiry

Directions: Combine what you have learned in this chapter with your own ideas to answer these questions.

1. Australia is four-fifths as large as the United States. If Australia were as close to Europe as the United States is, do you think it would have almost as large a population as the United States? Why or why not?
2. A week before Valentine's Day, a florist in Vermont can order summer flowers from a supplier in Sydney and have them in her shop in two or three days. Explain how changes in technology and communications have allowed Australians to develop a global economy.
3. Compare the ways that environment and relative location to Europe and Asia have influenced the development of Australia and of New Zealand.
4. Australia's earliest British settlers were convicts who had been sent to Britain's new colony. Why do you think most of these convicts became successful law-abiding colonists?

SKILLS

Using a Rainfall Map

Look at the rainfall map on page 358 to answer these questions.

1. What symbol on the map shows an annual rainfall of (a) under ten inches, (b) ten to twenty inches, (c) over sixty inches?
2. Does the most rain fall in the interior or on the coastal areas of Australia?
3. Does the island of Tasmania receive light or heavy rainfall?
4. Does the northeastern coast or the southwestern coast of Australia receive more rainfall?
5. The areas bordering Australia's dry interior generally receive about how much annual rainfall?
6. Notice that some of the coastal areas receive more than sixty inches of annual rainfall. Would these areas be mainly deserts, steppes, or tropical forests?

7. In the areas with twenty to forty inches of rainfall, would there be mainly deserts, grasslands, or rain forests?

8. In the areas with less than ten inches of rainfall, would there be mainly deserts, savannas, or rain forests?

Locating Australia's Mineral and Energy Wealth

Look at the map on page 367 to answer the following questions.

1. What two mineral resources found nearby help to explain why Sydney is the leading steel manufacturing city in Australia?
2. Lead, zinc, silver, copper, and uranium are found near which two Australian towns?
3. Large nickel deposits are found near what town?
4. What mineral resource is found near Perth and also near the northeastern tip of Australia?
5. Some of Australia's largest oil reserves are offshore near Melbourne. Are Australia's other chief offshore oil reserves near the northeastern coast or the northwestern coast?
6. Can mineral products be shipped by railroad from Broken Hill to Sydney?
7. Can manufactured products be shipped by railroad from Melbourne to Alice Springs?
8. What mineral resources are found on the island of Tasmania?

Vocabulary Skills

Directions: Number your paper from 1 to 11. Write the word(s) or phrase(s) from the list below that complete each sentence.

coral polyps	road warrior
coral reefs	savanna
leeward	secondary production
middle latitudes	stations
outback	tertiary production
postindustrial society	windward
primary production	

1. Over millions of years |||||||||| |||||||||| are created by small marine animals, called |||||||||| ||||||||||, that secrete lime.
2. A country that makes use of large numbers of service occupations such as computer programmers, doctors, and teachers is termed a |||||||||| ||||||||||.
3. In tropical and subtropical latitudes, where rainfall supports mostly grassland and few trees, a |||||||||| environment predominates.
4. Because the |||||||||| |||||||||| are more temperate, more people live there.
5. A |||||||||| |||||||||| is a huge double-decker Australian truck that carries cattle cross-country.
6. People whose occupations are mainly farming, ranching, and mining are said to be engaged in |||||||||| ||||||||||.
7. Tropical islands generally have two types of natural environments. On the |||||||||| side, heavy rainfall results in a very lush environment. On the |||||||||| side, less rainfall results in a drier environment.
8. People whose occupations involve converting raw materials into manufactured goods are said to be engaged in |||||||||| ||||||||||.
9. Australians and New Zealanders call sheep and cattle ranches ||||||||||.
10. Work that is based more on knowledge and information than on manufacturing is termed |||||||||| ||||||||||.
11. The inland region of Australia is a vast desert with thin grasses on its margins. Australians refer to this area as the ||||||||||.

THE PACIFIC WORLD IN TRANSITION

The Pacific Ocean covers a third of the Earth. This water world is larger than all the world's land areas combined. It is seventeen times larger than the United States. The Pacific Ocean reaches from Australia to the Aleutian Islands located between Alaska and Russia. Between Australia and the Hawaiian Islands, to the north and east, the Pacific Ocean is dotted with the islands and island archipelagoes of Melanesia, Micronesia, and Polynesia.

◄ *Economic change is transforming life on many islands in the Pacific, the world's largest ocean. What elements in this photograph suggest modernization?*

373

1. Islands of the Pacific World

IT'S A FACT The Pacific world has between 20,000 and 30,000 islands. No one knows how many there really are.

The Pacific World

The Pacific world includes tens of thousands of tropical islands, most of them located in the southern and western parts of the Pacific Ocean. Major powers colonized many of these islands over the course of three centuries. Spain, Britain, France, Germany, Japan, the United States, Australia, and New Zealand held islands in the Pacific prior to the end of World War II. Today, only France retains a colonial empire in the Pacific.

Independent countries are now emerging from these former colonies and United Nations Trust Territories (a **trust territory** is a territory under the authority of a nation or of the United Nations). These new countries are part of a region of strategic importance where the sea dominates the land, where places are isolated from one another, and where distances between them are enormous.

High Islands and Low Islands

Two types of islands are found in the Pacific world. **Volcanic islands** are the peaks of mountains that rise from the sea floor. These high islands have coral platforms, or **fringing reefs**, attached to their shores. Many volcanic islands also have coral **barrier reefs** that ring the high island and encircle a

The Pacific World

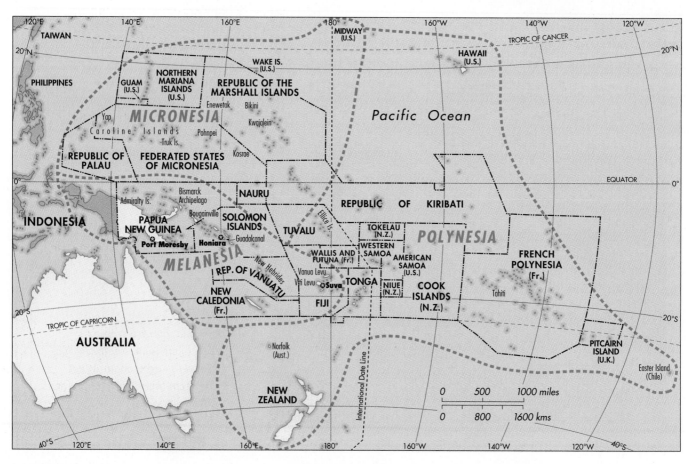

lagoon, which is a shallow body of water separated from the open sea by these reefs.

Low islands, or **atolls**, are coral barrier reefs that continued to build after the volcanic peak that they once surrounded sank beneath the sea, leaving a circle of low-lying coral reefs surrounding an empty lagoon.

Volcanic islands and atolls have very different natural environments. Volcanic islands are large, and their fertile soils and abundant fresh water provide a favorable environment for tropical crops such as sugarcane, yams (sweet potatoes), cassava (also called manioc), and taro (another starchy root).

Crops like bananas, coconuts, and breadfruit thrive on volcanic islands. Lush tropical forests offer forage for pigs and wild fowl. Nearby coral reefs provide an abundant fishing ground for fleets of local canoes and protect fishing fleets from storms. Volcanic islands are densely settled and have been occupied for thousands of years.

In contrast, atolls provide a precarious environment for living. These coral reefs rise only a few feet above sea level. Their soils are thin and infertile. The vegetation on atolls is mainly composed of salt-tolerant palm trees that provide food, clothing, and building materials for local inhabitants. Fresh water is scarce and must be collected in reservoirs and basins when it rains. Fishing provides most of the food.

Frequently, atolls are washed over by the wind-driven waves of huge tropical storms, or typhoons. Violent winds often exceed 100 miles per hour, and these winds push waves that flood low-lying areas. Palm trees and houses are swept away, and people must abandon these low islands because there is no high ground where they can take refuge.

⊕ REVIEW QUESTIONS

1. What is an atoll?
2. What is a volcanic island?
3. What is the difference between a fringing reef and a barrier reef?

⊕ THOUGHT QUESTIONS

1. Would you prefer to live on an atoll or a volcanic island? Why?
2. What is coral? Have you ever seen it? Can you describe it?

2. The Three Subregions of the Pacific

The three major subdivisions of the Pacific Ocean are Melanesia, Micronesia, and Polynesia. These areas encompass thousands of islands dotted across a vast ocean surface.

Melanesia

Melanesia, which is located in the southwest Pacific north and east of Australia and New Zealand, has more land area and more people than either Micronesia or Polynesia. Papua New Guinea, the eastern half of the island of New Guinea, anchors the western edge of Melanesia. Melanesia extends eastward some 2,500 miles across the Pacific to Fiji. It includes the Solomon Islands, Vanuatu (previously called New Hebrides), and the French possession of New Caledonia.

Papua New Guinea became an independent country in 1975, after being administered for more than a century by Britain and Australia. It is one of the poorest and least developed countries in the Pacific. Most of its 3.9 million people are farmers who make a meager living in tiny villages located deep in the rain forests that drape the volcanic slopes of their island. Port Moresby, the capital of Papua New Guinea, is growing rapidly and has a population close to 200,000. Today, the forest and mineral resources—copper, silver, and gold—of Papua New Guinea are attracting developers to this volcanic island, breaking the isolation of its people.

East of Papua New Guinea are the Solomon Islands, which have a population of 300,000. The Solomons gained independence from Britain in 1978. Palm and cocoa plantations are the basis of these islands' economy. Most people in the Solomon Islands are farmers, although the capital city of Honiara, on Guadalcanal Island, is now

IT'S A FACT More than 700 different languages are spoken in Papua New Guinea.

PLACE

HOW CORAL ATOLLS FORM

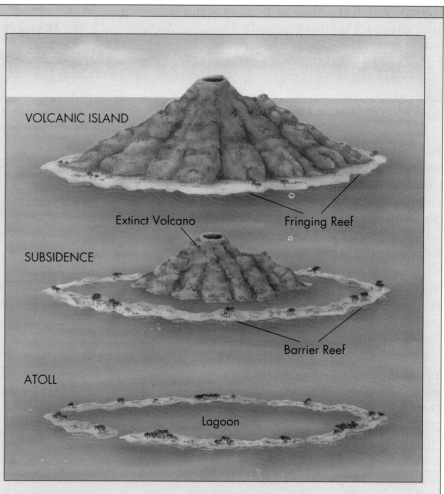

Stages in the Formation of Coral Atolls

Coral atolls are ring-shaped reefs surrounding empty lagoons. They look like necklaces strewn across the Pacific. In the 1830s, the British naturalist Charles Darwin proposed a theory to explain how atolls form—a theory that still stands today.

Noting that coral can grow only in the top few hundred feet of the ocean, Darwin suggested that atolls begin as fringing reefs surrounding an active volcanic island (top diagram). After the volcano dies, it begins to sink under its own weight. The coral, however, continues to grow and create a barrier reef offshore (middle diagram). Ultimately, the volcano sinks beneath the sea, leaving behind a ring of coral islands surrounding an empty lagoon—an atoll (bottom diagram).

Recently, geologists drilled into Pacific atolls, penetrating as much as 4,000 feet of volcanic soil and coral growth beneath the atolls. It turned out that Darwin was correct: atolls are tombstones marking the location of dead volcanoes.

QUESTIONS

1. What knowledge of animal life did Darwin use to help form his theory about atoll formation?
2. What causes the inactive volcanoes to sink into the ocean?

attracting migrants from the twenty-one major islands in this chain. Development of the timber resources of the Solomon Islands appears to hold promise for the future.

Vanuatu, which is located southwest of the Solomon Islands, gained its independence in 1980.

These rugged volcanic islands with a population of 200,000 support a mixed economy of cattle raising, farming, forestry, and fishing. Manganese mining is an important source of income. Tourists, particularly Australians, are bringing money to these islands.

The two remaining countries in Melanesia, New Caledonia and Fiji, are currently in turmoil. New Caledonia is a French possession. Its population of 200,000 is about equally divided between native Melanesians and French settlers. The Melanesians have demanded independence, and France has promised a referendum on the question in June of 1998. New Caledonia is the world's third largest producer of nickel, and also has important reserves of chrome, cobalt, and iron.

In Fiji, the population of over 700,000 is also divided into two groups. Native Melanesians form half of the total population, and the other half are Indians from South Asia who were originally brought by the British to Fiji to work on sugar plantations. A Melanesian coup d'état—a sudden, often violent change in government—overthrew a government run by Indians in 1987 and declared Fiji to be a republic. Their leader was elected prime minister of the country in 1992. Fiji's economy is based on sugar and coconut production, and a growing number of people who own these crop-bearing lands live in the capital city of Suva. Tourism is growing in importance.

Micronesia

North of Melanesia and east of the Philippines is Micronesia, which includes 2,000 or more tiny islands (*micro* in Latin means "small") strewn across an expanse of ocean as large as the United States. Many of these islands are atolls or low islands, whose people live by fishing. Farming is practiced only on the few volcanic or high islands in Micronesia. Originally populated by seafarers from Southeast Asia, Micronesia today has 380,000 people, who differ in race and culture from their Melanesian neighbors.

Until 1986, most of Micronesia was a United Nations Trust Territory administered by the United States. As a result of a vote, the people of this former trust territory now live in five independent countries and one commonwealth. The Caroline Islands extend for 2,000 miles from Palau near the Philippines to the island of Kosrae in central Micronesia. They are now two nations—the Republic of Palau, which became independent in 1994, and the Federated States of Micronesia.

IT'S A FACT The water area of Micronesia is as large as the United States, but its land area is smaller than Rhode Island.

The Republic of the Marshall Islands, located northeast of these countries, became a member of the United Nations in 1991. Still farther north, people voted to create the Commonwealth of the Northern Mariana Islands with rights similar to those of the people of the Commonwealth of Puerto Rico. Eventually, this commonwealth could become a state of the United States. On the eastern rim of Micronesia, the independent countries of Nauru and Kiribati gained independence in 1968 and 1979, respectively. Guam, the largest island in Micronesia, and Wake Island remain possessions of the United States.

Fishing on low islands and farming on high islands occupy most of the people in Micronesia, but other activities are important as well. Guam's economy is supported by a U.S. naval base, one of America's most important in the Pacific, that is on the island. Kwajalein Atoll in the Republic of the Marshall Islands is a practice target for missiles launched from California. Payment for military use of this atoll is an important source of income for the Marshall Islands. In the Marianas, sugar cultivation and tourism form the basis of the economy.

On the coral atoll of Nauru (its name means "nowhere"), the people have chosen to destroy their eight-square-mile island by mining the coral as phosphate and selling it as fertilizer to Australia and New Zealand. Its roughly 9,000 citizens pay no taxes and receive free education and medical care. The surplus money they earn goes into a trust account as a buffer against the day when their island will no longer exist.

Throughout Micronesia, a variety of new activities are springing up as the region's past isolation is broken.

Polynesia

Polynesia is located east of Melanesia and Micronesia in the heart of the Pacific Ocean. Polynesia has a large number of islands (*poly* in Latin means "many") and a total population of 1.9 million people. Its boundaries form a huge triangle extending from New Zealand to Hawaii to Chile's Easter Island.

Within Polynesia are many different tropical environments. Volcanic islands like Hawaii are covered by dense rain forests, have ample water, and support extensive sugar, pineapple, and palm plantations. In addition, hundreds of tiny low-lying atolls with a few palm trees and small fishing communities dot this ocean realm. The first settlers in this region were Polynesians, famous for their skill in navigating the Pacific in double-hulled, outrigger canoes.

Many of the islands of Polynesia are now independent, but others are owned by larger nations. Hawaii, the largest and most populous island group in Polynesia, became part of the United States in 1959. Midway, which consists of two small islands

north of Hawaii with a total population of 2,000, is governed by the United States through the U.S. Navy. American Samoa (with a population of 51,000) is a center for tuna fishing and canning in central Polynesia.

South of American Samoa, Tonga is a kingdom that gained independence from Britain in 1970. In Tonga, which has a population of 107,000, all land is held in reserve by the king and is distributed to young men when they become adults. Neighboring Tuvalu (previously the Ellice Islands) gained independence from Britain in 1978. Western Samoa became one of the world's smallest (1,093 square miles) independent countries in 1962.

Other Polynesian island groups are not independent states. New Zealand administers Niue (population of 3,000) and the Cook Islands (population of 21,500) as self-governing dependencies. Tiny Tokelau (population of 1,800) is a New Zealand ter-

▼ *Low islands, or atolls, dot the Pacific. If you lived on an atoll, what major hardships would you have to overcome?*

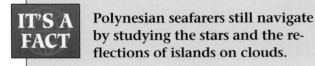

IT'S A FACT Polynesian seafarers still navigate by studying the stars and the reflections of islands on clouds.

3. Australia, New Zealand, and the Pacific Peoples

ritory. Far to the east, French Polynesia (population of 200,000), with five major island groups including Tahiti, is still governed by France, in spite of local protests against the testing of nuclear weapons. The French also administer the rain-soaked islands of Wallis and Futuna, which have a combined population of 10,000.

Far out in the Pacific, the fifty-five inhabitants of Pitcairn Island, all descendants of the mutineers of the H.M.S. *Bounty*, retain links with Britain. Easter Island, known for its mysterious stone statues, is a possession of Chile.

Most Polynesians (except those who live in Hawaii) either fish or raise bananas, breadfruit, cassava, yams, taro, and pineapples—the staple foods of Polynesia. These traditional ways of living are changing, however, as tourism grows in the region. The French and American military installations continue to demonstrate the strategic importance of this ocean realm.

To deal with forces of change, Polynesians have joined together in an association called the South Pacific Forum. The thirteen member countries of this organization discuss common problems relating to fisheries, tourism, transportation, and nuclear testing. The new nations and the dependent territories of Polynesia are facing the future together.

Europeans, Aborigines, and Maori

Most European settlers who came to Australia and New Zealand were convinced of the superiority of European culture. In Australia, native **aborigines** were viewed as lesser people, pushed off the best land, and finally housed on reservations.

▼ *The rich culture and heritage of Australia's aborigines are increasingly being appreciated throughout the world. Do you believe that the cultural treasures of native peoples will survive into the future?*

⊕ REVIEW QUESTIONS

1. Which Pacific people were especially known for their seafaring abilities?
2. What are the major crops grown on volcanic islands in the Pacific?
3. What are the chief islands of Melanesia? of Micronesia? of Polynesia?

⊕ THOUGHT QUESTIONS

1. How do the natural environments of volcanic islands and coral atolls differ?
2. What are the main economic activities in the islands of Melanesia?

In New Zealand, the more sophisticated **Maori**, a Polynesian people, were treated with greater respect. But soon after the Europeans arrived, fierce battles between the Maori and the Europeans led to the signing of a peace treaty that promised to respect the Maori's rights to land. In the years that followed, however, the Maori lost most of their land and many died of diseases introduced by Europeans.

Today, Australia and, to a lesser extent, New Zealand are struggling with problems of race relations similar to those found in many other countries. Aborigines and Maori have moved into urban centers, and racial tensions have erupted as these native peoples compete with Europeans for jobs, money, and education. Resentment is growing, along with demands for compensation for past losses. Neighboring Asian countries are watching to see how native peoples in Australia and New Zealand are treated. Racial tensions between Europeans and natives in these two countries echo across the Pacific Ocean.

> **IT'S A FACT**   One out of every five people in Australia was born on some other continent.

The Australian Aborigines

Australia's aborigines are a Melanesian people who lived as hunters and gatherers in Australia for thousands of years. When the Europeans arrived in 1787, some 300,000 of these aborigines were widely scattered across the continent. They lived in more than 500 territorial groups, each with its own language.

This aboriginal way of life was destroyed by hundreds of well-armed European ranchers, herders, and miners who occupied Australia's most fertile land in the mid-1800s. Most aboriginal tribes were pushed onto less desirable land, much like the Native Americans in the United States. Surviving on

▼ *This aboriginal child carrying a young kangaroo lives in the Australian outback. Increasingly, aborigines are giving up the traditional way of life and moving to cities. What factors do you think are drawing these people into the cities?*

Australian aborigines have 300 different names for the boomerang.

this marginal land was exceedingly difficult. By 1901, only 67,000 Australian aborigines survived.

Today, about 170,000 aborigines, many of mixed ancestry, live in Australia, mainly in those areas least attractive to Australians of European descent—in the Northern Territory, Queensland, and Western Australia. Although aborigines form only 1 percent of Australia's total population, relations with the aborigines are a national problem. As a recent prime minister noted, "Australia's treatment of her aboriginal people will be the thing upon which the rest of the world will judge Australia and the Australians."

In the last twenty years, the Australian government has attempted to correct the injustices of the past. Aborigines have been given the right to vote, to collect welfare benefits, and to own and occupy land on native reserves. In a symbolic gesture, the Australian government recently returned Ayers Rock, the world's largest rock, to aborigines who consider the caves at its base to be sacred. Moreover in 1993, aborigines were promised new laws that would accept claims to much of the land used by their ancestors, and compensation if that land is being farmed or mined.

The Maori of New Zealand

Sometime between 800 and 1000 A.D., Maori voyagers discovered New Zealand. Its two islands were far larger than other Pacific islands on which the Maori had settled. Moreover, New Zealand had a middle-latitude climate and a variety of environments open to human use.

The Maori lived primarily as hunters and gatherers in New Zealand, although they also tended gardens. When the British reached New Zealand, about 200,000 Maori lived there, mainly on North Island. They had a well-organized society with experienced leaders. Indeed, the Maori made a major contribution to the early European development of New Zealand.

▲ *The Maori of New Zealand are famous for their creative sculpture, as this example from Rotorua illustrates. Has discrimination been a greater problem for the Maori of New Zealand or the Australian aborigines?*

As the European population grew, however, conflicts over land led to war between the Maori and the Europeans. The Treaty of Waitangi in 1840 ended this war and guaranteed Maori rights to tribal lands. By 1900, however, the Maori population of New Zealand had fallen to 45,000.

Recently, Maori culture and society have experienced a revival. Improved living conditions, increased resistance to European diseases, programs of government help, legal protection of Maori land, and strong Maori leaders have made a difference.

Today, the Maori number some 300,000 and form 9 percent of New Zealand's population. Maori standards of living, life expectancy, and quality of life now compare favorably with those of other New Zealanders. This is a tribute to the mutual respect that has generally governed European-Maori relations, despite early injustices and some racial tension.

Moreover, New Zealand's highest courts have ruled in favor of Maori claims to land and fishing rights that extend back to 1840, the date of the Waitangi Treaty. By 1988, some Maori fishing claims off North Island had been declared valid. In 1993, a similar case on South Island led to compensation to the Maori for past injustice.

Many New Zealanders see this as the single biggest issue in their future, because the Maori have already filed claims that would include 60 percent of the country's land area and most of its offshore fisheries. The courts will be reviewing these claims well into the next century.

Discrimination in Early Immigration Policies

Very early, Australia and New Zealand recognized that their nations were much less crowded than neighboring Asian countries. For this reason, Australians and New Zealanders carefully controlled immigration. Their immigration policies reflected their view of their countries as British outposts in the shadow of Asia. In Australia these policies of exclusion were started after thousands of Chinese and Indians entered the country during the gold rushes in the 1800s. Popular resistance to these Asian immigrants led to a "White Australia" policy. A similar immigration policy existed in New Zealand. Immigrants of British origin were preferred, and Asians were excluded entirely.

These policies are now changing because Australia and New Zealand realize that they are Pacific Ocean countries. Their economies are becoming more closely tied to those of nearby Asian countries, particularly Japan.

Australia has relaxed its "White Australia" policy. Asian immigrants are now being admitted to Australia, along with thousands of Asian students. In 1990, half of all immigrants came from Asian countries. By 1992, however, a declining economy and high unemployment caused Australia to reduce considerably the total number of immigrants accepted.

New Zealand's immigration policies have also been liberalized. More than 200,000 of its Pacific neighbors from the Cook Islands, Tonga, and Fiji have become citizens of New Zealand.

⊕ REVIEW QUESTIONS

1. Australians and New Zealanders are now struggling with problems of race relations. How are they similar to those found in other countries?
2. In recent years, how have the immigration policies of Australia and New Zealand changed? With what results?

⊕ THOUGHT QUESTIONS

1. How was the aboriginal way of life destroyed in Australia? Where do most of the Australian aborigines live today?
2. What has caused the recent revival in Maori culture and society?

⊕ CHAPTER SUMMARY

In this chapter you have learned about the three island subregions of the Pacific world: Melanesia, Micronesia, and Polynesia. These subregions are composed of many islands and island groups, called archipelagoes composed of high islands (volcanic islands) and low islands (atolls). Most of the island groups were colonized by European powers during the last three centuries, thus their populations are composed of both native peoples and Europeans. Some of these archipelagoes have only recently become independent countries, and today they are struggling with problems of self-government and economic development. Depending on the natural resources found on the islands, the countries of these subregions have widely varied economies, ranging from fishing and farming to mining.

European settlement of Australia and New Zealand had an adverse impact on the aborigines and the Maori, the original inhabitants of those countries. Today, both governments are working to eliminate past practices of discrimination and exclusion and to compensate the aborigines and the Maori for land and fishing rights they lost to the Europeans.

EXERCISES

Select the Correct Ending

Directions: Find the correct ending for each of the following sentences and then write the complete sentences on your paper.

1. The Maori of New Zealand are (a) Polynesians (b) Europeans (c) Indians.
2. Both Australia and New Zealand (a) are south of the equator (b) have been colonies of more than one country (c) are heavily populated countries.
3. Sydney, Melbourne, and Perth all are (a) on Australia's east coast (b) in the southwestern part of Australia (c) port cities.
4. Volcanic islands usually (a) have fertile soils (b) are smaller than coral atolls (c) are poorer farming areas than coral atolls.
5. Papua New Guinea is in (a) Melanesia (b) Micronesia (c) Polynesia.
6. The most important products of rural Australia are wool, meat, and (a) corn (b) wheat (c) barley.
7. The most important resource in Nauru is (a) meat (b) phosphate (c) nickel.
8. New Zealand consists of (a) one island (b) two islands (c) four large islands and many smaller ones.

Which Occurred Later?

Directions: Notice that five pairs of sentences follow. An event described in one sentence in each pair occurred later than the event described in the other sentence. Write on your paper the sentence in each pair that describes the later event.

1. (a) The continent of North America was occupied by Europeans. (b) The continent of Australia was occupied by Europeans.
2. (a) The aborigines were widely scattered across the Australian continent. (b) Most aborigines lived in urban centers.
3. (a) The Maori standard of living compared favorably with that of other New Zealanders.

(b) The Maori lived mainly as hunters and gatherers.
4. (a) Australia and New Zealand developed closer relations with their Pacific neighbors. (b) Australia and New Zealand developed closer relations with Britain.
5. (a) The governments of Australia and New Zealand allowed no Asian immigrants to settle in their countries. (b) The governments of Australia and New Zealand allowed both Asian immigrants and students to be admitted to their countries.

Inquiry

Directions: Combine what you have learned in this chapter with your own ideas to answer these questions.

1. In what ways was the settlement of Australia similar to that of the Great Plains in the United States?
2. In Australia, the aborigines were pushed onto less desirable land by European settlers who thought they were bringing a higher form of civilization to the country. Do you think this was a satisfactory reason for the European settlers to take over the best land? Why or why not?
3. Why did the Europeans colonize the archipelagoes of the Pacific world? From a geographer's perspective, were they more interested in location (the need for bases from which to control Pacific Ocean shipping lanes), movement (trade with the native populations), or human-environmental interaction (access to the natural resources of the islands).

    In thinking about this problem, it would help to find out why the United States wanted control of the Philippine Islands after the Spanish-American War of 1898.

SKILLS

Using Encyclopedias

For many centuries, people have compiled information in encyclopedias. In 77 A.D., a Roman scholar named Pliny the Elder compiled an encyclopedia called *Natural History*. It provided a record of what the Romans of that day knew about many subjects.

Modern encyclopedias contain knowledge about a large variety of topics. In addition to multi-volume encyclopedias in book form, electronic encyclopedias are also available for use on computers.

Encyclopedia articles are arranged alphabetically by subject. To find the subject you want in a book encyclopedia, you must look at the letter or letters on the spines of the volumes. For example, if you want to read about automobiles, the volume you would need might be labeled *Arith–Aztec*.

Various subjects are named in List A, below. The letters on the spines of some encyclopedia volumes are shown in List B. On your paper, match the subject with the volume in which it appears by writing the correct letters from List B next to each number in List A.

List A

1. Formations caused by coral polyps
2. Refrigerated ships
3. The climate of New Zealand
4. Military bases on Guam
5. The major universities in Australia

List B

Arith–Aztec	Chinc–Czech
Gab–Gyro	Nao–Ney
Sav–Sound	

One helpful feature in many encyclopedias is a list of additional references for a subject. These references, usually found at the end of an encyclopedia article, tell the reader that the encyclopedia has other articles that contain information about the same subject. These helpful hints are called cross-references. Look at the list of possible cross-references that follows, and write on your paper the names of those that might pertain to Australia or New Zealand.

6. Moslem, *shogun*, Great Barrier Reef, sheep, Melbourne, Singapore.

Many encyclopedias also have yearbooks that update the encyclopedia by telling about things that happened in a recent year. If, for instance, you need information about events that occurred last year, look in the encyclopedia yearbook for that year. If you need to find out about something that happened last week or last month, however, you should turn to magazine and newspaper articles. Also, remember that encyclopedias and yearbook articles are generally brief and do not include all the information available on a subject.

Tell where you would look to answer questions on each of the following subjects. On your paper, write *e* for encyclopedia, *y* for yearbook, and *o* for other publications.

7. Did Australia export more wheat in 1985 or in 1984?
8. Is New Zealand a monarchy or a republic?
9. What is the size of Australia?
10. How extensive were Maori seafaring journeys?
11. Did Australian tennis players win the Davis Cup competition two months ago?
12. When was New Zealand colonized by the British?
13. Who was prime minister of Australia last year?
14. What is the capital city of New Zealand?

Vocabulary Skills

Directions: Look up the definitions of the following terms in a dictionary. Write the definitions on your paper. Indicate the parent language and etymology (root) of the word if it is given (example: platform—French *plateforme*, literally flat form).

1. aborigines
2. atoll
3. barrier reef
4. coup d'état
5. fringing reef
6. lagoon
7. Maori
8. trust territory
9. volcanic island

Human Geography

Directions: Each of the following sentences has three possible endings. Find the correct ending and then write the complete sentence on your paper.

1. Australia and New Zealand were explored late by Europeans because of their (a) rugged coastlines (b) tropical climates (c) remote location in the South Pacific.
2. Prisoners started a colony at Botany Bay in (a) Australia (b) New Zealand (c) both Australia and New Zealand.
3. The Maori are the native people of (a) New Zealand (b) Australia (c) all of the islands in the South Pacific.
4. In the 1700s and 1800s, the native people of Australia were (a) integrated into the European society (b) respected by the British (c) viewed as lesser people and pushed off the best lands by the British.
5. The first successful economic activity in Australia was (a) raising cattle (b) raising merino sheep (c) mining gold.
6. Australia and New Zealand became independent countries in the (a) 1500s (b) 1700s (c) 1900s.
7. The most important farm crop in Australia is (a) corn (b) cotton (c) wheat.
8. Refrigerated ships led to an increase mainly in (a) manufacturing (b) mining (c) sheep and cattle raising.
9. Australia's largest city is (a) Sydney (b) Melbourne (c) Auckland.
10. New Zealand's capital is (a) Auckland (b) Wellington (c) Christchurch.
11. In recent years, the Australian government has (a) tried to compensate the aborigines for past injustices (b) neglected the needs of the aborigines (c) treated the aborigines worse than in the past.
12. Today, the Maori people of New Zealand (a) generally share the high standard of living of the British population (b) make up less than 2 percent of the population (c) still are mainly hunters and gatherers.
13. Australia is the world's largest exporter of (a) beef and iron ore (b) wool and cotton (c) iron ore and tin.

Physical Geography

Are the following statements true or false? Write the true ones on your paper.

1. The only continent that is smaller than Australia is South America.
2. Australia is almost as large as the United States.
3. Australia is the world's flattest continent.
4. Australia is the world's rainiest continent.
5. Southern Australia is located in the middle latitudes.
6. Northern Australia is located in the tropics.
7. Australia and New Zealand are both huge storehouses of mineral and energy resources.
8. New Zealand is much more mountainous than Australia.
9. The builders of the Great Barrier Reef are coral polyps.
10. There is a long lagoon between the Great Barrier Reef and the mainland of New Zealand.

Writing Skills Activities

Directions: Answer the following questions in essays of several paragraphs each. Remember to include a topic sentence and several sentences supporting your main idea in each paragraph.

1. What are the most important geographical features of Australia and New Zealand? How did these features affect human settlement and the distribution of population?
2. How were the native people of Australia treated until recent times? What changes in race relations have occurred in the past twenty years?
3. Why cannot Australia and New Zealand support much larger populations than they have today?
4. Why do you think Australia and New Zealand are included among the countries of the technological world?

LATIN AMERICA

▲ Rio de Janeiro, framed by Sugar Loaf Mountain and Guanabara Bay, is one of the most beautiful cities in South America. It is one of the largest Portuguese-speaking cities of the world. Why is Portuguese spoken in Brazil?

UNIT INTRODUCTION

Latin America stretches from the southern border of the United States in North America to Cape Horn at the southern tip of South America. It includes Middle America, its mainland as well as the hundreds of islands in the Caribbean Sea. The large continent of South America, which lies south and east of Middle America, is also part of Latin America. In South America are the world's longest chain of mountains (the Andes), greatest river (the Amazon), driest desert (the Atacama), largest area of tropical rain forest (Amazônia), huge areas of nearly empty wilderness, and some of the fastest-growing cities in the world.

Latin America has many countries whose history and geography are very different from those of the United States. Three centuries of Spanish rule in Middle America and South America and Portuguese domination in Brazil shaped their development. Today, complex environments, rapid population growth, and political turmoil hinder the growth and prosperity of this vast region. Latin America's current problems are of deep concern to the United States, because they will play a significant role in our own future.

▼ *The mountain ranges of the Andes soar to elevations four miles above sea level, and stretch the entire length of South America. What forces of nature created these mountains?*

▼ *Maya pyramids, such as the Great Pyramid at Chichen Itza, have often been compared to the pyramids of Egypt. How does this pyramid compare with the photograph of Egypt's great pyramids located on page 520?*

UNIT OBJECTIVES

When you have completed this unit, you will be able to:

1. Describe the advanced civilizations that were created by Indian groups long before European settlers arrived in Middle America and South America.

2. Explain how Spain and Portugal influenced Middle America and South America during their three-hundred-year rule.

3. Describe how difficult environments—such as mountains, deserts, and rain forests— have contributed to the uneven distribution of people in mainland Middle America, the Caribbean, and South America.

4. Identify many areas of Latin America that are rich in natural resources but poor in per-capita income.

5. Explain how the population explosion in Latin America has greatly aggravated the region's problems.

SKILLS HIGHLIGHTED

- Using Special-Purpose Maps
- Using Population Maps
- Using Elevation Diagrams
- Vocabulary Skills
- Writing Skills

KEYS TO KNOWING LATIN AMERICA

1

Latin America, which is composed of Middle America (including Mexico, the countries of Central America, and the islands of the Caribbean) and South America, is more than twice the size of the United States.

2

Two-thirds of Latin America is located in the tropics.

3

Mountains, deserts, rain forests, and other difficult environments are found throughout most of Latin America.

4

Highly developed civilizations existed in the highlands of Mexico and Peru long before Europeans arrived in the New World.

5

Spain ruled Middle America and much of South America for three hundred years. Portugal ruled Brazil. Together, they deeply influenced the culture of Latin America.

6

Most Latin Americans are descendants of Indians, African slaves, and European immigrants.

7

People are unevenly distributed in Middle America and South America. In Middle America, most people live in the highlands or on the islands of the Caribbean; fewer along the coast. In South America, most people live near the coast, and the interior of the continent is mostly unoccupied.

8

Many Latin American countries export low-priced raw materials and import high-priced machinery and advanced technology.

9

Population is growing very rapidly throughout Latin America. Four of every ten Latin Americans are under fifteen years of age.

10

The cities of Latin America are growing in size. Mexico City may be the world's largest city.

▼ *This woman's dress and general appearance identify her as a native of Latin America. Does she live in a highland or lowland environment?*

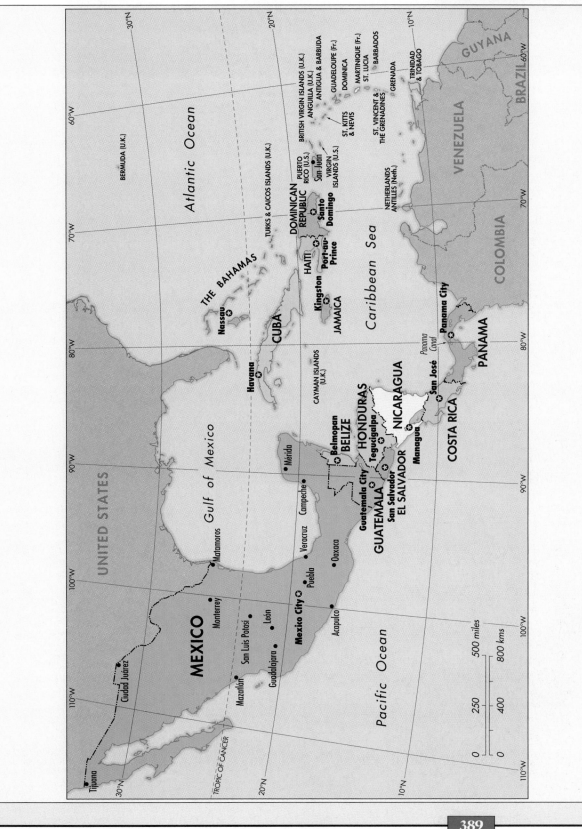

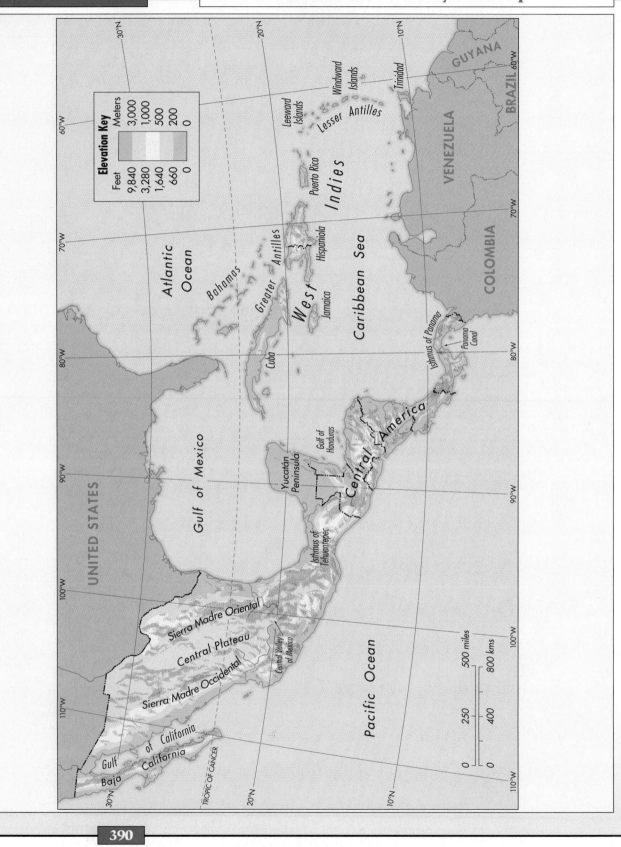

Elevation Key

Feet	Meters
9,840	3,000
3,280	1,000
1,640	500
660	200
0	0

30°N

20°N

10°N

60°W

70°W

80°W

90°W

100°W

110°W

GUYANA

BRAZIL 60°W

Trinidad

Windward Islands

Leeward Islands

Lesser Antilles

VENEZUELA

Puerto Rico

West Indies

70°W

Atlantic Ocean

Greater Antilles

Hispaniola

Caribbean Sea

COLOMBIA

Bahamas

Jamaica

80°W

Cuba

Isthmus of Panama

Panama Canal

Gulf of Mexico

Central America

Gulf of Honduras

90°W

Yucatán Peninsula

UNITED STATES

Isthmus of Tehuantepec

100°W

Sierra Madre Oriental

Central Plateau

Central Valley of Mexico

Pacific Ocean

Sierra Madre Occidental

110°W

Gulf of California

Baja California

TROPIC OF CANCER

20°N

10°N

30°N

100°W

110°W

500 miles

800 kms

250

400

0

0

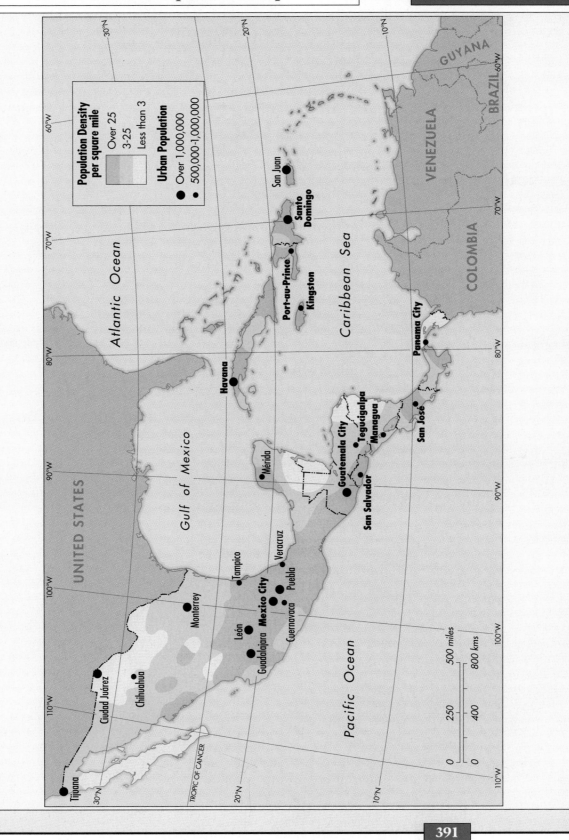

Population Density per square mile
- Over 25
- 3-25
- Less than 3

Urban Population
- ● Over 1,000,000
- • 500,000-1,000,000

GUYANA

BRAZIL

60°W

VENEZUELA

70°W

COLOMBIA

San Juan

Santo Domingo

Caribbean Sea

Port-au-Prince

Kingston

Panama City

80°W

Atlantic Ocean

60°W

70°W

80°W

Havana

Teguigalpa

Managua

San José

90°W

Mérida

Guatemala City

San Salvador

UNITED STATES

Gulf of Mexico

90°W

100°W

Tampico

Veracruz

Mexico City

Puebla

Monterrey

León

Guadalajara

Cuernavaca

100°W

Chihuahua

Ciudad Juárez

Pacific Ocean

500 miles

800 kms

110°W

250

400

Tijuana

TROPIC OF CANCER

30°N

20°N

10°N

0

0

30°N

20°N

10°N

30°N

20°N

10°N

South America Political Map

Caribbean Sea

NICARAGUA

COSTA RICA

PANAMA

Barranquilla
Cartagena
Maracaibo
Caracas
VENEZUELA
Ciudad Bolívar
Georgetown
GUYANA
Paramaribo
SURINAME
FRENCH GUIANA
(France)
Cayenne

Atlantic Ocean

Medellin

Cali
Bogotá
COLOMBIA

ECUADOR
EQUATOR
Guayaquil
Quito
Iquitos

Manaus
Santarém
Belém

0°

Fortaleza

PERU

BRAZIL

Recife

Machu Picchu
Cuzco

Lima

10°S

Salvador (Bahia)

La Paz
BOLIVIA
Cuiabá
Brasília
Arica
Sucre
Potosí
Goiânia

Belo Horizonte

PARAGUAY

Campinas
São Paulo
Santos
Rio de Janeiro
Curitiba

TROPIC OF CAPRICORN
Antofagasta

Tucumán
Asunción

CHILE

Pôrto Alegre

Córdoba

Valparaíso
Mendoza
Rosario
URUGUAY
Santiago
Buenos Aires
Montevideo

Atlantic Ocean

Concepción
ARGENTINA
Bahía Blanca

Valdivia

Pacific Ocean

Comodoro Rivadavia

| 0 | 250 | 500 miles |
| 0 | 400 | 800 kms |

Falkland Islands (U.K.)
(Malvinas)

Punta Arenas

Caribbean Sea

NICARAGUA

COSTA RICA

PANAMA

Isthmus of Panama

Lake Maracaibo

Orinoco River

Llanos

Cauca River

Magdalena River

Guiana Highlands

Rio Negro

EQUATOR

Gulf of Guayaquil

A m a z o n

B a s i n

Amazon River

River

Tapajós River

Atlantic Ocean

Andes

Ucayali River

Madiera

São Francisco River

Sertao

Tocantins River

Lake Titicaca

Altiplano

Mato Grosso Plateau

Brazilian Highlands

Pacific Ocean

Mountains

Atacama Desert

Paraguay River

Chaco

Paraná River

Gran Chaco

Paraná

River

Uruguay River

Andes Mountains

Mt. Aconcagua
22,887 ft.

Pampas

Río de la Plata

Atlantic Ocean

TROPIC OF CAPRICORN

Bio-Bio R.

Patagonian Plateau

Tierra del Fuego

Falkland Islands
(Malvinas)

Cape Horn

Elevation Key

Feet		Meters
9,840		3,000
3,280		1,000
1,640		500
660		200
0		0

▲ Mountain Peak

0	250	500 miles
0	400	800 kms

80°W   70°W   60°W   50°W   40°W

10°N

10°N

0°

10°S

10°S

20°S

20°S

30°S

30°S

40°S

40°S

50°S

50°S

100°W   90°W   80°W   70°W   60°W   50°W   40°W   30°W   20°W

South America Population Map

Country	Capital City	Area (Square miles)	Population (Millions)	Life Expectancy	Urban Population (Percent)	Per Capita GNP (Dollars)
Antigua and Barbuda	St. Johns	170	0.1	72	34	4,770
Argentina	Buenos Aires	1,068,297	33.5	71	86	2,780
Bahamas	Nassau	5,382	0.3	72	64	11,720
Barbados	Bridgetown	166	0.3	75	45	6,630
Belize	Belmopan	8,865	0.2	67	48	2,050
Bolivia	La Paz	424,162	8.0	61	51	650
Brazil	Brasília	3,286,475	152.0	67	76	2,920
Chile	Santiago	292,259	13.5	73	85	2,160
Colombia	Bogotá	439,734	34.9	71	68	1,280
Costa Rica	San José	19,575	3.3	76	45	1,930
Cuba	Havana	42,803	11.0	76	73	—
Dominica	Roseau	290	0.1	76	—	2,440
Dominican Republic	Santo Domingo	18,815	7.6	68	60	950
Ecuador	Quito	109,483	10.3	67	55	1,020
El Salvador	San Salvador	8,124	5.2	63	48	1,070
Grenada	Saint George's	131	0.1	69	—	2,180
Guadeloupe	Basse Terre	687	0.4	76	49	—
Guatemala	Guatemala City	42,042	10.0	63	33	930
Guyana	Georgetown	83,000	0.8	65	33	290

—Figures not available

Country	Capital City	Area (Square miles)	Population (Millions)	Life Expectancy	Urban Population (Percent)	Per Capita GNP (Dollars)
Haiti	Port-au-Prince	10,714	6.5	54	44	370
Honduras	Tegucigalpa	43,278	5.6	65	44	570
Jamaica	Kingston	4,243	2.4	74	52	1,380
Martinique	Fort-de-France	425	0.4	77	75	—
Mexico	Mexico City	761,602	90.0	70	71	2,870
Netherlands Antilles	Willemstad	302	0.2	74	—	—
Nicaragua	Managua	50,193	4.1	63	57	340
Panama	Panama City	20,761	2.5	72	53	2,180
Paraguay	Asunción	157,046	4.2	67	48	1,210
Peru	Lima	496,224	22.9	65	72	1,020
Puerto Rico	San Juan	3,436	3.6	75	74	6,330
St. Kitts-Nevis	Basseterre	139	0.04	68	49	3,960
St. Lucia	Castries	239	0.1	71	44	2,500
St. Vincent and the Grenadines	Kingstown	150	0.1	72	20	1,730
Suriname	Paramaribo	63,039	0.4	68	70	3,610
Trinidad and Tobago	Port of Spain	1,981	1.3	70	65	3,620
Uruguay	Montevideo	68,039	3.2	73	89	2,860
Venezuela	Caracas	352,143	20.7	70	84	2,610

—Figures not available

LATIN AMERICA: TWO CULTURES COLLIDE

History and environment shaped Latin America. Long before the Spaniards and Portuguese settled in what is now Latin America, highly developed civilizations existed in this region. Empires in Mexico and Peru controlled the lives of 75 million to 100 million people. These complex Indian civilizations were destroyed in the 1500s by the Spaniards; the Portuguese conquered Brazil.

For the next three hundred years, Latin American affairs were directed from Spain and

◄ *Discovered by Hiram Bingham in 1911, Machu Picchu was completely covered with jungle vegetation. There are many mysteries surrounding this ancient city still unsolved by historians and archaeologists. How were the buildings shown probably used?*

397

Portugal. European patterns of settlement, migration, land use, economics, government, and religion were transplanted to Spanish and Portuguese colonies in the New World. The slave trade brought millions of Africans—a third cultural group—to Latin America.

The legacy of these three centuries of Latin colonial rule remains a powerful force in both Middle America and South America, because when Latin Americans finally won their independence in the early 1800s, much of what they had inherited from Spain and Portugal persisted.

1. Latin American Landscapes

A Huge Region

Latin America, which is more than twice the size of the United States, is located south and east of our country and stretches 6,000 miles from northern Mexico to southern Argentina. The northern boundary of Latin America separates Mexico from the United States. The southern limit of Latin America is a cold, rockbound area called Tierra del Fuego, or "Land of Fire"—a name given by the early Spaniards because the Indians there treasured fire so much that they carried it with them from place to place.

Thirty-seven countries of many shapes and sizes make up Latin America. Brazil, Mexico, and Argentina are the region's three largest countries, and Brazil is nearly as big as the United States. The smallest country, the tiny island of Grenada in the Caribbean, has a land area of only 133 square miles.

Latin America can be thought of as two triangular-shaped areas, Middle America and South America. The two triangles join where Panama meets Colombia. The northern triangle of Middle America

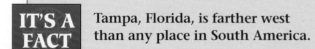

IT'S A FACT   Tampa, Florida, is farther west than any place in South America.

is made up of Mexico and the seven countries of Central America; it also includes the islands of the West Indies in the Caribbean. Middle America is widest in northern Mexico, and it tapers like a funnel to a width of only forty miles at the Isthmus of Panama. An **isthmus** is a narrow strip of land connecting two landmasses—in this case Middle America and South America.

The southern triangle of South America is seven times larger than Middle America and is divided into thirteen countries. Most of South America lies to the east of the United States. Lima, Peru, on the west coast of South America, in fact, is located further east than Miami, Florida.

Settlements in Latin America's Tropical Regions

The equator slices across South America from the mouth of the Amazon to the Ecuadorian Andes. Most of Latin America lies between the Tropics of Cancer and Capricorn. Like Africa, the other tropical continent, the average density of population in Latin America is generally quite low—fewer than fifty people per square mile.

Dense clusters of population, however, are found in favored locations. The east coast of South America is densely settled, as are the highlands of Middle America, the Caribbean, and parts of the Andes Mountains in South America. Locate these areas of dense settlement on the maps on pages 391 and 394. Notice how these regions' crowded settlement contrasts with the empty stretches of the Mexican north, the Amazon lowlands, and much of the Andes.

This uneven distribution of Latin America's current population of 460 million people is the result of two forces: (1) the nature of colonial settlement and (2) the different environmental opportunities and difficulties in each subregion.

The Human Geography of Latin America

Settlement history and physical environments explain a great deal about the human geography of Middle America and South America today. Within

Latin America, one-quarter of the land is mountainous, another quarter is covered with tropical swamps, and another tenth is barren desert. The coastal and upland centers that grew during the colonial period are now the dynamic cores of modern countries.

In Middle America—for example, in the highlands of Mexico and in the seven smaller countries of Central America—large estates called *haciendas* are worked by farmers who are *mestizos*, or people of mixed European and native Indian ancestry. By contrast, the islands of the Caribbean, a tropical lowland environment, is occupied by people of mixed African, Indian, and European heritage. Many of them work on **plantations**, which are large landholdings devoted to one crop. These differences in racial background and land use reflect early Spanish patterns of farming and their enforced use of Indian and African laborers to produce crops during the colonial period.

The larger and more diverse southern triangle of South America includes the Andean ranges that curve southward along the west coast of the continent. The Andean ranges are occupied by Indian peoples. The middle-latitude regions of the south and the huge nation of Brazil to the east with its densely populated coast and tropical interior are occupied principally by Europeans. In all of these regions, the human use of land and resources has varied. Overall, however, environmental use has reflected traditional patterns as well as current needs.

▲ *The Isthmus of Panama connects North and South America. The Panama Canal, which was completed in 1914 by the United States, has been the subject of political disagreement. Why is the canal important?*

Environments and Landscapes

Latin America stretches a full 80° of latitude from northern Mexico and the tropical islands of the Caribbean in the north to the clouded islands of Tierra del Fuego in the south. Two-thirds of the region lies in the tropics. Tropical rain forests near the equator give way to scrub forests, grasslands, and drylands as one moves into the subtropics north and south of the equator and upslope into the

◄

Tierra del Fuego is at the southern tip of South America. What other continent is close to this frozen area?

mountain environments of Middle America and the Andes of South America (see the elevation diagram on page 446).

The Tropical Rain Forests

The tropical rain forests, *selvas*, of Amazônia form the largest continuous mass of vegetation on the planet. Located astride the equator, Amazônia is a rolling plain extending 2.3 million square miles with constantly high temperatures, constant rainfall, and a continuous growing season. Moisture-laden trade

▼ *Trees in this rain forest in Venezuela grow to heights of more than 100 feet. Sunlight hardly penetrates the dense canopy. What global problems may be caused by the destruction of these rain forests?*

IT'S A FACT If the Amazon River and all the rivers that feed into it were placed end to end, they would encircle the globe four times.

winds from the Atlantic Ocean sweep over this lowland, dropping vast quantities of water (see diagram on page 403). This water streams down the 4,000-mile-long Amazon River to the Atlantic Ocean with such volume and force that the ocean is stained brown by silt as far as 200 miles offshore. Between November and June, the Amazon breaks over its banks, flooding surrounding plains 50 to 60 miles away. Like the Congo Basin of Africa, Amazônia until recently has been occupied mainly by native peoples practicing **shifting cultivation**, in which they plant crops in areas cleared of trees and then abandon these fields after several years.

Near the equator, the tropical rain forest consists of tall, closely spaced broadleaf evergreen trees, which compete for light and form a complex mass of vegetation. Many separate species of large trees are found in these forests. The tallest trees frequently reach more than 100 feet above the surface. A canopy of trees at 75 feet high forms the second layer of the tropical forest, usually blocking out sunlight below. At 50 feet, mainly shade-tolerant trees are found.

The nourishment for these rain forests, in which 3,000 or more plant species can be found in a single square mile, is provided mostly by leaves scattered on the ground and by the air. The soils beneath this layer of leaves are soaked by constant rainfall, and nutrients are swiftly washed away unless held in place by tree roots. When the surface is cleared of vegetation by farmers or loggers, these soils turn into iron-hard, sterile **laterites** (from the Spanish for "brick") in which hardly anything will grow.

The Effects of the Dry Season Away from the Equator

North and south of the equator, the tropical evergreen forest of the equatorial zone gives way to more open deciduous forests. Deciduous trees drop their leaves during the cool season because of a

DISCOVERING

REGION

THE VERTICAL LAYERING OF TROPICAL RAIN FORESTS

The tropical rain forests of Latin America often have vertical layering with three separate levels at 100 feet, 75 feet, and 50 feet competing for sunlight, as the illustration on this page shows.

The crowns of the tallest trees, such as the valuable mahogany, reach to heights of more than 100 feet. In fact, loggers often locate the mahoganies by having one man scale a mahogany tree that reaches above the 75-foot canopy who then directs ground crews to other mahogany trees nearby.

The 75-foot layer, by contrast, is made up of a dense, shade-producing canopy of broadleaf evergreen trees with smooth trunks. Below this, at the 50-foot level, a variety of shade-tolerant species of trees with slender trunks and narrow crowns compete for a place in the sun. Vines, some as thick as human torsos, drape tree trunks and branches throughout the tropical rain forests. Epiphytes and bromeliads, plants like orchids and Spanish moss that derive their moisture and nutrients from the air, seem to hang everywhere.

This vertical layering is most pronounced in the central Amazon, where only 1 percent of the sunlight reaches the ground. This dim light at the surface eliminates almost all undergrowth (see the illustration). The atmosphere is one of dappled shade, a damp cathedral-like twilight. Currently, rapid destruction of this complex ecosystem has become a global problem.

QUESTIONS

1. How do loggers sometimes locate the tallest trees for cutting?
2. Why is there so little vegetation on the floor of the rain forest?
3. Why are the soils of the rain forest infertile?

Tropical Rain Forest

more pronounced dry period and cooler temperatures. Here, the canopy of dense vegetation typical of the tropical rain forests is broken. Forests are scattered, undergrowth is thicker, and soils are more fertile.

In areas farther from the equator with longer dry seasons, scrub forests and grassland savannas appear on the margins of the Amazon. The fertile long-grass savannas of the Argentine *pampas* and the *llanos* of Colombia and Venezuela are occupied by great cattle and sheep ranches. In spite of their sterile, acid soils, the scrub woodlands of Paraguay and Brazil are increasingly being used for the raising of stock for hides and meat.

In Mexico, the much drier climates of the north support a short-grass prairie that is the basis of livestock ranching. Only in southern South America does this short-grass prairie region extend deep into the middle latitudes. Here, colder temperatures and higher elevations create the scrubby barren landscape of the Patagonian Plateau, the glacier-carved valleys of southern Chile, and the foggy islands of Cape Horn.

Highland Environments of Middle and South America

Mountains play a major role in the distributions of environments and people in Latin America. Mountains and high plateaus form the backbone of Mexico, the seven countries of Central America, and many of the islands in the Caribbean. In western South America, the Andes Mountains curve 4,000 miles southward from Venezuela to southern Chile, the longest mountain chain in the world. You may recall the Discovering feature, *Ring of Fire*, in Chapter 2 which describes the tectonic activity in this Pacific rim region.

Mountains change environments because they change climate. Temperatures drop as elevation increases so that the Andes Mountains have very cool temperatures. In mountainous areas, rainfall varies with the direction of the wind, increasing when moist air is forced to rise on windward slopes that face the wind and decreasing on leeward slopes that face away from rain-bearing air masses.

Moisture-bearing winds from the Atlantic Ocean provide much of the rainfall in tropical Latin America, particularly in South America. When these winds reach the eastern, or windward, slopes of the Andes Mountains, they are forced to rise. As they rise, the air cools, reaches its saturation point, and then rainfall pours down.

Vegetation changes when temperatures fall at higher elevations. Tropical trees are replaced by species that can grow in cooler temperatures. At higher elevations, these mountain forests give way to cloud forests, whose trees are covered with moss. Elfin forests of trees stunted by cold and wind are found still higher. Above 12,000 feet, tundra vegetation takes over.

Vegetation is also different on the western, leeward, slopes of the Andes. The moisture-laden air from the Atlantic Ocean drops its rain before climbing over these mountains, so that in Peru and northern Chile, no rain comes from either the Atlantic or the Pacific. (Refer to the Discovering feature on page 403 for a detailed description of this climate pattern.) As a result, the extraordinarily dry Atacama Desert is found there.

Latin America's Uneven Distribution of People

The variety of environments found in Middle and South America affects where people live. The rain forests and savannas of Amazônia are still largely unoccupied. Some people earn a living in valleys and basins in the Andes Mountains, but in general the highlands in South America are thinly settled. In Middle America, highlands have been more attractive zones of settlement. In both subregions, deserts discouraged most permanent settlement.

A majority of South Americans are concentrated at or near the coast, particularly on the east coast, which faces the Atlantic. Remember that the Spaniards and Portuguese built most of their cities and towns on coasts and left the interior of South America largely untouched. In Middle America, mineral finds in the highlands and tropical plantations on the coast led to a more even distribution of people. In both areas, patterns of settlement have been—and still are—strongly influenced by their environments and their historical development.

Environmental History

Much of the environmental history of Middle America and South America involved the quest for forested land that could be used for planting or grazing. Before the 1500s, the population of Middle and South America was relatively small, and pressure on the land was not intense except in selected areas of high civilization. In the realms of the Maya, Aztec, and Inca peoples, sophisticated systems of cultivation supported dense populations, but elsewhere settlement was light, people were few, and

REGION

MOUNTAINS CHANGE CLIMATE

In mountain regions, elevation influences climate, vegetation, and soils. In the mountains of Middle and South America, elevation is an important factor in human settlement and land use.

As elevation increases, air temperatures get cooler and the air's ability to hold moisture decreases. When air masses are forced upward by the slopes of mountain ranges, the air cools off and drops its moisture as rain or snow on windward slopes. This is called **orographic precipitation**. Once over the mountain peaks, the now drier air mass descends and warms up, decreasing the amount of rainfall on leeward slopes. These changes in temperature and rainfall determine the distribution of vegetation in major mountain ranges, and they also affect patterns of human settlement.

In South America, warm, humid trade winds from the Atlantic cross the Amazon Basin and climb the eastern, or windward, slopes of the Andes, as the diagram below shows. As this air cools, moisture condenses and falls as rain. Downpours are heaviest in the **cloud forests** (moss-covered tree forests); rainfall decreases at higher elevations in the **elfin forests** (where cold and wind stunts tree growth) and the tundra zone. On the high peaks of the Andes, rain freezes to snow.

The climate of the dry western, or leeward, slopes of the Andes is totally different. Here, descending air warms up and is able to hold more moisture, thus reducing the amount of rain that falls. On this side of the Andes, the Atacama Desert extends to the equator. This desert's only moisture comes from fog, which forms as morning dew when warm ocean air crosses a cold ocean current that flows along the coast.

QUESTIONS

1. Why do temperatures get cooler as elevations increase?
2. What does the term *orographic precipitation* mean?
3. Is it unusual to have a desert, such as the Atacama, so near a large body of water? Can you think of other deserts that are close to seas and oceans?

Mountain Climate

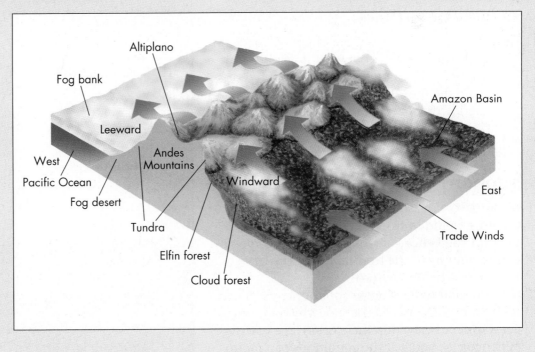

Altiplano

Fog bank

Leeward

West

Pacific Ocean

Fog desert

Andes Mountains

Tundra

Elfin forest

Cloud forest

Windward

Amazon Basin

East

Trade Winds

vast tracts of land in Middle and South America were relatively untouched.

For several thousand years, shifting cultivators cut and burned patches of forest land, abandoned the cleared fields when yields declined, and moved on to new settlement sites. Abandoned sites usually had time to recover before people returned.

Soil erosion was limited by the small size and scattered distribution of the clearings. The changes to vegetation were gradual but cumulative. Small groups of people armed with primitive tools and their own physical energy did alter vegetation, but we know from European eyewitnesses in local areas like the Central Valley of Mexico and Peru during the conquest period that their impact on dense forests was limited.

After the conquest, however, areas close to colonial cities were stripped of vegetation by charcoal burning and cattle grazing, but elsewhere the natural environment changed little. Deforestation and overgrazing were less intense in Latin America than in Africa, principally because animals played a more limited role in native economies (farmers in Latin America had domesticated the dog, llama, alpaca, and turkey) before cattle and sheep arrived from the Old World in the 1500s.

⊕ REVIEW QUESTIONS

1. How many countries are there in Latin America? Name the three largest countries. Is Middle America or South America larger?
2. Where are the most densely settled areas in Middle America? In South America?
3. What are the tropics, and how much of Latin America is located in the tropics?

⊕ THOUGHT QUESTIONS

1. Of the several reasons that few large towns or cities are found in tropical rain forests, which ones do you believe are important? Would you consider living in the tropical regions of Latin America? Why or why not?
2. How do mountains change environments? Since mountains are closer to the sun, why don't they get warmer as elevation increases? Have you experienced for yourself the conditions of a mountain environment? What was it like?

2. The Advanced Civilizations of the New World

The Mayas of the Yucatán Peninsula

Centuries before the Spanish conquest in the early 1500s, 12 million to 15 million people lived in

The Aztec, Maya, and Inca Realms

What three Indian civilizations are shown on this map?

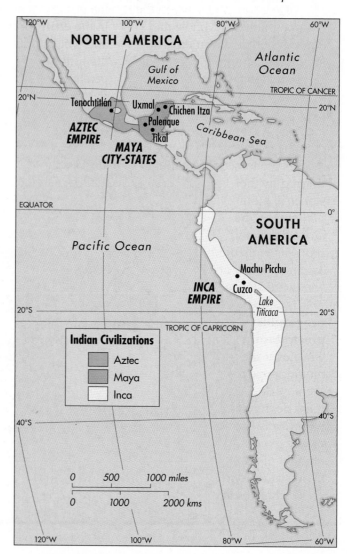

the temperate, fertile uplands of central Mexico. Another million people lived in the forested highlands of southern Mexico and Guatemala, and about 500,000 more in the tropical rain forests of the Yucatán Peninsula. The best known of these culture groups were the Mayas, who were one of the first groups of native Americans to create a highly developed civilization.

For hundreds of years, Maya farmers raised enough corn and other crops in the wet lowlands of the Yucatán Peninsula of Mexico and in neighboring Guatemala to support a large population. The Mayas built farms, religious centers, and roads in this region. The Spaniards paid little attention to the Mayas after the conquest because their lands held little gold and silver. Moreover, Maya civilization was already in decline by the 1500s, with many of its most spectacular buildings concealed beneath layers of forest vegetation.

In their religious centers, the Mayas built stone pyramids topped with temples. They explored the fields of astronomy and mathematics, developed a calendar, and kept records on stone slabs. Maya religion focused on the careful study of time and the stars.

About A.D. 900, the Mayas abandoned the forested lowlands of the Yucatán to move into the drier part of the peninsula and to highlands in the interior, for reasons that are still unknown. When the more powerful Aztecs moved southward into these highlands, much of this newly settled area was occupied by Maya groups.

The Aztecs of Central Mexico

The Aztecs established a civilization in the Central Valley of Mexico, where they had moved in the 1300s from somewhere in northwest Mexico.

▼ *The Mayas kept their astronomical calculations, mathematical records, and calendars in their religious centers. Why were these temples undiscovered for centuries?*

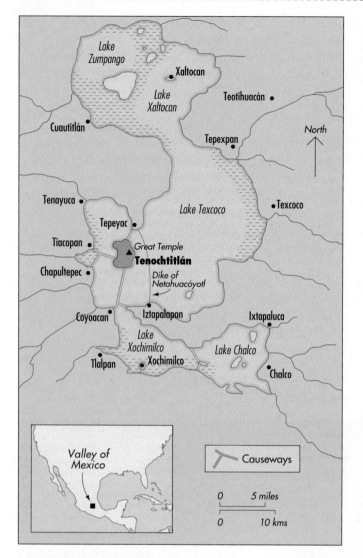

The Aztec Capital of Tenochtitlán

The Aztecs played a game like basketball in which they had to "shoot a basket" with either their hips, knees, or feet.

of **causeways**, or paved bridges, connected this city in a lake to the surrounding shores. In 1519, the city was described by Hernán Cortés as having "such excellence and grandeur . . . that in Spain there is nothing to compare."

The center of Tenochtitlán was the Great Temple, the largest building on the city's central plaza. This enormous structure was a 200-foot-high step pyramid topped with shrines. Surrounding the Great Temple were the palaces of the emperor and his nobles, large public buildings, and the houses of priests, administrators, artisans, and soldiers. Lightweight volcanic rock and wooden pilings supported these buildings on the muddy lake bed beneath. Canals formed the major avenues of the city, much like the canals of Venice do today.

More than a million Aztec farmers lived in the Central Valley of Mexico, a fertile basin thirty to forty miles wide whose major urban center was Tenochtitlán. In this balmy climate, the farmers grew three to four crops a year of corn, squash, and beans in gardens. In this way, Aztec farmers fed the large number of townspeople who lived in their capital city.

Additional wealth came to the Central Valley of Mexico from the vast empire conquered by the Aztecs. Each year, 7,000 tons of corn, 4,000 tons of beans and other foods, 2 million cotton cloaks, and precious goods like gold, silver, amber, and the feathers of beautiful birds were sent to Tenochtitlán.

Politically, the Aztec Empire was a military dictatorship. Its leaders did not try to integrate the varied peoples they conquered into Aztec society. Military outposts kept order and ensured that the conquered people paid tribute in the form of agricultural products or precious goods each year. The conquered, unhappy people of the Aztec Empire helped the Spaniards. Their discontent made it possible for Hernán Cortés, the *conquistador*, or conqueror from Spain, to defeat an empire of millions with a handful of soldiers in 1521.

On islands and sandflats in a lake in the Central Valley, they built a village of reeds that they called Tenochtitlán.

From the islands in this lake, the Aztecs extended their control over the entire 3,000-square-mile Central Valley of Mexico. In two centuries, they conquered surrounding lands and peoples. Their island village of Tenochtitlán grew into one of the most magnificent cities of its time and became the Aztec capital. Today it is called Mexico City.

The city of Tenochtitlán had an area of four square miles and a population of 250,000. A system

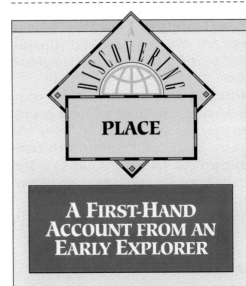

PLACE

A FIRST-HAND ACCOUNT FROM AN EARLY EXPLORER

Hernán Cortés sent a series of reports home to Spain to justify his conquest of the Aztec Empire. In his second such report, he described the city of Tenochtitlán.

The great city Tenochtitlán is built in the midst of this salt lake, and it is two leagues from the heart of the city to any point on the mainland. Four causeways lead to it, all made by hand and some twelve feet wide. The city itself is as large as Seville or Córdova. The principal streets are very broad and straight, the majority of them being of beaten earth, but a few and at least half the smaller thoroughfares are waterways along which they pass in their canoes.

Moreover, even the principal streets have openings at regular distances so that the water can freely pass from one to another, and these openings which are very broad are spanned by great bridges of huge beams, very stoutly put together, so firm indeed that over many of them ten horsemen can ride at once. Seeing that if the natives intended any treachery against us they would have every opportunity from the way in which the city is built, for by removing the bridges from the entrances and exits they could leave us to die of hunger with no possibility of getting to the mainland, I immediately set to work as soon as we entered the city on the building of four brigs, and in a short space of time had them finished, so that we could ship three hundred men and the horses to the mainland whenever we so desired.

The city has many open squares in which markets are continuously held and the general business of buying and selling proceeds. One square in particular is twice as big as that of Salamanca and completely surrounded by arcades where there are daily more than sixty thousand folk buying and selling. Every kind of merchandise such as may be met with in every land is for sale there, whether of food and victuals, or ornaments of gold and silver, or lead, brass, copper, tin, precious stones, bones, shells, snails and feathers; limestone for building is likewise sold there, stone both rough and polished, bricks burnt and unburnt, wood of all kinds and in all stages of preparation. . . .

Finally, to avoid prolixity in telling all the wonders of this city, I will simply say that the manner of living among the people is very similar to that in Spain, and considering that this is a barbarous nation shut off from a knowledge of the true God or communication with enlightened nations, one may well marvel at the orderliness and good government which is every-where maintained.

The actual service of Montezuma and those things which call for admiration by their greatness and state would take so long to describe that I assure your Majesty I do not know where to begin with any hope of ending. For as I have already said, what could there be more astonishing than that a barbarous monarch such as he should have reproductions made in gold, silver, precious stones, and feathers of all things to be found in his land, and so perfectly reproduced that there is no goldsmith or silversmith in the world who could better them, nor can one understand what instrument could have been used for fashioning the jewels. . . .

QUESTIONS

1. Who wrote this document? What was the purpose of writing this document?
2. When reading and evaluating a first-hand account, it is helpful to determine the writer's viewpoint. Identify some words or phrases in this document that give evidence of the writer's personal opinions.
3. How do you think a first-hand account by a resident of Tenochtitlán would differ from this account?

The Incas of the Andes Mountains

While Cortés was engaged in conquering Mexico, in the Andean highlands of South America a group of people called the Incas controlled a vast and complex empire. The Inca Empire extended for 3,250 miles from Colombia in the north to the Central Valley of Chile in the south. Most of the people in both the lowlands and highlands of this empire had been conquered by Inca armies.

The capital of the Inca Empire was Cuzco, a city of 100,000 people. It was located in a basin high in the Andes Mountains of Peru. Ten thousand miles of all-weather roads radiated outward from Cuzco, crisscrossing the Inca Empire. These highways spanned some of the most difficult landscapes in the Americas, and runners on the highways kept distant corners of the empire in constant touch with leaders in Cuzco.

The economic base of the Inca state was farming. Potatoes, corn, peppers, tomatoes, beans, squash, and a nutritious root called manioc were the major food crops. They were grown in irrigated valleys on the slopes of the dry Pacific coast, in mountain basins, and on terraced hillsides high in the Andes. Bird droppings, or **guano**, from offshore islands were used as fertilizer to increase crop yields. A well-planned system of canals and reservoirs channeled water to fields, and herds of the native **llamas**, which were used as pack animals, and **alpacas**, which were raised for their silky fleece, were grazed on pastures nearby.

One-third of the harvest was given to the community that produced it, one-third to Incan priests, and one-third to the ruler and his family. There was no private property. Work loads within each community were assigned according to family size. Surplus food was put in public storehouses and saved for distribution in drought years. The cycles of poverty and hunger that now occur in parts of Peru and Bolivia did not happen in Incan times.

⊕ REVIEW QUESTIONS

1. Where did each of these groups live: the Mayas, Aztecs, and Incas?
2. What kinds of tribute did the Aztecs receive from the peoples they conquered?
3. How did the Inca ruler keep in touch with the distant parts of his empire?

⊕ THOUGHT QUESTIONS

1. How do we know that the Mayas had a highly developed civilization?
2. The Aztecs did not integrate the peoples they conquered into their society. Why do you think this was or was not a good policy?

▶

Most of the golden treasures of Incas were melted down by the conquistadors, which makes those items remaining even more precious. The Incas used the lost wax method of casting many of their gold objects. Discover what this method involves by personal research.

3. The Clash of the New World with the Old

Spain and Portugal Claim a New World in the Americas

Spain and Portugal are located on the Iberian Peninsula, which juts out into the Atlantic Ocean. In the early 1500s, they were among the strongest countries in Europe.

Portugal, the smaller of the two Iberian nations, had developed maritime skills and technology quite early. Lisbon, the Portuguese capital, had one of the best harbors on the Atlantic Ocean. From this port, Prince Henry the Navigator sent out ships to explore the Atlantic coast of Africa, and in 1497 the explorer Vasco da Gama sailed around the tip of southern Africa and reached India. Trade with Asia made Portugal a very rich country.

In 1500, another Portuguese ship set out from Lisbon for India, but it landed on the shoulder of what is now Brazil. Its captain, Pedro Cabral, claimed the east coast of South America for Portugal. At first, this discovery did not interest the Portuguese because they were deeply involved with their rich Asian trade in spices and silk.

Spain, by contrast, was a much larger nation of 10 million people. In 1492, Queen Isabella of Spain agreed to finance Christopher Columbus in the first European venture westward across the Atlantic. Later that same year, the *Niña*, the *Pinta*, and the *Santa Maria* sailed across the Atlantic Ocean to find Asia. In October, Columbus landed on the island of San Salvador in the Bahamas, just southeast of Florida. The region was claimed for Spain, and three centuries of European domination began.

Spanish Settlements in the Caribbean

When Columbus returned to Spain to report that land lay only thirty-three days west of Europe, Spain immediately outfitted a large fleet with equipment, seed, and livestock. Some 1,500 settlers were recruited to colonize the large Caribbean is-

▲ *Christopher Columbus landed on San Salvador in the Bahamas on October 12, 1492. Why is Christopher Columbus viewed as the "discoverer of America" when other Europeans surely preceded him?*

land of Hispaniola. Spain had three goals in establishing this colony: (1) to gain the wealth of the Indies, (2) to convert native peoples to Christianity, and (3) to establish permanent settlements as bases for further Spanish exploration.

In 1493 the Spaniards built their first town, Isabella, on the northern coast of Hispaniola, the island now shared by Haiti and the Dominican Republic. Within a few years, 12,000 Spanish colonists migrated to the West Indies.

Broad powers were given to early colonial governors, the first and perhaps the worst of whom was Columbus. These governors collected taxes, administered law, and gave out land. They encouraged farming so the settlers could be fed. This was why horses, sheep, donkeys, cattle, and chickens were introduced into the New World. Crops like wheat, barley, melons, and cucumbers were also brought

IT'S A FACT Columbus made four voyages to the New World, but he never set foot in what is now the United States.

from Spain, as well as grapes, olives, citrus fruits, figs, and bananas. In exchange, the native Americans gave the Europeans beans, potatoes, corn, turkeys, sweet potatoes, chocolate, and tobacco among others.

Cortés Conquers Mexico

In 1518, Cortés landed on the Mexican coast. Unhappy subjects of the Aztec Empire joined his army and marched into the highlands to the Aztec capital of Tenochtitlán. The Spaniards and their native allies burned down the Aztec capital in 1521 and quickly extended their control throughout the Aztec realm. Indians were parceled out to work Spanish landholdings, and Catholic priests arrived to preach to the Indians and to set up missions. Mexico City was built on the ruins of Tenochtitlán. The rich gold, silver, and copper mining area on the Pacific slope of Mexico's western mountains was developed.

There was little resistance to the spread of Spanish rule, because deadly epidemics of smallpox, measles, and typhus—diseases brought by the Europeans—killed many thousands of Indians. Native Indians declined in number from an estimated 12 to 15 million at the time of the conquest to 2.5 million after a century of Spanish rule.

Pizarro Conquers Peru

Francisco Pizarro's conquest of Peru was similar to Cortés's conquest of the Aztec Empire. In 1531, Pizarro sailed southward along the Pacific coast of South America to find the wealthy civilization rumored to exist in the Andes Mountains; a year later, he confronted the ruler of the Inca Empire.

The Spaniards killed the Inca emperor and occupied the Incan capital city of Cuzco. They destroyed the city's temples and palaces, took control of the Inca Empire, and built the city of Lima on the coast

to connect Spanish Peru with Spain. Native Indian rebellions broke out during the next forty years, but Spanish control was never seriously threatened.

As in Mexico, the impact of smallpox and measles in the Andes was devastating. The population of the Inca Empire fell by a third. During this same period, Spanish settlers came to Peru to manage farmlands, organize laborers in the mines, work in newly established Spanish cities, and search for more gold and silver.

The Portuguese Move into Brazil

In 1494, Pope Alexander VI divided the unexplored worlds across the Atlantic into two separate regions—one Spanish and the other Portuguese—to avoid conflict between these two Catholic nations. The pope's line, or the Line of Demarcation as it was called, was drawn near what is now longitude 46° west. To eliminate competition and conflict, all

The Spanish and the Portuguese in the New World

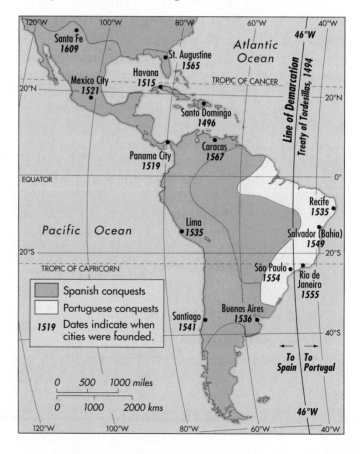

lands west of this line were granted to Spain; those east of it were granted to Portugal. At the time, no one knew that this meridian, or longitude, cut through the shoulder of modern Brazil, as the map on page 410 shows. Because Brazil was east of the line, it became a Portuguese-speaking colony.

By 1530, Portugal decided to settle what is now Brazil. It divided the Brazilian coast into fifteen *capitanias*, which were large tracts of land extending inland from the coast. Each *capitania* was granted to a wealthy family in Portugal in the hope that the family would use its personal wealth to develop the region. These families were given the right to establish towns, levy taxes, and control trade in their *capitanias*. In return they had to give one-fifth of their profits to the king of Portugal.

And so Portuguese colonies were started in northeastern Brazil. By 1550, colonists were exporting dyewood (such as Brazilwood for red and purple dyes), sugar, cotton, and tobacco to Portugal. They also raised grain and cattle for Europe. Soon they realized that huge profits could be made from growing sugar in Brazil, and sugarcane fields were established throughout the region. Because native Indians died by the thousands working in these fields, the Portuguese began to import enslaved African peoples to replace Indian laborers on the sugar plantations. Northeastern Brazil became the world's primary area of slavery and sugar production.

⊕ REVIEW QUESTIONS

1. What three goals motivated the Spaniards to colonize Hispaniola in the New World?
2. What major problems did they face in building colonies, and how did they solve them?
3. What area was conquered by Cortés? By Pizarro?
4. What were the *capitanias* in Brazil?

⊕ THOUGHT QUESTIONS

1. Why did the Indian populations of Mexico and Peru decline sharply in the first century of Spanish rule? Such contact has caused many experts to question whether space exploration is a good idea. Do you see the connection?
2. Why do you think Europeans called the Americas the "New World"? What would you have called it?

4. The Heavy Legacy of Colonial Rule

How Spain and Portugal Governed Their Colonies

In Spain, a small group of advisers called the Council of the Indies was established to govern the economic development of Spain's colonies in the New World. This council was made up of Spain's richest families. A second organization, the House of Trade, controlled all trade and immigration. Catholic missionaries, a third force, were sent to Latin America by the church. Spain governed its vast empire in the New World without serious challenge for three centuries by maintaining strict

▼ *Converting people to the Roman Catholic religion was a goal of Spain's colonial policy. This church in Cartagena, Colombia, is an example of what became known as colonial architecture. How does it compare with Maya temples?*

control of the economic and spiritual lives of Latin Americans.

Portuguese rule in Brazil was less organized than that of Spain. A Portuguese viceroy ruled the colony of Brazil from the coastal town of Salvador or Bahia, but Portugal was too weak to govern such a huge colony in the New World. Gradually, Brazil came to be run by rich plantation owners who lived in America.

Farming and Mining in Spain's American Colonies

Agriculture was the mainstay of Spain's New World economy. Because the value of agricultural exports to Spain was greater than the value of the precious metals sent to Spain, the Council of the Indies closely supervised land ownership and land use in the colonies.

Spanish soldiers, merchants, and favorites of the king were granted large tracts of land that were called *haciendas* in Mexico and *latifundios* in South America. Because of this royal policy, most land was owned by a few wealthy families. The majority of Latin Americans, a growing number of them *mestizos*, worked the land without owning it.

Commercial agriculture was introduced wherever possible to make money for Spain. Wheat, barley, olives, and grapes were planted on *haciendas* and *latifundios* in moderate climates. Sugar, tobacco, cacao, and indigo (blue dye) plantations were estab-lished in the Caribbean and along the coasts of the mainland in tropical regions.

Enormous herds of wild horses and cattle introduced by the Spanish flourished in the grasslands of northern Mexico and Argentina at either end of Latin America. A special way of life developed in these areas—that of the rancher-cowhand. Minerals also added to the wealth of Spain's New World colonies. Gold and silver were mined in Mexico, Colombia, Peru, and Bolivia.

Farming, cattle herding, and mining meant work, and through it all, the Spaniards refused to work with their hands. As a result, the entire economy of Spain's colonies ran on the muscle power of native Indian and African workers. Under this stress the number of native Indians dwindled. Ironically, it was a humanitarian effort to save the last of these Indians that led to large-scale African slavery in Spain's American colonies.

Africans in the New World

The Atlantic slave trade began when a cargo of Africans was shipped from the Guinea coast of Africa to Hispaniola in 1518. This event led to four hundred years of human brutality that numbed the hearts of all who took part in it. As it turned out, the Africans were better able to survive in the Americas. Native Indians lacked immunity to Old World diseases, but the death rate among African slaves

▼ *This engraving depicts a ship bringing enslaved persons to the United States from Africa before the Civil War. How would you describe their condition?*

MOVEMENT

THE ATLANTIC SLAVE TRADE

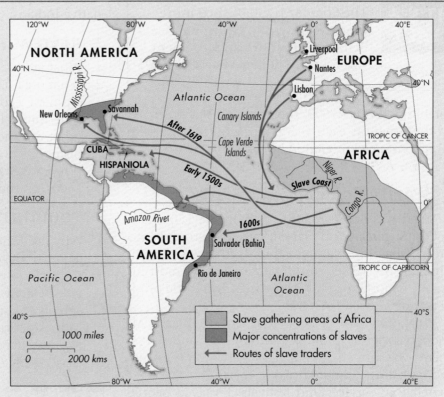

Europeans took captives from the west coast of Africa as early as the 1400s. By 1500, 3,500 enslaved people a year (a fraction of those starting the brutal journey) were shipped from West Africa to Portugal and Spain. They were put to work as domestic servants, field hands, and sailors. Africans also worked as enslaved laborers in fields and orchards in the Canary Islands and the Cape Verde Islands off the coast of West Africa.

The growing demand for workers in the huge colonies of the New World created the Atlantic slave trade; 11 to 12 million Africans were shipped to the New World in a period of four hundred years. The traffic reached its peak in the middle 1700s but did not decline until around the 1850s.

African slaves were better able to survive in the New World than native Indians. As a result, their value soared as did the profits from this trade in human lives.

Half of the enslaved were delivered to plantations in Brazil, and another third to the Caribbean. Nearly one of every twenty-five Africans arrived in the United States, a recorded 399,000 people. They altered the ethnic composition and enriched the cultures of virtually every New World society.

QUESTIONS

1. How many Africans were forcibly brought to the New World?
2. Approximately how many Africans were eventually brought to the United States?
3. Name at least three ways in which the cultures of the New World were enriched by the enslaved Africans. You may need to discuss this question with your families, friends, and classmates.

was only one-third as high as that among native Indians.

African slaves solved the chronic problem of labor shortages in the Americas, and demand for African slaves increased as more plantations were

IT'S A FACT From 1650 until about 1850, Africans outnumbered Europeans in the New World.

IT'S A FACT   The Spaniards established more than 250 towns in Latin America before the first British settlement was founded at Jamestown in 1607.

established. After 1600, thousands of slaves arrived in the Americas each year. The slave trade was the largest enforced migration in history.

Brazil's Economy and Culture

Sugar was Brazil's main export in the 1500s and 1600s. Sugarcane was raised, cut, and ground into brown sugar on plantations in northeastern Brazil. Then, in the 1700s, cotton became Brazil's most lucrative export crop. Tobacco, cacao, and cattle products were also sold abroad.

Plantation society took root first in Brazil. Although society was divided into masters and slaves, there was much interaction between the two groups. All the children played together and joined in religious services and festivals. Enslaved people had rights to earn money and inherit land. Laws protected them from cruel masters and permitted them to gain their freedom in various ways.

The Many Landscapes of Colonial Latin America

By 1600, Spanish and Portuguese colonial rule had created patterns of settlement and economic activity that would persist for two hundred years. In tropical areas closer to Europe—areas like the Caribbean—sugar, cacao, indigo, and tobacco plantations were established. Because land transport was costly, these crops, when grown on the mainland, were mainly grown on the Atlantic coast, and from there they could be shipped to Europe.

Food crops were planted near the densely populated silver and gold mining centers of Mexico and Peru to feed local populations. On the grasslands of Argentina, Uruguay, Venezuela, and Mexico, a ranching economy developed. Only the vast stretches of the Amazon, remote mountain areas of Middle America, and the southernmost part of

South America remained the undisturbed homes of native Indians.

⊕ REVIEW QUESTIONS

1. What are *haciendas* and *latifundios*?
2. What organizations in Spain controlled the government, economy, and spiritual life of Spain's American colonies?
3. What was the largest enforced migration in history?
4. What was Brazil's chief export in the 1500s and 1600s? In the 1700s?

⊕ THOUGHT QUESTIONS

1. Why did African slaves provide a solution to labor shortages in the Latin American colonies?
2. The economy of the Latin American colonies was based chiefly on sending raw materials to Spain and other European countries. Why do some Latin American countries still depend on the sale of raw materials for most of their income?

5. Independence in Latin America

Independence from Spain

Long distances made it difficult for the Spanish to preserve tight control of its New World colonies, although they managed to maintain their grip for three hundred years. During this period of colonial rule, a society developed in the New World that was very different from that in Spain. Between 1815 and 1825, the Portuguese and Spanish empires in the New World disintegrated.

By the 1800s, 23 million people lived in Spanish America. Spaniards from Spain formed the ruling class. Spaniards and other Europeans who were born in the Americas (known as *Creoles*) were the landowners, merchants, and business leaders of the region. Nearly half of the people in Spanish America were native Indians. A third were *mestizos*. The rest were Europeans, Africans, and *mulattos* (people of mixed African and European heritage).

Spain's New World colonists gradually felt more and more confined by Spain's control of their lives, just as Americans came to resent British rule. Local officials often ignored orders and laws from Spain. Spanish rulers came to distrust anyone born in the colonies, and American-born *Creoles* were bitterly resentful of this distrust. These *Creoles* led the struggles for independence in Spain's New World colonies in the early 1800s.

Spain tried to ease resentment in its colonies by relaxing trade restrictions, so that colonies could trade with one another. And although this new policy created a large class of *Creoles* made wealthy by increased trade with Spain, Spain's concessions were too few and came too late. By 1800, many Spanish Americans viewed their homelands as separate from Spain. Hard-fought revolts in Argentina, Mexico, and Chile led to the creation of three independent republics in the New World. The five countries of Venezuela, Colombia, Ecuador, Peru, and Bolivia became free under the inspired leadership of Simón Bolívar, a young *Creole* from Venezuela. Eighteen independent nations were established during the next twenty-five years.

By 1900, Spanish Latin America had crumbled into twenty independent countries. Today there are thirty-seven countries and a number of small colonies and dependencies in Latin America. No large, unified, powerful nation emerged after the wars of revolution against Spain in Latin America. Instead, there were many small and weak countries.

Democratic forms of government were not adopted in these countries. Autocratic rulers prevailed, and government by generals became common in Latin America. These generals, or **caudillos**, recruited armies in the countryside and attacked those *caudillos* who were already in power in the cities. The resources of many countries were and are squandered in these struggles for power.

Independence for Brazil

The exception is Portuguese Brazil, where one large colony became a single nation and achieved independence peacefully. In 1807, when Napoleon I of France tried to take over Portugal, the Portuguese royal family was able to escape to Brazil. They were warmly received in Rio de Janeiro, which immediately became the capital of both Brazil and Portugal.

After Napoleon was defeated in Europe, the Portuguese king returned to Portugal and left his son to govern Brazil. With the support of Brazil's leading citizens, Brazil became an independent monarchy under his rule. Brazil was the only country in Latin America to be ruled by a king after becoming independent.

Latin American Raw Materials for European Manufacturing

Economic patterns in Latin America changed little after independence. Export of raw materials in exchange for imported manufactured goods from Europe continued. Europe needed raw materials for its factories and food for its people, and Latin America wanted manufactured goods.

The British became the most important investors in Latin America. They loaned money to build railroads across the plains of Argentina and helped develop the wool industries of Uruguay and Argentina.

U.S. corporations also invested in Latin America. They developed the copper and nitrate deposits in Chile and drilled oil wells along the east coast of Mexico. The United Fruit Company made bananas a major crop in Central America. Latin American governments began to depend on revenues from these foreign-owned businesses.

Two Economies in One Country

Two very different economies existed side by side in many Latin American countries, resulting in a **dual economy**. A modern, technologically advanced economy was owned by a small group of foreign investors, wealthy landlords, mine operators, and plantation owners. These people sold Latin America's minerals and foodstuffs abroad. At the same time, most Latin Americans lived as poor farmers, untouched by the modern part of the economy and barely able to feed their families.

The best farmland in Latin America was used to grow crops for export. Colombia and Brazil became famous for their coffee. Argentina grew wheat and raised beef cattle for Europeans. Sugar, bananas, cacao, and cotton were grown on plantations in the Caribbean.

The modern part of the economy served the rich. Railroads opened up land away from the coast for new plantations, ranches, and mines. Modern technology used by mining companies increased mineral exports—tin from Bolivia, copper and nitrates from Peru and Chile, and oil from Venezuela and Mexico. Investments of money expanded these export industries.

A prime example of a dual economy developed in Argentina. By 1900, British money had built railroads and port facilities to increase exports of meat and grain to Britain. In exchange for this food, Britain sent coal and manufactured goods to Argentina. The sale of beef and grain benefited large landowners and merchants, but no industrial revolution improved the lives of a majority of the people of this largely undeveloped country who continued to live in traditional ways.

The Rich Still Own the Land

Centuries ago, Spanish and Portuguese rulers gave huge grants of land to colonists to encourage them to come to Latin America. Much of this land has remained in the hands of a small minority in each country. Wealthy landowners control large estates, while most Latin American farmers live on tiny plots of land or work as laborers on the large estates.

Today, half of the farmland of Latin America is held in large estates. This unequal pattern of land ownership divides the population of Latin America into the rich and the poor. Poverty in farming regions is a serious problem throughout most of Latin America. Attempts to break up large estates, which have been resisted by the wealthy, have resulted in only minor improvement in the lives of the people of much of this region.

IT'S A FACT An early Spanish land grant (now the King Ranch in Texas) is bigger than the state of Rhode Island.

▲ *Coffee is a major export crop of several Latin American countries. What countries probably import most of this coffee?*

The United States Enters the Picture

In the early 1900s, U.S. businesses began to invest money in Latin America. This money encouraged the production of exports, just as British investment had earlier. The United States also began to supply manufactured goods to Latin America in exchange for raw materials.

Approximately 65 percent of Mexico's imported goods come from the United States, while most

Caribbean countries and many in South America get 50 percent of their imports from our country. Only Argentina and Brazil import less than 40 percent of their manufactures and food from the United States. These goods are paid for by selling tropical food products and raw materials to the United States.

Development Plans Face Great Difficulties

Many Latin American countries are now trying to reduce their dependence on the export of raw materials and to develop higher standards of living for more of their people. The great cities of Latin America are becoming industrial centers, but rapid population growth, weak governments, land ownership by wealthy elites, political strife, and difficult environments are serious obstacles to economic development.

Few Latin Americans earn their livings in manufacturing. Three countries—Brazil, Mexico, and Argentina—produce three-quarters of all of Latin America's manufactured goods. These countries are becoming increasingly important manufacturing centers as the United States becomes more post-industrial. The least-developed Latin American countries, however, have virtually no industry and are among the poorest of the lesser developed countries.

Products from plantations, ranches, and farms are the principal exports in many Latin American countries. Some countries are implementing plans to grow a larger variety of farm products, but most still depend on one or two agricultural products, so that their economies are strongly affected by changes in world prices. Because the best land is used for export crops, Latin America does not grow enough food for its people. Latin America now must import one-tenth of its food.

⊕ REVIEW QUESTIONS

1. In the 1800s, what proportion of the people in Spanish America was native Indian? What proportion was *mestizo*?
2. Who were the *Creoles*?
3. Which five South American countries won their freedom from Spain under the leadership of Simón Bolívar?

⊕ THOUGHT QUESTIONS

1. Why were the *Creoles* unhappy with Spanish rule, and how did Spain try to ease their resentment?
2. How was Brazil's pathway to independence different from that of the Spanish colonies?

🌐 CHAPTER SUMMARY

The interplay of history and environment have shaped modern Latin America. Historically, Latin American civilization is the product of the collision of several cultures. One was the advanced civilizations shaped and sustained by native Indians centuries before the Europeans arrived. Another was the European civilization imposed by Spanish and Portuguese conquerors and colonizers over the past five hundred years. A third was the forced African migration to the New World.

European colonization destroyed native American civilizations and replaced them with European social structures, landholding patterns, economics, government, and religion that placed European descendants in positions of power and authority over Indians. Large numbers of Africans, brought to the colonies to work as slaves on the plantations and in the mines of Latin America, were ruled by Europeans. Many of Latin America's political and economic problems in the twentieth century can be traced to the effects of European colonization on these peoples.

Most of Latin America is tropical—that is, it lies between the Tropics of Cancer and Capricorn. It is generally thinly populated. Latin America can be described as two triangles: one encompasses Middle America which includes Mexico, Central America, and the Caribbean, and the other encompasses the countries of South America farther south and east. Climate and latitude have combined to create several environments that are common to both triangles. Tropical rain forests dominate in most areas near the equator. Scrub forests, grasslands, and drylands, in turn, extend both north and south of the rain forest. Elevation limits settlement in areas of Middle America and South America, as does access to areas of dense settlement.

EXERCISES

Mayas, Aztecs, and Incas

Directions: Some of the following statements apply to only one group of people native to America; others apply to two or three groups. On your paper write the name of the Indian group or groups to which each statement applies.

1. They kept records on stone slabs.
2. Tenochtitlán was their major city.
3. Cuzco was the center of their empire.
4. They lived in Mexico.
5. They had a highly developed civilization before the first European settlers arrived.
6. Bird droppings were used as fertilizer.

Find the Wrong Word

Directions: There is one wrong word in each of the following statements. Find the wrong word and then write the statement correctly.

1. Before the Spanish conquest, many Incas lived in Guatemala and the Yucatán Peninsula of Mexico.
2. The Aztecs controlled a vast South American empire.
3. Lima was built by Cortés in Spanish Peru.
4. One interest of the Spanish was to convert native Indians to the Protestant religion.
5. The Spanish landholding system gave the Spaniards much power over their Portuguese workers.
6. All lands east of the Line of Demarcation were granted to Spain.
7. Portugal divided the coast of Brazil into fifteen *haciendas*.
8. Mining exports from Spanish America brought more wealth to Spain than any other exports.
9. The Mayas built stone pyramids topped with beautiful gardens.
10. In 1497, Vasco da Gama sailed around the tip of southern Africa and reached Brazil.
11. The Spaniards built their first town in the Americas on the northeastern coast of Brazil.
12. A huge number of slaves from Asia were brought to the Spanish colonies to work in the fields and mines.
13. The northeastern part of colonial Brazil became Europe's primary source of tobacco in the 1600s.
14. After the Spanish and Portuguese conquests of the New World, only the Amazon region and southernmost South America remained the undisturbed home of the *Creoles*.
15. The Inca Empire extended southward from Colombia to the central valley of Peru.

America or Europe?

Directions: Some of the products and animals in the following list were found in America before the first Europeans arrived in the New World. Others were brought to America by the Europeans. On your paper, write "America" next to the name of each product or animal that was native to America; write "Europe" next to the name of each product or animal that was brought to America.

1. alpacas
2. bananas
3. barley
4. cattle
5. citrus fruits
6. corn
7. dyewood
8. grapes
9. horses
10. llamas
11. manioc
12. olives
13. potatoes
14. sheep
15. squash
16. sugarcane
17. tomatoes
18. wheat

Inquiry

Directions: Combine the information in this chapter with your own ideas to answer these questions.

1. The Aztec leaders did not try to integrate the varied peoples they conquered into Aztec society. Military posts kept order and ensured the payment of tribute each year. What problems did this policy create for the Aztecs?

2. The bureaucracy that controlled the Inca Empire was very efficient but allowed little personal freedom. Today, in some Latin American countries and in other parts of the world, personal freedom is greatly restricted. Explain why you do or do not believe that restricting personal freedom in modern countries leads to greater efficiency.

3. Explain why and how Latin American independence movements grew out of differences between Latin American *Creoles* and the Spanish colonial government. Compare the *Creoles'* feelings with those of British descendants in the North American colonies—feelings that led to the American War of Independence.

SKILLS

Using Special-Purpose Maps and Diagrams

A. Using the Map of the Aztec Capital of Tenochtitlán

The map on page 406 shows many interesting things about the Aztec capital and the surrounding countryside. Look at the map and answer these questions:

1. What is the name of the lake surrounding Tehnochtitlán?
2. Tenochtitlán was connected with the mainland by causeways. How does the map show these?
3. Using the scale on the map, estimate the length of the longest causeway.
4. Besides Tenochtitlán, other places are shown on this map. What symbol shows their locations?
5. The small inset map shows the location of what place in Mexico?

B. Using the Atlantic Slave Trade Map

Directions: Answer the following questions by studying the map on page 413. Look at the arrows on the map. Some of them show the major slave trade routes at different times in history.

1. In the early 1500s, where were most of the enslaved Africans sent? During this time period, why weren't enslaved Africans sent to the mainland of North America?
2. After 1619, some of the enslaved were sent to the southern part of what later became the United States. What were two of the southern cities where the enslaved Africans arrived?
3. When did the large slave trade to South America begin? What South American city received many enslaved Africans? In what country is this city?
4. Other arrows on the map begin at Liverpool, Nantes, and Lisbon. What do these arrows tell you about the slave trade?
5. Look at how the map shows areas where there were major concentrations of enslaved people. Notice that these areas are near the eastern coasts of South America and North America. Why do you suppose concentrations of enslaved Africans were not as large in the western parts of South America and North America?
6. The equator is almost parallel to the Amazon River as it crosses South America. Were the major concentrations of enslaved people in the Americas generally near or far from the equator?
7. The equator crosses Africa roughly halfway between the Congo and Niger Rivers. Was the Slave Coast of Africa near or far from the equator?
8. What does the location of the equator on the map suggest about the climate of the regions where large numbers of Africans lived?
9. What kinds of crops probably were raised in these regions?

C. Using a Mountain Climate Diagram

Directions: Use the diagram and text on page 403 to fill in the numbered blanks. Number your paper from 1 to 9, and beside each number write the term that belongs in each blank.

The diagram on page 403 shows that trade winds move in a direction from (1) |||||||||| to (2) |||||||||| . When the air masses are pushed upward by mountain ranges, the air cools off and drops its moisture as (3) |||||||||| or (4) |||||||||| on the (5) |||||||||| slopes. Once over the mountains, the now drier air masses descend and become (6) ||||||||||. They leave much less (7) |||||||||| on the (8) |||||||||| slopes. This diagram shows a rain forest east of the mountains and a (9) |||||||||| desert west of the mountains.

Vocabulary Skills

Directions: Match the numbered definitions with the vocabulary terms. Write each term next to the number of its correct definition on your paper.

1. A large estate in Mexico or Central America
2. An animal native to Latin America and used as a pack animal
3. Tropical savanna grasslands in Venezuela and Colombia
4. A domesticated hoofed mammal having a long, soft, silky fleece, related to the llama
5. An economy split into a small modern sector geared to export and a large traditional sector
6. Military leaders that ruled as dictators in Latin American countries
7. The tropical rain forests of Amazônia
8. Hard, compact, yellow-to-red soils that are poor for farming because of their low fertility
9. Composed of trees covered with moss and found on cloud-shrouded mountain slopes like the eastern Andes
10. A narrow strip of land connecting two larger landmasses
11. People of mixed European and native Indian ancestry in Latin America
12. The fertile, middle-latitude grasslands of Argentina
13. A raised earthen bridge across water or wet ground
14. Large tracts of land owned by members of the elite in Latin America
15. Term used in colonial Latin America for Spaniards and other Europeans born in the Americas
16. Bird droppings used as fertilizer to increase crop yields
17. Caused by the lifting of moist air over a mountain barrier
18. People of mixed African and European ancestry
19. A system of tropical farming in which clearings are cut in forests, planted for several years, and then abandoned
20. Trees at high elevations that are stunted by cold and wind
21. A leader in the Spanish conquest of Mexico and Peru
22. A large landholding usually devoted to one crop
23. Large tracts of land awarded by Portugal to noble families in the colonial period that extend inland from the coast of Brazil

alpaca	laterites
capitanias	*latifundios*
caudillos	llama
causeway	*llanos*
cloud forest	*mestizos*
conquistador	*mulattos*
creole	orographic precipitation
dual economy	*pampas*
elfin forest	plantation
guano	*selvas*
hacienda	shifting cultivation
isthmus	

THE LANDS AND PEOPLES OF MIDDLE AMERICA

Middle America is a region whose people are trying to improve their ways of living but are burdened by political institutions, economic patterns, and social customs from the colonial past. It is a storehouse of resources that can provide its citizens with higher standards of living in the future; yet it has drained many of its resources to benefit a few. It is also a political cauldron that could explode again at any time.

Remember that Middle America, because of its location and proximity next to our southern border, is a region of the

◄ *These people are weaving baskets in the highlands of Guatemala. Here, traditional crafts are important for home use as well as for sale to tourists. Are any of the goods used in your home made by hand? If so, by whom?*

developing world likely to influence the future of our country. Judge for yourself what may happen as you read this chapter and come to know Middle America. Will this region become a part of the technological world? Will it remain mired in poverty? Will it join the United States as a partner in a common future?

1. Regions and Subregions

Mainland and Rimland

Middle America is usually divided into two major subregions: the **mainland**, which includes most of Mexico and the seven smaller nations of Central America and (2) the **rimland**, which includes some coastal lowlands of the mainland and the islands of the Caribbean. The term *Central America* refers only to the seven small nations between Mexico and South America; the term *Middle America* refers to the entire mainland, as well as the Caribbean islands located between the United States and South America.

Mountains and plateaus sweep southward from the western United States through Mexico and Central America. Of the 156 million people who live in Middle America, three-quarters live on the mainland and 90 million of those live in Mexico. The seven countries of Central America have small populations. Guatemala is the largest with 10 million

people. With the exception of Belize, which was founded by the British, all were colonized by Spain.

Most people in mainland Middle America live in the highlands rather than along the coast. The capital cities of six of the eight countries of this region are located in the highlands. Locate these countries and capital cities on the map on page 389. Other large concentrations of population are found on the four largest islands of the West Indies called the Greater Antilles—Cuba, Hispaniola (Haiti and the Dominican Republic), Puerto Rico, and Jamaica.

The differences between mainland Middle America and the rimland are environmental and cultural. The highlands of the mainland from Mexico to Panama have temperate climates and are populated by people of Spanish and Indian background, many of them *mestizos*. Indians tend to be most numerous in southern Mexico and neighboring Guatemala. The heritage of the rimland, by contrast, is Spanish and African. People of African descent are dominant in the lowlands along the coasts of the Yucatán Peninsula, Belize, Costa Rica, and Panama. Here, and in the coastal lowlands of southern Mexico and Guatemala, tropical environments prevail. The ancestries of the people in Middle America reflect the historical pattern of the settlement of this area.

A mixture of peoples inhabits the West Indies, which include the four islands of the Greater Antilles located in the western Caribbean, as well as

Subregions of Middle America
Middle America has two major subregions: (1) the Mainland and (2) the Rimland.

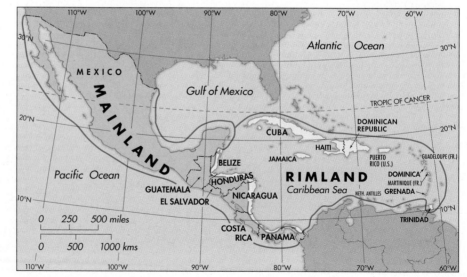

the many islands of the Lesser Antilles, which lie to the east and separate the Caribbean Sea from the Atlantic Ocean. A number of these islands were strongly influenced by European cultures: the Spanish in Cuba, the Dominican Republic, and Puerto Rico; the French in Haiti and a number of islands in the Lesser Antilles; and the British and Dutch on smaller islands in the Caribbean.

Patterns of Land Use

The mainland was a region of large *haciendas* many thousands of acres in size, owned by Spanish landlords of wealth and position, and worked by native Indian and *mestizo* laborers. *Haciendas* tended to be self-sufficient; they raised food crops such as corn, beans, and squash as well as cash crops for sale. They also kept domestic animals.

Usually, *haciendas* occupied the most fertile land in the highlands of Middle America. Poorer land was worked by peasants and sharecroppers who formed the vast majority of the population. The *hacienda* system relied on a large and dependable supply of cheap labor, which was available in the densely settled highlands of Mexico and Central America. The goal of the Spanish overlords was prestige not profit. The *hacienda* system was a way of life for aristocratic Spaniards, or *caballeros*, as they were called after the Spanish for "knight." This system of land use ensured poverty for most of the people in mainland Middle America, where land was the primary source of wealth.

Along the tropical coasts of Middle America and on low-lying Caribbean islands, a quite different system of land use, the plantation, was introduced into the region. Plantations were established primarily to produce a single cash or food crop like sugar or bananas for profit. On plantations, the need for labor was and still is seasonal, so that many people are out of work at least part of the year. Most plantations are foreign owned and operated, and their owners have chosen to locate in Middle America because of the availability of cheap labor and favorable environmental conditions.

⊕ REVIEW QUESTIONS

1. How many countries are located in Middle America? In Central America?

▲ *The Dominican Republic shares the island of Hispaniola with Haiti. Large sugar and coffee plantations are found in the lowlands and on lower mountain slopes. What European culture influenced the Dominican Republic? What European culture influenced Haiti?*

2. What are the two major regions of Middle America called?
3. What countries make up the Greater Antilles? Where are the Lesser Antilles located?

⊕ THOUGHT QUESTIONS

1. What racial groups are found in this region? How are these groups distributed in Middle America?
2. What are the major differences between a *hacienda* and a plantation? If you had to work on one or the other, which would you choose?

2. Resurgent Mexico

Patterns on the Land

Mexico is made up of a band of plateaus and mountains that sweep northwest to southeast into the seven Central American countries that link North America with South America. In northern Mexico, steep cliffs or **escarpments** 7,000 to 13,000 feet in elevation flank either side of the country's Central Plateau. This plateau is a mile above sea level, and even higher in its southern portions. Five hundred miles wide at its broadest in the north, it tapers to Mexico's southern border where the eastern and western *cordilleras*, or principal mountain ranges in a chain, converge.

Mexico, with an area of 756,066 square miles and a population of 90 million, is the largest and most significant country in mainland Middle America. The economic and social core of the country is located in the southern part of the Central Plateau at elevations of 6,000 to 7,000 feet.

This area has sufficient rainfall, a temperate climate, and rich volcanic soils. These environmental factors favored the early growth of dense agricultural settlement on Mexico's Central Plateau in **pre-Columbian** times (before the arrival of Europeans), as you have already learned. Currently, one-half of Mexico's people, primarily *mestizo* in ethnic background, are concentrated in this section of the Central Plateau, which is the core of highland Middle America.

As in the pre-Columbian era, intensive cultivation of corn, beans, and squash supports rural pop-

Mexico

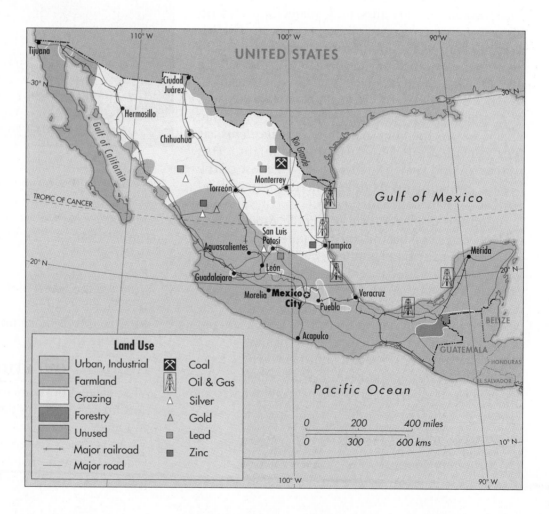

▲ *Commercial crops like cotton are of increasing importance in Mexico's rapidly diversifying economy. Why do you think it is advantageous for Mexico to grow cotton rather than import it?*

ulation densities of 200 to 300 people per square mile, some of the highest densities in Latin America. In especially favored locations, such as the Central Valley of Mexico, rural settlements blanket all but the steepest slopes with densities of 500 to 1,000 people per square mile, numbers rarely attained outside the rice-growing regions of Asia.

The Mexican North, the Mexican South

North of the Central Valley of Mexico, rainfall decreases to less than 20 inches per year. The densely settled fields of the Central Plateau give way to scattered oasis-like villages, ranches, and mining centers, and population densities fall to fewer than twenty people per square mile. In the states of San Luis Potosí and Zacatecas, discoveries of silver, lead, zinc, copper, and gold were early attractions for Spanish settlement. Now most of these mines are depleted.

Landscape and culture change south of the Central Valley of Mexico. Rainfall increases, and the upland escarpments here are covered with a mantle of trees that grade into the tropical rain forests of the Mexican south. Regional population densities are low—the entire south has a total population of only 10 million people, one-ninth of the inhabitants of all of Mexico.

In the state of Morelos, sugarcane and rice fields are found, adding new foods to the subsistence crops of Middle and Central America. **Subsistence crops** are grown primarily for family use and less for sale. Farther south, in Oaxaca, wheat, coffee, oranges, and tobacco are cultivated in the cooler piedmont uplands.

In the Chiapas Highlands, the agricultural economy is distinctly tropical, and the major commercial crops here are corn, cacao, and bananas. This pattern is repeated in the wetter part of the Yucatán, the lowland Caribbean peninsula that Mexico shares with Guatemala and Belize (population 200,000). Throughout southern Mexico, the people are native Indians, 50 to 80 percent of whom speak **indigenous** (native to the area) languages such as Maya in the Yucatán. Most of these native Indians are poor and powerless, although major oil discoveries in the Mexican south are rapidly changing traditional patterns of life in parts of this region.

⊕ REVIEW QUESTIONS

1. What is the core area of Mexico?
2. What is the pattern of rainfall in Mexico? Where is it driest? Wettest?
3. Where is population most densely concentrated in Mexico?

⊕ THOUGHT QUESTIONS

1. Mexico City is one of the largest cities in the world. What problems does this present to its people?
2. Of the various regions of Mexico, in which would you prefer to live? Give your reasons for this choice.

MEXICO CITY, THE CROWDED METROPOLIS

By many measures, Mexico City, which is located in the Central Valley of Mexico, is considered by urban geographers to be the world's largest city. When Cortés and his men first laid their eyes on the capital of the Aztec in November 1519, they were amazed at the splendor of Tenochtitlán, a city of 300,000. The shining grandeur that was Tenochtitlán is Mexico City today, a metropolis larger than any city in Europe or North America. It is at the center of a band of cities that stretches from Veracruz in the east to Guadalajara in the west.

The city's setting—at an elevation of more than one mile—is striking. Snow-capped mountains surround this highland basin that has been the focal point of mainland Middle America for centuries. But this superb human environment is rapidly becoming a grim urban setting for the millions of people who call Mexico City home.

Authorities do not know the true population of Mexico City, because about 1,000 migrants stream into this urban area each day and 1,000 new babies are added daily. Approximately 22 to 23 million people are thought to live in this enormous megalopolis.

The statistics of misery are abundant: 2 million of the city's people have no running water; 3 million have no sewage facilities. Six thousand tons of garbage lie uncollected every day; 100,000 deaths per year can be attributed to pollution.

The city operates on a gigantic scale. Local supplies of water are virtually exhausted, and increasing amounts must be piped long distances across the mountains and into the valley. Piping sewage out of the valley is an equally difficult problem. An estimated 550 shantytowns ring the city's fringe. Simply providing needed roads, schools, hospitals, and other city services consumes most of the city's resources.

Pollution is the most startling phenomenon to the visitor. Polluted air rushes through the air vents of approaching airplanes, making passengers acutely aware of what to expect. On the ground, the air

◀ *Mexico City is a huge urban center that sprawls for many miles across the floor of the Central Valley of Mexico. Note the smog dome that hangs over the city. What is a smog dome?*

often has a yellow or gray color—a cloud of pollutants resulting from the huge number of factories and cars that operate in this mountain-ringed basin.

Pollution controls are now being implemented, but the task is enormous. The city's 3 million cars burn fuel inefficiently because at an elevation of 7,350 feet the atmosphere has 30 percent less oxygen than locations at sea level. In the 1990s, pollution on bad days is one hundred times greater than the acceptable level. Athletes are warned not to train in the city's parks, and birds frequently die from the smog. The government twice has declared a state of emergency and closed factories. Workers are instructed to leave their cars at home one day a week. Nevertheless, with one-half of Mexico's total industry crammed into the narrow confines of the Central Valley of Mexico, it is questionable how effective these measures can be.

Mexico City is not simply a vast urban wasteland: it is a stylish and historic city with great cultural and economic power. Its social and environmental problems, however, threaten its future, and in Mexico City, which will have a population of 40 to 50 million by the year 2010, that future is now.

Bright colors adorn the stalls of this street market in Mexico City. Are street markets common in the United States? Why or why not?

Gondolas still transport goods and people on ancient Aztec canals, originally arteries of the Aztec capital city of Tenochtitlán, now a tourist attraction in Mexico City. What other city or cities are often associated with gondolas and canals?

QUESTIONS

1. What ancient city was on the present site of Mexico City?
2. What are the biggest challenges facing Mexico City today?
3. What does the author mean in the last sentence, ". . . that future is now"?

427

3. Mexico's Changing Economy

Changes on the Land

Centuries of Spanish rule had left most of Mexico's good farmland in *haciendas* so large in area that just 260 families owned four-fifths of Mexico's farmland. Much of this land lay idle, while nine of every ten Mexican farmers were *peones*, landless peasants who worked on the *haciendas*. After the revolution of 1910, the Constitution of 1917 redistributed nearly half of Mexico's agricultural land to groups of peasants who asked for it. The lands given to these communities were called *ejidos*, communal farms worked in common by 20 to 30 farm families. To prevent the reappearance of large landholders, *ejido* land could not be sold or rented.

The collective nature of these farms left them stagnant, because ownership rights were unclear. As a result, only one-tenth of Mexico's farmland is actually used for agriculture, and less than 5 percent is irrigated. Similarly, Mexico's 125 million acres of forest land are barely used.

A change in this situation in 1991 is revolutionizing the Mexican countryside. Three reforms have been implemented: (1) the government can no longer simply grant land to whomever asks, so that landowners large and small now have greater securi-

▼ *The production and assembly of automobiles and trucks in Mexico City is a vital industry. As the country modernizes, more industries are being created and more jobs are becoming available. Do you think this process will change the lives of the many unemployed and underemployed people in this huge urban center?*

ty; (2) clearly defined rights to private property have been restored; and (3) *ejido* land can be sold or rented with the consent of other members of the community. These reforms are expected to attract new, large-scale investment in Mexican agriculture. This is particularly important, now that the North American Free Trade Agreement (NAFTA) was signed and Mexico, the United States, and Canada have become a free trade area. Mexican grains, fruits, and vegetables are becoming important export products. This is not true in areas like Chiapas in the south where corn cannot compete with U.S. grain on world markets, as indicated by an Indian revolt on January 1, 1994—the day that NAFTA went into effect.

Industrial Progress

Along with these crucial changes in agriculture, major areas of Mexico's economy are being shifted from government ownership to private ownership. Throughout the country, banks and telephone companies are now private concerns. New opportunities are attracting investment from throughout the world.

Along Mexico's northern border with the United States, foreign-owned assembly plants called *maquiladoras* have been springing up since the late 1960s. In these plants, everything from auto parts and machine goods to small appliances and furniture are assembled from imported raw materials or component parts. The investors gain from Mexico's low wages, which in the automobile industry, for example, are one-eighth those in the United States. The Mexicans have gained many new jobs through this investment.

The greatest growth of *maquiladora* plants has been in Tijuana (which now has a population of over a million) across the border from San Diego and in Ciudad Juárez opposite El Paso. Although a relatively recent phenomena, these assembly plants already employ one-fifth of Mexico's industrial workers.

Mexico's vital government-owned oil industry is also opening up to foreign investment. Mexico now is the second largest oil-producing country in Latin America (after Venezuela) and is self-sufficient in energy. The major fields along Mexico's Gulf Coast between the cities of Tampico and Veracruz are one of the world's largest finds of oil and natural gas reserves since the North Sea discoveries in 1970. Mexico is joining with foreign companies in drilling dozens of offshore wells in the shallow waters of the Gulf. A major oil refinery is also under construction.

The North American Free Trade Agreement (NAFTA)

Nothing symbolizes the rapid and effective modernization of Mexico more than its entrance into the North American Free Trade Agreement. Mexico already conducts 80 percent of its trade with the United States. Reform leaders of Mexico have recognized that trade with the United States can provide the investment funds needed for a nation that will have 100 million people in the next decade. They have embarked on a series of dramatic and successful economic reforms. Despite the financial crisis of 1995, Mexico's future holds great promise.

REVIEW QUESTIONS

1. What is an *ejido*?
2. What are the *maquiladoras*? Where are they located?
3. Where is Mexico's oil industry located?

THOUGHT QUESTIONS

1. Why are Mexico's 1991 reforms in agriculture so important? What effects do you think they will have?
2. In the *maquiladoras*, workers in Mexico's automobile industry earn salaries that are one-eighth those in the United States. Do you think this shift in jobs will hurt the United States, or do you think both the United States and Mexico will benefit from this relationship? Consider such things as cost of living in each country, and other opportunities for employment and/or job retraining.
3. The North American Free Trade Agreement (NAFTA) went into effect on January 1, 1994. Will it have an impact on the economy of the United States in the future? As a class project, find out why some people in the United States opposed NAFTA and others supported it.

4. Turbulence in Central America

Lands and Peoples

Central America is composed of seven republics—El Salvador, Nicaragua, Guatemala, Honduras, Costa Rica, Panama, and Belize. This region forms a narrow funnel that connects Mexico with South America. The densely forested mountain core of Central America is made up of two volcanic mountain ranges. One of these ranges slices east-west across Guatemala, Honduras, and Nicaragua with peaks rising 11,000 to 13,000 feet above sea level. The second and less imposing mountain range runs northwest to southeast through Costa Rica and Panama.

Both of these mountainous areas are wet and densely settled. Roughly 80 percent of the 31 million people of Central America live in highlands above 2,000 feet. Most are of Indian, European, or *mestizo* descent. People of African heritage work on plantations on the Caribbean coasts of Panama, Costa Rica, Nicaragua, Honduras, and Belize.

In Central America, civil wars in the 1980s and early 1990s prevented any meaningful social or economic development. The worst conflicts broke out in Guatemala, El Salvador, and Nicaragua and devastated the economies of these countries. But no country in Central America completely escaped the **ripple effects** of these wars, one of which was the stream of refugees into nearby countries. Honduras suffered greatly from the wars going on in three neighboring countries at the same time.

Guatemala: Urban Concentration and Rural Poverty

Guatemala, just south of Mexico, has three major problems that are common in Central America: (1) a huge concentration of people living in one city, (2) large numbers of farmers whose standard of living is very low, and (3) a long and punishing civil war.

Central America

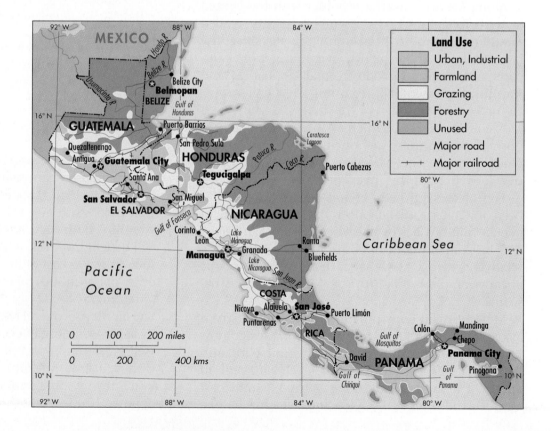

▲ *People of the highlands of Guatemala still do much of their farming by hand. What obstacles prevent the people from using modern farming methods?*

Once the heartland of the late Maya Empire, Guatemala has a location that could have protected it from Central America's wars but did not. Mexico forms its northern and western border, and in the east Belize and Honduras block much of the country from the Caribbean. Its primary orientation has been toward the Pacific Ocean, and its capital city is located in the western highlands. However, these geographical factors did not protect Guatemala from civil strife.

For thirty years, a civil war between communists in the countryside and military leaders in the capital of Guatemala City has devastated the country's economy. In the 1980s, the military engaged in a **scorched-earth policy** of extreme cruelty, destroying the land and driving peasants deep into the forested mountains. Guatemalan refugees also fled northward to Mexico and the United States. Finally, in 1993 some 50,000 of these people began to return home under a new but tentative agreement with the Guatemalan government.

Nearly one-fifth of Guatemala's 10 million people live in its capital, Guatemala City, where the country's best education and medical care can be found, and where there are more jobs. Heavy expenditures on weapons, however, have left few resources for these social needs. More than 60 percent of Guatemalans are poor farmers who raise corn, squash, and beans on small farms. Many of these people still speak Indian languages. Guatemala has lost 65 percent of its forests by rampant clearing, and war-time needs further devastated the countryside. Some peasants have moved to Guatemala City and live in slums, but many prefer to live in mountain villages.

A small class of wealthy people in Guatemala City, backed by the military, own and run the coffee, cotton, and sugar plantations that produce most of the country's wealth. These plantations are located along the Pacific coastal plain and in nearby highlands. Indian farmers come down from the mountains each year to earn extra money and relieve their poverty by harvesting plantation crops.

Guatemala and neighboring Honduras (population of 5.6 million) have lacked the resources to improve the lives of these rural people. Indeed, Honduras, which has provided shelter to thousands of refugees from neighboring countries, is Central America's least developed and poorest country.

El Salvador, a Country in Despair

Similar problems face the 5.2 million people of El Salvador. Civil war has torn the country apart for a decade. In the 1990s, it is now estimated that one-fifth of El Salvador's population has fled to the United States.

Although El Salvador is the smallest country in Central America—even smaller than Belize—its population is twenty-five times larger than that former British colony, previously called British Honduras. How El Salvador will support its people is unclear. Most of the country's money has been spent on weapons, life is insecure, and unemployment is high. A large population lives in an area with few resources and limited farmland, and many of them are taking flight to the capital city of San Salvador.

Most of the people of El Salvador are *mestizo* and Indian farmers who eke out a living on small plots of land. These people are subsistence farmers who

LOCATION

PANAMA, THE PATH BETWEEN THE SEAS

Panama was a crossing place for the early Spaniards. In the 1500s, Spanish *conquistadors* led mule trains laden with treasure across the Isthmus of Panama. By the 1880s, the advantages of cutting a canal through this isthmus were obvious. A canal would provide a shortcut that would reduce the sailing distance from the Atlantic coast to the Pacific by about 9,200 miles.

At that time, Panama was part of the South American country of Colombia. A French engineer attempted to construct a canal in Panama; however, thousands of laborers died of tropical diseases and his venture failed.

In 1903, the United States supported Panama's declaration of independence from Colombia. In return, Panama granted the United States control over the Canal Zone, a strip of land in central Panama that is ten miles wide and fifty miles long. In this area, they dug the Panama Canal, a project equivalent to digging a ten-foot trench fifty-five feet wide from New York to California.

In 1914, the Panama Canal was completed. Locks raise and lower ships eighty-five feet above sea level, making it possible for ships to make the fifty-mile passage across the mountains from coast to coast. The diagram on page 433 shows how these locks work. Since 1914, more than 600,000 ships have sailed across Panama, which is a "crossroads of the world."

By the 1970s, the Panama Canal was becoming outdated. Its 110-foot-wide locks were too narrow for new, larger ships. A U.S. commission studied twenty-five alternative canal routes before concluding that the existing canal should be widened.

In 1977, the United States made a new agreement with Panama, by which the Canal Zone was transferred to Panamanian control in 1979. The Panama Canal itself will become Panamanian property at the end of 1999.

This changeover has created great uncertainty. Without investment, maintenance, and skilled workers, the canal will deteriorate. Under the dictatorial rule of Manuel Noriega, the port facilities at the sea entrances began to crumble. Concerns about the safety and security of the canal, as well as the reliability of the Noriega government led to international economic sanctions and the U.S. invasion that ousted the regime from power in the late 1980s. For countries like Japan, which depend on the Panama Canal to ship goods to the East Coast of the United States and Europe, these events were worrisome.

Although there is general agreement that the canal should be widened and a new set of big-

---

raise squash, beans, and other vegetables to feed their families. Wealthy plantation owners once grew coffee and cotton on the best land in the highlands, but warfare has driven many landowners out of the country. Moreover, 93 percent of the forests of El Salvador have been cut down, causing erosion and flooding.

As in other countries of Central America, the difference in living standards between the wealthy minority and the poor majority has led to unrest and violence in El Salvador. Although eased in the early 1990s, these tensions are an active force in life in El Salvador.

Nicaragua, the Geographical Heart of Central America

Nicaragua, the largest country in Central America, is a troubled land of 4.1 million people. The Sandinistas achieved power in a bloody revolution

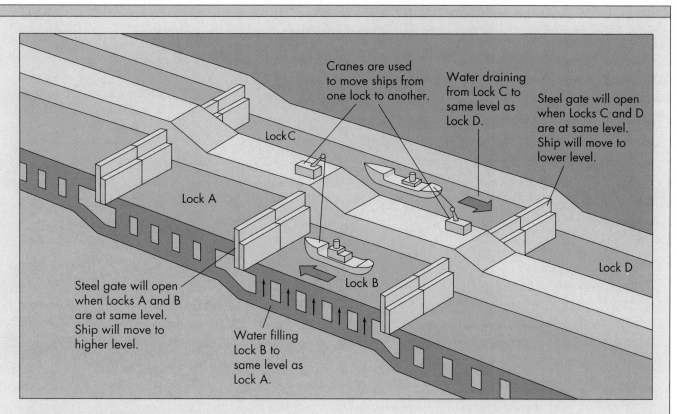

Cranes are used to move ships from one lock to another.

Water draining from Lock C to same level as Lock D.

Steel gate will open when Locks C and D are at same level. Ship will move to lower level.

Lock C

Lock A

Steel gate will open when Locks A and B are at same level. Ship will move to higher level.

Water filling Lock B to same level as Lock A.

Lock B

Lock D

Panama Canal Locks

ger locks created, how this would be paid for is not clear. Yet the canal's future looks bright. Almost all of the world's cargo ships being built in the 1990s can move through the locks, and one-third of them are specifically designed to make the passage. So in the immediate future, the Panama Canal will remain an important waterway in world trade and will provide badly needed wealth to the 2.5 million people of Panama.

QUESTIONS

1. Where is the Panama Canal and when was it completed?
2. What service does the Panama Canal provide that makes it so important to world shipping?

in 1979. This revolution set up the first communist government in mainland Middle America. This caused immediate concern in the United States and brought U.S. military and financial support to the *contra* rebels. Throughout the 1980s, the country was engaged in a violent civil war that pitted the Sandinista government against the U.S.– backed *contras*.

In 1990, in a United Nations-supervised election, the Sandinistas were themselves ousted by a coali-

tion government. Although the new government included Sandinistas in positions of power, insurrections still flare in Nicaragua in the middle 1990s, and little progress has been made in solving the

IT'S A FACT In parts of Panama, one can watch the sun rise over the Pacific and set in the Atlantic.

▶
The majority of Nicaragua's urban population lives in poor slum areas called barrios. *What are some problems that people there face on a daily basis?*

country's economic problems. Armed bands still control parts of northern Nicaragua. The government is deeply in debt, and the country's landscape has been badly damaged.

Strategically, Nicaragua is located at the core of Central America. The region's poorest country, Honduras, lies to the north. Its richest country, Costa Rica, is south of Nicaragua. The capital city of Managua is located on the rich volcanic soils of western Nicaragua, which has always been the center of the country's population.

The best land in Nicaragua was owned, until the 1980s, by rich landlords. They owned coffee *haciendas* in the cool highlands, and collaborated with foreign-owned companies in running cotton and banana plantations in the hot lowlands.

One of the Sandinistas' goals was to redistribute this land, but their efforts to increase production in the 1980s were unsuccessful. Land distribution did not provide better seeds, pesticides, fertilizers, and other elements essential to agricultural progress. Nicaraguans who had run the *haciendas* and plantations left the country, and few foreign investors would provide capital to a politically unstable coun-

try. An overwhelming majority of Nicaraguans were and still are poor subsistence farmers.

Most observers recognize that overwhelming inequalities between the rich and the poor in Nicaragua are the root of the turmoil that led to the revolution in Nicaragua in 1979. The ultimate solution will lie in economic development programs that will improve the lives of all Nicaraguans, but such programs cannot be carried out in an environment of strife, violence, and guerrilla warfare.

Costa Rica, a Prosperous Country of Small Farms

Costa Rica differs from all other countries in Central America. Despite its location between Nicaragua to the north and Panama to the south, Costa Rica has maintained a prosperous and successful democracy. It is the oldest continuous democratic state in all of Latin America. It has no standing army; the country spends its wealth on its people.

Many observers believe that Costa Rica's society is more democratic because this was the only coun-

IT'S A FACT   Costa Rica has more species of mammals and birds than the continental United States and Canada combined.

try in Latin America where Spaniards settled with their families rather than as single men. Most of this nation's 3.3 million people live in the cool highlands near the capital city of San José. Here, farmers raise coffee, sugarcane, and food crops on small farms and herd cattle for beef. In the lowlands, bananas are an important crop.

A large number of Costa Ricans own small farms. Wealth has been distributed among the people of Costa Rica, and this has resulted in the highest standard of living in Central America. Moreover, economic development and environmental protection have enabled Costa Ricans to benefit from the many products of its rain forests.

In contrast to other countries of Central America, most people in Costa Rica can read and write. San José is free of slums. A model of modern social and economic development, Costa Rica has fewer problems than other countries in Central America.

⊕ REVIEW QUESTIONS

1. How many countries are there in Central America? Which is the largest? Which has the most people?
2. What countries have been most affected by civil wars in Central America?
3. What are the principal highland and lowland crops grown in Central America?

⊕ THOUGHT QUESTIONS

1. In what ways does Costa Rica differ from other Central American countries? Do you believe these differences have benefited the people of Costa Rica or not?
2. Why is Panama called "a crossroads of the world"? What sequence of events has led to Panama's gaining control of the Panama Canal at the end of 1999? Do you think the agreement between Panama and the United States was a good idea or not?

5. Islands of the Caribbean

Islands and Peoples

Hundreds of islands stretch in a 2,500-mile arc east and south from the coast of Mexico to northern South America. The map on page 436 shows that these islands, the West Indies, circle the Caribbean Sea and separate it from the Gulf of Mexico and the Atlantic Ocean. The West Indies have a combined population of 35 million people. Nine out of ten people live on the four largest islands—Cuba, Hispaniola, Puerto Rico, and Jamaica. The largest and second-most populous of these islands, Cuba, lies only 100 miles south of Key West, Florida.

The topography of the islands of the West Indies is quite varied. Many islands, like Puerto Rico, Hispaniola, and Jamaica, are mountains that rise steeply from the sea. Others, like the Bahamas, are flat coral reefs built up from beneath the sea by coral polyps. Most of the people of the Caribbean live near the coasts at elevations below 1,000 feet. Tropical winds from the Atlantic keep these islands pleasantly warm the year around.

Two hundred years ago, many Caribbean islands were colonies of European countries. Spain ruled most of the islands, including Cuba, the eastern part of Hispaniola, and Puerto Rico. Britain owned Jamaica and a number of smaller islands. France controlled the western third of Hispaniola. The Dutch still own the Netherlands Antilles—including Aruba, Bonaire, and Curaçao located off the coast of Venezuela. Find these islands on the map on page 436.

Early on, Europeans established sugar plantations on the islands of the Caribbean. Sugar was worth its weight in gold in those days. Sugarcane could survive the frequent Caribbean hurricanes as well as most local plant diseases. Enslaved Africans were brought in to work on the sugar plantations, and

IT'S A FACT   The Caribbean exports a higher percentage of its population abroad than any other world region.

Islands of the Caribbean

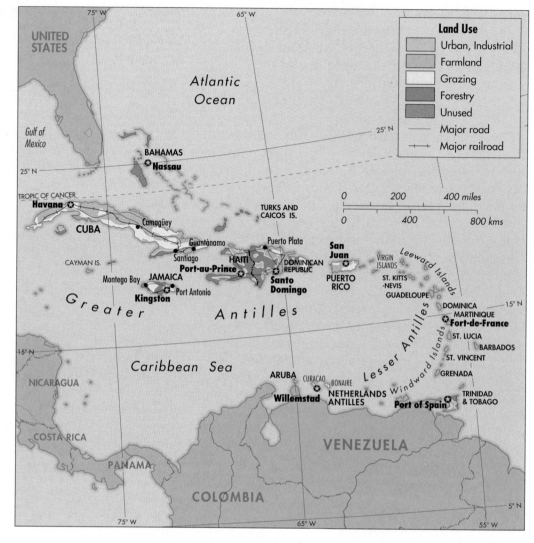

later to grow bananas. Today, the descendants of these enslaved people form a majority of the population on most Caribbean islands. About 65 percent speak Spanish as their first language, and another 20 percent speak French. Almost all of these former island colonies are now independent countries.

Cuba, a Declining Economy

Cuba has about 11 million people, a third of the entire population of the Caribbean. Most of this 800-mile-long island is gently rolling countryside punctuated by three small areas of low mountains. Cuba has fertile soils and a warm, rainy climate that

have helped make the island the world's largest producer of sugar.

The Spaniards developed Cuba's sugar economy long ago. When Cuba gained independence from Spain in 1898, U.S. corporations invested money in expanding the island's sugar production. Huge sugarcane plantations covered more than half of the good land on the island. Cuba became completely dependent on the sale of sugar, much of it sold in the United States.

The price of sugar on world markets began to fall as beet sugar in Europe and sugarcane grown in the United States provided new sources of competition. The fall of sugar prices brought economic hardship to Cuba. Also, after the sugar harvest, a quarter of Cuba's people were unemployed until the planting season rolled around months later. These social problems and the poverty they created for Cubans helped Fidel Castro and his followers to overthrow the island's dictator in 1959. For over thirty-five years, Castro has been the sole ruler of Cuba.

Castro's government confiscated the big sugar plantations, including those owned by U.S. citizens, and created a socialist society. Cane cutters began to work on large farms owned and controlled by the government. Castro's attempts to start new kinds of farming and new businesses were hindered by the unwillingness of most countries to invest in Cuba, which had become closely affiliated with the former Soviet Union.

This association also led the United States to sever relations with Cuba and to ban the sale of Cuban sugar in American markets. Soon Cuba was trapped in a trading relationship with the Soviet bloc in which Russian oil was exchanged for Cuban sugar. When the Soviet Union collapsed in 1991, Cuba's reliance on its communist allies for aid led to economic disaster.

Today, Cuba is still dependent on sugar, but revenues from sugar sold on world markets are small. The entire economy of the country is believed to have dwindled to half its former size. Cubans now ride bicycles rather than automobiles because gas and oil are not available. Factories are working shorter hours, unemployment is rising, and power blackouts in cities are common.

Castro's alliance with the former Soviet Union has led to economic disaster. Revenues from sugar cannot sustain its rapidly growing population. Poor relations with the United States have meant that Cuba lost the tourist business it had before Castro came to power. Food is now rationed, and as one observer notes, "Feeding a family has become a challenge."

Hispaniola: One Island, Two Countries

Hispaniola is the second largest island in the Caribbean. More than 14 million people live in the two countries that share the island, Haiti and

IT'S A FACT The two poorest countries in the Caribbean, Haiti and the Dominican Republic, also have been independent the longest.

the Dominican Republic. Hispaniola is smaller than Cuba and much more mountainous. Settlements of people are clustered along the narrow coasts and in valleys between steep mountain ranges. Climate and vegetation vary with elevation and with exposure to rain-bearing winds from the ocean.

Haiti, a former French colony, occupies the western third of the island of Hispaniola. French is still the official language of Haiti. Under French rule, Haiti became the richest colony in the Caribbean. The thousands of captives brought from Africa to work its sugar plantations greatly outnumbered the French colonists. In 1791 the slaves, led by Toussaint L'Ouverture, revolted, and by 1804 they had succeeded in throwing out the French. They also destroyed the sugar plantations and sugar mills. Haiti became the first colony in Latin America to win its independence, but the nation has had a troubled history, with brief periods of democratic leadership between long periods of dictatorship. Haiti accepted various forms of U.S. rule from 1905 to 1941, a time of stability and tremendous population growth. After thirty years of dictatorship by "Papa Doc" and then "Baby Doc" Duvalier, democratic elections were held in 1990. But after a short time in office, President Aristide was forced out in a military coup.

Today Haiti has a population of 6.5 million. Its main exports are coffee and some bauxite. Haitian farms are squeezed into narrow valleys, because 80 percent of the country is mountainous. Virtually all of the land has been stripped of forests, and over-cultivation is intensifying erosion. More than half of all Haitians are farmers, and population pressure on the land is intense. History, environment, and rapid population growth have contributed to making Haiti the poorest country in Latin America.

The eastern two-thirds of Hispaniola, the Dominican Republic, was ruled by Spain. Its capital, Santo Domingo, is the oldest European settlement in Latin America. Most of the Dominican Republic's 7.6 million people are *mulattos*. This country is twice as large as Haiti and is less mountainous, so scarcity of level land is not such a pressing problem. Large sugar and coffee plantations spread out over the lowlands and extend up the lower mountain slopes of the Dominican Republic. The main export is sugar. As is the case in many Caribbean countries, when the price of sugar falls on world markets, the economy and people of the Dominican Republic suffer.

Jamaica, the Third-Largest Island

Most of Jamaica is a rugged plateau. The island's best agricultural land is found on lowlands along the coast, where the British colonists started sugar and banana plantations, forcing native farmers to move inland onto the easily eroded plateau. Now 2.5 million Jamaicans live on an island where less than a quarter of the land can still be farmed.

Jamaica, however, has a major mineral resource—**bauxite**, the ore from which aluminum is made. This island nation produces one-fifth of the bauxite used in the world and has large reserves of bauxite ore. Jamaica is using money from bauxite mining to build factories and new businesses.

The Commonwealth of Puerto Rico—a Political Partner of the United States

Puerto Rico is correctly called the Free Associated State of Puerto Rico, because Puerto Rico is "associated" with the United States. Puerto Ricans are citizens of the United States and can move back and forth between their island and the United States. In addition, the United States has agreed that should Puerto Rico ever vote for complete independence, it

▼ *El Morro Fort in Old San Juan, Puerto Rico, follows the shape of the coastline. Notice that waves are seen only on one side of the island. Why is this?*

will be granted immediately. Puerto Rico also has the option of becoming the fifty-first state, a choice they turned down in 1993.

Puerto Rico is a rectangular-shaped plateau, 135 miles long and 35 miles wide. Only 15 percent of the land is cultivated, mainly along the coasts and in narrow mountain valleys. The windward northern coast is wet; the leeward southern coast is dry. Do you remember how wind direction affects rainfall in mountainous regions?

Sugar and tobacco plantations are found along the wetter northern and western coasts. Coffee is grown in the interior. Before 1940, large plantations occupied 85 percent of Puerto Rico's farmland, and most Puerto Ricans were poor plantation workers.

In 1945, after World War II, the plantations were broken up and the land was turned into small farms. Factories were built with the help of the United States; until 1993 they were not forced to pay any federal income taxes. Unfortunately, this economic growth in Puerto Rico has been offset by rapid population growth. In the last seventy-five years, the island's population has tripled to more than 3.6 million people.

Most Puerto Ricans are still farmers. Because little farmland is available, 40 percent of the population—a total of 1.4 million people—live in the capital, San Juan. Many Puerto Ricans leave the island and come to the United States to earn a living. About 2.6 million Puerto Ricans live in the United States today, half of them in New York City.

Puerto Rico's standard of living has been one of the highest in the Caribbean; however, the island's limited amount of usable land and its growing population are creating serious problems.

The Lesser Antilles

Tropical agricultural development is similar on the islands of the Lesser Antilles, which form a 700-mile-long arc across the Caribbean from Puerto Rico in the north to the coast of Venezuela. These islands are the peaks of a double line of submarine volcanoes. The outer ring is made up of low islands, old volcanoes, and limestone banks called the Leeward Islands; the inner ring of higher volcanic peaks is called the Windward Islands. These volcanic peaks meet with the islands of Trinidad, Bar-

bados, and the Netherlands Antilles near the coast of South America.

The largest of the low-lying or Leeward Islands are French Guadeloupe and independent Antigua. The mainstay of their economies is sugarcane grown on large plantations. The usual problems of poverty and unemployment characteristic of the Caribbean plantation system are found on these islands. St. Kitts and Nevis (British), also among the Leeward Islands, produce sugar.

On the higher Windward Islands, agriculture is quite varied. Forested mountains rise to between 3,000 and 4,000 feet in elevation. Each island has rich volcanic soils, little level land, and abundant rainfall. St. Vincent produces arrowroot, a starchy tuber; Dominica, limes; Grenada, cacao and nutmeg; Martinique (French), sugar; St. Lucia, bananas. On some islands, such as Grenada, small holdings dominate. On Martinique and St. Lucia, however, so much land is devoted to commercial estates that most food must be imported. About 1.7 million people live in the Lesser Antilles, nearly half on the two French islands of Guadeloupe and Martinique.

Approaching the South American coast, the Lesser Antilles join with a string of islands that includes the former British territories of Barbados, Trinidad, and Tobago, and the "A-B-C islands" of the Netherlands Antilles, Aruba, Bonaire, and Curaçao.

Barbados, which has a population of 300,000, most closely follows the Caribbean economic pattern: 95 percent of the island's exports are sugar, molasses, and rum. Population densities on Barbados reach levels of 1,400 per square mile. In Trinidad and Tobago, which were settled much later, a mixture of Africans (43 percent), South Asian Indians (37 percent), and Chinese (9 percent) were imported to work on sugar plantations in the nineteenth century.

Although sugar and rum are still the major agricultural exports of Trinidad, petroleum has been discovered on the southern third of the island, and Trinidad now produces and refines more than 100 million barrels annually. This provides an economic

IT'S A FACT Of the 2,700 or so islands in the Bahamas, only 30 are occupied.

balance rare in the Caribbean. The islands of Trinidad and Tobago have a population of 1.3 million.

In Aruba and Curaçao, petroleum refining is a dominant economic activity. Near-desert conditions on these islands limit agricultural development, and water is so scarce that distillation of seawater is required to maintain the population of 200,000. On all of these Caribbean islands, tourism is a growing industry.

⊕ REVIEW QUESTIONS

1. Where do nearly half of the people of the lesser Antilles live? Where are the Leeward and Windward Islands?
2. What are the four largest islands in the Caribbean?
3. Which island opted for a socialist government?
4. What is the principal crop grown in this region?

⊕ THOUGHT QUESTIONS

1. What are some of the reasons that so many people from the Caribbean migrate to other countries?
2. What Caribbean country is most closely linked to the United States? What is the nature of this association? If you were a citizen of that country, which political option would you choose?
3. Given the growing population, small land base, and few resources found in the Caribbean, how do you view its future? What steps might be taken to improve the standards of living in this region?

⊕ CHAPTER SUMMARY

Middle America has two subregions: the mainland and the rimland. The two subregions have problems and prospects in common as well as some that are unique to each subregion.

In mainland and rimland countries, colonial landholding practices have left enduring patterns on the land and have put most of the best land in the hands of a few wealthy families and left large numbers of poor families with little or no land. Investments by large American and other foreign-owned companies have helped to develop the resources of these countries, but this development has not improved the lives of the people. Economic inequalities have continued, and have led to political unrest and revolution in most mainland countries in this century. Four countries, Guatemala, El Salvador, Nicaragua, and Mexico are currently enduring the consequences of revolution and some civil war.

In Mexico, early attempts to settle landless peasants onto communal farms resulted in agricultural stagnation. Beginning in 1991, a new land reform policy was implemented that was designed to make farmers owners of the land and to increase investment in agriculture. Related to this policy of increasing investment, the Mexican government has entered the North American Free Trade Agreement (NAFTA) with the United States and Canada. Both policies are controversial in Mexico because they may in the short run benefit the wealthy at the expense of the poor, as the 1994 peasant revolt in the state of Chiapas in Mexico demonstrated.

Large plantations worked by enslaved Africans characterized many of the rimland countries of the Caribbean. On some islands land was more evenly distributed after independence, but on others colonial plantations remained. Many islands depend on a single crop, sugar, whose price fluctuates on the world market. Caribbean sugar must compete with sugar produced in countries around the world.

Many nations of both mainland and rimland Middle America have suffered from severe poverty and poor economies, as well as from political instability and war. All these conditions have led to mass migrations of Middle Americans to other countries in the hemisphere, especially to the United States.

EXERCISES

Record Setters

Directions: Write on your paper the answers to the following questions.

What is the . . .

1. largest city in Latin America?
2. oldest European settlement in Latin America?
3. oldest democratic state in Latin America?
4. largest island in the Caribbean Sea?
5. largest and most significant country in mainland Middle America?
6. smallest country in Middle America?

Which Country?

Directions: Write on your paper the answers to the following questions.

Which Middle American country . . .

1. is made up of coral reefs?
2. in the West Indies produces one-fifth of the world's bauxite?
3. has a higher standard of living than other Central American countries?
4. is called the "crossroads of the world" because of its major waterway?
5. has a densely populated Central Valley?
6. has huge oil and natural gas fields south of Veracruz?
7. has had its countryside and economy devastated by a thirty-year civil war?
8. is an island colonized by the French?
9. has its capital city more than a mile above sea level?
10. was once the heartland of the Maya Empire?
11. has one-fifth of its population living in its capital city?
12. shares an island with Haiti?

Which Direction?

Directions: Complete these sentences by writing on your paper the terms or phrases that belong in the blanks.

1. Cuba is |||||||||| of Mexico.
2. Haiti is |||||||||| of the Dominican Republic.
3. Panama lies |||||||||| of Colombia.
4. Honduras is |||||||||| of Nicaragua.
5. El Salvador is |||||||||| of Belize.
6. Mexico is |||||||||| and |||||||||| of Guatemala.

Inquiry

Directions: Combine the information in this chapter with your own ideas to answer these questions.

1. In the mainland part of Middle America, most of the large cities are in the highlands. Why is this so?
2. Settlement patterns are aspects of people-environment relations that have political and economic consequences, giving access to needed resources to some social groups and denying them to others. Explain how and why Spanish colonial settlement patterns led to political turmoil and revolution in Mexico and other Middle American countries after they became independent.
3. Movement, human migration, has been one of the most influential forces in history. Who moves where and why is also of interest to geographers because a place is partially the product of the kinds of people who live there. Costa Rica, for example, is thought to be a more settled and prosperous country because it was colonized by Spanish families rather than by single men. Why do you think families might have made a crucial difference in Latin America's colonies before and after independence?
4. Explain why Cuba's location is a major factor in the U.S. government's opposition to the communist government there.

SKILLS

Using Population Maps

Directions: The population maps on pages 391 and 394 show how population is distributed throughout Latin America. These maps do not show the names of countries, so you may want to refer to the political maps on pages 389 and 392 to help answer some of these questions.

1. What colors are used to show the regions that are the most densely settled and the least densely settled?
2. Find the boundary between the United States and Mexico. Is the area of Mexico that surrounds Ciudad Juárez densely or thinly settled?
3. Farther south is the Central Plateau of Mexico. In general, is it densely or thinly settled?
4. Is most of Middle America densely or thinly settled?
5. Are the largest islands in the West Indies densely or thinly settled?
6. Are the interior regions or the coastal regions of Middle America more densely settled? What do you think is the chief reason for this?
7. Is the northernmost region of Middle America densely settled? What do you think is the chief reason for this?

Now look at the map symbols on page 391 that show the size of cities in Middle America.

8. The largest black circle on the population map of Middle America shows cities of what size?
9. Which seven cities in Mexico have over 1 million people?
10. How many cities in Middle America have at least 500,000 people?
11. What are the three largest cities in the West Indies?
12. Which country in Middle America has the most large cities?
13. Are there more large cities along the eastern coast or the western coast of Middle America?

Vocabulary Skills

Directions: Complete the sentences by filling in the blank with the correct vocabulary term from the list. Write this term on your paper next to the number of the sentence it completes.

bauxite	*maquiladoras*
caballeros	*peones*
cordillera	pre-Columbian
ejidos	rimland
escarpment	ripple effect
indigenous	scorched-earth policy
mainland	subsistence crops

1. A line of steep cliffs rimming a plateau is termed an ||||||||||.
2. |||||||||| is a mineral used to make aluminum.
3. When several things are affected in turn by one thing, such as a recession resulting in jobs lost or in fewer sales of goods, it is called a ||||||||||.
4. After the 1910 revolution, the Mexican government redistributed the land and settled poor farmers on communal farms called ||||||||||.
5. Native cultures that existed before the Europeans encountered them are termed ||||||||||.
6. |||||||||| are foreign-owned assembly plants along the Mexican border with the United States.
7. Something that is native to a place or region is said to be |||||||||| to the place.
8. Peasant farmers in Latin America are called ||||||||||.
9. The Guatemalan government engaged in a |||||||||| designed to destroy everything in the region where the communist insurgents lived.
10. Crops that are grown for family consumption and not for sale are termed ||||||||||.
11. A |||||||||| is the principal mountain chain of a system of mountain chains.
12. Aristocratic Spanish colonists in Mexico called themselves |||||||||| from their word for "knights."
13. The coastal lowlands and the islands of the Caribbean make up a region of Middle America referred to as the ||||||||||.
14. The |||||||||| of Middle America includes most of Mexico and the seven countries of Middle America.

THE MANY LANDS OF SOUTH AMERICA

S outh America is a continent in change. It contains 305 million people, two-thirds of all Latin Americans. Today, the larger countries of this region are diversifying their economies and attempting to meet the demands of rapidly growing populations. Many are becoming urban nations as peasants leave the countryside for the cities. Other countries are burdened by social stress and political turmoil.

◄ *Bogotá, the capital of Colombia, has a population of over 7 million people. Pictured here is the international and banking area of the city, which is surrounded by highlands where several important export crops are grown. What are the main export crops of Colombia?*

443

Three Subregions

Topographically, South America has several very different variants. In general, the continent can be divided into three subregions: (1) the towering Andes Mountains, which sweep southward along the north and west coasts of the continent; (2) the temperate middle-latitude plains of the south; and (3) the vast expanse of Brazil's plateaus and lowlands.

The Andes Mountains rise in eastern Venezuela in the north and arc southward for more than 4,000 miles to the tip of South America. Throughout most of the Andes system, its mountain ranges and highland valleys form a band about 200 miles wide. The Andes broaden to a width of 400 miles in Bolivia, and they narrow to a width of only 20 miles in Chile. The countries of Venezuela, Colombia, Ecuador, Peru, Bolivia, and Chile share significant parts of the Andes Mountains.

The three countries of Argentina, Uruguay, and Paraguay occupy the middle-latitude plains of southern South America. Three important rivers drain these temperate plains—the Paraguay, the Paraná, and the Uruguay. These rivers join at the Rió de la Plata, an **estuary**, the sea-flooded mouth of a river valley, which is the main outlet to the world for these countries.

Brazil, as you know, is the largest country in South America, and occupies nearly half of the entire continent. It stretches 2,300 miles north-south and about the same distance east-west. Its size, diversity of resources, and variety of environments make it unique in Latin America.

Subregions of South America

South America has three major subregions: (1) the Andes Mountains, (2) the Middle-Latitude South, and (3) Brazil.

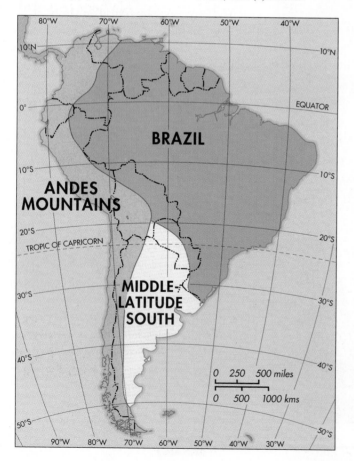

1. The Andes Mountains System

Environments and Peoples

The Andes Mountains form the spine of South America and are the longest mountain chain in the world. They rise from the waters of the Caribbean in eastern Venezuela and curve southward along the continent's west coast for 4,000 miles. Fertile valleys, basins, and plateaus reach 10,000 feet above sea level in the Andes. Individual snow-covered peaks in Ecuador, Peru, Bolivia, and Chile soar to 22,000 feet, a full 4 miles above sea level.

The six major countries in this subregion are Venezuela, Colombia, Ecuador, Peru, Bolivia, and Chile. Together these nations have a combined population of 100 million people, one-third of all South Americans.

In the Andean ranges, as in the higher mountains of Central America, land use varies with elevation. Four zones of human settlement and economy are frequently found. These are called the *tierra caliente*, *tierra templada*, *tierra fria*, and *tierra helada*—lowlands, foothills, highlands, and snow-covered peaks, as the diagram on page 446 shows.

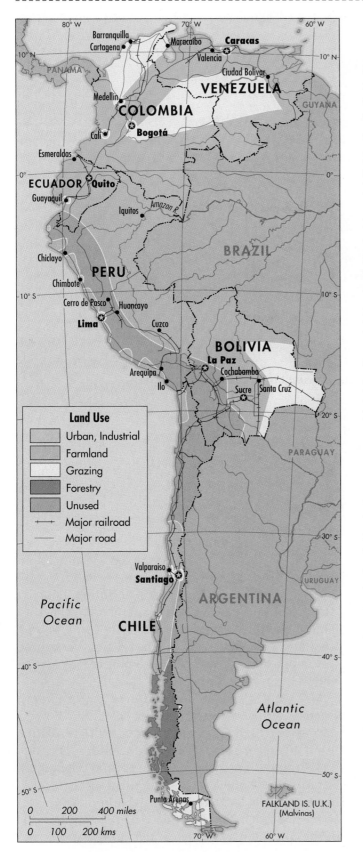

Land Use

- Urban, Industrial
- Farmland
- Grazing
- Forestry
- Unused
- —+— Major railroad
- —— Major road

The Andes Mountains System

The Eastern Slopes of the Andes

The Andes Mountains broaden to a width of 400 miles in Bolivia. In Chile, they narrow to only 20 miles. Throughout most of their course, however, these mountains form a 200-mile-wide barrier. These ranges are cut by fertile intermontane valleys and pocked with mountain-rimmed basins and plateaus. It is in these small niches that the people of the Andes, most of them indigenous peoples, live.

The eastern or windward slopes of the Andes are heavily watered, deeply dissected, and covered by tropical forests that extend eastward into the Amazon Basin as the diagram on page 403 illustrates. These tropical regions are poorly connected to the west, even though the eastern slopes of the Andes include half of the national territories of Colombia, Ecuador, Bolivia, and Peru.

The region is still inhabited by a thin scattering of native Indians, a few missionaries and pioneers, and more recently, oil exploration teams. Now, however, new roads are being pushed over the mountains to reach this resource-rich region.

The Western Slopes of the Andes

Tropical rain forests also cover much of the western slopes of the Colombian and Ecuadorian Andes, where moisture-laden winds from the Pacific drop more than 100 inches of rainfall each year. South of the Gulf of Guayaquil in Ecuador, however, an incredibly dry stretch of desert runs from the southern coast of Ecuador to the dune-covered Atacama Desert of northern Chile.

Along this desert coast, the only source of moisture is fog produced when air warmed by the Pacific Ocean passes over the cold northbound ocean current, the Humboldt Current, which runs parallel to the coast. Settlement here is limited to coastal oases dotted along streams that pour down from the Andes. Recently, Chilean engineers began to string nets across valleys near the coast to catch the fog in the form of dew and use this valuable water for cultivation.

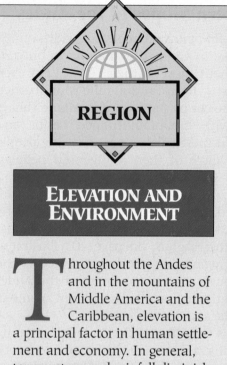

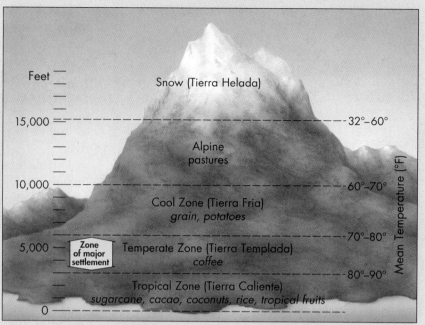

Elevation Diagram of Living Environments in the Andes

Labels within the diagram:
- Feet
- Snow (Tierra Helada)
- 15,000 — 32°–60°
- Alpine pastures
- 10,000 — 60°–70°
- Cool Zone (Tierra Fria) *grain, potatoes*
- 70°–80°
- 5,000 — Zone of major settlement — Temperate Zone (Tierra Templada) *coffee*
- 80°–90°
- Tropical Zone (Tierra Caliente) *sugarcane, cacao, coconuts, rice, tropical fruits*
- 0
- Mean Temperature (°F)

REGION

ELEVATION AND ENVIRONMENT

Throughout the Andes and in the mountains of Middle America and the Caribbean, elevation is a principal factor in human settlement and economy. In general, temperature and rainfall diminish with elevation.

Hot and humid coastal lowlands, the *tierra caliente*, usually have a plantation economy near streams, where bananas, cacao, sugarcane, and other commercial crops are grown. On the warm piedmonts above 2,500 to 3,000 feet, the *tierra templada*, dense populations are engaged in commercial cultivation of coffee and

Farther south, the Chilean Andes reach into the middle latitudes, producing a Mediterranean climate in the fertile Central Valley of Chile. Cold, wet, temperate forests stretch southward from the Central Valley toward Tierra del Fuego.

The Northern Andean Nations of Venezuela, Colombia, and Ecuador

In the northern Andes, the mountain chain splits into several parallel ranges that slice through Ecuador, Colombia, and Venezuela and create three distinct environmental zones: (1) tropical coastlines thinly populated by a mixed *mestizo* and African-descent population; (2) interior valleys and plateaus occupied by *mestizo* and indigenous peoples; and (3) eastern rain forests drained by the Orinoco and Amazon rivers, also thinly occupied by indigenous people.

These areas were the object of Simón Bolívar's early dream of a north Andean state to be called La Gran Colombia. Although they share common cultural and environmental characteristics, Venezuela, Colombia, and Ecuador have had different patterns of development.

Most of Venezuela's population of 20.7 million is located in the country's richest agricultural zone, the central highlands in the interior. Here plantations of cacao, sugarcane, cotton, and coffee support a large *mestizo* and European population. Caracas, the capital, with a population of 3.4 million people, is located at an elevation of 3,000 feet above sea level. The birthplace of Bolívar, it is one of the continent's leading centers of education.

sugarcane and subsistence cultivation of food crops.

At elevations above 6,000 feet, the cool uplands of the *tierra fria* are used for subsistence cultivation of grain and potatoes and, in places, for mining. Above 10,000 feet, just below the permanent snow of the *tierra helada*, alpine pastures predominate. The elevations and temperatures in the Andes reflect a widespread pattern of human habitation in the highlands of Latin America.

In addition to shaping the agricultural economy of the highlands, altitude in the Andes has a direct physical impact on the human body. Much of the population of the countries of the northern Andes lives at elevations of 1 to 2 miles above sea level. In Peru and Bolivia, large numbers live in intermontane valleys and on the *altiplano* (or high plateau) at elevations of between 10,000 to 13,000 feet. One of the largest mining centers of the region, Cerro de Pasco in Peru, is located at an altitude of 13,800 feet.

At about 10,000 feet, lower levels of oxygen and lower atmospheric pressure begin to have direct effects on animal and human biology. During the Spanish conquest, the movement of coastal people to mines at high elevations caused many deaths as well as long-term infertility. Similarly, highlanders who were accustomed to living at higher elevations frequently died and still do die at lower elevations from respiratory diseases.

Ultimately, the Spanish abandoned their plans to locate their capital at an elevation above 10,000 feet, possibly because European livestock became infertile, as did those Spaniards who ultimately settled in the region to manage the silver mines of the Incas. In any case, ethnic differences between the politically powerful Spaniards who live on the west coast of South America and the neglected Indians of the Andean highlands are a source of political tension and friction that have contributed to civil wars and revolution.

QUESTIONS

1. In what two ways does elevation generally affect climate?
2. What are the most important crops of the *tierra templada* plantations?
3. In what ways does elevation affect the workings of animal and human biology?

Although agriculture remains the principal economic activity, Venezuela's energy wealth makes it one of the richest countries in South America. With proven oil reserves of 60 billion barrels, Venezuela has the largest reserves of any country outside the Middle East, with the possible exception of Russia and Indonesia. It is also the second-largest supplier of oil to the United States, next to Saudi Arabia.

In the middle 1990s, Venezuela has begun to collaborate with foreign companies to tap residual oil reserves in the 6,500 oil wells that dot the 30,000-square-mile Maracaibo Basin. Efforts are also being made to develop the technology to extract oil from the vast tar deposits along the Orinoco River.

Oil capital in the 1970s and 1980s made it possible for Venezuela to launch an ambitious program of industrial diversification in the *llanos* (grasslands) of the Orinoco Basin. In the *llanos*, large deposits of

▼ *When oil prices on the world market dropped in the 1980s, the development plans of Venezuela suffered. As a result many people are poor and out of work, and they live in large slum areas. How do these living conditions compare with conditions before the arrival of Europeans?*

▲ *Thousands of oil wells have been drilled in the Maracaibo Basin of Venezuela. Lake Maracaibo is located at the northern tip of South America. How did Venezuela use the money earned from oil exports to benefit the population?*

iron at El Pao and Cerro Bolívar north and east of Ciudad Bolívar in the Orinoco Basin are being developed, new oil fields are being brought into production, and a multipurpose dam is being constructed on the Guárico River. Most of these projects have proved disappointing; moreover, they have left Venezuela deeply in debt.

Today, less than 5 percent of Venezuela's labor force is engaged in industry and its population in the 1990s is growing rapidly. But the Venezuelan economy has depended so heavily on oil for revenues that oil's large drop in market price is now causing considerable financial tension. Because the country's standard of living depends more on oil than on its productive capacity, Venezuela is facing low economic growth, lower standards of living, high levels of inflation and unemployment, and,

with the two coups it experienced in 1992, the real threat of political instability.

On the Guiana Plateau east of Venezuela the three small countries of Guyana, Suriname, and French Guiana raise sugar, rice, and coffee and produce bauxite. Their economies are static, their people quite poor.

In Colombia, the largest country in the northern Andes, the mountains divide into three separate chains that finger northward toward Panama and Venezuela and are separated by the valleys of the Cauca and Magdalena Rivers. Most of Colombia's 34.9 million people are scattered in fourteen separate valleys and basins dotting the highlands. Bogotá, the capital, with a population of 6 million people, is located at an elevation of 8,500 feet in the easternmost mountain range. Gold initially lured the Spaniards to Colombia, but cotton, sugarcane, cocoa, cattle, and coffee quickly became the country's most important economic products.

Two major regions of Colombia have never attracted significant populations: the forested slopes of the Pacific coast and the grasslands east of the Andes (the *llanos orientales*), which constitute 60 percent of the total area of Colombia. On the Caribbean coast, a stock-raising area, people are concentrated near large port cities, like Barranquilla and Cartagena, which are export centers for agricultural products grown in the highlands.

Coca, from which cocaine is derived, has emerged as a high-value crop that is replacing food-producing agriculture. Sent to the United States, the value of cocaine far exceeds anything else that highland peasants can grow. Often grown in Peru and Bolivia, and processed in Colombia, the revenues have enabled drug cartels to conduct terrorist warfare against the government to maintain their control of illegal drug trafficking.

In 1993, however, oil began to flow from a large field located about 100 miles east of Bogotá, providing the country with its largest economic bonanza ever. In five years, Colombia may be exporting a half million barrels a day. One uncertainty is the decade-long war between government forces and guerrilla forces who operate in this remote region. Whether the government of Colombia can use its new capital to regain control of its territory and build a stable economy is uncertain.

Ecuador is clearly divided into three separate environmental and cultural regions. On the narrow Ecuadorian coast lives a small population of native Indians, *mestizos*, and *mulattos* who produce virtually all the agricultural and mineral wealth of the country. Cacao, once the principal crop of the coast, has been replaced by large-scale production of rice and banana cultivation. All three products are exported through Guayaquil, Ecuador's largest city and most important port, with a population of 2 million.

In the Andean highlands, where most of Ecuador's 10.3 million people live, the population is predominantly native Indian. Here, population pressure is intense, and subsistence cultivation of corn, barley, wheat, and potatoes is the basis of agriculture. On the eastern slopes of the mountains, the vast tropical forests of the Amazon headwaters, where the Amazon River begins, are the homelands of the Jivaro Indian tribes. Their previously undisturbed existence has been disrupted by the discovery of oil in this region, one of the most remote

> **IT'S A FACT** | **Panama hats are made in Ecuador.**

parts of South America. In the 1990s, a considerable amount of oil is being shipped by a trans-Andean pipeline to the coastal port of Esmeraldas. Ecuador is now the second-largest producer of crude oil in South America.

The Mountain Republics of Peru and Bolivia

In Peru and Bolivia, the Andes commonly reach elevations above 20,000 feet. Two parallel ranges flank a broad, flat tableland at 10,000 to 14,000 feet, the *altiplano*, which is shared by Peru and Bolivia. The Peruvian *altiplano* is studded by jagged mountain blocks. In Bolivia, the *altiplano* forms a broad basin of interior drainage.

▼ *Ecuador's Chimborazo Volcano, with an altitude of over 20,000 feet, is part of the Ring of Fire, the volcanic chain that encircles the Pacific Ocean. What possible consequences would an eruption of this volcano have for the land and people of Ecuador?*

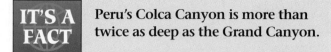

IT'S A FACT Peru's Colca Canyon is more than twice as deep as the Grand Canyon.

This upland plateau was the core of Inca civilization in Peru and Bolivia. Here, subsistence agriculture supported a relatively dense and prosperous pre-Columbian New World civilization. The dry, dune-covered coastal zone of Peru was also well developed by pre-Columbian peoples, whereas the Amazonian forests to the east were thinly settled.

The Spanish did not, however, establish their capital in the heavily populated uplands of Peru as they had in Mexico, Colombia, and Ecuador. Instead they founded Lima (which now has a population of 7 million) near the Pacific coast as a commercial port to link the wealth of the Andes to Europe. The social and economic gulf that developed between the *mestizo* population of the coast and the isolated native Indian population of the highlands exploded into guerrilla warfare in the 1980s and 1990s.

Modern Peruvian economic development has taken place in oases along its 1,500 miles of desert coast. European-owned sugar and cotton plantations are located in the fertile valleys with streams that cascade down from the Andean highlands through this desert. Roughly one-third of Peru's population, its richest agricultural land, and most of the country's manufacturing, fishing, and petroleum production, is found along the coast.

Until recently, a small European elite owned 90 percent of the cultivated land and the factories in which *mestizo* and *mulatto* populations labored. But native Indians have migrated toward coastal opportunities in large numbers in the last generation. Although land reform is only being discussed, the coast continues to be the economic heartland of Peru, and efforts at industrial diversification and agricultural expansion still focus on the capital city of Lima.

By contrast, the Andean highlands, where most of Peru's 22.9 million people live, and the Amazon forests beyond are virtually undeveloped—even unknown. Quechua-speaking Indians, the descen-

▼ *Bolivia, a landlocked country, shares Lake Titicaca, the world's highest navigable lake, with its neighbor Peru. Islands in this lake hold the ruins of what civilization?*

Iquitos, Peru, on the Amazon River is farther inland than any large port in the world.

dants of the Inca, cultivate terraced fields of wheat, barley, and potatoes in mountain valleys and on high plateaus; they also tend herds of sheep, llamas, and alpacas in mountain pastures.

Three areas of dense settlement are found in the highlands: (1) around Lake Titicaca, where the moderating climatic influence of the lake makes corn cultivation possible at high elevations; (2) near Cuzco, the old capital of the Inca Empire; and (3) at the mining complex of Cerro de Pasco, inland from Lima.

Desperately poor and short of land (only 2 percent of Peru is under cultivation), the people of the mountains—and particularly the native Indians—live outside the mainstream of Peruvian life.

This isolation has triggered one of the most persistent and bloody civil wars in South America. In the highlands, the Shining Path, a group of communist guerrillas, has won considerable sympathy from people who are alienated from the government on the coast. Poverty is extreme, and schools, hospitals, or any social service—except those provided by the guerrillas—nonexistent. Racial and cultural differences between the modern Peruvians of the coast and the native Indians of the highlands make the problem even more difficult.

In Bolivia, racial differences are worsened by an extremely complex natural environment. Trapped between the Atacama coastal desert and the Amazon rain forests, most of Bolivia's 8 million people live in small communities on the *altiplano*, in the gorges of streams flowing eastward to the Amazon, and in narrow mountain valleys. Over one-third of Bolivia is one mile or more high, and much of the rest is either rain forest or semiarid steppe. Until recently, Bolivia was a **landlocked** country, having lost its passage to the sea to Chile in 1879. In 1993, Peru agreed to give Bolivia a free trade zone in its port city of Ilo, providing better access for the free-flow of people, goods, and ideas.

The people of Bolivia, two-thirds of whom speak indigenous languages, are divided by race, language,

and local economy. The most significant concentration of population is found in the basin of Lake Titicaca, near the capital city of La Paz, which has a population of 1.3 million. Lake Titicaca is the highest large lake in the world, and La Paz is the highest capital at more than 12,000 feet in elevation.

Elsewhere, settlement is limited by climate and terrain, and primary settlement nodes are determined by the location of Boliva's mining economy (tin and other metals), which provides 80 percent of the country's exports. Petroleum production and some commercial agriculture has begun on the eastern slopes of the Andes. A program of land reform has broken the grip of landlords on farmland. But the limited natural resource base of Bolivia makes it one of the poorest nations on the continent. As in Colombia and Peru, poor farmers have increasingly turned to cultivation of coca, which is transformed into coca paste (from the leaves of the shrub) and then into cocaine.

Elongated Chile

The Andes Ranges converge in the south to form a narrow mountain spine that sweeps 2,500 miles toward the South Pole. This outlines the long, narrow Pacific republic of Chile. Despite the steep rise of the Andes from the sea, climate rather than terrain has influenced human settlement in Chile.

In northern Chile, the Atacama Desert extends 600 miles along the coast. Some agriculture and herding are found in the southernmost part of this coastal desert. But inland from the coast, rich nitrate deposits and copper ore mines are the vital contributions of the Chilean north to the national economy. Elsewhere, the barren landscapes of the north are virtually uninhabited.

The southern extreme of Chile, from the Bío-Bío River to the tip of Tierra del Fuego, is also thinly populated. It is a region with a cool, damp environment of forests and fjords. Until the last century, the Chilean south was physically and culturally outside the society and economy of the nation.

Punta Arenas, Chile, is the southernmost city in the world.

Today, forestry on the Pacific coast, sheep raising in high mountain valleys, and petroleum discoveries in Tierra del Fuego have integrated this region into the national economy.

The heartland of Chile lies between the northern deserts and the southern forests. Between these two regions lies the Central Valley, where some 65 percent of the nation's 13.5 million people live. With a Mediterranean climate and rich alluvial soils, the Central Valley is the most favored agricultural region of Chile. Centered on the capital, Santiago (population of 5.9 million), land in the Central Valley is held in large *haciendas,* called **fundos** in Chile, on which wheat and beef cattle are raised.

The Chilean population, unable to expand into a countryside of fenced ranchland, has been forced to migrate to industrial centers. Four-fifths of Chile's population is urban. After considerable political turmoil, the great estates have been confiscated, but national development plans aimed at the redistribution of agricultural land are still uncertain. Meanwhile, despite its fertile farmland, Chile still imports food. However, some agricultural products, such as fruit and wine, are now being shipped north during the Northern Hemisphere's winter months to provide fresh fruit "out of season." The political instability of the last two decades has undermined progress, but in the 1990s new efforts to lift people out of poverty are being initiated.

⊕ REVIEW QUESTIONS

1. How long and wide are the Andes Mountains? Where are they narrowest?
2. How has elevation influenced human settlement in the Andes?
3. Where is the Atacama Desert? Why is there a desert so close to the equator?

⊕ THOUGHT QUESTIONS

1. In what Andean countries has political turmoil created difficulties? Can you suggest any causes for this turmoil?
2. Which Andean countries rely on oil for much of their economic wealth? Has this reliance on oil as the main product been of benefit to these countries or not?

2. The Middle-Latitude South

The three temperate countries of Argentina, Uruguay, and Paraguay are part of a middle-latitude prairie that stretches from the foothills of the Andes Mountains eastward to the coast of the Atlantic Ocean. This plain is drained by the Paraná-Paraguay-Uruguay river system that flows into the estuary of the Río de la Plata, the region's outlet to

Middle-Latitude South

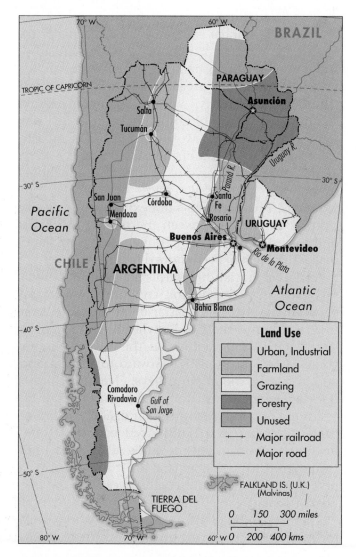

Land Use
- Urban, Industrial
- Farmland
- Grazing
- Forestry
- Unused
- Major railroad
- Major road

the world. Together these countries have a combined population of nearly 41 million.

Throughout most of the colonial period, these grasslands were occupied by indigenous hunting tribes, and the region was a backwater of the Spanish Empire. Thus separated from the rest of Spanish America, the middle-latitude South was peopled by immigrants from Europe during the nineteenth and twentieth centuries.

Argentina in Development

In Argentina, early colonization of native Indians, strong European immigration, and the economic development of the fertile grasslands of the *pampas* have created a unique land and people reminiscent of the American development of the Great Plains. Development, however, has been limited by a series of military dictatorships, a war with Britain over the Falkland (Malvinas) Islands in 1982, and a faltering economy.

Until just over 100 years ago, this region of temperate climate and rich loess soil supported a small population. After the 1880s, 7 million Europeans—nearly one-half of them Italians and one-third of them Spaniards—immigrated to Argentina. Native Indians were driven from the *pampas*, and cattle ranches were founded to feed the expanding populations of industrial Europe. European grasses, windmills, barbed-wire fencing, and six-shooters were vital to the settlement of this South American prairie. The range was parceled and fenced off, and wheat was sown. The same struggle between rancher and farmer that occurred earlier in the North American Great Plains occurred here. Soon a dense network of railroads—the best on the continent—connected the *pampas* with the now bustling port city of Buenos Aires.

The *pampas* are now the core of Argentina. Two-thirds of Argentina's population of 33.5 million people live on this grassland, which produces 80 percent of the nation's exports. Buenos Aires, with a population near 12 million, is a **primate city** which is much larger than any other city in the country. No other Argentine urban center approaches its size even though four-fifths of Argentina's population is urban. Market gardens are found nearest the city; intensive farming of wheat, fruit, corn, and flax farther out; and a purely cattle-ranching economy on the edge of the *pampas*.

◄

Argentina's capital city of Buenos Aires hosts a weekly flea market. Does your area have a swap meet or flea market? What are some advantages and disadvantages of such a market?

The three less-developed regions of Argentina—the scrub forests of the Chaco in the north, the Andean foothills in the west, and windswept Patagonia to the south—now contribute to the national economy. The Chaco is now a pioneer region where logging and cattle raising are supplemented by cotton, sugar, and tobacco. In the foothills of the Argentine Andes, sugar, grapes, and other fruit crops are grown. On the barren tablelands of Patagonia, which stretch 1,000 miles south from the *pampas* to the tip of the continent, population growth has been limited by political enmity between Chile and Argentina and by environmental constraints. In the middle 1990s, Chile and Argentina initiated talks to resolve territorial disputes in this remote region.

The La Plata Countries of Paraguay and Uruguay

The two smaller La Plata countries, Uruguay and Paraguay, are very different in environment, population, and social and economic development. Uruguay, with its mild climate, low-rolling terrain, and rich grasslands, is a buffer zone between the two large and powerful nations of Brazil and Argentina. Originally settled by the Portuguese, then taken over by Spain, Uruguay revolted against both countries and eventually became independent in 1825.

The introduction of sheep and immigration from Spain and Italy are keys to Uruguay's modern development. The country's 3.2 million people are mostly of European descent. Its economy is deeply engaged in animal husbandry. In fact, sheep and cattle outnumber people by ten to one in Uruguay, and about 70 percent of the territory of the country is in pasture. These grasslands produce the wool, hides, and meat that Uruguay exports.

In Uruguay, the expected social pattern of a wealthy landowning elite and poverty-stricken working class did not evolve. The large *haciendas* were complemented by many medium-sized and small farms on which wheat, flax, wine, and vegetable production has doubled in recent years. Enlightened government policies have provided social welfare to the poor. The Uruguayans, half of whom live in the primate city of Montevideo (population of 1.6 million), have the highest literacy rate, the lowest rate of natural increase, the best diet, and one of the highest standards of living of any South American country.

In Paraguay, society and economy followed a quite different course. The eastern third of Paraguay, with its rich soils, luxuriant grasslands, and gentle terrain, was settled by the Spanish. The western two-thirds of the country, the wilderness scrub forest known as the Chaco, was brought into the Spanish domain by Roman Catholic missionaries.

In the 1800s and 1900s a series of disastrous wars and revolutions devastated Paraguay, destroying the economy of the country. After the worst of these, the five-year War of the Triple Alliance against Brazil, Argentina, and Uruguay in the 1860s, Paraguay's population was cut in half. Experts estimate that only 28,000 adult males were left alive, and Paraguay had lost 55,000 square miles of territory. The country has never recovered from this catastrophe.

Currently, the only productive agricultural zone in Paraguay is located near the capital city of Asunción where cotton, tobacco, and market gardening are important. Throughout the remainder of Paraguay, extensive cattle ranches and slash-and-burn agriculture of cotton, corn, manioc, and beans are found. **Slash-and-burn agriculture** is the practice of clearing and burning trees off the land which is then cultivated for a short period of time. This meager economy barely supports a Paraguayan population of 4.2 million that is increasing very rapidly.

⊕ REVIEW QUESTIONS

1. What major river system drains the middle-latitude South?
2. Describe the *pampas* and how they are used.
3. What is slash-and-burn agriculture?

⊕ THOUGHT QUESTIONS

1. In what ways was the European settlement of Argentina similar to that of the American West? What factors have limited development in Argentina?
2. How is Uruguay's society different from that in other South American countries?

3. Brazil

IT'S A FACT   Brazil shares a common border with every South American country except Chile and Ecuador.

A Portuguese Domain

Discovered by accident in 1500 by the Portuguese explorer Pedro Cabral, Brazil is the fifth-largest country in the world and the largest in Latin America. Roughly 10 percent smaller than the United States, Brazil reaches 2,300 miles north-south and 2,300 miles from the Atlantic coast to its western boundaries. Largely because of Spanish indifference to the inner core of South America, Brazil became a Portuguese empire in the New World. As the only Portuguese-speaking country in the Western Hemisphere, Brazil is culturally isolated, and for the most part, Brazilians have been largely concerned with domestic affairs rather than foreign affairs.

Throughout most of Brazil's history, its people have been concentrated along the east coast. Today, Brazil has a population of 152 million that increases by nearly 3 million a year. United States and Euro-

pean investment is developing heavy industry. The Brazilian stock market is booming; exports, such as orange juice, are doing well worldwide. However, the pressure to expand into the interior has become the dominant theme in modern Brazilian life recently. A true twentieth-century frontier is central to the geography of modern Brazil.

In all of Latin America, Brazil is unequaled for the diversity of its environments and resources. Five basic regions are found in Brazil: (1) the old Northeast on the country's Atlantic "shoulder"; (2) the East, focused on the throbbing industrial heartland of São Paulo and the traditional capital, Rio de Janeiro; (3) the South; (4) the wilderness of the Central West, brought into the modern age by the construction of the new capital of Brasília; and (5) Amazônia, the world's largest drainage basin

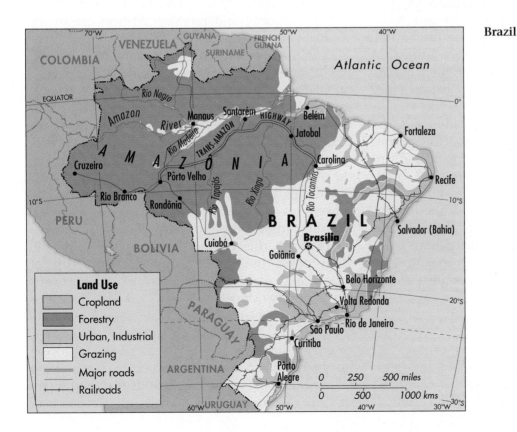

Brazil

and rain forest, a region now threatened by increasing human activity.

The Northeast

The old Northeast, surrounding the shoulder of Brazil, is the culture hearth of Portuguese America. It was here that European colonists with the forced labor of African slaves first established permanent settlements. The warm, rainy coasts of the Northeast are well suited to sugarcane cultivation, and the Northeast still produces one-third of Brazil's agricultural harvest. In some places, the fertile red soils of the coastal lowlands have been producing cane and cacao for 400 years. Recently, cotton has been introduced in the drier coastal areas and in the fringes of the upland interior.

The bulk of the Northeast's population of 44 million lives east of a line drawn across the shoulder from the northerly city of Fortaleza (population of 2 million) to the southerly city of Salvador (population of 2.3 million). Recife, long the most important settlement in the region, has a population of 2.5 million today.

Away from the coast, rainfall diminishes sharply and becomes erratic. Cycles of drought and flood characterize the interior and shape the lives of its people. The culture of Brazil was largely formed in the backcountry of the Northeast, in what Brazilians call their *sertao*. The *sertao* is a barren land covered by dryland grasses and thorny scrub.

The coast and the *sertao* are quite different. The cyclic nature of drought, worsened by deforestation in the *sertao*, makes this region overpopulated. In the coastal lowlands, the persistence of rigid social barriers, absentee landlords, and exhaustion of the soil have encouraged migration out of the Northeast to the more dynamic regions of contemporary Brazil. More than 3.5 million Brazilians have emigrated from the Northeast in the last generation to escape one of the most poverty-stricken areas in the hemisphere.

The East

Eastern Brazil makes up one-tenth of the land area of Brazil, but has nearly one-half the country's population. This is the most mature and diversified industrial region in Latin America. Four of every ten Brazilians live in densely settled countryside around the nation's two largest cities, São Paulo (population of 19 million) and Rio de Janeiro (population of 12 million).

In the 1880s, coffee estates (*fazendas*) were established on patches of fertile red soil in the hinterlands of São Paulo and Rio de Janeiro. These landholdings drew a stream of 2 million European immigrants, principally Italians, Portuguese, Spanish, and later Japanese into this region to labor on the coffee estates and on sugar, rice, and cotton farms along the coast and in the river valleys of the interior.

Shortly thereafter, in 1910, the discovery of rich mineral deposits north of Rio de Janeiro led to a flood of international investment in the region. However, Brazilians thought the resources should benefit Brazil directly, and development was delayed for a generation. Finally, to exploit the rich iron, manganese, and tungsten ores of the region, an integrated steel complex was constructed at Volta Redonda near Rio in the 1940s. Many others have been constructed since. The world's largest reserves of iron ore are found in Brazil, which is both the top exporter and producer of iron ore in the world.

Despite the creation of a new capital at Brasília in the 1960s, the coast remains the center of Brazil's population. São Paulo produces 40 percent and Rio 20 percent of all Brazilian manufactures. Very different cities from each other, São Paulo is a vast maze of skyscrapers, industrial plants, and residential suburbs spread over a large area. Rio de Janeiro, by contrast, is cramped between the sea and coastal mountains, and so it grows uphill, with the city broken up into different urban areas. Although stunning from a distance, Rio's beauty diminishes up close. In the middle 1990s, tropical rains still wash down the hillsides and bring torrents of shacks and garbage in their wake. Shantytowns cloak the city's slopes.

A basic problem for Brazil has been energy resources. Until recently, no oil had been located anywhere in Brazil. In the 1980s, the discovery of oil off the Northeast coast was encouraging, but in the 1990s, Brazil still imports half of its fuel, and its future plans include piping Bolivian oil thousands of miles to the industrial cities on the coast. Gasohol, a mixture of gasoline and alcohol used as a response to costly imported gasoline, was made from a variety of agricultural crops grown specifically for this purpose. Gasohol now accounts for more than 70 percent of the fuel used in Brazilian vehicles.

To counter high petroleum costs in the 1970s, a massive program of dam building was initiated. In southeastern Brazil, the vast hydroelectric potential of the upper Rio Paraná has been harnessed for the rapid industrialization and urbanization of the region. Enormous hydroelectric projects are underway along the Rio São Francisco in Bahia State in the Northeast. The giant Itaipú Dam on the Paraná, a joint Brazilian-Paraguayan project costing $20 billion, opened in 1982 as the world's largest hydroelectric project. Still, less than 15 percent of Brazil's hydroelectric potential has been tapped.

The Brazilian South

The Brazilian South, which supports one-sixth of the country's population, experienced a different pattern of colonization than elsewhere in Brazil.

The first European settlers filtered southward from São Paulo in the 1700s and 1800s and established a cattle- and sheep-raising economy like that in neighboring Uruguay. At the same time, gold seekers spread southward along the Atlantic coast and established fortified outposts along the coast.

In the 1800s and 1900s, these coastal centers became destinations for a wave of immigrants from Germany, Italy, and Eastern Europe. The first to penetrate the interior of the South were the Germans, who grew corn, rye, and potatoes, and raised pigs. Italian immigrants later extended the frontier, clearing the forests and planting vineyards deeper in the interior. Now extensive high-technology agriculture based on cultivation of wheat and soybeans for export is being introduced into the farmlands of the South. Brazil now earns as much from soybeans as from coffee.

As high-technology farming envelopes the Brazilian South, people with small farms are being driven out, and the South is beginning to bear the mark of a fundamental Brazilian problem—landlessness. At the same time as food crops such as beans and corn have been replaced by sugarcane for gasohol and by

◄

The Itaipu Hydroelectric Dam on the border of Brazil and Paraguay dams Rio Paraguay and produces much of the needed electricity for the area. What dam in the northwestern United States borders two states? What river is it on?

REGION

THE BOOM-BUST ECONOMY OF BRAZIL

Brazil has been historically adept at "harvesting the fruit without planting the trees," creating a boom-bust cycle that has prevented balanced economic growth in Latin America's largest country. Each of the "booms" brought great wealth to Brazil; most of it was lost in the "busts" that followed.

In the 1600s, a sugar boom led to the settlement of the Northeast; in the 1700s, the gold and mining boom in Minas Gerais State led to the establishment of Belo Horizonte. In the 1800s, rubber was harvested in the Amazon and coffee cultivation created wealth in the Paraíba Valley.

Today, the industrial complex centered in São Paulo and Rio de Janeiro is the focus of Brazilian life. The capital created by these series of booms was never invested in economic growth in Brazil, a major reason the country has not fully developed its resource base.

QUESTIONS

1. What four "bonus" products are described above?
2. What is meant by the term *capital* in the last paragraph?
3. What is meant by the phrase, "harvesting the fruit without planting the trees"?

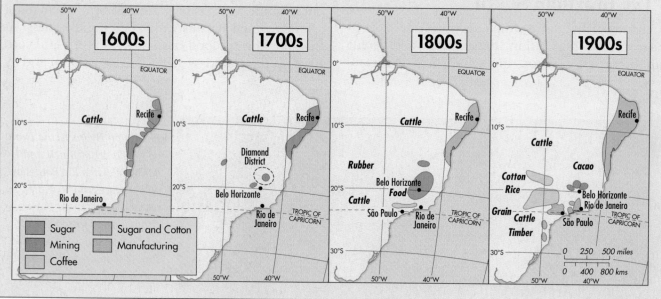

soybeans for export, Brazil still faces the fundamental problem of feeding its people in the 1990s.

The Central West

The Brazilian response to landlessness has been to extend the frontier inland. The "first front" of the Brazilian frontier in the middle 1960s extended into the Central West, whose surface geography is largely composed of exposed rock. Here, soils are poor, and laterites have been known to extend to a depth of ten feet. Vegetation tends to be a mix of savanna and scrub woodland known as the *campo cerrado*.

Despite these conditions, farming expanded throughout the region, but the mainstay of the Central West soon became livestock raising. The most dramatic event in the Central West was the implantation of a new capital in the wilderness of the *campo cerrado*, Brasília.

Symbol of the Brazilian desire (and need) to conquer the wilderness, Brasília was founded in 1959 as a planned, **forward capital**—an architectural design project intended to encourage settlement and to change the Brazilian perception of the interior. Despite its problems, Brasília has grown to a population of more than 2 million.

Amazônia

The vast Amazonian interior has been hailed as one of the world's last frontiers. The region's tropical environment stubbornly resisted all but the most determined efforts at permanent European settlement until recently. Settlement has traditionally been concentrated near the mouth of the Amazon River or far inland on patches of arable land directly along the Amazon and its many **tributaries**, the smaller streams that feed into the river system.

Belém (population of 1.2 million), gateway to the Amazon, is the focus for the development of the Eastern Amazon. Iron ore and hydropower are under development in the state of Pará. Inland, the 3,900-mile-long river passes by Manaus (population of 1.4 million), capital of Brazil's largest state, Amazonas. The Amazon continues another 2,900 miles beyond Manaus and is navigable as far inland as the frontier city of Iquitos, Peru (which has a population of 204,000).

The continuously hot, humid climate and infertile soils of the tropical rain forests that cover the Amazon basin kept it almost unoccupied by Europeans until modern times. Indigenous people collected rubber and Brazil nuts and practiced slash-and-burn agriculture. A brief rubber boom at the turn of this century was frustrated by inaccessibility, plant diseases, and labor shortages. The Ford (Fordlandia) and Goodyear rubber plantations are today in ruins, defeated in the end by the discovery of synthetic, petroleum-based rubber in the 1940s. The government's determination, however, that the empty heart of Brazil should no longer remain empty has led to the construction of a trans-Amazonian highway network linking the major regions of the Amazon.

It is south of the trans-Amazon Highway, however, where the bulk of current human development on the Brazilian frontier is taking place. With the construction of a paved highway heavily financed by the World Bank and the Inter-American Development Bank, lumber companies, gold miners, peasant settlers, and cattle ranchers have stampeded into this region. In the 1980s, immigrants arrived at a rate of 12,000 per year. In the 1990s, 12,000 arrive each month.

Poor people from the East and South are given a sum of money by city governments and sent west on a bus to colonize this region and reduce crowding in coastal cities. Few arrive with knowledge of the environment they are about to face in this diminishing wilderness. Indigenous groups retreating before the advancing line of settlement are in immediate danger of extermination.

▼ *This is a picture of the Amazon basin as seen from the space shuttle. What is the source of the smoke in the photograph?*

Vast areas of rain forest are being cleared. Developers are using wood to fuel new mining smelters. Forests are being cleared for farms and for pasture for cattle. However, farmland in this area of poor soils must soon be abandoned if fertilizer is not used. In addition, many decades are required before rain forests can reestablish themselves after being cut down. During the planting season, 7000 fires a day can be seen on satellite photographs of Amazônia. In the time it takes you to read this page, 400 acres of rain forest will disappear. Environmentalists are concerned that clearing the rain forests of Amazônia will cause erosion and might even change the climate of our planet.

Amazônia is Brazil's hope for the future. Mining and selling the minerals of Amazônia are seen as a way to pay off the country's soaring debts. Its farmland is needed to feed Brazil's growing population. Brazilians recognize the risk. As one popular saying goes, "Brazil is on the edge of an abyss but won't fall in because it is bigger than the abyss." Many geographers now disagree with this statement.

⊕ REVIEW QUESTIONS

1. What are the five regions of Brazil?
2. Describe the economy of the Northeast.
3. Which region in Brazil has the most people, the most industry, and the largest cities?

⊕ THOUGHT QUESTIONS

1. Do you think Brazil's plan to extend its frontier into the Amazon is wise? Why or why not?
2. Brazil imports half of its energy as well as some of its food. What does this tell you about the economy of the country?

 ## CHAPTER SUMMARY

Because of their geographic distribution and location, the countries of the South American continent have certain assets and liabilities. These countries can be grouped according to topography into three subregions: those of the Andes Mountains; those of the middle-latitude prairie; and Brazil, whose vast size comprises a single subregion of lowlands and plateaus.

In the northern Andes lie Colombia, Ecuador, and Venezuela. Venezuela has by far the richest reserves of natural resources, but both Venezuela and Colombia have oil deposits that promise a rich future, if they are well managed. Peru and Bolivia lie in the Andean highlands. Because Bolivia has been landlocked until recently, its population is concentrated in the high plateau known as the *altiplano*. Peru shares the *altiplano* with Bolivia, but its population and economic activities are concentrated on the Pacific coast. Chile, to the south, is also an Andean country. The heart of Chile's population lies in its Central Valley between the Atacama Desert in the north and the forests in the south, but economic activities take place in all three regions.

Argentina, Uruguay, and Paraguay occupy the middle-latitude prairie that stretches from the Andean foothills eastward to the Atlantic coast. The fertile *pampas* is the core of Argentina's population and economic activities, with areas to the north (Chaco) and south (Patagonia) as smaller economic centers. Uruguay's chief economic activity is animal husbandry, primarily sheep and cattle raising. Paraguay, having suffered devastating wars in the last century, is the poorest country in the region.

Brazil is the most diverse and complex region in Latin America. It has five broad subregions: the Northeast, the East, the South, the Central West, and Amazônia, which is the world's largest drainage basin and rain forest. While the Northeast is the culture hearth, each region is characterized by economic activities that make it unique. Brazil's efforts to develop its frontier in the interior have included building roads and encouraging settlement and exploitation of the resources of the rain forest region. (This was not so very different from the exploitation of the mid-latitude forests of the United States in the last century.) Because the rain forests are home to the greatest diversity of biological species in the world and because the burning of the rain forests may change the climate of the region and perhaps the world, environmentalists are concerned.

EXERCISES

Record Setters

Directions: Write on your paper the answers to the following questions.

What is the . . .

1. highest capital city in the world?
2. largest country in Latin America?
3. oldest capital city in South America?
4. largest tropical rain forest on Earth?
5. longest mountain chain in the world?
6. largest oil producer in South America?
7. world's largest producer of coffee?
8. highest large lake in the world?

Which Country?

Directions: Write on your paper the answers to the following questions.

Which South American country . . .

1. has a common border with every country in South America except Chile and Ecuador?
2. has a population that lives mainly on the *altiplano*?
3. has the mineral-rich valley of the Orinoco River?
4. shares the Itaipú Dam with Brazil?
5. has ten times more sheep and cattle than people?
6. has the nitrate-producing Atacama Desert?
7. exports most of its products through the port city of Guayaquil?
8. has more than 20 million people who live on the *pampas*?

Which Direction?

Directions: Fill in the missing words by writing the complete sentences on your paper.

1. Most of South America lies south and |||||||||| of the United States.
2. Middle-latitude regions are |||||||||| and |||||||||| of the equator.

3. The Andes Mountains stretch in a |||||||||| direction from Venezuela to Chile.
4. Iquitos, Peru, is a port that lies |||||||||| and |||||||||| of Rio de Janeiro, Brazil.
5. Buenos Aires, Argentina, lies |||||||||| and |||||||||| of São Paulo, Brazil.

Inquiry

Directions: Combine the information in this chapter with your own ideas to answer these questions.

1. In what ways is Brazil the most important country in Latin America?
2. In many ways, Venezuela is well situated economically with one of the richest natural resource bases in Latin America. Its oil reserves make its economy equal to many oil-producing countries of the industrialized world. Yet oil has also created many problems for Venezuela's economy. Describe some of the problems Venezuela has had developing its economy. If you were in charge of Venezuela's economy, what would you do to develop the country?
3. In what ways is Argentina a unique country? As you answer this question, think about the timing of the immigration of large numbers of German, Italian, and Spanish Europeans into Argentina in the 1900s. How would their impact differ from that of the Spanish Europeans who colonized other Latin American countries much earlier in the 1500s?
4. Compare the colonization of the La Plata countries, Uruguay and Paraguay. In what ways did their history differ from that of other countries to the north and east of the Rio de la Plata?
5. How have the locations of South American countries in relation to one another influenced the rate and timing of colonization? Have any of these factors had a lasting influence that is still felt today?

SKILLS

Using Elevation Diagrams

Directions: Use the diagram and text on pages 446 and 447 to fill in the following numbered blanks. Write the correct words on a separate piece of paper.

The diagram on page 446 shows that the zone of major settlement in the Andes is at an elevation that extends from about (1) |||||||||| feet above sea level to about (2) |||||||||| . In this diagram, the mean temperatures that are shown range from a high of about (3) |||||||||| degrees to a low of about (4) |||||||||| degrees.

Land above 10,000 feet is too cold for most crops, but some of it is used for (5) |||||||||| . Sugarcane and cacao are usually raised in regions that have (high or low) (6) |||||||||| elevations and (high or low) (7) |||||||||| temperatures. Coffee is generally raised at elevations that are (higher or lower) (8) |||||||||| than elevations where tropical fruits are raised. Grains and potatoes are generally raised at elevations that are (higher or lower) (9) |||||||||| than elevations where coffee is raised.

Vocabulary Skills

Directions: Match the numbered definitions with the vocabulary terms. Write each vocabulary term next to the number of its matching definition.

altiplano
campo cerrado
culture hearth
estuary
fazenda
forward capital
fundo
headwaters
landlocked

llanos orientales
primate city
sertao
slash-and-burn agriculture
tierra caliente
tierra fria
tierra helada
tierra templada
tributary

1. A city much larger than any other city in the country
2. The cool uplands of the Andes Mountains above elevations of 6,000 feet where grain and potatoes are cultivated
3. A large estate on which wheat and beef cattle are raised in Chile
4. The practice of clearing and burning trees in order to cultivate the land for a short period of time
5. A river or stream that flows into a larger river or stream
6. The area where a culture or civilization originates
7. Vegetation mix of savanna and scrub woodland that grows in the poor soils of the Central West region of Brazil
8. The area of the Andes Mountains above 10,000 feet where alpine pastures predominate
9. An area of barren land covered by dryland grasses and thorny scrub in the backcounty of northeast Brazil
10. A mountain-ringed high plateau in the Andes Mountains of Bolivia and Peru
11. Where a river begins
12. The warm foothills of the Andes Mountains; areas of dense population and commercial cultivation
13. The tidal mouth of a flooded river valley
14. An area surrounded by land with no direct access to the sea
15. Coffee estates established on patches of fertile soil in the hinterlands of São Paulo and Rio de Janeiro
16. The coastal lowlands surrounding the Andes Mountains
17. The sparsely populated grasslands located east of the Andes Mountains in Venezuela and Colombia
18. A planned architectural design project intended to encourage settlement and conquer the wilderness of the interior of Brazil

Human Geography

Directions: Three possible endings are given for each of the following sentences. Select the correct ending and then write the complete sentence on your paper.

1. The first high Indian civilizations in Latin America began (a) before the European settlers arrived (b) about the same time that the European settlers arrived (c) after the European settlers arrived.
2. The Incas developed a high civilization in (a) Mexico (b) Central America (c) South America.
3. Mexico City was built on the site of the former capital city of the (a) Mayas (b) Aztecs (c) Incas.
4. Spain and Portugal ruled Latin America for about (a) fifty years (b) one hundred years (c) three hundred years.
5. The *Creole* who helped liberate five South American countries was (a) Las Casas (b) Lisboa (c) Bolívar.
6. Most of the people in Mexico and Central America are (a) *mestizos* and native Indians (b) Europeans (c) Asians.
7. One of the largest groups of Latin Americans are descended from immigrants from (a) the United States (b) Great Britain (c) Africa.
8. In recent years the population of Latin America has been (a) declining (b) growing slightly (c) growing very quickly.
9. Most Latin Americans live (a) on terraced slopes in the Andes (b) near the coast (c) in the interior regions.
10. Latin Americans who are also citizens of the United States live in (a) Jamaica (b) Puerto Rico (c) Costa Rica.
11. The largest city in Latin America is (a) Rio de Janeiro (b) Buenos Aires (c) Mexico City.
12. The Latin American country with the largest population is (a) Brazil (b) Mexico (c) Argentina.
13. A country in Latin America that is controlled by communists is (a) Uruguay (b) Cuba (c) Venezuela.
14. Most Latin Americans are (a) Muslims (b) Catholics (c) Protestants.

Physical Geography

Directions: Tell whether each of the following statements is true or false.

1. Latin America is about the same size as the United States.
2. The Andes Mountains are near the eastern coast of South America.
3. About two-thirds of Latin America lies in the tropics.
4. The Amazon River is the longest river in Latin America.
5. The regions with the highest elevation in Mexico and Central America are along the coast.
6. The Atacama Desert is shared by Colombia and Venezuela.
7. The *pampas* is in Argentina.
8. The Orinoco River is the chief river of Colombia.
9. Argentina, Uruguay, and Paraguay are part of a middle-latitude grassland.
10. Quito is farther from the equator than Lima.
11. Cuba is the largest island in the Caribbean Sea.
12. South America's only landlocked country is Paraguay.

Writing Skills Activities

Directions: Answer these questions in essays of several paragraphs each. Remember to include a topic sentence and several supporting sentences in each paragraph.

1. How do we know that the Mayas, Aztecs, and Incas had highly developed civilizations?
2. Why are settlements of people unevenly distributed in Latin America?

UNIT
8

AFRICA SOUTH OF THE SAHARA

▲ *These sand dunes in Namibia are not part of the Sahara. What other large deserts are found on the African continent?*

UNIT INTRODUCTION The forty-nine countries of Africa south of the Sahara Desert are home to one-tenth of the world's population. Most of these countries were controlled by European colonial powers for much of this century, and did not gain independence until the 1960s. For the last thirty-five years, their leaders have attempted to build strong economies and to raise the standards of living of their people. With few exceptions, they have failed.

The people of Africa south of the Sahara are among the poorest in the world. Their lives are spent in a struggle for survival. Most are farmers and herders whose environments are deteriorating because of forest cutting, population pressure on the land, and the southward spread of desert-like conditions from the Sahara. The few areas or "islands" of modern development in Africa are burdened by boundaries that unnaturally divide ethnic groups, poor transportation networks, and raw-material-producing economies created during the colonial period. Food production is declining, population is growing, political turmoil is widespread, and many countries are bankrupt. Unless large-scale technical and financial assistance is given to this continent by the world's wealthier nations, Africa's decline will not be reversed.

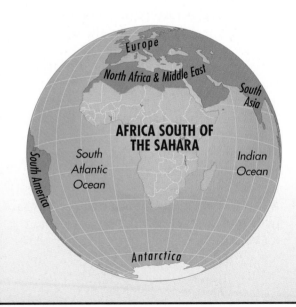

Most of the people in Africa are herders or farmers, like this woman fertilizing maize in Zimbabwe. Why is food production in Africa declining?

UNIT OBJECTIVES

When you have completed this unit, you will be able to:

1. Explain how the landforms, climates, and vegetation of sub-Saharan Africa have influenced the lives of the African people.

2. Explain why millions of Africans have a low standard of living, suffer from disease, and die young.

3. Identify the remarkable achievements of African kingdoms that existed many centuries before European settlers arrived in these lands.

4. Describe the effects of European colonization on the people of Africa.

5. State that although most African countries became independent in the 1960s, independence alone did not solve the most pressing problems of the African people.

This village sits peacefully amid the African savanna in Tanzania. What European country had a colony here, and what was it called?

SKILLS HIGHLIGHTED

- Using the Physical Map of Sub-Saharan Africa
- Using the *Readers' Guide*
- Deciding Whether to Use the *Readers' Guide* or the Library Catalog
- Understanding the Map of African Economic Migration
- Vocabulary Skills
- Writing Skills

KEYS TO KNOWING AFRICA SOUTH OF THE SAHARA

1

The continent of Africa is nearly three times the size of the United States and has about twice as many people.

2

Africa is a huge plateau fringed by narrow coastal plains. The highlands of East Africa are cut by the 4,000-mile-long Rift Valley.

3

Most of the climates of Africa south of the Sahara are tropical or subtropical. Rain forests, savannas, steppes, and deserts cover large areas.

4

Four of every five Africans are agriculturalists, shifting cultivators, settled farmers, or herders who live close to the land.

5

Kingdoms and empires have existed in Africa in and south of the Sahara region for thousands of years.

6

European powers divided Africa into colonies and ruled the continent for seventy years (1890–1960).

7

At independence, the new states of Africa south of the Sahara were burdened with colonial boundaries, transportation networks, and dual economies.

8

Islands of modern development in Africa south of the Sahara are found along the coasts of West Africa and South Africa and in mining and commercial farming areas in the interior. Migration to these islands of development for jobs is intense.

9

Most African countries export low-priced raw materials like minerals and commercial crops and import high-priced food, machinery, and technology.

10

Population is growing faster in Africa south of the Sahara than anywhere on Earth. Cities are exploding in size, and standards of living are falling.

◄ *These dredges in Port Nollath are used in the production of South Africa's major export commodity. What product is this?*

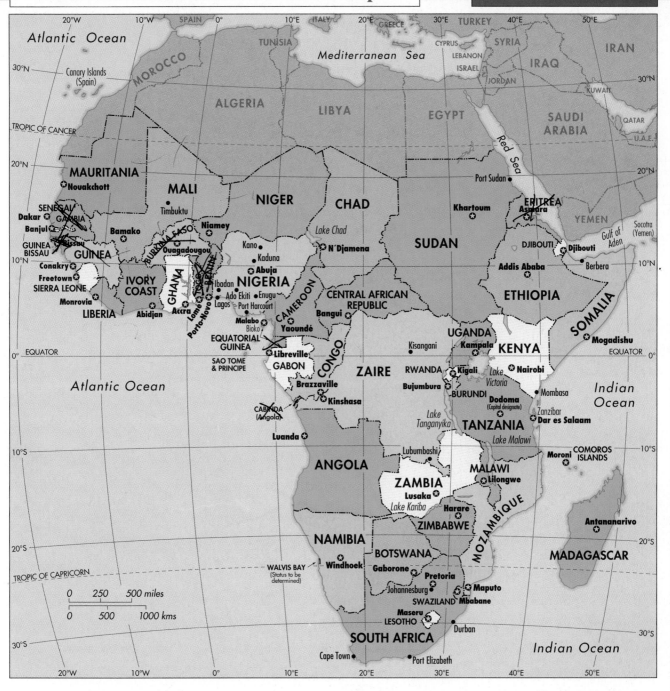

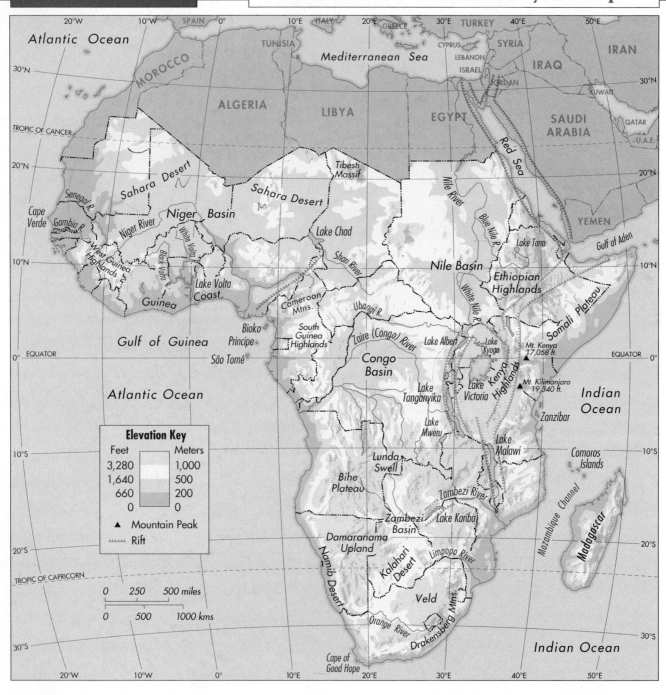

Atlantic Ocean

SPAIN

TUNISIA

Mediterranean Sea

ITALY

GREECE

TURKEY

CYPRUS

SYRIA

IRAN

LEBANON

MOROCCO

ISRAEL

JORDAN

IRAQ

SAUDI ARABIA

KUWAIT

QATAR

U.A.E.

TROPIC OF CANCER

ALGERIA

LIBYA

EGYPT

Red Sea

Sahara Desert

Tibesti Massif

Sahara Desert

Nile River

YEMEN

Gulf of Aden

Senegal R.

Cape Verde

Gambia R.

Niger Basin

Lake Chad

Blue Nile R.

Lake Tana

Niger River

White Volta R.

Shari River

Nile Basin

Ethiopian Highlands

West Guinea Highlands

Black Volta R.

Guinea

Lake Volta Coast

Cameroon Mtns.

Ubangi R.

White Nile R.

Somali Plateau

Gulf of Guinea

Bioko Principe

South Guinea Highlands

Zaire (Congo) River

Lake Albert

Lake Kyoga

Mt. Kenya 17,058 ft.

EQUATOR

São Tomé

Congo Basin

Lake Victoria

Kenya Highlands

Mt. Kilimanjaro 19,340 ft.

Indian Ocean

Atlantic Ocean

Lake Tanganyika

Zanzibar

Lake Mweru

Lake Malawi

Comoros Islands

Elevation Key

Feet		Meters
3,280		1,000
1,640		500
660		200
0		0

▲ Mountain Peak

⏣ Rift

Lunda Swell

Bihe Plateau

Zambezi River

Zambezi Basin

Lake Karia

Mozambique Channel

Madagascar

Damaranama Upland

Limpopo River

Namib Desert

Kalahari Desert

Veld

Orange River

Drakensberg Mtns.

Indian Ocean

Cape of Good Hope

0 250 500 miles

0 500 1000 kms

TROPIC OF CAPRICORN

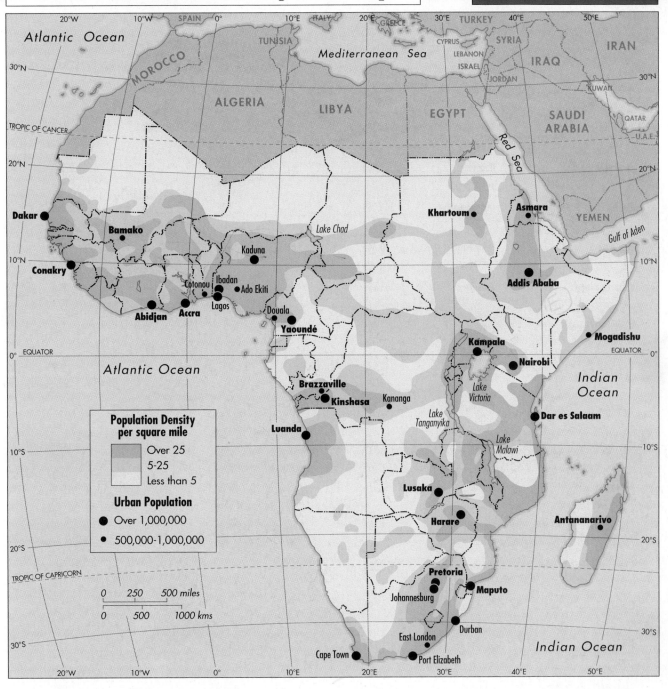

Population Density per square mile

- Over 25
- 5-25
- Less than 5

Urban Population

- Over 1,000,000
- 500,000-1,000,000

0   250   500 miles

0   500   1000 kms

Atlantic Ocean

Mediterranean Sea

SPAIN   ITALY   GREECE   TURKEY   CYPRUS   SYRIA   IRAN   LEBANON   ISRAEL   IRAQ   JORDAN   KUWAIT   SAUDI ARABIA   QATAR   U.A.E.   YEMEN

MOROCCO   TUNISIA   ALGERIA   LIBYA   EGYPT

Red Sea   Gulf of Aden

TROPIC OF CANCER

EQUATOR

Atlantic Ocean

Indian Ocean

TROPIC OF CAPRICORN

Indian Ocean

Lake Chad

Dakar   Bamako   Kaduna   Khartoum   Asmara
Conakry   Cotonou   Ibadan   Ado Ekiti   Addis Ababa
Abidjan   Accra   Lagos   Douala   Yaoundé
Mogadishu   Kampala   Nairobi
Brazzaville   Kinshasa   Kananga   Lake Victoria   Dar es Salaam
Luanda   Lake Tanganyika   Lake Malawi
Lusaka   Harare   Antananarivo
Pretoria   Maputo
Johannesburg   Durban   East London
Cape Town   Port Elizabeth

469

Country	Capital City	Area (Square miles)	Population (Millions)	Life Expectancy	Urban Population (Percent)	Per Capita GNP (Dollars)
Angola	Luanda	481,351	9.5	45	28	—
Benin	Porto-Novo	43,483	5.1	46	38	380
Botswana	Gaborone	231,803	1.4	61	25	2,590
Burkina Faso	Ouagadougou	105,869	10.0	52	20	350
Burundi	Bujumbura	10,745	5.8	52	6	210
Cameroon	Yaoundé	183,568	12.8	56	40	940
Cape Verde	Praia	1,556	0.4	67	33	750
Central African Republic	Bangui	240,533	3.1	47	47	390
Chad	N'Djamena	495,753	5.4	47	32	220
Comoros	Moroni	694	0.5	56	28	500
Congo	Brazzaville	132,046	2.4	54	41	1,120
Djibouti	Djibouti	8,494	0.5	48	81	—
Equatorial Guinea	Malabo	10,830	0.4	50	37	330
Eritrea	Asmara	46,080	2.7	—	—	—
Ethiopia	Addis Ababa	425,776	54.0	46	14	120
Gabon	Libreville	103,347	1.1	53	46	3,780
Gambia	Banjul	4,363	0.9	44	23	360
Ghana	Accra	92,100	16.4	55	32	400

—Figures not available

Country	Capital City	Area (Square miles)	Population (Millions)	Life Expectancy	Urban Population (Percent)	Per Capita GNP (Dollars)
Guinea	Conakry	94,927	6.2	42	26	450
Guinea-Bissau	Bissau	13,946	1.0	43	20	190
Ivory Coast	Abidjan	124,502	13.4	52	40	690
Kenya	Nairobi	224,961	27.7	62	24	340
Lesotho	Maseru	11,718	1.9	59	19	580
Liberia	Monrovia	43,000	2.8	54	44	—
Madagascar	Antananarivo	226,656	13.3	55	24	210
Malawi	Lilongwe	45,745	10.0	47	16	230
Mali	Bamako	478,764	8.9	45	22	280
Mauritania	Nouakchott	397,954	2.2	47	39	510
Mauritius	Port Louis	718	1.1	69	41	2,420
Mozambique	Maputo	309,494	15.3	47	27	70
Namibia	Windhoek	318,259	1.6	58	33	1,120
Niger	Niamey	489,189	8.5	45	15	300
Nigeria	Abuja	356,668	95.1	53	16	290
Réunion	Saint-Denis	969	0.6	73	73	—
Rwanda	Kigali	10,170	7.4	46	5	260
São Tomé and Principe	São Tomé	371	0.1	62	42	350

—Figures not available

Country	Capital City	Area (Square miles)	Population (Millions)	Life Expectancy	Urban Population (Percent)	Per Capita GNP (Dollars)
Senegal	Dakar	75,749	7.9	48	39	720
Seychelles	Victoria	108	0.1	68	50	5,110
Sierra Leone	Freetown	27,699	4.5	42	32	210
Somalia	Mogadishu	246,201	9.5	46	24	—
South Africa	Pretoria	471,444	39.0	64	56	2,530
Sudan	Khartoum	967,494	27.4	53	21	400
Swaziland	Mbabane	6,703	0.8	55	23	1,060
Tanzania	Dar es Salaam	364,900	27.8	52	21	100
Togo	Lomé	21,927	4.1	56	29	410
Uganda	Kampala	91,135	18.1	43	11	160
Zaire	Kinshasa	905,564	41.2	52	40	220
Zambia	Lusaka	290,583	8.6	46	49	420
Zimbabwe	Harare	150,803	10.7	56	29	620

—Figures not available

AFRICAN ECONOMIES AND SOCIETIES

One-tenth of the world's population, some 550 million people, live in the poorest of the world's culture regions—Africa south of the Sahara, or sub-Saharan Africa. Sub-Saharan Africa reaches 3,200 miles southward from the Sahara Desert to the Cape of Good Hope. At its widest point, Africa extends 4,600 miles from the Atlantic to the Indian Ocean. The world's second largest continent, it is three times the size of the United States.

Today, sub-Saharan Africa is divided into forty-nine countries. Most of these countries are small. Only four—Nigeria,

◀ *By many technological standards, Africa is the world's poorest continent. Do you recall from Unit 1 some of the general characteristics of developing countries?*

Ethiopia, Zaire, and South Africa—have populations of more than 30 million. Thirty-two African countries have fewer than 10 million citizens. Most of these countries are poor. The great variations in levels of development range from the destitution of countries like Somalia, Ethiopia, Tanzania, and Mozambique to the relative wealth of countries like Gabon, Botswana, and South Africa.

1. African Environments

The Plateau Continent

Africa is almost completely surrounded by two oceans and two seas. The Atlantic Ocean lies to the west and the Indian Ocean to the east; and to the north and northeast are the Mediterranean Sea and the Red Sea. The Sahara Desert stretches across the continent, separating North Africa from sub-Saharan Africa. Only at the narrow Isthmus of Suez is Africa connected to another continent.

Africa is a huge plateau fringed with narrow coastal plains. Nearly all of Africa is higher than 1,000 feet in elevation, and half of the continent is more than 2,500 feet above sea level. There are no great mountain chains in Africa like the Rocky Mountains of North America or the Andes of South America, but rugged lands are found in East Africa, where the volcanic peaks of Mount Kenya and Mount Kilimanjaro soar well above the highlands. In general, the African landscape is higher in the east and slopes gently to the north and west.

A series of rift valleys slices southward through the highlands of East Africa from the Red Sea to South Africa. A **rift valley** is a deep trench formed where large sections of Earth's crust drop between two parallel cracks or faults. The East African Rift System, which extends for 4,000 miles, cuts through the Ethiopian Highlands, splits into two trenches on either side of Lake Victoria, and runs southward parallel to Africa's east coast. Trace the course of the East African Rift Valley on the map on page 468. Along its course lie a series of lakes, the largest of which are Lake Tanganyika and Lake Malawi.

The edges of sub-Saharan Africa's plateau are marked by **escarpments**, or continuous lines of steep cliffs or slopes. The African escarpment stretches along Africa's east coast, rises to high elevations in the Drakensberg Mountains in South Africa, and sweeps clockwise around the Cape of Good Hope and northward along the west coast of Africa. Waterfalls and rapids are found where rivers cascade off the edge of the escarpment to the sea. These waterfalls on major rivers made it difficult for Europeans to penetrate the African interior in boats or canoes.

The African plateau is dented by a series of broad, deep basins or depressions where the great rivers of Africa laid down sediments whose weight caused the plateau's surface to sag. The Niger, Shari, and Nile rivers formed basins along the southern edge of the Sahara Desert. The Congo Basin covers much of central Africa, and the Zambezi River formed the Kalahari Basin of southern Africa. Locate these

Subregions of Africa South of the Sahara

Africa south of the Sahara has five major subregions: (1) West and Central Africa, (2) the Sahel, (3) the African Horn, (4) East Africa, and (5) Southern Africa.

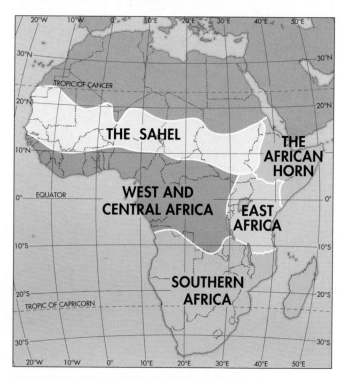

Mount Kilimanjaro is something of an oddity in Africa which has no continental mountain ranges like the Rockies or the Alps. How was Mt. Kilimanjaro formed?

basins on the map on page 468. Each of them is surrounded by an upland plateau.

Tropical and Subtropical Climates Cover Most of Africa

Because the equator runs through sub-Saharan Africa, most of this region's climates are either tropical or subtropical. The center of the continent is occupied by the Congo Basin, which is constantly hot and wet. Tropical rain forests that blanket the Congo and extend beyond the Congo Basin along the West African coast have lush vegetation but poor soils. As a result, much of Central and West Africa is only thinly inhabited. Only along rivers, near mining operations, and on the West African coast are there dense populations. Elsewhere, small farming and fishing communities are scattered in the rain forest.

Rainfall diminishes as we move away from the equator, and tropical rain forests give way to savanna grasslands. The savannas have a distinct dry season both north and south of the equator. Tall grasses are interspersed with low trees on either side of the Congo Basin. In East Africa, these thick grasslands support large herds of antelope, zebra, and giraffe that populate Africa's big game country. Throughout most of the African savanna, farming

and herding are major ways of living. Permanent settlement, however, is limited by the long dry season, uncertain rainfall, and poor soils.

Beyond the savannas, the dry season becomes longer. Vegetation thins out into low-grass steppes north and south of Africa's tropical core. These steppes give way to the parched, rainless lands of the Sahara Desert in the north and the Kalahari Desert to the south. Nomads graze camel and cattle herds on the steppe margins of these deserts. Farmers grow crops in desert **oases**, where the availability of water supports settled populations. In Africa, middle-latitude climates are found only near the southern tip of the continent.

Complex Environments

Africa has many complex environments. The lowlands of West and Central Africa receive too much rainfall, have poor soils, and are infested by **tsetse flies**, which kill cattle and cause sleeping sickness in people. The steppe and desert areas of northern and southwestern Africa have too little

IT'S A FACT   Between 800 and 1,000 languages and dialects are spoken in Africa.

rainfall. In the highlands of East Africa, drought hits once every four or five years. Overall, Africa has a great deal of unusable land, more than most other world culture regions.

These environmental problems directly affect African life because a majority of Africans live off the land. The population is growing rapidly, and the pressure of people on the land is intense, even though Africa has a lower population density than any continent except Australia. The scarcity of productive natural environments in Africa presents difficult problems for the goal of development to support more people in each subregion of the continent.

Five subregions of Africa are generally recognized: (1) the hot, wet rain forests of West and Central Africa; (2) the dry southern fringes of the Sahara, called the Sahel; (3) the varied landscapes of the African Horn; (4) the highlands of East Africa; and (5) the temperate lands of Southern Africa.

The Rain Forests of West and Central Africa

Tropical rain forests cover much of the Congo Basin as well as the coastal lowlands of West Africa. The Congo Basin, which is drained by one of the world's largest rivers, is a huge shallow saucer ringed by low hills. Locate the Congo Basin and the Zaire (Congo) River on the map on page 468. The hot, wet climate of West and Central Africa supports a luxuriant rain forest. Evergreen trees arch overhead to heights of 50, 75, and 100 feet above the surface and block sunlight from reaching the forest floor. Dense tangles of vegetation (called **jungle**) line river banks where sunshine does reach the surface.

Nearly a third of the rain forest in the Congo Basin has been cut down by farmers clearing land, ranchers creating grasslands, and timber companies harvesting valuable tropical hardwood trees like ebony and teak. The destruction of these forests is an issue of global concern.

These economic activities create environmental problems. Rain forest soils, already thin and low in plant nutrients, wash away once the forest cover is removed by farmers, ranchers, and timber companies. Because these forests take much longer to reestablish themselves than do woodlands in the middle latitudes, **reforestation**, the process of replacing damaged or destroyed forests, is extremely difficult. Much of the land in the Congo Basin, therefore, is now being damaged beyond repair, even though only 47 million people live in the rain forests of Zaire, Congo, and the Central African Republic—an area as large as the European continent.

Damage to rain forests is also a problem along the densely settled West African coast, where forests have been cleared to make space for palm, coffee, cacao, and rubber plantations, as well as for many small farms. Population pressure on cultivated land is increasing rapidly. More than half of the 138 million people of Sierra Leone, Liberia, Ivory Coast, Togo, Benin, Nigeria, and Cameroon live along the West African coast. Dense clusters of settlement ring port cities like Abidjan (Ivory Coast), Accra (Ghana), and Lagos and Port Harcourt (Nigeria). Recent oil discoveries in eastern Nigeria, Cameroon, and Gabon are now attracting even more people to the West African coast.

Northern Savannas

Along the northern and southern edges of the West and Central African rain forests, rainfall decreases. Savanna grasslands and woodlands replace

▼ *This dam on the Volta River holds back the waters of Lake Volta, Ghana, which covers much of the western half of this country. What are two important uses for dams?*

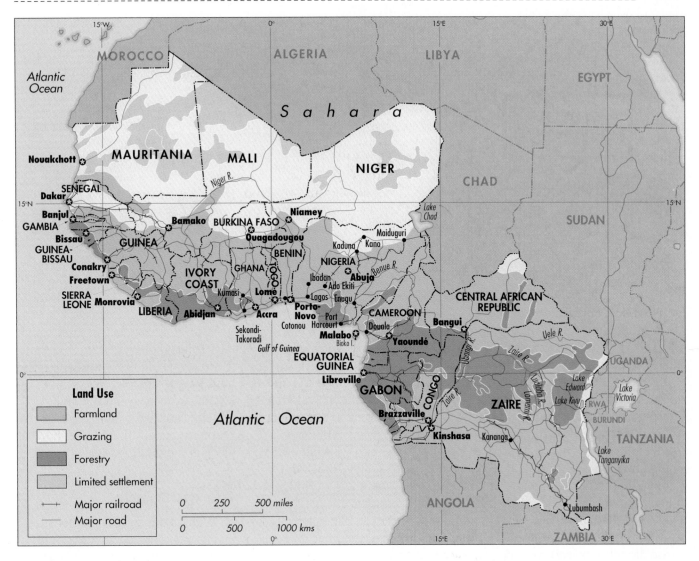

West and Central Africa

forests in both hemispheres because the winter dry season lengthens. Cotton and peanut farms are found in this densely settled area to the north that once supported ancient African empires. Cattle herding is an important activity south and east of the rain forest.

Still farther to the north, as one moves into the Sahel, savanna grasslands thin out into short-grass steppes. Overgrazing and cultivation of marginal soils already low in plant nutrients is gradually destroying what were once rich farming and grazing lands. As a result, many African farmers and herders are now leaving the land and moving south to West Africa's crowded port cities in search of work.

Drought in the Sahel

The Sahel is a 200- to 700-mile-wide belt of semi-arid land that stretches across the African continent from Senegal on the west coast to the Ethiopian Highlands in the east. The word *Sahel* means "shore," for the southern shore of the Sahara Desert. Rainfall in this subregion varies from twenty-four inches a year in the south to four inches a

PEOPLE-ENVIRONMENT

DROUGHT IS SHATTERING WAYS OF LIVING IN THE SAHEL

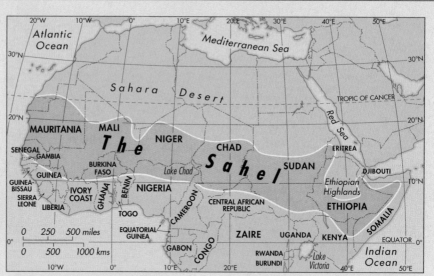

Desertification in the Sahel

In better times, herders drove cattle and sheep from pasture to pasture across the steppes of northern Mali, a country in the western Sahel. These nomads counted their wealth in animals. Then the rains failed and failed again. Water holes dried up and one by one, the animals died. In Mali alone, more than 100,000 nomads lost their entire herds. Their way of life was shattered.

Now millions of people from the drought-ravaged countries of the Sahel are crowding into refugee camps established in the wetter south. Torn from his homeland, one herder exclaims, "What is a nomad without animals? I tell you he is nothing."

Peasant farmers in the Sahel also face catastrophe. Crops have shriveled in their fields. With no seed left to plant, the farmers

have fled their homes. Entire villages have disappeared. As one government official in Mali notes, "Misery has come to live among our people."

Far to the east, in southern Sudan, before the drought hit in the late 1960s, the government started a program called "Freedom from Thirst." The government drilled 4,000 wells to provide people with water throughout an area as large as Kentucky. Thousands of families settled down with their herds near the wells. Farmland was cleared in the vicinity to grow food, but the crops soon exhausted the soil.

The desert began to spread in rings around the wells as animals ate everything in sight, causing "circles of death" to appear on

the landscape. People and animals stripped the land clean. Finally, everybody left the area, and the wells were abandoned. **Desertification** in the Sahel had killed the land.

QUESTIONS

1. What is the chief characteristic of a nomadic life style?
2. What are the "circles of death" referred to by the author in the last paragraph?
3. By drilling wells to provide water, the Sudanese government allowed the population in a particular area to grow larger than it would have normally. Which of the five themes of geography are involved in this situation?

year in the north. Farmers grow drought-resistant crops like millet and sorghum in the wetter south. Cattle, sheep, and camels are herded on thin grasses in the north.

Today, the 70 million people who live in the seven Sahel countries of Senegal, Mauritania, Mali, Burkina Faso, Niger, Chad, and Sudan face a tragic environmental crisis. The Sahara Desert is expand-

Africa was once home to rich and powerful nation-states, evidence of which is becoming more and more apparent as archaeological clues are located. What were some of these lost empires?

ing southward, drying up pastures and destroying the herds and farmland of millions of people.

The spread of desert-like conditions into semiarid areas is called **desertification**. Desertification is caused by human destruction of vegetation, a change in climate, or both. In the 1960s and 1970s, the drilling of water wells and setting up of new health programs in the countries of the Sahel improved living conditions. With more water and improved health conditions, population soon doubled in this subregion of Africa. Herds of cattle and sheep increased in the northern steppes, and new farms were planted in the wetter lands to the south. Population pressure on available land began to increase, which meant that overgrazing and soil erosion became more intense. Then, in the late 1960s, the Sahara Desert began its southward march. In the last thirty years, sand dunes have overrun an area in the Sahel as large as the state of Texas. Although the Sahara expands and recedes during periods of greater and lesser rainfall, generally the desert is still moving southward at a rate of four to five miles a year. The drought has also spread into the African Horn.

IT'S A FACT Food production per person has been falling in Africa for fifteen years.

The African Horn

The Horn of Africa juts out from East Africa into the Indian Ocean just south of the Red Sea. The African Horn is a jumble of hills, mountains, canyons, and valleys that slope eastward to dry lowlands along the Red Sea coast. It is shared by four countries—Ethiopia, Eritrea, Somalia, and Djibouti.

The Great Rift Valley slices southward from the Red Sea through the middle of Ethiopia. It runs between the densely settled western highlands of Ethiopia and Eritrea and the barren hills and parched lowlands of eastern Ethiopia, eastern Eritrea, Somalia, and the tiny coastal country of Djibouti.

Most of Ethiopia's 54 million people are farmers who live in the western highlands between the capital city of Addis Ababa (population of 1.8 million) and Lake Tana. To the east, nomads herd camels, cattle, sheep, and goats on sparse pastures. They also collect frankincense, a fragrant gum that has been used as incense for many centuries. The communist government of Ethiopia forcibly settled several million of these people in the south after the devastating drought in the late 1980s and in 1990. In addition, this radical regime is continuing to force farmers to relocate in tightly controlled government settlements, a process called **villagization**. This process is no longer occurring in Eritrea, a former province of Ethiopia that gained its independence from Ethiopia after a long war in 1993.

During much of the last thirty years, these highlands have been in the grip of the same withering drought that has afflicted the countries of the Sahel. This drought has devastated farmland and pastures throughout much of the African Horn, and it is spreading southward into northern Kenya. In the late 1980s, an intense drought severely affected the lives of 20 million people in this subregion and threatened nearly 10 million people with famine. This catastrophe triggered an international effort, including fundraising rock concerts, to provide money and food to these people. In the mid-1990s, an estimated 1 million people per year in the African Horn are still suffering from malnutrition.

International relief efforts to provide food to these countries have been disrupted by ethnic warfare in Ethiopia, Eritrea, and Somalia. The United Nations sent troops to supervise the distribution of food to the starving people of Somalia in 1993. Although initially welcomed, these troops were soon drawn into battles with local warlords who fought to retain their control of incoming food. Similarly, in nearby Sudan and in Chad, civil wars pit the Islamic peoples of the north against Christian and animist groups in the south (animist beliefs include faith in spirit powers often associated with animals, as well as ancestor worship). Food donated to the people of these countries frequently is not delivered or is used as a weapon of control. Poor roads and warring factions hinder most deliveries of humanitarian aid.

East Africa

South of the African Horn, a high plateau studded with volcanoes forms the roof of the African continent. This plateau slopes gently to the shores of the Indian Ocean in the east. The western border of this plateau is marked by one branch of the East African Rift System; a second branch cuts through the center of the plateau. Lake Victoria, the largest body of water in Africa, lies between these two trenches.

About 87 million people live in Kenya, Uganda, Tanzania, Rwanda, and Burundi, the five countries located south of the African Horn in the highlands of East Africa. Most of these people earn a living as farmers or herders. Their settlements are concentrated in the cool, green highlands and fertile lake basins where reliable rainfall and fertile soil are available. Dense populations are found where the land is free of tsetse flies. Tanzania is by far the largest of these countries, with an area greater than the other four combined, although tsetse flies are a problem over much of the country.

Nine-tenths of Kenya's 27.7 million people live on the well-watered lands along the shores of Lake Victoria and in nearby highlands around the capital city of Nairobi (population of 1.8 million). Coffee, tea, cotton, and sisal (a fiber used for making rope) are grown as commercial crops on small farms.

The African Horn and East Africa

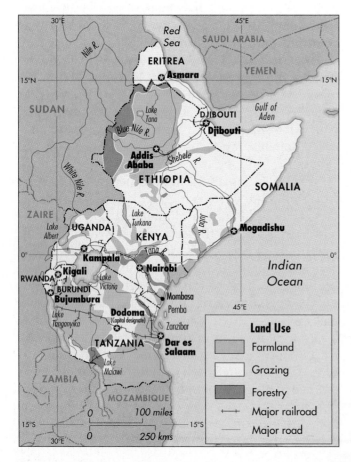

Land Use
- Farmland
- Grazing
- Forestry
- Major railroad
- Major road

IT'S A FACT Kenya has one of the world's highest rates of population growth.

Corn, wheat, and cassava (a starchy root crop, also called manioc) are important food crops. Elsewhere, population is thinly scattered. Drought has invaded northern Kenya, driving herders southward.

Kenya has magnificent scenery, but only one-sixth of the country receives enough rainfall to support settled farming. Its coast is lined with mangrove swamps except near the port city of Mombasa. The tree-studded, open grasslands of central and southern Kenya are great game preserves that are closed to farming. In the deep south, the Masai and other cattle herders live on the dry plains. With population in Kenya now growing faster than in any other country in the world, population growth has become a national crisis. More infants are surviving due to better health care, and on average Kenya's women have seven to eight children each. Life expectancy has risen from 39 years of age in 1960 to 62 years today.

People are clustered in favorable environments in the four other countries of the East African highlands. Most of Tanzania's 27.8 million people farm the fertile eastern shores of Lake Malawi and around Lake Victoria in the highlands. Coconuts and cloves are grown along the 500-mile-long coastal plain on the shore of the Indian Ocean. On this coast Dar es Salaam, with a population of 1.9 million, is Tanzania's major port. The western border of the country is lined by high mountains that dive steeply to the waters of Lake Tanganyika, which lies in the western branch of the East African Rift System. Much of the plateau and highlands of central Tanzania is covered with thornbush and savanna grasslands, and is infested with tsetse flies. Only a quarter of the country is cultivated.

Similar patterns of land use occur in Uganda, a landlocked country dependent on Kenya for access to the sea. At independence in 1962, Uganda was the largest producer of coffee in the British Commonwealth, and was a major exporter of cotton, tea, and sugar. In the 1970s, however, civil war badly damaged the economy. For a decade, Idi Amin, a cruel dictator, slaughtered his opponents, threw all Asians (who controlled much of the nation's commercial interests) out of the country, and ruined the economy. Recovery has been slow, foreign investment is limited, and the onslaught of AIDS has been severe. Most of the country's 18.1 million people live in the relatively small, fertile crescent of land that borders Lake Victoria or alternatively in the capital city of Kampala on the lake's shores.

▼ *A Kenyan is silhouetted before the skyline of Nairobi. Why are many Kenyans being forced to move south or into large cities like Nairobi?*

The tiny mountain countries of Rwanda and Burundi, with a combined population of 13.2 million, are located between Lake Victoria and smaller lakes at the bottom of the western branch of the African Rift Valley. Most citizens of these two densely populated countries are subsistence farmers and cattle herders. Coffee is the primary export product of both countries. Dense populations in these mountain nations are causing severe problems of deforestation, overgrazing, and soil erosion. In the 1990s, civil war in Rwanda and Burundi escalated into genocide between the two largest ethnic groups and devastated the economies of both countries.

Southern Africa

South of the East African highlands and the Congo Basin, a vast upland plain extends eastward across the continent from Angola to Zambia and southward to the Cape of Good Hope. The northern edges of this upland plain are covered by woodlands. To the south, these forests thin out into open parkland and short-grass steppes because rainfall decreases away from the equator. The Kalahari Desert covers most of Botswana, Namibia, and western South Africa. Locate the Kalahari Desert on the map on page 468, and see how "deserted" this area is on the population map on page 469. Khoikhoi peoples (Hottentots) and San (bushmen) live as hunters and gatherers in this dry land, cleverly adapting their way of life to desert conditions.

On nearby steppes, herding is practiced. Most of the area is very thinly peopled, except for highlands in Angola and Namibia. Inadequate rainfall and severe overgrazing of grasslands restrict farming and herding in southwestern Africa.

In southeastern Africa, the East African Rift Valley plunges southward between Mozambique and Malawi, its trench filled by Lake Malawi. In eastern South Africa, the Drakensberg Mountains separate the high plain of the interior from the densely settled coast of the Republic of South Africa. The Drakensberg Mountains run parallel to the east coast of South Africa. The escarpments of the Drakensberg Mountains form a boundary between the cattle and sheep ranches and grain farms of the interior grassland or **veld**, and the sugarcane, cotton, palm, and

Southern Africa

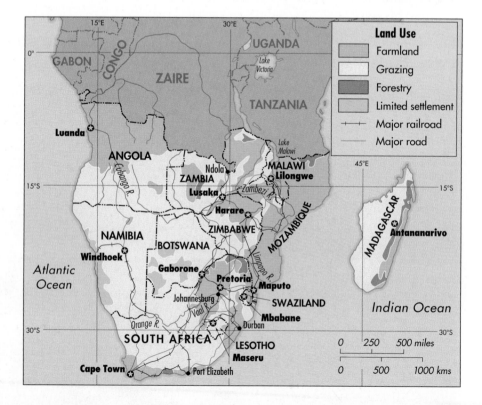

In spite of its beautiful scenery and hospitable people, for years South Africa was virtually isolated from the world community because of its racist policies. What term described that "whites only" policy?

banana plantations that line the eastern coasts of Mozambique and South Africa.

The forested highlands that extend from Zambia through Zimbabwe into South Africa are not densely settled. Many of the 19.3 million people who live in Zambia and Zimbabwe work in areas important for chrome, copper, gold, and diamond mining. But in most areas in these countries, the dry season lasts for more than half the year. Farmers are able to cultivate land only in the fertile lowlands of the Zambezi, Limpopo, and Orange Rivers, where water is available year-round. Cattle and sheep are grazed wherever disease-carrying tsetse flies are absent.

The tip of the African continent extends into the middle latitudes and has two different types of temperate climates. The southwest, near the city of Cape Town (population of 2.5 million), has a Mediterranean climate like southern California, with dry summers and wet winters. Vineyards, fruit orchards, and wheat farms spread across the southwest. The southeast has a humid subtropical climate with summer rainfall and moderate temperatures. Sugarcane, pineapples, bananas, and citrus fruit are grown in the southeast. Both areas are densely settled. Three of every four black South Africans are farmers, and many work on farms and ranches owned by white South Africans.

⊕ REVIEW QUESTIONS

1. Sub-Saharan Africa has what fraction of the world's population? It is divided into how many countries?
2. What are the five subregions of Africa south of the Sahara?
3. What is the chief type of vegetation in the Congo Basin?
4. Where do nine-tenths of Kenya's people live? What are the chief crops they produce?
5. What tragic environmental crisis is faced by the people who live in the Sahel?

⊕ THOUGHT QUESTIONS

1. How have economic activities in recent years created new problems in the Congo Basin? Do you think this area can support a dense population? How have environmental conditions affected the economies of tropical countries like Zaire?
2. Why is the subregion near the southern tip of Africa more productive than most other parts of the continent? What types of climate are found here? What crops are grown in the southwest? In the southeast? Are these environments similar to any that you know firsthand or have visited?

2. Farming and Herding in Africa

Environment and Population

Africa's environments help to explain many of the continent's current problems. Nearly 80 percent of the people of sub-Saharan Africa are farmers and herders who live close to the land. Small areas of sub-Saharan Africa did develop modern sectors in their economies—for instance, in commercial farming and mining—during the colonial period. Most Africans, however, still live in small groups whose lives are tied to the environment. Many of these ethnic groups maintain their own languages and beliefs and until very recently were isolated from other peoples.

Many Africans learned over time how to deal with the difficult environments of their continent. In West and Central Africa, they farmed soils with less than one-tenth the nutrients of the soils of Western Europe and the United States. In East Africa, they coped with low and uncertain rainfall. Animal disease and malaria cut off many areas of sub-Saharan Africa from permanent settlement. In the relatively few fertile environments, Africans built large kingdoms based on intensive farming and herding.

African farmers adapted to low soil nutrients in tropical areas by frequently moving their crops from areas where soil fertility was declining and crop yields were dwindling to new land that held greater promise. In dry regions, herders moved their cattle and sheep from one pasture to another from season to season to maintain larger herds. They also learned to avoid areas infested by tsetse flies and other disease-bearing insects. These were intelligent adaptations to Africa's environmental realities.

These types of subsistence farming and herding, however, require large amounts of land to support

IT'S A FACT   Tropical rain forests grow twenty times slower than middle-latitude forests.

small numbers of people. Now Africa's population is growing rapidly, which means that Africa's limited resource base must produce more food. Africa's subsistence economies—*shifting cultivation, settled farming,* and *pastoral nomadism*—cannot support large populations except in very favorable environments. Because there is only a small amount of fertile land, the incomes of Africans have declined by nearly 20 percent in the 1980s and 1990s. Poverty and hunger are increasing as population increases.

Shifting Cultivation in the Rain Forests

Farmers in the rain forests of West and Central Africa practice **shifting cultivation**, a form of agriculture in which farmers move from one area to another in response to environmental conditions. Most of these farmers live deep in the forest in small settlements. A dozen or so houses are surrounded by vegetable gardens, banana plants, palm, and fruit trees. Patches of land on nearby hillsides are cleared of trees and planted in grain, pumpkins, yams, and cassava. Tree crops are also important sources of food. These tropical farmers stay put as long as their fields produce enough food to feed the group—usually for four or five years. Then crop yields and soil fertility decline, and they must move to another part of the rain forest, clear new land, and begin the cycle again. The map on page 477 shows the rain forest areas of West and Central Africa where shifting cultivators are primarily found.

These farmers "shift," or move from place to place, because the soils of the rain forest cannot support permanent cultivation. In the first year, crop yields are high, then the crops begin to drain soils of their few nutrients. Heavy tropical rains begin to erode the cleared land, and forest undergrowth closes in wherever sunlight reaches the ground. At this point, these forest farmers must abandon their fields and move to a new area in the rain forest.

Forest clearings take twenty to thirty years to recover their soil fertility, so shifting cultivation requires a large amount of land to feed a small number of people. It is estimated that only fifteen

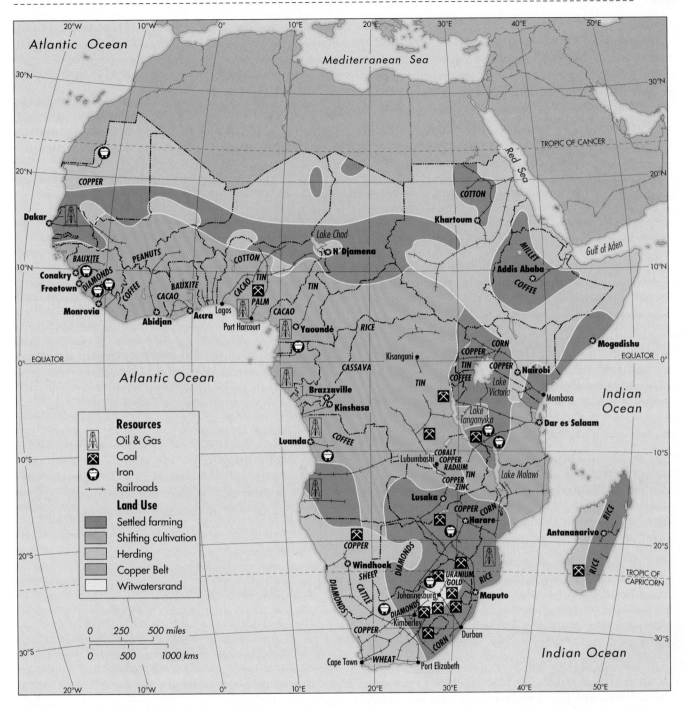

Economic Activities and Resources in Africa

people can be supported on one square mile of land. In recent years, however, population has grown rapidly, and a third of the rain forest in the Congo Basin has been cut down. Finding new areas suitable for shifting cultivation is becoming more and more difficult. Shifting cultivators are now forced to return to old clearings before soils have had time to recover their fertility. As a result, they raise fewer and poorer crops in these clearings, and many are being forced to leave the land.

Another form of shifting cultivation is practiced on the savanna margins of the rain forests. In the savannas, trees are cut down over a large area and then stacked on fields, where they are burned. The wood ash adds fertilizer to the soil. Crops like sorghum, millet, and cassava are grown for several years; then the soil is exhausted, and the farmers must move on. In places where cattle are raised, animal fertilizer is used to prolong the fertility of the soil, yet even in these areas, droughts, pests, and disease make shifting cultivation a hazardous way of living. As in the forests, large amounts of land are needed to support small communities.

Settled Farming in Africa

Settled farming is practiced in sub-Saharan Africa wherever fertile soils and reliable rainfall can support permanent settlement. These areas are located between the tropical rain forests of equatorial Africa and the dry lands of the Sahara Desert in the north and the Kalahari Desert in the south. Locate these transitional areas on the map on page 485. For the most part, settled farming is found in a belt stretching across interior West Africa south of the Sahel; in parts of the Ethiopian Highlands; in the fertile lake districts of East Africa; and in the tropical, subtropical, and middle-latitude environments of southern Africa.

Grains are the main food crops in settled farming areas in West Africa. Upland rice is grown in wetter parts of the savanna from Senegal to northern Nigeria. Where rainfall decreases in interior West Africa, corn, millet, and sorghum are the most important food crops. African villagers rarely raise both animals and crops, as they do in settled farming areas in other parts of the world. In some areas in West Africa, however, farmers hire herders to tend their cattle. And in parts of East Africa, groups of farmers and nomadic herders regularly exchange grain for meat and milk products. This arrangement provides farmers with animal fertilizer and herders with grain, balancing and enriching the diets of both groups.

In the settled farming areas of East and Southern Africa, combined crop and cattle raising is more common. Here, Africans use fertilizer and crop rotation to maintain the fertility of the land. They grow food crops and commercial crops and also keep herds of cattle and sheep. Coffee, tea, and cotton

are raised in the cooler highlands that stretch from Ethiopia to South Africa, and tropical commercial crops like bananas, coconuts, and sugarcane are grown on plantations near the coast. In South Africa, wheat farms and large cattle and sheep ranches are found throughout the interior. In much of South Africa, patterns of land use are similar to those found in other parts of the world settled by Europeans.

These areas of settled farming hold the key to Africa's future. Food production in Africa has fallen 20 percent in the last ten years as a result of the desertification of the Sahel, the destruction of the forests in Central and West Africa, and growing population pressure on farmlands in East and Southern Africa. Africa's agricultural environments are deteriorating. Only the areas of settled farming have environments that can support substantially larger numbers of people. Only here can crop yields be raised to meet the food needs of the continent.

African governments now recognize that improvements in farming are vital to the development of their countries. In many areas, village farmers have started local programs to conserve their land resources and use them more wisely. Such developments are among the most important changes taking place in sub-Saharan Africa today. If they become widespread across the continent, many rural Africans may have an improved standard of living, and a more stable and productive future may be ensured.

African Herders

A majority of African herders raise cattle on lands between the rain forest and the desert on the dry margins of areas of settled farming. In the Sahel, herders raise cattle, sheep, and goats in a belt extending from Senegal in the west to the African Horn on the east. Cattle keeping is a way of life throughout large parts of East Africa wherever rainfall is low. Herding is also practiced throughout the dry lands of the Kalahari Desert to the south, as the map on page 485 shows.

African cattle herders move their herds from one pasture to another throughout the year, a practice known as **pastoral nomadism**. In areas of low rainfall and sparse vegetation, some African herders

**PEOPLE-
ENVIRONMENT**

REVERSING AFRICA'S DECLINE

Village collection of firewood has contributed to environmental decline in Africa. Why?

I n Africa today, groups of farmers are joining together to fight hunger. Through self-reliance, they are determined to reverse Africa's decline.

In Burkina Faso, farmers have learned to dig underground storage pits that are cool enough to preserve crops like potatoes, which would rot above ground. Elsewhere, farmers are placing lines of stones along the contours of their fields to halt soil erosion.

Similar small-scale development projects are improving nutrition and living standards in many African villages. Some of these African farmers have joined together in an association that helps them to take control of their own lives. As one of their leaders exclaims, "If we want to end hunger, we can. . . . It all depends on us, on our motivation, our commitment."

Many of the local groups are led by women; they produce about three-quarters of Africa's food. In Kenya, one such woman, Wangari Maathai, decided to take action when the desert began to invade her life.

"I noticed springs that I knew as a child drying up," she says, "and I could see there was no longer firewood." So she founded the Green Belt Movement in Kenya by planting seven symbolic trees. Since then, 2 million trees have been planted.

Following her example, thousands of women throughout Kenya who are unwilling to see the earth stripped of its protection and color have established 670 community tree nurseries to ward off the desert and improve their villages. As Wangari Maathai notes, "If we do not take care of the environment, we will die."

 **QUESTIONS**

1. How would placing stones along the contours of fields halt soil erosion?
2. What is the Green Belt Movement in Kenya?
3. Why would the active involvement of African women in most small-scale improvement projects be necessary for the project's success?

must move as often as sixty times a year. In better-watered environments, herders migrate only two or three times a year—with the change of seasons. Large tracts of grazing land are required to support small numbers of people and animals.

Since 1900, Africa's cattle herders have suffered one disaster after another. Epidemics destroyed the cattle herds of the Masai and Kikuyu in East Africa,

forcing many of these people to settle in towns and villages. European immigrants who came to East Africa in the 1900s occupied the grasslands that had been pastures, preventing herders from returning to their nomadic way of life. Later the drought that began in the 1960s and 1970s destroyed the cattle herds of the Sahel, as you have already learned. In addition, population growth has caused

▲ *The Masai of Kenya consider cattle an important source of wealth. What problems are large cattle herds creating for the Kenyan environment?*

settled farmers to encroach on the grasslands in their search for more farmland. This expansion of village settlement has deprived herders of their best-watered, most reliable pastures. Also, the newly independent states in Africa have been eager to tax cattle, placing a further burden on herders. As a result, many of Africa's herders have been forced to abandon their way of life or to live in less productive environments.

⊕ REVIEW QUESTIONS

1. What percentage of the people of sub-Saharan Africa are farmers and herders?
2. What areas of sub-Saharan Africa have enough fertile soils and reliable rainfall for settled farming?
3. What are the main commercial crops grown in the farming areas of East and Southern Africa?
4. What are the major problems of African herders, and how do they cope with these problems?

⊕ THOUGHT QUESTIONS

1. Why do farmers who live in the rain forests of West and Central Africa usually move every four or five years? How could their condition be improved? Do you believe that the rain forests could support significantly larger populations? If

not, where will Africa's growing population live in the future?
2. Why do the areas of settled farming hold the key to Africa's future? Notice how limited the area of settled farming is on the map on page 485. Can this agricultural economy be made more productive? If so, how?

⊕ CHAPTER SUMMARY

Most of Africa is a vast plateau, rather like an inverted pie plate that is slightly tipped up on the east side. Escarpments ring the plateau, so that travel by rivers from the coast to the interior was difficult until recently. The environments of sub-Saharan Africa include savanna grasslands north and south of a rain forest belt in Central Africa, and in the south the Kalahari Desert. These African environments are difficult for the people who must live on the land. Central Africa has too much rainfall, poor soils, and vast regions infested by the disease-bearing tsetse fly. In the north and southwest, too little rain falls, and East Africa has periodic droughts.

Africa can be divided into five subregions: (1) West and Central Africa with hot, wet rain forests; (2) the Sahel, an arid region on the southern edge of the Sahara Desert; (3) the African Horn, a hilly region of varied landscapes; (4) East Africa, a fertile region with highlands and mountains; and (5) the temperate Southern Africa.

The majority of Africans can be classified as either farmers or herders. Among the farmers are those who practice shifting cultivation, periodically farming new fields, and those engaged in settled farming in areas where fertile soils and adequate rainfall will support it. African herders practice pastoral nomadism in arid and semiarid environments. They move their herds from one pasture to another throughout the year in order to feed larger herds.

These ways of living require large amounts of land to support relatively small numbers of people. With population growing rapidly, population pressure on the land is intense. As environments deteriorate under this growing pressure, more and more farmers and herders will be forced to move to Africa's cities.

EXERCISES

Matching

Directions: On your paper, write the term that matches each numbered phrase in the left-hand column.

1. The capital of Kenya
2. A farming practice in which farmers move from one field to another every three to four years
3. An important city in Nigeria
4. Cause disease in animals and humans
5. Waterfalls cascade off the edges
6. Partly caused by human destruction of vegetation
7. A country in the Sahel

escarpments
desertification
Mali
Lagos
Nairobi
shifting cultivation
tsetse flies

Find the Right Ending

Directions: Choose the correct ending to each of the following sentences, then write the completed sentence on your paper.

1. Sub-Saharan Africa stretches from the Sahara Desert to the (Mediterranean Sea, Cape of Good Hope).
2. Africa is the world's (largest continent, second largest continent).
3. An area of Africa that is constantly hot and wet is the (Congo Basin, Sahel).
4. Africa has a lower population density than any continent except (Europe, Australia).
5. In the late 1980s, many Ethiopians suffered from (floods, droughts).
6. There is a distinct dry season in (savanna grasslands, tropical rain forests).
7. The largest body of water in Africa is (Lake Malawi, Lake Victoria).

8. The country with the fastest growing population in sub-Saharan Africa is (Kenya, Tanzania).
9. Most of the people in the rain forests of West and Central Africa are (miners, shifting cultivators).
10. The chief exports of East Africa are coffee, tea, and (cotton, rice).
11. Nearly 80 percent of the people of sub-Saharan Africa are (city workers, farmers and herders).
12. In the 1980s and 1990s, the food production of Africans (declined by nearly 20 percent, rose by nearly 20 percent).
13. The economy of Uganda was ruined by the dictatorial polices of (Idi Amin, Wangari Maathai).

Inquiry

Directions: Combine the information in this chapter with your own ideas to answer these questions.

1. What programs have local groups launched in sub-Saharan Africa to cope with environmental problems and to improve living standards? Can you think of groups in your state, city, or town that are attempting to improve the environment?
2. Since 1900, what serious problems have confronted African herders? Do you believe they will ever be able to reestablish their control over Africa's grazing lands? Why or why not? What role will population growth play in this process?
3. What areas of sub-Saharan Africa do you think hold the most promise for economic growth in the next decade? Why have you selected these areas? What led you to exclude other areas?
4. Ask your school librarian to help you locate articles on the impact of population growth and farming on animal habitats in Africa. Discuss with the class the steps African governments are taking to protect these habitats. Does your local, state, or national government face similar issues?

CHAPTER 19

SKILLS

Using the Physical Map of Africa South of the Sahara

Directions: Turn to the map on page 468 and write the answers to these questions on your paper.

1. What is the name of the basin that is nearly in the center of Africa?
2. Is most of this basin covered by lowlands or mountains?
3. What deserts are in sub-Saharan Africa?
4. Are these deserts in sub-Saharan Africa larger or smaller than the Sahara Desert?
5. What is the largest lake in sub-Saharan Africa?
6. The Ethiopian Highlands border on what sea?
7. The Somali Plateau is near what ocean?
8. What is the northernmost lake south of the Sahara Desert on this map?
9. What two large lakes lie south of Lake Victoria?
10. Is the veld in South Africa or Central Africa?
11. Is Mount Kilimanjaro in West Africa or East Africa?
12. Is Mount Kilimanjaro located almost on the equator?
13. Does the Zambezi River empty into the Indian Ocean or the Atlantic Ocean?
14. Are the Cameroon Mountains near the Indian Ocean or the Atlantic Ocean?
15. What mountains are near the southern tip of Africa?
16. What two large rivers join to form the Nile River?
17. The Ubangi River flows into what river?
18. What large island is located off Africa's southeastern coast?
19. Is most of sub-Saharan Africa in the low latitudes or the middle latitudes?
20. Is most of sub-Saharan Africa in the zones of east longitude or west longitude?

Vocabulary Skills I

Directions: Match the numbered definitions with the vocabulary terms listed below by writing the term on your paper next to the number of its matching definition.

desertification	rift valley
escarpment	settled farming
jungle	shifting cultivation
oasis	tsetse fly
pastoral nomadism	veld
reforestation	villagization

1. The source of a disease that kills animals and makes people seriously ill
2. A deep trench formed when a large section of the earth's crust drops between two parallel cracks or faults
3. The open grassland that covers the central plateau in South Africa
4. The spread of desert-like conditions into semi-arid areas
5. A fertile watered area in the midst of a desert
6. Dense tangles of vegetation that line river banks in forested areas
7. In Ethiopia, the relocation of farmers in government-controlled settlements
8. A line of steep cliffs rimming a plateau
9. The reestablishment of woodlands damaged by human economic activities
10. A form of agriculture performed in one location where fertile soils and reliable rainfall can support it
11. The practice of moving cattle, sheep, or goats from one pasture to another in arid areas where vegetation is too sparse to support them in one place
12. A form of agriculture in which farmers move from one area to another in response to environmental conditions

Vocabulary Skills II

Directions: Write a paragraph about Africa south of the Sahara using at least five vocabulary terms.

TRADITION AND CHANGE IN AFRICA

L arge states and empires existed in Africa long before the continent was conquered by Europeans, but modern Africa has been shaped in large part by the economies and boundaries established during seventy years of European domination in the late 1800s and early 1900s. The forty-nine independent countries of Africa south of the Sahara have been struggling for thirty-five years to overcome the lingering effects of European rule.

◀ *This young Masai woman is wearing a traditional beaded necklace. What significance do you think this jewelry might have?*

1. The African Heritage

Discovering Africa's Past

Great nations rose and fell in Africa in and south of the Sahara long before Europeans entered this region. These civilizations, which were based on farming, herding, and long-distance trade, controlled large areas of the continent.

Many of these states were led by kings who were believed to have supernatural powers. Under their rule, city-states evolved into kingdoms and empires that governed the lives of millions of Africans. Because written records are sparse in Africa, the scale and grandeur of these early civilizations have only recently been discovered. Today we know that the story of African civilization goes back thousands of years and that Africans organized their societies and economies in many of the same ways as people did in other early centers of civilization.

African kingdoms sprang up in the Sahara (when it was less dry) and south of the Sahara—places where mineral resources were abundant, where the environment was especially productive, or where

African Kingdoms and Empires

Well-organized kingdoms and empires existed in Africa for thousands of years.

IT'S A FACT Africans conducted trade with India and China more than two thousand years ago.

access to world trade created wealth. These societies grew larger and more complex as Africans developed the skills to master the difficult environments of their large continent.

Africans settled in favored areas and became productive farmers, herders, metal workers, artisans, and merchants. Riches from these activities supported royal courts and capital cities of increasing splendor. By the 1800s, a majority of Africans lived in states ruled by kings. The map on page 492 shows the distribution and dates of some of the most important kingdoms and empires in Africa south of the Sahara.

The earliest and largest of these kingdoms arose in northeast Africa, in the grassland interior of West Africa, and in Central and Southern Africa. In each of these areas, states with opulent courts, finely constructed cities, and well-organized economies prospered. Here, Africans expressed their own cultures, religions, and ways of living. Later, Islamic and Christian beliefs spread southward across the Sahara Desert and became the ruling faiths of a number of African kingdoms. Most Africans, however, preserved their own ancient and highly developed cultures that explained the workings of the world around them and guided their daily behavior.

Kingdoms and Empires

About 800 B.C. a rich and powerful kingdom emerged at Kush in the upper (southern) reaches of the Nile Valley. Ideas, skills, and technology from the Middle East spread up the Nile River from Egypt into this part of northeastern Africa, which Egypt had ruled for centuries. The kingdom of Kush, therefore, played a major role in introducing the discoveries of early Middle Eastern civilizations into Africa south of the Sahara.

The people of Kush were mostly herders and farmers, but the merchants of Kush conducted trade over long distances. Important deposits of

iron ore existed in Kush, and nearby forests supplied wood for smelting the iron ore. The people of Kush gradually became skilled iron workers whose durable weapons and farming implements were highly valued over wide areas of Africa.

The kingdom of Kush became Africa's iron-mining and manufacturing center, and it retained its power for many centuries. Kings and queens ruled the kingdom of Kush in unbroken succession for nearly a thousand years. Its sprawling capital city at Meroe was adorned with stone temples, pyramids, and palaces whose ruins are among the great monuments in the ancient world.

The people of Meroe lived in a literate, well-ordered society that produced fine pottery, a distinctive style of art, and writings engraved in stone that cannot yet be understood. They wore cotton clothes from India and bought silk from China. Traders and ambassadors from Mediterranean countries to the north and African nations to the south exchanged goods and ideas at Meroe. Kush's influence extended deep into the heart of the continent. Conquered in A.D. 350 by Axum, a neighboring state in the Ethiopian Highlands, the kingdom of Kush with its splendid capital at Meroe was one of Africa's most important early civilizations.

Soon after the fall of Kush, empires based on control of international trade arose in the grassland interior of West Africa. Ancient Ghana, an empire located at the headwaters of the Senegal and Niger rivers, flourished between A.D. 700 and 1200. The people of Ghana knew how to make iron weapons, which they used to conquer neighboring groups of farmers and herders. Located on trade routes that connected the gold mines in the rain forests of West Africa with the copper and salt mines in the Sahara Desert, Ghana prospered by taxing the goods that moved north and south along these trade routes. Ghana also had major deposits of gold.

The capital city of ancient Ghana covered a square mile and housed 30,000 people. The kingdom had a well-developed bureaucracy, controlled a large population, and could field an army of 200,000 warriors—at a time when major European battles involved only small numbers of soldiers. The wealth of the king's court was legendary. Europeans called Ghana "the land of gold."

Ancient Ghana fell into decay in the 1200s, but other West African empires arose to control its rich trade in gold, copper, and salt. The most powerful of these states was Mali (A.D. 1200–1500), which managed to acquire both the southern gold mines and the northern salt deposits in West Africa, and thereby gained a monopoly over the richest trade network on the continent. This vast empire governed more than a half-million square miles of well-watered grasslands.

The fabled city of Timbuktu in Mali was an international center of learning and commerce. Located on the great bend of the Niger River, Timbuktu was the place where the camel met the canoe. Canoes laden with gold and slaves traveled up the Niger River to Timbuktu. These goods were exchanged for salt, copper, Venetian beads, and fine swords that had been carried southward from the Mediterranean by camel caravans crossing the Sahara Desert.

In time, Mali was conquered by the neighboring state of Songhai, which ruled this region and its trade between 1350 and 1600. Still later, the centers of trade in West Africa shifted eastward to the kingdoms of Kanem and Bornu and southward to the Guinea coast. These later kingdoms derived their wealth primarily from the slave trade.

Other African kingdoms arose in Central and Southern Africa. The kingdom of Kongo (A.D. 1400–1600), located south of the Congo River, was known for the prosperity of its royal court and capital city. Its wealth was based on an efficient system of taxing farmers. Other organized states existed in

▼ *The African city-state of Mali (A.D. 1200–1500) was home to a fabled city which was the center of learning and commerce. Traders from North Africa and Europe met with people of the south and exchanged, gold, copper, food, and other goods. Today the name of Mali's capital city carries connotations of the farthest, most remote spot on the face of the earth. What is this city called?*

▼ *Great Zimbabwe was one of the royal centers of a great civilization of Southern Africa. Huge walls still stand in mute testimony to the greatness of the Karanga people. How did they make these walls without mortar, concrete, or modern stone-working techniques?*

the lake regions of East Africa and on the southern margins of Africa's tropical rain forests.

The inland kingdom of Monomotapa (A.D. 1400–1800) was the most impressive civilization in Southern Africa. This kingdom of farmers and herders traded gold, iron, and ivory with Arab merchants on the East African coast in exchange for brass rods, shells, Chinese porcelain, and other luxury goods.

The capital city of Monomotapa, after which modern Zimbabwe is named, was renowned for its splendid architecture. The city housed a large population as well as the royal court, markets, and religious centers. The artisans of Great Zimbabwe were skilled stonecutters who constructed a spectacular stone palace ringed by an 800-foot-long wall of carefully fitted granite slabs thirty feet high and twenty feet thick. A million blocks, weighing 15,000 tons, were required for this wall alone.

Great Zimbabwe's stone ruins comprise 150 or more stone enclosures spread over sixty acres. The first European to find these ruins wrote that he picked up "bracelets of pure gold by the dozen." Monomotapa fell into decline after the Portuguese took over the ports of East Africa and asserted control over the gold trade of this highly-organized African state.

⊕ REVIEW QUESTIONS

1. The earliest and largest African kingdoms existed in what three parts of the continent?
2. What do we know about the cultural achievements of the early kingdom of Kush?
3. Why was Timbuktu described as the place where the camel meets the canoe?

⊕ THOUGHT QUESTIONS

1. How did the location of Ghana help this kingdom to prosper? Can you think of other places where relative location enhanced the economy of a city or region? What factors might give one city greater prospects of wealth than other cities?
2. Why can the inland kingdom of Monomotapa be considered the most impressive civilization in Southern Africa? On what was the wealth of this kingdom built? Why do you think the citizens of Zimbabwe named their country after the capital of this kingdom?

2. Early European Contacts with Africa

The Portuguese Sail Around Africa

For centuries Europeans had no direct contact with tropical Africa. A lively trans-Saharan trade in gold, ivory, and enslaved peoples from tropical Africa to the Mediterranean coast existed, and these trade goods were prized in Europe. The trade routes across the Sahara were controlled by camel nomads and by Muslim kingdoms on the Mediterranean coast of North Africa.

Then, in the 1400s, new sailing ships (caravels) enabled Europeans to bypass the Sahara and sail directly to the coast of tropical Africa. A visionary member of the Portuguese royal family, Prince Henry the Navigator, became convinced that well-constructed ships could sail down the West African coast and then eastward to the Indies. Previously, small, unsafe sailing ships had been caught in the trade winds that pushed them steadily to the south and west toward South America. No one knew how to navigate northward against the wind back to Europe. Because the sailing ships that Prince Henry developed were faster and safer, and could sail against the wind, they could carry the Portuguese southward along the West African coast and return northward along that same coast to Europe.

Portuguese exploration of the African coast gained momentum. By 1446, Portuguese sailors had reached the Cape Verde Islands off the great bulge of West Africa. They made contact with the kingdom of the Kongo in 1483, and five years later a Portuguese sea captain reached the Cape of Good Hope at the southern tip of Africa. The Portuguese discovered few natural ports or protected harbors along the west coast of Africa; therefore, most of their trade with Africans was carried on from ships anchored offshore at the mouths of rivers.

 IT'S A FACT | **Africa has a smaller coastal fringe than any other continent.**

During the next four hundred years, European merchants traded with African kingdoms along the coast of West Africa, but the African interior remained a mystery to the Europeans. The Portuguese, Dutch, British, and French built forts and trading posts at key points along the coast. Inland from these trade centers, African kings controlled the trade routes along which European metals, firearms, cloth, and food crops flowed into the African interior. These products were exchanged for gold, ivory, and slaves sent out from the interior to the coast.

Barriers Kept Europeans Out of the African Interior

Many obstacles prevented Europeans from entering the African interior for nearly four hundred years. Most rivers in Africa are blocked by rapids and waterfalls where they drop off the central plateau that covers the heart of the continent. Bypassing these waterfalls by canoe was difficult, so members of trading expeditions into the interior had to carry everything on their backs. This was especially difficult because the tropical rain forests of lowland Africa were infested with tsetse flies, which

killed animals and made people seriously ill. Europeans also risked battle with hostile African kings who often fought to protect their lucrative trade routes.

Perhaps the most serious barriers to the African interior were endemic diseases like malaria, dengue fever, and yellow fever. The continent came to be known as the "white man's grave," because between 40 and 60 percent of newcomers from Europe died during their first year in Africa. A song of the time warned sailors that "Forty go in for every one that comes out."

The Slave Trade Grows

Many African states had sold people as slaves to Europeans and Asians long before the Portuguese reached Africa. Most of these slaves were prisoners of war captured in local battles who were transported to the Mediterranean coast and sold into bondage to Europeans. Only after the development of European sailing ships did slavery become a major source of income for the kings of West African states. These kings controlled a network of trade routes linking the West African coast with other parts of Africa. European trading ships an-

▶ *Victoria Falls, the area surrounding the Zambezi River, is very near to five African nations. What are these countries?*

A slave could be traded for sixteen muskets or six hundred pounds of sugar in London, England, in 1700.

chored off the coast of the West African kingdoms unloaded their cargoes of trade goods, and filled their ships with slaves. The Europeans then sold these slaves in the newly established colonies of the New World where laborers were in high demand.

The Atlantic trade opened a huge market for slaves that were needed to work on sugar, tobacco, cotton, and coffee plantations in Brazil and the Caribbean. High death rates on these plantations meant that new cargoes of slaves were needed each year. African kingdoms, in turn, began to depend on the manufactured products of Europe. Exchanges between Europeans and Africans usually took place at trading posts at the mouths of the rivers of West Africa. European cloth, brass kettles, knives, axes, copper and iron bars, rum, and guns flowed through these trade centers into the most densely settled parts of Africa. New World crops, like corn and manioc, also were introduced into the African interior.

What amounted to a gun-slave cycle increased warfare in Africa. African kings bought guns and used them to enslave enemies. These captives were sold for profit, and the profit was then used to buy more guns. Distrust, fear, terror, and warfare increased in the African interior as the trade in guns and slaves reached deeply into the African continent.

The enslaved people were held in pens at coastal trading posts until European ships arrived to collect them and begin the six- to ten-week voyage across the Atlantic Ocean to the Americas. Between 1700 and 1850, an estimated 11 to 12 million slaves were shipped to the New World. Many more died on the brutal journey from interior Africa to the coast, in the slave pens at coastal trading posts, below deck on the notorious journey across the Atlantic, and in rebellions. Half of the slaves that crossed the Atlantic were sent to the islands of the Caribbean and about a third went to Brazil. Only one in twenty-five arrived in the United States. The slave trade was a despicable and degrading business that held itself to only one standard—that of profit and loss. It deadened the hearts of all who participated—traders, sailors, and slaves alike.

Explorers and Missionaries Enter Africa

For hundreds of years, Europeans who came to Africa stayed on the coast and rarely penetrated the continent, as you have already learned. Three great motives caused Europeans to venture into the African interior. These were curiosity, Christianity, and commerce.

European exploration began with the journey of the Scottish explorer James Bruce to the highlands of Ethiopia in East Africa in 1769. Although he failed in his attempt to discover the source of the Nile River, his writings fired the imagination of Europeans. Somewhat later, expeditions were sent from Europe to trace the course of the Niger River in West Africa. By 1825, a Frenchman succeeded in reaching the fabled city of Timbuktu. In East Africa, the famous explorations of Richard Burton and J. H. Speke solved the riddle of the Nile and proved that Lake Victoria was the river's ultimate source. On the basis of these explorations, European cartographers began to fill in the blank spaces on their maps of Africa.

In the late 1700s and 1800s, a humanitarian movement in Europe led to laws abolishing participation in the slave trade. Missionaries soon flooded into Africa to convert native peoples to Christianity, to introduce Western civilization into the region, and to eliminate the slave trade at its source. Believing that they had a duty to help Africans, these missionaries built many schools, hospitals, and churches in Africa.

By the 1850s, the discovery of quinine, an anti-malaria drug, made it safer for Europeans to live in the interior of Africa. Trading stations and diplomatic offices were set up throughout West Africa, and European trade depots dealing in palm oil, ivory, and rubber were scattered along the great

No motor vehicle reached the city of Timbuktu until after 1920.

MOVEMENT

DAVID LIVINGSTONE, MISSIONARY AND EXPLORER

David Livingstone embodied the three great motivations—curiosity, Christianity, and commerce—that lured Europeans to the interior of Africa. More than any other explorer-missionary, Livingstone worked to see Africa brought into the community of nations and alerted Europeans to the miserable living conditions of many Africans.

Livingstone was born into a poor Scottish family in 1813. Although he worked in a textile mill from age ten to age twenty-four, his burning desire was to become a medical missionary to China. He studied science and medicine in Glasgow, Scotland, and London, England. After he completed his studies, the London Missionary Society sent him to South Africa.

After eight years of missionary work in what is now Botswana, Livingstone mounted an expedition across the Kalahari Desert to explore the rich agricultural lands of east central Africa. For the next fifteen years, Livingstone tramped across Africa converting Africans to Christianity, exchanging remedies with African healers, and keeping detailed notes on the geography of these newly explored lands. He actually walked across the continent from Luanda, Angola, on the Atlantic coast to the mouth of the Zambezi River in modern-day Mozambique.

In 1865, Livingstone set out on his last journey to find the source of the Nile River and was lost to the outside world for five years. The American journalist, Henry Stanley, found him sick with fever in the village of Ujiji on the shores of Lake Tanganyika. After recovering his health, Livingstone continued his search for the source of the Nile. He died in a village near the headwaters of the Congo River in 1873. His African followers carried his body a thousand miles to Zanzibar, an island off the east coast of Africa. He is buried at Westminster Abbey in England.

Livingstone's explorations and writings attracted commercial interest to the basin of the Zambezi River, led to the end of the slave trade in East Africa, and inspired many other missionaries to establish hospitals in Africa. Rejecting what he called "the stupid prejudice against color," Livingstone was the first missionary to bring the sufferings of the African people to the attention of Europeans.

▼ *This engraving was made for Henry Stanley and depicted his meeting with Dr. Livingstone at Ujiji, Lake Tanganyika. Who was Henry Stanley?*

QUESTIONS

1. Although supported by the London Missionary Society, Livingstone is best known for what activity?
2. How do you think Livingstone helped put an end to the slave trade in East Africa?

rivers of West Africa. A similar mixture of exploration, missionary activity, and commercial ventures opened East Africa to Europeans in the 1800s.

As late as the 1880s, Africa was ruled by Africans, but the foundations for European political domination already had been established. In West Africa, the British and French occupied outposts along the Atlantic coast between the Senegal and Niger Rivers. Farther south, Henry Stanley was exploring the basin of the Congo River.

In South Africa, Dutch colonists, who had moved into this area in the 1650s, were clashing with British settlers, who had arrived in the early 1800s. The Dutch abandoned the coast to escape British influence, and moved into the interior of South Africa, which brought them into conflict with various African groups. Meanwhile, Cecil Rhodes, a British multimillionaire who controlled the diamond mines of South Africa, was working to set up a huge British colony in East and Southern Africa that would extend from Cape Town at the southern tip of the continent to Cairo, Egypt, in the north. After four centuries of marginal contact with Africans, Europeans were poised for the conquest of a continent.

⊕ **REVIEW QUESTIONS**

1. How did advances in sailing ships enable Europeans to sail south along the west coast of tropical Africa and then return to European ports?
2. What European and New World products flowed through coastal trading posts into the most densely settled parts of Africa?
3. Between 1700 and 1850, about how many slaves were shipped to the New World?
4. What did David Livingstone accomplish?

⊕ **THOUGHT QUESTIONS**

1. What obstacles prevented Europeans from entering the African interior for almost four hundred years? Do these obstacles still prevent people from the middle latitudes from settling in tropical areas? Would you consider working in a tropical country? Why or why not?
2. Why did the Atlantic slave trade open a huge market for slaves from Africa? What contributions have Africans made in building New World societies and economies?

3. Europeans Take Over Africa

European Countries Divide Up Africa

After four hundred years of contact with Europeans, Africans fell under direct European rule in only thirty years. Between 1885 and 1914, a number of European countries fought among themselves about subdividing Africa into European colonies. African resistance was fierce. The rebellions that occurred in virtually every part of Africa were crushed by better-equipped European armies whose weapons gave them a clear military advantage.

European countries wanted trade, natural resources, and access to Africa's wealth. They believed that they had to establish outposts wherever possible in Africa or else they might lose control of trade and resources of a region to some other European power. Boundaries of European colonies, therefore, were drawn to minimize clashes among these foreign powers; little attention was paid to the distribution of African kingdoms or ethnic groups. Africa became the last continent to fall under European domination.

The Impact of Colonialism on Africa

Because Europeans valued raw materials, they opened Africa to trade and settlement. Africans within the European colonies were forced to produce these raw materials. Farmers and herders were compelled to abandon traditional cultivation of food crops and animal herding and instead to raise commercial crops like palm oil, cacao, cotton, peanuts, coffee, and tea for export to Europe. Transportation systems, plantations, and mines were established and paid for by taxes on Africans. These taxes had to be paid in cash, and Europeans paid cash only to Africans producing commercial crops and working for them.

The colonial era lasted only seventy years, but during this brief period, many of modern Africa's problems were created. Africa was divided into colonies that later became the states of independent Africa, as the map on this page shows. Railroads and roads were built so that European soldiers could be moved from one place to another quickly to maintain law and order in areas of production, and also so that Europeans would be assured access to the food and mineral resources of the African continent. In the process of dividing the continent, African societies and cultures were badly damaged.

The European Colonies in West Africa

In 1850, fewer than one thousand Europeans lived in West Africa. They worked in trading posts and forts along the West African coast. Fifty years later, Europeans ruled seventeen different colonies in West Africa, which were populated by one-third of all the Africans living south of the Sahara.

By 1900, the French had carved out an empire in West Africa that was nine times the size of France. Seven French-controlled territories were joined together to form the very large colony of French West Africa, which was governed from the port city of Dakar in Senegal. French provincial administrators were sent out to villages and towns to maintain control of local African groups. An estimated 17 million people lived in this huge French colony.

French West Africa had few mineral resources, so the French forced Africans to grow commercial crops. Peanuts and cotton were planted on the grassy plains of the Senegal Valley and in the colony's interior. Along the tropical coast, cacao, palm trees, exotic hardwoods, coffee, and bananas were raised for export. Africans had to pay taxes to

The European Conquest of Africa, 1884–1914

The "scramble for Africa" brought the entire African continent, except for Liberia and Ethiopia, under direct European colonial rule in only thirty years. Boundaries were drawn between European spheres of influence. Little attention was paid to the distribution of African peoples or their ways of living.

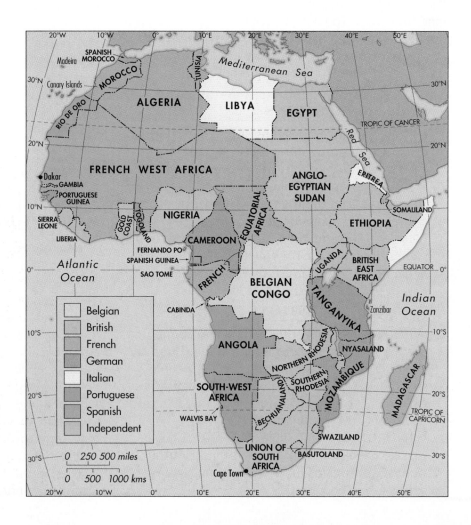

the French government. In French West Africa, however, people were scattered over an area larger than the United States, so many Africans were largely untouched by French colonial rule.

In contrast, the British colonies in West Africa were small, were more densely populated, and had important mineral resources. The Gold Coast (now Ghana) became the richest colony in West Africa. Railroads were built inland to transport shipments of gold, manganese, and bauxite ore to coastal ports for export to Europe. Cacao and palm oil plantations were established along the coast of the much larger colony of Nigeria, and peanuts were grown in the interior.

The British did not rule their colonies in West Africa from one central headquarters like the French did. Instead, they adapted their forms of local government to local conditions. The many minerals and crops exported from British colonies in West Africa made it possible for the British to raise money by taxing goods instead of people.

Other European powers also had colonies in West Africa. Germany imposed rigid discipline on its West African colonies of Togo and Cameroon. In Portuguese Guinea, little changed. The one area of West Africa that did not fall under European con-trol was the independent nation of Liberia, created in the 1820s by the United States as a home for freed slaves. Liberia retained its traditional agricultural ways of life until 1925, when a rubber company leased a million acres of Liberian forest land and established large-scale rubber plantations there.

The Tropical Colonies in Central Africa

The tropical environments of Central Africa posed different problems for the French, the British, and King Leopold II of Belgium, all of whom established colonies in this region. Central Africa's rain forests were thinly settled by small groups of people who hunted, fished, and engaged in subsistence farming. There were few Africans to raise commercial crops for export, no trade goods to tax, and no regions of dense settlement where taxes could be collected. As a result, colonial development was paid for by European companies who invested money in Central Africa in the hope of making large profits. Mineral and farming rights to large areas of the interior were given to these private companies, on condition that they build highways, railroads, plantations, and mines. These rights to

Plantations such as this one in the Ivory Coast were originally set up by Europeans in the 1800s. What crop is being produced here?

PLACE

"NEGRITUDE": THE QUALITY OF BEING AFRICAN

In the view of Africa's writers, poets, and politicians, black Africans share precious qualities—emotion, sensibility, and understanding—that, taken together, make up "negritude," the quality of being African. This is the basis of Aimé Césaire's passionate condemnation of colonialism. He writes:

They speak to me of progress and "accomplishments," sickness conquered, higher standards of living.

I speak of societies emptied of themselves, of trampled cultures, undermined institutions, confiscated lands, . . . annihilated artistic masterpieces, of extraordinary possibilities suppressed.

They throw up to me facts, statistics, the number of kilometers of roads, canals, and railways. I speak of thousands of men sacrificed . . . I speak of millions of men torn away from their gods; their land, their customs, their way of life, their livelihood . . . and their wisdom.

Source: Aimé Césaire, "On the Nature of Colonialism."

QUESTIONS

1. Who does "They" refer to in the first line of Césaire's excerpt?
2. Who does the "I" in the second paragraph of the excerpt refer to?
3. Is this excerpt written to appeal more to emotion or logic? Why?

---

land and minerals were known as **concessions.** Central Africa was largely developed by concession companies.

The Congo Free State was the personal domain of King Leopold II of Belgium. Beginning in 1886, he sold land concessions in the Congo to companies who agreed to build railroads into the interior. All lands not under cultivation were declared to be the property of the government. This had a disastrous effect on the native peoples of the rain forest, most of whom were shifting cultivators who moved from one area to another every few years. Now all new areas were owned by the government. Additional laws forced Africans to work for concession companies and to pay taxes. Armed police enforced a reign of terror in the region from 1889 to 1908.

During these nineteen years, half of the African population in the Congo died of forced labor, starvation, or rebellion. In 1908, Leopold II was forced to cede the Congo to the Belgian government.

The first products brought out of the Congo were ivory and rubber. Cotton plantations, palm groves, sisal farms, and mines were established on tracts of land granted to concession companies. Then air-filled automobile tires were invented, and the demand for rubber soared. African farmers were taken from their fields and villages and forced to tap rubber trees in the rain forest. The dreadful human conditions that existed under the concession companies improved after the Belgian government took over the colony, but Africans still worked for European profit.

The French colonies in Central Africa were also developed by concession companies, although the policies applied in French Equatorial Africa were less brutal than those in the Congo Free State. Africans worked on palm, coffee, and cotton plantations and harvested wild rubber. Here, as in the Congo, the food economy was disrupted because people could not grow food crops while working for the Europeans.

IT'S A FACT   African elephants are being wiped out by ivory poachers, particularly in Somalia and Tanzania; 40 percent have died in the last seven years.

The British, Germans, and Portuguese in East and Southern Africa

In the highlands of East Africa from Kenya southward to the Union of South Africa, the British found fertile land with a temperate climate that was sparsely populated by Africans. Initially, the British government attempted to have native Africans raise cash crops here. Then they encouraged immigration from Europe. In what is now Malawi, dense African populations occupied the best farmland for a time and restricted settlement by incoming European immigrants. In Uganda, in return for helping the British, the powerful Buganda people were given farmland on the shores of Lake Victoria that they promptly turned into profitable cotton farms.

European immigration to British East Africa increased when large tracts of land were taken from Africans and set aside for European settlers in Kenya and Northern and Southern Rhodesia (modern Zambia and Zimbabwe). Other immigrants came to

IT'S A FACT   Some 100 different languages and dialects are spoken in Tanzania, more than 300 in Nigeria.

work in the copper mines. By 1914, there were nearly 20,000 British settlers in East Africa.

Germans also encouraged immigration to their African colonies of Tanganyika (now Tanzania) on the east coast and South-West Africa (Namibia) on the west coast. Their goal was to create a German homeland in Africa that would provide new farmland for Germany's growing farm population and would produce raw materials for Germany's industries. In both colonies, German administrators brutally suppressed local peoples, levied heavy taxes on Africans, and forced African farmers to raise commercial crops for export.

In South-West Africa (now the Republic of Namibia), the Germans tried to settle Africans on the dry grasslands near the city of Windhoek in the

▼ *Kampala, the capital of Uganda, rests on the shore of Lake Victoria. Why do so many African places have non-African names?*

interior. Most Africans in this colony were part-time farmers and herders who moved about with their herds, a way of living that was well adapted to the area's dry climate. These people fiercely resisted settlement in German-controlled areas. As a result, the Germans declared all grazing land to be owned by the German colony and forbid Africans to raise cattle. After two years of warfare, two-thirds of the Africans in South-West Africa had been killed; the rest had either fled the colony or worked as laborers on German farms and ranches.

In Tanganyika, African resistance to German rule began as soon as the Germans bought the area from the sultan of the offshore island of Zanzibar in 1884. The Germans suppressed Swahili revolts along the coast, but in the interior, much larger African groups resisted German settlement. After five years of warfare, however, native settlements in Tanganyika were destroyed.

In the Portuguese colonies of Angola and Mozambique, conditions were no better. The Portuguese shipped some 3 million slaves from Angola to Brazil, seriously depopulating this dry, thinly inhabited land. In Mozambique, palm and cacao plantations were established in rain forests along the coast. When local African laborers began to die off, Indians from South Asia were imported into Mozambique to work on the estates of Portuguese colonists.

▲ *Soldiers return from the Anglo-Boer War in South Africa. The Dutch settlers (the Boers) used guerrilla tactics against the British who were coming to South Africa in ever-increasing numbers. What did the word Boer originally mean?*

The Boers and the British in South Africa

In South Africa, colonial rule took quite a different path. Europeans originally settled the coast of South Africa to establish trading stations that provided food and fresh water to ships sailing between Europe and Asia. The first of these stations was built by the Dutch in 1652 at Cape Town, on the southern tip of the continent. During the next hundred years, Dutch settlers increased in number and spread out from Cape Town; they established large wheat farms and vineyards in this fertile, temperate region. Slave laborers from West Africa, Malaysia, and the area around Cape Town worked the farms. As the Dutch population grew, the Dutch conquered and enslaved more and more Africans to work the land.

After a century in South Africa, the Dutch colonists (called **Boers**, or farmers) had developed their own distinctive rural culture. They spoke Afrikaans, a language simpler than the Dutch spoken in their original homeland. Their rigid religious beliefs included the right to own slaves.

In 1806, the British occupied Cape Town to protect their trade in the Indian Ocean, and by 1820, 5,000 British people had settled as town dwellers in what became known as the Cape Colony. Friction arose between the Boers and the British. The British, shocked by Boer cruelty to native Africans, gave legal protection to various tribes in the Cape Colony. When the British government forbade the Boers to own slaves in 1834, the Boers decided to move inland beyond the reach of British rule.

The Boers left the Cape Colony in ox-drawn wagons and traveled north across the Orange River in a march that became known as the Great Trek. This migration into the interior brought the Boers into direct conflict with Bantu-speaking Africans, who also were moving into the rich pasture lands of southern Africa's interior. The Boers and Bantus fought it out for control of these fertile lands. By 1850, the better-armed Boers conquered the Bantu peoples and set up two Boer colonies north of the Orange River—the Orange Free State and the Transvaal.

By now, the British and Boer colonies in South Africa had a combined population of 300,000 Europeans living among 1 to 2 million Africans. As the Europeans spread out, they seized all but the least fertile land from the Africans; one by one, African peoples were disarmed and destroyed. As a result, many Africans moved to Cape Town in search of work and food. With the discovery of diamonds at Kimberley (1867) and gold at Witwatersrand (1884) in Boer territory, many displaced Africans found jobs in the mines.

These great mineral discoveries, however, increased tensions between the British and the Boers. Thousands of foreign miners and adventurers flocked to gold-mining and diamond-mining areas in the Boer colonies. The Boers, already outnumbered by these intruders, feared that their culture and way of life would be destroyed. Conflict between the Boers and the British intensified and led to the Anglo-Boer War at the turn of the century, which ended in a British victory. The Union of South Africa, made up of the British and Boer colonies in South Africa, was formed in 1910.

The Union of South Africa Prospers While Africans Suffer

During the next fifty years, the Union of South Africa developed a thriving industrial economy. The gold fields of Witwatersrand supported a major mining and manufacturing zone centered at Johannesburg, the largest city in Southern Africa. A web of railroads was built across South Africa to connect inland industrial districts with port cities.

Some Africans participated in South Africa's economic development by providing cheap labor in the mines and factories of the industrial districts and by working on the cotton, rice, and sugar plantations in the east. Others were pushed onto **homelands**, which were reserves or reservations located in the least desirable areas of South Africa. Although Africans made up three-quarters of South Africa's population, their homelands covered only 13 percent of the land. However, Africans were forced by law to live either on these reserves, in slums in restricted areas near large cities (called **townships**), or on farms owned by whites.

In South Africa, the largest European population on the continent built a modern economy using Africans as cheap labor. Moreover, these descendants of Europeans—who have no European home country to return to—are determined to stay in Africa.

⊕ REVIEW QUESTIONS

1. When European colonies were established in Africa, what kinds of crops were African farmers forced to raise?
2. By 1900, how much larger was French West Africa than France? How many Africans lived in this French colony?
3. For nineteen years the Congo Free State was the personal domain of what individual?
4. The Union of South Africa was composed of colonies that had been settled by what two groups of Europeans?

⊕ THOUGHT QUESTIONS

1. How and why were the costs of colonial development paid for by European companies in the tropical environments of Central Africa? Can you think of an example in U.S. history where a similar system was used? Do you believe these systems were fair to farmers and herders in Africa and America?
2. What were the impacts of colonialism on Africa? Why were they more severe than in the United States, which was also a colony? Do you think the policies applied to Africa by Europeans during the colonial period have hindered economic development and social progress in Africa? Why or why not?

4. Africans Gain Independence

Independence Spreads Across Africa

Independence spread swiftly across tropical Africa in the years following World War II. The success of India and other countries in gaining independence encouraged African leaders to demand independence for their people. The once strong European colonial powers were exhausted by war and were eager to rebuild their own countries, and the costs of keeping their African colonies promised to be high. For these reasons, independence was achieved throughout most of Africa.

In 1955, only two African states—Liberia and Ethiopia—were already independent. A decade later, however, thirty-one new independent nations existed in Africa, seventeen of them created in 1960 alone.

Independence was first achieved in West Africa, where in 1957 the Gold Coast became the independent state of Ghana under the leadership of Kwame Nkrumah. By 1960, almost every European colony in West and Central Africa had won independence. East Africa soon followed, although independence came more slowly in areas of dense white settlement.

In the south, the Portuguese, after years of fighting against independence movements, withdrew from Angola and Mozambique in 1975. In 1965, the white minority government of Southern Rhodesia broke away from British rule and for the next fourteen years fought against the guerrilla forces of the Patriotic Front to remain an all-white government. The British helped work out the peace, and in 1980, Southern Rhodesia became the Republic of Zimbabwe with a multiracial government.

Namibia (formerly South-West Africa) was one of the last to gain independence. South Africa had taken the country by force from the Germans in 1915 and then was given supervision over South-West Africa by the World War I peace treaty. Despite efforts by black independence fighters, South Africa extended its laws

▲ *In 1955, Ethiopia was one of only two independent African nations. In the 1980s and 1990s, when United Nations relief efforts were trying to get food and aid to the Ethiopian people, Ethiopia was struggling for its identity amidst a civil war. What was the other independent nation in 1955?*

to South-West Africa in 1969. The United Nations refused to allow South Africa to incorporate the country and officially condemned South Africa for its actions. After several more years of fighting against black separatists, South Africa handed over limited powers in 1985. Multinational negotiations on the handling of elections and the form of the new government finally took place, and Namibia became independent in 1990.

IT'S A FACT African countries have nearly a third of the votes in the United Nations General Assembly.

In 1993, South Africa signed a draft constitution guaranteeing individual rights and free elections to all peoples within the Republic of South Africa.

Problems of Independence

Independence in Africa did not bring peace, prosperity, or an end to poverty. The boundaries of old colonies became the boundaries of the new independent states. Africa south of the Sahara remains divided into forty-nine different countries. Within many countries, tensions between different ethnic groups are strong, and few African leaders have experience in government, because whites controlled their continent for more than fifty years.

The problems of developing the resources of these newly independent African states were immense. Farmland, mines, railroads, and ports had been constructed to benefit the former European colonial rulers, not the African people. Production of minerals and agricultural products for sale in European markets was the main source of revenue in most countries; in fact, entire economies were dependent on the sale of these products. Virtually no investment had been made in the health, education, and welfare of the people. Africa south of the Sahara became independent, but it was the poorest and least developed culture region in the world.

⊕ REVIEW QUESTIONS

1. Between 1955 and 1965, how many new independent nations were formed in Africa?
2. When they achieved independence, the Gold Coast and Southern Rhodesia took what new names?
3. What major political event took place in South Africa in 1993?

⊕ THOUGHT QUESTIONS

1. Why is Africa south of the Sahara one of the poorest and least developed regions of the world? Had you been a leader of an African nation, what would you have done to improve the living conditions of your people?
2. How did Europeans justify their actions in Africa? Do you believe that Africans and Asians believe these justifications? What do you believe? Why?

5. Islands of Modern Development in Africa

Problems from the Colonial Period

In the late 1800s, European countries divided Africa into a number of colonies, within which they mined minerals and produced commercial crops for export to Europe. The Europeans established port cities to handle cargoes going in and out of Africa. Railroads were built to connect these ports with inland mining and commercial farming areas.

These colonial investments left Africans with three problems when the colonies became independent states in the 1960s. First, the boundaries of their countries were drawn to suit the needs and interests of Europeans, not Africans. Second, transportation systems within each country and between countries were not adequate to serve the needs of independent countries. Third, many African countries had dual economies with small islands of modern industrial development run by and for Europeans, alongside large traditional farming and herding populations.

African Countries Inherited Colonial Boundaries

The boundaries of most African colonies were straight lines or were drawn along rivers or mountain ranges, as the map on page 500 shows. These boundaries separated the territories controlled by different European powers. In drawing the boundary lines, little attention was paid to Africa's needs. As a result, fourteen African colonies were landlocked. Other colonies were absurdly small. Many culture groups were split apart, and peoples with different cultures were thrown together. Very few countries in sub-Saharan Africa (Cape Verde, the Comoros Islands, Lesotho, Swaziland, and Somalia) are composed of a single national group.

The leaders of newly independent states in Africa inherited countries that were very difficult to govern. Countries like Djibouti, Niger, and Burkina Faso had no real sources of income. Other countries

IT'S A FACT Africa has more countries than any other continent.

like Nigeria, Zaire, Uganda, and Burundi faced conflict between and within ethnic groups who wanted to govern themselves. Africans inherited countries whose people spoke different languages and practiced different faiths. Modern political movements in Africa are often rooted in differences among ethnic groups, resulting in rebellion and civil war.

Transportation Systems Reinforce Dual Economies

Inadequate transportation systems were a second obstacle to modern development in many African countries after independence. Europeans had built 50,000 miles of railroad track in Africa, but most rail lines connected coastal ports with interior mining and commercial farming areas. Few railroads or roads crossed colonial boundaries except to reach the sea. As a result, few of these trade routes connect neighboring African countries with one another. Each country is forced to modernize within the limits of its own boundaries.

So colonial boundaries and transportation networks reinforced the dual economies that most African countries inherited at independence. Heavy European investment in mines and commercial farms provided many African countries with their only source of income. Minerals still provide most of the wealth in countries like Zambia, Zimbabwe, and Zaire. Oil is the mainstay of the economies of Nigeria, Cameroon, and Gabon. Peanuts, palm oil, and cacao support many West African nations, and cotton, coffee, and tea are the most important export products of East Africa.

The human resources of African countries were also not developed during the colonial period. Less than one African adult in ten could read or write at independence, and few people worked for wages. Most Africans still live outside the islands of economic growth created by the colonial powers. As a result, modernization in Africa is restricted to two zones of economic activity: (1) coastal cities, and (2) interior mining and commercial farming.

Cities in Sub-Saharan Africa

One of the largest and most varied zones of economic development in sub-Saharan Africa is located in the twelve countries that line the coast of West Africa from Senegal to Cameroon. The cities of this coast have diversified industries that process the crops and minerals produced in their countries. This coastal zone is the most densely populated, highly urbanized, and economically developed area in sub-Saharan Africa except for South Africa. Some 160 million people live here, the majority in Nigeria.

Development began in the 1800s, when Europeans organized plantations along the coast of Nigeria to produce palm oil for soap. Other commercial crops soon followed. Peanut cultivation was started on the dry plains of Senegal and Gambia and in the interior of West Africa. Coffee and bananas became important products in Guinea and the Ivory Coast. Cacao was a major cash crop in Cameroon and Ghana. Later, rubber plantations were established in the rain forests of Liberia.

Minerals were also discovered in West Africa. Diamonds in Sierra Leone; bauxite (aluminum ore) in Guinea and Ghana; iron in Liberia; and petroleum in Nigeria, Gabon, and Cameroon contributed to West Africa's wealth. Railroads connected producing areas with the coast, where early European trading posts grew into large cities after independence.

Most large cities in West Africa are located on the coast or at railroad junctions in the interior. Coastal cities with more than a million residents include Dakar in Senegal, Conakry in Guinea, Abidjan in the Ivory Coast and Accra in Ghana. Lagos, with a population of more than 8 million, is the largest city in Nigeria. Yaoundé, the capital of Cameroon, has a million residents. Locate these cities on the map on page 469 and notice how different-sized dots are used to show cities of different sizes. Only Nigeria has sufficient development to support cities of more than a million people in the interior. These include Ibadan in western Nigeria and Kaduna in the north. In an effort to foster national unity, Nigeria has built a new capital city at Abuja in the center of the country. The capital was moved to Abuja from the large port city of Lagos.

These West African cities are growing rapidly. Many new city dwellers are migrants from the countryside, as the map on this page shows. As a result of this stream of immigrants, the gleaming skyscrapers that rise above the center of West Africa's largest cities are ringed by shantytowns, where hundreds of thousands of rural migrants live and seek work. Many are single men trapped between two worlds. They are driven from their villages by a shortage of land and opportunity and are drawn to the city by hopes of a job and a better way of life.

In sub-Saharan Africa more than a million people are found in the port city of Kinshasa in Zaire which serves the Congo Basin. Luanda is the capital of Angola. In East Africa and the African Horn,

Addis Ababa is the capital of Ethiopia and Dar es Salaam is the major port of Tanzania. The capital of Tanzania is in transition between Dar es Salaam and the interior city of Dodoma. Maputo is the port-capital of Mozambique. Interior capital cities include Nairobi in Kenya, Kampala in Uganda, Lusaka in Zambia, and Harare in Zimbabwe.

The industrial economy of South Africa supports five cities of a million or more—Johannesburg, Pretoria, Cape Town, Port Elizabeth, and Durban. Elsewhere, Africa's economy is too poor to support large cities. Sub-Saharan Africa is the least urbanized region in the developing world.

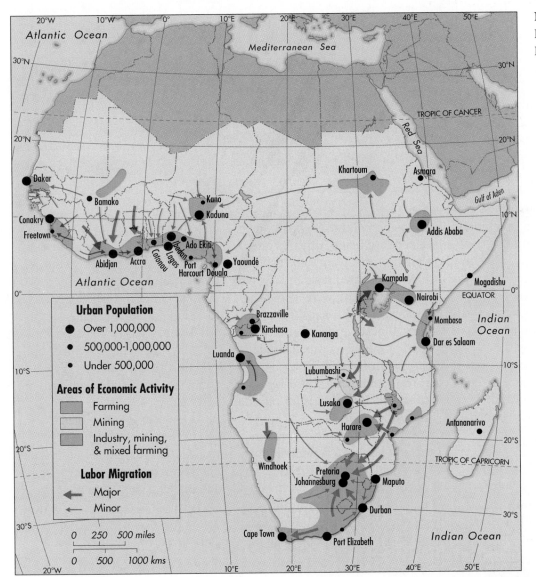

Migration to Africa's Islands of Modern Development

THE NEW POLITICAL GEOGRAPHY OF SOUTH AFRICA

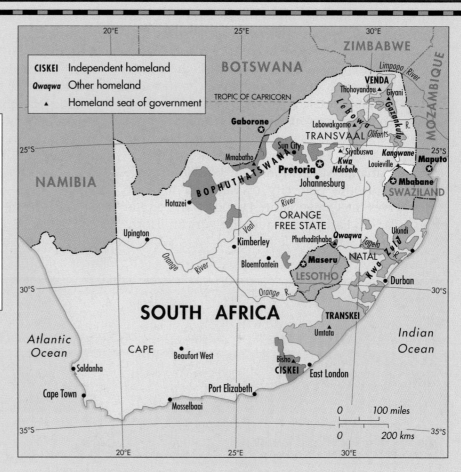

CISKEI — Independent homeland
Qwaqwa — Other homeland
▲ — Homeland seat of government

In 1948, a whites-only election transformed the politics and geography of South Africa. Black Africans were not allowed to vote. Dutch-speaking whites (called Afrikaners) took control of the country.

They began to carry out a policy of **apartheid**, or "apartness," in order to keep the white minority in control of the country. "Whites Only" signs appeared on buses and trains, and schools were segregated. Various semi-independent territories were established in South Africa.

Each ethnic group—whites, blacks, Asians, and Coloured (people of mixed blood)—was assigned its own territory, or homeland. Outside the homelands, members of each ethnic group had limited rights. The African majority was segregated from the European minority, which maintained strict control over most of South Africa.

In the late 1950s, 250 separate areas inside South Africa were set aside as homelands in which ten independent black African states, or **bantustans**, were to be established. Segregated residential areas surrounding cities were also designated as **townships** for the Asians, blacks, and Coloured.

Four bantustans were actually created—Transkei, Bophuthatswana, Venda, and Ciskei. As the map shows, homelands for black South Africans included about 13 percent of South Africa's land area. This land had no known minerals, no large cities, and few resources.

Many whites were determined to retain control of South Africa. The policy of apartheid stemmed from fears that in a democratic society the white minority would be swamped by the black African majority. Today, blacks make up 74 percent of South Africa's 39 million people; whites are 18 percent of the total population. These population pressures have made it increasingly difficult for white South Africans to retain the policies of racial discrimination they held in the past.

In the mid-1980s, protests against apartheid—in the form of work stoppages, strikes, sit-ins, and riots—became more intense. In reaction, the government in 1985 and again in 1986 imposed a state of emergency, giving the police special powers to arrest and detain people indefinitely. Freedom of the press and freedom of association were also curbed.

These actions in turn led the United States, the European Community, and the United Nations Security Council to impose sanctions on South Africa. Because of its apartheid policy and repressive measures, South Africa became an international outcast. Multinational companies doing business with South Africa were boycotted and many left the country. South African athletes were prohibited from competing in the Olympics and other international events.

Then South Africa took some of the actions the world had been urging. By the late 1980s, the government began gradually dismantling apartheid laws and loosening re-

This scene from a township near Cape Town is typical of the housing and general living conditions of many urban black South Africans. What has recently happened to segregation laws that dictated where people could and could not live?

strictions in the homelands. When F. W. de Klerk was elected president in 1989, he lifted the government's ban on the African National Congress (ANC) and other black African political parties. In 1990 Nelson Mandela, the leader of the ANC and a major symbol of black resistance, was released after twenty-seven years in prison. He has since played an important role in negotiations with the government on the formation of a majority-ruled government. Mandela and de Klerk were awarded a joint Nobel Peace Prize in 1993 for their efforts to bring peaceful change to South Africa.

Although much progress has been made, the tragic reality is that South Africa is now at a critical point. Many whites are willing to discuss alternative futures, but some Afrikaner groups are still digging in their heels, fighting the changes. And violence still erupts in South Africa, often between members of competing black groups with long histories of conflict and distrust. Some black leaders want a weak central

government and strong regional autonomy in order to maintain their power bases. Among these are the Zulu of the Inkatha Freedom Party, led by Mangosuthu Buthelezi, and the leaders of the four bantustans. In contrast, Mandela wants a strong central government to consolidate black African power.

The outlines of the new political geography that is emerging in South Africa became clearer late in 1993, when Mandela and de Klerk and other leaders signed a draft constitution that would abolish the homelands, guarantee fundamental rights for all, and establish free elections. In the new spirit of cooperation, two goals have been achieved: the elections of 1994, won by Nelson Mandela, and the writing of a permanent constitution. It seems likely that racial peace in South Africa will benefit all segments of its complex society.

QUESTIONS

1. What does the term *apartheid* mean?
2. What are the four major ethnic groups in South Africa?
3. What was the government's purpose in declaring states of emergency in 1985 and 1986?
4. According to the map, what is unusual about the location of the bantustans and homelands?
5. In your opinion, what is the major obstacle in South Africa to the peaceful transfer of power and participation in government of all peoples?

Mining Districts in Sub-Saharan Africa

Mining began in South Africa when diamonds were found near the city of Kimberley on the banks of the Orange River. Twenty years later, the world's greatest gold field was discovered in the Witwatersrand near Johannesburg, South Africa. Wealth from the Kimberley diamond mines provided money to develop the gold deposits of the Witwatersrand. South Africa's booming industrial economy was financed by the gold and diamonds of these mineral finds.

In the Witwatersrand, coal, precious minerals, and underground water are found next to one another. Gold is the area's most important mineral product, but uranium, platinum, nickel, iron, and manganese are increasing in importance. These mineral resources support a fifty-mile-long, twenty-mile-wide commercial and industrial district centered on Johannesburg that forms the economic core of South Africa.

The major problem in the Witwatersrand is finding laborers to work in its mines, steel plants, and factories. Workers are recruited from neighboring countries like Botswana, Mozambique, Lesotho, and Swaziland. Men from these countries are not allowed to bring their families with them when they work in South Africa. Although black Africans built the mineral economy of South Africa, black workers still earn much lower wages than whites in comparable jobs. This situation is now changing.

A second zone of economic development exists in the mining belt that sweeps northward from South Africa through Zimbabwe to Zambia and southern Zaire. Locate this mining zone on the map on page 509 and find its major mineral products. In the 1890s, white settlers moved into this area in search of another Witwatersrand. They found a rich mineral district stretching southward from Harare to Bulawayo. Gold was the most important mineral in the early days. Coal, iron, chrome, and nickel are Zimbabwe's most important mineral exports today.

Until Zimbabwe gained independence in 1980, black Africans earned much less than whites in this industrial area. Now salaries are equal.

Two other important mineral and mining districts straddle the border between Zambia and Congo. In Zambia, the huge ore resources of the Copper Belt were developed in the 1930s to satisfy growing world demand for copper. Seven towns in this mining area house 90 percent of Zambia's urban, industrial work force. As the country's only important export, copper provides half of all government revenue in Zambia.

Zaire's share of the Copper Belt was developed by Belgian concession companies. As in neighboring Zambia, copper is the most important mineral export. Today this district also produces tin and zinc, two-thirds of the world's cobalt, and almost all the world's radium.

⊕ REVIEW QUESTIONS

1. The effect of European colonial investments left Africans with what three problems when the colonies became independent states?
2. Modernization in Africa is restricted to what zones of economic activity?
3. South Africa's booming industrial economy was financed by discoveries of what two minerals?
4. What problems were caused when colonial boundaries became the boundaries of newly independent states in Africa?

⊕ THOUGHT QUESTIONS

1. Why is the coastal zone of West Africa the most densely populated, highly urbanized, and economically developed area in sub-Saharan Africa, except for South Africa? What economic activities support its people? Will this area sustain continued immigration from rural areas? What would you suggest this and other urban areas in the developing world do to meet the needs of incoming people?
2. Why did the whites establish a policy of apartheid in South Africa, and how was this policy implemented? How will the white minority deal with the shift to black rule in South Africa when it comes?

IT'S A FACT Africa has one-third of the world's potential hydroelectric power.

6. Africa's Growing Problems

More People Than Can Be Fed

Population is growing faster in Africa than in any other world region. Sub-Saharan Africa is home to 550 million people, twice as many as when this area became independent in the early 1960s. In most countries, population will double again in the next fifteen years. Families have an average of six children, compared to four in other developing regions. One of every seven children dies in the first year of life. How Africa's growing number of children will be housed, fed, and cared for in the years to come is an overwhelming problem in many countries.

Poverty is at the root of Africa's problems. In the 1980s and 1990s, Africa's rate of economic growth was smaller than in any other world region. Real incomes have fallen during this period. This means that most Africans have less to eat than their parents did. Malnutrition afflicts one of every five Africans, and many more Africans do not have enough protein in their diets to be healthy. Already, thousands have died in famines, as you have learned.

Poverty and poor diets make Africans vulnerable to disease. Malaria, carried by mosquitoes, is a killer of up to 1 million children a year and infects many adults in Africa. Sleeping sickness, transmitted by the tsetse fly, affects most of tropical Africa. Snails carrying a parasite that causes schistosomiasis, a severe blood disease, infest many lakes and streams. An estimated 20 million Africans are blind because of worms spread by black flies that breed in fast-moving rivers. Five million Africans suffer from leprosy. Over wide areas of the continent, medical care is absent or inadequate. The average African can expect to live to fifty-two years of age as compared with seventy-six in North America. Africa has the lowest life expectancy of any world region.

 IT'S A FACT One in eight human beings lives in Africa.

▲ *The level of poverty in Africa is extremely high. These homeless children do not face a bright future, but they have made it past a significant barrier in the life of all Africans. What percentage do not make it past their first birthday?*

The Low Level of Economic Development

Poverty hampers most attempts to increase economic development in Africa. Except in South Africa, most people live as subsistence farmers and herders. Governments rely on the export of minerals and commercial crops for the money needed to build schools, hospitals, factories, and plantations. This income from exports is falling. Prices of minerals like diamonds, gold, copper, cobalt, manganese, and platinum have fallen in the last twenty years. So has the value of commercial crops like coffee, cacao, peanuts, and palm oil. Efforts to build industrial economies to provide jobs for Africans have not reduced Africa's dependence on the export of these raw materials. Few Africans are now employed in mining and manufacturing—occupations that usually provide higher incomes.

Small-scale efforts to improve the environment and increase food production—efforts like those of Wangari Maathai in Kenya—hold some hope for the future. However, these local undertakings are dwarfed by widespread poverty. In Zaire, for example, nearly 60,000 miles of roads have fallen into disuse because the government does not have the money to repair them. The collapse of the road system means that farmers cannot transport their crops to market. As in many African countries, Zaire's leaders now must spend vital resources on importing food to feed their people. They have little money left to invest in the jobs and services that Africans so desperately need.

Faced with these economic realities, many African farmers and herders are seeking new lives in the cities. Many of these cities are ports or mining centers built during the colonial period, as you know. None were planned to house large populations. Thus, Africa's cities are not prepared to take care of tens of thousands of farmers pouring in from the countryside. Few urban centers have clean water supplies, sewage systems, or sufficient housing, hospitals, and schools to care for their rapidly growing populations. As a result, slums surround Africa's largest cities. People without jobs, unem-

ployment benefits, or social security must live in these shantytowns. And every year, more and more people migrate to Africa's cities. They have nowhere else to go.

Political Turmoil

Political turmoil has also seriously disrupted efforts at sustained economic development in sub-Saharan Africa. Nearly half of the countries, twenty-one in all, have experienced ethnic, religious, or civil wars that threatened the very fabric of these nations. Many African observers believe that political instability is a crucial contributor to Africa's ongoing poverty.

In the mid-1990s, democracy and economic development had been achieved in the tiny nation of Benin. By contrast, Nigeria was fluctuating between civil and military rule. The second largest and strongest economy in sub-Saharan Africa, the Republic of South Africa, was in transition from white rule to power sharing with the black majority in a form now being determined. In a remarkable speech before the Organization of African Unity shortly after his country gained independence from Ethiopia in May of 1993, the first president of Eritrea noted: "The African continent is today a marginalized actor in global politics and world economic affairs." In large measure, this is because political stability is fundamental to the development of the basic social services and economic activities most needed in Africa.

International Aid to Sub-Saharan Africa

Millions of dollars of loans and international aid have flowed into sub-Saharan Africa from private banks in the United States and Europe and from development organizations like the International Monetary Fund and the World Bank.

In the past, much of this money was spent unwisely and did little to build Africa's economies. One out of every two World Bank development projects failed in East Africa, compared with one failure in twenty in South Asia. Loans and aid were used to buy imported food and to build huge dams

▼ *In the United States, there are on average 2.5 children per family. In many developing nations, there are 4. Africa has the highest average number of children per family. What is the number?*

and steel plants, which were seen as symbols of modernization. Numerous changes in government increased corruption in many African countries. Also, production of minerals and commercial crops in other parts of the developing world glutted the market for Africa's exports and drove down prices.

Few countries can now pay back these loans, and many countries are bankrupt. Not enough money-making enterprises have been created. Bankers are now reluctant to provide more loans to Africa; to throw good money after bad, in their view.

Wealthy nations, however, are beginning to realize that Africa needs help and that without it, future tragedy will dwarf present suffering. New development strategies focus on self-help projects like those underway in Burkina Faso and Kenya.

Success is possible. Africa could feed its population and heal its environment if Africa's farmers are given the seed, tools, and techniques to implement a "green revolution" like that in South and Southeast Asia. Investment in small businesses could provide Africans with daily necessities, as well as jobs for the unemployed. Governments could reduce expensive imports, halt civil wars, and focus attention on their people's basic needs. All of this requires investment. Wealthy nations must now respond to Africa's crisis, for Africa's future will hinge on this ongoing investment.

⊕ REVIEW QUESTIONS

1. In what ways are African cities not prepared for the many rural people who are moving into them?
2. How much higher was the failure rate of World Bank development projects in East Africa than similar projects in South Asia?

⊕ THOUGHT QUESTIONS

1. Millions of Africans are stricken by diseases that are rarely found in North America. What are these diseases, and why do they affect so many Africans? How would you solve the problem of delivering health care to these people?
2. Why has the income from African exports decreased in the last twenty years? What kinds of economic activities would provide Africans with higher incomes? What skills would they have to develop and what investments made to make

this happen? What is your opinion of Africa's probable future?

⊕ CHAPTER SUMMARY

Location, movement, and human-environment relations have played important roles in Africa south of the Sahara. From very early times African states and kingdoms based on agriculture and trade flourished. These kingdoms were linked by long-distance trade routes to Europe and to Asia. The dark side of African trade was the trade in slaves. While the slave trade between Africa and Europe existed from a very early date, it became a major source of revenue for African kings during the colonization of the Americas.

European-African trade grew when Europe began to industrialize and to look for sources of raw materials and places to trade finished products. Europeans set up trading stations on the African coast and later subdued local tribes and claimed large sections of the African continent. By the end of the nineteenth century, all of Africa except Liberia and Ethiopia was under the direct control of Europeans.

While European colonization in Africa lasted only seventy years, the effects of colonialism linger in the form of (1) state boundaries that reinforce political and social unrest, (2) transportation systems that do not serve the needs of an independent country, and (3) a dual economy consisting of pockets of modern economic development surrounded by a traditional agricultural or herding economy.

Added to these effects of colonialism and imperialism was the political unpreparedness of most Africans to handle the problems of independent African countries. Chief among the serious problems created by this situation are a rapidly growing population with inadequate health care and nutritional resources; widespread poverty in the countryside; political turmoil in the form of ethnic, religious, or civil war; and environmental catastrophes such as drought and flood. Many of Africa's problems could be solved if political stability can be achieved and if wealthier countries begin to help African countries to invest in enterprises that meet their peoples' needs.

EXERCISES

Find the Right Ending

Directions: Write on your paper the correct ending to each of the following sentences.

1. Minerals provide much of the wealth of (Zimbabwe and Zambia, Uganda and Angola).
2. The chief exports of East Africa are coffee, tea, and (cotton, rice).
3. In the 1980s and 1990s, the incomes of Africans (declined by nearly 20 percent, rose by nearly 20 percent).
4. Minerals, particularly gold, provide much of the wealth of (Angola, South Africa).
5. Zaire's most important export is (cotton, copper).
6. South Africa's industrial economy was financed by (gold and diamonds, copper and uranium).
7. An important natural resource of Nigeria, Gabon, and Cameroon is (iron, petroleum).
8. A major cash crop in Senegal and Gambia is (peanuts, coffee).
9. An important mineral resource of Guinea and Ghana is (bauxite, iron).
10. A major cash crop in Cameroon is (cotton, cacao).

Match the City to the Country

Directions: Using the political map on page 467, match the cities in the second column to the countries in the first column. Write the matched names of the cities and countries on your paper.

1. Zimbabwe
2. South Africa
3. Mozambique
4. Zambia
5. Guinea
6. Nigeria
7. Ivory Coast
8. Senegal
9. Ghana

a. Dakar
b. Conakry
c. Kinshasa
d. Pretoria
e. Maputo
f. Dar es Salaam
g. Addis Ababa
h. Lagos
i. Kampala

10. Zaire
11. Angola
12. Kenya
13. Tanzania
14. Ethiopia
15. Uganda
16. Somalia

j. Accra
k. Harare
l. Luanda
m. Mogadishu
n. Abidjan
o. Lusaka
p. Nairobi

Inquiry

Directions: Combine the information in the chapter with your own ideas to answer these questions.

1. If you had been born many centuries ago, would you have preferred to live in Kush, Ghana, Mali, or Monomotapa? When and why?
2. What do you think was David Livingstone's single most important accomplishment in Africa?
3. In general, how was the European domination of colonial Africa similar to the European domination of colonial South America?
4. Compare the exploration of the African continent with the exploration of the American West by Lewis and Clark, or any explorer you choose. How were their experiences similar? How were they different? How would you compare these explorers to astronauts today?

SKILLS

Using the *Readers' Guide*

Directions: The *Readers' Guide to Periodical Literature* lists articles and stories that have appeared in magazines, or periodicals, during a specific period of time. Hardback volumes of the *Readers' Guide* include references to magazine articles and stories published in a particular year, such as 1994. Paperback volumes of the *Readers' Guide* list the most recently published articles and stories. Just as the library catalog helps you find certain books, the *Readers' Guide* helps you locate information in magazines.

The following is the way an article is listed in the *Readers' Guide*. Read the listing and then notice how each part of it is identified by the terms above and below the listing.

Title of Article   Author   Has Pictures

"Apartheid on the Ash Heap." J. Contreras. il por *Newsweek* V 122 p 48 N 29 '93

Name of Mag.   Vol. No.   Page No.   Date

Read these three listings of magazine articles and then answer the questions that follow.

"Another Lost Decade for Africa?" O. Ongwen. il *World Press Review* V 40 p 41 May '93

"African National Congress—A Welcome Mat for U.S. Business from Mandela to de Klerk." A. Borrus and A. Fine. il *Business Week* p 51–5 Jl 12 '93

"Mandela's Welcome Mat Starts Drawing Visitors." K.L. Alexander. il *Business Week* p 54–6 O 11 '93

1. What is the title of the article in *World Press Review*?
2. Who is the author of this article?
3. What is the volume number of the magazine in which this article appeared?
4. On what page does this article begin?
5. Is this article illustrated?
6. K.L. Alexander is the author of which article?
7. Did K.L. Alexander's article appear in *Reader's Digest* or *Business Week*?
8. Is this article illustrated?
9. On what page does this article end?
10. What is the date of the issue of the magazine in which this article appeared?
11. Who wrote "African National Congress—A Welcome Mat for U.S. Business from Mandela to de Klerk"?
12. Does the magazine in which this article appeared use volume numbers?
13. Are there three or five pages in this article?

Readers' Guide or Library Catalog

Directions: Tell whether you should look first in the *Readers' Guide* or in the library catalog to research each of the following subjects:

1. The history of Ethiopia
2. Economic developments last month in Zimbabwe
3. An ethnic conflict in an African country
4. The customs and folklore of Ghana
5. The life of David Livingstone
6. Recent archaeological findings in Kenya
7. The current political changes in South Africa
8. The life spans of various African animals

Understanding the Map of African Economic Migration

Directions: Turn to the map on page 509 and answer these questions:

1. What map symbol shows areas of farming?
2. What map symbol shows areas of mining?
3. What map symbol shows areas of industry, mining, and mixed farming?
4. What map symbols show migration of the labor force?
5. Does Dakar have more than 1 million people?
6. Does Bamako have more than 1 million people?
7. Is there a larger labor migration to Dakar or Bamako?
8. There is a large labor migration to the area near Lubumbashi. Is this a mining area or an area of industry, mining, and mixed farming?
9. What three coastal ports in South Africa have more than a million people?
10. Is South Africa only a farming area, or is it an area of industry, mining, and mixed farming?
11. Find the area that includes Abidjan, Accra, Ibadan, and Lagos. Which of these four cities are coastal ports?
12. Are these four cities in a mining area or a farming area?
13. Are more workers moving to the Johannesburg area or the Khartoum area?
14. Are the farmers who are moving to the Kano area coming mainly from the north or from several directions?

15. Addis Ababa and Nairobi are two of the largest cities in the interior of Africa. Are most of the workers who are moving to the areas around these cities seeking jobs on farms or in mines?

16. Many workers are moving to the area near Luanda. Is this a farming area or a mining area?

17. Is the chief activity in the area near Mombasa farming or mining?

18. In West Africa, are more labor migrants moving to coastal areas or inland areas?

Vocabulary Skills

Directions: Use context to determine the meaning of the italicized words in the following sentences. Write the letter of the correct definition on your paper.

apartheid   homelands
bantustans   quinine
Boers   townships
concessions

1. Black South Africans were forced to live in *townships* where city services and jobs were scarce.
   a. sites of the ancient African kingdoms of Kush and Kongo
   b. areas rich in minerals and other resources
   c. urban districts assigned to ethnic groups

2. The South African government in 1948 adopted a policy of *apartheid* in order to keep the white minority in control of the country.
   a. the practice of "apartness" or segregation
   b. the practice of training black Africans to be political leaders in South Africa
   c. the policy of deporting black Africans

3. According to law, black South Africans make decisions for themselves only in their own *bantustan*.
   a. the elected branch of congress in South Africa that represents black South Africans

   b. the labor unions to which all black South African workers belong
   c. a separate self-governing state established by whites for other populations in South Africa

4. When British settlers arrived in South Africa beginning in 1806, they were shocked by the cruelty of the *Boers* to black Africans.
   a. Dutch colonists who settled in South Africa
   b. Malaysian farm laborers
   c. Dutch Catholics who settled in South Africa beginning in 1900

5. The government of South Africa set aside racially segregated, semi-independent areas known as *homelands*.
   a. area where a person was born
   b. smaller state within South Africa
   c. neighboring countries that acted as a buffer between South Africa and other African nations

6. To promote the kind of economic development that was desired, colonial African governments offered *concessions* to European companies that would invest in Central Africa.
   a. the policy of placing heavy restrictions on companies investing in Central African colonies
   b. the policy of granting the rights to land and minerals to companies investing in Central African colonies
   c. the practice of agreeing to everything a company wanted to do in a colony

7. By the 1850s, the discovery of *quinine* made it safer for Europeans to live in the interior.
   a. a forest path by which one could travel into the interior of Africa
   b. a type of airplane that could easily cover the long distances required in Africa
   c. an antimalaria drug used extensively in the tropics where the disease is prevalent

Human Geography

Directions: In each group of three sentences, two sentences are correct. Write the correct sentences on your paper.

1. (a) French West Africa had few mineral resources, so the French rulers of this area forced Africans to grow commercial crops. (b) The British colonies in West Africa had important mineral resources. (c) In the 1890s, the Congo Free State had a democratic form of government.

2. (a) The British who settled in South Africa were called Boers. (b) The German government said that one of the reasons for owning African colonies was to provide land for Germany's growing farm population. (c) Tiny Portugal controlled the large colonies of Angola and Mozambique.

3. (a) Independence spread across tropical Africa in the years between World War I and World War II. (b) The only two independent African countries in 1955 were Ethiopia and Liberia. (c) When independent states emerged throughout Africa, few new African leaders had much experience in government.

4. (a) The income of the average African has more than doubled in the last twenty years. (b) The average African has a shorter life span than the average North American. (c) African society is less urbanized than European society.

Physical Geography

Directions: Match the sentence beginnings in Group A with the sentence endings in Group B. Then write the complete sentences on your paper.

Group A
1. Africa is a huge plateau
2. The East African Rift System
3. The Niger, Shari, and Nile rivers
4. Tropical rain forests
5. Vegetation thins out into low-grass steppes
6. The Sahel
7. Most of Tanzania's people live in
8. The dry climates of the Kalahari Desert

Group B
a. north and south of Africa's tropical core.
b. fertile lowlands near Lake Malawi and Lake Victoria.
c. blanket the Congo and extend along the West African coast.
d. prevail in Botswana and Namibia.
e. fringed with narrow coastal plains.
f. divides the Ethiopian Highlands and extends southward parallel to Africa's east coast.
g. is a semiarid area on the southern shore of the Sahara Desert.
h. formed basins along the southern edge of the Sahara Desert.

Writing Skills Activities

Directions: Answer these questions in essays of several paragraphs each. Remember to include a topic sentence and several sentences supporting your main idea in each paragraph.

1. What are the most distinctive physical features, including climates, of Africa south of the Sahara?
2. Describe the advanced civilizations that existed in sub-Saharan Africa before Europeans entered this region?
3. How did the European colonization of sub-Saharan Africa affect the lives of Africans?
4. What problems have the African people faced since achieving independence?
5. Why is sub-Saharan Africa not classified among the regions of the technological world?

THE MIDDLE EAST AND NORTH AFRICA

UNIT INTRODUCTION The region of the Middle East and North Africa reaches across North Africa and thousands of miles deep into the Asian interior. In North Africa, this region stretches along the southern shore of the Mediterranean Sea from Morocco to Egypt. In the Middle East, it includes the countries that line the eastern Mediterranean (an area that is also called the Levant), the nations of the Arabian Peninsula, and the northern and eastern highlands of Turkey, Iran, and Afghanistan.

The area of the Middle East and North Africa is a crossroads of cultures. Great civilizations and world religions developed in its deserts and rugged mountains. It has also been a crossroads of conflict, a place where people and civilizations have clashed for thousands of years. Today, this region is a focal point of world attention because of its large oil reserves, its internal wars, and the ongoing strife between Arabs and Israelis.

▲ *The three pyramids at El Gîza are considered to be one of the Seven Wonders of the Ancient World. Chephren's (Khafra's) pyramid appears to be the tallest, but is actually built on a small hill. It is also the only large pyramid that retains some of the original limestone covering. Why were these great structures built?*

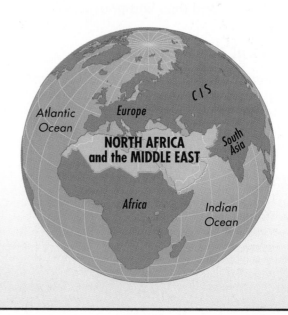

Atlantic Ocean

Europe

CIS

NORTH AFRICA and the MIDDLE EAST

South Asia

Africa

Indian Ocean

▲ Twenty-eight of the world's thirty-three giant oil fields are located in the Middle East and North Africa. The five largest oil-producing countries of that area are Saudi Arabia, Kuwait, Iran, Iraq, and Libya. Do you think it is cheaper to produce oil in the Middle East or in Texas?

▲ Istanbul, which was earlier called Byzantium and Constantinople, was once the capital of the Eastern Roman Empire as well as the Ottoman Empire. The Saint Sophia Basilica and the Blue Mosque are two of the city's landmarks. Istanbul is the only major city in the world which is located on two continents. Name the continents.

UNIT OBJECTIVES

When you have completed this unit, you will be able to:

1. Describe how environmental conditions have affected ways of life in the Middle East and North Africa.

2. Explain how the conquests of this region by various groups have helped shape the civilizations of the Middle East and North Africa.

3. Name three of the world's great religions that were born in this region.

4. Identify the problems and accomplishments of the Jewish people in Israel.

5. Explain why people throughout the world view the Middle East and North Africa as a region whose tensions could have worldwide repercussions.

SKILLS HIGHLIGHTED

- Using a Table
- Using a Population and Rainfall Map
- Using a Special-Purpose Map
- Studying the Map of a Single Country: Israel
- Using a Location-Movement Map
- Vocabulary Skills
- Writing Skills

KEYS TO KNOWING THE MIDDLE EAST AND NORTH AFRICA

1

The area of the Middle East and North Africa is nearly twice the size of the United States.

2

Deserts and mountains cover most of the Middle East and North Africa. The Sahara, the largest desert in the world (unless one considers Antarctica a desert), is located here.

3

Only 5 to 10 percent of the land in the Middle East and North Africa is farmed.

4

The Middle East and North Africa are a crossroads of cultures where three continents meet.

5

Judaism, Christianity, and Islam, which are the three great religions of the West, originated in this region.

6

Most people in the Middle East and North Africa are Muslims, believers in the Islamic religion. Two of every three speak Arabic as their native language.

7

Half of the more than 350 million people of this region live in the three countries of Egypt, Turkey, and Iran.

8

Vast reserves of oil exist in the Middle East and North Africa, but most of the twenty-one countries in this region are poor.

9

Political conflict in the eastern Mediterranean and the Arabian-Persian Gulf is of vital concern to the world.

▼ *Egypt, Saudi Arabia, and the Sinai Peninsula are areas in one of the driest deserts in the world. Only 5 to 10 percent of this land can be used for farming. Some of the water used for farming is obtained from desalinization plants. What is desalinization?*

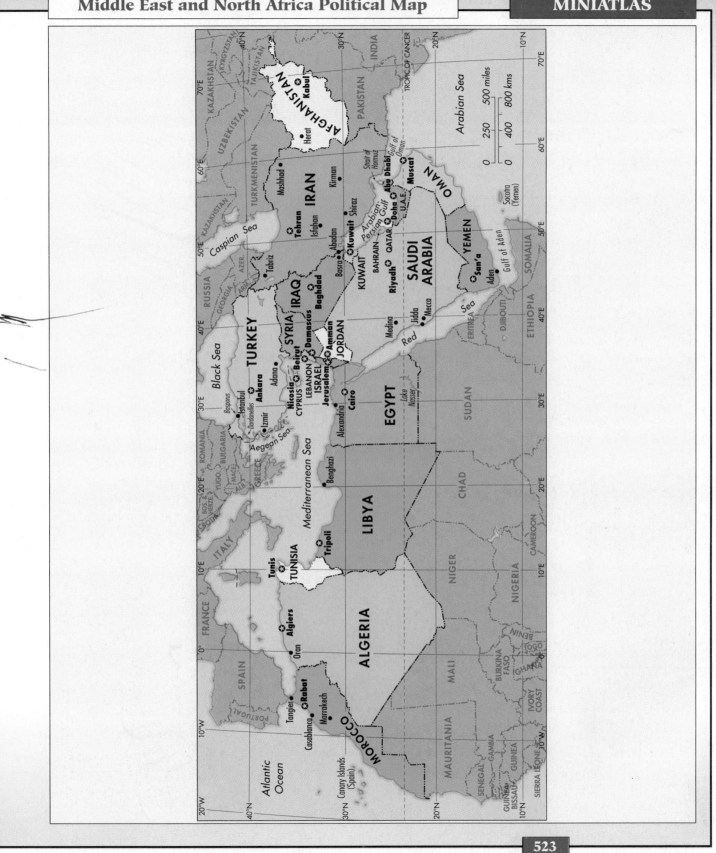

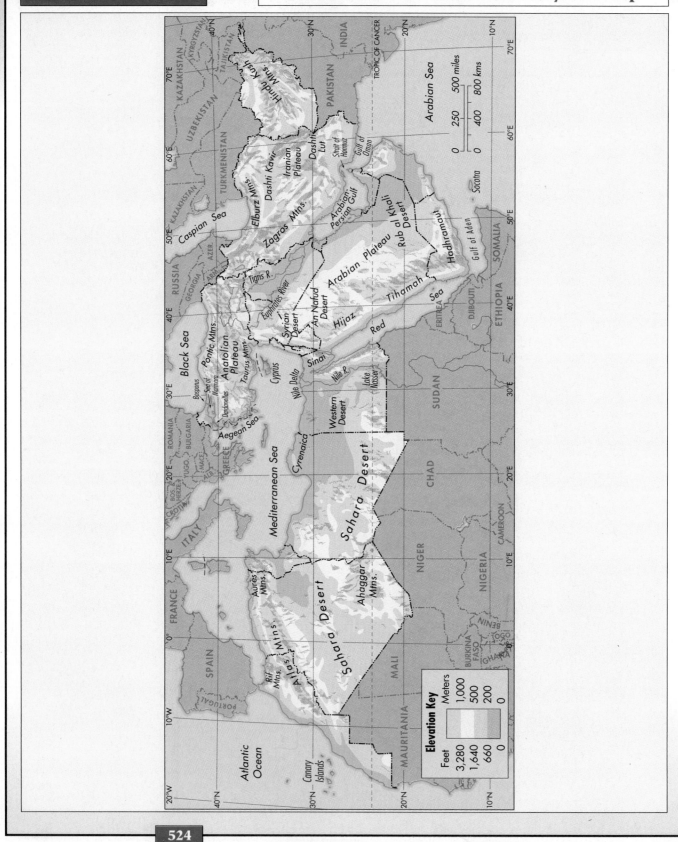

Atlantic Ocean

Canary Islands

Mediterranean Sea

Black Sea

Caspian Sea

Arabian Sea

Red Sea

Gulf of Aden

Gulf of Oman

Persian Gulf / Arabian Gulf

Strait of Hormuz

Aegean Sea

Sea of Marmara

Bosporus

Dardanelles

Nile R.

Lake Nasser

Tigris R.

Euphrates River

Sahara Desert

Western Desert

Syrian Desert

An Nafud Desert

Rub al Khali Desert

Atlas Mtns.

Rif Mtns.

Aures Mtns.

Ahaggar Mtns.

Cyrenaica

Sinai

Nile Delta

Hijaz

Tihamah

Hadhramaut

Arabian Plateau

Iranian Plateau

Anatolian Plateau

Pontic Mtns.

Taurus Mtns.

Zagros Mtns.

Elburz Mtns.

Hindu Kush

Dashti Kavir

Dasht-i Lut

Cyprus

Socotra

Elevation Key

Feet	Meters
3,280	1,000
1,640	500
660	200
0	0

500 miles
250
0

800 kms
400
0

TROPIC OF CANCER

SPAIN
PORTUGAL
FRANCE
ITALY
GREECE
ALB.
MACE.
YUGO.
BULGARIA
ROMANIA
CROATIA
BOS. & HERZE.
RUSSIA
GEORGIA
AZER.
ARM.
KAZAKHSTAN
UZBEKISTAN
TURKMENISTAN
KYRGYZSTAN
TAJIKISTAN
INDIA
PAKISTAN
ETHIOPIA
ERITREA
DJIBOUTI
SOMALIA
SUDAN
CHAD
NIGER
NIGERIA
CAMEROON
BENIN
TOGO
GHANA
BURKINA FASO
MALI
MAURITANIA

MOROCCO

20°W 10°W 0° 10°E 20°E 30°E 40°E 50°E 60°E 70°E

10°N 20°N 30°N 40°N

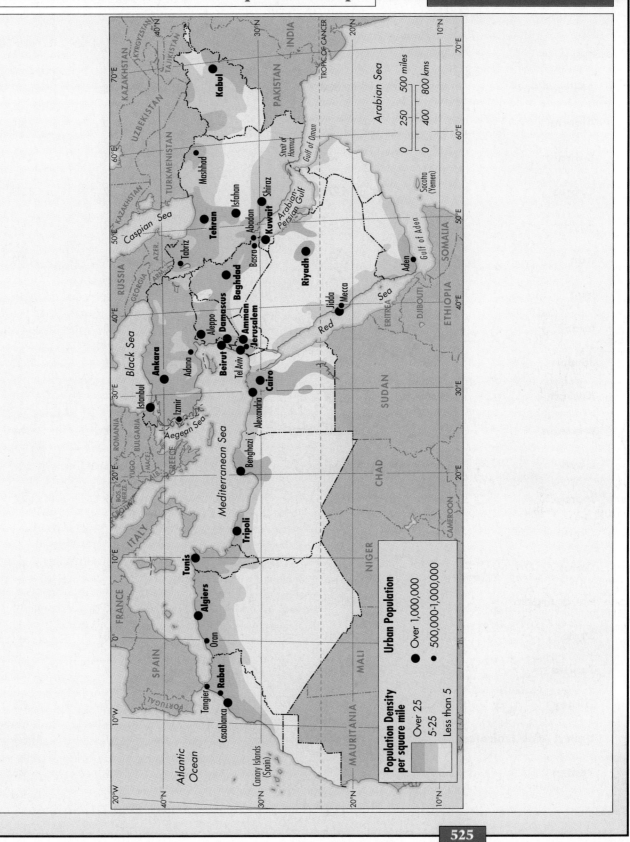

Atlas of the Middle East and North Africa Population Map. Map shows population density per square mile (Over 25, 5-25, Less than 5) and urban population (Over 1,000,000; 500,000-1,000,000).

Cities labeled: Kabul, Mashhad, Isfahan, Shiraz, Tehran, Tabriz, Abadan, Kuwait, Basra, Baghdad, Riyadh, Aleppo, Damascus, Amman, Jerusalem, Beirut, Tel Aviv, Ankara, Adana, Cairo, Istanbul, Izmir, Alexandria, Jidda, Mecca, Aden, Benghazi, Tripoli, Tunis, Algiers, Oran, Rabat, Tangier, Casablanca.

Seas and regions labeled: Arabian Sea, Caspian Sea, Black Sea, Mediterranean Sea, Aegean Sea, Red Sea, Gulf of Aden, Gulf of Oman, Strait of Hormuz, Arabian-Persian Gulf, Atlantic Ocean.

Countries labeled: KAZAKHSTAN, KYRGYZSTAN, TAJIKISTAN, UZBEKISTAN, TURKMENISTAN, INDIA, PAKISTAN, RUSSIA, GEORGIA, AZER., ARM., SOMALIA, ETHIOPIA, ERITREA, DJIBOUTI, SUDAN, CHAD, NIGER, MALI, MAURITANIA, CAMEROON, FRANCE, SPAIN, PORTUGAL, ITALY, GREECE, ROMANIA, BULGARIA, MACE., YUGO., BOS. & HERZE., CRO., Socotra (Yemen), Canary Islands (Spain).

Country	Capital City	Area (Square miles)	Population (Millions)	Life Expectancy	Urban Population (Percent)	Per Capita GNP (Dollars)
Afghanistan	Kabul	251,773	17.4	42	18	—
Algeria	Algiers	919,595	27.3	66	50	2,020
Bahrain	Manama	267	0.5	72	81	6,910
Cyprus	Nicosia	2,277	0.7	76	62	8,640
Egypt	Cairo	386,662	58.3	60	44	620
Iran	Tehran	636,296	62.8	62	54	2,320
Iraq	Baghdad	169,235	19.2	64	70	—
Israel	Jerusalem	8,473	5.3	76	90	11,330
Jordan	Amman	35,467	3.8	71	70	1,120
Kuwait	Kuwait	6,880	1.7	74	—	—
Lebanon	Beirut	4,015	3.6	68	84	—
Libya	Tripoli	679,362	4.9	63	76	—
Morocco	Rabat	275,117	28.0	65	47	1,030
Oman	Muscat	82,030	1.6	66	11	5,650
Qatar	Doha	4,247	0.5	71	90	15,870
Saudi Arabia	Riyadh	830,000	17.5	66	77	7,070
Syria	Damascus	71,044	13.5	65	50	1,110
Tunisia	Tunis	63,170	8.6	68	59	1,510
Turkey	Ankara	300,948	60.7	66	59	1,820
United Arab Emirates	Abu Dhabi	32,278	2.1	71	81	19,870
Yemen	San'a	203,850	11.3	46	29	540

—Figures not available

THE LANDS AND PEOPLES OF THE MIDDLE EAST AND NORTH AFRICA

The Middle East and North Africa is an area of deserts and mountains. Only about 350 million people live in this vast region, which is nearly twice the size of the United States. Half of them live in the three countries of Egypt, Turkey, and Iran. Settlement is keyed to the environment. Well-watered lands have many people; deserts and mountains have few people.

◀ *The Nile River Delta and the Nile River Valley stand in sharp contrast to the Sahara Desert. In ancient times several branches of the Nile emptied into the Mediterranean Sea; in modern times only two or three remain. What happened to the other branches?*

527

1. Patterns of Settlement in a Dry and Rugged Land

City, Village, and Tribe

For many hundreds of years, three ways of life have existed in this region—city life, village life, and tribal life. Cities are centers of government, trade, and religion. Middle Eastern cities, like Baghdad, Iraq, and Cairo, Egypt, are often located on rivers. Some, like Beirut, Lebanon, and Casablanca, Morocco, are on coasts. Others, like Tehran, Iran, and Kabul, Afghanistan, lie at the foot of snow-capped mountain ranges. Farming villages are clusters of mud, stone, or reed houses located along rivers, coasts, or in highland valleys—wherever water is available. Nomadic tribes live in the deserts and mountains, where they pasture herds of camels, horses, sheep, and goats.

In the past, caravan routes and roads linked cities to one another, following ancient trails from oasis to oasis. Many villagers lived near these cities and sold their grain and vegetables to urban dwellers in exchange for manufactured goods and a variety of services. Nomads were quite independent, but even they had to trade their wool, hides, meat, and milk products for goods they needed in the marketplaces of Middle Eastern cities and villages. The peoples of the cities, villages, and tribes in the Middle East were bound together by economics and often by blood ties for many centuries.

Growing Cities

In the past, cities in this region were centers of great culture and civilization. Many were originally centers of Greek and Roman learning. These cities were not large by modern standards, and less than one-fifth of all people in the region lived in cities. Famous cities like Cairo in Egypt, Baghdad in Iraq, and Constantinople (modern Istanbul) in Turkey had less than a half million residents in the 1800s.

Most of these cities were ringed by high walls, and inside the city, a palace or fortress was usually protected by still higher ramparts. Even in residential districts, houses had very high courtyard walls studded on top with pieces of glass. A maze of narrow alleyways ran through these quarters.

At the center of these old cities, two institutions were usually found. The first was an *Islamic* cathe-

Subregions of the Middle East and North Africa

The Middle East and North Africa has three major subregions: (1) the Northern Highlands,
(2) the Dry World, and (3) the Central Middle East.

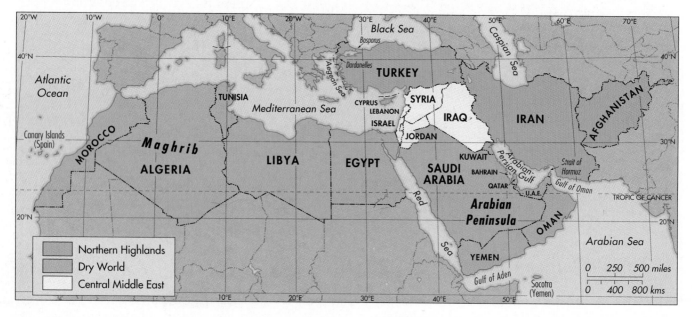

dral or **mosque**, a place of worship usually decorated with beautiful tiles. The second was the **bazaar**, the city's major shopping district. The mosque was a center of day-to-day life as well as religion. Here, *Muslim* (followers of Muhammad) religious leaders made decisions about family life, settled economic disputes, allocated drinking water, and managed the city's affairs. The nearby bazaar was equally busy. Covered stalls lined crowded lanes in which cloth sellers, carpet merchants, jewelers, and silversmiths engaged in wholesale and retail trade.

Today, these cities have attracted large numbers of people from the countryside. Roughly half of all people in the region are now urbanites, and twenty-two cities in this region have more than 1 million people, as shown here in the table. Rising birthrates and migration from the countryside have caused cities to grow rapidly. Many unemployed villagers have left their small settlements and sought jobs in the capital cities and seaports of the Middle East.

These large cities are quite different from the small Islamic cities of the past. Broad avenues cut through residential areas, and skyscrapers line wide new streets. Modern government offices, banks, department stores, and hotels have replaced the old inns where travelers stayed, the **caravanserais** in the old bazaars. New suburbs now spread far beyond the old city walls. As in many large cities in the developing world, a growing number of poor people who cannot find jobs live in slums.

Changing Villages

For centuries, most people in the Middle East and North Africa have lived in villages, raising wheat, barley, date palm, and other crops. These villagers make up at least half of the total population of the region. They supply grain and vegetables to city dwellers and nomads in exchange for products and services not available in their villages.

Water determines the location of villages in this region. Farming settlements are found wherever

Large Cities in the Middle East and North Africa

Capital Cities	Population (millions)
Cairo, Egypt	10.1
Tehran, Iran	7.5
Baghdad, Iraq	4.5
Algiers, Algeria	3.7
Tripoli, Libya	3.3
Ankara, Turkey	2.8
Riyadh, Saudi Arabia	2.6
Damascus, Syria	2.1
Tunis, Tunisia	2.1
Kabul, Afghanistan	2.1
Beirut, Lebanon	1.8
Kuwait City, Kuwait	1.5
Amman, Jordan	1.2
Seaports	**Population (millions)**
Istanbul, Turkey	7.8
Alexandria, Egypt	3.6
Casablanca, Morocco	3.3
Tel Aviv, Israel	2.3
Jidda, Saudi Arabia	1.5
Benghazi, Libya	1.1
Other	**Population (millions)**
Aleppo, Syria	1.9
Isfahan, Iran	1.8
Shiraz, Iran	1.5

▼ *An Algerian sells vegetables at a market in Ghardaïa. The country must import 20 to 25 percent of its food even though half its people live in villages. What does this tell you about the efficiency of agriculture in Algeria? The standard of living?*

water is available or can be secured by digging wells or using water-lifting devices. Dense rural populations are found along the coasts of the Mediterranean, Black, and Caspian Seas; in the few river valleys of the Middle East and North Africa; and in mountain areas where winter rains provide water. Locate these densely settled areas on the map on page 525.

Just as in the cities, village houses face inward to courtyards and are surrounded by high walls that provide privacy. The houses are built of mud, stone, reeds, or whatever building material is available locally. The villagers usually grow grain, tend a few goats and sheep, and weave carpets or other handcrafted textiles. Before Middle Eastern countries became independent, farmers often were **sharecroppers**, people who received a share of the harvest (usually one-fifth) in return for their labor. The land they worked and its water were usually owned by landlords who lived in cities.

However, in the 1950s and 1960s, in one country after another, village land owned by absentee landlords was redistributed to farmers. Egypt redistributed farmland after Gamal Abdel Nasser came to power in 1952. In Iraq, when the king was overthrown in 1958, large estates were seized by the new government and turned over to farmers. In

> ▼ *Many Egyptians live very much like their ancestors of 2,000 to 3,000 years ago. This farmer still uses a shaduf to lift water from the Nile to water his fields. What is the energy cost of lifting water this way?*

| IT'S A FACT | One of every six Egyptians lives in Cairo. |

Algeria, millions of acres of land abandoned by French colonists in 1962 were taken over by the new government. In Iran, the former Shah attempted to limit the amount of land owned by an individual landlord to a single village. In spite of these land distribution programs, the majority of farmers in this region are still very poor.

New land reform programs have recently brought better varieties of seed and new farming techniques to some areas. In northern Syria, machinery is being used to plant and harvest wheat and cotton. Some Egyptian farmers now produce high yields of wheat, corn, and cotton in the Nile Valley. In Israel, water has been piped to deserts, and the land has been replanted in gardens, orchards, and woodlands.

These areas are exceptions, however. Overall, the farmers of this region get crop yields that are less than a quarter of those produced by European farmers. Poverty, debt, and population pressure on the land are intense in the villages of the Middle East and North Africa.

Nomadic Tribes

Until the early 1900s, about 5 million nomads lived in this region. Masters of the deserts and mountains, they tended flocks of animals and controlled many settled villages. Governments often paid them money for the right to travel safely on overland routes of international trade in areas of difficult terrain controlled by the nomads.

Camel nomads, or **Bedouin**, live in the Sahara Desert of North Africa and in the deserts of the Arabian Peninsula. These areas are among the driest in the world and have little vegetation for feeding animals. Rainfall is so scarce here that some weather stations have never recorded a rain shower. Therefore, camel nomads stay on the edges of the deserts, if possible, or in oases where water is permanently available. Some move out from these base camps on 1,000-mile-long, nine-month migrations to pasture their herds.

MOVEMENT

THE REMARKABLE CAMEL, "SHIP OF THE DESERT"

The deserts of North Africa and the Arabian Peninsula are vast areas of sand and rock. For centuries, camel nomads with their one-humped dromedaries moved freely across these forbidding lands. Camels are marvels of biological engineering in their adaptation to desert life.

Camels are built to survive in deserts. Their eyes are protected from sand and dust by three sets of eyelids; they can close their nostrils during sandstorms. Camels have three stomachs and the ability to store food in a fat reserve in their humps. Their extremely long stride allows camels to forage over large, thinly vegetated areas, browsing as far as thirty miles in a day. Their jaws and teeth enable them to eat even the most prickly plants. Moreover, because they recycle nitrogen compounds in their bodies, camels thrive on low-protein desert shrubs and grasses that other grazing animals cannot eat.

Camels conserve water by using it up very slowly. Instead of perspiring in hot weather, camels simply get hot. They are able to drink water so salty that it would kill a human being. In addition, they can lose an amount of water equal to a quarter of their body weight, and one great drink can restore water to their tissues in less than twenty-four hours. Camels can go seven days without water in the hottest desert. If pasture is adequate, they can survive indefinitely without water. This is why camel nomads were able to rule the desert regions of the Middle East for so long.

Arabs call the camel "the milk-giving palm tree of the desert." Camel milk will stay sweet and fresh for as long as three months when refrigerated. The Israelis are now experimenting with turning camels into "desert cows." Camel meat is eaten in many countries, and fine camel wool is marketed in the cities of the region and throughout the world.

Camels are also excellent beasts of burden. They can carry loads up to 1,000 pounds a distance of 25 miles a day across sand or rock. Some racing camels can actually cover 100 miles a day. Their feet are as soft as foam rubber, and the soles of their feet spread out like pillows to give the animals balance on shifting sand dunes.

Camels are now disappearing from the region. So are camel nomads. Forced by many governments to settle down, tribal nomads now are abandoning a way of life that has lasted for thousands of years in the Middle East. They have no choice. The nomads' camels are being replaced by trucks.

▲ *Jordanian desert police patrol borders with their camels. In this war-torn area, the people face many political problems. Why are these police using camels rather than cars?*

❓ QUESTIONS

1. Name at least six ways in which the camel has adapted to a desert environment.
2. Besides transportation, in what ways does the camel provide for people of the desert?

IT'S A FACT

Bedouin in Arabic means "nomad."

Sheep and goat nomads graze their animals in the mountains, using upland meadows in summer and lowland pastures in winter. Because neither the uplands nor the lowlands can feed herds of sheep and goats year-round, nomadism is an intelligent adaptation to the changes of seasons in these dry and rugged environments. This type of nomadism is called **transhumance**. In the Zagros Mountains of Iran, the nomadic tribes once formed powerful confederations that ruled large areas. Powerful tribal groups also lived in mountains in Turkey, Afghanistan, and North Africa.

Modern airplanes, tanks, trucks, and jeeps now enable governments in the region to control their deserts and mountains. As a result, most nomads have lost their independence. In Iran, Reza Shah led army campaigns against mountain tribes after he took control of the country in the 1920s. In eastern Turkey, nomads must pay taxes to the central government. Nomadic groups in the central Middle East have been trapped in the wars between Arabs and Israelis. In Saudi Arabia, Bedouin nomads, who make up about half of the population, have fared somewhat better. The Saudi government has drilled wells in the desert and built oasis settlements for them.

⊕ REVIEW QUESTIONS

1. What is the population of the Middle East and North Africa, and in what three countries do half of the people live?
2. Villagers account for how much of the population of the Middle East and North Africa? What crops do they raise?
3. The densest rural populations are found in what areas in the Middle East and North Africa? Why have so many people settled in these regions?

⊕ THOUGHT QUESTIONS

1. What environmental problems prevent the Middle East from sustaining a large population?
2. Since the 1950s, what have the governments in Middle Eastern countries done to improve the living standards of many villagers?

2. The Northern Highlands: Turkey, Cyprus, Iran, and Afghanistan

The Middle East and North Africa can be divided into three major subregions: (1) the Northern Highlands, (2) the dry world of North Africa and the Arabian Peninsula, and (3) the Central Middle East, which is a center of turmoil at the crossroads of the Middle East.

Rugged Mountains and High Plateaus

The Northern Highlands include the three large countries of Turkey, Iran, and Afghanistan, as well as the small, disputed Mediterranean island of Cyprus. Most of the people of this region speak Turkish or Persian (Farsi), not Arabic. These nations have lofty mountain ranges and high plateaus.

The heart of Turkey is the Anatolian Plateau, which is bordered by the Pontic Mountains to the north and the Taurus to the south. In Iran, the Elburz Mountains to the north and the Zagros on the west frame Iran's desert center. In Afghanistan, the Hindu Kush Mountains occupy the middle of the country.

The climates of the Northern Highlands are wetter than other parts of the Middle East for two reasons. First, they lie north of the desert belt that stretches across North Africa and the Arabian Peninsula. Second, their high mountains force moisture-bearing air masses to rise and drop rain and snow. About one-third of all Middle Easterners live in the more humid environments of the Northern Highlands. Turkey and Iran are two of the three most populous countries in the entire region.

Turkey and Cyprus

Turkey is a large, mountain-rimmed peninsula about twice the size of California. It is bounded on three sides by water—the Black Sea on the north,

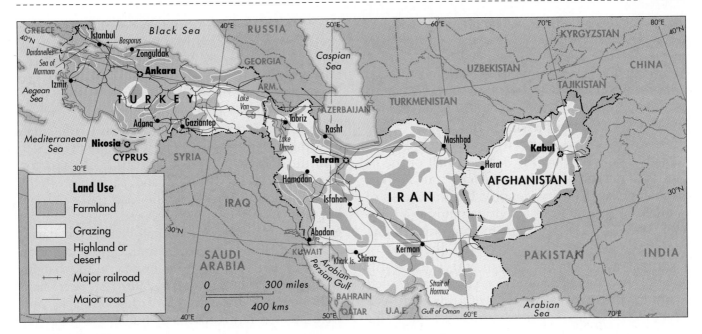

The Northern Highlands

the Aegean on the west, and the Mediterranean on the south. A small portion of Turkey extends into Europe across the Bosporus and the Dardanelles, the two straits that separate Europe from Asia. Istanbul, Turkey's largest city and most important seaport, spreads across both sides of this waterway.

Turkey has three environments. Its coasts are densely settled because they are well watered. These lowlands receive winter rain from air masses that flow westward along the Mediterranean Sea and reach Turkey. Water trapped in the mountains flows down valleys to the coast and supplies additional water for farming.

The center of the peninsula, the Anatolian Plateau, is Turkey's second environment. Bracketed by the Pontic and Taurus Mountains, the plateau spreads eastward to high mountains near the CIS and Iranian borders. These mountains are Turkey's third environment.

Turkey's most densely settled and fertile land lies in the coastal areas along the Black, Aegean, and Mediterranean Seas—in areas that receive plenty of rain in the winter. Here Turkish farmers grow cotton, tobacco, and wheat. Orchards and vineyards cover hills that slope to the sea. In contrast, the Anatolian Plateau is dry and sparsely settled, a barren land dotted by small mud villages of farmers and shepherds. Mountain areas in the east are occupied by the Kurds, a rebellious minority people who live in Turkey, Iraq, and Iran and are fighting for their own independent country.

Turkey is the only Middle Eastern country with large deposits of high-grade coal. It is located east of Istanbul and just south of the Black Sea at Zonguldak. Iron ore nearby has made it possible for the Turks to build a small steel mill here. Most industry in Turkey, however, is located in its largest cities, Istanbul and Ankara, the capital. These cities are the centers of change in Turkey today.

The country's population has more than doubled to 60.7 million people in the last forty years, and it continues to increase at a rate of a million people a year. Nearly half of the Turkish people are farmers, but only 7 percent of Turkey's land can be farmed because of difficult environments. As a result, villagers are migrating to cities in search of jobs. The Turkish government needs to expand its industry rapidly enough to provide urban jobs for its growing population. To improve its economy, Turkey has applied for membership in the European Community.

The rocky Mediterranean island of Cyprus, located forty miles off the southern coast of Turkey, is a source of bitter dispute between Greece and Turkey. Before 1974, Greeks made up 75 percent of the island's 700,000 people; the rest were Turks who lived in the mountains that cut across the island. When the Greeks of Cyprus demanded union with Greece in 1975, the Turks invaded the island. Turks in Cyprus then moved to the northern half of the island and declared it to be a separate state. Turkey has recognized this new country, but the Greeks of southern Cyprus do not accept its status as a separate state. Greece is reluctant to confront its more powerful neighbor to the east over the issue of Cyprus. Turkey's membership in NATO (the North Atlantic Treaty Organization) has caused the United States to be cautious in dealing with this issue. As a result, Cyprus has remained a divided island for nearly twenty years and is likely to remain so.

Troubled Iran

A central plateau covers most of Iran, much of which is desert or wasteland. High mountains ring the plateau on three sides. In the north, the Elburz Mountains separate the wet, forested coast of the Caspian Sea from the deserts of the central plateau. In the west and south, the rugged Zagros Mountains curve south and east for 1,400 miles along the Iraqi border and the Arabian-Persian Gulf. To the east lies Afghanistan. Locate Iran's mountains, central plateau, and borders on the map on page 524.

Iranians have experienced great difficulties in recent years. They are people who speak Persian (Farsi) and belong to a minority (Shi'ite) Muslim sect, which is quite different from the majority (Sunni) Muslim faith. In 1978, Shi'ite religious leaders rebelled against the government of the Shah of Iran, who was forced to flee the country in 1979.

An exiled religious leader, Ayatollah Khomeini, returned to Iran and set up a government under strict religious control. Alcohol was strictly forbidden, and women had to wear traditional dress, which covered them from head to toe. Many other cultural changes were initiated.

Within a year, Iran was attacked by its neighbor, Iraq, which believed it could benefit from the turmoil in Iran. This began a devastatingly destructive war between these two oil-rich countries that border the Arabian-Persian Gulf. The war lasted for eight years and produced more than a million casualties. Many industrial nations were concerned about protecting the shipment of oil in tankers out of the Arabian-Persian Gulf during this war.

Iran's population has increased from 10 million to 62.8 million in the last century. Additional crops to feed this increased population have been raised on Iran's few areas of fertile farmland, particularly in the north and northwest. Along the coast of the Caspian Sea, rice, tea, tobacco, grain, and citrus fruit are grown. South of these mountains, however, Iran is too dry to farm without irrigation. Oasis settlements fed by mountain streams are located on the margins of the central plateau. In the southwest, large dams were built to store water for irrigation. The dams were later severely damaged by war.

City populations have grown rapidly. The capital city of Tehran now has nearly 8 million residents. The cities of Isfahan and Shiraz have over 1 million people and Mashhad, Tabriz, and Abadan more than 500,000 residents, but no census is available. Villagers who have moved to the city in search of work are often forced to live in slums.

Oil is Iran's most important source of money to pay for badly needed development programs. The largest Iranian oil fields lie along the western slopes of the Zagros Mountains near the Arabian-Persian Gulf. This oil is piped to refineries in the city of Abadan and to deep-water offshore terminals like Khark Island built to receive modern supertankers. Many of these important facilities were seriously damaged during the Iran-Iraq War and are only now being rebuilt.

War-Torn Afghanistan

To the east of Iran lies another war-torn land, the country of Afghanistan, which was invaded by the Soviet Union in 1979. The center of landlocked Afghanistan is occupied by the four-mile-high Hindu Kush Mountains, the "roof of the world." Rivers fed by melting snows from this mountain range flow south into the deserts of Iran and Pakistan and northward into the grasslands of Central Asia.

◄

Kabul, Afghanistan's capital city, on the slopes of the Hindu Kush Mountains, was taken over by a Soviet invasion in 1979. After a 9-year war, the Soviet troops withdrew leaving approximately one-third of the people as homeless refugees. What other countries have serious refugee problems?

Villages are scattered along the rivers and on the outer slopes of the Hindu Kush Mountains. Wherever water is available, small patches of land are cultivated by the villagers. Before the Soviet invasion, more than 70 percent of Afghanistan's estimated 17.4 million people lived in such villages, growing wheat and barley and herding small flocks of sheep and goats. Another 15 percent were nomads who pastured sheep and goats in high mountain meadows in summer and returned to lowland valleys in winter.

Now, about a third of the people of Afghanistan are refugees, driven from their homes by war. They live in camps in neighboring countries, mainly in Pakistan and Iran, because civil war continues, even though Soviet troops have left Afghanistan.

Soviet troops occupied Kabul, the capital city, and other Afghan cities, but Afghan resistance forced them to pull out in 1989. The countryside remained in the hands of the Afghans throughout the war, even though villages were regularly bombed by Soviet helicopter gunships and other aircraft. Although Afghans belong to three different ethnic groups and speak many different languages, they were united in their Islamic religion and in

their resistance to the brutal Soviet occupation of their country.

In 1988, the Soviet Union gave up the attempt to set up a communist government in Afghanistan and withdrew its troops a year later. Since then several Afghan factions have struggled for power, continuing the destruction inflicted by the Soviet Union. Different parts of the country are controlled by various Afghan tribal leaders. Rebuilding the country after the suffering it has endured will take a very long time.

⊕ **REVIEW QUESTIONS**

1. For what two reasons are the climates of Turkey, Iran, and Afghanistan generally wetter than other parts of the Middle East?
2. What are Turkey's three environments?
3. In what ways are the Afghans divided, and in what ways are they united?

⊕ **THOUGHT QUESTIONS**

1. Why is the Turkish government now facing pressure to expand its urban industries?
2. What important events have occurred in Iran since 1978?

3. The Dry World: North Africa and the Arabian Peninsula

Settlement in the Dry World

The Sahara, the world's largest desert (except for Antarctica), is about 500 miles wide and stretches 4,500 miles across North Africa and the Arabian Peninsula. Many people think of Arabia and the Sahara as a vast area of shifting sand. In fact, only one-seventh of the Sahara's surface is covered by sand dunes. The world's largest sand desert is the Empty Quarter in southeast Arabia. Rock-strewn plateaus and plains cover the rest of this barren region we call the dry world.

The dry world of North Africa and the Arabian Peninsula is equal in size to the United States, but only 160 million people live in it. Settlements exist where water is found. Egypt, with a population of 58.3 million people, is the "gift of the Nile," a river that rises in central Africa and flows northward across the Sahara to the Mediterranean Sea.

Elsewhere in the dry world, people live wherever water is available. In places, mountain ranges intercept rain-bearing air masses and more than ten inches of rain falls in an average year. Along the coast of North Africa, for example, winter rains that fall in the Atlas Mountains and along the shores of the Mediterranean support a population of 65 million people. The map on page 537 shows how low the population is in areas of northwest Africa where rainfall is less than ten inches per year.

Because it rarely rains more than once a year on the Arabian Peninsula, this enormous area supports only 35 million people. Most Arabians live in the rugged highlands along the western and southern coasts of the peninsula.

The "Western Isle": Morocco, Algeria, Tunisia, and Libya

The Arabs call the three countries of Morocco, Algeria, and Tunisia the Maghrib or "western isle," because these countries are located far from the heartland of Islamic civilization. Along with Libya, the four countries of northwest Africa have a combined population of 68.8 million.

The Dry World

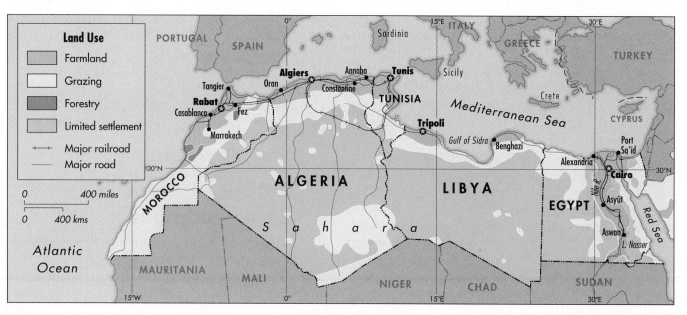

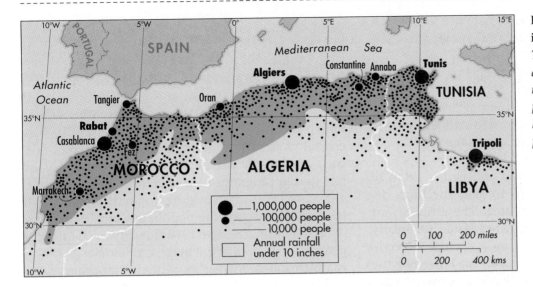

Population and Rainfall in North Africa

This map shows how rainfall affects where people live in northwest Africa. Notice how few people live in areas where less than ten inches of rain falls in most years.

The people of the Maghrib are clustered on the shore of the Mediterranean, and in highlands located between the sea and the desert. Most people live north of the Atlas Mountains, which rise as a barrier between the Sahara Desert and the Mediterranean Sea. In these mountains, Berber peoples live as farmers and herders. Only in the north, along the mountain-backed coast, is there enough rain to support large urban and farm populations.

Mountains and deserts dominate most of Morocco. The Atlas Mountains catch the rain-bearing winds of winter; their crests are usually covered with snow. Short rivers flow down from the mountains and water the coastal fringe, where the important

The Dry World

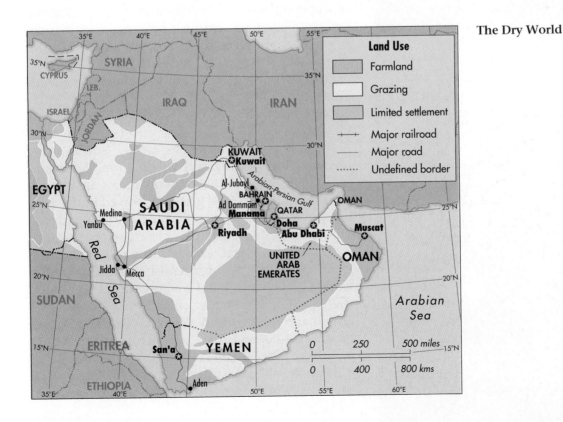

trading cities of Casablanca, Rabat, and Tangier are located. Many Moroccans are now moving to these rapidly growing cities.

Most of Morocco's 28 million people are farmers who cultivate the fertile lowlands between the Atlantic Ocean and the Atlas Mountains. Wheat, barley, citrus fruit, grapes, olives, and cork are grown. South and west of the Atlas Mountains lies the Sahara Desert.

Morocco annexed the Arab and Berber tribes of the Western Sahara in 1976 to gain the huge deposits of phosphate in that area. The ensuing war with local Arab and Berber tribes lasted for nearly fifteen years. It drained resources from Morocco's economic development programs and angered the neighboring state of Algeria. But Morocco now controls the Western Sahara and sells phosphate as fertilizer on world markets.

Farther east, Algeria also shows the environmental contrast between the well-watered Mediter-

IT'S A FACT Algeria is more than three times the size of Texas.

ranean coast, the high ranges of the Atlas Mountains, and the Sahara Desert to the south. The coastal plains and mountain ranges in northern Algeria are densely settled. The government of Algeria has started programs of land reform and agricultural expansion in these regions to provide more farm jobs for its fast-growing population of 27.3 million. As in Morocco, the goal is to slow down the flow of people from farming villages to cities. A strong Islamic fundamentalist movement that opposes the secular principles of the Algerian government makes these efforts more difficult.

Wealth from oil and natural gas is helping Algerians to pay for modernizing their country. Pipelines carry Algerian oil and natural gas from fields located deep in the Sahara to the city of Algiers and to Tunisian ports for export to Western Europe, Japan, and the United States.

Tunisia, a country of only 8.6 million people, is sandwiched between Algeria and Libya. One-third of the country is cultivated, and most of its fertile land is tilled by independent farmers. Melting snow from the mountains feeds rivers that flow to the coast. Along the coast, wheat, olive and citrus orchards, cork oak groves, and vegetables are grown. Tunis, the capital and main port of Tunisia, is located at the heart of the country's most fertile agricultural area.

In addition to prosperous farms, Tunisia gains revenue from its deposits of oil, iron, lead, and phosphate. Most of these raw materials are exported. Tunisia has avoided getting involved in the wars of the region, and its democratic approach to development is raising standards of living in the country.

In contrast, Libya, east of Algeria, is one of the most aggressive and extremist countries in the world. The country's 4.9 million people live along the Mediterranean coast, mostly in the cities of Tripoli in the west and Benghazi in the east. Between and beyond these islands of settlement, 93 percent of Libya is desert.

Without oil and natural gas, Libya would be one of the poorest countries in the world. However,

▼ *Dunes such as these in the Moroccan Sahara are relatively rare in the great deserts of this region. What type of surface covers most of the Sahara?*

> **IT'S A FACT**   **In parts of southern Egypt, it rains about once every fifty years.**

huge fields deep in the Sahara were discovered and developed in the 1950s and 1960s, and these oil revenues are used for Libyan economic development. Under the leadership of Colonel Muammar al-Qaddafi, Libya has become a difficult, unpredictable country on the world stage; it is known to support terrorist groups and fund terrorist activities around the world.

Egypt, "The Gift of the Nile"

Egypt is the most populous nation in North Africa. In 1900, there were 10 million Egyptians, most of them farmers. Today 58.3 million people are crowded along the banks of the Nile River, a narrow riverine oasis—one that is created where the Nile flows northward across the Sahara to the Mediterranean Sea. Too many people live on too little land in Egypt. Virtually all of Egypt's people are crammed into the narrow valley of the Nile River or on its delta. The valley of the Nile varies from 2 to 5 miles in width. The Nile delta, which is about 100 miles long and 125 miles wide, spreads north of the capital city of Cairo.

The sources of the Nile River are in the highlands of East Africa. Heavy rains in these highlands pour into the Nile each summer. Until the 1900s, when dams began to be built, these rains caused the river to rise and flood its banks between August and December.

For many hundreds of years, the flooding of the Nile provided the water needed to grow crops to feed the people of Egypt. A system of **basin irrigation** was used. Huge basins were built as holding ponds next to the river to trap the floodwaters, and this water was used to grow one or two crops of wheat and barley on the same land each year. As a bonus, the Nile flood deposited river silt, and a new layer of fertile topsoil was dropped in the valley, the basins, and on the delta when the flood receded. This renewed the fertility of Egypt's farmlands every year.

In the 1900s, the system of basin irrigation was changed. Low dams were built on the Nile to hold and store floodwater year-round. This system of **perennial irrigation** made farming possible twelve months a year. An Egyptian villager could now grow three or four crops each year on the same land. Cotton and corn became important crops. The amount of farmland and the number of jobs for Egypt's growing village population increased by 50 percent.

Then in 1970, the Aswân High Dam was finished. This huge dam is 365 feet high and more than 2 miles wide. It completely blocks the Nile River. Lake Nasser stretches out behind it for 300 miles. The Aswân Dam increased Egypt's farmland by a third, yet this is still not enough to feed Egypt's

◄ *The Aswân High Dam, located in Upper Egypt, has had a major impact on at least two countries, Egypt and the Sudan. Lake Nasser, which is the result of damming the Nile, displaced many people, and at least two major temples were moved to higher ground before the dam was built. The annual flooding pattern of the Nile was changed. What effects did this change produce?*

growing population. Moreover, the dam is now silting up, and other environmental problems are emerging from this major change in the Nile's flow.

Egypt's population has increased five times in this century, and its city populations have become enormous. Cairo, now the largest city in the region, has 10.1 million residents, and Alexandria has 3.6 million. Like other large cities in the developing world, Cairo and Alexandria are crowded, unhealthy places that are growing too fast. Many of these people moved to cities from villages because the available land was no longer able to support rural Egyptians. Only 4 percent of the country is cultivated. The average size of an Egyptian farm has decreased from 5.7 acres of cropland in 1990 to half that much today, because the growing population has made it necessary to subdivide the land.

Egypt faces serious problems in building industry to provide jobs for these people. Oil fields along the Gulf of Suez provide some revenue. And Egyptians working abroad sent money home until the 1990–91 conflict in the Arabian-Persian Gulf convinced the oil-producing Gulf States that they should hire workers from outside the region instead. Many Egyptian and other Middle Eastern workers have been replaced by Indians and Pakistanis who are viewed as politically reliable in the Gulf.

Most industry is located in Cairo and Alexandria. Textile and steel factories have been built, but Egypt imports more goods than it exports. Few sources of revenue are available to Egypt, yet the government has borrowed money to build its army and provide jobs and services to its people. Its determination to invest in military forces is a decision many developing countries have made. In Egypt, this choice affects the lives of virtually all of its citizens.

Countries of the Arabian Peninsula

The Arabian Peninsula is a huge desert platform bounded on three sides by the Red Sea, the Arabian Sea, and the Arabian-Persian Gulf. The peninsula is the largest in the world, nearly one-third the size of the United States, but it has a population of only 30 million people.

Most of the Arabian Peninsula receives less than three inches of rain each year; only the highlands

IT'S A FACT Saudi Arabia is the fifth largest country in Asia.

of Yemen and Oman at the southern corners of the peninsula get more than ten inches a year. Daily temperatures often rise above 100° (F). Vegetation is sparse, and less than 1 percent of Arabia is farmed. Vast stretches of this region are empty of people.

The Arabian Peninsula is carved up into a number of states, although nine-tenths of its area is ruled by the king of Saudi Arabia. More than half—17.5 million—of the people of the peninsula live in Saudi Arabia. The recently reunified nation of Yemen, at the southwestern corner, has 11.3 million people. Yemen, a mountainous country, receives enough rainfall to grow crops without irrigation, which explains its relatively large population. One of the smaller Arabian states is Oman, with 1.6 million people. Oman is located at the entrance to the Arabian-Persian Gulf.

In addition, a number of small, oil-rich kingdoms line the western shore of the Arabian-Persian Gulf. These include the United Arab Emirates (population of 2.1 million), Qatar (population of 0.5 million), the island nation of Bahrain (population of 0.5 million), and the desert city-state of Kuwait (population of 1.7 million). Locate these countries of Arabia on the map on page 523.

Saudi Arabia is the land where the Arabic language originated and the Islamic religion was born. This country is the keeper of the most sacred sites in Islam—Mecca and Medina—the places where Muslims believe that Mohammed received the revealed word of God. Pilgrimages bring some 2 million people to Mecca each year. Now the world's leading exporter of oil, this Islamic kingdom is trying to change from a nomadic society into an industrial country.

Most Saudi Arabians live in the rugged highlands that line the coast of the Red Sea in the west. The capital city of Riyadh, in central Saudi Arabia, is one of a number of oasis towns that rely on wells and springs for water. Settlement in the east, on the shores of the Arabian-Persian Gulf, is related to the development of the huge deposits of oil discovered there in the 1940s.

PLACE

AL-JUBAYL, SAUDI ARABIA: FROM SAND DUNES TO STEEL CITY IN TEN YEARS

◄ *The model city of al-Jubayl, built in the desert, even includes attractive play areas for children. How did Saudi Arabia pay for this new industrial center?*

Saudi Arabia has used some of its money earned from oil exports to change a patch of desert into the world's largest new industrial center—the steel city of al-Jubayl. Al-Jubayl, on the east coast of Saudi Arabia, is a clean, smog-free, industrial center with glistening hotels, shopping centers, mosques, and baseball fields. More than 100,000 imported trees line the boulevards of al-Jubayl. The trees are irrigated by treated water reclaimed from city sewers. A total of 250 million cubic yards of sand was moved to create space for new factories, shopping districts, residential suburbs, beaches and playgrounds, harbors, and 150 miles of roads. (About 10 cubic yards of sand will fill the average dump truck.)

Water is provided by **desalinization** plants that convert salt water into fresh water for use in homes and factories. Al-Jubayl already has 30,000 residents, and its population is expected to grow to 250,000 residents in the next century.

Al-Jubayl is intended to help Saudi Arabia become a country that produces manufactured goods for sale in world markets. At al-Jubayl, the Saudis already produce steel, fertilizer, plastics, and refined oil. Al-Jubayl has become a state-of-the-art industrial city.

QUESTIONS

1. In what two ways is al-Jubayl dealing with providing the water needed by the city?
2. Why is it important for Saudi Arabia to develop a "state-of-the-art" industrial city?

Oil refineries now line the Saudi coast and that of neighboring Bahrain. Petroleum has paid for free health, education, and social services for the Saudi people, as is the case to a lesser degree in the smaller neighboring countries of Oman, Qatar, Bahrain, and Kuwait. With extensive oil reserves, Saudi Arabia is the richest country in the Middle East.

⊕ REVIEW QUESTIONS

1. The dry land in North Africa and Arabia is equal in size to what country in the Western Hemisphere? How many people live in this region?
2. What is the largest city in the Middle East?
3. What is the largest peninsula in the world? Nine-tenths of its area is ruled by what country?

⊕ THOUGHT QUESTIONS

1. Why do you suppose that, for many centuries, Egypt has been called "the gift of the Nile"?
2. What has prevented Saudi Arabia, with its vast stretches of desert, from being a poor country?

4. The Central Middle East

Turmoil at the Crossroads

The Central Middle East occupies the area between Turkey and Iran in the north, Egypt in the southwest, and the Arabian Peninsula in the south and southwest. The countries of Syria, Lebanon, Israel, Jordan, and Iraq are located here.

Together, they have 45 million citizens. Iraq and Syria each have more than 10 million residents; the other three nations have less than 6 million.

These five countries, which have been caught up in political turmoil, civil war, and external war for most of this century, share a common history. They were **mandate** territories under British and French control after World War I; they gained independence during and after World War II. The League of Nations gave mandated territories, the former colonial possessions of Germany and Turkey, to Britain and France after World War I. Britain and France were to administer these territories until they were considered "ready" for independence.

Israel, the former British mandate of Palestine, was declared a nation by the United Nations in 1948. War with Jordan and Egypt erupted immediately. Israel fought wars for survival against its neighbors twice more in 1967 and 1973. Lebanon, a French mandate, disintegrated in 1975 into civil war, a conflict that was to last into the 1990s. Iraq was mired in a destructive war with Iran for eight years (1980–1988). Following its invasion of Kuwait in 1990, Iraq was devastatingly punished by a United Nations Coalition Force led by the United States. These political events have torn apart the economic and social fabric of life in much of the Central Middle East.

Israel, a Country Besieged

Few countries have experienced as much turmoil in recent times as Israel. Located on the shores of the eastern Mediterranean, Israel was carved out of the British mandate territory of Palestine in 1948. Since its formation, Israel has provided a refuge for Jews from all over the world, and Jews make up 85 percent of the Israeli population.

The Central Middle East

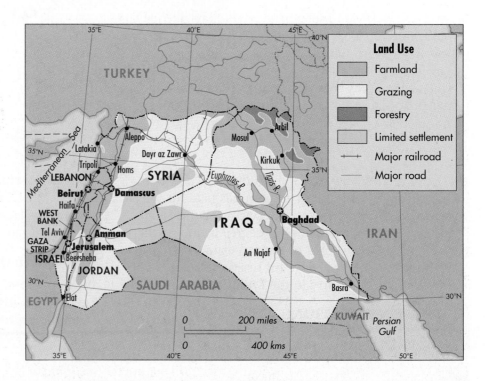

Tel Aviv, with 2.3 million citizens, is the largest city in the country. Population is dense on the well-watered coastal plain between Tel Aviv and the port city of Haifa. Fully 90 percent of all Israelis live in cities.

On farms that reach inland to the shores of the Sea of Galilee, citrus groves, vineyards, and market gardens thrive on winter rains. To the south lie the bleak wastes of the Negev Desert. A large pipeline, called the National Water Carrier, transports water from the Jordan River to the oasis town of Beersheba in the Negev. At Beersheba, irrigated agriculture uses the most modern techniques and crop combinations. Ninety-five percent of Israel's water goes to farming, industry, and domestic use. Israel has made more efficient use of available water than any country in the world.

Israel has few other resources to develop. The Dead Sea is a source of potash and other salts. Phosphates, gypsum, and copper are mined in the Negev Desert and exported through the southern port of Elat on the Red Sea. Human skills make services like diamond cutting important in the economy. But, still, the costs of maintaining a large modern army have forced Israel to rely on massive amounts of aid from the United States.

The country has had a difficult job in governing the Golan Heights, the West Bank, and the Gaza Strip, territories occupied by the Israeli army in 1967. (The Israelis eventually annexed the Golan Heights from Syria.) These occupied territories are populated by roughly 1.8 million Palestinians, and another 900,000 of Israel's 5.3 million citizens are Palestinians. Until recently, opinions differed in Israel about what occupied territories could be given up so the Palestinians could have a homeland, when they could be given up, and in exchange for what. After 45 years in which the daily life of the country has been shattered by terrorist bombs, guerrilla raids, internal strife, and hostile incidents between Israelis and Palestinian Arabs, a breakthrough has occurred. In 1993, the Israelis sat down with the PLO (Palestine Liberation Organization) to hammer out the details of a lasting peace.

▼ Tel Aviv is Israel's largest port and most important economic center. Israel has made more efficient use of its available water than any other country in the world. To what primary uses was this water put?

The Desert Kingdom of Jordan

East of Israel lies the kingdom of Jordan. Five-sixths of Jordan is a barren land used only by nomads. Most of Jordan's 3.8 million people live in the capital city of Amman and in farm villages that line the Jordan River Valley. Irrigation enables Jordanians to grow wheat, grapes, and citrus fruit along the banks of the river. Jordan must import food, even though a third of its people are farmers, because not enough of the land can be cultivated. Phosphates are the only other important resource.

Jordan has faced serious political problems since it gained independence in 1946. It has fought and lost wars with Israel. When Palestine was divided and Israel was created in 1948, a million Palestinian refugees fled into Jordan. In the 1967 war, Jordan lost the West Bank and its part of Jerusalem to Israel. Landlocked, resource poor, and overwhelmed by 2 million Palestinian refugees from lands held by Israel, Jordan faces tremendous problems.

Divided Lebanon

North of Israel is the Mediterranean country of Lebanon, another small state fighting to survive. In this fertile land, narrow coastal plains are backed by two mountain ranges separated by a wide, fertile valley. Most of Lebanon's 3.6 million people live in this interior valley and in cities and villages along the coast. Mediterranean crops like olives, grapes, and citrus fruits are grown in these areas.

Until 1975, Lebanon was the business and tourist capital of the eastern Mediterranean. Then, in 1975, civil war broke out. Fighting raged among Christians, Sunni and Shi'ite Muslims, a Muslim sect called the Druze, and Palestinian refugees. Invasions by the Syrian and Israeli armies further devastated Lebanon. The capital city of Beirut, once described as the Paris of the Middle East, became a battlefield. Rival militias still control different parts of the country, although fighting has died down.

At the root of the problem are centuries-old religious and cultural differences, as well as a wide gap in standards of living. Lebanon, once noted for the peaceful coexistence of peoples with different beliefs and ways of life, has been destroyed by civil war and foreign invasions.

Rural Syria

Unlike urban Israel and Lebanon, Syria is a nation of farmers. Syria is bordered by Turkey on the north, Iraq on the east, and Lebanon, Israel, and Jordan to the south. Its narrow coastal plain is backed by mountains along the Mediterranean, where crops can grow without irrigation. Much of northern Syria is mountainous, but in the south these highlands flatten into rainless desert.

Most of Syria's 13.5 million people live in farming areas along the coast, in interior valleys watered by rivers from nearby mountains, and on the inland plains of the Euphrates River. Syria's two largest cities, Damascus and Aleppo, are actually spring- and river-fed oases. Damascus claims to be the oldest continuously inhabited city on Earth. Many scholars believe that one of its winding streets, "The Street Called Straight," is described in the Bible.

Syria is another country in the Central Middle East whose resources are being drained by war. Syrian troops occupy much of war-torn Lebanon. An opponent of Israel, Syria does not recognize Israel's existence. Now negotiations with respect to the strategic Golan Heights annexed by Israel may lead to a peace agreement between these past enemies.

The River Valleys of Iraq

Iraq lies east of the lands bordering the Mediterranean. Iran is still further east and Saudi Arabia is to the south. The heart of the country lies in "Mesopotamia," the land between the Tigris and Euphrates Rivers where one of the world's earliest civilizations began.

These rivers rise in the mountains of eastern Turkey and flow southward for 1,000 miles into the Arabian-Persian Gulf. In some years, ten-foot floods pour down these rivers and spread out over the flat plains of Mesopotamia below Baghdad. Dikes and dams have been built on the rivers to control the flooding; however, drainage and irrigation are still difficult in this wheat- and barley-growing area. Where the Tigris and Euphrates join in the marshlands of the Shatt-al-Arab farther south, three-quarters of the world's dates are grown.

IT'S A FACT   Iraq is planning to rebuild the biblical city of Babylon as a tourist attraction.

Iraq's irrigated farmland can be expanded, and the country has the potential oil resources to pay for such expansion. Yet an increasing number of Iraq's farmers are moving to large cities like Baghdad, the capital. Plans to develop industry in these cities have been postponed because oil exports from Iraq's fields along the Arabian-Persian Gulf and northeast of Baghdad were damaged by Iranian bombing during the eight-year war in the 1980s.

Environmental problems, population growth, and the high costs of the war with Iran all lowered standards of living in Iraq. The Persian Gulf War between Iraq and a U.S.-led Coalition Force caused by Iraq's invasion of Kuwait has further damaged Iraq's economy and society.

Iraq, a country of 19.2 million people, has serious problems. Its invasion of neighboring Iran in 1980 led to an eight-year war (1980–1988) with Iran that consumed almost all of Iraq's national revenue. The country also lost an estimated 250,000 men and women during this period. Kurds, a non-Arabic people who live in northern Iraq, still threaten the country's oil fields there. Shi'ite Muslims, who share the Islamic faith of Iraq's enemy Iran, make up a majority of Iraq's population in the south. They are rapidly being destroyed by Iraqi troops as the marsh environment in which they have lived is drained.

Iraq's problems concern countries all over the world because of the fear that Iraq could hinder the flow of the oil resources of the Arabian-Persian Gulf region to the industrial world. This is why Iraq's invasion of the country of Kuwait sparked an immediate international response. As long as the country is led by Saddam Hussein, a ruthless tyrant, peace in the Arabian-Persian Gulf will probably be maintained by the armed presence of Western nations.

⊕ REVIEW QUESTIONS

1. What are the five countries in the Central Middle East?

2. What Syrian city claims to be the oldest continuously inhabited city on Earth?
3. The heart of Iraq lies between the valleys of what two rivers?

⊕ THOUGHT QUESTIONS

1. How has Israel made good use of its sources of water?
2. What are the causes of the devastation Lebanon has faced since 1975? How do you think these issues might eventually be resolved?

⊕ CHAPTER SUMMARY

The Middle East and North Africa make up a culture realm that is a vast and complex region embracing many countries. The majority of people in this realm have the same religion and religious heritage. Middle Easterners live as city dwellers, villagers, and tribal nomads.

The Middle East and North Africa can be divided into three subregions: the Northern Highlands, the dry world of North Africa and the Arabian peninsula, and the Central Middle East. The people of the Northern Highlands primarily speak a Turkic language or Persian (Farsi), those in the Arabian Peninsula and many countries of North Africa speak Arabic, some as their second language. The people of the Northern Highlands live in wetter mountainous and high plateau environments, compared to those on the Arabian Peninsula and North Africa or the Central Middle East whose cultures are shaped by arid environments.

The region is defined, in part, by religion. A majority of people in this region practice Islam, a religion that is both a faith and a way of living. Yet, here too, there are some differences. The majority of Muslims, as followers of Islam are called, follow the Sunni tradition of Islam, whereas the minority are Shi'ites. Only in Iran is the Shi'ite sect both the majority and the government. While warfare between Sunnis and Shi'ites is not common, this did happen in the Iran-Iraq War. War has also broken out repeatedly between neighboring Arab countries and Israel, which form the core of the third major subregion of this area, the Central Middle East.

EXERCISES

End the Sentences

Directions: Part A has ten sentence beginnings; Part B has ten sentence endings. Match the correct ending to each beginning, and then write the complete sentences on your paper.

Part A

1. Densely settled areas in the Middle East
2. About a third of all the people in the Middle East
3. In the 1950s and 1960s, much land owned by absentee landlords in the Middle East and North Africa
4. The Mediterranean island of Cyprus
5. In 1980, the oil-rich countries of Iran and Iraq
6. In 1979, military forces from the Soviet Union
7. The people of Algeria achieved independence after they
8. Under the rule of Colonel Muammar al-Qaddafi, Libya
9. In 1993, the Golan Heights, the West Bank, and the Gaza Strip were all territories
10. A large amount of oil from the Middle East

Part B

a. fought a long war against France.
b. is the source of a bitter dispute between the Turks and the Greeks.
c. controlled by Israel, but under negotiation.
d. are city dwellers.
e. went to war against each other.
f. is shipped out of the Arabian-Persian Gulf.
g. are places where water is available.
h. was redistributed to poor farmers.
i. has supported international terrorism.
j. invaded and conquered parts of Afghanistan.

Environments, Countries, and Cities

Directions: Each of the following sentences has three possible endings. Find the correct ending and then write the complete sentence on your paper.

1. For centuries, the Sahara and the desert area of the Arabian Peninsula have been peopled mainly by (a) sharecroppers (b) nomads (c) Islamic warriors.
2. The Northern Highlands of the Middle East include Turkey, Iran, and (a) Egypt (b) Morocco (c) Afghanistan.
3. The capital city of Iran is (a) Tehran (b) Tabriz (c) Istanbul.
4. The capital city that was seized by Soviet troops is (a) Tunis (b) Kabul (c) Damascus.
5. The country bounded by the Black Sea to the north, the Aegean to the west, and the Mediterranean to the south is (a) Turkey (b) Saudi Arabia (c) Israel.
6. An important Moroccan trading center is (a) Casablanca (b) Algiers (c) Baghdad.
7. The largest city in the region is (a) Istanbul (b) Jerusalem (c) Cairo.
8. Oman and Yemen are located in (a) North Africa (b) the Arabian Peninsula (c) Turkey.
9. The most sacred sites of the Islamic faith are (a) Jerusalem and Nazareth (b) Mecca and Medina (c) Tehran and Beirut.
10. In 1948, Israel was carved out of the British mandate territory of (a) Palestine (b) Syria (c) Lebanon.

Inquiry

Directions: Combine the information in this chapter with your own ideas to answer these questions.

1. What factors have made life difficult for Iranians during the last fifteen years? Do you believe that Iran has the resources to improve the living conditions of its people?
2. In most parts of the world, people who have depended on the export of a single product have been greatly handicapped. Why is this less true for the people who live on the eastern shore of the Arabian Peninsula?

SKILLS

Using a Table

Directions: Tables, like graphs, present information in a way that is easy to understand. They also show clearly the relationships between various facts. Look at the table on page 529 and answer these questions.

1. How many cities in the Middle East and North Africa have populations of more than 1 million people?
2. How many of these cities are capitals of countries?
3. What are the three largest cities in the Middle East?
4. How many cities have more than 2 million people?
5. How many cities have more than 3 million people?
6. How many cities have more than 6 million people?
7. Which cities in Egypt have more than 3 million people?
8. Which cities in Turkey have more than 2 million people?
9. Is Turkey's capital city the largest city in that country?
10. Does Morocco have any cities with more than 1 million people?
11. Does Saudi Arabia have any cities with more than 1 million people?
12. Is Damascus larger or smaller than Alexandria?
13. How much larger is Tehran than Baghdad?
14. How much larger is Istanbul than Ankara?
15. How much larger is Cairo than Tehran?

Using a Population and Rainfall Map

Directions: Look at the map on the top of page 537 and then write on your paper whether each of the following statements is true or false.

1. The most important generalization we can make based on this map is that the regions near the coast of North Africa have the most rainfall and the most people.
2. The southern part of Tunisia is the most densely settled part.
3. The area in Algeria that receives under ten inches of rain each year has more land but fewer people than the coastal area.
4. Most of the cities that are named on this map are seaports.
5. The interior of Libya is more densely settled than the interior of Morocco.
6. The two northernmost cities on this map are Rabat and Tangier.
7. The easternmost city shown on this map is Tripoli.
8. The westernmost city on this map is Casablanca.

Vocabulary Skills

basin irrigation	mandate
bazaar	mosque
Bedouin	perennial irrigation
caravanserai	sharecropper
desalinization	transhumance

Directions: Sometimes the meaning of an unfamiliar word is revealed by the context in which the word appears. Use the context in each of the following sentences to determine the meaning of the term in italics. Write on your paper the letter of the correct definition corresponding to the number of the sentence. Check your answers in this book or a dictionary.

1. The Nile River overflows its banks between August and December. To have water for spring planting, Egyptian farmers until 1960 employed *basin irrigation*.
    a. farmers capture floodwaters in shallow depressions
    b. farmers depend on water to remain in the soil of low-lying land
    c. the government constructs large dams to capture floodwaters

2. People in the Middle East and North Africa frequent the *bazaar* where many kinds of goods are sold.
   a. large museums where Egyptian and Mesopotamian antiquities are displayed
   b. a place where many strange and odd things are kept for people to see
   c. a traditional shopping district made up of rows of shops and stalls

3. *Bedouins* are tribal people who live in tents in dry regions and used to transport trade goods across the desert.
   a. local village merchants who travel the North African deserts
   b. camel nomads who live in North Africa and Arabia
   c. hunters and gatherers who live on the desert and trade their handicrafts in the local towns

4. Until recently every major trading center had at least one *caravanserai*, but now they are disappearing as modern buildings and transportation systems make them obsolete.
   a. an oasis maintained for the convenience of camel caravans
   b. a bazaar where camel caravans would bring their trade goods and where they would sell camels
   c. a type of inn with a large central court where camel caravans could be unloaded and their drivers could rest

5. Because some desert countries have easy access to the sea, they can provide water for drinking and irrigation by means of *desalinization*.
   a. a convenient method for pumping drinking water using gravity
   b. a method for converting sea water into fresh water
   c. a method for transporting fresh water using tanker ships

6. In 1920 the League of Nations made five countries—Syria, Lebanon, Israel, Jordan, and Iraq—*mandates* of either Great Britain or France.
   a. countries that owed large sums of money to Britain and France
   b. former colonial possessions supervised by another country on their road to independence
   c. members of the United Nations who are fully independent

7. *Perennial irrigation* is difficult in regions with seasonal rainfall.
   a. year-round irrigation
   b. irrigation during the dry season
   c. use of deep wells for irrigation

8. Devout Muslims can go to the *mosque* as often as they like, but they generally go every Friday.
   a. a traditional shopping district in Middle Eastern cities
   b. an entertainment district found in large cities
   c. a Muslim place of worship

9. Until recently, land in many Middle Eastern countries was owned by landlords and cultivated by *sharecroppers*.
   a. a type of landowner who grows only his share of crops
   b. a landless farmer who pays rent by giving the landowner a share of the harvest
   c. a farmer who shares his crops with his sons and daughters

10. Many sheep and goat nomads in arid regions practice *transhumance* because the land will not support a permanent pasture for their animals.
    a. the practice of tending grazing animals only part of the year and farming the rest of the year
    b. the practice of separating sheep from goats to lessen their impact on grazing land
    c. the practice of rotating grazing animals between upland summer meadows and lowland winter pastures

CHAPTER 22

CONFLICT AT THE CROSSROADS

The Middle East lies at the heart of the Old World, where three continents meet. Many different peoples have been drawn to this location at the junction of Europe, Africa, and Asia.

For centuries, this region has been a focal point of conflict. It still is. The most recent conflict followed Iraq's 1990 invasion of Kuwait, which led U.S. troops under United Nations authority to drive Iraqi forces out of Kuwait. This was the twenty-fifth war fought at the head of the Arabian-Persian Gulf over the centuries.

Yet no other culture region has produced so many world faiths. Millions of people world

◄ *The Treasury at Petra, in Jordan, is an unusual architectural achievement as it is built directly into the cliff. On top of the facade is the urn which, according to the locals, contains the treasure. Do you know what motion picture of the 1980s featured this temple?*

wide view the Middle East as their spiritual home. Three great world religions arose in this region. Judaism and Christianity were born in what is now the country of Israel. Islam began in the city of Mecca and spread out of the Arabian Peninsula over virtually all of the Middle East and North Africa. Islam unified the region under large empires, the last of which was the Ottoman Empire.

The Middle East and North Africa were largely isolated until the 1800s, when European trade, technology, and power spread into the region. By 1900, most people in North Africa and the Middle East lived under the direct or indirect rule of foreign countries, and the countries in the Middle East and North Africa did not win their independence until after World War II. The tensions and problems in this region today were created when the cultures of this region and those of Europe collided.

1. Where Three Continents Meet

Crossroads Location

The deserts and grasslands of this region extend 6,000 miles from the Atlantic shores of North Africa across the countries at the eastern end of the Mediterranean Sea deep into the heart of Asia. Desert oases, well-watered coasts, and fertile river valleys are the only densely settled areas in this region. The Middle East has played an important role in history because of its crossroads location.

The core of this region is the meeting place of three continents—Europe, Asia, and Africa. Here the Mediterranean, Black, and Red Seas reach deep into the land, as the map on page 524 shows. Overland trade and movement of peoples across the region must use land bridges, like the **Isthmus** of Suez, which connects Asia and Africa. Important sea routes between Asia and Europe pass through narrow **straits**, like the Bosporus and Dardanelles, which connect the Black and Aegean Seas. The people, ideas, and trade goods from Europe, Asia, and Africa that have met at this crossroads have over the centuries enriched the cultures of the Middle East.

Center of Creativity

This mixture of cultures led to great creativity. Middle Easterners were probably the first people to cultivate plants and to domesticate animals, as you have already learned. Soon after these discoveries in food production, civilizations sprang up in Mesopotamia, located between the valleys of the Tigris and Euphrates Rivers in modern Iraq, and in the Nile Valley in Egypt. Settlements also lined the banks of the Jordan River that flows from modern Syria southward along the border between Jordan, the West Bank, and Israel. Irrigation agriculture in these river valleys yielded rich harvests, which provided food for villagers and city dwellers.

More discoveries followed. Writing made it possible for Middle Easterners to keep records and to pass on knowledge from one generation to the next thousands of years ago. Legal codes were created to govern day-to-day life. Inventions like the wheel,

▼ *Mecca, in Saudi Arabia, is where Muhammad, the founder of Islam, was born, and is the holiest religious center of Islam. The Kaaba in the center of the crowd of pilgrims is the shrine to which all Muslims pray. Can you think of another place sacred to another religion in the Middle East?*

Arab scholars introduced the decimal system and Arabic numerals into Western Europe.

the plow, and metal tools made work easier. International trade with distant places led to a rich exchange of goods and ideas. The scholars of the Middle East became the foremost scientists of their day. For centuries during the Middle Ages, from the 700s to the 1700s, this region was one of the most advanced parts of the world. As a result, the Middle East had an enormous influence on the history and culture of Western Europe.

Cradle of Religions

Three major world faiths—Judaism, Christianity, and Islam—were born in the Middle East. Judaism and Christianity developed in the hills lining the eastern shore of the Mediterranean. The Muslim religion, Islam, started at Mecca, on the west coast of the Arabian Peninsula.

A basic principle of all of these three religions is **monotheism**, which means a belief in one God. These faiths share common roots and common places of worship, especially the city of Jerusalem. Today, more than one-third of the people in the world are members of one of these three religions.

⊕ REVIEW QUESTIONS

1. Which three continents meet in the Middle East, and which three seas reach deep into this land?
2. What three major faiths were born in the Middle East?
3. Why is Jerusalem sacred to these three faiths?

⊕ THOUGHT QUESTIONS

1. To what degree has each of the five major themes in geography affected life in the Middle East?
2. Some scholars believe that monotheistic religions developed in the Middle East because of its harsh desert climate. Do you agree? Why or why not?

2. Islam Awakening

The Origin and Spread of Islam

The Islamic religion was founded by the Prophet Muhammad, who was born in the small trading city of Mecca on the west coast of the Arabian Peninsula in A.D. 570. At the age of forty, he retired to the desert to meditate on the meaning of life. In the desert, he experienced a series of revelations about a single, all-powerful God—Allah—who was the creator of the universe. Muhammad returned to Mecca and announced that he had been chosen to be the "messenger of God."

Muhammad preached this new religion of **Islam**, a word which means both "peace" and "submission." Muhammad's message of one God, justice, and concern for all fellow Muslims caused some Meccan leaders to persecute him and his early followers, who were called **Muslims**. In the next ten years, after Muhammad moved to Medina in A.D. 622, many Arabs became Muslims. Muhammad unified most of the tribal groups of the Arabian Peninsula under the banner of Islam. He died in A.D. 632, having given Arabs a new religion and a new way of life.

After Muhammad's death, Muslim armies swept out of the Arabian Peninsula. In twenty-five years, these nomadic warriors conquered Syria, Egypt, Iraq, and Iran. By the middle of the 700s, Muslims ruled an empire that stretched from Spain in Europe to Pakistan in Asia.

During this "golden age of Islam" a civilization was created with its own religion, language, and laws. This civilization was driven by belief in the truth of Muhammad's message. Muslims came to live by a code of behavior based on Muhammad's teachings, and on the word of God as found in the holy book of Islam, the **Qur´an** or **Koran**.

The Five Pillars of Islam

The religious duties of Muslims involve five basic acts upon which they build their lives. These acts are called the "five pillars of Islam." The first pillar is the expression of faith stated in the Qur´an:

THE HOLY CITY OF JERUSALEM

The city of Jerusalem in the Middle East is sacred to Muslims, Christians, and Jews. Here, many believe people "can cup their hands to the wind and hear the voice of God."

For Muslims, Jerusalem is where Muhammad the Prophet mounted a winged horse at the Dome of the Rock and ascended to heaven. For Christians, the Son of God died on the cross and rose from the dead in Jerusalem. For Jews, Jerusalem is the capital city of the ancient Jewish kings. For some Jews, it is the place where the Ark of the Covenant (an ancient chest in which the scrolls of the Torah reside) will be found and, for most, the location of their great temple.

For centuries European maps showed Jerusalem as the center of the world, and many people believe that Jerusalem is closer to heaven than any place on Earth.

An especially sacred site for Jews in Jerusalem is the western or Wailing Wall, which many believe is what remains of King Herod's magnificent temple. This temple was destroyed by Roman soldiers in A.D. 70. when the Romans ordered the Jews into exile. This scattering of the Jews, the **diaspora**, spread Jewish people throughout Europe and Asia. In cities on three continents, the Jews kept their faith alive in **synagogues**, their centers of worship, often despite severe persecution.

Today, Jerusalem is Israel's capital. Conquered by Israel in the 1967 war with Arab countries, it was united under Jewish rule for the first time in eighteen hundred years.

At the time of partition, Jerusalem and nearby Bethlehem were envisioned by the United Nations as under international control because they were the locations of most of the shrines holy to Jews, Christians, and Muslims. Israel occupied the greater part of New Jerusalem; Jordan had control of Old Jerusalem and Bethle-

▼ *The Dome of the Rock, Islam's oldest and most sacred shrine in Jerusalem, stands not far from the remains of King Herod's Temple, now called the Wailing Wall. Jerusalem is sacred to three of the world's major religions. Name these religions.*

purposes, a dead issue. This was victory for the Israelis and a humiliation for the Arabs.

Since the 1967 war, the city of Jerusalem has been governed by Israel. Large-scale apartment complexes peopled by Israelis now ring the city, damaging the aesthetic appeal of this holy place, but providing the Israelis with proof "on the ground" of their permanent possession of Jerusalem. Israel's determination to maintain control of Jerusalem is one of the major obstacles to any peace agreement between Israel and its Arab neighbors.

By 1995, agreement was reached on less controversial areas like Gaza and Jericho, but because Jerusalem is sacred to both Arabs and Jews, neither country is willing to compromise. The holy city of Jerusalem— conquered by Arabs, crusaders, Turks, and British—remains at the center of the storm that swirls through the Middle East.

▲ *Jerusalem is a city in conflict. Incidents of violence between Palestinians and Israelis occur frequently. Here, as a Palestinian woman walks by, an Israeli Border Policeman frisks a young Palestinian man on a stairway where the week before a Jewish seminary student was stabbed to death. What is a Palestinian? An Israeli?*

hem. Jerusalem became a divided city. This issue remained on the United Nations agenda for the first two decades of Israel's existence, during which time both groups attempted to resolve this territorial partition in their own favor.

In the end, the perplexing issue of Jerusalem was decided by war. In 1967, the Israelis took over the city of Jerusalem, and two years later declared Jerusalem to be their capital, although few other countries regard it as such, and most maintain their embassies in Tel Aviv. Internationalization of the city became, for all intents and

❓ QUESTIONS

1. Why did medieval map makers place Jerusalem at the center of their maps? If you were to design a world map on the same principles, what would be at the center of your map?
2. What was the diaspora?
3. How was the issue of a "divided Jerusalem" settled?
4. Do you believe one nation should control Jerusalem?

IT'S A FACT   Muslims are forbidden to eat pork or drink alcohol.

"There is no god but God, and Muhammad is His Messenger." A person becomes a Muslim by making this declaration.

Prayer is the second pillar of Islam. Muslims pray five times a day—at dawn, noon, midafternoon, sunset, and bedtime. Religious leaders call believers to prayer at these times from **minarets**, tall graceful spires that rise beside Muslim mosques, or places of worship. The most important prayer meeting is held on Friday, when the entire Muslim community gathers to worship in the mosque.

The third pillar of Islam is almsgiving, which requires each Muslim to give part of his or her wealth to the poor. Today, the principle of almsgiving eases the lives of poor and sick people, many of whom live in countries with few hospitals, clinics, or orphanages.

Fasting during the month of Ramadan, the ninth month of the Muslim calendar, is the fourth pillar of Islam. During this month, no food or drink is consumed from dawn to dusk. In many Muslim countries like Algeria, Sudan, Iran, and Pakistan, breaking the fast is illegal. In no country is it acceptable.

The fifth pillar of Islam is the **pilgrimage** to Mecca, the most important religious center of Islam. Making this journey at least once in a lifetime is the duty of all Muslims, provided they can afford it. The pilgrimage unites Muslims from many different countries, who converge on Mecca from all over the world each year. In the 1990s, an estimated 2 million pilgrims from more than 70 countries make the pilgrimage to Mecca each year.

Islam Resurgent

Islam is now experiencing a major revival in the Middle East and North Africa. The number of pilgrims to Mecca has doubled in the last ten years. A political revolution led by religious leaders overthrew the Shah of Iran in 1979 and created an Islamic state in that country. In Kuwait, law codes are being revised to incorporate religious principles of

▲ *Muslims pray five times a day—at dawn, noon, midafternoon, sunset, and night. This Algerian Muslim stops to pray along the road. What underlying principle does this religion share with Christianity and Judaism?*

Islamic law. Both Sudan and Pakistan have declared themselves Islamic states. Egypt is changing its laws in response to demands from conservative Muslims. Algeria cancelled elections when it became clear that conservative Muslim parties would win.

Islam's awakening began as a spiritual movement but has increasingly become a political force. Fueled in part by the inability of political leaders to improve economies and raise standards of living, people are increasingly turning to their traditional faith. Some want to create Islamic states that do not allow Western influences, and are calling for a basic restructuring of their societies based on Islamic law. Each year, their strength and influence in the region have been growing.

⊕ REVIEW QUESTIONS

1. What are the five pillars of Islam?
2. What four countries have the largest Muslim populations?
3. What changes is Islam experiencing in this region?

⊕ THOUGHT QUESTIONS

1. Do you believe that religion can be a powerful source of change in society? Why or why not?
2. Why, in your opinion, is Islam the fastest growing religion in the world?

REGION

THE WORLD OF ISLAM

The world of Islam spreads far beyond the Middle East and North Africa. This religion has more than 950 million believers, or one of every six people on Earth. Islam is the fastest growing religion in the world.

Muslims live throughout Africa and Asia. They form a majority of the population in thirty-six countries and a near majority in four others. Indonesia is the largest Muslim country, with 126 million believers. In Pakistan there are 93 million, and India and Bangladesh each have 80 million. More than 50 million Muslims live in the independent republics that once formed the Soviet Union.

Throughout the world of Islam, religion is attracting many young people. In the past, Islam encouraged believers to be politically passive, but now some religious leaders are encouraging a revolutionary call to arms to implement God's law through the creation of Islamic states. From Morocco to Pakistan, fundamental Islam is a powerful spiritual force. This revival of Islam involves belief in the Islamic way of life and rejection of Western values.

QUESTIONS

1. Why are young people in Africa and Asia attracted to Islam?
2. An important part of Islamic revival is a rejection of what kind of values?
3. Which country has the largest number of Muslims? Can you think which countries in the Central Middle East are not Muslim countries?

The Distribution of Muslims in Africa and Asia

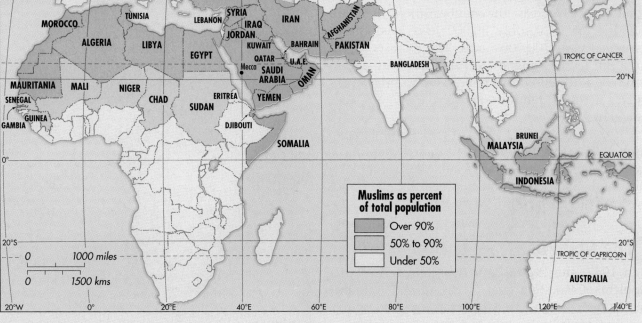

3. The Arabs and the Israelis

The Origin of Zionism

The struggle between the Arabs and the Jews in the Middle East actually began in the 1880s, when large Jewish communities were brutally persecuted in Russia and Eastern Europe. This persecution led to the birth of a political movement called **Zionism**. Zionists were devoted to the creation of a Jewish state as a political homeland for all Jews.

Zionists, convinced that safety from persecution could be guaranteed only in an independent Jewish state, looked for a place to establish one. They explored territories in Latin America and Africa, but no other place had the powerful emotional and historic appeal of Palestine, the original homeland of the Jews. In 1917, Zionists in Europe secured a commitment from the British government, which took over this land that same year, to support the "establishment in Palestine of a national home for the Jewish people." In 1920, Palestine became a British mandate.

Jewish Immigration to Palestine

Zionists realized that to create a country they would have to increase the number of Jews living in Palestine. In 1920, only 80,000 Jews lived there, compared with 700,000 Arabs. The Zionists urged European Jews to migrate to Palestine. They began to buy land there and to build Jewish settlements, and as a result, tensions between the Arabs and Jews grew.

In the 1930s, when a flood of Jewish immigrants from Nazi Germany greatly increased Palestine's Jewish population, the Arabs feared they would lose their land. Guerrilla warfare between the Arabs and Jews broke out just before World War II.

The State of Israel Is Created

In 1947, after World War II, Britain turned the problem of Palestine over to the United Nations. The United Nations devised a plan for dividing (or partitioning) Palestine into two separate countries—

one for the Jews, one for the Arabs. At the time, Jews made up 31 percent of the population and owned about 7 percent of the land in Palestine.

Under the United Nations plan, the Jewish state was to be given 56 percent of Palestine and the Arab state 43 percent. The holy city of Jerusalem

The State of Israel

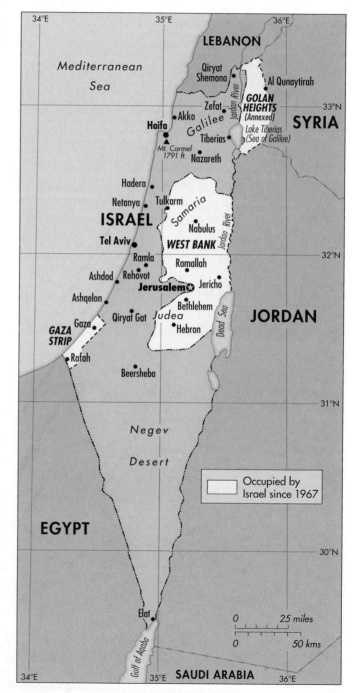

was to be administered as an open city by the United Nations. The Jews agreed to the United Nations partition plan, but the Arabs did not. Both sides prepared for war.

In 1948, the British left Palestine and the state of Israel was created by vote of the United Nations. The Arabs and the Jews were at war. After fourteen months of fighting, Arab armies were defeated, and the new state of Israel occupied nearly four-fifths of what had been Palestine.

During the 1948 war, 750,000 Arabs fled their homes and land in Palestine and became refugees. European Jews who survived World War II and Jews from Arab countries in the Middle East, North Africa, and the United States came to Israel and occupied the land and houses left behind by the Palestinians.

The Palestinians, the Arabs, and the Jews

This is how a half-century of bitterness and warfare between Arabs and Jews in the Middle East began. In the eyes of many Arabs, Israel is a colonial state created by Europeans. They see Israelis as foreigners who forced the Arab population of Palestine—the Palestinians—to leave their land. In the process, the Palestinians became a stateless, rejected people.

After the 1948 war, the surrounding Arab nations refused to absorb the 750,000 homeless Palestinian refugees. To do so, they believed, would be to recognize the reality of the state of Israel. As a result, about one-third of the region's 6.5 million Arab Palestinian refugees now live in United Nations camps in Syria, Jordan, Lebanon, and Gaza. Many Palestinians have lived their entire lives in these camps.

Hostilities have continued for more than two generations. The Palestinians express their frustration in border raids, terrorist activities, and uprisings. In most cases, Israel counterattacks to protect its citizens.

For the first time, in the early 1990s, Israelis and Palestinians agreed to sit at the same table and try to resolve some of their differences. The basic issue under discussion is "land for peace." Will the Israelis provide the Palestinians with enough land to satisfy their needs for a homeland? What security guarantees can the Palestinians provide to the Israelis if such land is given? Thus far, autonomy for some Arab areas, notably Gaza and the town of Jericho, have been agreed upon.

But many difficult issues remain. How much land will be given and where it will be located are matters of fierce debate among Palestinians and Israelis as well as between the two groups. Some obstacles seem insurmountable. Israel sees the Golan

◄
Violence erupts in Gaza, a small strip of land that houses large numbers of Palestinians in refugee camps. Here Israeli Border Police are attempting to disperse stone-throwing Palestinian youths who set fire to the tires. Why do Palestinians live in refugee camps like those in Gaza? How long have they lived in such places?

▲ *Oil is the most vital natural resource in the Arabian-Persian Gulf. Along its shores, long piers covered with oil pipelines reach far out into the Gulf. Here the water is deep enough for large oil tankers to be loaded directly from the pipelines. What factors make it cheaper to produce oil in the Gulf as compared with the United States?*

Heights, previously Syrian territory, as vital to its security. The Syrians may consider peace with Israel if Golan is returned to Syria. Jerusalem, in the words of both Palestinian and Israeli leaders, is nonnegotiable. Many Jewish settlements already exist in the occupied territories that the Palestinians want to control. Look closely at the map on page 556, and you will see how difficult the geographical situation is. Despite this, progress has been made on specific issues, the talks between Israel and the PLO go forward, and peace seems possible after 50 years of intermittent warfare.

⊕ **REVIEW QUESTIONS**

1. What is Zionism?
2. What major events occurred in Palestine in 1917? In 1948?
3. Where do many Palestinian refugees now live?
4. What are the occupied territories?

⊕ **THOUGHT QUESTIONS**

1. Why did the Zionists want an independent Jewish state, and why did they select Palestine as its site?
2. Why did the Jews agree and the Arabs disagree with the United Nations partition plan for Palestine?

4. Petroleum and Politics in the Arabian-Persian Gulf

Vast Reserves of Oil

About three-quarters of the known oil reserves outside the republics of the former Soviet Union (for which no reliable figures are available) are located in the Middle East and North Africa. Exploration teams have found large deposits of oil in the region, and each year more oil is discovered in the Middle East than is pumped out of its wells. Twenty-eight of the world's thirty-three giant oil fields are found in this region.

Wars in the Middle East are of worldwide importance because of the area's abundant oil reserves. Oil from the Middle East fuels the machines that run the economies of industrialized countries like the United States, Japan, and those of Western Europe. Events in this region—particularly in the Arabian-Persian Gulf, the heart of the Middle East's oil industry—affect the economy of industrialized and developing countries.

Oil Production Is Inexpensive

The five largest oil-producing countries in the Middle East are Saudi Arabia, Kuwait, Iran, Iraq, and Libya. Four of these states border the Arabian-Persian Gulf, as the map on page 559 shows. The economies of these countries are based on selling oil to industrial and industrializing countries. This is also true of smaller oil-producing countries like Oman, the United Arab Emirates, Qatar, and Bahrain. Jordan, Syria, and Lebanon indirectly benefit from Gulf oil by charging fees for petroleum piped across their countries to the Mediterranean coast. Locate these pipelines on the map on page 559.

The costs of producing oil in this region are lower than elsewhere in the world. A barrel of oil can be pumped to the surface for 6 cents in Kuwait and 35 cents in the oil-rich kingdoms of southeastern Arabia. These costs are much lower than comparable costs in the United States.

There are three reasons for this difference. First, oil is near the surface in the Middle East and flows freely while wells in the United States must be dug very deep. Second, an average well in the Arabian-Persian Gulf produces 4,500 barrels of oil a day, compared with less than 100 barrels a day in the United States, which means drilling costs per barrel are far less in the Middle East. Third, the oil fields in the Middle East are located near the coast, reducing transport costs to port refineries or loading platforms. In contrast, the United States has many expensive offshore wells and large oil reserves in the inland west and Alaska. As a result of all of these factors, the Middle East produces the least expensive crude oil in the world.

Many countries in the technological world depend on imported oil. Japan, for example, relies on imported oil for three-quarters of its energy needs. Western Europe imports more than half of its oil. After the price of oil skyrocketed 1,400 percent in the 1970s, the United States temporarily reduced its reliance on imported oil to 20 percent in the 1980s—by conserving energy and by encouraging

producers to find new oil deposits. But in the 1990s the United States imports nearly half of the oil it consumes, and this is a major factor in the U.S. trade imbalance with other countries. Industrialized and industrializing countries are likely to continue for some time to compete for the oil reserves of Middle Eastern countries despite a temporary glut in the world oil market.

The Oil-Rich Countries of the Arabian-Persian Gulf

Saudi Arabia is the largest producer of oil in the Arabian-Persian Gulf area. In the late 1940s, a deposit of oil 150 miles long and about 22 miles wide was discovered on Saudi Arabia's east coast. Its reserves are equal to the oil reserves of the entire United States. Oil from this enormous field and others farther north and beneath the shallow waters of the Arabian-Persian Gulf is pumped through

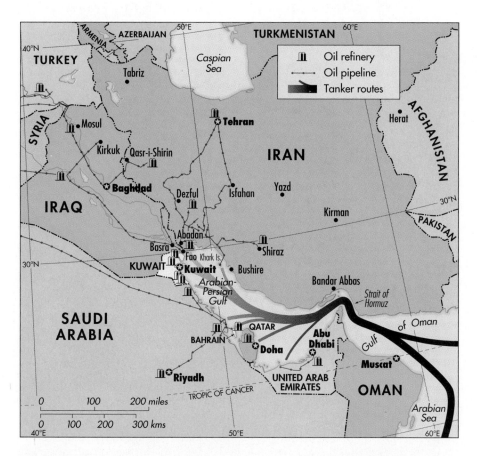

Oil Production in the Arabian-Persian Gulf

pipelines to local refineries on the Gulf coast or to tanker loading platforms in the Gulf itself. As you have learned, the kingdom of Saudi Arabia is spending large sums of oil money on development projects, desalinization plants, and industry, as well as military needs.

Iran is the Gulf's second largest oil producer. The country's largest oil fields, shown on the map on page 559, are located in the foothills of the Zagros Mountains just north of the Persian Gulf. Here, oil trapped in large reservoirs underground is pumped to the surface and piped to ports on the Arabian-Persian Gulf. Much of this oil goes to a massive tanker terminal on Khark Island off the coast of Iran north of Bushire. The rest is piped to the huge port refinery at Abadan. Both facilities were badly damaged during the 1980s war between Iran and Iraq.

Iraq's largest oil fields lie north of Baghdad and are connected by pipelines to Mediterranean ports in Lebanon and Syria. Important oil fields near the port city of Basra are still out of production since the refinery and port facilities at Fao were destroyed during the war with Iran and during the Gulf War.

Neighboring Kuwait, which had eight producing oil fields located near the coast, refined much of its crude oil into gasoline, jet fuel, and lubricating oil before the systematic destruction of the country's oil fields by Iraq. Kuwait's oil was protected by the military presence of U.S. forces in the Gulf during the Iran-Iraq War, but the country was devastated by Iraq's unexpected invasion in 1990. Kuwait is now constructing new pipelines to link with those of Saudi Arabia. These pipelines cross the Arabian Peninsula to the western port of Yanbu located 200 miles north of Jidda on the Red Sea coast.

▶

Pro-Iranian militants brave a torrential rain in Beirut to protest the actions of Saddam Hussein (the Iraqi leader pictured on the poster). Why would Iranians lead a protest against the Iraqi dictator? Describe relations between Iran and Iraq in the 1980s. What does the poster of Hussein suggest to you?

Oil is also produced in other small kingdoms that line the west coast of the Arabian-Persian Gulf. Life in countries like the United Arab Emirates, Qatar, Bahrain, and Oman has been transformed by oil revenues. Highways, airports, skyscrapers, and new cities dot the barren coasts of these countries. Now, however, these kingdoms are concerned with security and with recovering from problems created by the Iran-Iraq War and the Gulf War.

U.S. Concerns in the Arabian-Persian Gulf

The Arabian-Persian Gulf is a vital energy lifeline on which the United States, Western Europe, and Japan rely. About 40 percent of the world's oil is transported in tankers through the narrow mouth of the Gulf, the Strait of Hormuz. The Strait of Hormuz at its narrowest point is 21 miles wide. The depth of the main tanker channel varies from 250 to 700 feet. An estimated twenty-five to forty tankers carrying 6 to 9 million barrels of oil pass through this channel each day. The Iran-Iraq War demonstrated that tankers were vulnerable to attack and that the flow of oil could be in jeopardy. The United States, which has declared the Gulf to be an area vital to its interests, was deeply concerned about the threat to the oil supply.

The nations on the western shore of the Gulf— Saudi Arabia, Kuwait, Bahrain, Oman, Qatar, and the United Arab Emirates—have a total population of only 20 million people and about 140,000 armed troops. They were not able to defend tankers in the Gulf from attacks by Iran and Iraq.

The industrial nations reacted to this situation in two ways. First, naval forces from the United States and Europe moved into the Gulf to protect the tankers carrying oil. Second, huge reserves of oil were stockpiled. Japan has 150 million barrels of oil parked in tankers offshore, and the United States has stored a strategic petroleum reserve of 530 million barrels. A network of pipelines is being built to reduce dependence on the Strait of Hormuz, but the pipelines, of course, are still vulnerable to sabotage.

When Iraq invaded Kuwait in 1990 and threatened the oil-producing nations in the region, the United Nations and the United States again felt compelled to act. The magnitude of the multinational force led by U.S. forces that attacked Iraq and liberated Kuwait stressed the incredible importance of this area to industrial nations.

⊕ REVIEW QUESTIONS

1. What are the five largest oil-producing countries in the Middle East, and how many of them border the Arabian-Persian Gulf?
2. Where are the largest oil fields in Iran? In Iraq?
3. The Arabian-Persian Gulf is a vital energy lifeline to what parts of the world?
4. Why did the United States participate in the Gulf War in 1990–91?

⊕ THOUGHT QUESTIONS

1. Why are the costs of producing oil in the Middle East lower than costs in the United States?
2. How have the industrial nations of the world that depend on oil from the Middle East reacted to the hostilities that have broken out in this region during the last fifteen years?

🌐 CHAPTER SUMMARY

The Middle East is located where three continents meet. As a result, it has been a place of great creativity and great tension. Historically, the Middle East has been the crossroads of many people and many rich cultures, as well as the birthplace of three world religions—Judaism, Christianity, and Islam.

At the same time, the Middle East has been the scene of much conflict. In the twentieth century, some of the causes for conflict and war have been European imperialism, the founding of the state of Israel, and the tension between traditional and contemporary beliefs and lifestyles. This tension, for example, gave impetus to Islamic fundamentalism, a religious movement that rejects the Western emphasis on materialism in favor of traditional Islamic values.

The Middle East has some of the world's largest oil reserves, and some countries of this region are among the wealthiest oil producers in the world. Location—oil in the Middle East and the demand for oil in the technological world—has added to tensions and war in the Middle East.

EXERCISES

Fill in the Blanks

Directions: Find the missing word that best completes the sentence in the list below. Write the word on your paper next to the number of the sentence.

Africa	Iran
Asia	Iraq
Christianity	Islam
European	Israel

1. The core of the Middle East and North Africa is located where the continents of Europe, Africa, and |||||||| meet.
2. Morocco, Tunisia, Algeria, and Libya are on the continent of ||||||||.
3. The three major world religions that started in the Middle East are Judaism, Islam, and ||||||||.
4. Muhammad founded the religion of ||||||||.
5. Zionists were mainly responsible for creating the state of ||||||||.
6. |||||||| attacked Iran in 1980 precipitating eight years of warfare between the two countries.
7. The second largest oil-producing country on the Persian Gulf is ||||||||.
8. After World War I, most of the Middle East fell under the control of |||||||| countries.

Right or Wrong?

Directions: Some of the following sentences are true as written; other sentences are incomplete or wrong. Correct the wrong sentences, and then write only the corrected sentences on your paper.

1. The Middle East and North Africa are regions of deserts and grasslands.
2. For centuries, the region of the Middle East and North Africa was one of the most culturally advanced in the world.
3. Judaism, Christianity, and Islam are all monotheistic religions.
4. There are more Jews than Muslims in the Middle East.
5. Jerusalem was the birthplace of Muhammad.
6. By the middle of the 700s, the Muslims ruled an empire that stretched from Spain in the west to Pakistan in the east.
7. Islam is based upon five religious acts, called the "five pillars of Islam."
8. Today, the religious importance of Islam is fading as Muslims adapt to Western values.
9. The republics of the former Soviet Union have more than 50 million Muslims.
10. Istanbul is the modern name for Constantinople.
11. India has many millions of Hindus but only a few Muslims.
12. After the 1948 war, neighboring Arab nations refused to absorb the 750,000 homeless Palestinian refugees.
13. Many Jews from Nazi Germany migrated to Palestine in the 1930s.
14. The struggle between the Arabs and Jews in the Middle East began in Russia and Eastern Europe in the 1880s, when Jewish people there were persecuted.

Inquiry

Directions: Combine the information in this chapter with your own ideas to answer these questions.

1. How, in your opinion, do the "five pillars" of the Islamic faith affect the lives of Muslim people?
2. When the Turkish Ottoman Empire disintegrated, various parts of the empire became mandates of the United Nations. Was this wiser than giving the regions immediate independence? Give reasons for your answer.
3. Today, why is the Middle East a focus of world concern?

SKILLS

Using a Special-Purpose Map

Directions: The map on page 555 shows the percentage of Muslims in the populations of various countries. Use this map to do the following exercise. Find the two correct statements in each group of three statements. Then write the correct statements on your paper.

1. (a) Muslims make up a large percentage of the population in the Middle East. (b) Muslims make up a small percentage of the population in North America. (c) Muslims make up a large percentage of the population in North Africa.

2. (a) Most Japanese are Muslims. (b) Most Indonesians are Muslims. (c) Most Algerians are Muslims.

3. (a) There is a larger percentage of Muslims in the population of Egypt than in the population of Sudan. (b) There is a larger percentage of Muslims in the population of Mali than in the population of Mauritania. (c) There is a larger percentage of Muslims in the population of Libya than in the population of Chad.

4. (a) At least 90 percent of the people of Somalia are Muslims. (b) At least 90 percent of the people of Morocco are Muslims. (c) At least 90 percent of the people in the southernmost countries of Africa are Muslims.

5. (a) The holiest city in the Islamic faith is in Saudi Arabia. (b) Egypt's Muslim neighbor to the west is Libya. (c) Indonesia's Muslim neighbor to the south is Australia.

6. (a) Morocco is bounded on the north and south by countries that are mainly Muslim. (b) Afghanistan is bounded on the east and west by countries that are mainly Muslim. (c) Iran is bounded on the east and west by countries that are mainly Muslim.

Studying the Map of a Single Country: Israel

Directions: Look at the map on page 556 to do this exercise.

Select the correct term in parentheses in each of the following sentences, and then write the completed sentences on your paper.

1. The Negev Desert is in (northern, southern) Israel.
2. Israel's neighbors to the east are Syria and (Jordan, Egypt).
3. One of Israel's neighbors to the north is (Saudi Arabia, Lebanon).
4. Israel's nearest neighbor to the west is (Egypt, Syria).
5. The (Dead Sea, Mediterranean Sea) lies along Israel's eastern border.
6. A city that is almost in the center of Israel is (Nazareth, Jerusalem).
7. The Golan Heights region is nearer to (Syria, Egypt).
8. An Israeli port city is (Tel Aviv, Beersheba).
9. Gaza is nearest to (Syria, Egypt).
10. Haifa is (north, south) of Akko.
11. Israel lies along the coast of the (Mediterranean Sea, Indian Ocean).
12. Elat is (south, north) of Nazareth.
13. From north to south, the length of Israel is a little over (200 miles, 2,000 miles).
14. At its narrowest part, Israel's width is about (15 miles, 50 miles).
15. At its widest part, Israel's width is about (65 miles, 650 miles).

CHAPTER 22

Using a Location-Movement Map

Directions: Study the map on page 559. Notice that it has some political features, such as the names of countries and cities and the boundaries between countries. It also tells us about oil production and shipment in the Middle East. Look at the map symbols for an oil refinery, an oil pipeline, and tanker routes. Then answer the following questions.

1. Are most of the oil refineries within 400 miles of the Arabian-Persian Gulf?
2. Which country has an oil refinery near its border with Turkey?
3. Which country has an oil refinery near the Caspian Sea?
4. Do oil tankers leaving the Arabian-Persian Gulf pass through the Strait of Hormuz and the Gulf of Oman before reaching the Arabian Sea?
5. Is oil sent from the refinery near Tehran to the seaport of Abadan by pipeline or ship?
6. Do Saudi Arabia, Iran, and Iraq ship much of their oil through the Arabian-Persian Gulf?
7. Do the small countries of Kuwait, Bahrain, and Qatar ship oil through the Arabian-Persian Gulf?
8. Are there more oil refineries in eastern or western Iran?
9. Is Riyadh connected to the Arabian-Persian Gulf by pipeline?
10. Does Afghanistan ship oil by pipeline to the Arabian-Persian Gulf?

Vocabulary Skills

Directions: Match the numbered definitions with the vocabulary terms by writing the term next to the number of its matching definition.

diaspora   pilgrimage
Islam   Qur´an (Koran)
isthmus   strait
minaret   synagogue
monotheism   Zionism
Muslim

1. A believer in the Islamic religion
2. The holy book of Islam
3. Religion founded by the Prophet Muhammad
4. A narrow stretch of water lying between two land masses
5. A journey undertaken to visit a religious place or shrine
6. A meeting place for Jewish worship and religious instruction
7. The scattering of Jews outside of Palestine
8. A tall spire from which Muslims are called to daily prayer
9. A political movement devoted to the creation of a Jewish state
10. Belief in only one God
11. A narrow strip of land connecting two land-masses

Human Geography

Directions: Select the correct ending to each of the following sentences and then write the complete sentences on your paper.

1. The religion of Islam started on the west coast of (a) Turkey (b) Palestine (c) the Arabian Peninsula.
2. Jerusalem is a sacred city for Jews, Christians, and (a) Buddhists (b) Hindus (c) Muslims.
3. The central shopping districts in ancient Middle Eastern cities were called (a) bazaars (b) minarets (c) mosques.
4. For many centuries, camels have played a major role in the lives of (a) city dwellers (b) village farmers (c) nomadic tribes.
5. The number of people in the Middle East and North Africa is about (a) 170 million (b) 350 million (c) 580 million.
6. Half of the people in the Middle East and North Africa live in the three countries of Egypt, Turkey and (a) Morocco (b) Iraq (c) Iran.
7. Before 1974, the largest group of people on the island of Cyprus were (a) Greeks (b) Turks (c) Arabs.
8. More than half of the people in the Middle East and North Africa are (a) nomads (b) city dwellers (c) villagers.
9. The religious leader who seized control of Iran after the Shah was deposed was (a) Kemal Ataturk (b) Gamal Abdel Nasser (c) Ayatollah Khomeini.
10. About 90 percent of the people live in cities in (a) Israel (b) Afghanistan (c) Tunisia.
11. In recent years, fighting has raged between Christians, different Muslim sects, and Palestinian refugees in (a) Saudi Arabia (b) Lebanon (c) Libya.
12. Most Iranians speak (a) Russian (b) Persian (c) Turkish.
13. Turkey has applied for membership in (a) the CIS (b) the European Community (c) OPEC (Organization of Petroleum Exporting Countries).

Physical Geography

Directions: On your paper, write *True* for the following statements that are correct and *False* for the statements that are not correct.

1. Less than 15 percent of the Middle East and North Africa is cultivated.
2. Wheat, barley, and dates are major crops in the Middle East.
3. Crop yields in the Middle East are generally higher than crop yields in Europe.
4. The Sahara is the world's largest desert, except for Antarctica.
5. Winter rains help account for the relatively dense population between the Atlas Mountains and the Mediterranean Sea in North Africa.
6. The Nile River begins in East Africa and flows northward to the Mediterranean Sea.
7. Most of the Arabian Peninsula receives less than three inches of rain each year.
8. Two landlocked countries are Jordan and Afghanistan.
9. Oil from the Middle East is mainly consumed by countries in that area.
10. The Strait of Hormuz is important mainly because some of its water is channeled to dry farming areas.

Writing Skills Activities

Directions: Answer the following questions in essays of several paragraphs each. Remember to include a topic sentence and several supporting sentences in each paragraph.

1. What environmental factors have helped shape the lives of people who live in the Middle East and North Africa?
2. Why is the Middle East called the "cradle of religions"?
3. What major problems have confronted Israel since its independence?
4. Why is the Middle East and North Africa one of the critical regions in the world today?

CHINA

UNIT INTRODUCTION For forty-five years, the Chinese people have been told what to do and what to think under Communist rule. Recently, Communist leaders have introduced elements of a capitalist economic system into some regions of China, and the economy is booming. Their efforts to modernize the country, however, are still based on rigid political control of the Chinese people. They have achieved the remarkable feat of lowering birthrates in a population of more than 1 billion people by implementing strict controls on the number of children couples may have.

Can the Chinese achieve the prosperity enjoyed by their fellow Chinese in Taiwan and Hong Kong, centers of capitalist enterprise in East Asia? Will a combination of communist political control and capitalist economic principles enable the Chinese to achieve a higher standard of living? Nearly one-fifth of all the people on Earth live in China, so the answers to these questions will be a major factor in shaping our world.

▲ *Riding a bicycle is the primary method of transportation in China. As they begin to shift to motorcycles, what effect will this have on the environment of Beijing?*

Hong Kong is the last important Chinese port under British control. In 1997, Hong Kong will become a part of the People's Republic of China. What are some advantages and disadvantages that might occur in Hong Kong after this change in political status?

▼ *This schoolboy in China appears to be wearing a uniform. Why would you expect to see students in China wearing uniforms, and not expect to see most students in the United States wearing uniforms?*

UNIT OBJECTIVES

When you have completed this unit, you will be able to:

1. Identify the main geographical features of China, a country about the size of the United States but in which only 4 percent of the land can be farmed.

2. Explain the important changes that occurred in China when its long isolation from the West ended.

3. Describe how ways of life in China were changed when the Communists gained control of the country and how, in recent years, communist policies are being altered to encourage private enterprise.

4. Identify crucial problems that China, with more than 1 billion people, faces in meeting the needs of its people.

SKILLS HIGHLIGHTED

- Using the Map of China's Resources and Industrial Areas
- Using an Almanac
- Using a Graph That Shows Population Growth
- Vocabulary Skills
- Writing Skills

KEYS TO KNOWING CHINA

1

China is about the same size as the United States.

2

China is geographically divided into two parts. Outer China is a sparsely settled region of high mountains, plateaus, steppes, and deserts. Agricultural China is where 95 percent of the Chinese people live.

3

China has three great river systems—the Huang (Yellow), Chang Jiang (Yangtze), and Xi Jiang.

4

China has a population of more than 1 billion people, the largest national group in the world.

▼ *One of the most important archaeological finds of this century are the thousands of terra-cotta warriors, which have guarded the tomb of the first emperor of China for 2,000 years, near Xi'an, China. What does this tell you about the length and depth of Chinese culture and history?*

5

Two-thirds of the Chinese people are farmers, but only 4 percent of China's land can be cultivated.

6

China has been under Communist rule since 1949. In the late 1970s, China began to open its doors to the outside world.

7

China is a land of enduring traditions. Communist and Confucian values frequently come into conflict.

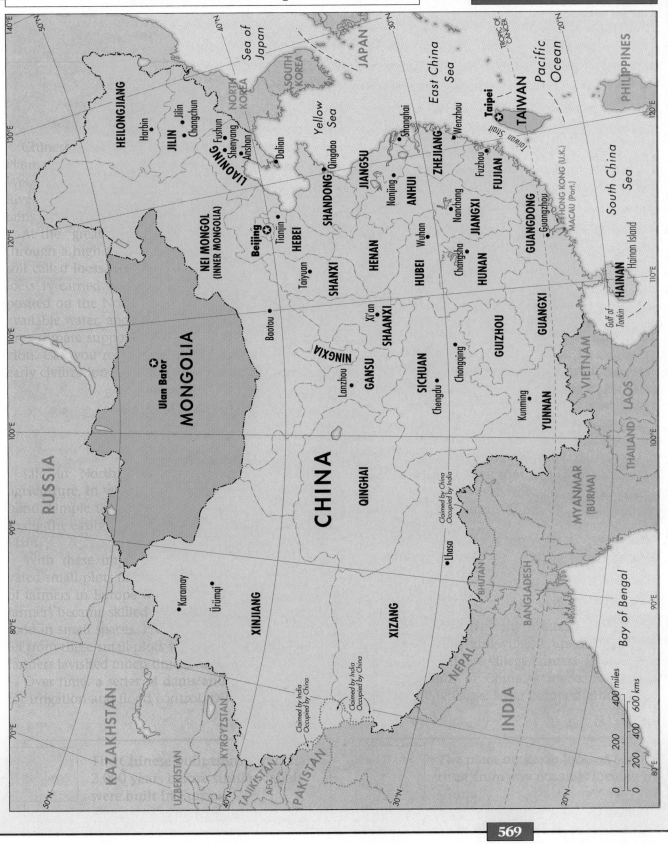

140°E
50°N
N
Sea of Japan
40°N
30°N
JAPAN
TROPIC OF CANCER
20°N
PHILIPPINES

HEILONGJIANG
Harbin
JILIN
Jilin
Changchun
Fushun
Shenyang
Anshan
Dalian
NORTH KOREA
SOUTH KOREA
Yellow Sea
Shanghai
East China Sea
Wenzhou
TAIWAN
Taipei
Pacific Ocean

LIAONING
130°E
Beijing
Tianjin
HEBEI
SHANDONG
Qingdao
JIANGSU
Nanjing
ANHUI
ZHEJIANG
Nanchang
Fuzhou
FUJIAN
Taiwan Strait

120°E
NEI MONGOL (INNER MONGOLIA)
Taiyuan
SHANXI
HENAN
Wuhan
HUBEI
Changsha
JIANGXI
HUNAN
GUANGDONG
Guangzhou
HONG KONG (U.K.)
MACAU (Port.)
South China Sea

Baotou
Xi'an
SHAANXI
GUIZHOU
GUANGXI
Hainan Island
HAINAN
Gulf of Tonkin
110°E

MONGOLIA
Ulan Bator
NINGXIA
Lanzhou
GANSU
SICHUAN
Chengdu
Chongqing
Kunming
YUNNAN
VIETNAM
LAOS
THAILAND

RUSSIA
100°E

CHINA
QINGHAI
Claimed by China Occupied by India
MYANMAR (BURMA)

90°E
Karamay
Ürümqi
XINJIANG
Lhasa
XIZANG
Claimed by India Occupied by China
Claimed by India Occupied by China
NEPAL
BHUTAN
BANGLADESH
Bay of Bengal
INDIA
90°E

80°E
KAZAKHSTAN
KYRGYZSTAN
UZBEKISTAN
TAJIKISTAN
AFG.
PAKISTAN
30°N

70°E
40°N
20°N
400 miles
600 kms
200
400
200
400
0
0
80°E

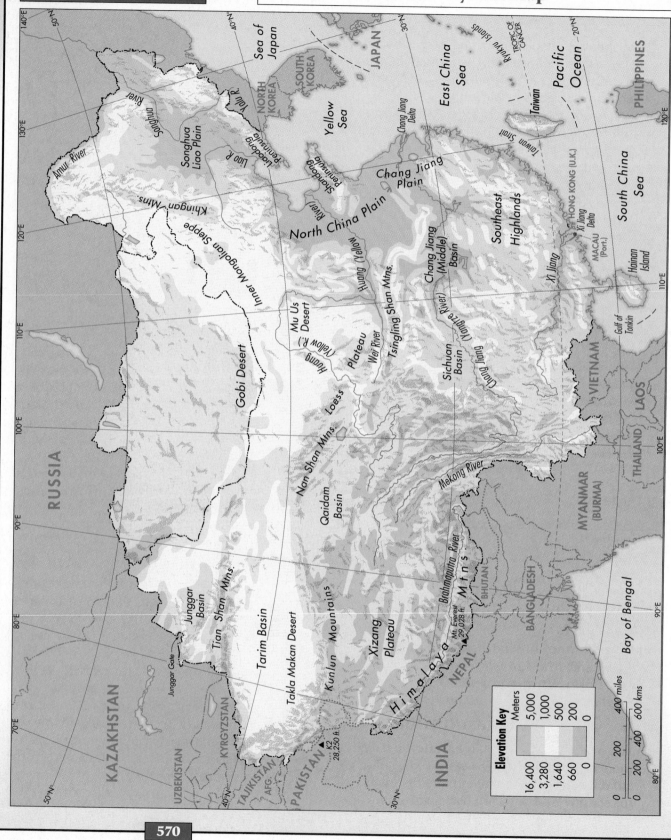

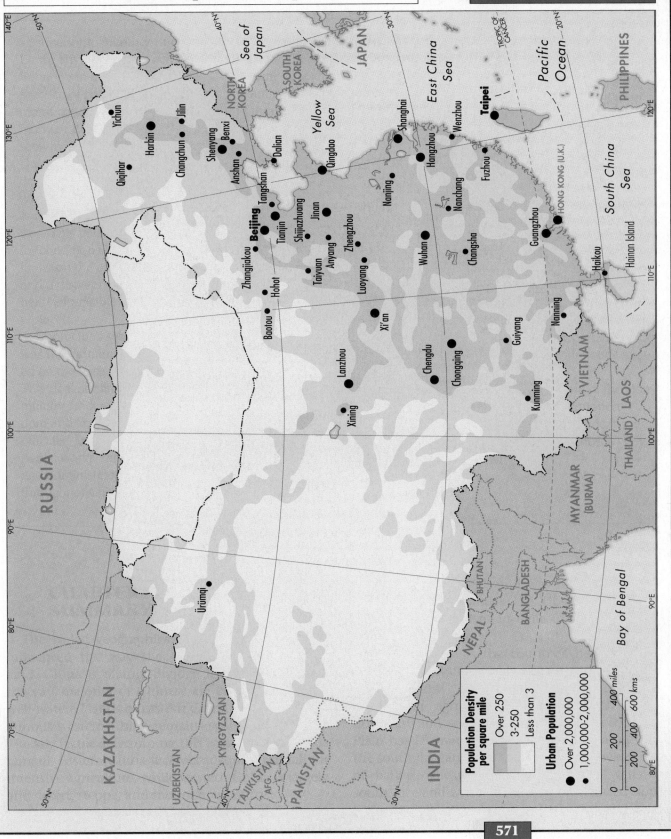

Sea of Japan

30°N

TROPIC OF CANCER

20°N

NORTH KOREA

SOUTH KOREA

JAPAN

East China Sea

Pacific Ocean

PHILIPPINES

Yellow Sea

Yichun

Jilin

Harbin

Changchun

Shenyang

Benxi

Qiqihar

Anshan

Dalian

Tangshan

Shijiazhuang

Qingdao

Shanghai

Taipei

Wenzhou

Hangzhou

Fuzhou

Zhangjiakou

Beijing

Tianjin

Jinan

Nanjing

Nanchang

Baotou

Hohot

Taiyuan

Anyang

Zhengzhou

Wuhan

Changsha

HONG KONG (U.K.)

South China Sea

Luoyang

Guangzhou

Xi'an

Haikou

Hainan Island

Lanzhou

Chengdu

Chongqing

Guiyang

Nanning

VIETNAM

LAOS

Xining

Kunming

MYANMAR (BURMA)

THAILAND

Ürümqi

RUSSIA

KAZAKHSTAN

UZBEKISTAN

KYRGYZSTAN

TAJIKISTAN

AFG.

PAKISTAN

INDIA

NEPAL

BHUTAN

BANGLADESH

Bay of Bengal

Population Density per square mile

Over 250

3-250

Less than 3

Urban Population

Over 2,000,000

1,000,000-2,000,000

400 miles

600 kms

200   400

200   400

0

0

AT A GLANCE

Country	Capital City	Area (Square miles)	Population (Millions)	Life Expectancy	Urban Population (Percent)	Per Capita GNP (Dollars)
China	Beijing	3,705,390	1,178.5	70	26	370
Hong Kong	Victoria	402	5.8	78	—	13,200
Macau	Macau	8	0.4	79	97	—
Mongolia	Ulan Bator	604,250	2.3	65	57	—
Taiwan	Taipei	12,456	20.9	74	71	—

—Figures not available

ENVIRONMENTS AND PEOPLES OF CHINA

The People's Republic of China is deciding how to best use its resources to meet the needs of its people. After forty-five years of Communist rule, the Chinese are debating priorities. How much emphasis should be placed on heavy industries like steel and shipbuilding? How much should be placed on light industries that produce consumer goods like clothing and appliances? How can China feed four times as many people as the United States in a country that has only the same amount of farmland? How far should China open its society

◀ *Originally started by Ch'in Shih Huang Ti in the second century B.C. to keep out invaders from the north, the Great Wall is now one of the great tourist attractions in China. What class of people do you think participated in the building of the Great Wall?*

to foreign technology, investment, and trade? What combination of communist and capitalist principles are best suited to modernize China?

The future of China is uncertain. On a limited amount of farmland, the Chinese have the largest population in the world. Whatever path China follows in its effort to modernize, solutions to China's problems must be found.

1. A Vast Country with Varied Regions

Where the Chinese Live

In a country roughly equal in size to the United States, China has the largest population of any country in the world. Much of the country is densely settled, but large areas of China are thinly settled, as the population map on page 571 shows. Why is this so? The answer to this question is essential to understanding the geography of China.

The most important geographical division in China is between the sparsely settled, dry lands and mountain regions of Western China and the densely settled, well-watered lands of eastern China. The geographical line that separates these two regions follows the twenty-inch rainfall line that cuts across China from northeast to southwest. West of this line, low rainfall limits farming so that the population is small. East of this line, rainfall increases and so does population.

These two regions are Outer China and Agricultural China. Look at the map of the subregions of China below. Now find each region on the physical map and the population map on pages 570 and 571. How would you describe the relationship between low rainfall and low numbers of people? Between rugged land and low population? Between level land and dense settlement?

Outer China, a Harsh, Mountainous Land

Outer China includes over half of the land area of the country, but it is sparsely populated. The

Subregions of China
China has three major subregions:
(1) Outer China, (2) Agricultural China,
and (3) Northeast China. Within
Agricultural China rice is dominant in
the south, wheat in the north. The
Northeast is China's most important
industrial area.

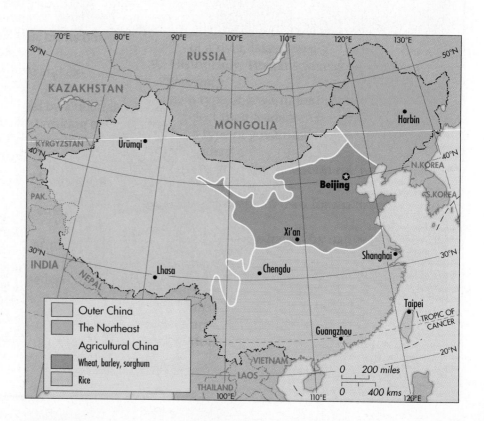

population map shows that most of Outer China has fewer than three people per square mile. Terrain and climate explain this very low density of human settlement.

The terrain in Outer China is as dramatic as any landscape in the world. Towering mountain ranges, many soaring to elevations of 20,000 feet, dominate the western frontiers of this region. The Himalaya Mountains have the highest mountains on Earth. Ringed by these soaring mountains is the three-mile high plateau of Xizang (or Tibet), a remote and isolated land.

To the north of the Xizang Plateau are two dry basins ringed by mountains. North and east of these basins are the cold, windswept plains of Inner Mongolia. These plains provide the only easy land route into Agricultural China from western Asia. The Chinese began to build the Great Wall along this frontier during the Han period (207 B.C. to A.D. 220), one of the magnificent periods in China's long history. Their goal was to keep the nomads of Inner Mongolia and other "barbarians" from attacking settled areas in Agricultural China.

The climates of the mountains, high plateaus, basins, and plains of Outer China are harsh. Cold, dry winds sweep out of Central Asia across this entire region in the winter. Only in the summer do air masses from the Pacific Ocean reach this deep into China. Usually, these moisture-bearing air masses bring enough rainfall for limited farming in the basins and plains of Outer China, but sometimes there are long droughts. As a result, Outer China is a hostile and difficult environment—a vast wilderness about the size of India.

Agricultural China, Rich in Farmland

Agricultural China, which lies to the east of the twenty-inch rainfall line, is one of the most densely settled regions in the world. Most of Agricultural China has more than 250 people per square mile, and virtually all of China's large urban areas are located in this region. Here, 95 percent of the Chinese are crowded into the industrial cities of the Northeast and onto the plains of China's three main rivers—the Huang (Yellow), the Chang Jiang

▲ *Since there are few automobiles and trucks in China, many people still use old-fashioned methods of transport for their goods. How does this basket method of carrying goods compare with methods you may have observed? Is it efficient, or would it require more trips to carry the same amount of goods?*

(Yangtze), and the Xi Jiang. Locate China's three major rivers on the map on page 570. Find the symbol for areas with more than 250 people per square mile on the population map on page 571. Notice that the symbols for large cities are scattered throughout these areas.

Latitude, or the distance north or south of the equator, has a strong influence on the climates of Agricultural China. Just as in the eastern United

IT'S A FACT An average Chinese farm is 300 times smaller than an average American farm.

States, climates are colder in North China and warmer in South China. Rainfall tends to be lower in the north, and heavy rainfall and warm temperatures are found in South China.

Overall, however, Agricultural China has temperatures about twenty degrees cooler than comparable locations on our East Coast. Bo Hai, the body of water just east of Beijing, which is at the same latitude as Washington, D.C., is often covered by ice in winter. Temperatures at Charleston, South Carolina, are usually twenty degrees warmer than those in Shanghai at the same latitude. Only southernmost China is free from frost year-round.

These temperature differences affect the crops that farmers grow at different latitudes in Agricultural China. Study the map of China's subregions on page 574. Notice the line that cuts across Agricultural China and separates crop regions. Hardy crops like wheat, barley, and sorghum are grown north of this line, and rice is raised south of this agricultural boundary. As the climate becomes warmer and wetter toward the south, rice fields cover most of the farmland. In the deep south, the warm, humid climate allows farmers to grow two or three rice crops in a single year on the same land.

⊕ REVIEW QUESTIONS

1. What is the most important geographical division in China? What are the country's two major subregions called?
2. About how many people per square mile are there in Outer China? In Agricultural China?
3. In Agricultural China, what are the chief crops grown in the north? In the south?

⊕ THOUGHT QUESTIONS

1. How does rainfall help account for the fact that Agricultural China is so densely settled?
2. Why can it be said that Outer China has a difficult environment for farming?

2. Outer China

China's Frontier Lands

Outer China is occupied by less than 4 percent of China's people, even though it covers a huge territory about the size of India. In Outer China, most people live in areas that are sheltered from wind and cold, in nomadic camps located near pastures, and in scattered oases. Little food is produced in Outer China; however, recent mineral and energy discoveries are being developed.

Two major problems have slowed Chinese attempts to modernize this region. The first is Outer China's harsh environment. The cold, dry climate and rugged topography make living conditions difficult. The second is the hostility of the region's indigenous non-Chinese peoples to unwelcome Chinese immigration and settlement in Outer China.

Xizang, the Land of the Tibetans

The Xizang Plateau is a large, cold, treeless land surrounded by snow-covered peaks. It is one of the most remote areas on Earth. About 2 million Tibetans lived here in relative isolation until the Communist takeover of mainland China in 1949. The Tibetans were farmers and herders who raised barley to feed themselves and tended herds of **yak**, large long-haired oxen.

The Tibetans were governed by a **theocracy**, which is a state ruled by religious leaders. Buddhist monks, headed by their god-king, the **Dalai Lama**, ran the country from the capital city of Lhasa in eastern Xizang.

In 1950, Chinese troops invaded Xizang and drove monks out of their 6,000 fortress monasteries, destroyed shrines, and brought the land under government control. Roads, mines, and factories are being built throughout Tibet by the Chinese government. Chinese efforts to mold the Tibetans into practicing communists have failed, however, despite the fact that 200,000 Chinese troops occupy Xizang.

◀ *The Himalaya Mountains include the highest peaks on Earth. Mt. Everest, considered to be the highest mountain on Earth, has an elevation of 29,028 feet. Why do you think climbers need to take oxygen and warm clothing with them when they climb high in the Himalayas?*

The Mountain-Ringed Basins of Outermost China

Between Xizang and Mongolia, at the outermost reaches of China, are the mountain-rimmed Tarim and Junggar Basins. The map on page 570 shows that these basins are separated from one another by the towering peaks of the Tian Shan Mountains. This basin region is twice the size of Texas, but its population is smaller than that of New York City.

The center of the Tarim Basin is a dune-covered desert. Oasis settlements are located at the foot of the mountains where qanats, underground water channels, supply irrigation water for farming. The vegetation is too sparse to graze animals in the Tarim Basin. In the cooler Junggar Basin to the north, some rainfall enables short grasses to grow so that nomads are able to graze their animals. In earlier times, the great trade routes connecting the Orient with Europe ran through this region.

Most of the people now living in these basins are Turkic-speaking Muslims. Their culture is similar to that of their Muslim neighbors across the border in Afghanistan and Russia.

The Chinese government has encouraged Chinese people to move into the Tarim and Junggar Basins in order to tighten Chinese control over this border region by assimilating local Muslims into Chinese society. They also want a buffer zone on the Chinese-Russian frontier, because clashes between Russian and Chinese troops here have been frequent. A buffer zone is a neutral region situated between two or more powerful states. Trace the long border between China and Russia and Mongolia on the map on page 569.

Today nearly half of the people in the region are Chinese immigrants. They live in towns in the Junggar Basin, where farm production has quadrupled. They also live in Karamay, a large petroleum center near the Russian border that is located on the third largest oil field in China.

IT'S A FACT | The Tarim and Junggar Basins combined are larger than all of Western Europe.

TIBET, LAND OF THE SNOWS

For many centuries, Tibet was a land of legend accessible only to the most determined explorers. Its natural barriers discouraged visitors. On three sides, Tibet was ringed by some of the highest and most impassable mountains in the world: the Himalaya to the south, the Karakoram to the west in Pakistan, and the Kunlun to the north. Only in the east was Tibet open to penetration.

The Chinese exerted some influence on Tibet from the early 1900s onward, but in the 1950s, it was from the east that the Chinese came in force to obliterate one of the world's most distinctive cultures.

Tibet is nearly half a million square miles in extent. Its northern half is almost uninhabitable, although a few nomads and hunters still live in its forbidding landscapes, where they herd, skin, and sell animals in mountainous areas three miles high. Most of Tibet's 2 million people live in the south where milder climates and accessible water enable village farmers to grow barley, wheat, and vegetables in protected valleys.

The origins of the Tibetan people are unknown. Reliable records trace them back to A.D. 600, at about the same time that Buddhism arrived in Tibet. The daily lives of Tibetans are affected by their belief in and practice of Tibetan Buddhism. The center of this faith is the city of Lhasa (literally, "the Ground of the Gods"). Despite lives of great hardship, Tibetans created a society of wisdom, patience, devotion, and faith.

The Chinese invasion of Tibet in 1950 had two goals: (1) to "raise" Tibetan culture (much like the colonizing countries attempted to do in the 1900s) and (2) to make Tibet a part of communist China, with its farmland organized into cooperatives. The Chinese later decided to use Tibet, one-quarter of all of China, as a nuclear waste dump. The Tibetan people have been deprived of the basic freedoms of speech, religion, and movement. Their uprisings bring swift repression by Chinese troops.

For the Dalai Lama, it has been a cruel destiny. Only fifteen years old when the Chinese first invaded, he was forced to flee to India in 1959. He was awarded the Nobel Peace Prize in 1989 for working for international peace, as well as for calling world attention to Tibet's plight. Now the leader of a disappearing people, he describes himself this way:

> "Dalai Lama" means many different things to different people. To some it means that I am a living Buddha, the earthly manifes-

▲ *The Dalai Lama, Tibet's Buddhist religious leader, led the government until 1959 when he escaped into exile in India. The Tibetans still seek independence from China. What are some ways they have tried to inform the world of their plight?*

▲ *Potola Palace, in Lhasa, Tibet, is the Buddhist "Fortress of the God." Tibet, nicknamed "The Roof of the World" has been remote from contact during much of its history. How would this relative isolation have influenced the development of this country?*

tation of Avalokiteshvara, Bodhissatva of Compassion. To others it means that I am a "god-king." During the late 1950s it meant that I was a Vice President of the Steering Committee of the People's Republic of China. Then when I escaped into exile, I was called a counterrevolutionary and a parasite. But none of these are my ideas. To me "Dalai Lama" is a title that signifies the office I hold. I myself am just a human being, and incidentally a Tibetan, who chooses to be a Buddhist monk.

China has encouraged Chinese people to settle in all parts of Tibet; the Chinese now outnumber Tibetans in their own land. The Chinese send young Tibetan boys to various parts of China for training. As one noted expert put it, "In ten or twenty years Tibetan culture will be something you can only read about in the library."

Source: Tenzin Gyatso (Dalai Lama). *Freedom in Exile: The Autobiography of the Dalai Lama.* New York: Harper/Collins, 1990.

❓ QUESTIONS

1. Describe your reaction to the way in which the Dalai Lama describes himself.
2. What do we know of the origins of the Tibetan people? How would you describe their society before the Chinese moved into Tibet in large numbers in the 1950s?
3. Note the comment: "In ten or twenty years Tibetan culture will be something you can only read about in a library." Do you think this is happening to other culture groups? If so, where are they located? Do you think it is important to preserve these culture groups? Why or why not?

New roads, railroads, and pipelines now connect this once inaccessible region with Agricultural China. These links to the country's western frontier have been carefully planned to discourage any Russian military aggression.

Inner Mongolia, Another Chinese Frontier

Inner Mongolia is a huge, rolling plain stretching from the Gobi Desert in the west to the Great Wall of China in the east. The climate here is harsh. Bitterly cold winds sweep across the Mongol plains in winter. Summer temperatures are searing, often exceeding 100° F. In the north and west are sand dunes and gravel deserts. In the east, where there is a bit more rainfall, short grass covers the rolling countryside. Mongol herders and Chinese farmers have fought for these grasslands, just as ranchers and farmers did in our Great Plains. In China, however, this conflict has continued for hundreds of years.

Today, more than 90 percent of the estimated 18 million people in Inner Mongolia are Chinese. Many who live on the plains north and west of the Great Wall are descendants of Chinese pioneers who settled these grasslands. Near the great bend in the Huang (Yellow) River, recent Chinese immigrants cultivate land irrigated by new dams and canals.

New industrial centers have been built in Inner Mongolia, as in other parts of Outer China. Roads and railroads connect these cities with Agricultural China. The largest of these cities is Baotou, built near iron mines on the great bend of the Huang River. As you would expect, Baotou is becoming an important steel-manufacturing center.

The million or so Mongols of Inner Mongolia have been pushed north and west toward the edge of the Gobi Desert and the borders of Mongolia. The Chinese have tried to force these nomads to settle down as farmers, yet many Mongols still herd goats and live in felt tents, or yurts, like those used by their ancestors. Mongolia is a thinly populated state that has acted as a buffer between China and Russia.

⊕ REVIEW QUESTIONS

1. Before 1949, what was the way of life of most Tibetans? How were they governed?
2. Why can't animals be grazed in most of the Tarim Basin? Why can they be grazed farther north in the Junggar Basin?
3. Why did the Chinese government encourage Chinese people to move into the Tarim and Junggar Basins?

⊕ THOUGHT QUESTION

1. Why have Chinese leaders had great difficulty in forcing the Tibetans and Mongols to conform to communist practices?

▶

This Mongol rider still lives in a traditional yurt even though the Chinese government has attempted to settle Mongols as farmers on collective farms or force them to live in apartment buildings and work in cities. Can you name other areas in the world where central governments have attempted to impose their majority culture on minority groups like the Mongols?

3. Agricultural China

China's Great Resource Base

You have learned that 95 percent of China's population lives in Agricultural China. Almost all of these people are clustered in the lowlands of China's three great rivers and in the industrial Northeast. Find again and retrace on the map on page 570 China's three great rivers: the Huang, the Chang Jiang, and the Xi Jiang.

Agricultural China produces almost all of the food for this nation of more than a billion people. It has China's largest cities, as well as most of China's new industrial centers. It is here that the Chinese farm the land, live as urbanites, and develop their natural resources to build a modern country.

One-Fifth of the Chinese Live on the North China Plain

Most of the plain of the Huang River is covered by farms. The fertile, windblown soils of the plain are planted in grain crops like winter wheat, barley, and millet. Winter wheat is planted in the fall and grows through the winter, as it does on our Great Plains. Also as in parts of the Great Plains, droughts are common in North China.

In the past, farmers on the North China Plain have also had to deal with floods. Flooding is a constant threat because silt is carried downstream and deposited in the bed of the Huang. **Silt** is a very fine-grained soil laid down in the bed of a river, lake, or sea. Silt buildup here has caused the Huang River to rise a bit each year, so the Chinese have built **levees**, or low earthen embankments, along its banks to hold the floodwaters. The river now flows in channels that are as much as twenty feet above the surrounding land. Any break in a levee can cause disaster. Do you understand why the Huang is often called "China's sorrow"?

In spite of these problems, the North China Plain is a farming area with a population of 210 million people. Its agricultural wealth feeds the great cities of the North China Plain. Find Beijing, the capital of China, on the map of China's resources and in-

dustrial areas on page 584. Also find the port city of Tianjin. Beijing has more than 9 million people and Tianjin about 7 million.

These two cities are located at the center of one of China's largest industrial areas. The industries in this area are based on important iron and coal deposits, as this map also shows. Identify the symbols for coal and iron on the map, and trace the broad arc of coal deposits that sweeps across the North China Plain. Do you see the city of Taiyuan on the map? Why do you think this city has become a center of steel production and heavy industry?

The Rich Plain and Inland Basins of the Chang Jiang

The 3,964-mile-long Chang Jiang (Yangtze River) is the third longest river in the world. Large ships can sail up the Chang Jiang for 700 miles, and smaller ships can reach still deeper into the Chinese interior. One of the most important transportation routes in China, the Chang Jiang serves very productive farming and industrial areas along its course.

▼ *The Chang Jiang, one of the longest rivers in the world, is one of the most important transportation routes in China. What types of ships and boats do you think sail on this river? Would you expect to see such vessels on the Mississippi River?*

The huge port city of Shanghai lies near the mouth of this water highway. Shanghai is one of the largest cities on the continent of Asia. Some 50 million people live in its hinterland, the area that the city serves. Shanghai handles more than half of China's foreign trade and is an industrial center with many different kinds of factories and businesses. Find the industrial area spreading out around Shanghai on the map on page 584. Notice that iron and coal deposits are nearby.

The lower Chang Jiang Plain stretches 700 miles along the Chinese coast and extends inland for several hundred miles. On the lower Chang Jiang Plain the important agricultural transition takes place between northern wheat and southern rice. Farm production of both crops on this plain is immense.

The Middle Basin, away from the coast, has frequently been called the rice bowl of China. Wuhan is the largest city in this basin. Here is another Chinese industrial area that contains one of China's major steel complexes.

Now find the province of Sichuan on the map on page 569. The highest and innermost basin of the Chang Jiang is in Sichuan, about 1,000 miles away from the coast. Sichuan has a population of approx-

> **IT'S A FACT** Food raised in the basin of the Chang Jiang feeds one out of every fifteen people on Earth.

imately 100 million people, most of whom are farmers. The protected location of the Sichuan Basin gives it a milder climate than that of Shanghai. Terraced slopes are planted in rice and wheat, and tea and mulberry groves grow on steeper slopes. Chongqing, the largest urban center in Sichuan, has a population of more than 6 million and is still growing. Its industries are supported by the iron, petroleum, and natural gas resources of the surrounding area.

The Delta of the Xi Jiang

A third important river lowland in Agricultural China is the delta of the Xi Jiang. Although the delta is only two-thirds the size of Connecticut, 15 million people live there. Population pressure on farmland is intense. Hill slopes have been carved into stepped terraces, and canals and ditches distribute water to every bit of level land. Rice is the

The limestone (karst) formations in Gulin are a unique and beautiful background for the many rice paddies in this area of southern China. How might these mountains have been formed? Where does limestone change topography in the United States?

IT'S A FACT China supports four times as many people as the United States on about the same amount of farmland.

most important crop, but sugarcane, silk, and fruit are also produced in the warm climate of South China.

Find Guangzhou (Canton) on the map on page 584. Many Chinese Americans are descendants of immigrants from this area, because Guangzhou was the first Chinese seaport to be opened to foreign trade (in 1834). Guangzhou has a population of about 6 million, as does the neighboring British territory of Hong Kong. The map shows that Guangzhou, like large port cities on China's other major rivers, is the center of an industrial area.

The population of Guangzhou and its surrounding plain is Chinese, but in the interior hills there are about 15 million non-Chinese people. They speak a variety of Thai languages and cling to their own cultures; however, they are facing the same pressures to become Chinese as the Muslim minorities of Outer China.

The Industrial Northeast

The Northeast, formerly called Manchuria, is the major industrial region in China. It is bordered on the north and east by Russia and North Korea, as the map on page 574 shows. Most of the 100 million people of the Northeast are concentrated in the valley of the Liao River. This lowland has a short growing season, severe winters, and uncertain rainfall; about one-third of the valley is farmed, and irrigation systems are being extended. The main crops are spring wheat, barley, corn, and soybeans.

Of far greater importance in the Northeast, however, are the heavy industries, most of which are located near deposits of coal, iron, and other mineral resources. The major industrial centers are Shenyang, Fushun, and Anshan. Shenyang has a population of 6 million people. For its major steel-manufacturing industry, it gets coal from Fushun and iron from Anshan. A modern transportation network links these cities. Find this industrial area on the map on page 584.

Recognizing that the Northeast is a key region in China's drive to become a modern industrial nation, Chinese leaders are now devoting special attention to developing industries and modern

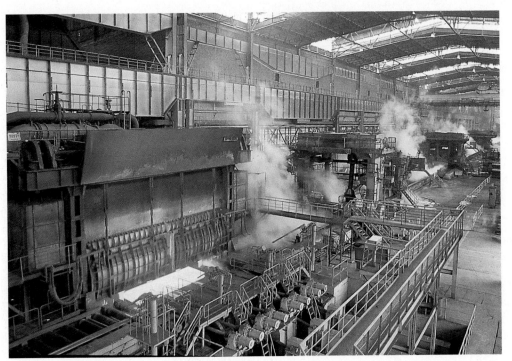

◀

Japan is one of the several countries that has helped China with modernization by entering into a "joint venture" agreement. This iron/steel complex is an example. How do you think a "joint venture" would work?

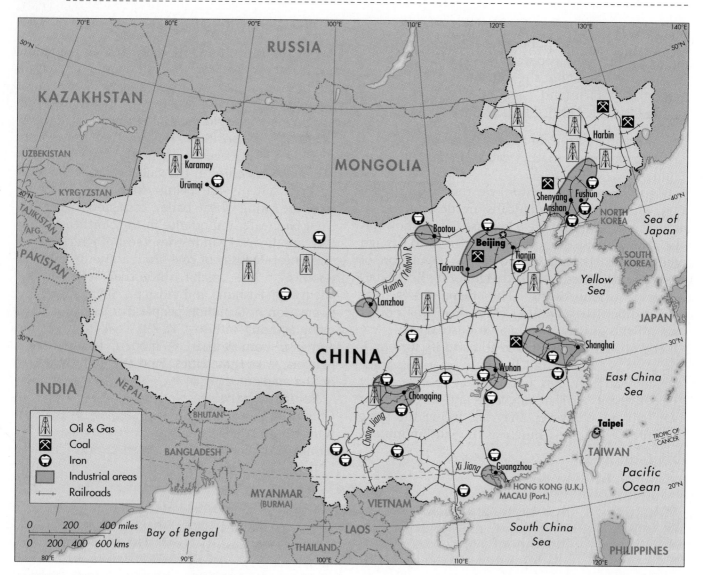

China's Resources and Industrial Areas

technology here. As a result, an industrial complex is developing in China's Northeast, similar to the one that stretches from Chicago to Pittsburgh in our own country.

⊕ REVIEW QUESTIONS

1. Most of China's huge population is clustered in the industrial Northeast and in the lowland basins of what three rivers?

2. What is the capital of China, and how large is it? What port city is located near the capital?

3. Why is the Chang Jiang one of the most important transportation routes in China?

⊕ THOUGHT QUESTIONS

1. Shanghai has a population density seven times higher than that of New York City. What problems do you think are caused because Shanghai is such a crowded city?

2. Why is the Northeast the major industrial region of China?

4. The Settling of the Land

Chinese Civilization Began in the North

Chinese civilization began on the North China Plain some 3,500 years ago in the area where the Wei River intersects the Huang River. The Huang River flows eastward across China from the mountains of Tibet to the Yellow Sea.

At the "great bend" in the Huang, the river cuts through a high plateau composed of fertile, yellow soil called loess. Each year, a huge quantity of this loess is carried downstream by the river and deposited on the North China Plain. These rich soils, available water, and a moderate (though somewhat dry) climate support millions of farmers in this region. Can you recall another river valley where an early civilization arose?

The Chinese Developed a Garden Agriculture Unlike That of Europe

Life in North China was based on garden agriculture, in which most farm work was done by hand. Simple tools like hoes and spades were used to dig the easily tilled loess soils of the North China Plain.

With these implements, Chinese farmers cultivated small plots of land, much smaller than those of farmers in Europe who used plows. But Chinese farmers became skilled in growing large amounts of food in small spaces. Production of wheat and millet from these small plots was high because Chinese farmers lavished much time and attention on them.

Over time, a series of dams and dikes were built for irrigation and flood control. Dikes are low walls of earth and stone used to close off or channel water. These water control systems protected villages and fields from raging floodwaters, and by making irrigation water available, reduced the threat of famine from drought. This efficient system of irrigation turned former fields and forests into farmlands. Control of water provided security and well-being to the people who lived on the North China Plain.

The Chinese Move Southward

These better living conditions in North China led to population growth. The Chinese began to move southward into the basins, plains, and delta of the Chang Jiang and ultimately as far south as the delta of the Xi Jiang in South China. As they moved, the Chinese absorbed local people into their culture or drove them out of the lowlands and into the hills. Chinese farmers, after all, wanted level land that was suited to their garden style of farming.

Chinese farmers were lured to the south as early as the Han period by the promise of new land, just as pioneers in the United States were drawn westward. They also fled southward to escape the frequent floods and droughts that destroyed their farms. Their belief that southern China was a fertile land was correct—even today, South China produces about 60 percent of the total food supply of modern China.

How the Chinese Organize the Land

Market towns, generally located within walking distance of the farming villages, are nerve centers in the world of Chinese village farmers. In the densely settled lowlands of China, a market town serves about twenty villages; fewer in mountainous areas.

 IT'S A FACT   The Chinese built transport canals 2,000 years before similar canals were built in Europe.

IT'S A FACT   The place on Earth located farthest from any ocean is located in China.

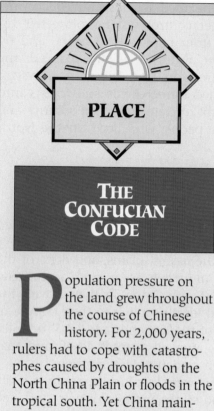

PLACE

THE CONFUCIAN CODE

Population pressure on the land grew throughout the course of Chinese history. For 2,000 years, rulers had to cope with catastrophes caused by droughts on the North China Plain or floods in the tropical south. Yet China maintained its cultural and territorial continuity. This was largely because the principles put forth by its most famous philosopher and teacher-scholar, Confucius (about 551–479 B.C.), were adopted and practiced as a state cult from the Han period (207 B.C. to A.D. 220) to the present.

Confucian teachings embody three basic concepts that were of supreme importance in the organization of life and landscape in China. First, the Chinese ruler was considered accountable for

human conditions. Second, imperial administrators were selected on the basis of competence. Third, the Chinese people were encouraged to value order, continuity, and social cohesion above the pursuit of personal gain. Although imperfectly practiced, these teachings defined the appropriate behavior of the three most important social groups in China: the royal court, the imperial administrators, and the peasants.

The first Confucian principle held the emperor personally accountable for the welfare of his people. The emperor was viewed

◀ *This inscribed relief of Confucius (about 551–479 B.C.), China's most celebrated philosopher, is based on traditional legends and myths about his appearance. What principles of action and belief did Confucius propose for the government and people of China?*

Almost every person in China's 900,000 villages takes advantage of the goods and services offered in these "hometowns." Whatever a villager needs—soap, matches, oil, incense, a barber, a marriage broker, a tailor, or a scribe—can be found there. These market towns are middle-sized places in an organized settlement system made up of villages, towns, and cities. The products of village farmers are sold to merchants in market towns. Town merchants then resell these products to city merchants. In return, manufactured goods and specialized services are provided by city merchants to town merchants for resale to village farmers. The market town, therefore, is the exchange point between China's cities and its villages.

⊕ REVIEW QUESTIONS

1. What is loess? In what part of China is it deposited?
2. What structures were built to control floods and provide irrigation water in North China?
3. Why did the Chinese occupy the plains of the Chang Jiang and the delta of the Xi Jiang?

⊕ THOUGHT QUESTIONS

1. What advantages did North China have that encouraged Chinese civilization in that region?
2. Describe the settlement system by which the Chinese organized the land. Can you see any similarities to the settlement system of the United States?

as superior to all other rulers on Earth. His task was to achieve harmony in China through the practice of good government and moral conduct. Evidence of harmony or disharmony could be seen in living conditions in China. Drought, flood, famine, and disorder were signs that the emperor was not fulfilling his duty. Peace and prosperity suggested that he was. This personal responsibility for conditions in the country set limits on the emperor's power, and if disasters occurred, he was held responsible and sometimes replaced.

Confucius's second principle was that administrators be selected on the basis of competence. The ideal administrators were not military leaders, but men of wisdom and learning. These "scholar-officials" levied taxes, supervised public works, ran police forces, and administered law courts throughout China. They were se-lected by an elaborate recruitment scheme, the imperial examination system. An exhausting series of examinations attracted hundreds of thousands of student-scholars from the lower reaches of society. Success quotas on some examinations were less than 3 percent, but the rewards of the system were so great that all over China young men spent as many as twenty years honing their knowledge of Confucian teachings.

Finally, because agriculture was the leading source of wealth in China, peasants were encouraged to value submissiveness to authority, respect for tradition, reverence for the past, prudence, caution, and moderation—Confucian virtues that contributed to conservatism, order, and harmony in society. Harmony exists when, as the proverb stated "Superior men diligently attend to the rules of propriety, and men in an inferior position do their best."

QUESTIONS

1. What three basic concepts are embodied in the teachings of Confucius? Would you describe these teachings as "good" or "bad," "traditional" or "modern"? Why?

2. How did people determine whether or not the emperor was practicing good government?

3. Should we hold our presidents responsible for floods, hurricanes, and earthquakes? Note the difference between our view that these are natural events and the Confucian view that they are evidence of mismanagement.

4. What do you think of the Confucian virtues mentioned above? Should Americans attempt to practice these? Why or why not?

CHAPTER SUMMARY

The varied geographical features of China have influenced the way the Chinese people use the land. China is mainly divided into east and west; the vast majority of Chinese are crowded onto the well-watered lands of eastern China and dry, mountainous western China remains sparsely populated. The key factor determining this east-west division is rainfall. Eastern China receives adequate rainfall for intensive agriculture, while western China is largely arid desert, steppe, and mountain country.

The mountainous Xizang Plateau, formerly Tibet, has been occupied by the Chinese govern-ment since the 1950s and was reorganized under communism. Still today the Tibetans, led by the Dalai Lama who lives in exile in India, resist the Chinese.

Eastern China is the country's heartland base, with its largest cities, industries, seaports, and agricultural lands.

A major agricultural division exists between northeast and southeast China. The North China Plain has less rainfall and cooler temperatures than the south. Irrigation systems are used to water crops of barley, wheat, and millet. In the water-rich south, rice and wheat are grown on terraced slopes. Chinese civilization began in the north and gradually spread southward.

EXERCISES

True or False?

Directions: On your paper, write *true* for the following statements that are correct and *false* for the statements that are incorrect.

1. The plateau of Xizang is a vast, treeless land where about 2 million Tibetans lived in relative isolation before 1949.
2. Inner Mongolia has a moderate climate and enough rainfall to be a rich wheat-producing region.
3. Many dikes and levees have been built in China to control water and reduce the danger from droughts.
4. The southwest is China's most important industrial area.
5. China's densest population is located in Outer China.
6. China's highly mechanized plow agriculture is the most productive in the world.
7. Villages are so remote from market towns that selling local crops is a major problem.
8. The climate of North China is suitable for large wheat crops.
9. The climate of the Tarim Basin is suitable for large orchards.
10. The climate of South China is suitable for large rice crops.

Cities

Directions: Number your paper from 1 to 8. Next to each number, write the letter in front of the word (or words) in Part B that matches the correct phrase in Part A.

Part A

1. The first Chinese seaport opened to foreign trade
2. The largest city in the Middle Basin, which is called China's "rice bowl"
3. One of the largest cities in Asia, it handles more than half of China's foreign trade
4. The main steel-making city in the Northeast
5. The largest new industrial center in Inner Mongolia
6. A major city in the Sichuan region
7. The large port city near Beijing
8. The chief city of the Tibetans in eastern Xizang

Part B

a. Baotou
b. Chongqing
c. Guangzhou
d. Lhasa
e. Shanghai
f. Shenyang
g. Tianjin
h. Wuhan

Inquiry

Directions: Combine the information in this chapter with your own ideas to answer these questions.

1. For hundreds of years the Tibetans were organized into a theocracy, which is a state ruled by religious leaders. What do you believe are the advantages and disadvantages of theocracies?
2. Much of China is troubled by droughts and floods. Is the region in which you live troubled by either or both of these natural hazards? If so, what has been done in your community to deal with these problems?
3. Use the *Readers' Guide to Periodical Literature* in your school or public library to find articles on China's policies in Tibet and their effects on the Tibetan population and environment. Organize a class debate based on your research.
4. Look at the regions of Outer China and Agricultural China on the maps in the miniatlas. As you learned in this chapter, the Chinese government has encouraged migration of Chinese farmers and laborers to Outer China. Imagine what it might be like for a Chinese teenager to move with his or her family from Agricultural to Outer China. Compose a journal entry that might be written by such a teenager describing one day's journey.

SKILLS

Using the Map of China's Resources and Industrial Areas

Directions: Study the map on page 584 and answer these questions.

1. Are the industrial areas of China generally located near the coast or in the country's interior regions?
2. Are there more large areas of industrialization in China's northeastern region or southwestern region?
3. Which country is located nearer to China's most industrialized region—North Korea or India?
4. Beijing is China's capital city. Is it also in the middle of an industrial area?
5. What city in southern China is both a port and the center of an industrial area?
6. Look at the map symbols for iron and coal. Are some of the industrial areas of China located near deposits of iron and coal?
7. Why do these same industrial areas probably manufacture steel products?
8. Find the map symbol for oil and gas. Are the oil and gas sources in China located almost entirely in one region, or are they spread through various parts of the country?
9. Are there any oil and gas sources near China's border with Russia?
10. Notice the map symbol for railroads. Do railroads connect Shanghai with Beijing?
11. Do railroads connect Beijing with Anshan? With Wuhan?
12. Is the railroad route from Shanghai to Ürümqi longer or shorter than the railroad route from Shanghai to Guangzhou?
13. Is the railroad route from Chongqing to Fushun longer or shorter than the railroad route from Chongqing to Lanzhou?
14. Do railroads connect Beijing with India and Pakistan?

Using an Almanac

Directions: The quickest way to find answers to many questions is to turn to an almanac, which is a reliable, accurate reference book that contains many facts. Two widely used almanacs are *The World Almanac and Book of Facts* and *The Information Please Almanac*. New editions of these almanacs are published each year.

To find a particular topic in an almanac, you must use the index. An almanac usually places its index at the front instead of the back of the book.

The following is part of an index as it might appear in an almanac. Look carefully at this list of items, or entries, dealing with China. The first entry (pages 321–322) shows where to find a brief overview about China. It includes such subjects as the country's population, ethnic groups, languages, religions, geography and topography, type of government, economy (including industries and crops), communications, transportation, education, and health statistics. It also includes brief highlights from China's history. The other entries show where to locate other types of information in the almanac.

Use these almanac index entries to tell on what page or pages you would look to find the answers to each of the following questions.

1. How many Chinese emigrated to the United States last year?
2. Has China's trade with the United States increased in the last five years?
3. What percentage of the world's petroleum production comes from China?

4. What are the chief religions of China?
5. What recent events have occurred in U.S. relations with China?
6. How many cargo ships does China have?
7. What are the major crops raised in China?
8. Does China have large deposits of coal and iron?
9. What is the size of China's air force?
10. How many languages are spoken in China?
11. What is the population of China's largest cities?

Vocabulary Skills

buffer zone   qanats
Dalai Lama   silt
dikes   theocracy
garden agriculture   yak
levees   yurt
loess

Directions: Use clues in the context of the sentences to determine the meaning of the italicized words. Write the letters of the best answers on your own paper.

1. China has maintained a *buffer zone* on its border with Russia.
   a. an area where each could trade and exchange goods with the other
   b. China and Russia were separated by a neutral territory to prevent fights from breaking out
2. Tibetans living in Tibet and in exile in India revere the *Dalai Lama.*
   a. a sacred Buddhist statue in Tibet
   b. the god-king of the Tibetans
3. For centuries, the Chinese have been experts at building *dikes.*
   a. earthen walls that surround a farmhouse or village
   b. low walls of earth and stone used to control and channel water

4. Chinese farmers engage in intensive *garden agriculture.*
   a. a type of farming in which intense labor is invested in small plots of land
   b. a practice of growing flowers together with crops to keep away pests
5. The Chinese have built *levees* along the banks of the Huang River in North China.
   a. earthen embankments that protect surrounding areas from flooding
   b. canals to channel river water to irrigate croplands
6. Brought by winds from the Gobi Desert, *loess* covers the North China Plain.
   a. a windblown fungus
   b. a fertile, windblown soil
7. In the Tarim Basin *qanats* are essential to farmers.
   a. underground water channels that supply water for farming
   b. earthen embankments that protect surrounding areas from flooding
8. The Huang (Yellow) River got its name because its waters carry so much *silt.*
   a. a type of yellow algae native to rivers in China
   b. a very fine-grained soil deposited in the bed of a river, lake, or sea
9. Historically, Tibetans were governed by a *theocracy.*
   a. a type of government in which the people elect their leaders
   b. a type of government ruled by religious leaders
10. Tibetans are dependent on the *yak* for many resources.
    a. a fibrous plant used for making mats and bags
    b. a large, long-haired ox native to Tibet and Mongolia
11. Mongol nomads take their *yurts* with them when they move.
    a. circular, felt tents
    b. large, long-haired oxen native to Tibet and Mongolia

CHINA, ASIA'S AWAKENING GIANT

In 1949, after nearly forty years of warfare and civil turmoil, the Chinese Communists defeated the Chinese Nationalists, who fled to the offshore island of Taiwan. Now the Communists, led by Mao Zedong, were ready to revolutionize life in the world's largest society. They had three goals: (1) to change Chinese agriculture, (2) to build a modern industrial state, and (3) to eliminate traditional Chinese philosophy and practice.

◄ Shanghai is China's largest city. It also is an international city, reflecting both Chinese and European ideas. Why might the city have built this pedestrian bridge? Why not a freeway overpass?

1. Fifty Years of Communism

The Communists Reorganize Agriculture

The Communists realized that China's farmers would have to grow more food to provide the money needed for building industries and importing modern technology. More wheat, rice, and other foods had to be produced, and more land had to be brought under cultivation. More efficient techniques of farming, better protection against floods and droughts, and a fairer distribution of the harvest were needed.

Breaking the power of China's landlords, the people who had owned much of China's farmland, was accomplished quickly. In three years, the Communists took the land away from the landlords and gave it to the farmers.

Several different ways of redistributing and reorganizing the land were tried, all under the direction of Communist planners, but these methods did not work well. Therefore, China's agricultural land and farmers were organized into large **collective farms**, on which all land was held in common by the group. All workers were paid on the basis of work points. By 1960, 90 percent of China's rural population lived on collective farms.

Communist Failures in Agriculture

The production results were disappointing; the government invested little money in improving farm machinery. Instead, the Communists plunged available resources into heavy industry. During the 1950s, food production increased by only 2.6 percent a year, making barely enough to feed China's rapidly growing population. Trade between city

dwellers working in industry and village farmers became one-sided. Harvests went to the city but little came back to the farms in exchange.

The Communists also underestimated the importance of environmental conditions to success in farming. Many of the tree-planting projects that planners put into action failed because of drought. Irrigation schemes to increase farm production were poorly planned. Many newly built **reservoirs** and ponds in North China dried up when there was a drought. Wells sunk on the drier, western edges of Agricultural China also failed.

As things got worse, Chinese farmers became more and more uncooperative. They resisted the Communist policy of taking their crops. They began to eat more, leaving less to be taken by the government, and as rural food consumption increased, so did population.

The Communists did manage to change the organization of rural society, but they did not solve the age-old problem of food shortages. China doubled its output of grains during the first thirty years of Communist rule. In the 1980s, however, more than 100 million Chinese were still hungry because of population growth, too great an emphasis on grains, and limited investment in agriculture.

The Development of Industry

The Chinese Communists were more successful in their efforts to develop industry, mainly because this is where they put their money. About 40 percent of the country's budget was invested in the production of coal, steel, electric power, and oil; the result was increased industrial production. By 1958, China became the third largest coal producer in the world, behind the United States and the former Soviet Union.

The Communists also tried to distribute new industries more evenly across China. Before 1949, factories in China's port cities manufactured three-quarters of China's industrial products. Chinese leaders decided to locate new industries in the interior. Two-thirds of the industries built during the first ten years of Communist rule were located away from the coast in the less-developed provinces of the west and northwest.

IT'S A FACT   On Arbor Day in China, each of its 1 billion people is expected to plant a tree.

China clearly had the resource base—the coal, iron, petroleum, and minerals—needed to build a modern industrial country. A network of roads and railroads was built to connect new industries with raw materials and cities throughout the country. China reported a 10 percent increase in industrial production each year, but experts think these figures are higher than the actual production.

Mao's Cultural Revolution Fails

Within fifteen years, Mao Zedong's goal of a new communist society in China was threatened by widespread discontent. Many Chinese leaders did not agree with all of Mao's policies. Farmers were unhappy, and students also complained. In the face of this opposition, Mao launched a "Cultural Revolution" in 1966 to eliminate his enemies. In the process, he nearly destroyed the country.

Universities were closed and college examinations were abolished. **Red Guards**, most of them inexperienced students, were encouraged to harass "disloyal" public officials. Gangs of Red Guards were ordered to purge "bad elements" in Chinese society. Millions of educated Chinese from the cities were sent to the countryside to work at farming. The turmoil created by these disruptions paralyzed China, and violence broke out in the nation's cities and villages. Millions of Chinese supported Mao; millions more opposed the violence and chaos of his Cultural Revolution.

Opposition to the Cultural Revolution began to surface. In spite of higher total agricultural production, China's farmers were no better off twenty years after the communist revolution than they had been before. And now there were millions more of them because of population growth. The Red Guards that Mao sent to the villages were young city dwellers with little or no knowledge of life on a collective farm. In the cities, harassment of leaders in industry, government, and business weakened the economy. These problems haunted China until Mao's death in 1976.

A New Chinese People, the Primary Goal of Mao Zedong

The third important goal of the Communists under Mao was to create a new social order. Traditional beliefs that elevated a class of people with more education and training over another class with less education were to be eliminated. More class-equal communist principles were to be introduced. To accomplish this, Communist leaders set up countrywide programs of education and health care.

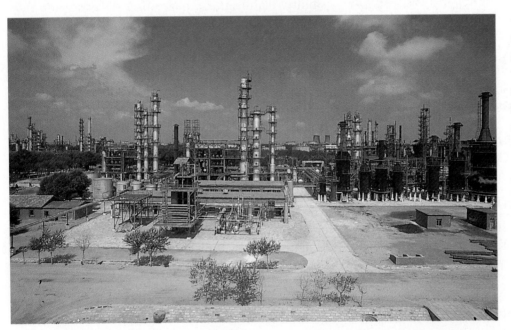

◄

China is believed to have the richest coal deposits in the world. Now petroleum engineers are discovering oil and natural gas deposits in the Northeast and in the interior. How will these discoveries help the economy and environment of China?

After ten years of Communist rule, 40 million people were attending elementary classes and another 31 million were studying part-time. Enrollment in primary and secondary school had quadrupled. In twenty years, more than 90 percent of the school-age children of China were receiving an education. The old practice of educating only the best qualified of China's young people was replaced by education for everyone.

At the same time, Communist leaders attempted to improve health conditions in China's villages. As late as the 1930s, two-thirds of the people of South China suffered from malaria. Typhus and plague, the primary diseases of poverty, were also widespread throughout China. To combat these diseases, Communist planners began campaigns of mass inoculation and improved sanitation; they built thousands of medical clinics in China's villages. They also tried to exterminate mosquitoes and rats, two major carriers of disease.

The goal was to transform Chinese society. Before 1949, China had an urban ruling class living in comfort at the expense of poor village farmers. The Communists wanted to make the quality of life as high in China's villages as it was in its cities. China has succeeded in reaching this goal to a greater degree than most developing countries. The availability of services in many villages is one reason urbanization has been slower in China than in other developing countries, although part-time workers are moving into the booming area on China's coast in search of additional off-season income.

⊕ REVIEW QUESTIONS

1. In 1949, the Communists set up a program to transform Chinese society. What were their three goals?
2. How were the large collective farms organized?
3. In the first ten years of Communist rule, where were two-thirds of China's new industries built?

⊕ THOUGHT QUESTIONS

1. Why did the Communist reorganization of Chinese agriculture have disappointing results?
2. Why were the Communists more successful in developing industry?
3. Why did Mao Zedong's Cultural Revolution nearly destroy the country?

2. The Four Modernizations

New Leaders, New Plans

After Mao's death in 1976, China began to pursue a new path to modernization called the "Four Modernizations." Many of Mao's most cherished policies were abandoned. China's new leaders recognized that after thirty years of Communist rule, China was still a poor, developing nation with much of its population working in farming. That was why new policies were needed.

Their goals, called the Four Modernizations, were (1) to make agriculture productive enough to support China's more than 1 billion people and to release workers from farming to work in industry, (2) to diversify China's industrial base with a new emphasis on light industry and consumer goods, (3) to modernize the army, and (4) to import foreign science and technology to make China a first-rate world power by the year 2000.

A New Approach in Agriculture

The changes made in agriculture are still in place today. Collective farms were subdivided into smaller units with more local independence and responsibility. Planting and harvesting decisions are now made at the village level. Traditional one-family farms are being revived in some regions. The government also allows farmers to carry on a certain amount of private cultivation and trade. Each farm family can work a small plot of land for itself and sell the crops from these fields in newly opened free markets, where the farmers can set their own prices and even make a profit. Privately owned land now makes up a growing percentage of the cultivated area of China.

Mao's policy of taking money and crops out of agriculture to pay for industry has also been changed. Government factories now produce farm machinery and fertilizer for China's farmers. In addition, the government has raised the price of grain 20 percent and lowered the amount of grain that farmers are forced to sell to the state. This means that those who work harder receive higher rewards

Water buffalo help plow a rice paddy. Although horses and mules do much of the work in other areas of China, the water buffalo does most of the heavy work in the south. Why do you think this is true? Is it related to cropping patterns?

and that more land has been planted in profitable commercial crops like cotton, jute, tea, and sugar.

So far, the results have been astonishing. They demonstrate how individual rewards motivate people. Farmers are better off. Food production has increased in spite of China's ancient problem with floods and droughts. China is exporting significant quantities of corn and soybeans to neighboring Asian countries.

Changes in Industry

The command economy of Communism, in which all economic decisions were made by government planners in Beijing, has been revised. Decision-making powers have been shifted to the local level, so that local businesspeople have more say in buying raw materials and selling their manufactured goods. The **profit motive**, or desire to make a profit, has been introduced into factory management. Under the new system, each business is assigned a **quota**, or an amount of goods or services to be produced. After this quota is filled, all additional manufactured goods can be sold directly

to consumers at market prices. Workers are paid bonuses each month if they work longer hours or produce more.

The government is working to raise standards of living by producing more energy, better transportation, additional housing, and more consumer goods wanted by the people. All factories, except those making armaments, are expected to make a profit. These reforms have introduced elements of the capitalist system into the Communist Chinese economy, resulting in a more mixed economy.

⊕ REVIEW QUESTIONS

1. What major change in policy increased farm production in China?
2. What is a quota?

⊕ THOUGHT QUESTIONS

1. Do you believe that the new Chinese policy of the Four Modernizations will succeed in raising the standard of living in China? Why or why not?
2. Why would the profit motive increase farm production?

3. China's Booming Economy

The Chinese economy has been growing substantially each year for the last decade, at a rate significantly higher than that of most other countries in the world. China is now the leading exporter of inexpensive products like sweaters, toys, and shoes. Foreign investment has been pouring into the country. Indeed, the United States now has a greater trade deficit with China than with any other nation except Japan. This means that more Chinese products are sold in the United States than U.S. products in China.

New Politics, New Economy

The new policies of China's leaders appear to be modeled after the record of the "**four tigers**"—

DISCOVERING

LOCATION

CHINA'S LARGEST CITY— SHANGHAI

More than 13 million people make Shanghai China's largest city and most crowded urban space. It has a population density seven times

▶

Nanjing Road, in Shanghai, is one of the busiest shopping streets in China. The stores along this road have the most modern, and best selection of goods. What type of goods would you expect to find in these stores? How would this have changed in the last decade?

Hong Kong, Singapore, South Korea, and Taiwan. Each of these countries experienced a recent and rapid economic miracle without having to deal with significant political disruption. The Chinese aim is to build a society like theirs, one with strict political control but relatively free economic opportunity. In other words, Chinese leaders are willing to eliminate restrictions on prices and companies, while maintaining strict control over the media and any political activity by private citizens. In June of 1989, the Chinese government took extreme measures against hundreds of students demonstrating in Tiananmen Square for greater political liberties.

If this policy succeeds, the impact on the world will be enormous. China, with more than a billion people as well as a nuclear arsenal, will be the dominant power in Asia and the Pacific. Even Chinese villagers, who make up a vast majority of the Chinese people, now have a stake. For the first time in many generations, village families are able to send a child to school, build a brick home, and eat meat occasionally. Never before has such a large portion of humanity risen from poverty so rapidly.

---

higher than that of New York City, five times that of London and three times that of Tokyo. Average living space in Shanghai is about forty-five square feet, or one nine-by five-foot room, per family. Shanghai's residents live with air and water pollution.

Shanghai has the most traffic accidents in the country. Its public transportation system attempts to cope with 12 to 13 million commuters on a busy day, and nearly 5 million bicycles and motorcycles clog its streets. Crime is becoming a major problem.

Shanghai was opened to foreign trade in 1842, when the Treaty of Nanking made the city a treaty port. In a section of the city called the International Settlement, British, French, and Americans operated concessions outside the control of China. With its excellent location at the mouth of the Chang Jiang, China's most important waterway, Shanghai soon grew into China's main port city. The foreign concessions lasted until World War II, and before the Communists came to power in 1949, half of China's trade passed through Shanghai. Silk, tea, and ores were the main exports. Food, manufactured goods, oil, steel, and chemicals were major imports.

In an effort to redistribute industry, the Communists forced more than 1 million skilled Shanghainese to migrate to other parts of China. Many more of these city people fled to Hong Kong or Taiwan to escape Mao's policies. Despite these losses, Shanghai still supplies almost half of all of China's manufactured goods and is a major center of heavy industry. One-sixth of all government taxes are collected in Shanghai. But its economy is growing more slowly than more favored cities in the southeast like Guangzhou.

Shanghai, however, still has China's most educated work force, and with its many universities, research institutes, museums, and theaters, Shanghai is the country's largest center of education and art. The city was President Chiang Kai-shek's political center under the Nationalists; it also was the birthplace of the Chinese Communist party.

QUESTIONS

1. The density numbers comparing Shanghai and New York City suggest that Shanghai has a high level of crowding. How do you think crowding affects people? Do you think it affects people from the U.S. differently than people from other countries? How would it affect you?

2. Describe Shanghai's site and situation. Is it similar to those of Paris and London described in Unit 2? Why or why not?

3. We tend to associate pollution with technological rather than developing nations. Does China fit this pattern? If not, what does that mean in terms of future pollution?

▶

Shanghai produces almost half of all China's manufactured goods. What type of product is this factory worker making? Why is Shanghai's location so important to its growth?

The Distribution of Economic Growth

The industrial revolution that is transforming China is most evident along the southern and eastern coasts of China. Cities from Wenzhou, 250 miles south of Shanghai, to Guangzhou, at the mouth of the Xi Jiang opposite Hong Kong, are experiencing explosive economic growth. In part this is because of the **Special Economic Zones (SEZs)**, which are areas set up along the China coast to attract foreign investment by providing special tax benefits. More likely, this industrial growth is a result of access, ability, and freedom from government regulation.

In China's interior, peasants derive fewer benefits from the economic boom on the coasts. They have benefited from land reform because privately owned parcels of land can now support a family. Yet in many remote rural areas, people still live in huts or caves, and are connected to modern China by footpaths. They have little access to the major changes that are transforming the world's largest nation.

There are reasons to be optimistic about China's economic future. Many experts see similarities between what is happening in China and what has happened in Japan, South Korea, Taiwan, Singapore, and Hong Kong. Moreover, traditional Chinese values stress education, thrift, and savings—key elements of modernization.

Even China's strict birth control policy (discussed in the next section) is seen as giving China an edge over countries like India, where so much wealth must be invested in a growing population.

The Price of Progress

The incredible economic leap that China has made is not without cost. Transportation networks, energy supplies, and basic infrastructure (the basic framework of public works) have not kept up with the needs of the new China. China is the world's leading producer of coal, the country's primary fuel; not surprisingly, it is now contributing to global warming faster than any other country. Railroads are clogged and so are highways. Movement of goods, particularly among the major coastal cities, is alarmingly slow.

Moreover, urban pollution is now a serious hazard. Beijing, China's capital, is suffocated with a blanket of hazardous chemicals and particles almost as intense as those that pollute Mexico City. Like Mexico City, Beijing is backed up against a half-ring of mountains that traps and holds pollutants. Despite the Chinese effort to build a "green great wall" by planting trees across North China to reduce pollution, economic growth has outpaced city facilities. In Beijing, sewage systems do not exist, so untreated sewage and industrial wastes flow freely into the sea. In addition, the city is running out of fresh water.

Shanghai, China's largest city, suffers even more from pollution because of neglect during Mao's rule. Heavy industry spews emissions into its sky. Shifting from bicycles to motorcycles and from burning wood to burning coal has intensified the process. In many Chinese cities, air pollution is seven to eight times worse than that in New York City. Many Chinese wear surgical masks when they go outside, particularly in the wintertime when the coal dust is at its worst. Coal-burning electric plants do not provide enough energy. Unpolluted water is in short supply. Indeed, 85 percent of China's cities are short of clean water, and in the countryside, only one in seven villagers has access to safe drinking water.

⊕ REVIEW QUESTIONS

1. How fast has the Chinese economy been growing over the last decade?
2. What does the term "four tigers" mean?
3. What is a Special Economic Zone (SEZ)?
4. What is China's primary source of energy?

⊕ THOUGHT QUESTIONS

1. Do you think that China's booming economy poses an opportunity or a threat to the United States? Why do you think as you do?
2. What has been the price of economic progress in China? Do you think that it has been worth the price or not?

4. More Than 1 Billion People

One of every five people on Earth lives in the People's Republic of China. In the late 1980s, the Chinese government estimated China's population at 1,087,000,000—slightly more than a billion people. The Chinese population at that time was still growing at a rate of 15 million more people a year.

Land and People

China's agricultural base is not growing, because little can be done to increase the amount of China's agricultural land. Virtually every square foot of **arable** land (suitable for growing crops) is already in use. In China, 7 percent of the world's farmland feeds more than a billion people and will have to feed even more. About one-tenth of China's farmland can grow two crops a year. This means that the area of China that is farmed is about equal to the area farmed in the United States.

Given the huge Chinese population, less than one-half acre of farmland is available per person in China. Also, China has been adding an amount equal to the U.S. population to its total population once every seventeen years. Virtually every expert has doubted that this rate of population growth could be controlled, because one-third of China's people are under fifteen years of age and have yet to marry and have children. But the experts have been wrong.

Early Approaches to Population Control

The first Communist policy designed to solve this overwhelming population problem was to increase the amount of farmland. Millions of acres of marginal farmland were brought under cultiva-

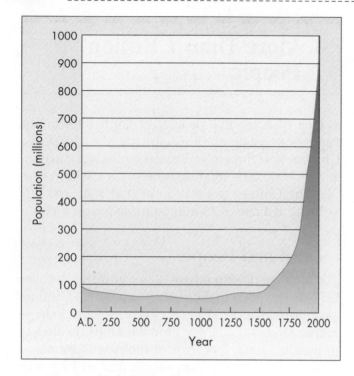

Population Growth in the People's Republic of China

The amount of grain produced per person after twenty years of Communist rule was not much higher than that in the 1930s, and food shortages were widespread.

Under China's new leadership, as you have learned, private control of farmland has replaced government control of the system of land organization. As a result, agricultural production has greatly increased. China now exports food; previously it was a major food importer. Was this enough to meet the needs of China's growing population?

Harsh Measures Reduce Births

In the early 1980s, a "one-couple, one-child" policy was adopted. Severe measures were employed to force Chinese couples to limit their number of children. Resentment and resistance to the government's interference in private lives was intense. Nevertheless, a nationwide crackdown by family planning authorities not only quelled resistance, it also led to a stunning decline in the birthrate.

Couples are now given "pregnancy permits" that restrict them to one or two children. Any deviation from this quota is severely punished by reducing food coupons, turning off water, or pulling down houses. Family planning targets are set for local planning officials; if they fail to meet their target, they are fined. As a result, Chinese women now

tion in Outer China and on the dry fringes of Agricultural China. After this policy failed, an effort was made to increase production on existing agricultural land by organizing China's farmers into large collective farms. Neither policy was very successful.

Ancestor worship and respect for the elderly are two of the important beliefs of the people in China. How does this picture reflect one of these ideas? Do people respect the elderly in the United States?

have about the same fertility rate as women in Great Britain and the United States.

The Dilemma of People and Land

Chinese leaders recognize that if a one-child policy is imposed on the Chinese people now, the nation's population will stabilize at about 1.2 billion and then fall to 700 million during the next 100 years. If, however, each couple has three children, China will have a population of 4.26 billion 100 years from now, a population about 1 billion less than the entire world today.

The current economic boom in China may support a strong and vibrant economy, but it is too early to tell. The sheer size of China's population and economy is staggering. Most are villagers and income varies from region to region. China's per-capita gross national product is estimated at $400 per year in some parts of the country and at $4,000 per year in others.

Millions of new jobs will have to be created each year for the 300 million Chinese under the age of fifteen and the 650 million under the age of thirty. The real test of China's new economic reforms will come in China's villages. Some 800 million Chinese still live in these villages, one-third of all the farmers in the world. Today, as in the past, the fate of China lies in the relationship between land and people.

⊕ REVIEW QUESTIONS

1. The Chinese population is now growing at a rate of how many new people a year?
2. About how much farmland per person is available in China?
3. How has China's "one-couple, one-child" policy affected the birthrate of the country?

⊕ THOUGHT QUESTIONS

1. In the early 1980s, why did the Chinese government adopt the "one-couple, one-child" policy? What would you have done if you had been the political leader of China?
2. About one-third of China's population is under fifteen years of age. What does this statistic suggest about the future problems the country must face?

5. Taiwan, Another Chinese Country

An Industrial Island off the Coast of East Asia

The island nation of Taiwan is shown on the map below. It is located 125 miles off the southern coast of mainland China. After the Communists took over mainland China in 1949, the Nationalist Chinese retreated to this island. At that time, most observers expected Communist armies to capture Taiwan. Instead, the U.S. Navy protected the Nationalists in Taiwan during the 1950s and 1960s.

Taiwan, Another Chinese Country

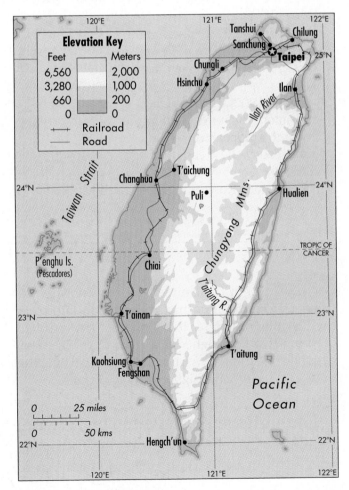

Elevation Key

Feet	Meters
6,560	2,000
3,280	1,000
660	200
0	0

+—+ Railroad
——— Road

PLACE

HONG KONG: A PORT BUILT BY OUTSIDERS

Until the 1700s, Asia was ruled by empires whose great cities were located in the interior. These were land-based empires. When the Europeans arrived by sea, they built trade bases on the coasts of Asia. These ports were Western outposts on the rim of Asia.

In the 1800s, ripples of modernization began to spread inland from the European centers of trade. Today, these early trading cities, like Bombay, Calcutta, Singapore, Hong Kong, and Shanghai, are among the largest urban centers in Asia.

In the 1700s, foreign traders in China were restricted to the port city of Guangzhou (then called Canton). But imperial China was forced to turn over Hong Kong to Great Britain after it lost the First Opium War in 1842. A large number of other Chinese port cities were later opened to Western influence.

As the power of the Qing dynasty declined, Western powers established footholds on the Chinese coast. The Chinese gradually granted them rights to mine minerals and gave land for railroads to foreigners. These outsiders also gained almost complete control over Chinese trade. Europeans, however, were never able to penetrate the interior of China in large numbers. Foreign ports in China remained European islands on the edge of an alien and resistant nation.

Today Hong Kong is a 400-square-mile British crown colony, the last important Chinese port under Western control. It is a bustling center of capitalist trade and business on the doorstep of mainland China. It has one of the finest deep-water, natural harbors on China's coast. The city includes the island of Hong Kong, the nearby Kowloon Peninsula, and adjacent areas on the mainland of China called the New Territories.

Hong Kong thrives on international trade. The city has nearly 6 million people, virtually all of whom are Chinese, and one of the highest standards of living in Asia. It is, however, one of the most crowded cities on Earth.

Hong Kong's future is uncertain. In 1997, the 100-year British lease on Hong Kong expires and the island becomes part of Communist China. China and Great Britain have agreed that the capitalist system of business and trade practiced in Hong Kong will be continued when the city becomes a part of Communist China. There is no guarantee, however, that this will happen.

With aid from the United States, Taiwan has become a successful industrial country also known as the Republic of China. Today, the island's standard of living is much higher than that of the People's Republic of China. In Asia, only the people of Hong Kong and Japan have higher annual incomes.

The Chinese Settlement of Taiwan

In the 1500s, Chinese farmers and fisherfolk from the mainland began to move to Taiwan. These Chinese immigrants gradually pushed the original inhabitants of Taiwan deep into the island's interior mountains. Today, about 250,000 of these Southeast Asian people still live in some 240 villages in the mountains of central Taiwan.

In the late 1600s, 2 million more Chinese fled to Taiwan to escape the invading armies of China's Qing (Manchu) dynasty. Gradually, Chinese culture and economy were firmly planted on the island. But in 1895, Taiwan was turned over to the Japanese after China was defeated in a war with Japan.

The future of Hong Kong will be determined by the leaders of China at that time.

QUESTIONS

1. How did Hong Kong's relative location change in the 1800s? Did its absolute location change?
2. What factors do you believe have enabled the people of Hong Kong to achieve one of the highest standards of living in Asia? Can you think of any small countries that have achieved high standards of living?
3. What political change will occur in Hong Kong in 1997? Can you speculate on what the future of Hong Kong might hold after 1997?

▲ *The harbor of Hong Kong is one of the most beautiful and busiest in the world. This port was built by foreigners, and was an important early trading city. Does this look like a modern Western city or a traditional Chinese city to you?*

The Japanese began a large-scale program of economic development.

The Economic Development of Taiwan

The map on page 601 shows that the Tropic of Cancer, 23½° north latitude, runs through Taiwan. In North America, this line passes through central Mexico. Taiwan has a climate similar to that of southern China. It is subtropical in the north and tropical in the south. Rainfall is heavy; however, it varies from year to year, because Taiwan is lashed by typhoons in the summer at the rate of three every year. **Typhoons** (or as we call them, hurricanes) are huge tropical storms with high winds and heavy rain.

In many parts of the island, climate is influenced by topography, because the island is quite mountainous. In the coastal lowlands of western Taiwan, Chinese immigrants cleared away tropical woodland, reclaimed swamps, and introduced their garden system of intensive farming. Irrigated rice fields spread along the coast. Sugarcane, tea, and bananas were planted inland in more hilly areas.

The Japanese extended the irrigated area and increased the production of rice and sugar. During their occupation of Taiwan, which began in 1895, the island became an important producer of food and raw materials for Japan.

The Japanese built hydroelectric plants on rivers that plunge down the slopes of the Chung Yang Mountains to the sea. These plants provided cheap electric power for small industries. The Japanese also linked together the towns of the west coast by roads and railroads. Taiwan's economy grew substantially.

The Japanese planned to make Taiwan a permanent Japanese territory. They taught Japanese in local schools, and their language became the language of business and trade. These plans ended when Taiwan was returned to China after Japan's defeat in World War II in 1945. Then, four years later, Taiwan became a country run by the 2 million members of the defeated Nationalist armies that arrived on the shores of the island when the Communists took over mainland China.

Taiwan Today

The Chinese Nationalists initiated a program of modernization in Taiwan. Two key policy decisions were made: (1) a land reform program was started to distribute farmland to Taiwanese farmers and (2) factories were built to produce the textiles and fertilizers that had previously been imported. Both policies were successful. Taiwan, with a population of nearly 21 million crammed onto an island a bit

▼ *The island of Taiwan has long been agriculturally important, and has recently become the twentieth largest industrial nation in the world. Why is the future of Taiwan uncertain?*

IT'S A FACT Taiwan has a greater surplus of foreign exchange than any country in the world including the United States and Japan.

larger than Maryland, now produces 85 percent of its food supply. It also has become the twentieth largest industrial producer in the world.

Taipei, the country's capital, is a modern city of 3.4 million people. It has 5 million automobiles, traffic jams, pollution problems, and one of Asia's highest standards of living. Industries in Taiwan's cities produce light machinery, bicycles, electronic components, textiles, and shoes—products they export to the United States, Europe, and the Middle East. Taiwan even trades with mainland China through the free port of Hong Kong.

An Uncertain Future

Today, Taiwan is no longer recognized as a country by many nations in the world. In 1979, the United States severed diplomatic relations with Taiwan and gave formal diplomatic recognition to the People's Republic of China on the mainland. However, Taiwan is still linked to the United States by defense treaties.

The current policy of the United States is to keep things as they are and follow a **two-China policy.** Mainland China until very recently has been committed to the "liberation" of Taiwan by invasion. The Nationalists on Taiwan have declared that they would never deal with the Chinese Communists.

Meetings of high-level delegates from China and Taiwan in 1993 suggested that a new policy of cooperation was at hand. But, by 1996, veiled threats of invasion from the mainland erased these hopes. As the People's Republic of China expands its role in world affairs, relations with Taiwan are becoming more complicated. How negotiations can continue between two countries—each of which claims to be the legitimate government of the other—is uncertain, but a start has been made.

REVIEW QUESTIONS

1. What role did the United States play in preventing the Chinese Communists from conquering Taiwan in the 1950s and 1960s?
2. Is rainfall in Taiwan usually heavy or light? What tropical storms lash the island in summer?
3. What are the chief crops grown in Taiwan?
4. What are Taiwan's main exports?

THOUGHT QUESTIONS

1. What two key policy decisions did the Nationalists make after Taiwan became their refuge? How do we know that these policies have been successful?
2. What is unusual about the political relationship of the United States with Taiwan?

CHAPTER SUMMARY

In 1949, the Chinese Communists embarked on one of the most daring social and economic experiments in history—to transform China from a country destroyed by war into a modern progressive society. Initially, Mao Zedong, the leader of the Communist revolution, sought to radically revolutionize Chinese society, to create a new Chinese people. This attempt largely failed after Mao's death in 1976.

New, more pragmatic goals were established. Called the Four Modernizations, they were intended to make China a world power by the year 2000. The government has implemented policies that mix free-market and controlled economies in a way that has proved successful for several other Asian countries on the Pacific Rim. The results thus far have been remarkable. But China's huge population of 1 billion people is a major threat to the country's future prosperity. Although the government has been largely successful in controlling population growth, the problem is not yet completely solved.

The island of Taiwan is sometimes called the "other China." The Chinese Nationalist government, defeated by the Communists, has transformed Taiwan into one of the major economic powerhouses of the Pacific Basin. Taiwan, South Korea, Hong Kong, and Singapore are known as the "four tigers" of Asia.

EXERCISES

Select the Correct Missing Information

Directions: Write the letter and word or phrase that properly completes each sentence on your own paper.

1. The Chinese ||||||||||, led by Mao Zedong, defeated the |||||||||| in 1949.
   a. Socialists
   b. Nationalists
   c. Communists
2. As part of their land reform policy, the Communists took land away from the |||||||||| and gave it to the ||||||||||.
   a. merchants
   b. farmers
   c. landlords
3. Farmers working on collective farms in the 1950s found that their food production ||||||||||.
   a. increased faster than the population
   b. barely kept pace with population growth
   c. decreased relative to the population
4. Under the Four Modernizations policy, each business must |||||||||| before it can then make a profit.
   a. fulfill its state quota
   b. meet its expenses
   c. accurately anticipate market demand
5. Today, the gap between rural and urban living standards is ||||||||||.
   a. increasing
   b. narrowing
   c. staying about the same

Inquiry

Directions: Combine the information in this chapter with your own ideas to answer these questions.

1. Analyze Mao Zedong's three goals to revolutionize Chinese society. How successful were the policies created to achieve these goals? Explain why some policies succeeded and some failed to achieve their goals.

2. What are the Four Modernizations and what are their goals? Why do you think that the Four Modernizations policies are succeeding where Mao's policies failed?
3. Ask your librarian to help you locate newsmagazine articles on the transfer of Hong Kong from a colony of Great Britain to a Chinese city. Describe some of the problems associated with the transfer and explain how Hong Kong residents feel about becoming Chinese citizens.

SKILLS

Using a Graph That Shows Population Growth

Directions: Look at the population graph on page 600 and answer these questions.

1. What does the vertical axis indicate?
2. Each number next to this line shows an increase of how many million people?
3. What does the horizontal axis indicate?
4. Each number next to this line shows an increase of how many years?
5. Did China's population increase or decrease between the years 250 and 1000?
6. In which 250-year period has China's population had the greatest growth?
7. A graph can dramatize facts and make the reader more aware of their impact. What situation in China is dramatized by this graph?

Vocabulary Skills

Directions: On your paper, write two or three sentences that explain what each of these terms means.

arable	Red Guards
collective farm	reservoir
"four tigers"	Special Economic Zones (SEZs)
profit motive	two-China policy
quota	typhoon

Human Geography

Directions: Finish each sentence by writing the words that belong in the blanks on a piece of paper.

1. About 95 percent of the Chinese people live in the region called |||||||||| China.
2. Chinese civilization began in |||||||||| China.
3. The occupation of most of the Chinese people is ||||||||||.
4. The philosopher who established many of China's long-held traditions was ||||||||||.
5. After their defeat in the First Opium War, the Chinese were forced to give the island of |||||||||| to the British.
6. When the Nationalists were driven from mainland China in 1949, they took refuge on the island of ||||||||||.
7. In the 1950s the Communists attempted to revolutionize Chinese society by pursuing three goals. They were: (a) ||||||||||, (b) ||||||||||, and (c) ||||||||||.
8. The Communist government took China's farmers off of their land and put them on large |||||||||| ||||||||||.
9. Faced with widespread discontent with his reform policies, Mao Zedong launched the |||||||||| ||||||||||.
10. China's new leaders launched the Four Modernizations. They had four goals by which they hoped to modernize China by the year 2000. They are (a) ||||||||||, (b) ||||||||||, (c) ||||||||||, and (d) ||||||||||.
11. By allowing businesses to sell the products they produce in excess of the state quotas at market prices, the government has introduced the |||||||||| |||||||||| into the Chinese economy.
12. As a result of this rapid growth, China's cities suffer from urban |||||||||| and ||||||||||.
13. To control the rate of China's population growth, the government adopted a |||||||||| policy in the early 1980s.
14. Taiwan's topography is quite ||||||||||, and its climate is |||||||||| in the north and |||||||||| in the south.
15. The two policy decisions the Nationalists made in their efforts to modernize Taiwan were (a) |||||||||| and (b) ||||||||||.

Physical Geography

Directions: Two of the statements following each number are correct. Write these correct statements on your paper.

1. (a) China shares a border with Russia. (b) China shares a border with Vietnam. (c) China shares a border with South Korea.
2. (a) Much of China's best farmland is on the North China Plain. (b) China has much more cultivated farmland than the United States. (c) China has about as much cultivated farmland as has the United States.
3. (a) The Huang River cuts through a high plateau of fertile loess. (b) Because of its unusually deep channel, the Huang seldom rises above its banks and rarely floods the countryside. (c) The Huang has been called "China's sorrow."
4. (a) Mountain ranges dominate the western frontier of Outer China. (b) Typhoons often strike the Xizang Plateau. (c) The plains of Inner Mongolia are generally cold and windswept.
5. (a) Beijing is the major ocean port city of North China. (b) Shanghai is a port city near the mouth of the Chang Jiang. (c) Guangzhou is an important port city on the Xi Jiang.

Writing Skills Activities

Directions: Answer the following questions in essays of several paragraphs each. Remember to include a topic sentence and several sentences supporting your main idea in each paragraph.

1. Explain the statement, "China is geographically diverse." Give specific examples in your explanation.
2. What drastic changes did the Communists make after they took over China?
3. What are the major problems China is facing today? How are these problems being handled?

SOUTH ASIA

▲ *The many styles of architecture in Bombay, India, reflect diverse cultural, religious, and national influences. Can you name some of the styles and influences shown in this photograph?*

UNIT INTRODUCTION South Asia includes the six countries of India, Pakistan, Bangladesh, Sri Lanka, Nepal, and Bhutan. These countries have a combined population of more than 1 billion people, nearly one-fifth of all the people on Earth. A majority of their citizens are very poor, and most are village farmers. All of these countries are struggling to modernize their economies, to raise the standards of living of their people, and to build stable governments. The results have been mixed.

Some geographers believe that the future of South Asia is already determined. They think that poverty, population growth, inefficient farming, slow growth in industry, and rapid urbanization must lead to chaos. Other experts think that South Asia will overcome these problems and achieve a higher level of modernization in the near future.

South Asia is such a complicated region that predictions about it have usually been wrong. As you read about this huge region and its many peoples, try to judge for yourself what the future may bring.

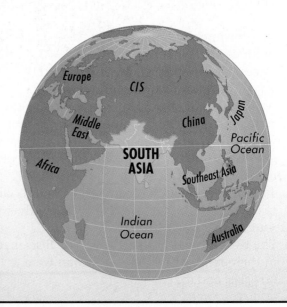

▼ Many people of India still live in small villages and do most of their chores by hand. The woman in this picture wears a traditional form of dress. What is this called?

SKILLS HIGHLIGHTED

- Making Generalizations
- Using a Physical Map
- Using a Political Map
- Differentiating between Fact and Opinion
- Vocabulary Skills
- Writing Skills

UNIT OBJECTIVES

When you have completed this unit, you will be able to:

1. Describe the geographical features of South Asia and how this region's monsoon rains greatly affect crop production and standards of living.

2. Identify the significant role that religion has played in the history of South Asia.

3. Describe the changes brought to South Asia by British colonialism.

4. Explain the conditions that have led to extreme poverty throughout most of South Asia.

▶

The people of India buy much of their food and their necessities daily in open street markets. How does this contrast to shopping patterns in your neighborhood?

609

1

South Asia is half the size of the United States but has four times as many people.

2

South Asia has five major subregions: the Himalaya Mountains, the Indus Valley, the Ganges Plains, the Deccan Plateau, and the tropical coasts.

3

South Asia is politically divided into six countries: India, Pakistan, Bangladesh, Sri Lanka, Nepal, and Bhutan.

KEYS TO KNOWING SOUTH ASIA

4

India is the world's largest democracy.

5

Indian civilization, which is 3,000 years old, is based on Hinduism and the caste system.

▼ *Rain is plentiful in the forested mountain slopes in Sri Lanka, where tea plantations flourish. Tea is one of the two main exports of this island nation. What is the other export?*

6

Hinduism, Islam, and British colonialism are the three major cultures that have influenced South Asia.

7

There is great conflict, often violent, between different political, religious, language, and ethnic groups in South Asia.

8

More than 1 billion people live in South Asia, and population is growing rapidly.

9

Three of every four South Asians are farmers. Most of the region's villages are crowded.

10

Some major obstacles to progress in South Asia are population growth, inefficient farming, slow rates of industrialization, poorly developed resource bases, and conflicting cultures.

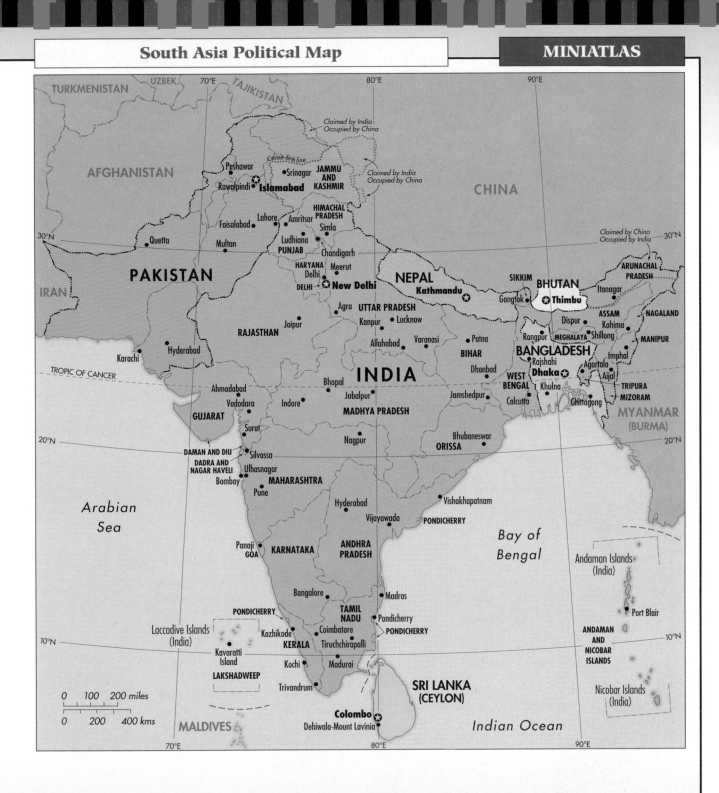

TURKMENISTAN   UZBEK.   70°E   80°E   90°E

TAJIKISTAN

AFGHANISTAN

Claimed by India
Occupied by China

Cease-fire line

Peshawar
• Srinagar   JAMMU
AND
KASHMIR   Claimed by India
Occupied by China   CHINA

Rawalpindi ✪Islamabad

HIMACHAL
PRADESH

30°N   Claimed by China   30°N
Occupied by India

Faisalabad Lahore Amritsar
Simla   ARUNACHAL
PRADESH
Quetta   Ludhiana Chandigarh
Multan   PUNJAB   SIKKIM BHUTAN   Itanagar

IRAN   HARYANA Meerut   Gangtok ✪Thimbu
Delhi   NEPAL   ASSAM   NAGALAND
PAKISTAN   DELHI ✪New Delhi   Kathmandu✪   Dispur • Kohima
Rangpur MEGHALAYA Shillong • MANIPUR
Agra UTTAR PRADESH   BANGLADESH
Jaipur Kanpur • Lucknow   • Rajshahi   Imphal •
RAJASTHAN   Patna •   ✪Dhaka   Agartala •
Hyderabad   Allahabad • Varanasi   BIHAR   Aijal •
Karachi   Ahmadabad   Bhopal •   Dhanbad •   TRIPURA
•   WEST Khulna   MIZORAM
TROPIC OF CANCER   Vadodara   Jabalpur • INDIA   BENGAL   Chittagong •
•   Indore •   Jamshedpur •   MYANMAR
GUJARAT   MADHYA PRADESH   Calcutta •   (BURMA)
Surat •
20°N   Nagpur •   Bhubaneswar   20°N
DAMAN AND DIU   • Silvassa   •
DADRA AND   Ulhasnagar   ORISSA
NAGAR HAVELI   • Bombay   MAHARASHTRA
Pune •

Vishakhapatnam
Arabian   Hyderabad •   •
Sea   Vijayawada   Bay of
• PONDICHERRY   Bengal   Andaman Islands
Panaji •   ANDHRA   (India)
GOA KARNATAKA   PRADESH
Madras   •Port Blair
Bangalore • •   ANDAMAN
PONDICHERRY   TAMIL   AND
Laccadive Islands   NADU • Pondicherry   NICOBAR
(India)   Kozhikode • Coimbatore • PONDICHERRY   ISLANDS
10°N   • Tiruchchirapalli   10°N
Kavaratti KERALA
Island   Kochi • • Madurai   Nicobar Islands
LAKSHADWEEP   (India)
• Trivandrum   SRI LANKA
0 100 200 miles   (CEYLON)

0 200 400 kms   ✪Colombo
MALDIVES   Dehiwala-Mount Lavinia •   Indian Ocean

70°E   80°E   90°E

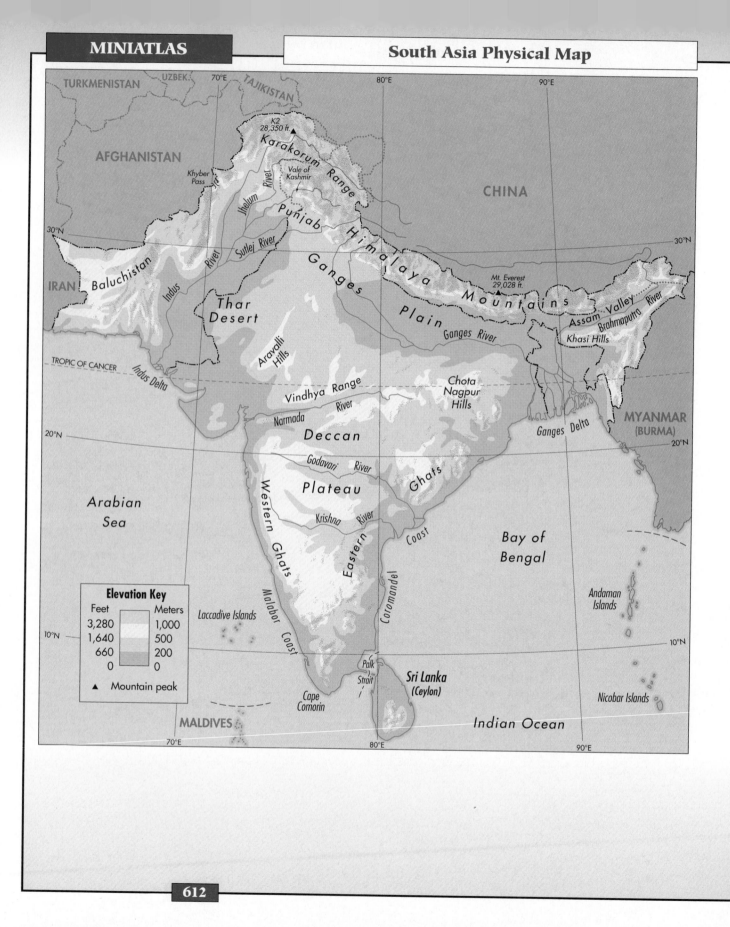

TURKMENISTAN   UZBEK.   70°E   TAJIKISTAN   80°E   90°E

AFGHANISTAN   K2
                               28,350 ft.
                          Karakorum Range
                    Khyber   Vale of   CHINA
                    Pass   Kashmir
                                         Punjab
30°N   Jhelum   Himalaya   30°N
         IRAN   River   Mountains
                    Baluchistan   Sutlej River   Ganges   Mt. Everest
                           Indus   Plain   29,028 ft.   Assam Valley   Brahmaputra River
                                   Thar   Khasi Hills
                                   Desert   Ganges River
TROPIC OF CANCER   MYANMAR
         Indus Delta   Aravalli   (BURMA)
                                         Hills   Chota
                                                                   Nagpur
                              Vindhya Range   Hills
20°N   River   20°N
                              Narmada   Ganges Delta
                                        Deccan
                                                                        Ghats
         Arabian   Godavari   River
         Sea   Plateau
                              Western   Krishna   River   Bay of
                              Ghats   Bengal
                                        Eastern
                                        Ghats
                                                    Coast
                                                                                                        Andaman
Elevation Key   Islands
Feet   Meters   Malabar
3,280   1,000   Laccadive Islands   Coast
1,640   500   Coromandel
660   200
0   0   10°N
10°N
▲ Mountain peak   Palk   Sri Lanka
                                             Strait   (Ceylon)   Nicobar Islands
                              Cape
                              Comorin
                    MALDIVES   Indian Ocean
                         70°E   80°E   90°E

612

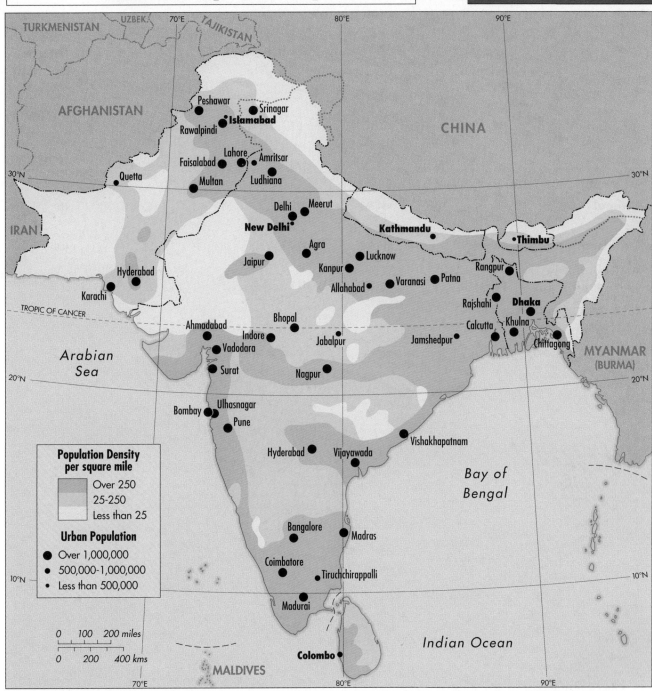

TURKMENISTAN
UZBEK
TAJIKISTAN
70°E
80°E
90°E

AFGHANISTAN

CHINA

Peshawar
Srinagar
•Islamabad
Rawalpindi
Lahore
Amritsar
Faisalabad
30°N
Quetta
Multan
Ludhiana

IRAN

Delhi
Meerut
New Delhi
Kathmandu
•Thimbu
Agra
Lucknow
Jaipur
Rangpur
Hyderabad
Kanpur
Karachi
Allahabad
Varanasi
Patna
Rajshahi
Dhaka
TROPIC OF CANCER
Khulna
Ahmadabad
Bhopal
Calcutta
Chittagong
Indore
MYANMAR
Vadodara
Jabalpur
Jamshedpur
(BURMA)
Arabian
Surat
Nagpur
Sea
20°N

Bombay
Ulhasnagar
Pune

Vishakhapatnam

Bay of
Bengal

Population Density
per square mile

Over 250
25-250
Less than 25

Hyderabad
Vijayawada

Urban Population

● Over 1,000,000
• 500,000-1,000,000
• Less than 500,000

Bangalore
Madras

Coimbatore
Tiruchchirappalli

0 100 200 miles

0 200 400 kms

Madurai

Indian Ocean

Colombo •

MALDIVES

Country	Capital City	Area (Square miles)	Population (Millions)	Life Expectancy	Urban Population (Percent)	Per Capita GNP (Dollars)
Bangladesh	Dhaka	55,598	113.9	53	14	220
Bhutan	Thimbu	18,147	0.8	49	13	180
India	New Delhi	1,269,340	897.4	59	26	330
Nepal	Kathmandu	54,363	20.4	54	8	180
Pakistan	Islamabad	310,402	122.4	56	28	400
Sri Lanka	Colombo	25,332	17.8	71	22	500

—Figures not available

LAND AND PEOPLE IN SOUTH ASIA

South Asia is separated from the rest of Asia by mountains and oceans, as the map on page 612 shows. This huge peninsula, shaped like a kite, is about half the size of the United States. Over the centuries, many kingdoms and societies arose here, and these ancient cultures still form the basis of daily life in the countries of South Asia. Differences of religion, language, and ethnic group are critical to the lives of South Asians.

Although geographic barriers often isolate and unify a country, South Asia has many different cultural groups in spite of the mountains that separate it from the rest of Asia. Hindu civilization developed in South Asia around 3,000 years ago. Muslims entered the region

◀ *This elderly Indian prepares to have his tea. Why might he be sitting on the ground?*

615

through mountain passes in the northwest more than 1,000 years ago.

Together Hindus and Muslims produced the brilliant Mogul civilization, which ruled this region from the 1500s until the mid-1700s, when the British East India Company began to take over. Europeans controlled South Asia for 200 years until it achieved independence in 1947. Given this long history, traditional ways of thinking and living are very strong influences in South Asia today.

1. The Subregions of South Asia

The Summer Monsoon, Lifeblood of Farming in South Asia

In most regions of South Asia, whether or not crops ripen depends on rainfall from the summer monsoon—a current of moist, warm air from the ocean that blows across the subcontinent between June and September. **Monsoon** is the Arabic word for "season." It is called a **subcontinent** because it is large and separated from the rest of Asia by mountains and seas. Geographers define a monsoon as any air current that blows steadily from the same direction for weeks or months at a time. The climate of South Asia is vitally affected by two monsoons. One monsoon blows cold, dry air from the Central Asian interior in the winter season onto the

subcontinent and the other blows warm, moist air from the Arabian Sea and the Indian Ocean inland in the summer season. The maps below show the directions of the two monsoons.

The wet **summer monsoon** blows inland from the ocean. It pushes warm, moist ocean air over South Asia, drenching the region with heavy rain from June to September. The **winter monsoon** flows in the opposite direction, blowing outward from Central Asia to the ocean from October to March. It is a current of dry air that produces little rain as it moves across the subcontinent.

The summer monsoon drops between 80 and 90 percent of India's total rainfall between June and September. Summer monsoon rains are the lifeblood of agriculture, except where farmland is irrigated. Good monsoons, where the rain falls at the right time and in the right amount, bring bountiful harvests in Pakistan, India, Bangladesh, and Sri Lanka. Poor monsoons, where the rains come too early or too late and it rains too little or too much, mean food shortages, famine, and the need to buy expensive grain on world markets.

Both the timing of the summer monsoon and the amount of rain that falls are crucial to farmers. Plowing the land and sowing the basic food crops in each area of the subcontinent begin with the rains. If the monsoon is delayed, or if too little rain falls, crop yields are low. A good monsoon means that India's production of food grains may be 20 percent higher than in a year with a bad monsoon. One of the worst monsoons of the century occurred in 1987, leading to great stress in the villages of India.

The Winter Monsoon

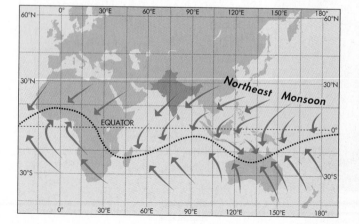

The Summer Monsoon

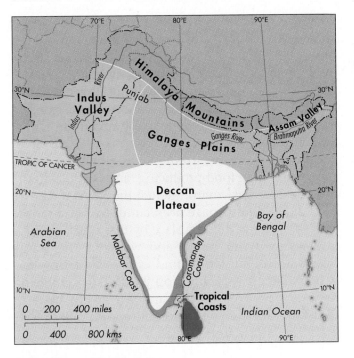

Subregions of South Asia
South Asia has five major subregions: (1) the Himalaya Mountains, (2) the Indus Valley, (3) the Ganges Plains, (4) the Deccan Plateau, and (5) the Tropical Coasts.

▲ *Heavy monsoon rains flood the streets of India's capital, New Delhi. How do the monsoons affect urban and rural life in India?*

The arrival of the summer monsoon, with its torrential rains, is a time of joy and celebrations. For South Asians, happiness is a cloudy day. As you read about the subregions of the subcontinent—(1) the towering Himalaya Mountains in the north, (2) the Indus Valley, (3) the Ganges Plains, (4) the dry Deccan Plateau, and (5) the tropical coasts—remember that, for most South Asians, farming is a gamble on the rains. Their well-being depends on the summer monsoon.

The Mountain Wall of the Himalayas

The Himalayas form a mountain wall up to four miles high along the northern borders of Pakistan and India. *Himalaya* means "house of snow." Although there are passes through the Himalayas in the northwest, these mountains are an impressive barrier.

Like most mountain regions, the Himalayas are thinly settled, except for the province of Jammu and Kashmir in the northwest. Here, some 7 to 8

million people live in the famous Vale of Kashmir and its well-known city of Srinagar. India now controls most of Jammu and Kashmir's broad valleys and fertile farmland, although the region is populated principally by Muslims and is still claimed by Pakistan. The area is an ongoing source of conflict between Pakistan and India.

Southeast of Jammu and Kashmir in the Himalayas lie the small mountain kingdoms of Nepal and Bhutan. These mountain hideaways, sandwiched between India and China, have been refuge areas for centuries. Nepal, with a population of 20.4 million, is the larger of the two—it is roughly the size of Georgia in the United States. The fertile Kathmandu Valley southwest of Mount Everest, the highest mountain in the world, is the population center of Nepal. Hinduism and Buddhism (another Indian religion) blend in this country, whose environments climb from fertile lowlands to the peaks of the Himalayas. Although a new constitution in 1990 provided for multi-party elections,

IT'S A FACT   In 1992, 32 people reached the top of Mount Everest, the world's tallest mountain, in a single day.

demonstrations for democracy and against the king of Nepal are still disturbing this small country in the 1990s.

The tiny highland kingdom of Bhutan has less than a million people. Although India conducts the country's foreign affairs, the Buddhist people of Bhutan are self-sufficient in food and have retained their culture through isolation. The fertile land in the country has attracted many immigrants from Nepal, and this has led to serious clashes between the two countries in the early 1990s.

The mountains of the eastern Himalayas are split by the Assam Valley of India. The Brahmaputra River flows through this wide corridor on its way to the Bay of Bengal. Rice paddies and fields of jute (from which rope is made) cover the valley floor. Tea plantations dot nearby hills. Recent immigration from other parts of India and Bangladesh has increased the population of Assam to nearly 20 million, and violent clashes have broken out between local people and the new immigrants. India has actually considered building an enormous fence to keep Muslims from Bangladesh out of Assam.

Pakistan, Gift of the Indus River

Water from melting snows pours down from the Himalayas into five rivers. These rivers flow into the Indus River, a 2,000-mile-long waterway that is the heartland of the Muslim state of Pakistan. The northern part of this area is known as the Punjab, which means "land of the five rivers." It is a fertile land where cotton, wheat, and rice are grown. Lahore, the Punjab's major city, has 4.6 million people; not far from Lahore is the newly built capital of Islamabad. In the south, the Indus River widens, and a second important rice-farming area extends east of Karachi, a huge port city of more than 8 million citizens.

These two major farming areas along the Indus River are vital food-producing areas in Pakistan; both face serious environmental problems today. In the Punjab, forests have been cut down. As a result, soil is no longer held in place by tree roots, and silt is blocking water canals and wells. In the south near Karachi, salt left behind from evaporating irrigation water is reducing crop yields. Both problems need to be solved quickly and effectively, because Pakistan—an agricultural country—is now faced with the need to import food.

The Ganges Plains of India and Bangladesh

The 200-mile-wide Ganges Plain is densely settled and very productive farmland. It stretches 1,500 miles from the dry Punjab in the west to the humid plains of Bangladesh in the east. Rainfall increases as one moves eastward down the Ganges. Crops change, farming villages increase in number, and cities are larger.

▶ *Mt. Chogolisa, Pakistan, is part of the Himalaya Mountain Range, which divides India and Pakistan from the rest of Asia. How might this natural barrier have affected the development of India and Pakistan?*

The Punjab, located in both Pakistan and India, is a huge plain made up of soil deposited by rivers flowing down from the Himalayas. It forms a **divide** (line or zone of higher ground) between the two great river systems of South Asia—the Indus of Pakistan to the west and the Ganges of India and Bangladesh to the east.

The Punjab is dry; only ten inches of rain fall each year. Crops are irrigated with water from wells or from dams on streams that are part of the Indus River system of Pakistan or the Ganges of India. Settlements are located wherever there is enough water.

Rainfall gradually increases as the Ganges Plain extends eastward to the cities of Delhi and Agra (site of the beautiful Taj Mahal). Wheat is the main food crop, and sugarcane is the most important commercial crop. Irrigation is used because of the possibility of a poor summer monsoon. In this area of dense population, a number of famous cities, such as Lucknow, Kanpur, and Varanasi (Banaras) line the banks of the Ganges.

Downstream on the Ganges, rainfall increases to seventy inches per year near the Bangladesh border. Because of higher rainfall, rice is grown here. Sugarcane is still an important commercial crop, and in the wetter regions, jute.

In the easternmost part of the Ganges Plain, the Ganges River joins the Brahmaputra River on an enormous flat plain before flowing into the Bay of Bengal. On this plain, in the Indian province of Bengal, with its huge city of Calcutta, and in Bangladesh, rainfall is heavy and often causes large-scale flooding. Most of the plain is farmed, and four-fifths is planted in rice.

The Deccan Plateau

The Deccan Plateau, an enormous triangle of forests, fields, scrub, and low mountain ranges, covers most of the Indian subcontinent. It is separated from the Ganges Plain by forested mountain ranges on the west and by a rugged plateau on the north and east.

Not surprisingly, the Deccan Plateau of India is sparsely populated as compared with the rest of the region. Farming depends almost entirely on the summer monsoon. Rainfall usually reaches a badly needed sixty inches per year in the northeast. It de-

IT'S A FACT A third of the world's poorest people live in India.

creases rapidly to twenty to thirty inches per year in the south and west, leaving farmers in a very risky position.

The center of the Deccan Plateau is heavily eroded and deforested. Some areas of this plateau have rich volcanic soils, however. And where water is available, cotton is grown commercially on modern, mechanized farms.

The Tropical Coasts of South Asia

The Malabar Coast on the west and the Coromandel Coast on the east are separated from the Deccan Plateau by two long escarpments called the Western Ghats and Eastern Ghats. The term *ghats* means "steps." The coastal areas are tropical with fertile soils and plenty of rain. They are carpeted with rich fields; in the foothills, rubber, tea, pepper, and coconuts are planted.

The climate becomes less tropical in the northern sections of these coasts, although rice remains the most important crop. On the west coast, cotton is grown near the city of Bombay, a major port with more than 12 million people. On the east coast, Madras, with a population of 5.9 million, is an important port. The eastern coastal plain widens into the densely populated delta of the Ganges and Brahmaputra Rivers in India and Bangladesh.

⊕ REVIEW QUESTIONS

1. How many people live in South Asia? Do most South Asians live in cities or villages?
2. What are the months of heaviest rainfall in South Asia, and what is the chief cause of this rainfall?
3. How wide and long is the Ganges Plain? How much does the amount of rainfall vary between the western and eastern parts of this plain?

⊕ THOUGHT QUESTIONS

1. What are some of the problems South Asia has in common with other developing areas?
2. Why is the summer monsoon much more important to South Asians than the winter monsoon?

2. Hinduism: Culture and Faith in South Asia

Early Cultures and Religions

Most of South Asia is part of a subcontinent. The subcontinent is bordered by the Arabian Sea to the west, the Bay of Bengal to the east, and the Indian Ocean to the south. The Himalayas, the world's highest mountain range, form a wall along the northern border of the subcontinent, as you have learned.

Over thousands of years, many different peoples entered the Indian subcontinent and settled on its plains and in its river valleys. Some came by land through high mountain passes in the northwest, and others came by sea. As a result, Indian civilization did not develop in isolation, but instead benefited from a rich heritage from many cultures. Indeed, the people of India made important discoveries in science, mathematics, architecture, and philosophy. Perhaps India's greatest contributions to the world's knowledge have been in the realm of religion.

Indo-European (Aryan) Invaders Bring Hinduism to South Asia

Hinduism began when Indo-European (Aryan) tribes invaded India from the northwest between 1500 and 1000 B.C. They settled down as farmers and herders on the plains near the headwaters of the Indus and Ganges Rivers. In the next 500 years, the Aryans organized kingdoms in the Ganges River Valley and gradually spread southward over the subcontinent. The Aryans conquered the original inhabitants of India, the Dravidians, and drove them southward. The priest class of the Aryans, the *Brahmans,* spread Hinduism throughout India.

IT'S A FACT   The Hindus were the first people to use the zero. They called it the *Sunya* (the void).

Hinduism as a Religion

Hinduism is a religion and a way of organizing society into classes, or castes. Today, Hinduism is practiced by more than 600 million people. Within Hinduism, some people worship a single god and others worship several gods. Most Hindus, however, accept the following five beliefs:

1. *A single unifying spirit, called* **Brahma***, is permanent in an everchanging world.* The soul of every creature is part of this spirit, and the thousands of gods and goddesses in Hinduism are the varied faces of Brahma.

2. *All life is sacred.* Many Hindus will not injure or kill an animal; many more eat only vegetables. The cow is especially sacred to Hindus. In India today there are thousands of hospitals and "rest homes" for cows.

3. *Reincarnation, or rebirth, is a basic principle of life.* Hindus believe that, when a person dies, he or she is reborn. They believe that birth, life, death, and rebirth are cycles that go on and on. People are released from these cycles only when they achieve union with the universal spirit, Brahma.

4. *The law of* karma *holds each person responsible for all of his or her deeds in all past lifetimes.* A good life is rewarded by rebirth into a better one; an evil life leads to rebirth as a lowly creature. The law of *karma* explains to Hindus why each person and group in Indian society is higher or lower in status than others.

5. *Each individual is born with a sacred duty in life.* This duty cannot be changed by becoming educated or gaining wealth and power. Self-improvement has had little meaning in Hindu society.

Caste Determines Everyone's Place

Hindu society is divided into five broad social groupings. *Brahmans* are the priests, teachers, and students who perform religious ceremonies in Hindu India. *Kshatriyas* provide the warriors, political leaders, and rulers of Indian society. *Vaisyas* are farmers, tradespeople, and merchants. *Sudras,* who form a majority in Hindu society, are the servant class and have a lower status. Beneath them are the *Harijans* or Untouchables.

These broad groups in Hindu society are broken into several thousand small subgroups called castes. Each caste forms an exclusive group; its members eat together, live together, marry one another, dress the same way, and usually work at the same occupation. Can you imagine such a highly divided society in the United States?

Until recently, the caste system was the basis of law in India. Now there are laws against discrimination toward Untouchables, who form about one-sixth of India's total population. In the 1990s, these laws, which guaranteed Untouchables a substantial number of government jobs, triggered caste warfare in India. After 2,000 years, traditions are not easily changed in South Asia.

⊕ REVIEW QUESTIONS

1. What do we call the Indo-European people who brought Hinduism to India?
2. What oceans and mountains border the subcontinent of South Asia?
3. What are the five main beliefs of Hinduism?

⊕ THOUGHT QUESTIONS

1. Why is India's caste system undemocratic?
2. Hindus who follow the teachings strictly do not eat any animal products. How would this both help and hurt the country?

▲ *This Indian beggar girl is a member of one of the Hindu Untouchable castes. Here, she is plying her trade near the entry to the Ranthambhore National Park, an area in Rajasthan State, set aside to protect tigers in 1989. The park attracts tourists and a nearby lake draws pilgrims, hence the beggar's location. What is the caste system? Who are the Untouchables?*

◄ *A religious celebration honoring one of the thousands of Hindu gods and goddesses passes down a Calcutta street. The five men process down the street carrying a pot, a small tree, and two fringed umbrellas. Is Hinduism a polytheism or a monotheism? What is Christianity? Judaism?*

PLACE

THE TAJ MAHAL: SYMBOL OF INDIA, MONUMENT TO LOVE

Most great buildings are monuments to religion, to power, or to wealth. The Taj Mahal at Agra, in northern India, is the most beautiful structure ever dedicated to love. It has a special story.

▲ *The Taj Mahal, at Agra in northern India, is the national symbol of India. It is an interesting mixture of Hindu and Muslim styles of architecture. Why was it built?*

3. The Muslims Invade South Asia

Muslim Armies Reach India

Arabs, buoyed by the new faith taught by their Prophet Muhammad, came out of the Arabian Peninsula and conquered many lands, as you learned in Chapter 22. These Arab Muslims reached the western borders of India in the early 700s. They set up a Muslim kingdom in the Indus Valley in what is now Pakistan and converted many people to Islam.

In the 1000s, a second invasion of India was launched by Muslims based in Afghanistan. Afghan armies poured through the northwestern passes onto the Ganges Plain and pillaged cities, looted temples for treasure, and took prisoners by the thousands. Their goal was plunder, but as a result of these raids, a Muslim province was established south of the Himalayas on the upper reaches of the Ganges River.

Mogul Rule Lasts for Three Hundred Years

The third great Muslim invasion of India occurred in the 1500s when the Moguls, descendants of the Mongols, established a center of power in north India. Soon thereafter, under the brilliant leadership of Akbar, Mogul rule was extended southward over most of the peninsula. Akbar was the genius who first brought together Hinduism and Islam and created a civilization based on the

In the early 1600s, Emperor Shah Jahan, one of the Mogul rulers of India, married a twenty-one-year-old woman who adopted the name Mumtaz Mahal, meaning "Exalted of the Palace," and became known for her generosity and understanding.

During their nineteen years of marriage, the emperor and his wife were constant companions. When Mumtaz Mahal died suddenly in 1629 while delivering their fourteenth child, the emperor was grief stricken. His hair turned gray in a few months. He vowed to build a monument to their love, a building more exquisite than anything the world had ever seen. That monument is the Taj Mahal.

He recruited 20,000 of the world's finest artisans and crafts-people from as far away as Greece and Italy. For twenty-two years these workers labored to capture Shah Jahan's love in white marble and precious stones.

Today the Taj Mahal is India's national symbol. As a brilliant mixture of Hindu and Muslim architectural styles, it joins together the two great civilizations that shape life in South Asia. But air pollution is now threatening to destroy this exquisite seventeenth-century shrine. Its glistening white marble is cracking and turning yellow and gray because of the pollutants in Agra's air. In 1982 the government ordered the two hundred or so iron foundries in the city to close until they installed pollution control equipment; in 1993 this process began.

QUESTIONS

1. When and why was the Taj Mahal built?
2. What effect is pollution having on this building? Can you think of another place in the United States or Europe where buildings are threatened by pollution?
3. The Taj Mahal was built during the Mogul Period when Muslims and Hindus cooperated. Now conflicts between Muslims and Hindus are common. What do these facts suggest to you in terms of creating a productive society?

blending of Hindu and Muslim cultures. The Mogul Empire (1526–1857) was one of the most glorious empires in the history of the world.

Mogul rulers supported new and graceful Indian-Islamic styles in art and music; their splendid architecture still stands in India's cities. The Moguls made detailed surveys of farmland in the countryside and developed intelligent, fair systems of land ownership and land taxation that still influence farming in Pakistan, India, and Bangladesh. Under Mogul rule, the population of South Asia in 1650 increased to 150 million, one of every four people on Earth at that time.

In the late 1600s, however, Mogul rule began to decline. Warfare broke out between Muslims and Hindus, and the costs of these conflicts drained the empire's treasury. Millions were killed or died of starvation and disease. Gradually, the empire was shattered by war. The pieces were picked up by strangers from Europe, British traders who seized the chance to make India theirs.

REVIEW QUESTIONS

1. Two groups of Muslims invaded South Asia, one in the 700s and the other in the 1000s. Where did each group come from, and where did each group settle?
2. What two cultures were blended together under Mogul rule?
3. Why did Mogul rule begin to decline in the late 1600s?

THOUGHT QUESTION

1. What evidence indicates that the Moguls had an advanced civilization?

4. The British in South Asia

The British Come to India as Traders

The great Mogul emperor Akbar allowed the British to enter India as traders in the 1600s when the country was rich and self-sufficient. The wealth in its national treasury was eighty-five times larger than Britain's. Indian goods were of better quality than those made in Europe. By comparison, Britain was an underdeveloped country and appeared to be no threat to India.

During its first fifty years, the British East India Company set up small trading posts at Bombay on the west coast of India, Madras in the south, and Calcutta in the delta of the Ganges River to the east. Locate these cities on the map on page 611. From these small footholds on the western, southern, and eastern edges of the subcontinent, the British ultimately gained control of India when Mogul rule crumbled.

The British East India Company Takes Over India

When fighting between Hindus and Muslims began to threaten law and order, the British East India Company hired troops to guard local areas that produced the cotton, tea, and other products that the company exported to Britain. While various armies fought battles with each other across the subcontinent, the British traders tried to protect their crops.

Gradually, the British East India Company took over Indian territory. It started in 1757, when a local Mogul ruler seized Calcutta, killed a number of the British, and locked others into a small, airless room called the "black hole of Calcutta." The British East India Company reacted in anger and drove the Mogul ruler out of the rich province of Bengal in the Ganges delta. In the years that followed, the British took over more and more Indian territory, and by 1857, three-fifths of India was under the direct control of the British East India Company.

India's Farms and Cities Change Under Company Rule

The British East India Company established new economic policies that caused great hardship for India's farmers and city dwellers. The company raised taxes on farmers, and when farmers could not pay, tax collectors were allowed to take land away from the farmers. This is how moneylenders in India's cities became owners of 40 percent of India's agricultural land.

The British East India Company also required that taxes be paid in cash rather than in crops. This forced India's farmers to grow crops that they could sell for cash. The best farmland on the subcontinent, therefore, was no longer producing food crops like wheat, millet, and barley. Land was planted in cotton for Britain's textile mills, in indigo for blue dye, in opium for sale to China, and in jute for rope. In the hills, tea and coffee plantations were established because these products were in great demand in Europe. South Asia quickly became an agricultural colony run by the British East India Company.

The populations of South Asia's cities also dramatically changed under the British East India Company's control. Tiny coastal towns like Calcutta, Madras, Bombay, and Karachi grew into large port cities, because trade with Europe was now big

▼ *Regiments of bagpipers march in the Republic Day Parade. What clues does the military dress provide about the history of this country?*

| IT'S A FACT | Queen Victoria of England assumed the title of Empress of India in 1877. |

business on the subcontinent. Older inland cities declined in size, influence, population, and wealth.

India as a British Colony

A century of British East India Company rule in India led to conflict in 1857. Native troops (Sepoys) fought British soldiers to a standstill for a year. The British called it the Sepoy Rebellion: Indians call this war the "First War of Independence." Do you understand why? After the rebellion failed, Britain decided to make India a colony of the British Empire directly run by the British government.

New and higher taxes were levied on India's farmers to pay for the 60,000 British troops now permanently housed in India and also for the railroad system Britain built in India to move its troops from place to place. Farmers were driven deeper into debt.

Meanwhile, India's population began to grow because of the introduction of Western medicine and sanitation. In the 1870s there were 250 million Indians. By the 1920s, India had 320 million people, a 28 percent increase. In the villages, this meant that less land was available for farm families. Nearly one-fifth of India's farmers owned no land at all. Almost one-half supported their families on an acre of land (an area smaller in size than a football field). As a result, hunger became widespread in India.

Nor were there enough jobs in the cities for those who lost their land. The British established some industrial plants but discouraged the full-scale development of industry in India's cities just as they had in North America centuries earlier. South Asia's primary role in the British plan was to produce commercial agricultural crops for export to Britain, so there were few factories in the cities to provide jobs for India's growing population. Having insufficient industry for the population made India poorly prepared to become a growing economy when the goal of independence was achieved in 1947.

REVIEW QUESTIONS

1. How did the Indian cities of Bombay, Madras, and Calcutta start?
2. By 1857, how much of India was under the direct control of the British East India Company?
3. What happened to the best farmland on the Indian subcontinent after the British East India Company gained control of it?

THOUGHT QUESTIONS

1. The British brought to India such improvements as new roads, railroads, Western medicine, and better sanitation. Why, then, did the Indians suffer great hardships when India became a colony of the British Empire?
2. Why did the British discourage the development of industry in India, and how did this affect the Indian people?

CHAPTER SUMMARY

The subregions of South Asia are the Punjab and the Indus River Valley in the northwest, the Himalaya Mountains in the north, the Ganges Plain and the Ganges River Valley in northern India, the Deccan Plateau in the south, and the tropical east and west coasts.

South Asia is a subcontinent that is home to six countries and many cultures. Like Southwest Asia, South Asia's location has made it a crossroads. Many different peoples arrived by land and by sea and settled the plains and river valleys of the subcontinent. The Indo-European Aryans were followed by Muslim merchants and then Europeans. South Asia's heritage has made it a region of great creativity and tension. The people of South Asia have given birth to two world religions, Hinduism and Buddhism. They have also produced a rich legacy of philosophy, science, literature, and architecture.

Among other Europeans, the British came to India to trade but ultimately took control of the country. By taxing the farmers and encouraging the production of export crops instead of food production, the British made India an agricultural colony. When Indians finally won their struggle for independence in 1947, they were poorly prepared to become an independent country.

EXERCISES

Two Correct Sentences

Directions: Two sentences in each group of three sentences are correct statements. Write the numbers of the correct statements on your paper.

Group A

1. The Himalayas form a mountain wall along the northern borders of Pakistan and India.
2. Because of heavy rainfall, the Himalayas are farming areas.
3. There are passes through the Himalayas in the northwest.

Group B

4. The mountains of the eastern Himalayas are split by the Assam Valley.
5. The population of the Assam Valley has grown rapidly in recent years.
6. Fields of corn and potatoes cover the floor of the Assam Valley.

Group C

7. The Punjab means the "land of the five rivers."
8. Bombay is the major city in the Punjab.
9. Cotton, wheat, and rice are grown in the Punjab.

Supply the Missing Word

Directions: Number your paper from 1 to 7. Find the missing word needed to complete each of the following sentences. Then write the complete sentences on your paper.

Bombay	Kashmir
Indus	Islamabad
Lahore	Madras
Karachi	

1. About 7 to 8 million people live in the Vale of ||||||||||.
2. The chief city in the western Punjab is ||||||||||.
3. The chief port in northwest India is ||||||||||.
4. The chief port in southeast India is ||||||||||.
5. The economic base of Pakistan is irrigated agriculture on the fertile plain of the ||||||||| River.
6. The capital of Pakistan is ||||||||||.
7. Pakistan's chief port city is ||||||||||.

Matching

Directions: Number your paper from 1 to 7. Then match the numbered terms in Part A with the letters of the descriptions in Part B.

Part A

1. British
2. Mogul Empire
3. Taj Mahal
4. Brahmans
5. Dravidians
6. Sudras
7. First War of Independence

Part B

a. the large servant caste of India
b. ruled the Indian subcontinent from the 1500s through the mid 1700s
c. helped Britain decide to take over India from the British East India Company and make India a colony
d. people conquered by the Aryans
e. spread Hinduism throughout India
f. ruled the Indian subcontinent from the mid-1700s to the middle 1900s
g. a brilliant mixture of Hindu and Muslim architectural styles

Which Direction?

Directions: Each of the following sentences has a correct word or phrase and an incorrect word or phrase in parentheses. Write on your paper the correct word or phrase in each sentence.

1. Pakistan is (east, west) of India.
2. Bangladesh is (east, west) of India.
3. Sri Lanka is (south, north) of India.
4. The Arabian Sea lies to the (east, west) of India.
5. The Himalayas form a wall along the (southern, northern) border of the Indian subcontinent.

Which Came First?

Directions: In each of the following pairs of sentences, two events are mentioned. On your paper write the letter of the event that occurred first.

1. (a) Most of India was controlled by the British East India Company. (b) India became a colony of Great Britain.
2. (a) The Indian subcontinent was invaded by Muslims from Afghanistan. (b) The Indian subcontinent was invaded by the Indo-Europeans (Aryans).
3. (a) The Taj Mahal was built. (b) The First War of Independence started.
4. (a) Hinduism came to India. (b) Islam came to India.

Inquiry

Directions: Combine what you have learned in this chapter with your own ideas to answer these questions.

1. Many Hindus will not kill animals. How does killing animals relate to *karma* and reincarnation?
2. How might Hindus view self-improvement in this life compared to the way it is viewed in American society?
3. How did location and movement in the forms of migration and trade influence the Indian subcontinent from earliest times to the 1800s?
4. Look at the map of monsoon wind patterns in your book on page 616. How might Muslim traders have used them to transport their goods between, for instance, Damascus and Bombay?

SKILLS

Making Generalizations

Directions: A generalization is a general statement or summary based on various facts. It is a logical, correct conclusion drawn from the facts. Look at the following facts. Then look at the generalizations that might be based on those facts.

Number your paper from 1 to 3 and next to each number write the letter or letters of the generalizations that can be supported by the facts listed above it. There may be *more than one* correct generalization for each group of facts.

GROUP 1

Facts
- Important discoveries in science have come from India.
- India has contributed major ideas in the realm of religion.
- Indian architects have designed beautiful buildings.
- Indian mathematicians have made significant discoveries.

Possible Generalizations
a. Indians have made important contributions in the fields of science, mathematics, architecture, and religion.
b. India has led all other countries in developing new information.
c. Indians have helped to advance the world's knowledge.

GROUP 2

Facts
- Members within each caste in Hindu society live and work together.
- In the caste system there is fear of contamination by contact with a *Harijan* (or Untouchable).
- The Brahmans belong to the highest caste.
- Even though there are now laws against discrimination toward *Harijans* or Untouchables, these people still are generally avoided.

Possible Generalizations
a. *Brahmans* will eat with Untouchables.
b. Untouchables are never seen by people of other castes.
c. In spite of the nondiscrimination laws in modern India, there is still much prejudice against the Untouchables.

CHAPTER 25

GROUP 3

Facts

- After India officially became a British colony in 1858, new and higher taxes were levied on India's farmers.
- The British opened up new areas of India for the cultivation of commercial crops such as cotton, jute, and tea.
- Under British rule, proportionally less land was planted in food crops, so there was more hunger among the Indian people.
- The population of India grew larger, partly because the British introduced Western medicine and sanitation.

Possible Generalizations

a. The British wanted India to produce mainly crops that could be sold, rather than crops that would feed the Indian people.

b. Indian farmers suffered from higher taxes and less food after their land became a British colony.

c. Britain treated India much more harshly than it did its other colonies.

Vocabulary Skills

Matching

Directions: Match the numbered definitions with the vocabulary terms. Write the correct term after the number of its matching definition.

Brahma	monsoon
Brahmans	reincarnation
caste	Sepoys
divide	subcontinent
ghats	*Sudras*
Harijans (Untouchables)	summer monsoon
karma	*Vaisyas*
Kshatriyas	winter monsoon

1. A topographical feature that separates two river systems
2. The lowest social group in Hindu society
3. Hindu law that holds each person responsible for all of his or her deeds in past lifetimes
4. A single unifying spirit that is permanent in an ever-changing world
5. A way of organizing Indian society in which each social group eats together, shares the same spiritual duties, and often practices the same occupation
6. Any air current that blows steadily from one direction for weeks or months at a time
7. The servant class in India
8. A large landmass that is connected to and protrudes from a continent
9. A current of moist, warm oceanic air that blows into South and Southeast Asia between June and September
10. A current of dry continental air that blows out from the Indian subcontinent between September and June.
11. The warrior and landowner caste in Hindu society
12. The Hindu belief that when a person dies, he or she is reborn
13. A long cliff or escarpment that divides India's coastal plains from the Deccan Plateau
14. The farmer and merchant class in Hindu society
15. Native troops of India
16. The social class in Hindu society that includes priests and teachers

DEVELOPMENT, DISUNITY, AND CHANGE IN SOUTH ASIA

South Asia, a region about half the size of the United States, has four times as many people. More than 1 billion human beings live in India, Pakistan, Bangladesh, Sri Lanka, and the mountain states of Nepal and Bhutan. Most of the people in these countries are villagers who are closely tied to the land.

At independence, India faced three major problems: (1) forging a nation out of diverse cultures, (2) easing village poverty, and (3) creating jobs in industry. These have proved to be formidable tasks for the world's largest democracy.

◀ *In some villages and cities, aspects of trade and travel have not changed much over the years. What old methods of transportation do you see in this crowded outdoor market in Old Delhi, India?*

Conflict between Muslims and Hindus is intense and exploded in the early 1990s. Antagonism between Hindu castes is also widespread. Village populations have increased so that many farmers have been forced to migrate to South Asia's growing cities. But not enough jobs in industry have been created. In the end, the future of the region will depend upon the solution to these three problems. At this time, there are few hopeful signs.

1. Independence and the Struggle for Unity in South Asia

A Divided India Gains Independence

Indians had protested British rule through riots, strikes, and marches for many years, but unrest against British oppression increased after World War I. The most beloved of India's nationalist leaders was Mohandas K. Gandhi, whose millions of followers called him **Mahatma**, the "Great Soul." Gandhi worked to free India from British authority, but he rejected violent tactics. He also dreamed of freeing India from poverty, illiteracy, and injustice.

Britain finally was forced to withdraw from South Asia and granted independence to India in 1947. Muslim leaders, however, demanded that two areas of India become a separate Muslim country to be called Pakistan. Rioting broke out between Hindus and Muslims, and many people were killed. Despite Gandhi's plea for unity, two independent countries—Hindu India and Muslim Pakistan—were carved out of the subcontinent.

India, Pakistan, and Bangladesh

The new country of Pakistan was made up of two regions located very far away from one another—West Pakistan in the northwest corner of the subcontinent and East Pakistan in the delta of the Ganges. Muslims formed a majority in both regions.

After this division, millions of Hindus and Muslims found themselves on the wrong sides of the new political borders, so Hindus migrated to India, and Muslims left India to live in East and West Pakistan. Tens of thousands died in the chaos. Since 1947, when these events occurred, three wars have been fought between Pakistan and India. In 1972, East Pakistan became a Muslim country called Bangladesh, separate from West Pakistan which was renamed Pakistan.

Unity and Diversity in India

India has a great variety of cultures. Fourteen major languages and 1,652 separate dialects are spoken in India. No common language exists in the country. Fostering a sense of national unity, therefore, is a problem. The Indian government has had to face one divisive crisis after another. For example,

▼ *As Mahatma Ghandi struggled for a peaceful separation of India from Great Britain, he emphasized the use of certain strategies against the British. What were two of these strategies?*

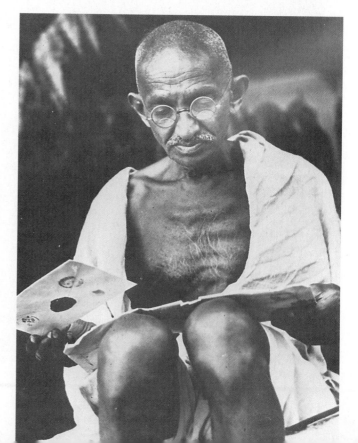

SATELLITES MAY SOLVE SOME PROBLEMS IN INDIA

It may seem strange that a poor developing country like India is investing $500 million in a sophisticated space satellite program. However, because India is a huge country with many different language groups, space satellites may be the most effective way of solving communication problems. Commercial satellites also appear to be the cheapest and most efficient means of providing information to India's estimated 700,000 villages.

In the 1980s, the United States and India cooperated in launching the world's first triple-function commercial satellite into stationary orbit over India. This is the only Indian commercial satellite capable of handling television broadcasting, telecommunications, and weather forecasting.

In early tests, community radio and television sets were installed in 2,400 Indian villages. More than 1,300 hours of entertainment and information were broadcast in four languages to

▲ *In Delhi, young teachers volunteer to teach the children of quarry workers who are Untouchables to read and write. Classes are held at night to avoid the summer heat. Do you think this kind of interaction would have occurred in the early 1900s? In what ways do you think a telecommunication satellite might affect education in India?*

an estimated 200,000 viewers throughout India.

Now in operation, this satellite may prove to be a great success. Its weather forecasting could provide early warning of storms headed for India's east coast. The ability to report the advance of the rain-bearing summer monsoon with greater accuracy would benefit many farmers. In addition, commercial satellites are a practical way to bring education and entertainment to India's widely

scattered villages. Newly imported satellite dishes could open India to the world and may also help to unify the nation.

QUESTIONS

1. India has about 700,000 villages. Why would satellite communications be more practical than visits by experts in conveying information to villagers?

2. India is a country where very traditional technology and very modern technology are found in different regions and sectors of the economy. Can you think of other developing countries where this is true?

3. If you ran India's commercial satellite, what kinds of programs would you consider to be most important? List several topics or programs you would put on the air.

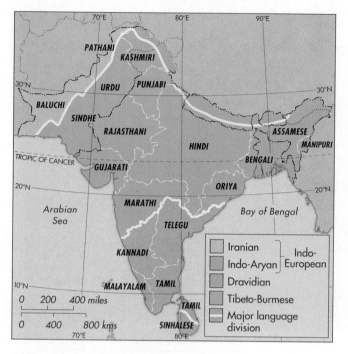

Major South Asian Languages

IT'S A FACT Chess, dice, and polo were first played in India.

Gandhi in 1984, continues today. Similar demands for separate states have come from tribal groups in the states of Assam in the east, in Bihar and Orissa west of Calcutta, and in other parts of the Deccan Plateau.

These cultural differences in language, religion, and race prevent the central government from creating a strong and unified India. In the 1980s, religious and linguistic disputes led to frequent riots. In the 1990s, clashes between Muslims and Hindus are common, and caste disputes break out sporadically. The force of these outbursts of anger from different groups challenges the authority and abilities of the central government to create a unified country.

For fifty years, India had been held together by **secularism**, a practice of not connecting the government to one religion, and **democracy**, a government system in which leaders are directly elected by the people and majority vote prevails. Now Hindu extremists and Muslim fundamentalists battle in the streets challenging both ideas. Moreover, the government decision to reserve a large number of all government jobs for Untouchables has brought the country to the brink of caste warfare. These ancient hatreds run deep, but they must be contained if Indians are to remain one people. The struggle for national unity is clearly a major task facing India in the coming decade.

different language, religious, and tribal groups have demanded separate states.

During their long struggle against British rule, Indians used English as their means of communicating with one another because each region of their country had a different language. But an independent India needed a national language of its own. This created a major challenge.

The Indian government chose the northern Indian language of Hindi as the country's "official language." This choice was quite unpopular. Dravidian-speaking people in southern India were very unhappy, as were speakers of other languages on the Deccan Plateau and in northern India. In the south, competition between language groups led to the splitting of Madras into two separate states based on language. Can you imagine this happening in some of our larger states, like Texas, California, or New York?

In the north, the Sikh people (another clearly defined Indian minority group) have pressed the central government for their own state in the Punjab for the last thirty years. This conflict, which led directly to the assassination of Prime Minister Indira

⊕ **REVIEW QUESTIONS**

1. When did India become independent?
2. What geographical problem did Pakistan face when it first gained independence from India?
3. What did the Indian government choose as the official language of India? Why was this choice unpopular?

⊕ **THOUGHT QUESTIONS**

1. How do regional religions and ethnic loyalties threaten the political unity of India?
2. What alternative to selecting Hindi as the national language might the Indian government have selected?

2. Feeding India's Growing Population

Food and People in India

Today, India has an estimated 897.4 million people. Each month India's population grows by more than 1 million people, and each year it increases by a population nearly equal to that of Australia. Population geographers expect that India will have more than 1 billion people by the year 2000. India may very well surpass China as the world's most populous nation within twenty-five years.

To feed this growing population, India must produce 1.5 million more tons of food every year. This poses tremendous problems. There has been some progress in agriculture, as you will learn. However, the needs of more than a million new children each month drain the Indian economy. Needs for new hospitals, houses, nurses, doctors, factories, jobs—the essential elements of living—must be met before funds can be invested in the development of natural resources. This is why population growth has hindered India from raising its standard of living.

The problem reached crisis proportions in India's villages twenty years after independence. A series of poor summer monsoons combined with more people living off crops from the same amount of farmland caused food shortages. Rationing was put into effect. The United States sent millions of tons of free food to ease the crisis.

Indian planners had to face the unpleasant fact that little progress had been made in agriculture. Rice yields were the same as they had been in the 1920s, when India's population was half as large. Village farmers were deeply in debt to moneylenders and landlords. Clearly a new strategy to feed India's people was needed.

IT'S A FACT Five times in the last fifteen years the summer monsoon has failed to produce enough rain.

The Green Revolution

The strategy that the Indian government adopted to increase total food production was called the **green revolution**. The plan involved three parts: (1) new "miracle" seeds and modern farming techniques were introduced, (2) these improvements were concentrated in irrigated areas where the water supply was reliable, and (3) farmers were encouraged to take part in the program by government guarantees of high prices for their crops. The subregion in India best suited to the green revolution was the Punjab in northern India. Locate this area on the map on page 617.

New high-yielding strains of wheat and rice were used. These "miracle" seeds do very well when chemical fertilizers are used and enough water is available. By the early 1980s, the miracle seeds were producing one-third of India's total grain crop.

The core area of this agricultural revolution, as noted above, was in the Punjab, where 80 percent of the land was planted in miracle wheats. Here, wheat yields doubled overall. In some newly irrigated areas, they increased ten times. Rice yields didn't increase as much along the tropical Malabar and Coromandel coasts and in southern India.

▼ *Simple tools and machines are used to harvest and winnow wheat in rural Rajasthan State in western India. How do you think these farming methods affect productivity? What is productivity?*

**PEOPLE-
ENVIRONMENT**

THE GREEN REVOLUTION INCREASES INDIA'S FOOD SUPPLY

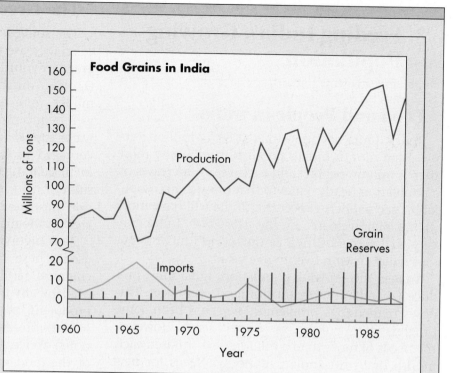

This graph illustrates several interesting and vital facts about the quality of life in India in the last twenty-five years. To see this, you must identify what the two jagged lines and the vertical bars represent. The top line (in blue) represents the amount of food grains produced in India each year. The bottom line (in orange) represents the amount of grain India imported each year. The vertical bars indicate how much grain India had in reserve each year. What can you discover about Indian agriculture from this graph?

First, grain production in the 1960s was so low that India had to import food. Notice how the bottom line representing imports went up and the top line representing food production went down. The graph tells you that 1966 was a very bad year in India's villages.

Second, the green revolution steadily increased total grain production during the 1970s and

1980s. Note that the top line gradually rises to reach about 150 million tons in the 1980s. Imports (the bottom line) drop to zero; with large harvests they are no longer needed. Grain reserves (the vertical bars), which were small in the 1960s, increase, indicating that India has a surplus.

Finally, note that production of food grains in India (the top line) is jagged and that production varies greatly from year to year. This is because the summer monsoon still affects Indian food production. Notice the effects of the bad monsoons of 1981 and 1987 on food production and reserves.

Can you see how useful graphs are in summarizing information? Can you see how learning to read graphs will give you insight into what is happening?

QUESTIONS

1. Why is India's agriculture so dependent on the monsoon rains? Has irrigation changed that? Are there agricultural areas in the United States where irrigation is essential? Name one or two.

2. What effect has the green revolution had on India's food production? What requirements had to be met before this "revolution" could be introduced? Where was it most successful in India?

3. Given India's rapidly growing population, do you think the green revolution is a permanent solution to India's food situation? If not, what would you do either about population growth or about agriculture?

The green revolution—a breakthrough in food production—prevented widespread famine in India, but only a small number of India's farmers have benefited. Today, wheat and rice crops cannot yet be grown on the three-quarters of India's farmland that is still not irrigated.

Indian Agriculture Today

India's food production is growing slightly faster than its population. This means that in the late 1980s and 1990s, famine no longer threatens the Indian countryside. However, conditions in rural India are still quite difficult. In India's villages, 120 million farmers own no land. An additional 240 million work on tiny patches of land. These poverty-stricken villagers make up half of the population of India, where three of every four people earn their livings directly from the land. Yet farm products make up only 40 percent of the national wealth, and wheat, rice, cotton, and sugar yields are very low compared with yields in other countries.

The future of Indian agriculture is hard to predict. Standards of living are extremely low in India's villages. The amount of government investment in farming has never been adequate. At present, a few Indians have highly productive farms, but many villagers are very poor. Most farmland is still owned by large landowners, and worked by these villagers. Despite impressive gains in food production, India's farming is still among the most inefficient in the world. For this reason, hunger is common in India.

⊕ REVIEW QUESTIONS

1. By how much does India's population grow each month? What is it expected to be by the year 2000?
2. What are the three major elements of the green revolution? To which area of India was the green revolution best suited?
3. How many Indian farmers own no land? How many own tiny patches of land?

⊕ THOUGHT QUESTIONS

1. Why were the new strains of wheat introduced in the green revolution called "miracle" wheats?
2. Why hasn't the green revolution solved India's food shortage problem?

3. Cities and Industry in the Republic of India

City Populations Are Growing Quickly

In the 1990s, one of every four Indians lives in a city. This may seem like a small number, but remember that the total population of India is nearly 900 million. This means that 225 million people live in India's cities, almost as many as the total population of the United States, which has 258 million people. The crisis in Indian cities is becoming even more serious: because Indian villagers are moving to cities, the rate of increase of urbanites is twice the rate of India's total population growth.

India's largest cities are also the fastest growing. Calcutta and Bombay, once small British trading ports, are now giant cities with populations of more than 10 million each. New Delhi, India's capital, now has 8.9 million people. India's three largest cities continue to increase in size despite government efforts to limit their growth.

▼ *Ticketless travelers "hitch" rides on the top of a train in India. Why do you think so many people ride the trains?*

CLOSE-UP

CALCUTTA: CITY OF THE DREADFUL NIGHT

🌐

▲ *Calcutta, India is one of the poorest and most overcrowded cities in the world. What are some of the problems people in Calcutta face daily?*

Nowhere are India's urban problems more pressing than in its largest city, Calcutta, where 11.9 million people are crowded into 490 square miles. Located on a wet, flat delta plain on a **tributary** of the Ganges River some 80 miles from the coast, Calcutta is India's largest seaport and most productive manufacturing center. (A tributary is a small river that feeds into a larger river.) Calcutta's cosmopolitan population is drawn from virtually every region, religion, and language group on the Indian subcontinent.

In the central business district along the banks of the Hooghly River, an estimated half-million pedestrians, a number of cows, and 40,000 vehicles cross the Howrah Bridge each day. Traffic

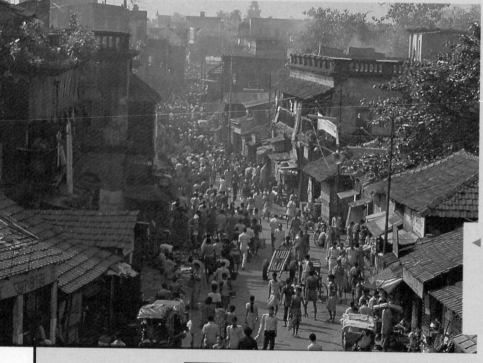

◀ *Pedestrian shoppers fill this narrow street in Calcutta, India's most crowded city. This is a city suburb, not a slum. What name is given to shantytowns or slums in Calcutta?*

636

jams choke movement; power blackouts are frequent. Residential apartment complexes and slums surround the central city, and higher-class suburbs lie to the south. Three-quarters of the population live in overcrowded tenement districts. Many live in houses made of unbaked brick.

More than half the families in Calcutta live in a single room, with an average of only 30 square feet of living space for every family member. A quarter of Calcutta's population lives in slums, locally known as **bustees**, in houses made of cardboard cartons, scraps of wood, and tin cans. In these slums, each water tap serves twenty-five to thirty people, and each latrine serves some twenty people. Garbage is not collected.

An estimated half-million Calcuttans live and sleep in railroad stations, on temple grounds, and on sidewalks. Another 100,000 beggars scavenge food from garbage heaps. In the *bustee* quarters of Calcutta, cholera and other waterborne diseases are common. The city has between 35,000 and 40,000 lepers in eight colonies around the city. Epidemics are frequent and disease is commonplace. Everywhere a formidable array of beggars, day laborers, and pieceworkers scour the streets. In the words of one prominent writer, "Calcutta is like a stone astride my chest."

An underlying problem in Calcutta is that industrialization is not proceeding rapidly enough to employ its growing population. Further, Calcutta's housing supply,

▲ *Hinduism, the religion practiced by the majority of people of India, grew from the beliefs of Aryan tribes who dominated India from 1500 to 100 B.C. What element in this picture reflects one of the major tenets of Hinduism?*

waterworks, electrical system, and other facilities are not sufficient to cope with the city's rapid growth.

Before 1980, no new water supply system had been built in a century. The last major road was paved in 1930; the last major sewer line was built in 1896. One Indian poet describes the human despair of Calcutta: "Blind herds in the bazaar; Tin-roofed hovels, wormy lanes; Meshes of bones, pits of eyes; People rotting, unable to die."

QUESTIONS

1. Why would anyone build a large city on a swamp? Why did this happen in Calcutta? Can you think of other large cities built in very difficult environments?

2. Can you imagine what it would be like growing up in one of Calcutta's *bustees*? Try, by listing on your paper what you would do on one typical day from the time you got up until the time you went to bed.

3. What is the connection between industrialization and population growth? How do people survive without a job in India which has no social security system?

In 1947, the average life expectancy in India was 32 years. Today it is 59.

Urban Problems Are Intense

India's difficulty in providing jobs for its growing city populations is creating an urban crisis in one of the world's most rural countries. When people don't have jobs, every aspect of life becomes uncertain. Many people live in shantytowns called bustees. Crowded conditions create severe problems for public health and transportation.

Urban leaders are trying to deal with these problems. New houses are being built at a rate of 200,000 a year in India's major cities. However, to meet the needs of growing city populations, houses are needed at a rate eight times higher than this. As a result, immigrants fleeing the poverty of India's villages pour into city slums without houses. Conditions of life are desperate.

Unless Indian industry can create jobs for this rising tide of village refugees, the human suffering caused by unemployment, appalling living conditions, and grinding poverty will increase.

India Has Ample Resources for Industry

India does have the energy resources needed to build a strong industrial nation. Coal is India's main source of industrial energy. Deposits are found in many areas of the peninsula, particularly in the hill country west of Calcutta. Petroleum from the Assam Valley in the far northeast and from Gujarat, north of Bombay, provides one-third of India's petroleum needs. Major offshore discoveries in the Arabian Sea near Bombay promise future large-scale production.

India has equally abundant mineral resources for industry. The country has large reserves of high-grade iron ore. The major iron-producing area is west of Calcutta, and new iron ore fields in a number of areas are being developed to produce iron ore for export. India also has important deposits of alloy metals used in making steel. Locate these resources on the map on page 639.

Industrial Regions Are Located in Cities or Near Resources

India's important industrial areas are located either near these energy and mineral resources or in the country's largest cities. India's largest center of heavy industry is located near the steel city of Jamshedpur. This industrial region is 150 miles west of Calcutta in an area with large coal and iron deposits. Steel production in this industrial region, however, has been disappointing. Power failures, labor strikes, and high transportation costs have forced India to import large amounts of steel in spite of its great potential as a steel producer.

Other Indian industrial regions are found in large cities. Calcutta produces chemicals, processed farm products, and cotton textiles. In the south, Bangalore, which grew up as a cotton textile center inland from the British port of Madras, has many factories. Bangalore now has cement plants, light industry, machine tool plants, and food-processing plants.

The manufacturing region near Bombay in the west has had a similar pattern of development; it began as a cotton textile center and has diversified into other industries, particularly petrochemicals based on offshore oil and gas fields.

Industrial Growth Is Slow

India has the resources to become an industrial nation. The pace of industrialization increased in the 1980s, but in the 1990s only 15 percent of all Indians are employed in manufacturing. Not enough industrial jobs have yet been created.

Better coordination of industries and transportation than India's leaders have provided is essential to India's future. Production suffers from frequent strikes, power shortages, aging manufacturing plants, and deteriorating railroads. Also, most industries are state owned and hopelessly inefficient.

These conditions mean that industry is failing to keep pace with the needs of India's growing population. But India's leaders have now launched a program of industrial expansion, and in 1992 they began to shift from a government-controlled to a market-oriented economy in order to spur economic growth.

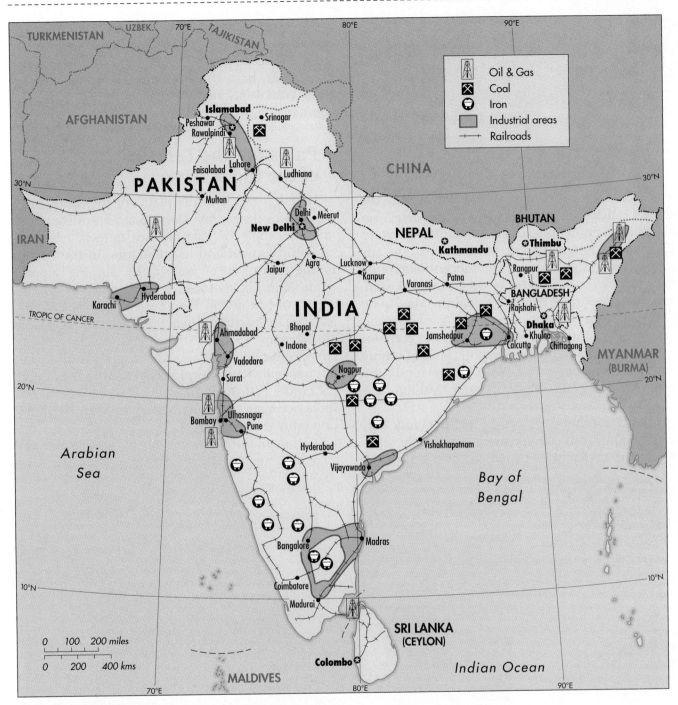

Industrial Regions and Resources in South Asia

⊕ REVIEW QUESTIONS

1. How many people live in India's cities? What are the three largest cities in India?
2. What are *bustees*?
3. What important energy and mineral resources does India have?

⊕ THOUGHT QUESTIONS

1. India is one of the world's most rural countries. Why, then, is it facing an urban crisis? Discuss both jobs and housing.
2. What conditions have caused India's industrial production to suffer?

4. South Asia's Two Islamic Nations: Pakistan and Bangladesh

The Creation of Two Islamic Countries

In 1947, the subcontinent was divided on the basis of religion into two separate countries: Muslim Pakistan and Hindu India. Muslim Pakistan included two quite different areas located on the western and eastern sides of the subcontinent. One was called West Pakistan, the other East Pakistan. They were separated from one another by 1,000 miles of territory belonging to India. West Pakistan was located in the Indus Valley, which was converted to the Islamic faith by invading Muslims more than 1,000 years ago. East Pakistan, previously the Indian province of Bengal, became Muslim 600 years later.

Distance and differences in culture and language made it difficult for these two widely separated parts of Pakistan to remain one nation. In 1971, when a huge tidal wave killed 500,000 people in East Pakistan, West Pakistani relief efforts were halfhearted. Resentful East Pakistanis, aided by Indian troops, declared their independence from West Pakistan in 1972 and called their new country Bangladesh.

Pakistan, the Land of the Indus River

Farming on the fertile plains of the Indus River depends on irrigation. More than half of the 122.4 million people of Pakistan are farmers who live close to the river and its tributaries. In the Punjab in the north, wheat and cotton farming support dense farm populations. Here, the ancient Muslim capital of Lahore has long been a center of Islamic learning in Pakistan, where 98 percent of the people are Muslims. Modern Lahore is a rapidly growing city of 4.6 million. Islamabad, a new forward capital, was built for political reasons in the north near the border with Kashmir, which Pakistan claims but India occupies. It is still a small city with a population estimated at 525,000.

In the south, the Indus plains widen near the coast and support a dense farming population that

▶ *Wheat fields grow in northern Pakistan on the fertile plains of the Indus River. What mountain range overlooks this wheat field?*

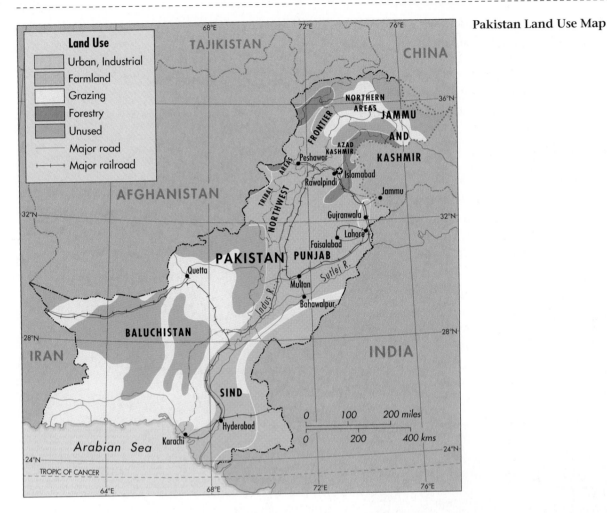

Pakistan Land Use Map

grows rice. Pakistan's largest city, Karachi, with a population of 8.2 million, is the country's most important center of industry and trade.

Farming along the Indus River has not kept pace with population growth, so Pakistan now imports food. Crop yields are low, and traditional methods of farming are still used by most farmers. Many villagers are very poor, just as in India. Meanwhile, population has soared, and the need for economic development and modernization has become intense.

Economic Development in Pakistan Is Limited

Pakistan launched a program of industrial development to provide jobs for its rapidly growing population, but this program has not been able to achieve its goals. Pakistan has no large deposits of iron or coal. Its oil production is also limited, although new oil and gas finds may reduce Pakistan's dependence on imported oil in the future. Moreover, wealth in Pakistan is concentrated in the hands of a small elite.

To cope with limited resources, Pakistan decided to build a huge steel mill at Karachi that would use imported coal and iron. This project is proving to be very expensive and not very profitable. Pakistan's important textile industry, based on cotton grown on the Indus Plain, is also becoming more and more inefficient. Government threats to take over private factories have caused mill owners to postpone programs of modernization.

Pakistan's economic future is not bright. The country is plagued by problems familiar in South Asia—a rapidly growing population, slow progress in agriculture, and slow development of industry. Pakistan now depends on loans from international organizations and other countries to maintain its relatively low standard of living. In 1990, however, the United States suspended aid to Pakistan because of the military overthrow of its elected government.

Bangladesh, One of the World's Most Densely Populated Lands

Many major rivers flow through Bangladesh and empty into the Bay of Bengal. Parts of the country are flooded much of the year. It is the world's eighth most populous country. Here, some 113.9

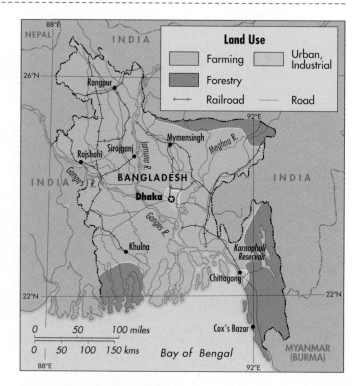

Bangladesh Land Use Map

▼ *The main Friday market in downtown Dhaka, the capital of Bangladesh, conveys some idea of why Bangladesh is considered one of the most overcrowded countries in the world. Note the large numbers of fruit and vegetable stalls and the baskets customers will use to take their purchases home. What do you know about the relationship between land and population in Bangladesh? About the country's economy?*

million people live in an area the size of the state of Alabama.

In this overcrowded country, 85 percent of the population depends on farming, but food production is failing to keep pace with population growth. Except for some natural gas, Bangladesh has no important resource other than land and no major industry other than jute manufacturing. The Muslim Bengalis of Bangladesh are among the poorest people on Earth. Malnutrition affects 60 percent of the population. Many people are ill fed and impoverished; they have less to eat now than they had twenty years ago.

A Farming Country That Imports Food

Bangladesh is the world's largest importer of rice, even though rice is its most important crop. The country's survival depends on the efforts of many countries to provide aid. For twenty-five years, the United States repeatedly gave thousands of tons of grain and cooking oil to Bangladesh to avert famine. This food was diverted into city markets, particularly in the capital of Dhaka, a rapidly growing city of 4.5 million, while the villagers of Bangladesh faced starvation.

No solution to these problems is in sight. In the villages, half of the people are landless laborers. Land distribution and land reform are blocked by large landowners. Floods periodically wash over this flat country. In the last century, fifty major cyclones, or destructive windstorms accompanied by floods, have drowned hundreds of thousands of villagers. The latest occurred in 1991.

⊕ REVIEW QUESTIONS

1. How large are the populations of Pakistan and Bangladesh?
2. What percentage of the people of Bangladesh are farmers?
3. What is the chief crop of Bangladesh? Is this crop exported or imported?

⊕ THOUGHT QUESTIONS

1. Why is Pakistan less able to develop industries than India?
2. Why does Bangladesh have to depend on international aid to survive?

5. Sri Lanka, a Divided Island

An Island Divided by Two Cultures

Sri Lanka, a former British colony called Ceylon, is a teardrop-shaped island located twenty-two miles off the coast of southern India across the Palk Strait. In this island country, 17.8 million people live in a mountainous land that is smaller than South Carolina. Sri Lanka shares with other nations of South Asia the difficulties of feeding a growing farm population on limited cultivated land while trying to build a diversified economy.

Sri Lanka's cultural mix, however, is quite distinctive. Sometime in the distant past, an Indo-European (Aryan) people, the Sinhalese, migrated to Sri Lanka from northern India. They were Buddhists, and they established an irrigation-based farming civilization on the northern half of the island.

Then Dravidian-speaking Hindu Tamils, who differ in language and religion from the Buddhist

◀

The government of Sri Lanka is building large-scale irrigation systems so that in areas of infrequent rainfall the rice paddies will be able to produce every year. What was the former name of Sri Lanka?

Sinhalese, later migrated to the island from southern India. Gradually the Tamils settled the mountainous central and southern section of the island as well as the Jaffna Peninsula in the north. Their numbers increased in the 1800s, when the British imported more Tamils from India to work on newly planted tea and rubber plantations in central and southern Sri Lanka.

In recent years anger between these two ethnic groups has exploded into violence. The Sinhalese, who make up about 70 percent of the population, have refused to agree to Tamil demands for a separate state. Further, the government of Sri Lanka deported a million of the "plantation" Tamils back to India. Guerrilla warfare between the two groups was halted when Sri Lanka invited India to send troops to quell the Tamils in 1987. A temporary and fragile peace was achieved, but violent incidents continue to occur. The ongoing conflict has hindered efforts to modernize Sri Lanka.

An Agricultural Economy

The southwestern quarter of Sri Lanka receives plenty of rain and is densely settled. In the 1800s, the British built roads from the coast into the wet forested mountain interior of this region and on the slopes established tea and rubber plantations. These two plantation crops are Sri Lanka's main exports, the former accounting for one-quarter of all Sri Lankan exports. The capital city of Colombo (population of 2 million) is the port for these exports and also the main industrial center.

▼ *Elephants are used for heavy labor in Sri Lanka. What other countries also use elephants?*

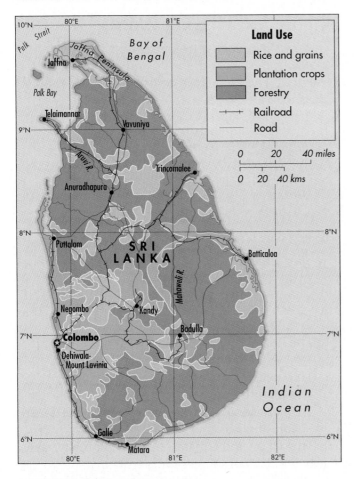

Sri Lanka Land Use Map

In the lowland north and along the southeastern coast, unreliable rainfall causes problems for farmers. In most years, rainfall is sufficient to grow rice, but when the summer monsoon fails, dry years occur. The government is building large-scale irrigation systems to make water supplies more dependable. As a result Sri Lanka is now able to grow enough rice for its own consumption, but still must import petroleum and industrial goods.

The government of Sri Lanka is encouraging foreign businesses to establish new industries, and it has set up a free-trade zone in the capital city to attract foreign trade. But the country's population is growing at the rate of 1 million people every four years, and conflict between the Tamils and Sinhalese still divides the country. These are the overwhelming difficulties Sri Lanka faces in paying its debts and providing higher standards of living for its people.

⊕ REVIEW QUESTIONS

1. Where is Sri Lanka located, and what is its population?
2. What two large groups of settlers migrated to Sri Lanka? Which group dominates the country today?
3. What are Sri Lanka's two chief exports?

⊕ THOUGHT QUESTION

1. How have tension and violence between two ethnic groups handicapped Sri Lanka?

⊕ CHAPTER SUMMARY

Following India's independence in 1947, its cultural, religious, and ethnic diversity proved to be sources of problems. Indian Hindus and Muslims split into two countries: Hindu India and the geographically divided country of West and East Pakistan. In 1972, East Pakistan broke away from West Pakistan and became Bangladesh. Today, the secular democratic government that Indian leaders hoped would bridge religious and other cultural differences is breaking down as ethnic groups demand independence, and Muslim and Hindu extremists attack one another. In Sri Lanka as well, the Hindu Tamil minority is fighting a guerrilla war against the Buddhist Sinhalese majority to press its demands for a separate Tamil state.

The Green Revolution helped India feed its people, but that is only one problem of many. Today, the Indian government continues to struggle to ease village poverty, to create basic services in its rapidly growing and overcrowded cities, and to develop industry to provide jobs and investment. To the west, Pakistan has fewer resources than India and many of its same population, food, and job problems. In the east, Bangladesh has no resources and is one of the most crowded and poorest nations on Earth. South Asia has the natural and human resources to improve the living standard of the people, but the obstacles the governments need to overcome are great.

CHAPTER 26

EXERCISES

Two Correct Sentences

Directions: Two sentences in each group are correct statements. Write on your paper the numbers of the correct sentences next to the letter of each group.

Group A
1. The Green Revolution has been limited to areas that do not need irrigation, because rainfall in these areas is usually plentiful.
2. The seeds used in the Green revolution respond well to chemical fertilizer when planted in areas with reliable water supplies.
3. Wheat yields increased more than rice yields as a result of the Green Revolution.

Group B
1. Most people in Pakistan are very poor.
2. Most people in Bangladesh are very poor.
3. Most people in Sri Lanka are not poor because their country exports large amounts of tea and rubber.

Group C
1. India has a single culture and language, Hindi, which gives it great potential for unity.
2. The demands of two different language groups in southern India resulted in Madras being divided into two states.
3. The biggest problem facing India in this decade is to maintain national unity in the face of cultural differences.

Group D
1. Much of India's farmland is owned by large landowners.
2. Government investment in farming and the Green Revolution have solved India's food problem permanently.
3. A few Indians have highly productive farms, but many villagers are very poor.

Supply the Missing Word

Directions: Number your paper from 1 to 10. Choose from the list the words missing from the sentences. Then write the complete sentences on your paper.

Assam Valley
Bangalore
Bombay
Calcutta
Hindus
Karachi

Muslims
New Delhi
Punjab
Sinhalese
Tamils

1. Pakistan's largest city and most important industrial center is ||||||||||.
2. The capital of India is ||||||||||.
3. |||||||||| is India's most productive manufacturing center, but it is also a city with miserable living conditions.
4. The Buddhist |||||||||| make up 70 percent of Sri Lanka's population.
5. The majority of Pakistanis are ||||||||||.
6. The majority of Indians are ||||||||||.
7. The subregion in India best suited to the Green Revolution was the ||||||||||.
8. The |||||||||| and Gujarat, north of Bombay, are two of India's sources for petroleum.
9. |||||||||| and |||||||||| both began as cotton textile processing centers and have grown to become industrial cities.
10. Hindu |||||||||| are an ethnic minority in Sri Lanka that is demanding a separate state.

Inquiry

Directions: Combine the information in this chapter with your own ideas to answer the following questions.
1. South Asia has been the scene of much tension and fighting between Hindus and Muslims. In what other areas of the world has religious strife occurred in recent years?
2. How is it possible that even though India has made impressive gains in food production in recent years, widespread hunger is still a problem?

3. India appears to have the natural resources and industrial potential to improve its people's standard of living. What are some of the major problems or limitations preventing the country from reaching that potential?
4. In what ways is Calcutta an example of the problems cities of developing nations have providing services to a rapidly growing urban population?
5. Explain how Bangladesh's location has brought difficulties for its people.

SKILLS

Using a Physical Map

Directions: Use the physical map of South Asia on page 612 to tell whether each of the following statements is true or false. Write *True* or *False* on your paper next to the correct sentence number.

1. Most of South Asia lies between 10 and 30 degrees north latitude.
2. Most of South Asia lies between 70 and 90 degrees west longitude.
3. The Khyber Pass is in the Eastern Ghats.
4. The Indus River begins in the Himalayas and flows to the Arabian Sea.
5. The Ganges River flows parallel to the Indus River.
6. The Ganges Plain is south of the Deccan Plateau.
7. The Coromandel Coast is on the Bay of Bengal.
8. The Malabar Coast is on the Arabian Sea.
9. The interior regions of the Indian subcontinent are generally lower than the areas along the northern borders.
10. Sri Lanka is separated from India by the Palk Strait.
11. Most of Sri Lanka is higher in elevation than most of Baluchistan.
12. The Himalayas extend over a distance of more than 800 miles.

Using a Political Map

Directions: Use the political map of South Asia on page 611 to match these countries with their capital cities. Write the name of each country followed by its capital city.

1. Pakistan   a. Colombo
2. India   b. Dhaka
3. Nepal   c. Thimbu
4. Sri Lanka   d. New Delhi
5. Bangladesh   e. Islamabad
6. Bhutan   e. Kathmandu

Differentiating Between Fact and Opinion

Directions: You need to know the difference between facts and opinions when you study geography or any other subject. *Facts* are things known to be true or known to have really happened. *Opinions* are beliefs that may be true but are not substantiated by positive knowledge or proof. On your paper, write "fact" or "opinion" next to the number of each of the following statements.

1. Calcutta has many *bustees*.
2. India is larger than the combined areas of Pakistan and Bangladesh.
3. All villagers are poor farmers who want to move to the cities.
4. The summer monsoon is a current of warm, moist, oceanic air that blows across the subcontinent between June and September.
5. Monsoons provide the best way to bring sufficient water for farming.
6. It would be more enjoyable to climb Mount Everest in the Himalayas than to climb the Matterhorn in the Alps.
7. India now controls most of Kashmir, but it should belong to Pakistan.
8. Nepal and Bhutan are small mountainous countries sandwiched between India and China.

9. The Deccan Plateau is sparsely populated, so living conditions there would be more comfortable than in crowded cities like Bombay.

10. India's population will be over 1 billion by the year 2000.

11. Major offshore petroleum discoveries in the Arabian Sea near Bombay will make India self-sufficient in energy in the 1990s.

12. It was a good idea for East Pakistan to become the separate country of Bangladesh.

Vocabulary Skills

Directions: Number your paper from 1 to 8. Write the word from the list below on your paper that completes the sentence.

alloy
bustees
cyclone
democracy
green revolution
Mahatma
secularism
tributary

1. |||||||||| is an honorific term meaning "the Great Soul." It is reserved for people with great spiritual powers like Mohandas K. Gandhi.

2. Some minerals are smelted and refined directly from ores; others are blended with other metals to form an ||||||||||.

3. Indian leaders have found it necessary to practice |||||||||| , that is not recognizing any one religion as an expression of government, in order to have a government that both Hindus and Muslims will support.

4. India's government is officially a |||||||||| because its leaders are elected directly by the people.

5. In order to feed India's growing population, the country's leaders adopted a plan to use new "miracle" seeds and modern farming techniques. Because of its hoped-for success in India and many countries in Asia, it was called the ||||||||||.

6. Many Indian cities are overcrowded and few jobs are available, the poor are crowded into shantytowns called ||||||||||.

7. A |||||||||| is a powerful, very destructive windstorm, like a hurricane. In this century, Bangladesh has experienced fifty such windstorms that have been accompanied by floods.

8. A |||||||||| is a small river that feeds into a larger river.

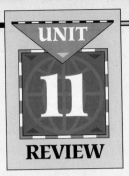

Human Geography

Directions: Each of the following sentences has three possible endings. Select the correct ending and then write the complete sentence on your paper.

1. South Asia is half the size of the United States but has (a) a smaller number of people (b) twice as many people (c) four times as many people.
2. The total population of South Asia is about (a) 500 million (b) 900 million (c) 1.5 billion.
3. South Asia's population is (a) growing rapidly (b) growing slowly (c) declining slightly.
4. The largest religious group in India is the (a) Muslims (b) Hindus (c) Buddhists.
5. The dominant religion in Bangladesh is (a) Islam (b) Hinduism (c) Buddhism.
6. The largest religious group in Sri Lanka is the (a) Muslims (b) Hindus (c) Buddhists.
7. Most South Asians live in (a) villages (b) towns and small cities (c) large cities.
8. Indian civilization based on Hinduism is about (a) 1,000 years old (b) 3,000 years old (c) 5,000 years old.
9. One of the most densely populated countries on Earth is (a) Bangladesh (b) Pakistan (c) India.
10. The Mogul Empire controlled most of South Asia from the (a) 800s to the 1200s (b) 1500s to the 1700s (c) 1800s to the 1900s.
11. The caste system developed from (a) Hinduism (b) Islam (c) British colonialism.
12. One of the lowest groups in the caste system is the (a) *Brahmans* (b) *Sudras* (c) *Kshatriyas*.
13. The Taj Mahal was built by (a) a Mogul ruler (b) a Hindu ruler (c) the British East India Company.
14. When the British East India Company took over most of India, it (a) lowered the farmers' taxes (b) allowed taxes to be paid in crops rather than cash (c) increased the amount of farmland used to produce cotton, indigo, tea, and opium.
15. The proportion of India's population that lives in cities is: (a) one-tenth (b) one-quarter (c) two-thirds.

Physical Geography

Directions: There is a correct term and an incorrect term in the parentheses in each of the following sentences. Write the completed sentences on your paper, using only the correct terms.

1. The Ganges River flows through India and (Pakistan, Bangladesh).
2. The plains of the Indus River provide farms for many people in (Pakistan, Bangladesh).
3. The largest island country in South Asia is (Nepal, Sri Lanka).
4. Pakistan has (many, few) natural resources.
5. The heaviest rainfalls in South Asia usually occur during the (summer, winter) monsoon.
6. The Himalayas form a mountain wall along the northern borders of India and (Sri Lanka, Pakistan).
7. On the northern plain of Pakistan, five rivers join to form the (Indus, Ganges) River.
8. The Punjab is an important farming area that has (heavy, light) rainfall.

Writing Skills Activities

Directions: Answer the following questions in essays of several paragraphs each. Remember to include a topic sentence and several sentences supporting your main idea in each paragraph.

1. In what ways does the rainfall pattern in South Asia differ from the rainfall patterns of other regions you have studied?
2. What role has religion played in the history of South Asia?
3. What conditions cause Pakistan and Bangladesh to need foreign economic aid?

SOUTHEAST ASIA

UNIT INTRODUCTION Southeast Asia extends from Myanmar (Burma) on the Asian mainland to the islands of Indonesia in the Pacific Ocean. This is a distance equal to that between Boston, Massachusetts, and Seattle, Washington. Southeast Asia is a region of incredible human and environmental diversity. Many different peoples live in the ten countries located in this region.

Today, these countries face many of the same problems found elsewhere in the developing world. Population is growing faster than farmland. As villages become more crowded, many farmers are moving to the rapidly growing cities of Southeast Asia. Plans for development vary from one country to another. Some countries are handicapped by limited resources; others have great resource potential. But many countries in Southeast Asia also have been torn apart by war, have unstable governments, and include different culture groups. Raising the standards of living of the 460 million people who live in this region will be a difficult task.

▲ *Indonesia is a country made up of over 13,000 islands. Many are uninhabited, but the warm climate and abundant rain make possible large harvests of crops like rice, tea, rubber, sugar, and tobacco on other islands. What is the woman in this photograph doing?*

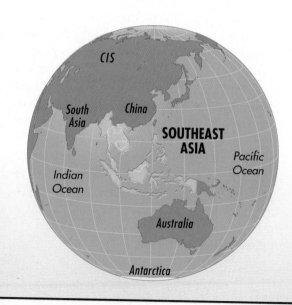

▲ Despite the wealthy appearance of cities like Jakarta, Indonesia, nearly 60 percent of Indonesians live in complete poverty. Recent discoveries of oil and gas fields have stimulated the economy, however. Is oil a dependable revenue raising export?

SKILLS HIGHLIGHTED

- Using the Maps in the Miniatlas
- Verifying Generalizations
- Vocabulary Skills
- Writing Skills

UNIT OBJECTIVES

When you have completed this unit, you will be able to:

1. Describe the important geographical features and tropical climates of Southeast Asia, and how the people of this region have adapted to their environments.

2. Explain how Indians, Chinese, and Muslims who invaded Southeast Asia affected the culture and economy of this region.

3. State that almost all the lands of Southeast Asia became colonies of European countries, and describe how the period of colonial rule influenced the lives of the local peoples.

4. Identify the ways some Southeast Asian countries have achieved more developed economies.

5. Relate the reasons some Southeast Asian countries have failed to make sufficient economic progress to lift the living standards of most of their people.

▲ Once called "the Venice of the East," Bangkok, Thailand, is experiencing major growth problems. Many canals have been paved over to create streets. This has destroyed the natural drainage system of the city. What problems will this cause?

KEYS TO KNOWING SOUTHEAST ASIA

1

Southeast Asia is fragmented into many peninsulas and islands.

2

Southeast Asia has a tropical environment; the equator cuts across much of this region.

3

Forested mountains cover much of mainland Southeast Asia and many of its volcanic islands.

4

Southeast Asia is a meeting place of peoples and cultures because of its crossroads location.

5

Indians and Chinese influenced Southeast Asia in the first millennium A.D. Muslims and European traders entered the region in the 1400s. European colonialism dominated Southeast Asia in the 1800s.

6

Southeast Asia is divided into ten countries with a combined population of about 460 million.

7

Nearly two out of three Southeast Asians earn their living directly from the land.

8

Three types of farming are practiced in Southeast Asia: shifting cultivation, wet rice farming, and commercial farming.

9

Population is growing rapidly in Southeast Asia. Population pressure on farmland is intense. Port cities are exploding in size because of migration from the countryside.

10

Development plans in Southeast Asia vary from country to country. The export economies of Thailand, Malaysia, the Philippines, Singapore, and Brunei are prosperous. Political disruption and war have hindered economic growth in Myanmar (Burma), Laos, Cambodia, Vietnam, and Indonesia.

▼ *Young Buddhist men of Southeast Asia are expected to spend several years as monks, searching for enlightenment. In many places, Buddhism adopts some Hindu beliefs, such as reincarnation. What does reincarnation mean?*

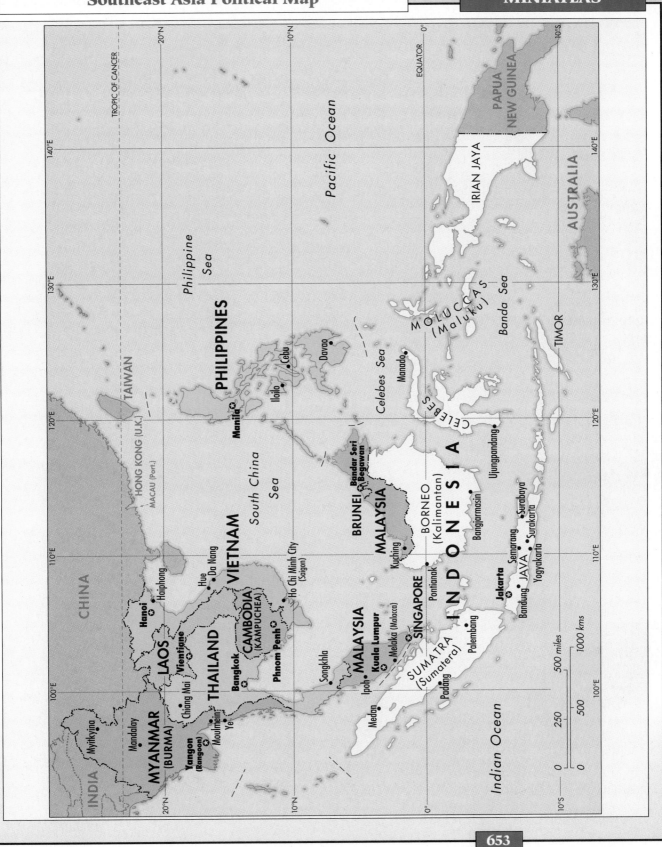

TROPIC OF CANCER

20°N 10°N 0° 10°S

PAPUA NEW GUINEA

IRIAN JAYA

Pacific Ocean

AUSTRALIA

140°E

130°E

Philippine Sea

MOLUCCAS (Maluku)

Banda Sea

TIMOR

TAIWAN

PHILIPPINES

Davao

Cebu

Iloilo

Manila

CELEBES

Celebes Sea

Manado

Ujungpandang

HONG KONG (U.K.)

MACAU (Port.)

120°E

BRUNEI Bandar Seri Begawan

MALAYSIA

Kuching

BORNEO (Kalimantan)

Banjarmasin

Surabaya

Semarang Surakarta

Yogyakarta

INDONESIA

CHINA

South China Sea

VIETNAM

Hue Da Nang

Haiphong

Hanoi

Ho Chi Minh City (Saigon)

LAOS

Vientiane

CAMBODIA (KAMPUCHEA)

Phnom Penh

THAILAND

Bangkok

Chiang Mai

Songkhla

SINGAPORE

Pontianak

JAVA

Jakarta

Bandung

MALAYSIA

Kuala Lumpur

Ipoh Melaka (Malacca)

SUMATRA (Sumatera)

Palembang

Padang

Medan

INDIA

MYANMAR (BURMA)

Myitkyina

Mandalay

Yangon (Rangoon)

Moulmein

Ye

110°E

100°E

Indian Ocean

20°N 10°N 0° 10°S

1000 kms

500 miles

500

250

0 0

Southeast Asia Physical Map

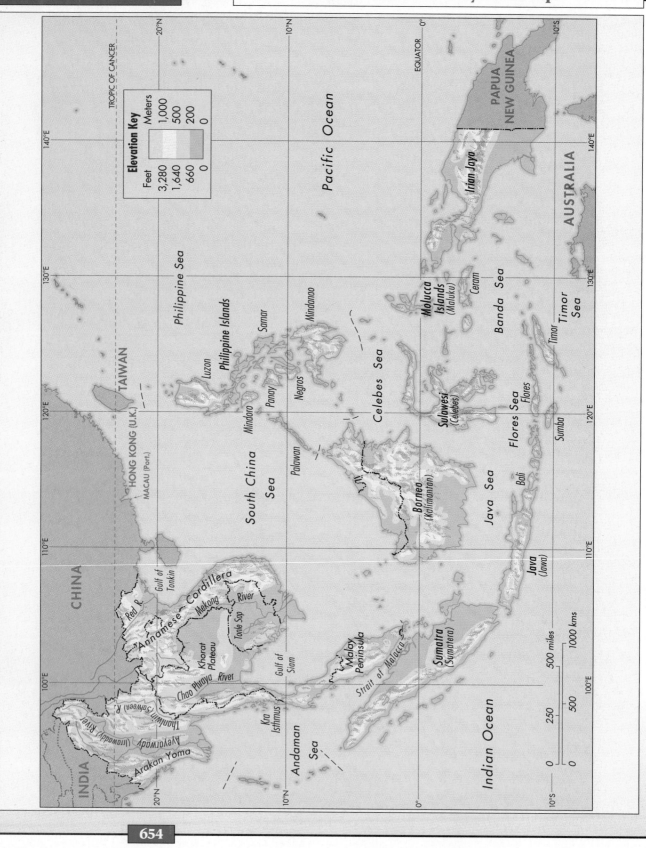

Elevation Key

Feet	Meters
3,280	1,000
1,640	500
660	200
0	0

TROPIC OF CANCER

20°N

10°N

0° EQUATOR

10°S

140°E

130°E

120°E

110°E

100°E

Pacific Ocean

PAPUA NEW GUINEA

Irian Jaya

AUSTRALIA

Philippine Sea

Philippine Islands

Luzon

Samar

Mindanao

Mindoro

Panay

Negros

Palawan

TAIWAN

HONG KONG (U.K.)

MACAU (Port.)

CHINA

South China Sea

Celebes Sea

Moluccan Islands (Maluku)

Ceram

Banda Sea

Timor Sea

Timor

Sulawesi (Celebes)

Flores Sea

Flores

Sumba

Borneo (Kalimantan)

Java Sea

Bali

Java (Jawa)

Red R.

Gulf of Tonkin

Annamese Cordillera

Mekong River

Tonle Sap

Khorat Plateau

Chao Phraya River

Thanlwin (Salween) R.

Ayeyarwady (Irrawaddy) River

Arakan Yoma

Kra Isthmus

Gulf of Siam

Malay Peninsula

Strait of Malacca

Sumatra (Sumatera)

Andaman Sea

Indian Ocean

INDIA

20°N

10°N

0°

10°S

0 250 500 miles
0 500 1000 kms

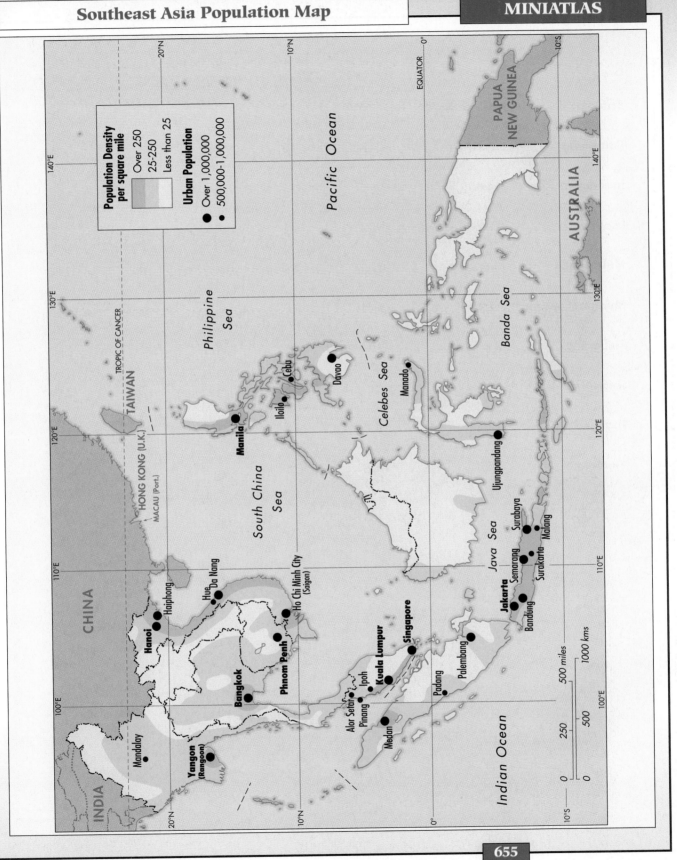

Population Density
per square mile
Over 250
25-250
Less than 25

Urban Population
● Over 1,000,000
• 500,000-1,000,000

CHINA

INDIA

TAIWAN

HONG KONG (U.K.)

MACAU (Port.)

Mandalay

Yangon
(Rangoon)

Hanoi

Haiphong

Hue
Da Nang

Ho Chi Minh City
(Saigon)

Bangkok

Phnom Penh

Kuala Lumpur

Ipoh

Alor Setar

Pinang

Padang

Singapore

Medan

Palembang

Jakarta

Bandung

Semarang

Surakarta

Surabaya

Malang

Ujungpandang

Manado

Davao

Cebu

Iloilo

Manila

Philippine
Sea

South China
Sea

Pacific Ocean

Celebes Sea

Banda Sea

Java Sea

Indian Ocean

PAPUA
NEW GUINEA

AUSTRALIA

EQUATOR

TROPIC OF CANCER

20°N

10°N

0°

10°S

20°N

10°N

0°

10°S

100°E

110°E

120°E

130°E

140°E

100°E

110°E

120°E

130°E

140°E

500 miles
250
500
0

1000 kms
500
0

Country	Capital City	Area (Square miles)	Population (Millions)	Life Expectancy	Urban Population (Percent)	Per Capita GNP (Dollars)
Brunei	Bandar Seri Begawan	2,228	0.3	71	59	—
Cambodia	Phnom Penh	69,900	9.0	50	13	200
Indonesia	Jakarta	735,355	187.6	59	31	610
Laos	Vientiane	91,429	4.6	50	19	230
Malaysia	Kuala Lumpur	127,317	18.4	71	51	2,490
Myanmar (Burma)	Yangon	261,216	43.5	58	24	—
Philippines	Manila	115,830	64.6	64	43	740
Singapore	Singapore	224	2.8	74	100	12,890
Thailand	Bangkok	198,456	57.2	68	19	1,580
Vietnam	Hanoi	127,243	71.8	64	20	—

—Figures not available

CHAPTER 27

ENVIRONMENTS, PEOPLES, AND ECONOMIES

Southeast Asia is a region rich in resources and has a population of only 460 million people. Conditions for economic development would seem to be favorable in the region. However, rapidly growing populations, warfare, and limitations on farmland have hindered economic growth in many Southeast Asian countries.

◄ *Rice was one of the first crops grown in Southeast Asia and is still a major food crop of Indonesia today. Why does rice grow so well in Indonesia?*

1. Environments of Southeast Asia

Mountains, Valleys, and Rivers on the Mainland

Southeast Asia is a huge peninsula that juts southward from the southeast corner of Asia. Just to the north of this region, the Himalaya Mountains bend sharply to the south, their ranges reaching down like fingers into Southeast Asia. They separate the great river valleys of mainland Southeast Asia from one another.

These mountain ranges are highest on the eastern and western edges of mainland Southeast Asia. In the west, a wall of towering mountains in Myanmar (Burma) forms the boundary between mainland Southeast Asia and India. In the east, another high Himalayan mountain range, the Annamese Cordillera, curves southward for hundreds of miles through Laos along the border with Vietnam. Between these two high ranges, low mountains and

hills are located along the boundary between Myanmar and Thailand and stretch southward down the length of the Malay Peninsula. Other hill ranges lie between Thailand and Cambodia to the southeast. These mountains encourage north-south movement and hinder east-west communication.

On the map on page 654, trace the great rivers of mainland Southeast Asia. They flow to the sea along the valleys that separate these mountain ranges. The river valleys are narrow in the mountainous north and broaden gradually to the south. In these river valleys and on the region's deltas and coastal plains, 40 percent of the people of Southeast Asia—about 185 million people—earn their living.

Climate and Vegetation on the Mainland

Temperatures are high throughout most of Southeast Asia because this region is located near the equator. The climate, like that of India, is a product of the summer and winter monsoons. In summer, the southwest monsoon blankets the mainland with dense clouds that bring heavy rain-

Subregions of Southeast Asia
Southeast Asia has two major subregions:
(1) Mainland Southeast Asia and
(2) Island Southeast Asia.

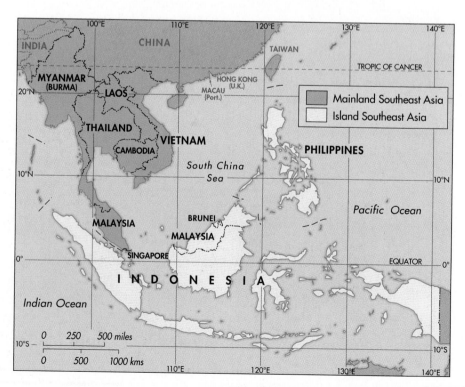

fall. In winter, the dry northeast monsoon brings little rain to Southeast Asia, except where mountain slopes face the wind. Vegetation and soils reflect these variations in rainfall.

Southeast Asia's **monsoon rain forests** do have a dry season; this explains why they have deciduous as well as evergreen trees. These forests cover most of the highlands in mainland Southeast Asia. Because tree species tend to grow together in groups, precious woods are easier to harvest in these forests than in tropical rain forests where species are more scattered.

Also, trees are spaced farther apart in the monsoon rain forests of Southeast Asia than in tropical rain forests. Sunlight reaches the forest floor, which encourages the growth of ground cover that enriches soils each year. Therefore, when monsoon forests are cut down, crops can be planted in these richer soils for much longer periods of time than in tropical rain forests.

Island Environments in Southeast Asia

Southeast Asia includes thousands of islands, which are scattered in chains and clusters that extend 2,500 miles from the Malay Peninsula eastward into the Pacific Ocean. The largest of these are the Indonesian islands of Sumatra (Sumatera), Borneo (Kalimantan), and Java (Jawa), and the Philippine islands of Luzon and Mindanao. Locate them on the map on page 654.

Together, Southeast Asia's islands have a total land area of about 850,000 square miles. Surrounding seas cover an area four times larger. About 275 million people, 60 percent of all Southeast Asians, live in this island world. As you would expect, fishing, trade, and other activities related to the sea play an important role in the daily lives of many of these people.

Climates of Island Southeast Asia

The equator runs directly through Sumatra, Borneo, and the Sulawesi (Celebes) Islands, all of which are in Indonesia. Most Southeast Asian islands are hot and wet year-round. They sometimes suffer severe tropical storms, called **typhoons** after the Chinese *tai fung*, meaning "big wind." These storms, which we call hurricanes, have winds that blow at rates of more than 100 miles per hour. They inflict great damage on exposed coastal areas.

◄

Most Laotians live and work in the valleys of the Mekong River. What type of work do most people in Laos do?

▲ *Dance is an important tradition in Indonesia that portrays episodes from the past or conveys spiritual and cultural values. This* barong *dancer is performing a ceremony to drive evil spirits from his village. Can you think of any Native American dances that express community needs?*

▲ *These colorful puppets from Jakarta are a popular and honored form of entertainment. Puppet theatre is an art form which evolved from elaborately staged dances. Puppet shows are part of a country's oral tradition. What does the term* oral tradition *mean?*

Rainfall varies from place to place in the islands of Southeast Asia, but most islands receive from 60 to 100 inches of rain a year. This is enough rain to support rain forests similar to those of Amazônia in Brazil.

Only islands on the northern and southern margins of Southeast Asia have long dry seasons and monsoon rain forests like those of the mainland.

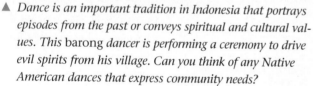

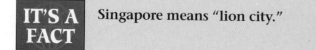
IT'S A FACT | **Singapore means "lion city."**

The northern Philippines and eastern Java have dry seasons that last from four to five months. On many mountainous islands, the windward and leeward sides of the mountains have very different climates and types of vegetation.

Volcanic Islands in Southeast Asia

Many of Southeast Asia's islands have volcanic mountains at their centers, with tropical rain forests blanketing their slopes. Short rivers flow downslope

IT'S A FACT The 1883 volcanic eruption on the Indonesian island of Krakatoa was so violent that the shock wave was felt 2,200 miles away in Australia.

into dense swamps and marshes that line the shore. Most islanders live on deltas at the mouths of these rivers or on narrow coastal plains that have been cleared for settlement.

Volcanoes form the mountain core of many islands in Southeast Asia and the Pacific, including our Hawaiian Islands. Volcanoes are openings in the Earth's crust; lava, or melted rock, flows through these openings. Some active volcanoes continuously spew forth steam, gas, and ash from their centers. Lava often bubbles in volcanic craters. Large eruptions are infrequent, however.

Many islands in Southeast Asia, as you already know, are part of the "Ring of Fire," a string of volcanic islands that stretch from Sumatra and Java eastward through the Sulawesi and Molucca (Maluku) Islands.

On some of these islands, volcanic debris forms deep, dark, rich soils that support populations as dense as the fertile river valleys of mainland Southeast Asia. That is why more than 125 million Indonesians can live on the small island of Java.

⊕ REVIEW QUESTIONS

1. Why are temperatures high throughout most of Southeast Asia?
2. What percentage of this region's people live on islands?
3. Where are volcanic islands found in Southeast Asia?
4. In which direction do the mountain ranges of mainland Southeast Asia tend to run?

⊕ THOUGHT QUESTIONS

1. In what ways are the monsoon rain forests of Southeast Asia different from tropical rain forests?
2. Although volcanoes can cause serious damage, they may have positive effects, too. What are these good effects?

2. Tropical Farmers in Southeast Asia

Three Kinds of Farming

Most Southeast Asians are farmers who live off the land. Three different systems of farming are found in this tropical region. **Shifting cultivation** supports small communities of people living in the highlands of Southeast Asia. **Wet rice farming** sustains dense farming populations in the river valleys, deltas, and coastal plains, where there is abundant water for irrigation. **Commercial farming**, found throughout Southeast Asia, is the growing of cash crops like rubber, coffee, sugarcane, and coconuts for sale on local and world markets.

Shifting Cultivation

Rugged land, dense forests, and poor soils discourage permanent farming settlements in the highlands of Southeast Asia. For these reasons, farmers practice a style of farming called shifting cultivation, which is well adapted for small populations living in tropical mountain environments. It begins when farmers cut small clearings on forested slopes and burn the fallen trees in these clearings. The ashes from these fires add nutrients to the soil and increase its fertility for a short time. Farmers plant food crops in these temporary clearings and live in shelters nearby.

After three or four years, the food crops drain nutrients from the soils in the clearings, and crop yields begin to decline. Farmers are forced to abandon the clearings and move to new locations in the forest, where the process of clearing the land, burning trees to ashes, and planting crops in new clearings is repeated. The need to shift locations every few years is the reason this farming system is called shifting cultivation.

The hill peoples who practice shifting cultivation in Southeast Asia have a simple technology. Hoes, digging sticks, and machetes are their main farming tools. The variety of crops grown, however, reflects their sophisticated knowledge of tropical environments.

PEOPLE-ENVIRONMENT

DIFFERENT TYPES OF RURAL SETTLEMENT

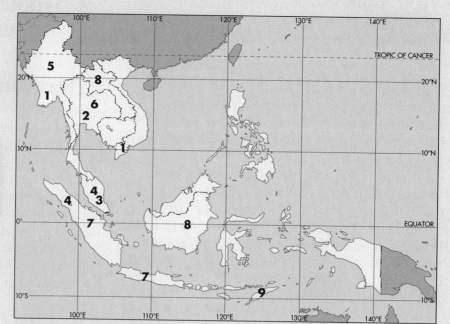

Agglomerated (Linear)

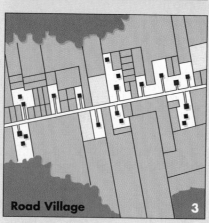

River Village   1

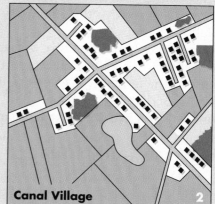

Canal Village   2

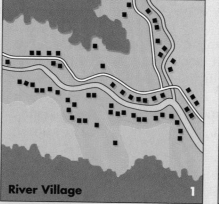

Road Village   3

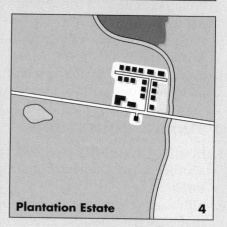

Plantation Estate   4

Major components of rural areas are built-up villages, farm buildings, and fields. These components are arranged with differing degrees of clustering. The distribution and pattern of dwellings varies with the style of agriculture, local traditions, and environmental conditions.

Geographers distinguish two types of rural settlement, depending on the arrangement of houses and other structures on the land. They are agglomerated settlements and dispersed settlements. **Agglomerated settlements** are clusters of dwellings called hamlets or villages, depending on size. All have a central concentration of buildings surrounded by the gardens, fields, and pastures of the settlement. Different patterns of concentration in agglomerated settlements take different shapes on the landscape. The two most common shapes of agglomerated settlements are **nuclear villages** where dwellings are clustered around a common center, and

linear villages where dwellings follow a road, path, or river. **Dispersed settlements**, by contrast, are composed of isolated farmsteads often far removed from neighbors.

In Southeast Asia, agglomerated settlements are the most common form of farm settlement, as the illustration shows. Linear agglomerated settlements occur throughout the region along rivers (1), canals (2), and roads (3), and on the plantation estates of Malaysia and Sumatra (4).

Other villages are nuclear, like the walled settlements of central Myanmar (5) and the less concentrated but still nuclear villages of central Thailand (6). Two smaller types of nuclear settlements are the clumped hamlets of Java and Sumatra in Indonesia (7) and the hamlets of shifting cultivators in the highlands of mainland Southeast Asia and Borneo (8).

Dispersed settlements composed of widespread individual farmsteads occur occasionally on rented rice fields in the river lowlands of Southeast Asia. Only in Timor (9), however, is dispersed rural settlement common.

Source: Adapted from: Colin S. Freestone, *The Southeast Asian Village: A Geographic, Social, and Economic Study.* London: George Philip and Son, 1974, pp. 6–7.

QUESTIONS

1. What are the three main parts of rural areas?
2. Hamlets or villages are also known by what geographic term?
3. The isolated farmsteads of the American Great Plains would be an example of which type of settlement pattern—agglomerated or dispersed?

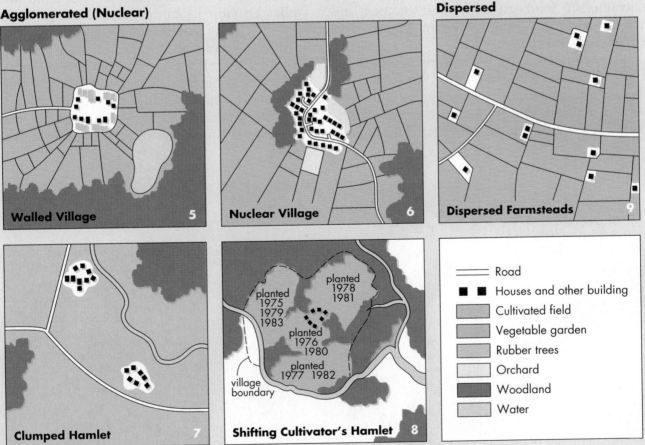

Agglomerated (Nuclear)

Walled Village 5

Nuclear Village 6

Dispersed

Dispersed Farmsteads 9

Clumped Hamlet 7

Shifting Cultivator's Hamlet 8

planted 1978 1981
planted 1975 1979 1983
planted 1976 1980
planted 1977 1982
village boundary

— Road
■ ■ Houses and other building
Cultivated field
Vegetable garden
Rubber trees
Orchard
Woodland
Water

About 400 different plants are grown in forest clearings in Southeast Asia. As many as 100 crops may be planted in a single field. Cassava, yams, sweet potatoes, and tree crops are the most important foods grown by shifting cultivators in this region.

Even small groups of shifting cultivators need large areas of tropical forest to support their way of life. They must move to new areas of uncleared land every few years. They supplement their farming with hunting and gathering, which require even more undisturbed land where wild animals and plants can be found.

Shifting cultivators live, therefore, in the remote highlands of Southeast Asia. They are found in the mountains of Vietnam and Laos, in the highlands of eastern Thailand, in the mountains of Myanmar, and along the rugged eastern side of the Malay Peninsula. In island Southeast Asia, they live in the interior of Borneo, Irian Jaya (western New Guinea), and much of Sumatra. There are 30 million shifting cultivators in Southeast Asia. Most of them are members of ethnic groups who are minorities in the countries in which they live.

Today, shifting cultivators are being pushed deeper into the recesses of the forests. Rice farms and commercial plantations are expanding into the foothills, reducing the amount of forest land suitable for shifting cultivation. Modern medical care is increasing the population of these mountain farmers. As a result, growing numbers of shifting cultivators are being forced to live on less and less land.

Wet Rice Farming

Rice is the preferred food of most Southeast Asians. In wet rice farming, water sources are controlled to provide the correct amount of water. More than half of the farmland of every country in the region, except Malaysia, is planted in wet rice. The amount of land planted in rice has doubled, and rice production has tripled in the last fifty years. Modern drainage, irrigation, and water control systems have turned swampy and often-flooded deltas into productive rice lands. Thailand and Vietnam export rice. Malaysia imports rice, and until recently Indonesia was the world's leading importer of rice.

Wet rice is grown in small, flooded fields surrounded by low embankments that hold water while the crop is grown. Once built, these embanked fields, or paddies, are used year after year.

A reliable, controlled source of water is essential for wet rice cultivation. Natural rainfall or river flooding provides this water in some parts of Southeast Asia. In many areas, however, irrigation systems are needed. Large reservoirs store the water, and miles of banked canals and channels transport

▶

Most people in Southeast Asia eat about a pound of rice per day. Rice is very nutritious and provides calories for energy. In some areas, rice is being harvested by machine, but the men in this photograph separate rice by traditional methods. Which countries in Southeast Asia export rice?

it to and from the fields. In densely settled rice-producing areas, villages often look like tree-ringed islands rising above carpets of flooded rice fields. In some places, rice is grown on terraces, level stepped fields that have been dug into hillsides.

Rice yields in Southeast Asia used to be one-third to one-half those of Japan, Taiwan, and the United States. In the 1980s, however, chemical fertilizers, pesticides, and new varieties of high-yielding, fast-growing rice seed were introduced into Southeast Asia.

This green revolution has raised rice yields in those parts of Myanmar, Vietnam, Indonesia, and the Philippines where reliable water control systems, chemical fertilizers, and pesticides are available. In much of Southeast Asia, however, water control systems are not efficient, and farmers cannot afford to buy fertilizers and pesticides. Therefore, the green revolution has been confined to the best irrigated and richest lowland areas in the region.

Southeast Asia now produces twice as much rice as it did thirty years ago. However, the population of Southeast Asia has also doubled during these thirty years. The green revolution has postponed large food shortages for a time, but the problem of feeding rapidly growing populations is still a major concern of many governments in Southeast Asia.

Commercial Farming

For centuries, native plants like pepper, cloves, and nutmeg were grown in Southeast Asia and sold on world markets. Today, commercial crops are still grown on some of the best farmland in the region. An estimated 20 million acres of land are planted in the most important commercial crops in Southeast Asia: rubber, coconuts, oil palms, coffee, tea, and tobacco.

Natural rubber is the most important commercial crop in Southeast Asia, which produces 85 percent of the world's supply. Rubber makes up one-third of all commercial farm exports in every Southeast Asian country except the Philippines, where coconuts are the most important commercial crop.

Rubber trees are grown on large plantations and small farms in Southeast Asia. Because this forest plant needs little care, can grow on sloping as well as flat land, and can be harvested year-round, rubber is grown in many areas in Southeast Asia. There are large commercial rubber plantations on the Malay Peninsula, in southern Vietnam, and on the Indonesian island of Sumatra.

Elsewhere in Southeast Asia, coffee, tea, sugarcane, and oil palms are grown. The spices that first lured Europeans to this region are still raised in the Molucca (Maluku) Islands. Changes in demand for these commercial crops on world markets often cause problems for the people who raise them. The development of synthetic rubber, for example, drastically reduced demand for natural rubber. The price of coffee often goes up or down, depending on total world production. To reduce their dependence on a single crop, many countries are raising a greater variety of crops.

The countries of Southeast Asia, however, cannot improve living conditions in the countryside by improvements in agriculture alone. More and more villagers are migrating to the region's growing cities, because farming settlements already are crowded. New job opportunities must be created in the cities if these refugees from the countryside are to earn a living and Southeast Asians are to achieve a higher standard of living.

⊕ REVIEW QUESTIONS

1. How many shifting cultivators are in Southeast Asia, and in what remote areas do they live?
2. What portion of the farmland of every country in Southeast Asia, except Malaysia, is planted in wet rice?
3. What are the most important commercial crops in Southeast Asia?

⊕ THOUGHT QUESTIONS

1. Why do shifting cultivators move their locations every few years?
2. What parts of Southeast Asia have profited from the green revolution, and why has it not helped the entire region?

3. Cities in Southeast Asia

The Growth of Cities in Southeast Asia

Port cities first became centers of economic activity in Southeast Asia when Europeans set up trading centers on the coast, and later millions of Chinese moved into the region to conduct urban trade and business. Inland towns grew up in the tin-mining areas of Malaysia, along railroad lines, and in areas where plantation crops were grown. Yet because trade was the main source of wealth in Southeast Asia, port cities grew much faster than inland centers.

When these cities began to grow, Southeast Asia had little manufacturing. Europeans preferred to ship raw materials out of Southeast Asia and to have them processed in factories in Europe. Also, Southeast Asia did not have significant deposits of coal and iron, two basic resources that fired up the factories in the industrialization of Europe.

IT'S A FACT Borneo has an area more than triple the size of Great Britain; Sumatra is more than double Britain's size.

As a result, the largest cities in Southeast Asia had few factories and few manufacturing jobs, and Europeans and Chinese immigrants controlled foreign trade. Southeast Asians who moved to the cities to find work ended up living in urban slums. Jobs were not available for Southeast Asia's growing population.

Large Metropolitan Centers

In the 1980s and 1990s, Southeast Asia's port cities have grown rapidly. Indeed, some countries in Southeast Asia are now dominated by a single huge metropolitan center. Yangon (Rangoon) in Myanmar has a population of 2.5 million people. Bangkok, in neighboring Thailand, holds 6.3 mil-

Penang, an island town, has the oldest European settlements in Malaysia. In this medical facility, workers are inspecting latex medical equipment. Penang is also home to one of the world's largest insect farms. Can you guess which insect?

lion people. Kuala Lumpur, the capital of Malaysia, has a population of more than a million. Phnom Penh, Cambodia's capital, is also a city of more than a million. The city-state of Singapore at the tip of the Malay Peninsula has 2.8 million people. Locate these important Southeast Asian cities on the map on page 655.

Three Southeast Asian countries have more than one large metropolitan center. In Vietnam, Ho Chi Minh City (formerly Saigon) in the south is the country's largest city, with 3.9 million people. Hanoi, the country's northern capital, has a population of 1.1 million and its port, Haiphong, has more than a million residents. The capital of populous Indonesia, Jakarta, has 8.3 million people. Two-thirds of Indonesia's 187.6 million people, however, live on the small island of Java, which has three other cities of more than a million; Manila, in the Philippines, is a metropolitan center of more than 10 million people, and Davao on the southern island of Mindanao has grown to more than a million. Only landlocked Laos and the tiny sultanate of Brunei do not have a city with more than a million people.

These fast-growing Southeast Asian cities are now becoming industrialized. The construction of food-processing mills, metalworking plants, textile factories, oil refineries, and service industries are beginning to provide badly needed urban jobs. But so far, manufacturing jobs are being created too slowly to meet the demands of growing populations.

Urban Problems in Bangkok, Thailand

Bangkok is located on a bend of the Chao Phraya River in the most densely populated area of Thailand. In the early 1970s, Bangkok had a population of 2.3 million. By 1993, the number of people in the city had swelled to 6.3 million—one of every ten people in the country. If the present rate of population growth continues, Bangkok's 597 square miles of flat, swampy land will be clogged with 14 million people in ten years. The city is already larger than central Los Angeles and also has more smog.

▼ *Rush hour traffic is bumper-to-bumper in most Southeast Asian cities as this photograph of Bangkok, Thailand, illustrates. Given your knowledge of Bangkok, how have streets and automobiles damaged life in that city? What role have the streets played? The automobiles?*

IT'S A FACT The city of Bangkok, Thailand, is sinking and might be below sea level in about fifteen years.

This incredible rate of growth has created major problems in Bangkok. In the past, canals branched off in all directions across the city, and Bangkok was called "the Venice of the East." During the last thirty years, however, most of these canals were paved over to provide streets, destroying the city's natural drainage system.

The people of Bangkok dug 11,000 wells beneath the city to get clean drinking water. When this water was drained from the subsoil, Bangkok began to sink at a rate of three inches per year. The city now is submerged under three feet of water during the rainy season each year.

Basic human services in Bangkok have not kept up with population growth. Entire sections of the city have no sewage removal system, and piped water is delivered to less than half of its people. Because housing is in short supply, many people in Bangkok live on boats. Some 600,000 automobiles jam the streets of Bangkok, creating traffic jams that strangle the city. Air and water pollution in Bangkok are bad enough to endanger health.

In spite of these problems, tourists regularly fill the 12,000 luxury hotel rooms in Bangkok. Western-style skyscrapers, office buildings, and condominium complexes rise at the city's center. Outside the inner city, 1 million Thais live in Bangkok's many slums. The city needs to build 400,000 new houses a year to satisfy its growing population, but only one-fifth of these are being built.

⊕ REVIEW QUESTIONS

1. What is the largest city in Indonesia, in the Philippines, and in Thailand?
2. What are the three largest cities in Vietnam?
3. What problems occurred when most of the canals in Bangkok were paved over to provide streets? These problems are examples of which theme of geography?

⊕ THOUGHT QUESTIONS

1. Why do you think that Europeans developed cities on the coasts of the countries of Southeast Asia?
2. Why was there little manufacturing in Southeast Asia during the colonial period?

🌐 CHAPTER SUMMARY

Geographically, Southeast Asia includes a large peninsula and many islands on the southeastern fringe of the Asian mainland. Most countries on the mainland are divided by mountains, ranges of the high Himalayas. Southeast Asia is home to 460 million people living mostly in the habitable river valleys, along the coasts, and on islands. Southeast Asia is rich in natural resources, but until recently has had limited means to develop them.

Three kinds of farming are practiced in Southeast Asia. In tropical forests in the mountains, shifting cultivation is common. Farmers clear trees and burn them, then plant crops in the soil that has been enriched by the ash. In a few years, the fertility of the soils declines, and farmers move on to other forest areas and repeat the process. Wet rice farming is practiced on flooded fields and terraces. The water sources are river floods, rainfall, or irrigation. Commercial farming produces cash crops for export, including rubber, coconuts, palm oil, coffee, tea, and tobacco.

Almost all Southeast Asian countries have to contend with pressures of geographic limits on farmland and growing populations that are pushing many people into the cities. However, large cities, such as Bangkok, do not have the infrastructure to absorb all these people. This means the cities suffer from major environmental problems, and the diminishing quality of life their people must endure.

EXERCISES

Cities and Countries

Directions: Match the city with the country in which it is located. On your paper, write the name of each city after the name of the country.

1. Indonesia
2. Myanmar (Burma)
3. Vietnam
4. Thailand
5. Malaysia
6. Philippines

a. Bangkok
b. Manila
c. Jakarta
d. Ho Chi Minh City (Saigon)
e. Yangon (Rangoon)
f. Kuala Lumpur

Complete the Sentences

Directions: Choose the correct ending for each sentence and then write the complete sentence on your paper.

1. The mountains that reach southward into mainland Southeast Asia are ranges of the (a) Urals (b) Carpathians (c) Himalayas.
2. There are many volcanoes in (a) Indonesia (b) Thailand (c) Burma.
3. One of the chief crops raised by shifting cultivators is (a) tobacco (b) yams (c) coffee.
4. The green revolution has produced the most successful results in increasing the yield of (a) rice (b) bananas (c) coconuts.
5. The most important commercial crop in Southeast Asia is (a) tea (b) rubber (c) rice.
6. Most of the large cities in Southeast Asia are (a) inland from the coast (b) high in the mountains (c) coastal ports.
7. A landlocked country in Southeast Asia is (a) Thailand (b) Laos (c) Malaysia.

True or False?

Write *true* on your paper next to the numbers of the correct statements and *false* next to the numbers of the incorrect statements.

1. The mountains of mainland Southeast Asia have hindered east-west communication more than north-south communication.
2. Because Southeast Asia is a hot tropical region, most of its people live on highland plateaus and mountainsides.
3. The monsoon rain forests of mainland Southeast Asia have a dry season.
4. More than half of all Southeast Asians live on islands.
5. There are more tropical rain forests than deserts on the islands of Southeast Asia.

Inquiry

Directions: Combine the information in this chapter with your own ideas to answer these questions.

1. Why does shifting cultivation on a large land area support so few people?
2. Compare the two most common types of agglomerated settlements with dispersed settlements. Reread the information on wet rice farming. What advantages would agglomerated settlements have for wet rice farmers compared to dispersed settlements?
3. If you decided to live in Southeast Asia, what location would you choose? Why?

SKILLS

Using Three Maps in the Miniatlas

Directions: In this activity, you will use the political, physical, and population maps on pages 653–655. You often will need to study more than one map to find an answer.

Each of these groups of sentences has three statements. Two of the statements in each group are true. Write these statements on your paper.

1. a. Thailand is larger than Cambodia.
   b. Laos is larger than Myanmar.
   c. Indonesia is larger than Vietnam.
2. a. Myanmar borders India and China.
   b. Vietnam borders China, Cambodia, and Laos.
   c. Thailand borders Indonesia and Borneo (Kalimantan).
3. a. Yangon and Singapore are capital cities.
   b. Ho Chi Minh City and Mandalay are capital cities.
   c. Jakarta and Phnom Penh are capital cities.
4. a. The Khorat Plateau is in Indonesia.
   b. Luzon is in the northern part of the Philippines.
   c. The Red River flows through Vietnam.
5. a. Myanmar is northeast of Brunei.
   b. Vietnam is northeast of Singapore.
   c. Sumatra is southwest of the Philippines.
6. a. Vietnam has a longer coastline than Cambodia.
   b. Laos has no coastline.
   c. Java is a larger island than Sumatra.
7. a. Vietnam is more densely populated than Laos.
   b. Cambodia is more densely populated than Thailand.
   c. Java is more densely populated than Borneo (Kalimantan).
8. a. All areas in Malaysia have more than 250 persons per square mile.
   b. All areas in Java have more than 25 persons per square mile.
   c. Some of the islands in the Philippines have more than 250 persons per square mile.
9. a. The northwestern part of Myanmar is a mountainous region.
   b. Mountains are found near the southern and western coasts of Sumatra.
   c. Sulawesi (Celebes) is a flat land without any mountainous regions.
10. a. The northwestern part of Myanmar is less densely settled than the southeastern part.
    b. The coast of Vietnam is more densely settled than the interior.
    c. All parts of Sumatra are more densely settled than Java.

Vocabulary Skills

Directions: Match the numbered definitions with the vocabulary term. Write the term on your paper after the number of its matching definition.

agglomerated settlement
commercial farming
dispersed settlement
linear village
monsoon rain forest
nuclear village
shifting cultivation
terrace
typhoon
wet rice farming

1. Level steps dug into hillsides to create fields and increase the area under cultivation
2. Isolated farmsteads removed from neighbors
3. Dwellings clustered around a common center
4. Burning off the forest cover to create farmland that can be cultivated for three or four years
5. Houses set out along the side of a roadway or path
6. Sprouting rice in beds and then transplanting the shoots to flooded fields
7. Clusters of houses surrounded by gardens, fields, and pastures
8. Severe tropical storm with high winds that can cause great damage
9. The growing of crops like rubber, coffee, sugar, and coconuts for sale
10. A tropical forest that has a dry season and is composed of deciduous as well as evergreen trees

CULTURE AND ECONOMIC CHANGE IN SOUTHEAST ASIA

Southeast Asia is a crossroads of peoples and cultures. People in this region speak many different languages, practice a variety of faiths, and come from different ethnic backgrounds.

The region's location on important sea routes and its natural resources have brought colonizers and traders from many parts of the world to this crossroads.

◀ *This beautiful photo of Phi Phi Island, Thailand, suggests some of the great beauty and diversity of Southeast Asia. How many climate environments can you identify in this picture?*

1. Indian and Chinese Influences in Southeast Asia

Early Farmers in Southeast Asia

The early peoples of mainland Southeast Asia invented a farming system that was well suited to warm, tropical environments. Southeast Asians were the first farmers to grow tropical crops like yams, taro, breadfruits, bananas, and coconuts. They also kept pigs and chickens as domesticated animals.

Very early, most of Southeast Asia's farmers lived in villages surrounded by orchards, vegetable gardens, and fish ponds. By 1000 B.C., these farmers were growing wet rice on terraced hillsides and in river valleys. Wet rice, as you know, is a very productive crop when grown in irrigated fields. These crops and animals provided an ample food supply for Southeast Asia's villagers long before Indian and Chinese influences led to the organization of these village farmers into city-based states ruled by kings.

The Indians Come to Southeast Asia

Indians were the most important outside influence on most of mainland Southeast Asia in the first millennium. Indian merchants sailed eastward, exploring the sea route between India and China. They sought gold, tin, teak, ebony, camphor, ivory, and most important, spices. These merchants set up trading posts on the Myanmar coast, along the shores of the Malay Peninsula, at the mouth of Thailand's Chao Phraya River, in the Mekong River Delta, and on the Pacific coast of Vietnam where Chinese influence was strong. Indian trading posts were strung out like beads along the entire coast of mainland Southeast Asia.

Indian priests and monks introduced Indian religions, literature, languages, and art into Southeast Asia. The Hindu way of life, with its caste groupings, spread throughout the region. Buddhism, another Indian religion, also spread eastward. Indians also introduced their system of government to this region. City-based coastal kingdoms were established throughout Southeast Asia.

All of these coastal kingdoms traded widely, built irrigation systems to improve crop yields, and constructed elaborate palaces for their rulers. These kingdoms lasted for nearly 1,000 years. The most famous of them was that of the Khmer people in the lower Mekong Valley. The people of this kingdom built the world's largest religious structure, the huge temple complex at Angkor.

The Chinese Influence

In time, these coastal kingdoms were replaced by large, highly organized inland states located in river valleys. There were two reasons for this shift from the coast to interior valleys. First, another wave of peoples from China moved southward into these valleys in the 1100s and 1200s and founded inland kingdoms that eventually conquered those on the coast.

Second, the river valleys of mainland Southeast Asia had a larger amount of fertile farmland than coastal areas. Dense populations were fed more easily in the river valleys, so these areas became major centers of population. One kingdom in northern Vietnam was part of the Chinese Empire for almost 1,000 years.

Chinese writing, systems of government, and Confucian thought prevailed throughout this Vietnamese kingdom. Chinese methods of water control were used to prevent floods and to store water for irrigation. Under Chinese rule, the plain of the Red River in northern Vietnam became the most densely settled place in mainland Southeast Asia. The Vietnamese people adopted virtually all of China's culture except its language.

When this Vietnamese kingdom became independent of China in A.D. 939, the north Vietnamese pushed southward into the delta of the Mekong River. Chinese influence remained strong after Vietnamese independence, although the Vietnamese cherished their new found freedom, and conflicts between the two countries were frequent.

⊕ REVIEW QUESTIONS

1. What crops did the early farmers of Southeast Asia raise?
2. What two Indian religions were introduced into Southeast Asia?

⊕ THOUGHT QUESTIONS

1. Why do you think Southeast Asia has been attractive to people for many centuries?
2. Why were the coastal kingdoms that grew out of Indian trading posts replaced by inland kingdoms in river valleys after about 1,000 years?

2. Traders and Rulers from the West

Muslims in Southeast Asia

A third wave of new ideas and ways of living swept into Southeast Asia in the 1300s and 1400s, when Muslim traders from northwestern India introduced Islam into this region. These Muslim merchants, who dealt in the trade that flowed between Europe and China through Southeast Asia, set up trading posts that soon became religious centers. The Islamic faith spread throughout Indonesia

▼ *Angkor, which had a very sophisticated irrigation system, was the capital of the early Khmer civilization. Later, Buddhists temples were built, the largest being Angkor Wat. What does the term* wat *mean?*

and the Philippines. Port cities on the Malay Peninsula and the islands of Sumatra and Java were converted to Islam. Indonesia remains predominantly Muslim.

The most important of these Muslim trading centers was Malacca on the southwest coast of the Malay Peninsula. This port city was located on a narrow strait on a major international sea route that passed through Southeast Asia. Find it on the map on page 677.

Large Chinese junks sailed southward to Malacca on the northeast monsoon that blows to the south in the winter. At Malacca, Chinese traders exchanged goods with Persian, Arab, and Indian traders who arrived on the southwest monsoon that blows to the east in the summer in Southeast Asia. In Malacca's magnificent bazaar, warehouses were crammed with gold, cotton, silk, tea, tin, precious woods, and spices.

The Portuguese Take Over the Spice Trade

When merchants in Western Europe learned of the riches of Malacca, they were especially interested in spices like pepper, cloves, and nutmeg that grew in Southeast Asia. By the time these goods reached Venice, Italy, their prices were so high that only the wealthiest Europeans could afford them. These Southeast Asian crops "spiced up" the monotonous meals—including salt-cured meat—that many Europeans ate at the time. That is why the Portuguese wanted to seize control of the spice trade of Southeast Asia.

In 1511, a Portuguese fleet attacked and captured the port of Malacca. The Portuguese then tried to

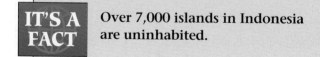

take over the Moluccas (Maluku) or Spice Islands in Indonesia, where pepper, nutmeg, cloves, and mace were grown. The Portuguese did manage to capture a few islands in Indonesia, and their profits were enormous.

The Spaniards in the Philippines

The Portuguese soon had European competitors for this wealth. During the next fifty years, the Spaniards tried unsuccessfully to loosen the Portuguese grip on Indonesia. Finally, they moved northward into the Philippines and established a fortress settlement at Manila on the island of Luzon.

▼ *This early print shows pepper harvesting in the Molucca Islands (Spice Islands). Various European powers tried to gain a monopoly over Southeast Asia's spice trade. Why were Europeans so anxious to control the pepper trade?*

LOCATION

THE FABLED PORT OF MALACCA, WHERE THE MONSOONS MEET

Malacca, a Portuguese sailor wrote in the 1500s, "is the richest seaport with the greatest number of merchants and abundance of shipping that can be found in the whole world." For centuries Malacca was a meeting place for seafarers from around the world. Arabs and Persians brought rosewater, carpets, tapestries, and incense to Malacca. From Japan and China came silk, porcelain, gold, and salt. Southeast Asian traders offered camphor, tin, bird feathers, spices, and precious woods for sale in the city's bazaar. Indian, Chinese, Japanese, Arab, Persian, Portuguese, Dutch, and British merchants and adventurers were all drawn to Malacca in search of wealth through trade or plunder.

Geography, specifically location, was responsible. Malacca is located on the narrow Strait of Malacca between the mainland of Southeast Asia and the island of Sumatra. The main sea route that links the Indian Ocean with the South China Sea passes through this strait.

Malacca is also located near the equator, where the winter and summer monsoons meet (see page 616). Chinese **junks** (large seagoing ships) and sailing ships from the northern reaches of Southeast Asia rode the northeast monsoon into Malacca in the winter. After unloading their cargoes of silk, pottery, and spices, these same vessels were filled with precious goods from India and the Middle East and sailed back to their home ports on the southwest monsoon in summer.

European, Indian, and Middle Eastern merchant ships rode the same monsoon winds to and from Malacca. They filled their holds with cargoes of spices, silks, and other riches from the East.

Two factors contributed to Malacca's great wealth: (1) the city's location where the monsoons meet and (2) its situation on a world trade route at the mouth of a river where sailing ships could find shelter. Indians, Arabs, Portuguese, Dutch, and British fought for control of the city over the centuries.

Today, Singapore has replaced Malacca as the chief center of trade. Singapore is a **city-state** at

▲ *Although once an important center of trade, today Malacca, on the Malay Peninsula, is no longer of strategic importance. Many people live on boats on canals, which are also major arteries of transportation. Can you think of a European city where canals play a similarly important role?*

the tip of the Malay Peninsula. A city-state is a large urban center that functions as an independent nation. Singapore, like the old port of Malacca, owes much of its wealth to geography.

QUESTIONS

1. What was the primary reason for Malacca's success as a chief trading port?
2. What role did the summer and winter monsoons play in encouraging trade?
3. What city now occupies the position that Malacca held for so many centuries?

Manila became an important Southeast Asian trading city. It was the Asian center for Spanish trade across the Pacific Ocean from Acapulco, Mexico. At Manila, Mexican silver was exchanged for silk and porcelain from China.

Filipinos who lived near Spanish settlements soon converted to Christianity and adopted Spanish systems of landholding and government. For the next 300 years, the Philippine Islands were Spanish territory.

The Dutch East India Company

Portuguese and Spanish interests in Southeast Asia were soon challenged by the Dutch, who formed a merchant company called the Dutch East India Company. The goal of this company was to win the riches of Indonesia and the Philippines for the Netherlands. The Dutch captured Malacca from the Portuguese in 1641 and also established their own port city at Batavia (now Jakarta) on the north coast of Java. From these two naval bases, the Dutch were able to control much of Southeast Asia's ocean trade for the next 200 years.

Like the Portuguese, the Dutch tried to gain a monopoly over the spice trade of Southeast Asia. Gradually, however, they attempted to take over the producing areas where the spices and other tropical products were grown. The Dutch expanded production of sugar, tobacco, palm oil, and pepper on Java. They also introduced new export crops like coffee to Southeast Asia.

⊕ REVIEW QUESTIONS

1. The Muslims spread the Islamic faith to what parts of Southeast Asia?
2. Which European country conquered the Philippine Islands in the 1500s?
3. From what two ports were the Dutch able to control most of Southeast Asia's ocean trade for nearly 200 years?

⊕ THOUGHT QUESTIONS

1. Why did Malacca become a center of trade between people from eastern and western Asia?
2. Why did Manila become a Southeast Asian trading center?

3. European Rule in Southeast Asia

The British and the French Move In

The British East India Company, which controlled India in the 1800s, began to expand into Southeast Asia. The company had already established the great port city of Singapore at the tip of the Malay Peninsula. It soon took over the functions of the old port of Malacca on the Strait of Malacca, through which most of the trade between Europe and Asia flowed.

The British established colonies in Myanmar (then called Burma) and the Malay Peninsula in the 1850s and also occupied the northern part of the large island of Borneo (Kalimantan). This string of colonial territories stretching from India through Southeast Asia guarded Britain's major trade lines to East Asia.

The French seized Cambodia from Thailand in the 1860s. Despite resistance, the French managed to establish colonial rule throughout all of Vietnam. They rounded out their colonial empire in mainland Southeast Asia by taking over Laos from Thailand in 1899. With the French in mainland Southeast Asia and the British, Dutch, and Spaniards in the part of Southeast Asia that is on islands, the European influence began to change the economic life of this region in the early 1900s.

Europeans Change the Economy of Southeast Asia

Two technological advances made it easier for European countries to control their colonies in Southeast Asia. First, in 1867, the Suez Canal opened a faster sea lane between the Mediterranean Sea and Southeast Asia, so that European ships no longer had to sail around Africa to reach Asia. Second, steamships replaced sailing ships, so that the speed and reliability of transportation and communication increased. This faster shipment of goods between Europe and Southeast Asia encouraged Europeans to invest in commercial activities in Southeast Asia.

Within fifty years, Europeans transformed the economic life of Southeast Asia. Port cities replaced inland settlements as the most important centers of business and trade. The little town of Saigon (now Ho Chi Minh City) in southern Vietnam grew into a large and busy port. The villages of Phnom Penh in Cambodia and Rangoon (now Yangon) in Myanmar also grew into large seaports. The important city of Kuala Lumpur in Malaya (now Malaysia) started as a tin-mining camp. Jakarta (then Batavia) in Indonesia and Manila in the Philippines became great cities where small trading settlements had existed earlier.

On the coasts, modern systems of water technology turned marshlands and swamps into productive fields of irrigated rice. The huge deltas of the Ayeyarwady (Irrawaddy), Chao Phraya, and Mekong Rivers began to produce vast amounts of rice.

Elsewhere, cash crops were grown for export and new mines were developed. The Dutch raised tobacco on Sumatra and coffee, tea, and sugar in Java. On other islands in Indonesia, they planted tea and spices and opened up tin mines. On the Malay

IT'S A FACT   Southeast Asia got its first rubber trees from London about 100 years ago.

Peninsula, the British turned jungles into plantations of rubber trees. In the French colonies of mainland Southeast Asia, coffee, tea, and rubber plantations were started. In the Philippines, sugar, tobacco, copra, and hemp were grown for sale on world markets. Huge amounts of rubber, sugar, rice, tin, and oil began to move from Southeast Asia through the Suez Canal to Europe.

At the same time, the introduction of improved medical and sanitation practices led to population growth in Southeast Asia. Within a little more than 100 years, the population of Southeast Asia increased six times. Many Southeast Asians found jobs in the region's huge port cities, and others worked in the rice fields on the coast, on plantations, and in mines managed by Europeans. Chinese and Indian immigrants also moved to Southeast Asia, attracted by job possibilities.

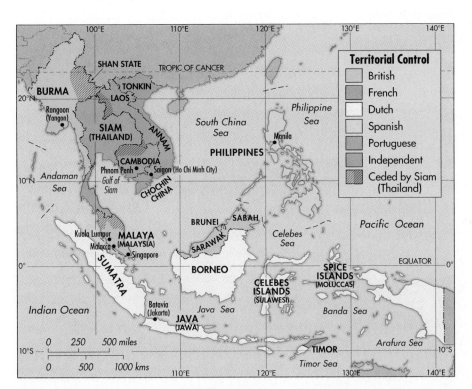

Colonial Rule in Southeast Asia

Independence After World War II

Discontent with European colonial rule grew rapidly in Southeast Asia as the gap between rich people and poor people increased. Soon Southeast Asians began to demand greater control over their own destinies. Then in 1941, the Japanese invaded and took over much of Southeast Asia. Japan lost these conquered lands when it surrendered to the United States in September 1945.

Independence came to much of Southeast Asia after World War II. In 1946, the United States fulfilled its promise and granted independence to the Philippines. One country after another in Southeast Asia became independent. Myanmar gained independence in 1947. Indonesia, Laos, and Cambodia became independent in 1949. The Federation of Malaysia, which included Malaya, Singapore, and British territories on the island of Borneo (Kalimantan), gained independence in 1963. In 1965, Singapore withdrew from the federation of Malaysia.

The French, however, decided to fight to keep their Southeast Asian colony in Vietnam. After years of guerrilla warfare against the independence movement organized by Communist leader Ho Chi Minh, the French were defeated in 1954. A result of the 1954 Geneva Conference held after this war was the provisional division of Vietnam into two parts, north and south, until national elections could be held. The leader of the south, Ngo Dinh Diem, refused to hold elections and declared the south an

| IT'S A FACT | Some 300 different languages and dialects are spoken in Indonesia. |

independent state, which Communist guerrillas tried to overthrow. During the conflict that followed, the Soviet Union and China supported the North Vietnamese, and the United States supported South Vietnam. North and South Vietnam were united by force in 1975, when Communist North Vietnam occupied the south.

REVIEW QUESTIONS

1. What colonies were established in Southeast Asia by the British after 1850?
2. The French established footholds in what parts of Southeast Asia in the 1800s?
3. What were the chief cash crops raised in each of these regions: Indonesia, the Malay Peninsula, and the Philippines?

THOUGHT QUESTIONS

1. What two technological advances made it easier for European countries to control their colonies in Southeast Asia?
2. What helped cause Southeast Asia's huge population growth in the last 100 years of the colonial period? Where did Southeast Asians find jobs?

▶ *Forty percent of the world's rubber comes from Malaysia. The trees are grown on large plantations, need little care and can be harvested all year long. What North American forest product is also harvested by tapping trees?*

4. Strategies for Development in a Troubled Region

Troubles in Southeast Asia

Independence did not erase the effects of colonial rule in Southeast Asia. Young governments throughout Southeast Asia still seek to unify people of varied cultures, languages, and ways of living within their boundaries. Civil wars within countries and wars between Southeast Asian countries have created chaos.

The main goal of most leaders in Southeast Asia is to raise standards of living, but rapid population increases and the explosive growth of cities hinder economic progress. Efforts are being made to diversify industry, to improve educational opportunities, to provide better living conditions, and to create more jobs. Accomplishing these objectives is very difficult in a region still suffering from the effects of war.

Various strategies are being used. Several Southeast Asian countries have adopted communism as a pathway to economic development. Laos and Viet-

Southeast Asia Resource Map

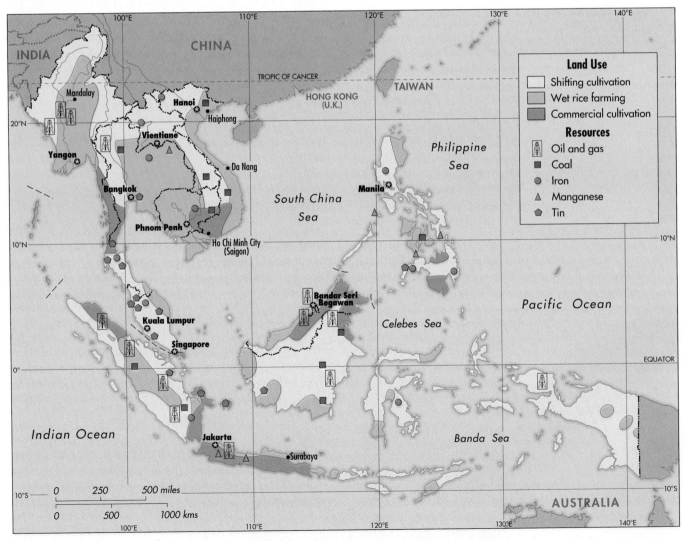

Land Use	
	Shifting cultivation
	Wet rice farming
	Commercial cultivation

Resources
- Oil and gas
- Coal
- Iron
- Manganese
- Tin

nam now have Communist governments. In 1993, Cambodia elected a new coalition government under United Nations' supervision. Other Southeast Asian countries do not want communism. Among these countries, Indonesia, Malaysia, the Philippines, Thailand, and Singapore have formed the Association of Southeast Asian Nations (ASEAN), an organization that encourages cooperation among its members and independence from all foreign countries.

Two Strategies for Development

The larger countries of Southeast Asia share many similar problems. Their people require more jobs, better social services, and higher incomes. The need for rapid economic development in this region is urgent.

Two different strategies to achieve economic development have been tried in Southeast Asia. The first, adopted by a group of outward-looking countries, has used Western-style planning to achieve economic growth. The countries' plans include increasing their involvement in the world economy and participating in regional organizations. Included in this group of countries are Thailand, the Philippines, Malaysia, Brunei, and the city-state of Singapore (a **city-state** is an urban center that is an independent country). All have experienced high rates of economic growth in the last twenty years.

In contrast, a second group of Southeast Asian countries has been damaged by civil war and internal divisions. Political problems have blocked economic growth. Myanmar, Vietnam, Laos, Cambodia, and Indonesia are members of the inward-looking group. With the exception of Indonesia, which has been developing rapidly during the 1990s, the rates of economic growth in these countries have been sluggish.

Thailand

Thailand is a country about the size of Texas that occupies the central part of mainland Southeast Asia and extends southward along the Malay Peninsula, as the map on page 653 shows. The valley of the Chao Phraya River is Thailand's core area. This river rises in the mountains of northern Thailand and flows southward across a broad lowland plain before passing through the city of Bangkok on its way to the Gulf of Thailand.

Most of Thailand's 57.2 million people live in the valley of the Chao Phraya and along its tributaries. Villages spread out along the banks of the rivers and blanket the lowland plains. This is the "rice bowl" of Thailand.

Most Thais earn their living directly from the land by working in farming, forestry, and mining. Thailand was the largest exporter of food in Asia, but the value of these exports has been overtaken by tourism and textiles. Rice, rubber, tin, and teak (a hard, beautiful, tropical wood) are other important export products of Thailand. Most local manufacturing industries in Thailand process farm and forest products, and many of the forests in the country have now been cut down. Petroleum, machinery, and manufactured goods are its leading imports.

Thailand's economy has grown more than 7 percent, twice the world average, each year during the last ten years. This growth has been enhanced by Thailand's tolerance for diversity and its measured approach to economic development. But a 1991 military coup brought its political stability into question.

Thailand faces several problems in maintaining this rate of economic growth. Oil imports are expensive, so Thailand is building a huge oil refinery south of Bangkok to cut energy costs. In addition, an oil field has been found in northern Thailand. In the late 1980s, natural gas from the Gulf of Thailand began to be pumped to Bangkok through Asia's longest undersea gas pipeline.

The country's huge forests are being leveled. Twenty years ago, forests covered more than half of Thailand. Today, forests cover only 30 percent of Thailand's land area. There is very little new farmland available to support Thailand's growing village population. More dams are needed to prevent flooding of existing farmland, to provide irrigation water to farmland in the dry season, and to generate hydroelectric power for industry.

The Thai government plans to continue its outward-looking policy of exporting raw materials while building the country's industry and agriculture. Investments in farming include efforts to develop new cash crops like jute and sugarcane. New industrial zones are being set up to encourage small industries throughout the country. Thailand is beginning to create manufacturing and service jobs for its growing city populations.

The Philippines

There are an estimated 7,107 islands in the Philippines. The two largest are Luzon and Mindanao, as the map on page 654 shows. This archipelago is actually a chain of half-submerged mountains located in the South China Sea about 600 miles off the coast of the Asian mainland. The total land area of the Philippines is about as large as that of Italy and slightly smaller than that of Japan.

The central plain of Luzon, the most densely populated area in the Philippines, is where Manila, the country's capital and most important port and one of Southeast Asia's largest cities, is located. One of every six Filipinos lives in Manila.

About 64.6 million people live in the Philippines, and population is growing rapidly. Because much of the country is mountainous, farmland is in short supply. Volcanic peaks rise to elevations of nearly 10,000 feet. Hillsides are terraced and planted in rice. New land is being brought into cultivation on the island of Mindanao, but population pressure on available farmland is intense. The country's wealth is unevenly divided.

Many Filipinos are village farmers who have little or no land and whose lives are barely touched by modern ways of living. Thousands more poor Filipinos are crowded into the slums that ring Manila. This widespread poverty was one of the reasons for the 1986 political upheaval that forced President Ferdinand Marcos to flee after more than twenty years of running the Philippines. Since then, political disruption, the fiery volcanic eruption of Mt. Pinatuba, and the elimination of U.S. military bases

◄

Manila, the capital city of the Philippines, has one of the best natural harbors in the world. The people of the Philippines speak two languages, the first being the national language, Filipino. What is the second language?

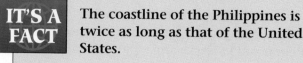

in the Philippines have had a serious impact on lives in the country.

As in Thailand, the economy of the Philippines is based on products from the land. Sugar, coconut oil, bananas, tropical woods, nickel, iron, and copper are the country's most important exports. Petroleum and machinery are its major imports. In the Philippines, however, manufacturing is growing rapidly to create a more diversified economy than that of Thailand. Textile, shoe, electric appliance, and electronics factories produce and assemble consumer goods. These industries benefit from the Philippines' low labor costs, which have attracted Japanese and U.S. investment to the islands. The opening of new copper, nickel, and chromite mines promises new jobs in metal-processing industries.

In addition, the United States introduced the green revolution into the Philippines twenty years ago in an effort to relieve the poverty of local farmers. High-yielding varieties of rice did prove successful where farms were able to supply fertilizers and pesticides and to control water. None of these measures have succeeded in reducing poverty in the Philippines, however. An estimated 70 percent of all Filipinos live in poverty, mainly because a wealthy elite in the Philippines owns or controls nine-tenths of the best farmland. Whether or not new political regimes in the mid-1990s will alter these conditions is difficult to predict.

The Federation of Malaysia

Malaysia is a divided country composed of a collection of former British colonies. It includes the southern half of the Malay Peninsula and much of the northern coast of the island of Borneo (Kalimantan). In 1963, these territories, along with Singapore, joined the Federation of Malaysia. Two years later, Singapore became an independent city-state.

Most of Malaysia is a country of mountains and lush rain forests. A majority of the nation's 18.4 million people live on a narrow coastal plain on the west side of the Malay Peninsula, where the federal capital of Malaysia, Kuala Lumpur, is located.

Malaysia has a variety of cultures. About 44 percent of the people are Malays, early settlers of this land. Another third are Chinese. The rest are Hindu Indians and members of hill tribes, many of whom live in Sarawak and Sabah, the two major Malaysian states on northern Borneo.

Malaysia is one of the richest in resources of all the countries in Southeast Asia. Its economy is based on exports of rubber, tin, palm oil, tropical woods, and petroleum from offshore fields. Malaysia now accounts for half the world's export of tropical timber. Chinese merchants and foreign investors have encouraged rapid growth in manufacturing.

Heavy machinery still must be imported, but Malaysians now produce their own textiles, electronics, and consumer goods. The chemical, electrical appliance, and transport equipment industries are booming. Standards of living are rising, and new jobs are being created. Relations between the Chinese and the Malays have improved, and Malaysia's government continues to encourage economic growth.

The Sultanate of Brunei

Brunei, a former British territory, became independent in 1984. It is ruled by a sultan, and has had a colorful history. Brunei was a haven for pirates, then was ruled by a British adventurer, and came to world attention when oil was discovered there in 1929. Located in the northeastern part of the island of Borneo, Brunei is nearly split in two because long ago the ruler of neighboring Sarawak took over a river valley that runs through the country's middle.

Brunei is smaller than Delaware, has a population of 300,000, and is rich because of its oil and gas reserves. The sultanate owes everything to oil. Its people have free medical services, pay no taxes, and enjoy a high standard of living.

Myanmar (Burma)

Myanmar is a Texas-sized country with a population of 43.5 million. The country was once called the "Golden Land" because of its diversity of natur-

IT'S A FACT Elephants feed for eighteen hours a day and sleep standing up.

al resources, including petroleum, minerals, rivers for irrigation and hydroelectric power, and good farmland. But in spite of Myanmar's wealth of resources, it now ranks among the poorest countries in the world.

Mountains and hills ring Myanmar's two rivers, the Ayeyarwady (Irrawaddy) and the Thanlwin (Salween). These rivers flow from the mountainous north to the sea and form the heart of Myanmar. The most important river is the Ayeyarwady, on whose banks Myanmar's two largest cities are located. Mandalay, in the center of this elongated country, is an important railroad hub. Yangon (Rangoon), on the delta of the Ayeyarwady, is Myanmar's main seaport and national capital. It is also located at the center of the country's largest rice-producing area.

Myanmar's poverty is the result of a thirty-year government policy of self-imposed isolation. In the early 1960s, the government closed the country to foreigners, expelled local Indian merchants, and took over all farmland and businesses. This "Burmese way to socialism" was based on the belief that the country's natural resources would provide great wealth once foreign business influence was removed. Companies left the country, foreign investment stopped, and economic growth came to a halt. This policy has been a total failure.

Faced with a growing population, the Burmese economy is going to ruin. Uprisings disrupt life in the mountains, and the incomes of the Burmese are

▼ *In this photograph, school children are in canal boats in Myanmar (Burma). Lack of government services is a great problem in this country. What clues to identifying a primary occupation are provided in this photograph?*

SINGAPORE, "THOROUGHFARE TO THE EAST"

In 1819, a young British merchant named Stamford Raffles came to Singapore on a week's business trip and bought the island for 33,200 Spanish dollars. He soon turned the island city into what novelist Joseph Conrad called "the thoroughfare to the East." Raffles made Singapore an exchange center for the world's ships, a role that the nearby city of Malacca had played in earlier times.

Singapore became a wealthy and dynamic trading center. Immigrants from China, India, and In-

▲ *Outdoor restaurants fill Bugis Street in Singapore, attracting diners from all over the city. Do you believe banning automobiles in some parts of downtown American cities might help their economies? Or would this just increase traffic problems?*

donesia came to Singapore as settlers, traders, and laborers. Today, Singapore is an independent republic whose 2.8 million people, three-quarters of them Chinese, are crowded into an area of less than 240 square miles.

Singapore is located at the tip of the Malay Peninsula. Only sixty miles from the equator, the island

◄ *Pedicabs are a common mode of transportation in Singapore. Do you think these could be used in the central business districts of some large American cities? If so, what would be the advantages?*

▲ *Singapore's port, the second busiest in the world, services over 300 ships per day. Singapore's location as a trade center links what large culture regions?*

is on the vital sea route linking Europe and the Middle East with East Asia. Today, Singapore is one of the busiest ports in the world, serving 300 or more ships a day. Singapore has used its advantageous location to become an important trade center for Southeast Asia, Japan, and other industrial countries.

Oil from the Middle East is sent to Singapore for refining and then to countries throughout the region. Rubber, tropical woods, rice, and spices flow through Singapore to destinations around the globe. Singapore is also a diversified manufacturing center. Its bustling shipyards, machine shops, food-processing plants, and high-tech industries have led many to call this island republic a "new Japan."

QUESTIONS

1. What other famous island in history was purchased as a trading post?
2. What administrative status does Singapore hold?
3. Why has the description "new Japan" been applied to Singapore?

falling. Muslims and tribal groups like the Karen are being systematically persecuted.

The military that has run the country since 1991 has decided to create a high-tech city for 4 million people in the forests south of Yangon (Rangoon), where a half-million people have been forcibly moved. This area has no local resource base and few educated engineers or computer specialists. Most people still work in farming and fishing. The generals remain in power, financed by the ruthless cutting of forests (mainly teak) and the export of gems. Economic and industrial growth in isolated Myanmar must overcome the results of thirty years of government neglect.

Vietnam

Vietnam extends southward along the east coast of mainland Southeast Asia for more than 1,000 miles. Two clusters of population exist along this coast: one is at the mouth of the Red River in northern Vietnam, and the second is on the delta of the Mekong River in southern Vietnam. Between these two lowlands, in central Vietnam, mountains plunge directly into the sea and form a barrier to communications between north and south. The interior of the country is mountainous.

In the 1960s, the two densely populated lowlands were the heartlands of separate states: North Vietnam and South Vietnam. These two states became one country when Communist North Vietnam took over all of Vietnam after twenty years of destructive warfare.

Today, the destruction of this war, the cost of maintaining a huge army, and economic mismanagement have seriously damaged the quality of life in Vietnam. The standard of living, which is much lower in the north than the south, is still falling. Vietnam, with 71.8 million people, is now the second most populous country in Southeast Asia.

Northern Vietnam has coal and iron deposits and a variety of other mineral resources that together support an industrial complex near Hanoi. Most industry here is state owned, very inefficient, and runs at a loss. The main food-raising area in the north is nearby on the plains of the Red River, where a network of canals carries water to and from large expanses of rice fields. In the south, the largest area of rice production in Southeast Asia is found in the delta of the Mekong River. Southernmost Vietnam once exported large quantities of rice.

The industrial north and the agricultural south together would seem to provide a solid base for future development. Yet today even Vietnamese officials expect that it will take many years to regain prewar levels of industrial and agricultural production. War destroyed the coal-based industries of the north and badly damaged the rice fields of the northern and southern lowlands. Chemical defoliants used during the war did great damage to the natural vegetation as well as cultivated fields. Instead of rebuilding after the losses, the present government spent much of its income on military equipment. By the mid-1990s after the proven economic failure of communism, it was beginning to shift to a market economy based on the export of rice and oil.

But the Commonwealth of Independent States (CIS) no longer sends aid to Vietnam, and the country is at odds with China over some offshore islands in the South China Sea. The United States has resumed diplomatic relations with Vietnam and, along with Japan and Korea, is now interested in investing in the country as its communist-run government strives for a more open economic policy.

Laos and Cambodia

Neighboring Laos and Cambodia are also victims of the French-Indochinese War that engulfed mainland Southeast Asia following World War II. This war destroyed the cultural landscape, or the human and natural patterns on the land, of these already poor countries for nearly thirty years.

Laos is the smallest country on mainland Southeast Asia, and the 4.6 million people of this landlocked country are among the poorest. Most of Laos is rugged country draped with monsoon rain forests. Mountain ridges run parallel to the valleys in which flow the Mekong River and its tributaries, or smaller rivers that feed into it. In these valleys, most Laotians live and work as rice farmers.

The country has no seacoast, no railroads, and only a few roads; rapids interrupt river traffic moving southward on the Mekong River. War has hindered the development of the country's farmland, forest, and mineral resources. Some tropical wood, coffee, and tin are exported, but not enough to lift Laos out of poverty. In the 1990s, Laos trades primarily with neighboring Thailand to the west. Its major export product is hydroelectric power. By world standards, Laos is one of the world's poorest countries.

Cambodia, an almost round country of 9 million people, spreads across a broad plain through which the lower Mekong River flows. This plain is the most densely settled area in Cambodia. In its center, a gigantic freshwater lake called Tonle Sap, or "Great Lake," floods to more than three times its normal size during the wet season and shrinks in the dry season. This annual change in lake size renews the soil and deposits rich silt on the lake's shores, where rice is grown.

In 1975, Communists won a civil war in Cambodia and changed the country's name to Kampuchea. These Communists, called the Khmer Rouge, devastated the country's economy. In an incredible act of cruelty, they drove huge numbers of Cambodians from the cities into the countryside to try to create a classless agricultural society. An esti-

mated 2.5 million Cambodians died; tens of thousands more fled to refugee camps in neighboring Thailand.

Vietnam invaded Cambodia in 1978 to control the Khmer Rouge, and 700,000 Vietnamese settled there. A new communist government was set up, and a large part of the country was controlled by the Vietnamese army until the early 1990s. In 1991, four major Cambodian factions, among them the Khmer Rouge, agreed to hold elections. By 1993, the United Nations, with an army of 21,000 in Cambodia, monitored elections and helped create a new coalition government.

Island Indonesia

The estimated 13,667 islands in Indonesia make up a land area about one-fourth the size of the United States. These islands reach east-west across three time zones and 3,000 miles of seas and oceans, about the same distance as between New York and Los Angeles. Indonesia supports 187.6 million people, making it the most populous country in Southeast Asia. Only 6,000 of its islands are inhabited. Two of every three Indonesians are

◀ *Wheat, here harvested by Indonesian women, is also an important crop grown on Java (Jawa). Here, wheat is being picked by hand. How is wheat harvested on the Central Plains of the United States and Canada?*

crowded onto the small island of Java. The government is now forcibly shipping millions of people from Java to more remote islands in an effort to reduce overcrowding, a policy that has provoked considerable world criticism.

Indonesia's year-round warmth and abundant rain enable the country to produce large quantities of coffee, tea, rubber, sugar, and tobacco. Oil deposits have been developed on Sumatra, Borneo, and Java. Despite these rich resources, Indonesia is one of the less developed, more inward-looking countries in Southeast Asia.

This island country has been unable to develop its rich resources since it became independent in 1945. During the 1950s and 1960s, the Indonesian economy was badly mismanaged because of corruption and government preoccupation with internal security and political problems. Its violent treatment, since 1975, of the 600,000 Christians who live on East Timor—punctuated by a major incident in 1991—has provoked worldwide criticism. Little effort or money has gone into programs of economic development.

Now there is reason for hope. By 1993, Indonesia's economy was growing twice as fast as the world average. Huge offshore petroleum and natural gas fields provide revenues for expanding farming and building industry. The government still

▼ *Rice farming on the fertile volcanic soils of central Java (Jawa) provides the food to support two of every three Indonesians who live on this island. Do you understand why population is so dense on Java as compared to other Indonesian islands? (Note smoking Merapi Volcano in the background.)*

estimates that 15 percent of all Indonesians live in poverty, but in the late 1980s and early 1990s that percentage has been falling.

Indonesia is now self-sufficient in food, an encouraging change from twenty years ago when the country was the world's leading importer of rice. Imports of machinery, iron and steel, and chemicals are paid for by oil revenues. About 40 percent of the nation's income comes from petroleum, and much effort has been devoted to diversifying the country's economy. Its major export products are textiles and wood products. With one of the richest resource bases in the developing world, Indonesia holds more promise economically than most nations.

⊕ REVIEW QUESTIONS

1. What are the Philippines' most important exports?
2. What independent city-state is one of the busiest ports in the world?
3. What transportation problems handicap Laos?
4. The islands in Indonesia stretch across how many time zones?

⊕ THOUGHT QUESTIONS

1. Why has Myanmar, which was once rich, become one of the poorest countries in the world?
2. Why has the standard of living been falling in Vietnam since the communists took over that country?

⊕ CHAPTER SUMMARY

Many cultural influences have left their mark on Southeast Asia and form the basis for some of the current problems of the individual countries of this region. Early in human history, Southeast Asia was inhabited by farmers who developed systems of wet rice agriculture in river valleys and on terraced hill-sides. Later, Indians sailed on the summer monsoon from India to Southeast Asia seeking trade. They introduced the religions of Hinduism and Buddhism to Southeast Asia, faiths which are practiced in many countries of the region today. At the same time and later, the Chinese brought many cultural influences overland, particularly into Vietnam, including writing, systems of government, and social ethics. In the 1300s and 1400s, Muslim traders from northern India sailed the trade routes that pass through Southeast Asia to China. Islam was spread throughout Indonesia, the world's most populous Muslim country, and the Philippines. Relative location contributed to the growth of the fabled port of Malacca, which in those days was a major international seaport, similar to today's Singapore. In the 1500s, the Portuguese and Spanish became dominant powers in the region, followed by the Dutch, French, and English. Initially, these Europeans wanted access to the spice trade and to other subtropical and tropical natural resources of Southeast Asia. Later, they conquered the kingdoms that controlled this trade. The Europeans colonized the region to exploit raw materials of value in industrializing Europe. World War II and its aftermath brought independence to the region.

Today, the nations of Southeast Asia can be categorized as outward-looking countries that seek trade and economic ties with industrial countries and inward-looking countries that reject external models of development and try to find their own paths. Some, like Vietnam, are undergoing a transition from an inward- to an outward-looking attitude.

Vietnam, Laos, and Cambodia are still wrestling with the economic, social, and environmental problems that followed the end of two decades of civil war. At the same time, the city-state of Singapore is a rapidly developing economic power in the Pacific Basin. The future success of the remaining Southeast Asian countries, however, is still not certain.

EXERCISES

Arrange in Chronological Order

Directions: Following each number are statements about three events, but they are not in the order in which the events occurred. Write the statements on your paper in the correct chronological order.

1. (a) European colonial rule came to Southeast Asia. (b) Traders and missionaries from India and China strongly influenced Southeast Asia. (c) Muslims introduced new ideas and ways of living in Southeast Asia.
2. (a) The Spanish established a fortress settlement at Manila. (b) The Portuguese captured the port of Malacca. (c) The Dutch captured the port of Malacca.
3. (a) Vietnam was a kingdom controlled by the Chinese Empire. (b) Vietnam became a communist country. (c) Vietnam was a colony of France.
4. (a) The United States owned the Philippine Islands. (b) Spain owned the Philippine Islands. (c) The Japanese conquered the Philippine Islands.
5. (a) Indonesia became an independent country. (b) World War II occurred. (c) Indonesia became a Dutch colony.

Find the Missing Words

Directions: Following the sentences below is a list of missing words. Find the word missing from each sentence and then write that term on your paper.

1. Southeast Asia is positioned astride the natural sea routes between the |||||||||| Ocean and the lands of the Pacific.
2. The Indians introduced the religions of Hinduism and |||||||||| to Southeast Asia.
3. Before 939 A.D., Chinese influence in Southeast Asia was most strongly felt in what is now the country of ||||||||||.

4. In the 1300s and 1400s, the religion of |||||||||| spread throughout Indonesia, the Malay Peninsula, and other parts of Southeast Asia.
5. European merchants in the 1500s were especially interested in obtaining |||||||||| at the trading center of Malacca.
6. The city of |||||||||| became the Asian center for trading Spanish goods that came from Acapulco, Mexico.
7. The Dutch established the city now known as |||||||||| on the north coast of Java.
8. The British East India Company established the port city of |||||||||| at the tip of the Malay Peninsula.
9. In 1867, the Suez Canal opened a faster sea lane between the |||||||||| Sea and Southeast Asia.
10. The chief cash product of the Malay Peninsula was ||||||||||.
11. In the 1800s, Myanmar (Burma) became a colony of ||||||||||.

Britain	Mediterranean
Buddhism	rubber
Indian	Singapore
Islam	spices
Jakarta	Vietnam
Manila	

Inquiry

Directions: Combine the information in this chapter with your own ideas to answer these questions.

1. Why did discontent with European colonial rule grow rapidly in the 1900s?
2. Even though Indonesia has one of the richest resource bases in the developing world, it is only now beginning to prosper. Why is this so?
3. Singapore is a small independent republic with less than 3 million people, yet it is one of the richest states in Southeast Asia. Why?

SKILLS

Verifying Generalizations

Directions: On page 627 you learned how to make a generalization based on supporting facts. Here you will learn how to verify, or prove, the truth of generalizations by identifying supporting statements called *proofs*. Here is an example of a generalization followed by a series of statements. See if you can find those statements that are correct proofs because they support the generalization.

Generalization:

The local peoples of Southeast Asia invented their own system of farming that was well suited to warm, tropical environments.

Possible Proofs:

a. The early farmers of Southeast Asia grew rice on terraced hillsides and in river valleys.
b. Missionaries from India and China introduced new religions in Southeast Asia.
c. There are many thousands of islands in Southeast Asia.
d. Southeast Asians were the first farmers to grow yams and breadfruits.
e. Southeast Asians were the first farmers to grow bananas and coconuts.

The correct proofs in the preceding example are statements **a**, **d**, and **e**. Now find the correct proofs for the following generalizations. Number your paper from 1 to 4, and next to each number write the letters of the correct proofs.

1. Generalization:

Malacca became one of the world's chief centers of trade.

Possible Proofs:

a. Malacca is close to European ports.
b. Malacca is located on a major international sea route that passes through Southeast Asia.
c. Muslims converted many people on Sumatra and Java (Jawa) to Islam.
d. At this port city, Chinese traders exchanged goods with Persian, Arab, and Indian traders.
e. The spices that could be obtained at Malacca commanded high prices in Europe.

2. Generalization:

The Dutch played an important role in the history of Southeast Asia.

Possible Proofs:

a. The Dutch tried to gain a monopoly over the spice trade.
b. The Muslims and the Portuguese wanted to control the spice trade.
c. The Dutch gained control over many islands that today are part of Indonesia.
d. The Dutch expanded production of sugar and tobacco on Java.
e. The Dutch established the port city of Jakarta (Batavia).

CHAPTER 28

3. Generalization:
For several reasons, the population of Southeast Asia increased tremendously in the last 100 years of colonial rule.

Possible Proofs:

a. Many millions of Europeans migrated to Southeast Asia.

b. Chinese and Indian immigrants moved to Southeast Asia, attracted by job possibilities.

c. Southeast Asia has the world's most comfortable climate.

d. Improved medical and sanitation practices were introduced.

e. Large numbers of Southeast Asians studied in the United States.

4. Generalization:
Today, economic progress is being hindered in Southeast Asia.

Possible Proofs:

a. Recently, there have been civil wars and wars between Southeast Asian countries.

b. Southeast Asia is still the scene of rapid population growth.

c. The Association of Southeast Asian Nations (ASEAN) encourages cooperation among its members.

d. Southeast Asia is experiencing explosive urban growth.

e. Efforts are being made to increase educational opportunities in Southeast Asia.

Vocabulary Skills

Directions: Number your paper from 1 to 4 and complete each sentence by writing on your paper the correct vocabulary term that belongs in each blank.

> archipelago
> city-state
> cultural landscape
> junks

1. The Chinese have maintained sea trade with Southeast Asia for centuries by means of large, seagoing ships called ||||||||||.

2. A ||||||||||, such as Singapore, a large urban center that is itself a country. This means that the city is the highest level of government and the people are citizens of that city.

3. Broadly defined as the panoramic view you see from a hill or mountain, in geography a |||||||||| includes human-built phenomena like cities, villages, and neighborhoods as well as natural phenomena like mountains, hills, or plains.

4. The Philippines is an example, like Japan, of a group or chain of islands called an ||||||||||.

Human Geography

Directions: In Part A you will find the beginnings of ten sentences. The endings to these sentences are in Part B. Choose the correct ending to each sentence and then write the complete sentence on your paper.

Part A

1. Southeast Asia is a region divided into
2. Nearly two out of three Southeast Asians
3. Southeast Asia has long been a meeting place of peoples because of
4. European colonialism prevailed in Southeast Asia
5. Muslims entered Southeast Asia
6. Indians and Chinese first influenced Southeast Asia
7. Countries in Southeast Asia with prosperous economies based on exports are
8. Wars, isolation, or political dissension have hindered economic growth in
9. Port cities in Southeast Asia are growing fast because of
10. After many years of warfare, the two Vietnams became

Part B

a. its crossroads location.
b. in the first millennium.
c. Myanmar, Laos, Cambodia, Vietnam, and Indonesia.
d. in the 1400s.
e. ten countries with a total population of 460 million.
f. earn their living directly from the land.
g. migration from the countryside.
h. in the 1800s.
i. one country under communist control.
j. Thailand, Malaysia, the Philippines, Singapore, and Brunei.

Physical Geography

Directions: Choose the correct term in each set of parentheses, and then write the complete sentence on your paper.

1. Southeast Asia extends from (Laos, Myanmar) on the west to the eastern islands of Indonesia.
2. The largest archipelagoes of Southeast Asia include the islands of Indonesia and (Malaysia, the Philippines).
3. Because it is located astride the equator, Southeast Asia has a (middle-latitude, tropical) environment.
4. Forested (mountains, plains) cover much of mainland Southeast Asia and many of its islands.
5. Singapore is at the southern tip of the (island of Sumatra, Malay Peninsula).
6. Indian and Chinese merchants established important trading posts in the (river valleys, mountain plateaus) of mainland Southeast Asia.
7. European merchants were attracted to Southeast Asia because of its (gold and silver, spices).
8. Monsoon rain forests differ from tropical rain forests because they experience a (dry, wet) season.

Writing Skills Activities

Directions: Answer the following questions in essays of several paragraphs each. Remember to include a topic sentence and several supporting sentences in each paragraph.

1. What are the chief geographical features of Southeast Asia?
2. What influences did the Indians, Chinese, and Muslims have on the culture and society of Southeast Asia?
3. What effect did the colonial period have on the lives of the Southeast Asian people?
4. Why have some Southeast Asian countries, such as Thailand and Malaysia, enjoyed relatively prosperous export economies?
5. Why has economic progress been hindered in some Southeast Asian countries, such as Vietnam, Laos, and Cambodia?

Glossary

aborigines   native people of Australia

absolute location   exact position of a mountain, river, or city on Earth's surface

acid rain   air pollution produced by sulfur oxide from factories and nitrogen oxides from auto emissions that combine with water vapor in the atmosphere

agglomerated settlement   dwellings clustered together and surrounded by gardens, fields, and pastures

agribusiness   a huge farm under corporate management that raises, processes, and sells food products nationwide

alloy   mixture of metals

alluvial plain   a level or gently sloping surface formed by sediments deposited by streams

alpaca   domesticated hoofed animal with long, silky fleece

altiplano   high plateau in the Andes

apartheid   policy of racial segregation adopted by South African government to keep the white minority in control

aquifer   porous layer of rock in which water is held and through which it moves

arable   suitable for growing crops

archipelago   group or chain of closely connected islands

assembly line   mass-production technique in which machines, equipment, and workers are arranged in line so that one operation follows another

atmosphere   envelope of gases—mainly nitrogen, oxygen and carbon dioxide—that encloses Earth

atmospheric pressure   weight of air

atoll   a low, ring-shaped island of coral that once surrounded a volcanic peak that sank beneath the sea

axis   imaginary line that passes through the center of the planet from pole to pole

azimuthal projection   map projection that measures equal distance from its central point to any other point on the map

bantustans   separate self-governing states established by whites for other racial and ethnic populations in South Africa

barrier reef   coral formation that rings high volcanic islands

basin   relatively level lowland surrounded by higher land

basin irrigation   a system in which huge basins hold floodwater, which is then used to grow one or two crops a year on the same land

bauxite   aluminum ore

bayou   slow-moving, sluggish stream

bazaar   Middle Eastern and North African city's main shopping district

Bedouins   camel nomads, a tribal group in the deserts of North Africa and Arabia

Bessemer process   a method of injecting oxygen into iron that purifies and strengthens the iron into steel

biodiversity   the number of species that exists in a given habitat and on the planet

biotechnology   making use of living organisms to manufacture products, such as drugs

Boers   Dutch colonists who settled in South Africa

boreal forests   taiga (northern forests) in subarctic climates

Brahma   a single and permanent unifying spirit, in the Hindu religion

Brahmans   the social class in Hindu society that includes priests and teachers

buffer state   a small country or region located between two larger, more powerful countries

bustees shantytowns in city slums of India

caballeros Spanish for "knights," name for aristocratic Spanish colonists

campo cerrado savanna and scrub woodland mixed

canal artificial waterway

capital money and materials used to produce more wealth

capitalism a social and economic system with open competition in a free market, and with private or corporate ownership of property and the means of production

capitanias large tracts of land in Brazil granted to wealthy families in Portugal

caravanserai inns in bazaars of North Africa and the Middle East, where travelers with camel caravans could rest

caravel small sailing ship of the 1400s and 1500s that could sail in any direction

carbonation a process in which weak acids in rainfall may dissolve rocks such as limestone

cartogram map on which the size of areas is based on a measure other than space—population, for instance

cartographer mapmaker

caste class in Hindu society

caudillo military dictator in Latin America

causeway paved bridge across water or wet ground

chaparral scrub vegetation in Mediterranean climate, also called maquis

chemical weathering process of breaking down rock surfaces by the action of heat, moisture, and oxidation

chernozem fertile soil that is rich in humus

chlorofluorocarbons (CFCs) family of chemicals used in consumer products; one of the main destroyers of Earth's ozone layer

circle of illumination imaginary line that separates the lighted from the darkened half of Earth and that moves as Earth rotates on its axis

city-state urban center that is an independent country

climate the average of weather conditions over a period of years

cloud forest trees covered with moss and found on cloud-shrouded mountain slopes

coastal plain long, relatively flat land, created when deposition occurs along a coastline

collective farm farm on which workers pool their land and labor under government supervision

collectivization system in which all the land is owned by the group and is under government control

command economy system in which government planners decide what is produced, how much and where it is produced, and what industries receive resources

commercial farming growing crops on a large scale to sell on world markets

communism social and economic system based on common ownership of property and of the means of production

compass rose an ornamental symbol showing all four major compass points

computer cartography creation of base maps on which data can be plotted, analyzed, and retrieved

concessions rights to land and minerals given to private companies investing in colonies

conformal projection map projection that shows the true shape for limited areas

conic projection cone-shaped map projection useful in depicting a hemisphere or smaller part of a globe

coniferous type of evergreen tree that holds its leaves year-round

conquistador conqueror from Spain

continental climate climate in interior of land masses, with long cold winters and short hot summers

contour lines on a topographic map, lines that connect points of equal elevation

coral polyps tiny sea creatures that live in a protective shell of lime secreted from their stomachs; builders of coral reefs

coral reef formations in the Pacific, built up by the secretions of tiny sea creatures called coral polyps

cordillera principal mountain range in a chain

coup d'état   a sudden, often violent change in government

creoles   Spaniards and other Europeans born in the Americas

crust   band of solid rock at the Earth's surface that floats on the mantle

cultural landscape   human-made and natural patterns on the land

culture   the learned behavior of a society or nation, including the people's knowledge, faith, laws, languages, technology, and ways of living

culture hearth   area where a country's culture or civilization began

culture region   large area of the world unified by a common culture

cyclone   a powerful and destructive windstorm like a hurricane

cylindrical projection   map projection in which every straight line is a line of true direction; invented by Mercator

czar   absolute ruler in Russia before the Communist revolution of 1917

Dalai Lama   the god-king of the Tibetans

deciduous   a type of tree that drops its leaves in the dry season

deforestation   rapid destruction of tropical and high-latitude forests

delta   flat lowland made up of sediments dropped by a river at its mouth

demand economy   economic system in which the marketplace or the demand for goods determines what will be produced and how much workers will be paid

demilitarized zone (DMZ)   a "no man's zone" that divides the Korean Peninsula and is patrolled by troops from the two Koreas and the United Nations

democracy   government system whose leaders are directly elected by the people and in which majority rules

deposition   the laying down of rock, sand, and silt picked up and moved by water, ice, wind, and ocean currents

desalinization   conversion of salt water into fresh water for drinking and irrigation

desert climate   climate near the Tropic of Cancer that is hot and dry year-round

desertification   the spread of desert-like conditions into semiarid areas

diaspora   the scattering of Jews outside of Palestine

dike   low wall of earth and stone used to channel water

dispersed settlement   arrangement of isolated farmsteads

divide   topographical feature that separates two river systems

DMZ (demilitarized zone)   a "no man's zone" that divides the Korean Peninsula and is patrolled by troops from the two Koreas and the United Nations

dome   large formation created by underground magma that warps Earth's surface

domesticated   adapted for use by and made dependent upon humans

dominion   a self-governing nation of the British commonwealth other than the United Kingdom that acknowledges the British monarch as chief of state

double cropping   the growing of two crops on the same field in one farm year

Dravidians   original inhabitants of India

dry farming   techniques in low-rainfall areas that involve conserving water by plowing but not planting the land every year

dual economy   a country divided between a wealthy modern sector geared to export and a large and poor traditional sector

easterlies   semipermanent belt of winds flowing from subpolar low-pressure regions to the middle latitudes

economic autonomy   independence of a country to set its own production goals and standard of living without interference by a more powerful country

ejidos   communal farms given to peasant farmers in land redistribution after Mexican Revolution in 1910

elfin forest   trees at high elevations that are stunted by cold and wind

emissions pollutants discharged into the air by factories and vehicles

enclave a territory occupied by a group of people belonging to one country or ethnic group but located within another country

environment the surrounding conditions within which an individual or organism lives; on Earth, its fundamental elements are land, air, and water

equal-area projection map projection that shows area in exact proportion to its reality on Earth

equator an imaginary line that marks the midway point between the North and South Poles

equatorial low a wide belt of low pressure around the equator created by solar energy heating Earth in the tropics

equidistant projection map projection that shows distance accurately from one or two points

equinox the time twice a year (about March 21 and September 23), when the vertical rays of the sun are directly over the equator

erosion the breaking down and movement of rock particles by running water, ocean currents, wind, or ice

escarpment line of steep cliffs rimming a plateau

estuary an area of the sea at the mouth of a river valley

ethnic group a group of people that shares beliefs, language, and culture and often lives in a particular territory and has a sense of national unity

European Community (EC) 12 European countries that cooperate on trade and economic matters

fall line imaginary line connecting points where highlands meet a coastal plain, at which rivers from highlands drop suddenly as waterfalls

farm belt an area where the climate encourages the cultivation of a single commercial crop

faulting the pushing and pulling of rock masses against one another, causing one mass to ride up over or slide below the other; caused by tectonic pressures

fazenda coffee estate in Brazil

fjord long U-shaped valley carved out by glaciers and partly filled by the sea

folding the compression of rock into a series of folds by the movement of plates in Earth's crust

forward capital an architectural design project in Brazil intended to encourage settlement in the interior

"four tigers" the countries of Hong Kong, Singapore, South Korea, and Taiwan, each of which has experienced an economic success without major political disruption

fringing reef coral platform attached to the shores of high volcanic islands in the Pacific

fundo large estate in Chile where beef cattle and wheat are raised

garden agriculture farming with most work done by hand on small plots

geography the study of Earth

geothermal energy heat from Earth's interior that surfaces in the form of hot springs and geysers

ghat a long cliff or escarpment

glacier river or sheet of ice that scrapes the soil off the land as it moves

global warming the increase in temperature of Earth's atmosphere, caused by pollution and carbon dioxide in the atmosphere

green revolution a government strategy to increase total food production

greenhouse effect the trapping of solar energy inside the atmosphere and the oceans by carbon dioxide

gross national product (GNP) a measure of the total value of all goods and services produced by a country in a given year

growing season the period between the last frost in spring and the first frost in fall

guano bird droppings used as fertilizer

gyres the circular flow of ocean currents carrying warm water poleward from the tropics and cool water back to the tropics

hacienda large estate in Central America and Mexico

Harijans **(or Untouchables)** the lowest social group in Hindu society

headwaters where a river begins

heartland place of origin and fullest expression of a culture

heavy industry   the production of heavy goods like steel and machinery

hemisphere   half of the globe

high latitudes (or polar regions)   belts located north of the Arctic Circle and south of the Antarctic Circle; they have six months of daylight and six months of darkness

high pressure   atmospheric condition created when air is cooled, loses its ability to hold moisture and becomes dense and heavy

highland climate   climate region with high rainfall and cool temperatures at higher elevations

hill   elevated landform with more gentle slopes and less relief than mountains

hinterland   region that serves an urban area with its natural resources

homelands   reserves set aside for black South Africans, located in the least desirable areas

human resources   the skills and knowledge of people that are used to create new technology

humid continental climate   climate region with hot summers and rich agriculture in the southern margins of the middle latitudes in the Northern Hemisphere

humid subtropical climate   temperate climate region with mild winters and hot, wet summers

humus   organic matter in soil

ice cap climate   climate region located around the poles; its temperatures are never above freezing

inanimate energy   energy derived from nonliving sources like coal or oil

indigenous   native to a place or region

industrial triangle   areas of the French northeast, the Benelux countries, and the German Ruhr with a concentration of heavy industry

inner core   the innermost portion of Earth, made of a 1,560-mile-wide ball of iron and nickel

intermontane   a landform region located between two mountain ranges

Inuit   native Canadian people who hunt whale and seal

irrigation   artificial watering of crops

Islam   religion founded by the Prophet Muhammad

isthmus   a narrow strip of land connecting two larger landmasses

juche   national economic self-reliance, a Korean concept

jungle   dense tangle of vegetation

junk   large Chinese seagoing ship

karma   Hindu law that holds each person responsible for his or her deeds in all past lifetimes

keys   sandbars, reefs, and small islands that line the coast

Koran (or Qur'an)   holy book of Islam

Kshatriyas   social class in Hindu India that includes warriors, political leaders, and rulers

lagoon   a shallow body of water separated from the open sea by barrier reefs in the Pacific

land breeze   created when cooler air from land flows toward the sea

land hemisphere   the half of Earth that contains most of its land area

landlocked   surrounded by land with no direct access to the sea

laterite   hard, compact, yellow-to-red soils with low fertility

latifundios   large tracts of land given to the Spanish elite in South America

latitudes   east-west lines used on globes and maps to show distances north and south of the equator

lava   liquid rock forced out of the ground onto the surface by volcanic activity

lava flow   the forcing out of magma through cracks in Earth's crust

leeward   the sheltered side of a mountain or landform, facing away from the rain-bearing air masses

legend   key that explains information on a map

levee   low earthen embankment along a riverbank to hold floodwaters

limestone   rock that dissolves in water and leaves behind caverns

linear village a type of agglomerated settlement in which houses are laid out along rivers or roads

literacy the ability to read and write

lithosphere a zone on Earth that includes the crust and the uppermost layers of the mantle

llama animal native to Latin America used as a pack animal

llanos tropical savanna grasslands in Venezuela and Colombia

llanos orientales grasslands east of the Andes

loess fine and fertile windblown soil

longitudes north-south lines used on globes and maps to show distances east and west of the Prime Meridian

low latitudes (or tropics) the belt between the Tropic of Cancer and the Tropic of Capricorn

low pressure atmospheric condition created when air is warmed and expands, becomes lighter, and rises over a landmass

loxodrome a line of true compass direction on certain map projections

magma partly melted, white-hot rock inside the layer called the mantle

Mahatma "Great Soul", name for Mohandas K. Gandhi and others with great spiritual powers

mainland a major subregion of Middle America, Central America, most of Mexico, and the islands of the Caribbean

makoto sincerity and fidelity to a person or a company in Japan

mandate former colonial possession supervised by another country on its road to independence

mantle a partly melted, white-hot inner layer of rock between Earth's crust and its core

Maori Polynesian people native to New Zealand

map key legend that explains map information

map projection grid of lines projected onto a geometrical surface

map scale line or bar that marks out how many inches or centimeters on the map equal how many miles or kilometers on Earth's surface

maquiladoras foreign-owned assembly plants along Mexico's border with the United States

maquis scrub vegetation found in Mediterranean climate; also called chaparral

marine west coast climate climate region with cool winters and warm summers

maritime located on or close to the sea

maritime climate climate region near oceans; temperature is more moderate and rainfall more plentiful than in the interior of landmasses

maritime technology systems for navigating oceans

mass emigration movement of people leaving a country with low-wage jobs and few opportunities to settle in industrial countries

mass movement the spontaneous downhill sliding of large amounts of material, such as a rock slide or mudflow

mass production large-scale processing of raw materials into products

mechanical weathering the breaking down of rock through exposure at the surface to wind and water

Mediterranean climate climate region on the west coast of continents between 30 and 40 degrees latitude; winters are mild and wet and summers hot and dry

megalopolis a continuous belt of cities that have expanded and merged together

meridian line of longitude that runs north-south on a globe or map

meseta dry, rugged central plateau in central Spain

mestizos people of the Americas who are of mixed European and native Indian ancestry

metropolis a large sprawling city with several small towns and suburbs

Middle America region that consists of Central America, the mainland of Mexico, and the Caribbean islands

middle latitudes (or temperate regions) two belts located between the Tropic of Cancer and the Arctic Circle and between the Tropic of Capricorn and the Antarctic Circle

minarets tall spires beside mosques, used to call Muslims to prayer

mirs communal villages in pre-Communist Russia

modernization the process by which a country tries to improve its standard of living

Moguls Muslim invaders of India in the 1500s

monotheism belief in one God

monsoon an air current that blows steadily from the same direction for weeks or months

monsoon rain forest tropical forest with a long dry season and both deciduous and evergreen trees

mosque Muslim place of worship

mountain a landform high in elevation, with steep slopes and great relief, formed by volcanism, folding, and faulting

mountain glacier river of ice that moves down a mountain valley to the sea, removing soil as it goes

mulatto person of mixed African and European heritage

multinational corporation large company that operates in more than one country

Muslim believer in the Islamic religion

narrows narrow passages

natural resources materials supplied in nature and available for human use

navigable deep and calm enough for ships to pass; describes a body of water

nonrenewable not able to be restored or replenished once it is used

nonrenewable resources materials such as coal, oil, and minerals that are on the planet in limited amounts and that cannot be replaced when used up

North Atlantic Drift an ocean current that carries water from the Gulf of Mexico to warm the southern British Isles

nuclear village an agglomerated settlement in which dwellings are clustered around a center

oasis a fertile, watered area in the midst of a desert

orbit the fixed path that Earth follows as it moves around the sun

orographic precipitation moisture dropped as rain or snow on windward slopes of mountains when cooled air masses are forced upward by a mountain barrier

outback a flat, stony, desert region in the interior of Australia

outer core layer of Earth made of hot molten iron and nickel inside the mantle

oxidation the binding of oxygen to a mineral to produce an oxide; a kind of chemical weathering

ozone a toxic chemical at lower levels of the atmosphere; in upper atmosphere, forms a protective shield that blocks most ultraviolet light from reaching Earth's surface

paddies embanked plots of irrigated land on which rice is grown

pampas fertile, middle-latitude grasslands of Argentina

panhandle narrow strip of land attached to a larger piece of land

parallels lines of latitude

pastoral nomadism African herders' practice of moving their cattle from one pasture to another in arid areas

peninsula land area that projects into a body of water

peones landless peasant farmers in Latin America

per-capita GNP a country's per-person gross national product

perennial irrigation a system of low dams built on the Nile to hold and store floodwaters year-round

perestroika reorganizing of the Soviet system

permafrost permanently frozen subsoil found in polar regions

petrochemicals products made from oil

piedmont foothills that are a transition between mountains and plains

pilgrimage journey to a religious place or shrine

place a particular city, town, or area with distinctive physical and human characteristics

plain low-lying level area, sometimes gently rolling

plantation large landholding devoted to one crop

plate tectonics theory that Earth's crust is made up of plates that are constantly drifting apart and sliding together; the study of large-scale movements of Earth's crust

plateau level highland that rises above surrounding areas

plates huge, rigid, but moving slabs that form Earth's crust

podzol poor and highly acidic soils of the taiga

polar highs high-pressure air cells that lie among the Arctic and Antarctic Circles

polar regions (or high latitudes) belts located north of the Arctic Circle and south of the Antarctic Circle; they have six months of daylight and six months of darkness

polder land reclaimed from the sea by construction of dikes

pollution contamination of air, water, or land by chemicals, gases, and other materials as a result of by-products of human activity

postindustrial economy an economy based on services, information, and high technology rather than production or manufacture of raw materials

prairie middle-latitude treeless grasslands in the interior of North America and Asia

pre-Columbian describes cultures that existed before the Europeans encountered them in the Americas

precipitation condensed droplets of water vapor that appear as dew, rain, snow, sleet, or hail

primary production work in which people make a living directly from the land and produce raw materials

primate city largest urban center in a country or region

Prime Meridian represented on a map as zero degrees longitude; runs from the North to South Pole through Greenwich, England

production quota the amount of goods assigned to be produced

profit motive desire to make a profit

province political subdivision in Canada

qanats underground water channels used for irrigation for farming in China

quinine antimalaria drug

quota the amount of goods or services required to be produced

Qur'an (Koran) holy book of Islam

rain shadow region in the lee of mountains that receives less rainfall than the region windward of the mountains

Red Guards Chinese students in the Cultural Revolution who were encouraged by Mao Zedong to harass public officials and others who were considered bad elements in Chinese society

reforestation the process of replacing damaged or destroyed forests

regions areas of Earth's surface that have distinctive characteristics

reincarnation rebirth

relative location the position of a place or point on Earth's surface in relation to other locations

relief variation in the elevation, shape, and forms of Earth's surface

renewable resources materials found in nature that can be used, replenished, and reused

reservoir a storage place for water

revolution movement of Earth around the sun; contributes to the seasons of the year

rice cycle planting, growing, and harvesting of rice plants

rift valley deep trench formed where large sections of Earth's crust drop between two parallel cracks or faults

rimland subregion of Middle America made up of the Caribbean islands and the coastal lowlands on the mainland

ripple effect a result from one event that spreads to affect or influence other events

road warrior double-decker truck with three trailers that carries cattle across Australia to urban markets

rotation turning of Earth on its axis, causing night and day

samurai warrior class that led efforts to modernize Japan in the nineteenth century

samurai code a way of living for a certain class in Japan that emphasizes discipline, loyalty, sacrifice, and selfless labor

savanna grazing land with fertile soil and fields of tall grasses dotted with bushes and trees

scientific method a process of observation and experimentation that includes recording the procedures for others to test the results

scorched-earth policy cruelly destroying land so that the people driven out cannot return, as in Guatemalan civil war

scrub short thorny trees found on margins of savannas

sea breeze air from the ocean that flows inland when air above land warms and rises

secondary production manufacturing and processing industry that converts raw materials into manufactured goods

secularism a practice of not connecting the government to any one religion

sedimentary rock rock composed of layers of sediment originally deposited under water

selvas tropical rain forests of Amazonia

Sepoys native soldiers in India

sertao a barren, thinly peopled land covered by dryland grasses and thorny scrub

settled farming type of agriculture practiced wherever fertile soils and reliable rainfall can support permanent settlement

shantytown section of a city where poor people live in makeshift shelters

sharecropper landless farmer who owes a large share of the harvest to the landlord

shield a massive block of very cold, very hard rock

shifting cultivation a form of agriculture in which farmers plant crops in cleared areas and then abandon the fields after several years, when the crops decline

silicon raw material used to make computer chips

silt dissolved soil carried to the mouth of rivers, where it is deposited, forming deltas

siltation the depositing of fine-grained soil in the bed of a river, lake, or sea

sinkhole lake hollow melted out of limestone that then fills with water

site the actual location of a place

situation the relative location of a place; the relationship between a location and its surroundings

slash-and-burn agriculture practice of clearing and burning trees off the land, which is then cultivated for a short period of time

slum crowded area of a large city where poor people live

smog dome a canopy of polluted air that settles over an urban area

solstice time when the vertical rays of the sun reach their northernmost or southernmost limit and the sun is directly above the Tropic of Cancer or the Tropic of Capricorn

sovereign independent, having supreme power

soviet committee chosen by the Communist party as a local governing council in the former Soviet Union

Special Economic Zones (SEZs) areas along the China coast that provide special tax benefits to foreign investors

standard of living the quality of life of people, based on such measures as income and ownership of material goods

staple a basic item of food or the main commodity grown in an area

station sheep-raising ranch in Australia

steppe climate climate region on the edges of deserts with short grasses and limited rainfall

steppe semiarid grassland found near deserts

strait narrow stretch of water lying between two landmasses

strip mining method of mining coal that removes surface layers of soil to expose coal deposits near the surface

subarctic climate continental climate with cold winters

subcontinent   large landmass that projects from a continent and is bordered on three sides by oceans; India, for example

subpolar low   semipermanent belt of low-pressure air cells

subsidy   government grant that keeps the price of a commodity higher than the product would sell for on world markets

subsistence crops   crops grown more for family use and less for sale

subtropical highs   zones of high pressure near latitudes 30 degrees N and 30 degrees S

Sudras   servant class in Hindu social system

summer monsoon   a current of moist, warm air that blows inland from the ocean, drenching South Asia and Southeast Asia with heavy rain from June to September

sun belt   band of southern and western states that extends from Florida to California

synagogue   center of worship for Jews

taiga   northern forests in subarctic climate

technology   methods, tools, and knowledge that people use to obtain products they need and want

technopolis   "science city" devoted to the manufacture of high technology components

temperate forest   area of deciduous and coniferous trees in middle-latitude environments

temperate regions (or middle latitudes)   two belts located between the Tropic of Cancer and the Arctic Circle and the Tropic of Capricorn and the Antarctic Circle

tenant farmer   worker on the land of a landlord

terrace   level stepped field dug into hillsides for growing crops

tertiary production   work based on knowledge and information rather than the manufacturing of raw materials

theocracy   government ruled by religious leaders

tierra caliente   hot and humid coastal lowlands where commercial crops are grown

tierra fria   cool highlands of the Andes Mountains above 6,000 feet

tierra helada   climate region that has permanent snow, in the Andes Mountains above 10,000 feet

tierra templada   foothills of the Andes Mountains above 2,500 to 3,000 feet; commercial cultivation and dense population are features of this region

topographic map   map that shows landforms by using contour lines to illustrate elevation

topography   physical features of landforms

townships   slums for black, Asian, and colored South Africans in restricted areas near large cities

trade winds   air masses from high-pressure areas near latitudes of 30 degrees N and 30 degrees S that flow toward the equator

transhumance   a practice of sheep and goat herders in dry environments who graze their animals in upland meadows in summer and lowlands in winter

tributary   smaller stream that feeds into a river system

tropical forest   mixed evergreen and deciduous forest in parts of the tropics with a dry season

tropical monsoon climate   climate region that is hot year-round but has a dry season in winter

tropical rain forest   luxuriant evergreen forest found in the tropics where rainfall is abundant and there is no dry season

tropical rain forest climate   hot and wet climate region located at or near equator

tropical savanna   grassland with a long dry season in winter and a wet season in summer; consists of tall grass interspersed with groves of trees

tropical savanna climate   climate region that is hot year-round but has a winter dry season

tropics (or low latitudes)   the belt between the Tropic of Cancer and Tropic of Capricorn

trust territory   a territory under the authority of a nation or the United Nations

tsetse fly   insect that kills cattle and causes sleeping sickness in people

tundra   vast, frozen, nearly treeless plains of the Arctic regions of Europe and Asia

tundra climate   climate region with freezing cold winters and short summers

two-China policy   the U.S. foreign policy position toward Taiwan and the People's Republic of China

typhoon   severe late summer windstorm on Asian coasts; in the Americas, called a hurricane

Untouchables (or *Harijans*)   the lowest social group in Hindu society

urban   describing a city or town

urbanites   people who live in cities

Vaisyas   Hindu social class that includes farmers, tradespeople, and merchants

veld   open grassland in central plateau of South Africa

villagization   practice of forcing farmers to relocate to government-controlled settlements

volcanic island   mountain peak that rises from the seafloor in the Pacific

volcanism   the outpouring of molten rock onto the surface of the land through cracks in Earth's crust

water hemisphere   the half of Earth that contains most of its water area

weather   the temperature and rainfall conditions of any place on a given day

weathering   the slow breaking down of rock into finer particles through chemical and mechanical means

westerlies   belt of winds that flow from subtropical high-pressure regions to the middle latitudes

wet rice farming   practice of growing rice in small flooded fields surrounded by low embankments

windward   facing the direction from which the wind is blowing

winter monsoon   a current of dry continental air that flows outward from the land to the ocean from October to March, producing little rain across the subcontinent

xerophyte   plant adapted to drought

yak   large, long-haired oxen native to Tibet and Mongolia

yurt   circular, felt tent used by Mongol nomads

Zionism   a political movement devoted to the creation of a Jewish state and a homeland for all Jews

Pronunciation
Guide

Please see page 712 for a pronunciation key.

A·ba·dán (ä′bə dän′, ab′ə dan′)
Ab·i·djan (ab′ə jän′)
Ab·i·lene (ab′ə lēn′)
Ab·kha·zi·a (äb käz′ē ə)
A·bu·ja (ä bōō′jä)
Ac·a·pul·co (ä′kə pŏŏl′kō, ak′ə-)
Ac·cra (ə krä′)
Ad·dis A·ba·ba (ad′is ab′ə bə)
Ad·e·laide (ad′′l ād′)
A·dri·at·ic (ā′drē at′ik)
Ae·ge·an (ē jē′ən, i-)
Af·ghan·i·stan (af gan′i stan′)
A·gra (ä′grə)
Ak·ko (ä kō′)
Al·ba·ni·a (al bā′nē ə, -bān′yə)
A·lep·po (ə lep′ō)
A·leu·tian (ə lōō′shən)
Am·man (ä′män′, ä män′)
A·mu Dar·ya (ä mōō′ där′yä)
A·mur (ä mŏŏr′)
An·a·to·li·a (an′ə tō′lē ə)
An·des (an′dēz′)
An·dor·ra (an dôr′ə)
Ang·kor (aŋ′kôr′)
An·ka·ra (aŋ′kər ə, äŋ′-)
An·ti·gua (an tē′gwə)
An·til·les (an til′ēz′)
Ant·werp (an′twɥrp′)
Ap·en·nine (ap′ə nīn′)
Ar·al (ar′əl)
Ar·me·ni·a (är mēn′yə, -mē′nē ə)
A·ru·ba (ə rōō′bə)
As·sam (a sam′, as′am)
A·sun·ción (ä sōōn syôn′)
A·ta·ca·ma Desert (ä′tä kä′mä)
Auck·land (ôk′lənd)
Ax·um (äk′sŏŏm)
Az·er·bai·jan (äz′ər bī jän′, az′-)
A·zores (ā′zôrz′, ə zôrz′)

Bagh·dad (bag′dad, bäg däd′)
Ba·hi·a (bə hē′ə; *Port* bä ē′ə)
Bah·rain (bä rān′)
Bai·kal (bī käl′)
Ba·ku (bä kōō′)
Ba·lu·chi·stan (bə lōō′chə stan′, -stän′)
Ba·ma·ko (bä mä kō′)
Ban·ga·lore (baŋ′gə lôr′)
Ban·gla·desh (bäŋ′glə desh′, baŋ′-)
Bao·tou (bou′dō′)
Bar·ba·dos (bär bā′dōs, -dōz)
Bar·ce·lo·na (bär′sə lō′nə)
Bar·ents Sea (bar′ənts, bär′-)
Bar·ran·quil·la (bä′rän kē′yä)
Ba·sel (bä′zəl)
Bas·ra (bus′rə, buz′-)
Ba·ta·vi·a (bə tā′vē ə)
Beer·she·ba (bir shē′bə, ber-)
Bei·jing (bā′jiŋ′, bā zhiŋ′)
Bei·rut (bā rōōt′; bā′rōōt′)
Bel·a·rus (bel′ə rōōs′)
Be·lem (bə len′)
Bel·grade (bel′grād′, -gräd′)
Be·lize (bə lēz′)
Be·lo Ho·ri·zon·te (bä′lō hôr′ə zän′tē; *Port* be′lô rē zôn′te)
Be·ne·lux (ben′ə luks′)
Ben·gha·zi (ben gä′zē, beŋ-)
Ber·ing Sea (ber′iŋ, bir′-)
Bhu·tan (bōō tän′)
Bi·har (bi här′)
Bo·go·tá (bō′gə tä′)
Bo Hai (bō′hī′)
Bo·liv·i·a (bə liv′ē ə)
Bo·phu·tha·tswa·na (bō′pōō tät swän′ə)
Bos·po·rus (bäs′pə rəs)
Bo·tswa·na (bät swä′nə)
Brah·ma·pu·tra (brä′mə pōō′trə)
Bra·sí·lia (brä zē′lyä; *E* brə zil′yə)

707

Bra·ti·sla·va (brä´ti slä´və)
Bre·men (brem´ən; *Ger* brä´mən)
Bru·nei (broo nī´)
Bu·da·pest (boo´də pest´)
Bue·nos Ai·res (bwā´nəs er´ēz)
Bu·kha·ra (boo kär´ə)
Bu·la·wa·yo (boo´lə wä´yō)
Bul·gar·i·a (bəl ger´ē ə, bool-)
Bur·ki·na Fa·so (boor kē´nə fä´sō)
Bur·ma (bʉr´mə)
Bu·run·di (boo roon´dē)
Bye·lo·rus·sia (bye´lō rush´ə)

Cal·e·do·ni·a (kal´ə dōn´yə, -do´nē ə)
Can·ber·ra (kan´bər ə, -ber´ə)
Car·ib·be·an (kar´ə bē´ən, kə rib´ē ən)
Car·pa·thi·an (kär pā´thē ən)
Car·ta·ge·na (kär´tə hä´nə, -tə jē´-)
Cas·cade (kas kād´)
Cau·ca (kou´kä)
Cau·ca·sus (kô´kə səs)
Cel·e·bes (sel´ə bēz´, sə le´bēz´)
Cer·ro de Pas·co (ser´ō dä päs´kō)
Cha·co (chä´kô)
Chao Phra·ya (chou´ prä yä´, -prī´ə)
Cher·no·byl (cher nō´bəl)
Chi·a·pas (chē ä´pəs)
Chi·le (chil´ē; *Sp* chē´le)
Chong·qing (choong´chiŋ´)
Cis·kei (sis´kī´)
Ciu·dad Bo·lí·var (syoo däd´ bō lē´vär´)
Ciudad Juá·rez (hwä´res´)
Com·o·ros (käm´ə rōs´)
Co·na·kry (kän´ə krē´; *Fr* kô nȧ krē´)
Con·stan·ti·no·ple (kän´stan tə nō´pəl)
Cor·o·man·del (kôr´ə man´dəl)
Côte d'A·zur (kōt dȧ zür´)
Co·to·nou (kô tô noo´)
Cri·me·a (krī mē´ə, krə-)
Cro·a·tia (krō ā´shə)
Cu·ra·çao (kyoor´ə sō´, koor´ə sou´)
Cuz·co (koos´kō)
Czech·o·slo·va·ki·a (chek´ə slō vä´kē ə)
Czech Republic (chek)

Da·kar (də kär´, däk´är)

Dan·ube (dan´yoob)
Dar·da·nelles (där´də nelz´)
Dar es Sa·laam (där´ es sə läm´)
Da·vao (dä vou´)
Dec·can (dek´ən)
Del·hi (del´ē)
Dhak·a (dä´kə, dak´ə)
Dji·bou·ti (ji boo t´ē)
Dnies·ter (nēs´tər)
Do·nets'k (dô nyetsk´)

Ec·ua·dor (ek´wə dôr´)
Eir·e (er´ə)
E·lat (ā lät´)
El·be (el´bə, elb)
El·burz (el boorz´)
Elles·mere (elz´mir´)
El·lice (el´is)
Er·i·tre·a (er´ə trē´ə)
Erz·ge·bir·ge (erts´gə bir´gə)
Es·to·ni·a (es tō´nē ə)
E·thi·o·pi·a (ē´thē ō´pē ə)
Eu·phra·tes (yoo frāt´ēz)
Eur·a·sia (yoo rā´zhə; *chiefly Brit*, -shə)

For·ta·le·za (fôr´tə lä´zə)
Fu·ji (foo´jē)
Fu·ku·o·ka (foo´koo ō´kə)
Fu·shun (foo´shoon´)
Fu·tu·na (fə too´nə)

Ga·bon (gȧ bō*n*´)
Gan·ges (gan´jēz)
Gen·o·a (jen´ə wə)
Gha·na (gä´nə)
Ghats (gôts, gäts)
Go·bi (gō´bē)
Gö·te·borg (yö´tə bôr´y´)
Gre·na·da (grə nä´də)
Gua·da·la·ja·ra (gwäd´´l ə här´ə; *Sp* gwä´thä lä ha´rä)
Gua·dal·ca·nal (gwäd´´l kə nal´)
Gua·de·loupe (gwä´də loop´)
Guam (gwäm)
Guang·zhou (gwäŋ´jō)
Gua·te·ma·la (gwä´tə mä´lə)
Guay·a·quil (gwī´ä kēl´)

Gui·a·na (gē an′ə, -ä′nə)
Guin·ea (gin′ē)
Gu·ja·rat (goo′jə rät′)
Guy·a·na (gī an′ə, -än′ə)

Hai·fa (hī′fə)
Hai·phong (hī′fäŋ′)
Hai·ti (hāt′ē)
Ham·burg (ham′bərg; *Ger* häm′boorkh)
Ha·ra·re (hä rä′rē)
Heb·ri·des (heb′rə dēz′)
Her·ze·go·vi·na (hert′sə gō vē′nə)
Hi·ma·la·yas (him′ə lā′əz, hi mäl′yəz)
Hi·ro·shi·ma (hir′ə shē′mə, hi rō′shi mə)
His·pan·io·la (his′pən yō′lə)
Hok·kai·do (hō kī′dō)
Hon·du·ras (hän door′əs, -dyoor′-)
Hon·shu (hän′shoo′)
Hoogh·ly (hoog′lē)
How·rah (hou′rə)
Huang (hwäŋ′)

I·ba·dan (ē bä′dän′)
I·be·ri·a (ī bir′ē ə)
In·chon (in′chän′)
I·o·ni·a (ī ō′nē ə)
I·qui·tos (ē kē′tôs)
Ir·ra·wad·dy (ir′ə wä′dē, -wô′-)
Is·fa·han (is′fä hän′)
Is·lam·a·bad (is läm′ə bäd′)
Is·ra·el (iz′rē əl, -rā-; *also* iz′rəl)
Is·tan·bul (is′tan bool′, -tän-; -bool;
    Turk is täm′bool)

Jaff·na (jaf′nə)
Ja·kar·ta (jə kär′tə)
Jam·shed·pur (jum′shed poor′)
Ja·va (jä′və, jav′ə)
Ja·wa (jä′və)
Jid·da (jid′ə)
Jo·han·nes·burg (jō han′is bɵrg′, yō hän′is-)
Ju·neau (joo′nō′)
Jung·gar (zhooŋ′gär′)
Jut·land (jut′lənd)

Ka·bul (kä′bool′)

Ka·la·ha·ri (kä′lä ha′rē)
Ka·li·man·tan (kä′lē män′tän′)
Kam·chat·ka (käm chät′kə)
Kam·pa·la (käm pa′lə)
Kam·pu·che·a (kam′poo chē′ə)
Kan·pur (kän′poor′)
Ka·ra·chi (kə rä′chē)
Ka·ra·gan·da (kä′rə gän′də)
Ka·ra·ko·ram (kä′rä kôr′əm, kar′ə-)
Ka·ra Kum (kä rä′ koom′; *E* kar′ə-)
Kat·mai (kat′mī′)
Kath·man·du (kät′män doo′)
Ka·zakh·stan (kä′zäk stän′)
Ken·ya (ken′yə, kēn′-)
Khar·toum (kär toom′)
Khy·ber Pass (kī′bər)
Ki·ev (kē′ef′, -ev′)
Kil·i·man·ja·ro (kil′ə män jär′ō)
Kin·sha·sa (kēn shä′sä)
Kir·i·bati (kir′ə bas′)
Ki·ta·kyu·shu (kē′ta kyoo′shoo)
Ko·be (kō′bā′; *E* kō′bē)
Kow·loon (kou′loon′)
Kra·kow (kra′kou′, krä-; -kō; *Pol* krä′koof)
Kua·la Lum·pur (kwä′lə loom poor′)
Kun·lun (koon′loon′)
Kush (kush)
Kuz·netsk (kooz netsk′)
Kwa·ja·lein (kwä′jə lān′)
Kyo·to (kē′ōt′ō)
Kyr·gyz·stan (kir′gi stan′)
Kyu·shu (kyoo′shoo′)

La·gos (lā′gäs′, -gəs)
La·hore (lə hôr′, lä-)
Lan·zhou (län′jō′)
La Pla·ta (lä plä′tä)
Lat·vi·a (lat′vē ə)
Leip·zig (līp′sig, -sik)
Len·in·grad (len′in grad′)
Le·so·tho (le soo′too, le sō′tō)
Lha·sa (lä′sə)
Liao (lē ou′)
Liao·dong (-dooŋ′)
Li·ber·i·a (lī bir′ē ə)
Liech·ten·stein (lik′tən stīn′; *Ger* liH′tən shtīn′)

Lille (lēl)

Lim·po·po (lim pō´pō)

Lith·u·a·ni·a (lith´oo ā´nē ə, lith´ə wā´-)

Loire (lə wär´; *Fr* lwả*r*)

Lu·an·da (loo än´də, -an´-)

Lu·bum·ba·shi (loo´boom bä´shē)

Lu·sa·ka (loo sä´kä)

Lux·em·bourg (luk´səm bɜrg´; *Fr* lük sän boo*r*´)

Lyon (lyōn)

Maas·tricht (mäs´tri*H*t)

Mac·e·do·ni·a (mas´ə do´nē ə, -dōn´yə)

Mag·da·le·na (mäg´dä le´nä)

Ma·lac·ca (mə lak´ə)

Ma·la·wi (mä´lä wē)

Ma·lay (mä´lā´)

Ma·lay·sia (mə lā´zhə, -shə)

Ma·lu·ku (mə loo´koo´)

Ma·na·gua (mä nä´gwä)

Ma·naus (mä nous´)

Man·chu·kuo (man choo´kwō)

Man·chu·ri·a (man choor´ē ə)

Ma·pu·to (mə poot´ō)

Mar·a·cai·bo (mar´ə kī´bō; *Sp* mä´rä kī´bô)

Ma·ri·an·a (mer´ē an´ə, mar´-)

Mar·seille (már se´y´; *E* mär sā´)

Mar·seilles (már sā´, -sālz´)

Mar·ti·nique (mär´tə nēk´)

Mash·had (mə shäd´)

Mau·ri·ta·ni·a (môr´ə tā´nē ə, -tān´yə)

Mec·ca (mek´ə)

Me·kong (mā´käŋ´, -kôŋ´)

Mel·a·ne·sia (mel´ə nē´zhə, -shə; *Brit*, -zē ə)

Mel·bourne (mel´bərn)

Mer·o·ë (mer´ō ē´)

Me·sa·bi Range (mə sä´bē)

Mes·o·po·ta·mi·a (mes´ə pə tä´mē ə)

Meuse (myooz; *Fr* möz)

Mi·cro·ne·sia (mī´krə nē´zhə, -shə)

Mi·nas Ge·rais (mē´nəs zhi rīs´)

Min·da·na·o (min´də nou´, -nä´ō)

Minsk (minsk)

Mo·ga·di·shu (mō´gä dē´shoo)

Mo·ja·ve (mō hä´vē)

Mo·luc·cas (mō luk´əz, mə-)

Mom·ba·sa (mäm bä´sə, -bas´ə)

Mon·go·li·a (mäŋ gō´lē ə, män-; -gōl´yə)

Mon·te·ne·gro (mänt´ə nē´grō, -neg´rō)

Mon·te·vi·de·o (mänt´ə və dā´ō)

Mont·ser·rat (mänt´sə rat´)

Mo·re·los (mô re´lôs)

Mo·zam·bique (mō´zəm bēk´)

Mu·nich (myoo´nik)

Mur·mansk (moo*r* mänsk´)

Myan·mar (myun´mä, -mär)

Na·ga·sa·ki (nä´gə sä´kē)

Na·gor·no-Ka·ra·bakh (nä gôr´nō kär´ä bäk´)

Na·go·ya (nä´gô yä´)

Nai·ro·bi (nī rō´bē)

Na·mib·i·a (nə mib´ē ə)

Nantes (nän*t*; *E* nänts, nants)

Na·u·ru (nä oo´roo)

Neg·ev (neg´ev´)

Ne·pal (nə pôl´, -päl´)

Ne·va (nē´və)

Ne·vis (nē´vis, nev´is)

Nic·a·ra·gua (nik´ə rä´gwə)

Ni·ger (nī´jər)

Ni·u·e (nē oo´ā)

Nizh·ny Nov·go·rod (nēzh´nē nôv´gu rət)

Oa·xa·ca (wä hä´kä)

O·khotsk (ō kätsk´; *Russ* ô khôtsk´)

O·man (ō män´)

O·ri·no·co (ôr´ə nō´kō)

Pa·ki·stan (pak´i stan´, pä´ki stän´)

Pa·lau (pä lou´)

Pal·i·sades (pal´ə sādz´, pal´ə sādz´)

Pap·u·a (pap´yoo ə)

Pa·rá (pä rä´)

Par·a·guay (par´ə gwä´, -gwī´; *Sp* pä rä gwī´)

Pa·ra·í·ba (pä rä ē´bä)

Pa·ra·ná (pä´rä nä´)

Pat·a·go·ni·a (pat´ə gō´nē ə, -gōn´yə)

Pel·o·pon·ne·sus (pel´ə pə nē´səs)

Pen·nine (pen´īn´, -in)

Pe·ru (pə roo´)

Phnom Penh (pə näm´pen´, näm´pen´)

Pit·cairn (pit´kern)

Port Mores·by (môrz´bē)

Prague (präg)
Pre·to·ri·a (prē tôr´ē ə, pri-)
Pun·jab (pun jäb´; pun´jäb, -jab)
Pun·ta A·re·nas (po͞on´tä ä re´näs)
Pu·san (po͞o´sän)
Pyong·yang (pyuŋ´yäŋ´)
Pyr·e·nees (pir´ə nēz´)

Qa·tar (ke tär´, kä´kär´)
Qui·to (kē´tō)

Ra·bat (rə bät´; *Fr* rȧ bá´)
Ra·leigh (rô´lē, rä´lē)
Ran·goon (ran go͞on´, raŋ-)
Re·ci·fe (rə sē´fə)
Rey·kja·vík (rā´kyə vēk´, -vik´)
Rhine (rīn)
Rhône (rōn)
Ri·ga (rē´gə)
Ri·o de Ja·nei·ro (rē´ō dā´ zhə ner´ō; *Port* rē´o͞o də zhə nā´ro͞o)
Ri·yadh (rē yäd´)
Ruhr (ro͝or; *Ger* ro͞o´ər)
Rwan·da (ro͞o än´də)

Saar (sär, zär)
Sa·bah (sä´bä)
Sa·hel (sä hel´)
Sa·kha·lin (sä´khä lēn´; *E* sak´ə lēn´)
Sal·ween (sal wēn´)
Sam·ar·qand (sam´ər kand´)
San Lu·is Po·to·sí (sän´ lwēs´ pô´tô sē´)
São Pau·lo (sou͝n pou´lo͝o)
Sap·po·ro (sä´pô rô´)
Sa·ra·wak (sə rä´wäk)
Sas·katch·e·wan (sas kach´ə wän´, -wən)
Schuyl·kill (sko͞ol´kil)
Seine (sān; *Fr* sen)
Sen·e·gal (sen´i gôl´, sen´ə gəl)
Se·oul (sōl; *Kor* syö´o͞ol´)
Shang·hai (shaŋ´hī´, shäŋ´-)
Sha·ri (shä´rē)
Shatt-al-A·rab (shat´əl ä´räb)
Shen·yang (shun´yäŋ´)
Shi·ko·ku (shē´kô ko͞o´)
Shi·raz (shē räz´)

Si·chuan (sē´chwän´)
Sic·i·ly (sis´ə lē)
Sri Lan·ka (srē läŋ´kə)
Sri·nag·ar (srē nug´ər)
Sta·vang·er (stä vaŋ´ər)
Su·ez (so͞o ez´, so͞o´ez)
Su·la·we·si (so͞o´lä wä´sē)
Su·ma·tra (so͞o mä´trə)
Su·ri·name (so͝or´i näm´, so͝or´i nam´; *Du* so͝or´ə nä´mə)
Su·va (so͞o´vä)
Syr·a·cuse (sir´ə kyo͞os´, -kyo͞oz´)
Syr Dar´·ya (sir där´yä)

Ta·briz (tä brēz´)
Tae·gu (tī´go͞o´, tī go͞o´)
Tai·pei (tī pā´)
Tai·wan (tī wän´)
Tai·yu·an (tī´yo͞o än´)
Ta·jik·i·stan (tä jik´i stan´, -stän´)
Taj Ma·hal (täzh´ mə häl´, täj´-)
Tal·linn (täl´in)
Tan·gan·yi·ka (tan´gən yē´kə)
Ta·rim (tä rēm´, dä-)
Tash·kent (tash kent´, täsh-)
Tas·ma·ni·a (taz mā´nē ə, -mān´yə)
Tau·rus (tô´rəs)
Tbi·li·si (tə bi lē´ sē´; tu´bi lē´sē; tə bil´i sē´)
Teh·ran (te rän´, tə-; -ran´)
Thai·land (tī´land´; -lənd)
Thames (temz; *for 3* thāmz, tāmz, temz)
Tian·jin (tyen jin´)
Tian Shan (tyen shän´)
Ti·bet (ti bet´)
Ti·er·ra del Fue·go (tē er´ə del fwā´gō)
Ti·gris (tī´gris)
Ti·jua·na (tē wän´ə, -hwän´ə; *Sp* tē hwä´nä)
Tim·buk·tu (tim´buk to͞o´)
Ti·mor (tē´môr´, tē môr´, tī´môr´)
Ti·ti·ca·ca (tit´i kä´kə; *Sp* tē´tē kä´kä)
To·ba·go (tō bā´gō, tə-)
To·go (tō´gō)
To·ky·o (tō´kē ō´)
Ton·ga (täŋ´gə)
Ton·le Sap (tän´lä säp´, -sap´)
Trans·kei (trans kā´, -kī´)

711

Trans·vaal (trans väl´, tranz-)
Trond·heim (trän´hām´)
Tu·ni·si·a (t͞oo nē´zhə, -zhē ə; ty͞oo-)
Tu·rin (t͞oor´in, ty͞oor´-)
Turk·men·i·stan (tərk men´i stan´, -stän´)
Tu·va·lu (t͞oo´və l͞oo´)

U·ban·gi (y͞oo baŋ´gē, -bäŋ´-; ͞oo-)
U·gan·da (y͞oo gan´də, -gän´-; ͞oo-)
U·kraine (y͞oo krān´, -krīn´; y͞oo´krān)
U·ral (y͞oor´əl)
U·ra·nus (y͞oor´ə nəs, y͞oo rān´əs)
U·ru·guay (y͞oor´ə gwā´, -gwī´, ͞oor´ə-;
   Sp ͞oo´r͞oo gwī´)
Ü·rüm·qi (͞oo´r͞oom´chē´)
Uz·bek·i·stan (͞ooz bek´i stan´, -stän´)

Va·nua·tu (vän´wä t͞oo´)
Va·ra·na·si (və rän´ə sē´)
Ven·e·zue·la (ven´ə zwā´lə, -zwē´-;
   Sp ve´ne swe´lä)
Ver·sailles (vər sī´; *Fr* ver sä´y´)
Ve·su·vi·us (və s͞oo´vē əs)
Vil·ni·us (vil´nē əs´)
Vis·tu·la (vis´cho͞o lə)

Vla·di·vos·tok (vlad´i väs´täk; *Russ* vlä´di vôs tôk´)

Wind·hoek (vint´ho͝ok)
Wit·wa·ters·rand (wit wôt´ərz rand´)
Wu·han (w͞oo´hän´)

Xi·zang (shē´dzäŋ´)

Ya·kutsk (yä ko͝otsk´)
Ya·lu (yä´l͞oo´)
Yan·gon (yan gôn´)
Yang·tze (yaŋk´sē)
Ya·oun·dé (yä ͞oon dä´)
Yem·en (yem´ən)
Yo·ko·ha·ma (yō´kə hä´mə)
Yu·ca·tán (y͞oo´kä tän´; *E* y͞oo´kə tan´)
Yu·go·slav·i·a (y͞oo´gō slä´vē ə, -gə-; -släv´yə)

Za·ca·te·cas (sä´kä te´käs)
Zag·ros (zag´rəs)
Za·ire, Za·ïre (zä ir´)
Zam·be·zi (zam bē´zē)
Zan·zi·bar (zan´zə bär´)
Zim·ba·bwe (zim bä´bwā´, -bwē´)
Zur·ich (zo͝or´ik)

Pronunciation Key

English Sounds

Symbol	Key Words	Symbol	Key Words
a	**c**at	b	**b**ed, du**b**
ā	**a**pe	d	**d**ip, ha**d**
ä	**c**ot	f	**f**all
		g	**g**et, do**g**
e	t**e**n	h	**h**elp
ē	m**e**	j	**j**oy
		k	**k**ick, **q**uit
i	f**i**t	l	**l**et, bott**le**
ī	b**i**te	m	**m**eat
		n	**n**ose
ō	**go**	p	**p**ut
ô	**a**ll, **or**	r	**r**ed
oo	l**oo**k, p**u**ll	s	**s**ee
͞oo	t**oo**l, r**u**le	t	**t**op, ca**tt**le
oi	**oi**l, t**oy**	v	**v**at
ou	**ou**t, pl**ow**	w	**w**ish, **qu**ick
		y	**y**ard
u	**cu**p	z	**z**ebra
ʉ	t**u**rn		
ə	**a**go, **a**gent,	ŋ	ri**ng**
	penc**i**l, at**o**m		
	foc**u**s	ch	**ch**ain, ar**ch**
		hw	**wh**ere
ər	p**er**haps, moth**er**	sh	**sh**e, mo**ti**on
		th	**th**in, tru**th**
	cattle (kat´´l)	*th*	**th**en, fa**th**er
	cotton (kät´n)	zh	mea**s**ure

Foreign Sounds

The following special symbols are used to represent non-English sounds in various foreign languages:

å   pronounced as intermediate between (a) and (ä)

ë   approximated by rounding the lips as for (ō) and pronouncing (e)

ö   approximated by rounding the lips as for (ō) and pronouncing (ā)

o   any of a range of sounds between (ô) and (u)

ü   approximated by rounding the lips as for (ō) and pronouncing (ē)

kh   approximated by placing the tongue as for (k) but allowing breath to escape in a stream, as in pronouncing (h)

H   formed as placing the tongue as for (sh) but with the tip pointing downward to air friction against the forward part of the palate

n   indicates that the preceding vowel sound is nasalized; that is, the vowel is pronounced with breath passing through both the mouth and the nose; the letter *n* is not pronounced

r   any of various trilled (r) sounds, produced with rapid vibration of the tongue or uvula

Spanish Equivalencies

Spanish Equivalencies for Important Geographical Terms

Aborigine—aborigen

Absolute location—sitio absoluto

Acid rain—lluvia ácida

Agglomerated settlement—población aglomerada

Alloy—aleación

Alluvial plain—llanura aluvial

Apartheid—política de segregación racial (esp. en Africa del Sur)

Aquifer—roca acuífera

Arable—fétil

Archipelago—mar poblado de islas

Assembly line—línea de montaje

Atmosphere—atmósfera

Atmospheric pressure—presión atmosférica

Atoll—atolón

Axis—eje

Azimuthal projection—proyección acimut

Barrier reef—barrera de coral

Basin irrigation—irrigación por una cuenca

Basin—cuanca

Bauxite—bauxita

Bayou—canalizo

Bazaar—bazar

Bedouin—beduino

Bessemer process—proceso de Bessemer

Biodiversity—biodiversidad

Biotechnology—biotecnología

Boers—agricultores holandeses; boeres

Boreal forest—bosque boreal

Buffer state—nación parachoques

Buffer zone—zona amortiguador

Capitalism—capitalismo

Caravanserai—posada para caravanas

Caravel—carabela

Carbonation—carbonización

Cartographer—cartógrafo

Caste—casta

Causeway—carretera elevada

Chemical weathering—acción corrosiva química de los elementos naturales

Chlorofluorocarbons (CFCs)—clorofluorocarburos

Circle of illumination—círculo de iluminación

City-state—ciudad-estado

Climate—clima; zona meteorológica

Cloud forest—bosque de nubes

Coastal plain—llanura costera

Collective farm—granja cooperativa

Collectivization—colectivización

Command economy—economía de mando

Commercial farming—agricultura comercial

Communism—comunismo

Compass rose—símbolo de una brújula

Computer cartography—cartografía por computadora

Concessions—concesiones

Conformal projection—proyección concordante

Conic projection—proyección cónica

Coniferous—conífero

Conquistador—conquistador

Continental climate—clima continental

Contour lines—líneas acotadas

Coral polyps—pólipos de coral

Coral reef—arrecife de coral

Coup d'etat—coup d'etat

Creole—criollo

Crust—corteza

Culture hearth—hogar cultural

Culture region—región cultural

Culture—cultura

Cyclone—ciclón

Cylindrical projection—proyección cilíndrica

Czar—zar

Deforestation—desforestación

Delta—delta

Demand economy—economía en demanda

Democracy—democracia

Deposition—deposición

Desalinization—desalinización

Desert climate—clima desértico

Desertification—desertificación

Diaspora—dispersión de un pueblo homogéneo

Dikes—diques; terraplenes

Dispersed settlement—población dispersada

Divide—divisoria

DMZ (demilitarized zone)—zona desmilitarizada

Dome—cúpula

Domestication—domesticación; aclimatación

Dominion—dominio

Dravidians—dravidianos

Dry farming—agricultura seca

Dual economy—economía dual

Easterlies—viento del este

Economic autonomy—autonomía económica

Elfin forest—bosque de duende

EC (European Community)—comunidad europea

Emissions—emisiones

Enclave—enclave

Environment—ambiente

Equal-area projection—proyección de área-igual

Equator—ecuador

Equatorial low—depresión ecuatorial

Equidistant projection—proyección equidistante

Equinox—equinoccio

Erosion—erosión

Escarpment—escarpa

Estuary—estuario

Ethnic group—grupo étnico

Fall line—línea de caída

Farm belt—región agrícola

Faulting—fracturándose (falla)

Fissures—fisuras

Fjord—fiordo

Flat—plano

Folding—plegamiento

Foward capital—capital delantero

"Four tigers"—"cuatro tigres"

Fringing reef—arrecife orlado

Garden agriculture—agricultura del jardín

Geothermal energy—energía geotérmica

Ghat—escalera a orillas de un río

Glacier—glaciar

Global warming—calentamiento global

Green Revolution—Revolución Verde

Greenhouse effect—efecto invernadero

Gross national product, (GNP)—producto nacional bruto

Growing season—temporada de cultivación

Gyres—círculos; espirales

Harijans (Untouchables)—intocables

Headwaters—cabeceras

Heartland—zona de importancia estratégica

Heavy industry—industria pesada

High latitude—latitud alta

High pressure—presión alta

Highland climate—clima de terreno montañoso

Hill—cerro

Hinterland—interior (de un país)

Human resources—recursos humanos

Humid continental climate—clima continental húmido

Humid subtropical climate—clima subtropical húmido

Humus—humus

Ice cap climate—clima casquete polar

Inanimate energy—energía inanimada

Indigenous—indígena

Indo-Europeans (Aryans)—indoeuropeos (arios)

Industrial triangle—triángulo industrial

Infrastructure—infraestructura

Inner core—médula interior

Intermontane—situado entre montañas

Irrigation—irrigación

Isthmus—istmo

Jungle—selva

Junk—trastos viejos

Karma—karma

Key—cayo; isleta

Lagoon—laguna

Land breeze—brisa de tierra

Land hemisphere—hemisferio de tierra

Landlocked—cercado de tierra

Landlord—propietario

Landscape—paisaje

Laterite—laterita

Latitude—latitud

Lava flow—flujo de lava

Leeward—sotavento

Legend—leyenda

Levee—ribero; bordo

Limestone—piedra caliza

Linear village—pueblo linear

Literacy—alfabetismo

Lithosphere—litosfera

Longitude—longitud

Low latitudes—latitudes bajas

Low pressure—presión baja

Loxodrome—línea loxodrómica

Mainland—tierra firme; continente

Mandate—mandato

Mantle—manto

Map key—explicación de mapa

Map projection—proyección de mapa

Map scale—escala de mapa

Marginal soils—tierras colindantes

Marine west coast climate—clima marítima de la costa del oeste

Maritime technology—tecnología marítima

Maritime—marítimo

Mass emigration—emigración en grupo

Mass production—producción en masa

Mechanical weathering—acción corrosiva mecánica de los elementos naturales

Mediterranean climate—clima del Mediterráneo

Megalopolis—megalópoli

Meridian—meridiano

Metropolis—metrópoli

Middle latitudes—latitudes medias

Minarets—alminares

Mirs—mires

Modernization—modernización

Moguls—mogoles

Monotheism—monoteísmo

Monsoon rain forest—monzón de selva tropical

Monsoon—monzón

Mosque—mezquita

Mountain glacier—glaciar de las montañas

Mountain—montaña

Mulatto—mulato

Multinational—multinacional

Muslim—musulmán

Narrows—estrecho

Natural resources—recursos naturales

Nonrenewable resources—recursos no renovables

North Atlantic Drift—terreno de acarreo atlántico del norte

Nuclear village—pueblo nuclear

Orbit—órbita

Orographic precipitation—precipitación orográfica

Outer core—núcleo exterior

Oxidation—oxidación

Ozone—ozono

Pampas—pampas

Panhandle—mendigar

Parallel—paralelo

Pastoral nomadism—nomadismo pastoral

Per-capita GNP—per cápita

Perennial irrigation—irrigación perenne

Permafrost—permafrost

Petrochemical—producto petroquímico

Piedmont—región al pie de las montañas

Pilgrimage—peregrinaje

Plain—llanura

Plantation—plantación

Plate tectonics—placa tectónica

Plateau—meseta; antiplanicie; antiplano

Plates—placas

Podzol—podsol

Polar high—alto polar

Pollution—contaminación

Post industrial—posindustrial

Postindustrial society—sociedad posindustrial

Prairie—pradera

Pre-columbian—precolombino

Precipitation—precipitación

Primary production—producción primaria

Primate city—ciudad primada

Prime Meridian—primer meridiano

Production quota—producción cuota

Profit motive—motivo de ganancia

Provinces—provincias

Quinine—quinina

Quota—cuota

Qur'an (Koran)—Corán

Rain shadow—sombra lluviosa

Red Guards—Guardias Rojas

Reforestation—repoblación forestal

Reincarnation—reencarnación

Relative elevation—elevación relativa

Relative location—localización relativa

Renewable resources—recursos renovables

Reservoir—embalse

Revolution—revolución

Rice cycle—ciclo de arroz

Rift valley—valle por falla

Rimland—región periférica

Ripple effect—efecto de rizo

Road warrior—guerrero de la carretera

Rotation—rotación

Savanna—sabana

Scientific method—método científico

Scorched-earth policy—sistema de tierra quemada

Scrub—depurar

Sea breeze—brisa marina

Secondary production—producción secundaria

Secularism—laicismo

Sedimentary—sedimentario

Seismic—sísmico

Sepoy—cipayo; soldado indio

Settled farming—agricultura estable

Shanty town—villa miseria

Sharecropper—aparcero

Shield—escudo

Shifting cultivation—cultivación que cambia

Silicon—silicio

Silt—Cieno; légamo

Sinkhole lakes—lagos sumideros

Site—sitio; emplazamiento

Situation—puesto; situación

Slash and burn agriculture—agricultura de acuchillar y quemar

Slum—barrio bajo

Smog dome—domo de smog

Solstice—solsticio

Sovereign—soberano

Soviet—soviético

Special Economic Zones, (SEZ's)—Zonas Económicas Especiales

Standard of living—nivel de vida

Staple—producto básico; materia prima

Station—estación

Steppe climate—clima estepa

Steppe—estepa

Straits—estrechos

Strip mining—explotar una mina a cielo abierto

Subarctic climate—clima subártico

Subcontinent—subcontinente

Subpolar lows—depresiones subpolares

Subsidy—subsidio

Subsistence crops—cosechas subsistentes

Subtropical high—altura subtropical

Sulfuric acid—ácido sulfúrico

Summer monsoon—monzón veraniego

Sun belt—estados del sur/suroeste de EE.UU.

Synagogue—sinagoga

Taiga—taiga

Technology—tecnología

Technopolis—tecnópoli

Tectonic forces—fuerzas tectónicas

Tenant farmer—agricultor arrendatario

Terrace—terraplén

Tertiary production—producción terciaria

Theocracy—teocracia

Topography—topografía

Trade winds—vientos alisios

Transhumance—transhumación

Tributary—afluente

Tropical forests—bosques tropicales

Tropical deciduous forest—bosque caduco tropical

Tropical rain forest climate—clima de selva tropical

Tropical rain forest—selva tropical

Tropical savanna climate—clima sabana tropical

Tropical savanna—sabana tropical

Tropics—trópicos

Trust territory—territorio bajo fideicomiso

Tsetse fly—mosca tse-tsé

Tundra climate—clima tundra

Typhoon—tifón

Urban—urbano

Urbanite—habitante de una ciudad

Veld—veldt

Volcanic island—isla volcánica

Volcanism—volcanismo

Weather—tiempo

Weathering—acción corrosiva de los elementos naturales

Westerlies—viento del oeste

Wet rice cultivation—cultivación de arroz mojado

Windward—barlovento

Winter monsoon—monzón invernal

Xerophyte—xerofita

Yurts—yurtas

Zionism—Sionismo

Index

Persian Gulf War, 545
Perth, 363
Peru, 402, 444, 449–50
    climate of, 402
    colonization of, 410
    Incas in, 450
Peter the Great, 177, 179, 181
Petrochemical products, 345
Petroleum, 156–57
Philadelphia, 28–30, 249
Philippine Islands, 377, 659
    economy of, 680, 681–82
    Spanish influence in, 674, 676
Phnom Penh, 667, 677
Physical map, 189
Piedmont, 255
Pilgrimage, 554
Pitcairn Island, 379
Pittsburgh, 250
Pizarro, Francisco, 410
Place, as geographical theme, 29–30
Plains, 48, 49
Plantations, 399
Plate tectonics, 41, 42–43
Plateau, 263
Plateaus, 48, 49
PLO (Palestine Liberation
        Organization), 543
Po River Valley, 134
Podzol soils, 176
Poland, 221, 222
    creation of, 225–26
    industry of, 228, 229
    resistance to collectivization in,
        231
Polar climates, 61
Polar highs, 56
Polar regions, 20
Polder, 128, 129
Polo, Marco, 87
Polynesia, 357, 373, 377–79
Pontic Mountains, 532, 533
Pope John Paul II, 225
Population
    in developing world, 96
    in technological world, 92
Port Elizabeth, 509
Port Harcourt, 476
Port Moresby, 375
Portugal, 132–34

colonization of Africa, 495–96,
        506
colonization of South America,
        410–12, 455–56
    in Southeast Asia, 674
Postindustrial economy, 243–44
Postindustrial society, 366
Prague, 221
Prairie Provinces, 298, 302–3
Prairies, 66
Pre-Columbian era, 424
Precipitation, 54, 57
    orographic, 403
Pretoria, 509
Primary production, 363
Primate city, 453
Prime Meridian, 20–21, 219
Prince Edward Island, 298, 300–1
Prince Henry the Navigator, 86–87,
        409
Production, primary, secondary,
        and tertiary, 363, 366
Production quota, 186
Profit motive, 195, 595
Provinces, 297
Puerto Rico, 377, 422, 423, 435,
        438–39
Punjab, the, 618, 619
    agricultural revolution in,
        633–35
    statehood for, 632
Pusan, 344

Q
Qanats, 577
Qatar, 540, 541, 558, 561
Qing (Manchu) dynasty, 602
Québec, 296, 301, 304
Queensland, 363, 366, 381
Quinine, 497
Quota, 595
Qur'an, 551

R
Rabat, 538
Raffles, Stamford, 684
Rain forests. *See* Tropical rain forests
Rain shadow, 57
Red Guards, 593

Red River (Vietnam), 686
Red Sea, 474
Reforestation, 476
Regina, 303
Region, as geographical theme,
        33–34
Reincarnation, 620
Relative location, 28–29
Renewable resources, 81
Republic of the Marshall Islands,
        377
Republic of Palau, 377
Research Triangle Park, 254
Resources, 80
    human and natural, 81
Revolution, of Earth, 17–18
Reykjavik, 123–24
Rhine River, 112, 124, 127, 161, 223
Rhodes, Cecil, 4
Rhône, 125
Rice cycle, 336–37
Rift valley, 474
Riga, 199
Rimland, 422
Ring of Fire, 42, 661
Rio de Janeiro, 386, 456
Río de la Plata, 452–53
Ripple effects, 430
Riviera, the, 126
Riyadh, 540
Road warriors, 360
Robinson projection, 26–27
Rochester, 251
Rockefeller, John D., 278
Rocky Mountains, 244, 259, 261–63
Roman alphabet, 225
Roman Catholicism, 225
Romania, 202–3, 221, 223, 225, 226
    oil fields of, 228, 229
Rome, 135
Ross Ice Shelf, 364
Rotation, of Earth, 17–18
Rotterdam, 147
Rubber, farming of, 665
Ruhr Valley, 127, 143, 154, 156
Russia, 148, 166, 174, 198, 373, 583.
        See also Commonwealth of
        Independent States; Union
        of Soviet Socialist
        Republics

Uzbekistan, 179, 205–7

V

Vaisyas, 620
Vale of Kashmir, 617
Van der Grinten projection, 26–27
Vanuatu, 375, 376
Varanasi (Banaras), 619
Vatican City, 116–17
Vegetation, 63
    climates and soils and, 63–68
    scrub, 65
Veld, 482
Venda, 510
Venezuela, 401, 444, 446–48
    agriculture in, 447
    oil deposits in, 416, 447
Venice, 134
Victoria (Australia), 363
Vienna, 130, 147
Vietnam, 658, 664, 667, 686
    division of, 678
    economy of, 680
    French control of, 678
    invasion of Cambodia, 687
    war in, 686
Villagization, 479
Virgin Lands plan, 212–13
Vistula River, 222
Vladivostok, 208
Volcanic islands, 374–75, 660
Volcanism, 41–43
Volga River, 177

W

Waitangi treaty, 381, 382
Wake Island, 377
Wales, 115
Wallis Island, 379
War of the Triple Alliance, 454
Washington, 267
Washington, D.C., 248, 249–50
Water hemisphere, 113, 114
Water pollution, 204–5, 338–39
Watt, James, 88–89
Weather, 53
Weathering, 44–45
Wei River, 585
Wellington, 351
Wenzhou, 598

West Africa, 476–77, 484–86
    colonies in, 500–1
    exploration of, 495–96
    resources and settlement in,
        508–9
West Bank, 543, 550
West Germany, 226. *See also*
        Germany
West Indies, 422, 435
West Pakistan, 630, 640
West (U.S.), 261–68
Westerlies, 56
Western Australia, 363, 381
Western Europe, 104–35, 139–40,
        221, 222
    in 1900, 147–48
    agriculture in, 159–60
    air pollution in, 160–61
    British Isles and northern
        Europe, 115–20
    climates of, 112, 124
    construction of railroads in,
        142–43
    economic development in,
        152–59
    emigration from, 146
    growth of cities in, 147–48
    Industrial Revolution in,
        140–41
    industrial triangle of, 143, 154,
        156
    Mediterranean, 132–36
    microstates of, 116–17
    as peninsula continent, 112
    population growth of, 147,
        158–59
    role of the sea in, 113–13
    science and technology growth
        in, 143–45
    shift to high technology in,
        157–58
    subregions of, 113
    water pollution in, 161–62
    world trade from, 145–46
    World War I in, 148–49
    World War II and, 111
    World War II in, 149–51
Western Ghats, 619
"Western Isle," 536–39
Western Samoa, 378

Wet rice farming, 661, 664–65
Windhoek, 503–4
Windward slopes, 57, 369
Winnipeg, 303
Winter monsoon, 616
Winter solstice, 20
Witwatersrand, 512
World
    human, 80
    population of, 85
World War I, 148–49
World War II, 149–51

X

Xerophytes, 65
Xi Jiang, 575
Xi Jiang delta, 582–83, 585
Xizang Plateau, 576
Xizang (Tibet), 575

Y

Yanbu, 560
Yangon (Rangoon), 667, 677
Yaoundé, 508
Yekaterinburg, 208
Yellow Sea, 585
Yeltsin, Boris, 197, 198
Yemen, 540, 541
Yucatán Peninsula, 404–5, 422
Yugoslavia, 187, 222, 224, 225, 227
    ethnic war in, 227
    resistance to collectivization in,
        231
Yukon, 267
Yurts, 580

Z

Zacatecas, 425
Zagros Mountains, 532, 534, 560
Zaire, 474, 508, 509
Zaire River, 476
Zambesi River, 498
Zambia, 482, 483, 503, 512
Zanzibar, 498, 504
Zimbabwe, 482, 503, 512
Zionism, 556
Zonguldak, 533
Zurich, 130

Photo Credits

734